THE ROUTLEDGE HANDBOOK OF ENVIRONMENTAL JUSTICE

The Routledge Handbook of Environmental Justice presents an extensive and cutting-edge introduction to the diverse, rapidly growing body of research on pressing issues of environmental justice and injustice. With wide-ranging discussion of current debates, controversies, and questions in the history, theory, and methods of environmental justice research, contributed by over 90 leading social scientists, natural scientists, humanists, and scholars from professional disciplines from six continents, it is an essential resource both for newcomers to this research and for experienced scholars and practitioners.

The chapters of this volume examine the roots of environmental justice activism, lay out and assess key theories and approaches, and consider the many different substantive issues that have been the subject of activism, empirical research, and policy development throughout the world. The *Handbook* features critical reviews of quantitative, qualitative, and mixed methodological approaches and explicitly addresses interdisciplinarity, transdisciplinarity, and engaged research. Instead of adopting a narrow regional focus, it tackles substantive issues and presents perspectives from political and cultural systems across the world, as well as addressing activism for environmental justice at the global scale. Its chapters do not simply review the state of the art, but also propose new conceptual frameworks and directions for research, policy, and practice.

Providing detailed but accessible overviews of the complex, varied dimensions of environmental justice and injustice, the *Handbook* is an essential guide and reference not only for researchers engaged with environmental justice, but also for undergraduate and graduate teaching and for policymakers and activists.

Ryan Holifield is an Associate Professor of Geography at the University of Wisconsin-Milwaukee. His research interests include environmental justice policy and practice, social and political dimensions of urban environmental change, and stakeholder participation in environmental governance.

Jayajit Chakraborty is a Professor of Geography in the Department of Sociology and Anthropology, and Director of the Socio-Environmental and Geospatial Analysis Lab at the University of Texas at El Paso. His research interests are located at the intersection of hazards geography, health geography, and urban geography, and encompass a wide range of environmental and social justice issues.

Gordon Walker is Professor of Environment, Risk, and Justice in the Lancaster Environment Centre, Lancaster University, UK. His research focuses on environmental justice, sustainable energy transitions, and the dynamics of energy demand. Recent books include *Environmental Justice: Concepts, Evidence, and Politics* (Routledge 2012) and *Energy Justice in a Changing Climate* (2013).

THE ROUTLEDGE HANDBOOK OF ENVIRONMENTAL JUSTICE

Edited by

*Ryan Holifield, Jayajit Chakraborty
and Gordon Walker*

Routledge
Taylor & Francis Group

LONDON AND NEW YORK

First published 2018
by Routledge
2 Park Square, Milton Park, Abingdon, Oxon OX14 4RN

and by Routledge
52 Vanderbilt Avenue, New York, NY 10017

First issued in paperback 2020

Routledge is an imprint of the Taylor & Francis Group, an informa business

British Library Cataloguing in Publication Data
A catalogue record for this book is available from the British Library

Library of Congress Cataloging in Publication Data
Names: Holifield, Ryan, editor.
Title: The Routledge handbook of environmental justice / edited by Ryan Holifield, Jayajit Chakraborty and Gordon Walker.
Other titles: Handbook of environmental justice
Description: Milton Park, Abingdon, Oxon ; New York, NY : Routledge, 2018. | Includes bibliographical references and index.
Identifiers: LCCN 2017005742| ISBN 9781138932821 (hardback : alk. paper) | ISBN 9781315678986 (ebook)
Subjects: LCSH: Environmental justice. | Environmental justice—Case studies. | Environmental justice—Research—Methodology.
Classification: LCC GE220 .R68 2018 | DDC 363.7—dc23
LC record available at https://lccn.loc.gov/2017005742

ISBN 13: 978-0-367-58112-1 (pbk)
ISBN 13: 978-1-138-93282-1 (hbk)

Typeset in Bembo
by Keystroke, Neville Lodge, Tettenhall, Wolverhampton

CONTENTS

FIGURES

TABLES

CONTRIBUTORS

Troy D. Abel's research and teaching interests focus on the dynamic tensions of environmental science and democratic politics in the fields of environmental justice, information disclosure and climate governance. He is an Associate Professor of Environmental Policy at Western Washington University's Huxley College of the Environment.

Aaron Aber graduated in 2016 with a B.S. in Environmental Science and Policy from the University of Maryland. From 2015 to 2016, he conducted research under Dr Sacoby Wilson, focusing on the health impacts of natural gas drilling in Western Maryland, urban health disparities in Washington, DC, and community-based environmental justice research methods.

Rhuks T. Ako, PhD (Kent), is an independent consultant and Senior Fellow (Natural Resources, Energy and Environment) at OGEES Institute, Nigeria. He has published widely on topics including environmental justice, environmental human rights, and natural resource conflicts, including the monograph *Environmental Justice in Developing Countries: Perspectives from Africa and the Asia-Pacific*.

Alison Hope Alkon is Associate Professor of Sociology at the University of the Pacific, where she co-founded their MA program in Food Studies. She is the author of *Black, White and Green: Farmers Markets, Race and the Green Economy* and co-editor of *The New Food Activism* and *Cultivating Food Justice*.

Ruchi Anand is an Indian-born academic living in France. She has authored or contributed to several books in the field of international relations and environmental politics and teaches international relations in France and the US. She authored *Self-Defense in International Relations* and *International Environmental Justice: A North–South Dimension*.

Isabelle Anguelovski is a social scientist trained in urban and environmental planning (PhD, MIT, 2011). Her research is situated at the intersection of sustainability planning, socio-spatial inequalities, and development studies. She is ICREA Research Professor at the Universitat Autònoma de Barcelona affiliated both with the ICTA and IMIM research institutes.

Kimberly L. Barrett is an Assistant Professor of Criminology in the Department of Sociology, Anthropology, and Criminology at Eastern Michigan University. Her research interests include green criminology, environmental crime and justice. Kimberly's recent work has been published in the *British Journal of Criminology*, *Journal of School Violence*, and *Deviant Behavior*.

Pratyusha Basu is an Associate Professor in the Department of Sociology and Anthropology and Director of the Asian Studies program at the University of Texas at El Paso. Her research focuses on development and inequalities in India and Kenya and has recently been supported by a Fulbright research award.

Karen Bell is a researcher at the University of Bristol and Teaching Fellow at Keele University, focusing on environmental justice; environmental and social equality; poverty and social exclusion; and the voluntary and community sector. She is currently working on an ESRC Future Research Leaders project to compare redistributive with market-based environmental transition paradigms.

Derek Bell is Professor of Environmental Political Theory at Newcastle University, UK. His research interests are in environmental and climate justice at scales ranging from the local to the global. He is the co-editor of *Environmental Citizenship* (MIT Press 2006) and *Justice and Fairness in the City* (Policy Press 2016).

Karen Bickerstaff is Associate Professor in Human Geography at Exeter University, UK. She has worked extensively on environmental justice, energy systems and equity and has written widely on the topic of energy justice – co-editing a recent collection on 'Energy Justice in a Changing Climate'.

Patrick Bond is a Professor of Political Economy at the University of the Witwatersrand in Johannesburg, and honorary professor at the University of KwaZulu-Natal Centre for Civil Society. His political ecology books include *Politics of Climate Justice* (UKZN Press 2012), *Looting Africa* (Zed 2006) and *Unsustainable South Africa* (Merlin 2002).

Christopher G. Boone is Dean of the School of Sustainability at Arizona State University. His research contributes to ongoing debates in sustainable urbanization, environmental justice, vulnerability and global environmental change. He has a PhD from the University of Toronto and was a post-doctoral fellow at McGill University.

William M. Bowen is a regional scientist who serves as Professor and Director of the PhD Program in Urban Studies and Public Affairs in the Maxine Goodman Levin College of Urban Affairs at Cleveland State University. He has studied environmental justice and related topics for nearly thirty years.

Anna Livia Brand is an Assistant Professor in the Department of Landscape Architecture and Environmental Planning at the University of California, Berkeley. Her research focuses on the intersections of race and redevelopment. Anna received her PhD in Urban Planning from MIT, her Master's from UNO, and her Bachelor's from Tulane.

Robert J. Brulle is a Professor in the Department of Sociology at Drexel University. His research focuses on environmental politics and critical theory. He is the author of three books and over seventy articles in these areas. He is also the recipient of the 2016 Frederick Buttel Distinguished Contribution award.

Geoffrey L. Buckley is Professor of Geography at Ohio University. His research interests include environmental history; urban parks and sustainability; environmental justice; and the evolution of mining landscapes. His articles have appeared in the *Annals of the American Association of Geographers, Geographical Review, Cities and the Environment* and *Urban Ecosystems*.

Michael Buzzelli, BA (Hons.), MA, PhD, M.Ed, is Associate Professor at the University of Western Ontario. After completing graduate work at McMaster University, Michael held academic appointments at UBC and Queen's University and has been a visiting scholar at the University of Glasgow and the University of Bologna.

Jason Byrne is Associate Professor of Urban and Environmental Planning at Griffith University, Australia. Jason's research addresses environmental (in)justices associated with urban greenspace access and use (e.g. parks, community gardens and street trees). A key element of his research is how urban greening may help adapt cities to climate change.

Mary L. Cadenasso is Professor of Plant Sciences and an urban and landscape ecologist at UC Davis. She is a founding member of the Baltimore Ecosystem Study, a research program studying the city as an integrated social-ecological system. She has co-authored four books and more than 70 peer-reviewed articles and 30 book chapters.

Jayne Carrick is a PhD student in Politics at Newcastle University, having previously worked in the UK's renewable energy and construction sectors. Her research interests are in participatory decision making and environmental justice. She is currently investigating methods to promote engagement and improve environmental and energy policy making.

Kuei-Tien Chou is Professor and Director of the Graduate Institute of National Development and also Director of the Risk Society and Policy Research Center at National Taiwan University. He leads the young researchers in RSPRC engaging in transboundary risk research. Recently he has dedicated himself to cosmopolitan studies of climate governance.

Eric Chu is Assistant Professor of Urban Studies in the Department of Geography, Planning, and International Development Studies at the University of Amsterdam. His research is on the politics of governing climate change adaptation, resilience, and development in cities.

Timothy W. Collins is a Professor of Geography and Director of the Institutional Development Core of the NIH-funded BUILDing SCHOLARS centre at the University of Texas at El Paso. His research focuses on environmental justice; health disparities; social vulnerability to risks, hazards and disasters; urban governance; and Hispanic/Latino populations.

Philip Coventry is a PhD researcher in the Department of Geography and Environmental Science at the University of Reading, funded by the Economic and Social Research Council. His PhD project focuses on the role of path dependence in the development of climate finance and capacity building policy in the UNFCCC.

Rosie Day is a Senior Lecturer in Human Geography at the University of Birmingham, UK. Her research centres on environmental and energy justice in diverse, international contexts. She collaborates on and supervises a number of projects funded by the UK Research Councils and overseas governments.

Séverine Deguen, after receiving a PhD in biomathematics (1999), became a researcher and a teacher in biostatistics in the Department of Environmental and Occupational Health in the School of Public Health (EHESP) in Rennes (Brittany). Her research deals with the role of environmental exposure in social health inequalities, the Equit'Area project.

Catalina M. de Onís is an Assistant Professor in Willamette University's Department of Civic Communication and Media. Her research examines environmental, climate, and energy (in)justices from a rhetorical perspective. Her scholarship appears in *Environmental Communication: A Journal of Nature and Culture*, *Women's Studies in Communication* and *Women & Language*.

Daniel Faber is Professor of Sociology at Northeastern University and Director of the Northeastern Environmental Justice Research Collaborative. His latest book is *Capitalizing on Environmental Injustice: The Polluter-Industrial Complex in the Age of Globalization*. Dr Faber is on the Board of Coming Clean, a national collaborative of 200+ environmental organizations.

Mei-Fang Fan is Professor at the Institute of Science, Technology and Society, National Yang-Ming University. She holds a doctoral degree in Environment and Society from Lancaster University. Research interests include environmental justice and citizenship; local knowledge and expertise; and risk and disaster governance.

Richard Filčák is a senior researcher at the Slovak Academy of Sciences. His main research area is environmental and social policy development in the transitional countries of Central and Eastern Europe, with a particular focus on poverty and social and territorial exclusion leading to exposure to environmental risks and vulnerability.

Sheila R. Foster is University Professor and the Walsh Professor of Real Estate, Land Use and Property Law at Fordham University. She is also the Faculty co-Director of the Fordham Urban Law Center. Professor Foster is the author of numerous publications on land use and environmental law and justice.

Louise Francis is a researcher at University College London (UCL) and a practitioner of citizen science and community mapping programmes. She is a co-founder and Managing Director of the social enterprise 'Mapping for Change'. Her research interest is in the role that participatory mapping, VGI and citizen science play in generating positive environmental and social change.

Greta Gaard is Professor of English and Coordinator of the Sustainability Faculty Fellows at University of Wisconsin-River Falls. Author or editor of six books and over eighty articles, Gaard's most recent volume is *Critical Ecofeminism*. Her creative nonfiction eco-memoir, *The Nature of Home*, was translated into Chinese and Portuguese.

Colleen George completed her PhD in the School of Environment and Sustainability at the University of Saskatchewan. Her research interests include social innovations toward sustainability, particularly in areas related to environmental governance and policy. As an interdisciplinary scholar, Colleen is interested in governance models working toward holistic definitions of sustainability.

Kian Goh, RA, PhD, is Assistant Professor of Urban Planning at UCLA. She researches urban ecological design, spatial politics and social mobilization in the context of climate change and global urbanization. She received a Master of Architecture from Yale University, and a PhD in Urban and Environmental Planning from MIT.

Aaron Golub is an Associate Professor in the Nohad A. Toulan School of Urban Studies and Planning at Portland State University. His teaching and research interests include the social impacts of transportation, planning for alternative transportation modes, and the history of transportation in the United States.

Donna Green is a senior research scientist in the Climate Change Research Centre, and an Associate Investigator in the Australian Research Council Centre of Excellence for Climate Systems Science. She conducts research leadership in the areas of climate extremes and health; climate impacts on Indigenous people; and environmental justice and air pollution.

Sara E. Grineski is a Professor of Sociology and Director of the Research Enrichment Core of the NIH-funded BUILDing SCHOLARS centre at the University of Texas at El Paso. Her research focuses on environmental justice issues related to children, environmental health, and Hispanic/Latino populations.

Ulrika Gunnarsson-Östling is a researcher at KTH – Royal Institute of Technology, Stockholm, Sweden. Her research focuses both on how sustainability is understood and put into practice in planning and policy making, and on how this practice can be altered in order to promote more socioecologically just developments.

Muki Haklay is a Professor of Geographic Information Science at University College London (UCL). He is Co-Director of UCL Extreme Citizen Science group, and a co-founder of the social enterprise 'Mapping for Change', which provides services in participatory mapping and citizen science. His research focuses on public access to environmental information, participatory mapping, and citizen science.

Krista Harper is Associate Professor of Anthropology and Public Policy at the University of Massachusetts Amherst. Her articles and her books *Wild Capitalism: Environmental Activism and Postsocialist Political Ecology in Hungary* (2006) and *Participatory Visual and Digital Methods* (with Aline Gubrium, 2013) present her research on environmental justice mobilizations.

Leila M. Harris is an Associate Professor at the Institute for Resources, Environment and Sustainability and in the Institute for Gender, Race, Sexuality and Social Justice at the University of British Columbia. Her work examines social, cultural and political-economic and equity dimensions of environmental and resource issues.

Megan Heckert is an Assistant Professor in the Department of Geography and Planning at West Chester University. She received her PhD in Urban Studies from Temple University in 2012. Megan's teaching and research interests are in spatial analysis, urban greening, and environmental justice.

Donna Houston is a Senior Lecturer in the Department of Geography and Planning at Macquarie University, Sydney, Australia. Her current research explores environmental and climate justice in multispecies worlds; political ecologies of urban nature and postindustrial transformation; and critical and creative geographies of activism, memory-work and place-making.

Alex Karner is an Assistant Professor in the Community and Regional Planning Program at the University of Texas at Austin. His research aims to assess the social equity, environmental and public health implications of transportation projects and plans by using emerging data sources and developing new, open source methods.

Heike Köckler is Professor for Place and Health at Hochschule für Gesundheit, Bochum and Dean of the Department of Community Health. She works in the field of environmental justice with a focus on procedural environmental justice, the potential of environmental and urban planning for more environmental justice and environmental justice indicators.

Stephen H. Linder is a Professor in the Division of Management, Policy and Community Health, and Director of the Institute for Health Policy at the University of Texas School of Public Health. His research interests focus on policy studies, social theory, risk assessment, and climate change and health.

Jonathan K. London studies and supports environmental justice through participatory research, rural community development, and community-engaged planning. He is Associate Professor in the Department of Human Ecology, Faculty Director of the Center for Regional Change, and Chair of the Community Development Graduate Group at the University of California Davis.

Michael Lukas is Assistant Professor in the Department of Geography at the University of Chile. He works on the political economies and political ecologies of urban growth in Latin America.

Michael J. Lynch is a Professor of Criminology and associated faculty member in the Patel School of Global Sustainability at the University of South Florida. His research primarily addresses green crime and justice, corporate crime and its control, and environmental justice. He is the author/editor of twenty books and more than 140 articles and book chapters.

Juliana Maantay is Professor of Urban Environmental Geography; GISc Program Director; and Director of the Urban GISc Lab at Lehman College, City University of New York (CUNY). Formerly an urban planner and policy analyst, she has been active in environmental health justice research and advocacy for more than 25 years.

Andrew Maroko is an Assistant Professor at The CUNY Graduate School of Public Health and Health Policy and Associate Director of the Urban GISc Lab at Lehman College. His research interests are in the examination of health inequities, exposures, accessibility, and environmental justice in a spatial framework.

Karel Martens is an international expert on transport and justice. He has authored numerous publications on the topic, including a book titled *Transport Justice: Designing Fair Transportation Systems*. Martens holds the Leona Chanin Chair at the Technion – Israel Institute of Technology, which he combines with a position at Radboud University (Netherlands).

José Mayorga is Professor in Urban Sociology at Pontificia Universidad Javeriana and Professor in GIS at Universidad de los Andes. He is a sociologist, MA in Urban and Regional Planning, specialist in Urban Law and specialist in Applied Statistics.

Scott McKenzie is a PhD student at the University of British Columbia's Institute for Resources, Environment and Sustainability. His work considers the relationship between the environment, governance and law.

Anders Melin is Associate Professor in Ethics at Malmö University, Sweden and his main field of study is environmental ethics. He is currently the leader of a research project on energy and justice.

Jeremy Mennis is a Professor in the Department of Geography and Urban Studies at Temple University. He received his PhD in Geography from Pennsylvania State University in 2001. Jeremy's teaching and research interests are in geographic information science, environmental justice and behavioural health.

Karrina Nolan is a descendant of the Yorta Yorta people. She is a community educator, organiser and facilitator. She has led programs and campaigns on women's rights, globalization and environmental justice with a focus on First Nations peoples. She consults on a range of climate justice, resource extraction and building power and capacity related issues.

Chukwumerije Okereke is a Professor in Environment and Development at the Department of Geography and Environmental Science, University of Reading, UK. His research interest is in the ethical and political economy dimensions of global climate governance.

Damilola S. Olawuyi, LL.M (Calgary), LL.M (Harvard). DPhil (Oxford), is Associate Professor of Petroleum, Energy and Environmental Law at the HBKU Law School, Qatar and Chancellor's Fellow at the Institute for Oil, Gas, Energy and Environment (OGEES Institute), Nigeria. Dr Olawuyi has published more than 30 peer-reviewed articles, books and reports on climate justice, extractive resource governance and sustainable development.

Gwen Ottinger is Associate Professor in the Department of Politics and the Center for Science, Technology, and Society at Drexel University. She is co-editor of *Technoscience and Environmental Justice: Expert Cultures in a Grassroots Movement* and author of *Refining Expertise: How Responsible Engineers Subvert Environmental Justice Challenges*, which won the 2015 Rachel Carson Prize from the Society for Social Studies of Science.

David N. Pellow is the Dehlsen Chair and Professor of Environmental Studies and Director of the Global Environmental Justice Project at the University of California, Santa Barbara. His teaching and research focus on environmental and ecological justice in the US and globally.

Phaedra C. Pezzullo is an Associate Professor in the Department of Communication at the University of Colorado Boulder. Her books include *Toxic Tourism: Rhetorics of Travel, Pollution, and Environmental Justice*; *Environmental Justice and Environmentalism* (co-edited with Ronald Sandler); *Cultural Studies and the Environment, Revisited* and *Environmental Communication and the Public Sphere* (co-authored with Robert Cox).

Laura Pulido is a Professor in the departments of Ethnic Studies and Geography at the University of Oregon. Her research interests include race, environmental justice and political activism. She is the author of numerous books, including *Black, Brown, Yellow and Left: Radical Activism in Los Angeles* and *Environmentalism and Economic Justice: Two Chicano Struggles in the Southwest*.

Andrea Ranzi is an environmental epidemiologist at the regional agency for prevention, environment and energy of the Emilia-Romagna region, Italy. His research activity focuses mainly on methods of exposure assessment in environmental epidemiology, health effects of air pollution from different sources and health impact assessment.

Vivek Ravichandran is a junior public health science major and global poverty minor, enrolled at the School of Public Health, Class of 2018 at the University of Maryland College Park. His primary research interests include analysing health disparities and planning interventions to combat and eliminate inequalities that may persist in society.

Maureen G. Reed is Professor and Assistant Director of the Graduate School of Environment and Sustainability at the University of Saskatchewan. With students and partners, she has created a "PROGRESS" lab dedicated to understanding and improving practices of governance for resilience, and environmental and social sustainability. She has worked and learned with biosphere reserves for about 15 years.

Lauren Rickards is Senior Lecturer and co-leader of the Centre for Urban Research's research programme on Climate Change and Resilience at RMIT University in Melbourne, Australia. A Rhodes Scholar, Lauren's research examines different sites, meanings and implications of resilience, particularly in terms of urban–rural relations and the Anthropocene.

Glenn Robinson is a transportation researcher in the Institute for Urban Research at Morgan State University and a University of Maryland, National Smart Growth Center affiliate. His research in environmental justice in transportation focuses on understanding the distribution of transportation system improvement inequities on low income and minority communities.

Lucy Rodina is a PhD candidate at the University of British Columbia's Institute for Resources, Environment and Sustainability. Her work examines water governance, inequality and resilience in urban contexts, particularly in the Global South, using a case study from Cape Town, South Africa.

Marcela Salgado is a sociologist and PhD candidate in Social Sciences at Universidad de Chile. Her main fields of study are socionatural disasters and socio-environmental conflicts. She is currently a post-graduate researcher in a socio-environmental project dealing with conflicts in forested territories of Chile and a researcher in the Centre of Study on Vulnerabilities and Socio-natural Disasters (CIVDES), both at Universidad de Chile.

David Schlosberg is Professor of Environmental Politics in the Department of Government and International Relations at the University of Sydney, and co-Director of the Sydney Environment Institute. David's current work includes environmental justice in adaptation and resilience planning, environmentalism and everyday life, and theoretical implications of the Anthropocene.

Ken Sexton is a Professor in the Division of Epidemiology, Human Genetics and Environmental Science at the University of Texas School of Public Health. His research interests involve human exposure to environmental hazards, health risk assessment and risk-based decision making.

Sameer H. Shah is a PhD student at the University of British Columbia's Institute for Resources, Environment and Sustainability. His work explores the relationships between agricultural livelihoods, water security and governance, with particular emphasis on South and South-East Asia regions.

Alana Shaw is a recent graduate of the Integrative Conservation and Geography program at the University of Georgia, where she also earned a graduate certificate in Native American Studies. Her doctoral research focused on the decision making processes surrounding the Arctic offshore drilling program and its effects on the Iñupiat of Alaska's North Slope.

Diane M. Sicotte is an Associate Professor in the Department of Sociology at Drexel University, where she researches environmental injustice and inequality. She is author of *From Workshop to Waste Magnet: Environmental Inequality in the Philadelphia Region* (Rutgers University Press 2016). Her current environmental justice research involves natural gas infrastructure.

Tamara Steger is an Associate Professor in the Department of Environmental Sciences and Policy at Central European University (CEU) in Budapest, Hungary. She is the founder of the Environmental Justice Program (est. 2003) of the Environmental and Social Justice Action Research Group (ACTJUST) at CEU.

Mark Stephan is an Associate Professor of Political Science at Washington State University. His research is in the areas of climate change governance, environmental justice and environmental policy in US states. Currently he is working on a study on climate change governance that intends to understand state–local policy connections.

Marianne Sullivan is an Associate Professor of Public Health at William Paterson University in New Jersey, USA. Prior to her academic position she served as an epidemiologist at Public Health – Seattle & King County. She has published on a range of topics including community based participatory research, mining/smelting and children's health, and community and occupational lead exposure.

Åsa Svenfelt, PhD, is a researcher in Environmental Strategies Research and Reader in Sustainability and Futures Studies at KTH – Royal Institute of Technology, Sweden. Her research is focused on systematic analysis of sustainability, consumption and social-ecological justice. She applies future studies and backcasting for exploring environmental policy and long-term planning.

Julie Sze is Professor and the Chair of American Studies at UC Davis. She is also the Founding Director of the Environmental Justice Project for UC Davis' John Muir Institute for the Environment. She has authored two books, and written or co-authored 40 peer-reviewed articles and book chapters.

Leah Temper, PhD, is a scholar-activist specializing in ecological economics and political ecology at the Autonomous University of Barcelona. She is Director of the Global Environmental Justice Atlas (www.ejatlas.org) and Project Leader of ACKnowlEJ (Activist-academic Co-production of Knowledge for Environmental Justice), a project analysing the alternatives born from resistance to extractivism.

Leire Urkidi is Assistant Professor at the University of the Basque Country (Department of Geography) and holds a PhD in Environmental Studies. Her main fields of research are mining conflicts, environmental justice, environmental movements and ecological debt. Recently, she has been participating in projects about energy transitions and gender justice.

Alexis Vásquez is an Assistant Professor in the Department of Geography at the University of Chile. His research interests lie in the area of landscape planning, multifunctional green infrastructure and geospatial analysis, ranging from theory and design to implementation.

Pavithra Vasudevan is a PhD candidate in Geography at the University of North Carolina, Chapel Hill, USA. Her dissertation research is a collaborative performance ethnography exploring racial capitalism in a Southern aluminium town. Pavithra serves on the Planning Committee of the North Carolina Environmental Justice Network.

Mariana Walter holds a PhD in Environmental Sciences. She is a researcher at the Institute of Environmental Sciences and Technologies (ICTA) of the Autonomous University of Barcelona. Her research addresses political ecology, ecological economics and resource extraction conflicts in Latin America.

Kyle Whyte holds the Timnick Chair in the Humanities and is Associate Professor of Philosophy and Community Sustainability at Michigan State University. His research seeks to address moral and political issues concerning climate policy and Indigenous peoples and the ethics of cooperative relationships between Indigenous peoples and climate science organizations.

Nicole J. Wilson is a PhD candidate at the University of British Columbia's Institute for Resources, Environment and Sustainability. Their work examines Indigenous peoples' relationships to water and water governance in the context of settler colonialism and environmental change, with an emphasis on Arctic and sub-Arctic regions of North America.

Sacoby Wilson is an Assistant Professor with the Maryland Institute for Applied Environmental Health, University of Maryland-College Park. He has over fifteen years of experience as an environmental justice scientist. He uses community-based participatory research (CBPR) including community–university partnerships to study and address environmental injustice and health disparities.

Kerri Woods is Lecturer in Political Theory at the University of Leeds, and author of *Human Rights* (Palgrave Macmillan 2014) and *Human Rights and Environmental Sustainability* (Edward Elgar 2010).

Lindsey Wright is a junior at the University of Maryland, College Park. She is currently pursuing a dual degree in Environmental Science and Technology and Government and Politics with plans to attend law school upon graduation. Her primary research interest is sustainable natural resource use.

ACKNOWLEDGEMENTS

The editors gratefully acknowledge the authors who contributed to this volume, and we also thank Egle Zigaite, Sarah Gilkes, Daniel Bourner, and Andrew Mould at Routledge for their encouragement, patience, and assistance, and Helena Power for her diligent copyediting. Finally, we would like to thank our families, loved ones, and colleagues for supporting us through the lengthy process of putting this handbook together. Ryan would like to give special thanks to Max, Baz, and Kristin for their inexhaustible patience.

1

INTRODUCTION

The worlds of environmental justice

Ryan Holifield, Jayajit Chakraborty and Gordon Walker

Introduction

Although some three decades have passed since the term *environmental justice* began mobilizing activists and making headlines – along with related terms like *environmental racism*, *environmental equity*, and *environmental inequality* – the conditions it names remain just as relevant today. During the period of time when we were compiling the chapters for this handbook, the US city of Flint, Michigan attracted international media attention for its lead-contaminated drinking water (also see the special issue of *Environmental Justice*, volume 9, issue 4). If the concepts of environmental justice and environmental racism had faded somewhat from North American public discourse and consciousness in recent years, the Flint water crisis brought them decidedly – and distressingly – back in. Flint has a majority African American population and a notoriously high rate of poverty, and the toxic burden of the lead pollution fell disproportionately on low-income African American children and their families. This distributive injustice was compounded by egregious procedural injustice, as residents' concerns about the highly corrosive water were repeatedly dismissed by state officials, several of whom subsequently received indictments for misconduct in office, tampering with evidence, and other criminal offences. Moreover, the Flint water crisis is by no means an isolated incident or an anomaly. As this book goes to press, the rallying cry of environmental justice can be heard in protests of the Dakota Access Pipeline in the Standing Rock Sioux Reservation (North Dakota, US), in the closure of London's City Airport by activists from Black Lives Matter UK, in the Ecuadorean president's call to institute an international court of environmental justice, and in countless lower-profile community struggles throughout the world.

Environmental problems, from water and pollution to biodiversity loss and global warming, have the capacity to affect all of us. However, as the Flint water crisis and other contemporary struggles so starkly remind us, they do not affect us all equally, or in the same ways. Nor do we have equal power to decide solutions to these problems, or to take the necessary action to solve them. This unequal and differentiated positioning, which typically places the heaviest environmental burdens upon marginalized, disadvantaged, and less powerful populations, forms the central premise of the problem of environmental injustice and the hope for environmental justice as its solution.

The aim of this handbook is to present a broad overview of the diverse, rapidly growing body of research on issues of environmental justice and injustice, both to introduce new scholars to the field and to help orient the agendas of experienced researchers. From its origins in grassroots activism and engaged sociological scholarship, primarily in the US, environmental justice (EJ) research has generated what is now a vast, multi-disciplinary literature encompassing a wide range of issues and politics in countries throughout the world. Initially focused on environmental hazards and pollution, the scope of EJ activism and research has now expanded to encompass almost everything that is unsustainable about the world, including rampant industrialization, resource depletion, energy use, consumption patterns, food systems, access to environmental amenities, and public policies that adversely affect minority, Indigenous, and low-income communities, other vulnerable groups such as disabled, immigrant, and linguistically isolated populations, as well as future generations.

Although this handbook is intended for a wide variety of audiences, from academics and students to practitioners, policymakers, and activists, we need to be clear from the outset that it is not a handbook of political tactics, community-organizing strategies, legal solutions, or best practices for public policy. These topics would no doubt require handbooks of their own, with a rather different set of authors. The purpose here, in keeping with the objectives of the Routledge handbook series, is more modest and narrowly focused: to present an introduction, guide, and reference to current debates, controversies, and questions in academic EJ research.

Of course, as EJ scholarship has spread to countries and regions throughout the world (Walker & Bulkeley 2006; Carmin & Agyeman 2011; Schroeder et al. 2008; Reed & George 2011), this academic literature has now become so large and diverse that no single handbook could cover it in its entirety. Although we have sought to present the topic with as much breadth as feasible, constraints of space and time make it impossible to be exhaustive and comprehensive. In addition, EJ continues to evolve as a concept and as a research agenda, and to some extent this handbook will inevitably be limited to presenting a snapshot of the field at one particular point in time. As we have put the handbook together new papers have appeared that seek to challenge received wisdoms in EJ research and push in new directions – see, for example, Velicu and Kaika (2015); Jamal and Hales (2016); and Mennis et al. (2016). However, despite these limitations, we challenged our authors not only to introduce and critically assess the current state of the art, but also to set an agenda for the future. In our view, the *Handbook* also makes it clear that the present moment is particularly rich and important for EJ research. For one thing, it is no longer a new field; it has had time to develop and evolve, in multiple contexts and directions. EJ scholarship now engages with and contributes to a wide array of social, political, legal, ethical, and geographical theories. It draws on and helps advance an increasingly sophisticated and diverse collection of methodological approaches. Finally, it addresses a growing variety of substantive issues in an expanding range of geographic settings.

The objective of the *Handbook* is to provide a critical introduction to these theories, approaches, issues and contexts, with a focus on what makes each distinctive. For example, quantitative, qualitative and mixed methodological approaches of various kinds, as well as different disciplinary or transdisciplinary perspectives, allow us to ask and answer distinctive questions and provide different insights about EJ. The same goes for different substantive issues; for instance, such problems as toxic waste, air pollution, depletion or degradation of water resources or threats to biodiversity – each present distinctive injustices, and for each the meaning of EJ may look slightly different. Moreover, as much prior work has now shown, the meanings and dimensions of EJ have undergone changes as the concept has travelled to different places. Theorists have argued that among the key distinguishing features of EJ are its plurality and multivalence (Schlosberg 2007; Walker 2012), and we have sought in the *Handbook* to emphasize and elaborate this remarkable pluralism. This extends to our roster of contributing authors, which includes social scientists, natural scientists, humanists and academics from law, public health, planning and other professional disciplines, recruited from almost every continent.

In the remainder of this chapter, we begin by briefly situating EJ research with respect to its historical origins and key premises. We follow this with an introduction to the structure and themes of the volume, closing with a call for renewed commitment to research dedicated to both critically engaging with environmental injustice, and advancing its eradication.

Situating environmental justice research

Origins

In the US, the most widely recognized landmarks of the pioneering stage of environmental justice research were published during the 1980s, prompted by instances of grassroots activism. These include sociologist Robert Bullard's (1983) classic study of the distribution of solid waste sites among communities of colour in Houston, Texas, as well as a regional study by the US General Accounting Office (1983) to follow up on issues raised in historic protests against the siting of a hazardous waste dump in Warren County, North Carolina. Such work set the stage for a national-scale investigation by the United Church of Christ's (UCC) Commission on Racial Justice (1987), which found the presence of hazardous waste facilities and toxic waste sites to be significantly associated not only with lower household income and housing values, but above all with higher percentages of people of colour. The UCC analysis both inspired new studies – some of which challenged its findings and others which supported them – and helped bring environmental racism and injustice to the federal agenda by the early 1990s (see, e.g., Bryant & Mohai 1992).

Since that time, the research literature on EJ has grown and diversified tremendously, and it has become the central focus of entire journals, such as *Environmental Justice* and *Local Environment*. Within the academy, EJ research has never been confined to a single discipline, and as a result it has invited diverse methodological and theoretical approaches and perspectives. In the US, sociologists such as Robert Bullard, Bunyan Bryant, Paul Mohai, and Beverly Wright played central roles both in setting the research agenda and applying this research to influence public policy (Cable et al. 2005). A crucial part of this agenda – to which geographers, economists, legal scholars, political scientists, historians, public health scientists, urban and regional planners, ecologists and others have also made key contributions – has always been the quantitative and spatial analysis of distributive EJ. Other sociologists, later joined by anthropologists, environmental historians, geographers, political scientists, and scholars from fields such as communication, rhetoric, literature, women's studies and gender and sexuality studies, ethnic studies, and American studies, produced some of the pioneering studies of EJ as a social movement and discourse (e.g. Čapek 1993; Taylor 2000). Qualitative approaches to research, including ethnographic, archival, and cultural studies methods, have been particularly prominent in this tradition. Interdisciplinary and transdisciplinary projects have also become increasingly important, involving collaborations both among academic disciplines and between academics and non-academics. Accompanying this methodological pluralism has been a remarkable geographic diffusion of EJ research, which, as this *Handbook* makes evident, has now become common in countries throughout the world.

The meanings of environmental justice

The terms *environmental justice* and *environmental injustice* have always resisted straightforward definition (Holifield 2001; Schlosberg 2007; Phillips & Sexton 1999; Sze & London 2008).

This is in part because they serve a variety of functions; for example, they can serve as descriptive terms for observable or measurable states of affairs, as normative terms condemning present conditions and naming desired outcomes for the future, or as political terms deployed to name substantive problems and antagonists, mobilize activism, and justify advocacy for particular policies and laws (Walker 2012). Initially associated most prominently with inequitable distributions of waste and pollution, the terms have also come to encompass a wide variety of substantive problems, struggles and aspirations. The classic document *Principles of Environmental Justice*, a product of the 1991 First National People of Colour Environmental Leadership Summit in Washington, DC, exemplifies this broad scope, addressing issues ranging from protection from hazardous waste and nuclear testing to self-determination and opposition to military occupation. As EJ terminology has spread spatially and evolved temporally, it has taken on new political meanings, embracing still more issues and aspirations depending on the particular contexts (Walker 2009; Schroeder et al. 2008; Sikor 2013). With this expansion of meaning, the question of "justice for whom" has also become more complicated. In some settings, such as politics and policy in the US, the language of EJ has traditionally focused on racialized minority groups and low-income populations. But the term's diffusion in other settings has raised questions about whether other groups of people might also be considered victims of environmental injustice – and indeed, whether EJ and injustice are also terms that should be extended to nonhumans (see, e.g., Schlosberg 2007).

In addition to engaging with its multiple definitions, scholars have increasingly addressed and theorized the multidimensionality of the justice in environmental justice. In addition to *distributive* justice, which remains a key focus of much quantitative and spatial analysis, a growing body of literature now attends to *procedural* and participatory justice, justice as *recognition*, and justice as *capabilities*, as well as the interrelations among these dimensions (Schlosberg 2007; Walker 2012).

Structure and rationale of the volume

The process of editing this handbook reminded us that it is impossible, even in a large volume, to do full justice to environmental justice. We had neither an infinite word count nor an infinite timeline to work with, and the authors we invited or considered had constraints of their own. We sought to capture as much of the breadth of the field as possible, but we realized early in the process that there would inevitably be omissions. One reviewer of the book proposal, for example, suggested that the volume include the following, whether as chapters or as topics: "weapons production and war, biotechnology, nanotechnology, nuclear power, green energy and biofuels, peak oil and oil vulnerability, green infrastructure, mining, farm chemicals, obesity and food deserts, natural hazards (e.g. earthquakes), urban agriculture and community gardens, industrial ecology, asbestos, medical technology, bio-prospecting and coal seam gas/shale oil." Some of these issues feature in the book; others unfortunately do not. All of these topics are worthwhile and important, but some of them are still emerging as EJ issues – perhaps in a few years they will be essential to a new volume or edition.

The same goes for regions of the world. As hard as we tried to represent every corner of the globe, there was not enough room to include all of the places in which EJ activism and scholarship are taking place. In addition, in some locations – above all, the US – EJ is a well-established field with numerous active scholars, while in others it is still in its infancy. We regretted, for example, that we could not include chapters devoted to Pacific Island states – facing unique threats to their existence in climate change – to the former Soviet states of northern Asia, or to the countries of northern Africa, the Middle East, and western Asia (see, e.g., Kuletz 2002; Omer & Or 2005; Agyeman & Ogneva-Himmelberger 2009). Even within continents or regions represented in the *Handbook*, we had to exclude far more than we wanted; sub-Saharan Africa, for example, has seen a blossoming of EJ scholarship in recent years to which this volume unfortunately cannot do justice (see, e.g., Myers 2008; Tschakert 2009; Margai & Barry 2011; Makene et al. 2012; Martin et al. 2014; Kubanza & Simatele 2016). In other cases, such as Latin America and South Africa, excellent book-length collections have already been produced (McDonald 2002; Carruthers 2008). For these reasons, readers should view the sections and chapters as representative rather than comprehensive.

The volume consists of four main parts. Part I begins with a chapter by Laura Pulido, who combines a reflection on her personal history with EJ with an argument for increased attention to the implications of changing racial formations for the efficacy and potential of the movement. It then works through key theories and approaches informing EJ scholarship, covering different perspectives on the dimensions of EJ, theories that bring specific insights into the production of injustice, and analytical frameworks that have been applied to revealing the tactics and strategies of activist work.

The first section within Part I presents discussions of a range of theories relevant to the analysis of EJ as a movement and phenomenon. Two chapters focus on dimensions of *social movement theory*, as applied to analysis of EJ movements. Dianne Sicotte and Robert Brulle present a broad overview of the different varieties of social movement theory that have informed EJ scholarship, from the field's origins to the more recent emergence of transnational and climate justice movements. David Pellow then focuses on one framework within social movement theory, arguing for a model of *political opportunity structures* more attentive to race and gender, the role of the nonhuman, and opportunities beyond the state. The following subsection introduces a set of contrasting approaches to explaining and addressing environmental injustices. William Bowen assesses the prospects of *rational choice theory*, an approach prominent in economics and related disciplines, arguing that formal approaches to rationality focused on utility maximization must be replaced in EJ analysis by explicitly normative approaches more attentive to the complexity of human behaviour. Daniel Faber presents an analysis of EJ based in *radical political economic theory*, which undergirds much EJ research in other social science disciplines, focusing on the ways in which capitalism constrains the mobility of disempowered populations in ways that make them the most vulnerable to ecological hazards. Greta Gaard considers the wide-ranging contributions of *feminist theory* to EJ analysis and practice, from the under-recognized importance of women's movements in the history of environmentalism to the present-day need within the climate justice movement to identify vulnerabilities associated with multiple, intersecting axes of difference – gender, sexuality, race and ethnicity, age, ability, and others. Gwen Ottinger then introduces *science and technology studies* (STS) approaches to EJ; these not only reveal ways that practices and infrastructures of conventional scientific knowledge production condition and constrain efforts to attain EJ, but also illuminate the potential of alternatives, such as community-based science and scientist-activism, to address environmental injustices more comprehensively and democratically.

The second half of Part I focuses on different dimensions of EJ as a normative concept in theory and practice, building and expanding on David Schlosberg's (2007) framework and considering how these dimensions relate to each other. Derek Bell and Jayne Carrick begin with an analysis of the complexities of *procedural* EJ in both theory and professional practice, proposing a framework for "assessing the relative injustice of actual environmental institutions". Kyle Whyte follows with a chapter devoted to *recognition* as a dimension of EJ, arguing that it calls for deeper consideration of resilience as a criterion for justice. An understanding of justice that has received less attention in EJ scholarship to this point is the *capabilities approach* (as developed by Amartya Sen and Martha Nussbaum), and Rosie Day argues that this can provide a rich framework for making explicit links between environmental issues and multi-dimensional human wellbeing. Next, Sheila Foster considers how equality in environmental decision-making – which in the US neither antidiscrimination law nor environmental regulation has adequately addressed – might be better served by analysis of *vulnerability*. As Kerri Woods contends in the subsequent chapter, the rising prominence of environmental issues in the theory and practice of human *rights* is also transforming our understanding of relationships between human and nonhuman nature with important implications for EJ. The final chapter in the section, by Ulrika Gunnarsson-Östling and Åsa Svenfelt, considers how justice figures in different sustainability discourses, and how the just sustainability framework provides one way to integrate EJ and sustainability.

Questions of method and methodology – the broad range of considerations about how to collect, analyse, interpret and apply data – have been central to EJ research throughout its history, and these constitute the focus of Part II. The chapters in this section consider quantitative, qualitative and mixed methods, and collectively they address many of the greatest challenges for conducting quantitative EJ research: addressing the distinctive statistical complexities associated with spatial data, disentangling causal processes underlying patterns of injustice, characterizing relationships between environmental hazards or amenities and their human health impacts, establishing appropriate relationships between researchers and communities, and implementing transdisciplinary partnerships. The first three focus on spatial and statistical methods for analysing geographic distributions of people and environmental risks. Jayajit Chakraborty begins the section by tracing the evolution of quantitative approaches to assessing environmental exposure and risk, highlighting both their increasing sophistication and the key challenges that remain. One of the persistent challenges in quantitative EJ analysis is estimating population counts and demographic characteristics in environmentally burdened or deprived areas, and the next chapter by Juliana Maantay and Andrew Maroko describes and compares the strengths and limitations of the most prominent estimation techniques. Another methodological challenge is that spatial data violate the assumptions of conventional statistical techniques; this has required the development and application of an array of distinctive spatial statistical methods in EJ research, reviewed and assessed in the subsequent chapter by Jeremy Mennis and Megan Heckert.

Other questions in EJ research have lent themselves to qualitative or mixed methods, the focus of the next four chapters in Part II. To address questions about the processes underlying present-day patterns of environmental inequality, EJ researchers have turned to archival, ethnographic, or longitudinal methods of historical analysis, critically evaluated by Christopher Boone and Geoffrey Buckley. With respect to the experience and politics of environmental injustice, Catalina de Onís and Phaedra C. Pezzullo contend that ethnographic methods have been of particular importance, because of both their ethical commitment to dialogue between researchers and participants and their attention to embodiment and experiential knowledge. Donna Houston and Pavithra Vasudevan introduce approaches to EJ grounded in cultural studies, focusing on the analysis of storytelling as a central practice in EJ activism and scholarship. The challenges of using mixed methods are perhaps at their most acute in interdisciplinary and transdisciplinary EJ research, as Jonathan London, Julie Sze and Mary Cadenasso demonstrate in their critical reflection on the strengths and difficulties associated with different approaches to collaboration across disciplines.

The last three chapters in Part II review innovations in some of the most important methods for assessing and helping to address health risks and other burdens facing disproportionately affected communities. One such method is cumulative risk assessment (CRA); as Ken Sexton and Stephen Linder show, CRA carries much promise for providing a more rigorous basis for policy decisions than traditional risk assessment, but it remains a framework very much in development. Other methods aim explicitly to transfer greater control over the research process to affected communities themselves, and the chapter by Sacoby Wilson, Aaron Aber, Lindsey Wright and Vivek Ravichandran assesses the potentials and pitfalls of a range of models for community-based participatory research. To close out the section, Muki Haklay and Louise Francis examine how two of these models – participatory geographic information systems (GIS) and citizen science – can be combined and applied in addressing community concerns about environmental justice and sustainability.

Part III shifts the focus to the many different substantive issues that have been the subject of EJ activism, empirical research, and policy development. Although all of these chapters deal with cross-cutting themes, we have arranged them loosely into three groups. The first group of chapters addresses some of the key environmental problems at the heart of concerns about environmental injustice, also considering their relationships with policy and practice. Troy Abel and Mark Stephan begin by introducing a framework that relates waves of research on toxic and hazardous waste disparities with the development of public policy in the US. Michael Buzzelli follows with a discussion of air pollution as an EJ issue, reflecting on the complications associated with both research design and the development of appropriate policy. Leila Harris, Scott McKenzie, Lucy Rodina, Sameer Shah, and Nicole Wilson provide a wide-ranging review of approaches to EJ concerns surrounding freshwater, considering disparities in water quality and access along the axes of gender and indigeneity, as well as the intersections of race, income, and other categories. In the next chapter, Timothy Collins and Sara Grineski find that flood hazards complicate the assumptions and approaches of conventional distributive EJ research in surprising ways, and they introduce a conceptual framework to attend to the distinctiveness of risks and inequities associated with flooding. This section concludes with a chapter by Philip Coventry and Chukwumerije Okereke on the justice dimensions of what has become the most prominent international environmental issue of our time – global climate change – examining the intertwined ways in which they have developed in scholarship, activism, and policy.

The second group within Part III highlights some of the economic activities associated with environmental injustice, dealing with aspects of production, consumption, and distribution. First, Leire Urkidi and Mariana Walter lay out a framework for the study of EJ and large-scale metal mining, reviewing the complexities of its biophysical, economic, and political aspects. The following chapter, by Karen Bickerstaff, considers the relations of distributive, procedural, and recognition justice dimensions within low carbon energy systems, focusing on the controversial case of nuclear power. Next, Alex Karner, Aaron Golub, Karel Martens, and Glenn Robinson provide a critical overview of EJ and transportation, emphasizing inequities in participation in planning processes, exposure to environmental burdens, and access to benefits and amenities. Alison Alkon traces the roots of food justice, which links EJ's emphasis on the inequitable distribution of benefits and harms with alternative agriculture's critique of industrial food systems and food scholarship's critical reflections on the links between food and identity. A final cross-cutting economic theme considered here is environmental crime, and Michael Lynch and Kimberly Barrett emphasize scholarship examining how criminal activity that harms nonhumans has intersected with conditions of environmental injustice for humans.

As the EJ movement has evolved and diversified, its focus on the inequitable impacts of waste and pollution has broadened to encompass a host of environmental "goods" as well, including amenities like parks and greenspace, along with efforts to protect and preserve biodiversity and wilderness. This theme unites the last three chapters in Part III. Jason Byrne reviews scholarship on the EJ dimensions of urban parks and greenspace, considering not only dimensions of activism and policy but also intersections with climate injustice, food insecurity, and ecosystem services. In a closely related chapter focused on planning and community development, Isabelle Anguelovski, Anna Brand, Eric Chu, and Kian Goh examine injustices associated with urban greening – such as ecological gentrification – and discuss emerging cases of resistance to inequitable approaches to urban sustainability. Finally, moving outside the urban realm, Maureen Reed and Colleen George consider EJ with respect to nature conservation and wilderness preservation, focusing on injustices and inequities associated with the establishment of biosphere reserves.

Part IV addresses the diverse geographic contexts of EJ, beginning with discussions of global and transnational concerns and themes and closing with a series of chapters focused on particular regions or countries. Among the key actors shaping patterns of environmental injustice at the global scale are multinational corporations, and Ruchi Anand begins the section with a discussion of these corporations and the distinctive legal and institutional context in which they operate. Leah Temper follows with a chapter about the global diversity of EJ activism, considering a set of common themes that both unite contemporary struggles and link them with historical environmental movements. Among the most important transnational EJ movements are movements organized around indigeneity, the shared focus of the next two chapters. A chapter by Alana Shaw provides an overview of Indigenous EJ scholarship, activism, and policy in the rapidly changing transnational region of the Arctic. Of course, Indigenous EJ activism also takes on unique valences in particular settings; in their contribution, Donna Green, Marianne Sullivan, and Karrina Nolan examine the issues distinctive to mining initiatives in Aboriginal Australia.

The remaining chapters, reflecting a range of different approaches, address research on EJ as it has emerged in other locales beyond the term's North American origins. The first group highlights three broad themes of significance to Latin America: borders, international alliances, and urbanization. The boundary between the US and Mexico, which by many definitions constitutes the northern border of Latin America, has become paradigmatic in the study of EJ issues distinctive to border regions in an era of globalization, and Sara Grineski and Tim Collins analyse this region using the lens of world systems theory. The broader region of Latin America and the Caribbean has recently hosted one of the world's most remarkable international alliances, among a set of countries with governments committed to the advancement of socialism; Karen Bell's chapter critically assesses the strengths and limitations of this alliance's policy approaches to achieving EJ. Finally, Latin America and the Caribbean has become one of the most urbanized regions in the world, and the chapter by Alexis Vásquez, Michael Lukas, Marcela Salgado, and José Mayorga reviews the growing scholarship on urban EJ issues in Chile – also one of the world's most urbanized countries.

In the second group, the chapters focus on either a single country or small groups of countries, highlighting the distinctive EJ issues that arise with their positions in a global economy. The first chapter, by Rhuks Ako and Damilola Olawuyi, reviews the unique politics of EJ surrounding oil production in the Niger Delta in Nigeria, where competing claims to justice have frequently been intertwined with violent conflict. This is followed by Patrick Bond's examination of South Africa and how involved 'sub-colonial' processes have served to produce and sustain environmental injustice through into the post-Apartheid period. The section closes with a look at one of the world's wealthiest countries, Australia; David

Schlosberg, Lauren Rickards, and Jason Byrne argue that its unique positioning, history, and circumstances have made it one in which place attachment has become particularly salient to EJ conflicts and struggles.

The final group of chapters is dedicated to Eurasia, the physiographic continent in which a large majority of the world's population lives; here the approach is to consider how diverse regions within the landmass are linked by shared EJ issues and concerns. Pratyusha Basu argues that in South and Southeast Asia, attention to EJ has been a product primarily of vibrant social movement struggle, and that this struggle has been defined heavily by divisions between growing urban areas and rural areas that remain heavily populated. Mei-Fang Fan and Kuei-Tien Chou examine EJ in several of the countries of Pacific East Asia, emphasizing how the uneven transition to democracy among these states has conditioned the dynamics of EJ politics in the region. Heike Köckler, Séverine Deguen, Andrea Ranzi, Anders Melin and Gordon Walker examine discourses of EJ as they have unevenly emerged in Western Europe, reviewing research that has focused on both patterns of distributional inequalities at national and regional scales, and public participation within the context of the Aarhus Convention. In the closing chapter, Tamara Steger, Richard Filčák, and Krista Harper consider the post-socialist context of Central and Eastern Europe, where dynamics of right-wing resurgence have intensified EJ issues for Roma communities.

Conclusion

The discourse of environmental justice continues to transform and evolve, and it is our hope that this *Handbook* will help set research agendas for the next generation of EJ scholarship. Thanks to the pioneering efforts of the first generation of EJ scholars in the 1980s and 1990s – themselves taking continual inspiration and cues from the tireless work of grassroots activists – we now have a rich and rigorous theoretical, methodological and empirical foundation from which to build.

Nonetheless, the persistence and proliferation of environmental injustices throughout the world demonstrate that tremendous challenges remain. For example, although research helped propel EJ to the US federal policy agenda in the early 1990s, and although the work of scientists played a central role in bringing attention to the environmental injustice of contaminated water in Flint, Pulido's chapter is a sobering reminder that producing knowledge of environmental injustices has too often fallen short in helping rectify them. As important as it has been to develop and refine normative and analytical theories of EJ, we must continue asking how we might apply these theories, not only to the shaping of laws and policies, but also to the growth of "best practices" in movement activism. We should continue striving to make our research methods ever more sophisticated, but we must not lose sight of the ultimate objective to create more fair and equitable environmental relations. Although we need to continue attending to the classic concerns of the EJ movement, we must continue listening as the discourse expands to encompass new issues. Finally, we must turn our attention not only to the contexts in which EJ discourse has only recently emerged, but also to the new transnational connections which EJ is enabling activists and policymakers to make. To put it bluntly, the goal for the future should be a world in which the achievement of environmental justice has made this handbook obsolete.

References

Agyeman, J. & Ogneva-Himmelberger, Y. (2009). *Environmental justice and sustainability in the former Soviet Union*. Cambridge, MA: MIT Press.

Bryant, B. & Mohai, P. (1992). *Race and the incidence of environmental hazards: A time for discourse*. Boulder, Colorado: Westview Press.

Bullard, R.D. (1983). "Solid waste sites and the black Houston community." *Sociological Inquiry*, vol. 53, pp. 273–288.

Cable, S., Mix, T., & Hastings, D. (2005). "Mission impossible? Environmental justice activists' collaborations with professional environmentalists and with academics." In Pellow, D.N. & R.J. Brulle (eds.), *Power, justice and the environment: A critical appraisal of the environmental justice movement*. Cambridge MA: MIT Press, pp. 55–76.

Čapek, S.M. (1993). "The environmental justice frame – a conceptual discussion and an application." *Social Problems*, vol. 40, pp. 5–24.

Carmin, J. & Agyeman, J. (2011). *Environmental injustice beyond borders: Local perspectives on global inequities*. Boston: MIT Press.

Carruthers, D.V. (2008). *Environmental justice in Latin America: Problems, promise, and practice*. Cambridge, MA: MIT Press.

Holifield, R. (2001). "Defining environmental justice and environmental racism." *Urban Geography*, vol. 22, pp. 78–90.

Jamal, T. & Hales, R. (2016). "Performative justice: new directions in environmental and social justice." *Geoforum*, vol. 76, pp. 176–180.

Kubanza, N.S. & Simatele, D. (2016). "Social and environmental injustices in solid waste management in sub-Saharan Africa: a study of Kinshasa, the Democratic Republic of Congo." *Local Environment*, vol. 21, pp. 866–882.

Kuletz, V. (2002). "The movement for environmental justice in the Pacific Islands." In Adamson, J., M. Evans, and R. Stein (eds.), *The environmental justice reader: Poetics, politics, and pedagogy*. Tucson: University of Arizona Press, pp. 125–142.

Makene, M.H., Emel, J., & Murphy, J.T. (2012). "Calling for justice in the goldfields of Tanzania." *Resources*, vol. 1, pp. 3–22.

Margai, F.M. & Barry, F.B. (2011). "Global geographies of environmental injustice and health: a case study of illegal hazardous waste dumping in Cote d'Ivoire." In Maantay, J. and S. McLafferty (eds.), *Geospatial Analysis of Environmental Health*. Dordrecht: Springer Netherlands, pp. 257–281.

Martin, A., Gross-Camp, N., Kebede, B., McGuire, S., & Munyarukaza, J. (2014). "Whose environmental justice? Exploring local and global perspectives in a payments for ecosystem services scheme in Rwanda." *Geoforum*, vol. 54, pp. 167–177.

McDonald, D.A. (2002). *Environmental Justice in South Africa*. Cape Town: University of Cape Town Press.

Mennis, J., Stahler, G., & Mason, M. (2016). "Risky substance use environments and addiction: a new frontier for environmental justice research." *International Journal of Environmental Research and Public Health*, vol. 13, pp. 607; doi:10.3390/ijerph13060607.

Myers, G.A. (2008). "Sustainable development and environmental justice in African cities." *Geography Compass*, vol. 2, pp. 695–708.

Omer, I. & Or, U. (2005). "Distributive environmental justice in the city: differential access in two mixed Israeli cities." *Tijdschrift voor economische en sociale geografie*, vol. 96, pp. 433–443.

Phillips, C.V. & Sexton, K. (1999). "Science and policy implications of defining environmental justice." *Journal of Exposure Analysis and Environmental Epidemiology*, vol. 9, pp. 9–17.

Reed, M.G. & George, C. (2011). "Where in the world is environmental justice?" *Progress in Human Geography*, vol. 35, pp. 835–842.

Schlosberg, D. (2007). *Defining environmental justice: Theories, movements and nature*. Oxford: Oxford University Press.

Schroeder, R., Martin, K.S., Wilson, B., and Sen, D. (2008). "Third World environmental justice." *Society & Natural Resources*, vol. 21, pp. 547–555.

Sikor, T. (2013). *The justices and injustices of ecosystem services*. London: Earthscan.

Sze, J. & London, J.K. (2008). "Environmental justice at the crossroads." *Sociology Compass*, vol. 2, pp. 1331–1354.

Taylor, D.E. (2000). "The rise of the environmental justice paradigm: injustice framing and the social construction of environmental discourses." *American Behavioral Scientist*, vol. 43, pp. 508–580.

Tschakert, P. (2009). "Digging deep for justice: a radical re-imagination of the artisanal gold mining sector in Ghana." *Antipode*, vol. 41, pp. 706–740.

United Church of Christ (1987). *Toxic waste and race in the United States*. New York: United Church of Christ.

US General Accounting Office (1983). *Siting of hazardous waste landfills and their correlation with racial and economic status of surrounding communities*. Washington DC: USGAO.

Velicu, I. & Kaika, M. (2015). "Undoing environmental justice: re-imagining equality in the Rosia Montana anti-mining movement." *Geoforum*, http://dx.doi.org/10.1016/j.geoforum.2015.10.012

Walker, G. (2009). "Globalising environmental justice: the geography and politics of frame contextualisation and evolution." *Global Social Issues*, vol. 9, pp. 355–382.

Walker, G.P. (2012). *Environmental justice: Concepts, evidence and politics*. Abingdon: Routledge.

Walker, G. & Bulkeley, H. (2006). "Geographies of environmental justice." *Geoforum*, vol. 37, pp. 655–659.

PART I

Situating, analysing and theorizing environmental justice

2

HISTORICIZING THE PERSONAL AND THE POLITICAL

Evolving racial formations and the environmental justice movement

Laura Pulido

Minorities and the environment

In 1985 I was working on a master's thesis in geography at the University of Wisconsin. Seeking to bridge my interests in Chicana/o Studies and environmentalism, my thesis examined farmworkers' experiences and attitudes towards pesticides in California's San Joaquin Valley using a survey. I struggled to weave together these disparate intellectual traditions, as no one had previously done so. Exacerbating the situation was the fact that I had never heard of environmental justice (EJ). In fact, the concept was just developing in North Carolina at the time and was relatively localized (this was before the internet). Having scoured the library for research that addressed the relationship(s) between people of colour and environmental issues, the most robust literature I found, if one could call it that, explored park usage patterns of various ethnic groups. Unsure what to call my field of study, I framed it as, "Minorities and the Environment".

By the late 1980s the situation had changed remarkably. I had completed my MS and transferred to the Urban Planning programme at UCLA and the idea of environmental justice was spreading rapidly. My dissertation research, building on my thesis, investigated how working class Mexican Americans understood and mobilized around environmental issues. Looking back, I cannot believe my good fortune. Within a few short years my position had changed from being an intellectual outlier to being on the cutting-edge of a new field, environmental justice.

In this chapter I reflect on my engagement with EJ scholarship as well as the literature's minimal attention to racial formation. I first discuss my personal history with EJ, and then argue that ignoring the changing nature of the US racial formation and its implications for the EJ movement is a major oversight. This is because such shifts influence what the EJ movement is (un)able to accomplish. I focus in particular on the degree to which the EJ movement has been able to improve the environments of vulnerable communities, given larger structural changes. While researchers have widely acknowledged the political economic changes associated with neoliberalism, race has been treated as a stable playing field. Although my chapter focuses on the US, hopefully such an analysis will prove useful to other places. Not only do we live in a global racial formation (da Silva 2007; Winant 2002), but racial dynamics are geographically distinct. Accordingly, more nuanced studies of race at multiple scales are needed.

Environmental justice

In the late 1980s Los Angeles witnessed several high-profile environmental justice campaigns involving African Americans and Latinas/os (Pulido 1997). I joined the Labor/Community Strategy Center (LCSC), which challenged the oil refineries in the harbour area and focused on air pollution more generally (Mann 1991). The LCSC's roots were in labour and consequently adopted a class-based anti-racist analysis of air pollution, which resonated deeply with me. Collectively, we developed a framework that privileged power and informed my dissertation. It was incredibly exciting to be involved in both building a grassroots movement *and* in building a new field of scholarship. At the time there were only two Chicana/o Studies scholars working on environmental issues, Devon Peña and myself. Although Peña studied very different places and environments, Chicanas/os in the Upper Rio Grande, I learned a great deal from him and was inspired by his intellectual creativity and courage (Peña 1999).

In reviewing my intellectual trajectory, I cannot separate my early scholarly development from the larger EJ movement. Not only did the movement provide a language and framework that I previously lacked, but it gave me energy, confidence, and power. This was important because it was difficult to be in Chicana/o Studies and do environmental work in the 1980s. Most colleagues considered it relatively unimportant compared to labour, poverty, and immigration. The EJ movement affirmed my efforts and offered a sense of belonging.

But the EJ movement also pushed me intellectually. Specifically, it enabled me to engage more directly with other communities of colour. Through the EJ movement I met African Americans, Asian Americans and Native Americans – people and communities that I may not have otherwise met. This had a profound effect on me politically and intellectually and drove my research towards explicitly comparative and relational ethnic studies (Pulido 1997; 2006; Kun and Pulido 2013).

Within environmental justice I have been particularly drawn to two questions. First, how are race and racism conceptualized within the literature, policy and movement (Pulido 2000)? And second, how do working class communities of colour within the US respond to and mobilize around environmental problems (Pulido 1996)? I was drawn to the first issue because I was poorly trained on the subject of race and desperately needed to understand it. In truth, I think few social scientists understood race well at the time (this was just prior to the social construction of race migrating from the humanities into the social sciences), but that did not preclude many researchers from conducting analyses on the subject. For me, this was a huge issue in need of interrogation. How could we possibly move forward if we were not clear on one of the fundamental processes creating environmental injustice? Second, through both my activism and scholarship I came to believe that real social change only comes from below. I admit that I was initially attracted to oppositional struggles through largely romantic ideas (Pulido 2008). While the romance has long since faded, I am more convinced than ever that power concedes nothing without struggle, and that for all their messiness and disappointments, social movements, including massive shifts in political consciousness, are the *only* ways to create meaningful change.

I drifted away from EJ for a number of years, but my passion for the topic, as well as the urgency of climate change, and a new concern with domination (versus resistance), lured me back. As I began to re-immerse myself in the world of EJ politics and scholarship, I began questioning the movement's ability to actually improve the environmental conditions of vulnerable communities. Two things triggered this interest. First was the story of Exide Technologies, a Los Angeles area battery recycling facility whose regulatory failure was breathtaking. The facility had been illegally contaminating over 100 000 primarily working-class Latinas/os for decades (Pulido 2015). At roughly the same time, I attended an EJ workshop at UCLA at which Lisa Jackson, then director of the Environmental Protection Agency (EPA), was present. Frankly, I was appalled at the discussion. Nobody was seriously challenging Jackson given the egregious kinds of violations that were occurring. Juxtaposing these two events forced me to reconsider the efficacy of the larger movement. Certainly the EJ movement has accomplished a great deal over the past decades, especially blocking new hazardous projects and the expansion of existing ones, but how has the movement dealt with *existing* pollution?

Has the EJ movement improved environmental quality?

To my knowledge, there is no clear answer to this question because no one has systematically studied it. Obviously, it is beyond the scope of this chapter to attempt a full answer, but as a first step, I turn to two bodies of literature. The first considers the policy and legal accomplishments of the EJ movement (also see Chapter 12). I have elaborated on this literature more fully elsewhere (Pulido et al. 2016). Here, I review briefly four relevant registers. The first one consists of EJ lawsuits based on the Equal Protection clause of the 14th amendment. To date eight lawsuits have been filed. All have failed. A second register is Title VI complaints. According to the EPA website, as of January 2014, 298 Title VI complaints had been filed. My analysis of the website indicated that only one has been upheld (see also Mank 2008; Gordon and Harley 2005). This is a success rate of .3 per cent. A third arena of activity is the actual implementation of Executive Order (EO) 12898. The EO is supposed to compel all federal agencies to incorporate EJ considerations into their activities. In 2003 the Civil Rights Commission examined the extent to which four critical agencies had done so: the EPA, Housing and Urban Development, the Department of Transportation, and the Department of Interior. It concluded that all four had largely failed to do so. Subsequent studies by Guana (2015) and Noonan (2015) provide more detailed analyses of precisely how environmental agencies fail to address EJ concerns. The fourth area is scholarship that explores the (in)equitable nature of environmental enforcement. While the overall literature is mixed (Ringquist 1998; Atlas 2001), there is compelling evidence that it is not equitable (Konisky 2009; Konisky and Reenock 2013; Lynch et al. 2004; Mennis 2005; Lavelle and Coyle 1992), especially where low-income Latinas/os are concerned.

The second body of literature that is key to assessing the efficacy of the EJ movement is research that analyses the political culture of the movement itself. Though I have always considered the movement to be politically diverse, fractures are increasingly visible. Carter (2016), in his article "Environmental Justice 2.0?", suggests that the EJ movement has become far less oppositional and is now focused on "quality of life" issues, such as parks and other amenities. While I believe that Carter understates the significance of current industrial pollution and environmental violations, he is correct in pointing out the less than oppositional nature of large strands of the movement. Perkins (2015) has documented how California activists, through disruptive tactics, initially achieved access and eventually became part of the state. However, this has not led to meaningful change in existing environmental conditions, largely because of neoliberal politics. Other scholars have begun documenting co-optation and the extent to which the state, at multiple levels, has increasingly favoured industry (Lievanos 2012; London et al. 2013; Kohl 2014; Harrison 2015; Pulido 2015). In short, this small but critical literature suggests that by participating in state-led processes, the EJ movement has lost much of its oppositional content and character.

There is no doubt that neoliberalism has been the primary force behind these shifts (see Heynen et al. 2007, more generally). Faber (2008, also Chapter 6) has detailed how state agencies have been captured by industry, in what he calls the "pollution industrial complex". The power of this complex is evident in EPA appointments, its interpretation of regulations and rule-setting, its priorities, and the utter failure of laws like the Toxic Substances and Control Act (Silbergeld et al. 2015). Holifield (2007) has argued that EO 12898 is classic "roll-out" neoliberalism in that it replaces substantive regulatory power with superficial and/or voluntary policies. Indeed, EO 12898 encourages community empowerment, citizen participation, and data collection – with no real teeth or enforcement mechanism. More recently, Harrison's (2015) study of EJ grant programmes shows the extent to which state bureaucrats have internalized neoliberal ideas regarding EJ and actively discipline activists by funding *non*oppositional projects, preferring instead those that emphasize individual behaviour modification and/or working towards solutions with industry. This research is vitally important as it enables us to discern precisely how neoliberal shifts have been implemented. Missing, however, are detailed analyses of how race and neoliberalism inform each other. Lisa Duggan (2003) has argued that it is impossible to separate race from neoliberalism. As a central power relation in the US, racism creates a textured landscape that capital can exploit in order to accumulate greater power and profits. Thus, it is hardly surprising that racism is deployed to facilitate various economic projects (see, for example, Gilmore 2007; López 2014; Inwood 2015; Pulido 2016). Indeed, this is the essence of racial capitalism – race as a structuring logic of capitalist processes (Robinson 2000). Given the importance of racism to economic relations, we must be cognizant of how the racial formation has shifted under neoliberalism and its implications for EJ.

The EJ movement and shifting racial formations

To clarify how the racial formation has shifted I draw on the work of Jodi Melamed. She argues that instead of conceptualizing the US state as breaking with white supremacy after World War Two (Winant 2002), we should see it as adopting various anti-racisms. Such forms of racial hegemony have "enabled the normalizing violences of political and economic modernity to advance and expand" (Melamed 2011: 5). Specifically, she identifies three eras of state anti-racisms: racial liberalism (1940s–1960s), liberal multiculturalism (1980s–1990s), and neoliberal multiculturalism (2000–present). According to Melamed's framework, the EJ movement arose during liberal multiculturalism and continued into neoliberal multiculturalism. So, for example, the 1980s saw the proliferation of multiculturalism not only in the literary canon, but also in environmental politics. Indeed, my first book can be seen as one such example (Pulido 1996), as it illuminated the unique relationships that working class Chicanas/os had to environmental issues. What is important about multi-culturalism for our purposes is that it arose as a "counterinsurgency . . . against the materialist anti-racisms of the 1960s' and 1970s' . . ." (Melamed 2011: 93). In short, it was meant to "disable . . . effective antiracism . . ." (ibid: 92). It is vitally important to understand the 1980s: this decade contains both the last vestiges of the energy of the Civil Rights Movement

(CRM) – which I believe gave rise to the EJ – and the seeds of its backlash. It is a decade of transition (Johnson 1991).

Environmental justice and the civil rights movement

The deep connection between the early EJ movement and the CRM has been well documented (Foster and Cole 2001: chpt. 1). The CRM is present in the EJ movement's origin story in Warren County, North Carolina in 1982 when the NAACP assisted local residents fighting a dump and contaminated soil (McGurty 2007). The struggle in Warren County attracted significant attention from African American leaders who had cut their teeth on the CRM, including Benjamin Chavis, Walter Fauntroy, Leon White and Fred Taylor. By 1992 EJ legislation had been introduced by John Lewis, a Democrat congressperson from Georgia. This effort clearly built on earlier civil rights strategies: force the federal government to recognize the problem and demand amelioration. Although legislative efforts never gained traction, President Clinton did issue Executive Order (EO) 12898 in 1994, as previously mentioned.

It is hardly surprising that EJ leaders, especially African Americans, would draw upon conventional CRM strategies. Not only were they steeped in the culture of the CRM, but they had achieved significant change via the state, including the Civil Rights Act (1964), the Voting Rights Act (1965) and the Equal Employment Opportunity Act (1972). Such accomplishments were the result of racial liberalism and contributed to a change in public attitudes and practices. EJ was fuelled by the last embers of the CRM. Already by the 1980s civil rights leaders encountered decreasing levels of support and growing opposition. Indeed, the Bakke decision, challenging affirmative action in university admissions, came in 1978. That EJ united two of the great movements of the twentieth century, environmentalism and civil rights, and refashioned them in a new and politically productive way was a strategic move that partially countered growing opposition to anti-racism. Not only did EJ provide evidence of a whole new realm of inequality, but it also created a bridge to mainstream environmentalism, however rocky at first. These were critical steps in bolstering and expanding anti-racist forces. Nonetheless, it was precisely the twin pillars of EJ, anti-racism and environmentalism, which would be subject to attack in an increasingly neoliberal era.

The resurgence of the white nation and neoliberal multiculturalism

As the US transitioned from liberal multiculturalism to neoliberal multiculturalism, the moral authority of the CRM was replaced by explicitly racist discourses and projects, such as Black criminalization, the prison industrial complex, the war on drugs, and the criminalization of immigrants. Numerous scholars have detailed the precise economic function of such projects (Gilmore 2007; López 2014; Duggan 2003). Though many racial projects enable neoliberalism, racism cannot be reduced to economic imperatives. A good example of this is the resurgence of the white nation (Bjork-James 2015). The white nation has been instrumental in eroding support for anti-racist initiatives and therefore is essential to understanding the larger racial landscape under which EJ activists labour. Moreover, while the white nation facilitates neoliberalism by rationalizing growing inequality, it is in no way contained by it. As I write, Republican presidential candidate Donald Trump's racist comments against Muslims and Mexicans attract ever more voters – much to the chagrin of the Republican Party.

The white nation refers not just to white racism, but a racism linked to a presumed ownership of the nation (Gerstle 2002). While the US has historically been defined by white nationalism, as seen, for example, through citizenship (López 1996) and territorial expansion (Horsman 1981), it was seriously challenged by the CRM and more recently by widespread demographic change. Together, these shifts have resulted not only in fewer whites, but whites increasingly feeling that they are losing control of *their* country (Huntington 2004). This, in turn, has led to a massive white backlash that has fundamentally altered the national racial formation. The resurgence of the white nation has become an entrenched part of contemporary political culture and is evident in the daily headlines, including, for example, opposition to Central American immigrant minors; Republican efforts to limit voting rights; the rise of the Tea Party and its desire to "take back" the country; and "birthers" who insist that Barack Obama was not born in the US. Indeed, the current fractures in the Republican Party are in no small part fuelled by the anxiety, anger, and activism of the white nation, in addition to economic insecurity. While some suggest that the anger is a misdirected response to economic precarity, they are only partially correct. At its root such thinking assumes that racism is an aberration or an epiphenomenon to the economy, rather than a formative feature of the US social formation (Dunbar-Ortiz 2014; Smith 2012).

This latest resurgence of the white nation is typically located in the 1980s, but it actually began in the 1950s with the Republican Party's Southern Strategy. The Southern Strategy, which developed in response to the CRM, sought to revitalize the party by appealing to southern white racism. As the Democratic Party became increasingly associated with racial integration, the Republican Party saw an opportunity to realign national electoral politics. By 1981 Lee Atwater was able to look backwards and explain the movement's rhetorical evolution:

> You start out in 1954 by saying, 'Nigger, nigger, nigger.' By 1968, you can't say 'nigger,' that hurts you. Backfires. So you say stuff like forced bussing, states' rights and all that stuff. You're getting so abstract now [that] you're talking about cutting taxes, and all these things you're talking about are totally economic things and a byproduct of them is [that] blacks get hurt worse than whites … we are doing away with the racial problem one way or another. You follow me.
>
> *(quoted in Woods 2017: 232)*

The Southern Strategy accomplished three important things in terms of EJ's historical context. First, it legitimized the tremendous anxiety and fear that whites, particularly white southerners, felt in response to racial integration. Second, it introduced a new, 'coded' way of talking about race (López 2014). And third, it assembled a broad coalition that went far beyond the South (Inwood 2015). According to Woods (2017):

> The national Republican party's Southern Strategy crafted an alliance composed of plantation blocs, major business sectors, white flight suburbs, rural whites, working class white urban Catholics, several unions, and White Citizens Council leaders. The new electoral alliance also supported a restructuring of urban, rural, regional, gender, and class relations. The institutions of national and global capitalism were also reformed under the banners of privatization, deregulation, and neoliberalism.

The ballast of the Southern Strategy has been long and powerful and has been instrumental in distracting attention from growing levels of economic insecurity associated with neo-liberalism, especially among whites, as well as endless war (De Genova 2012). Neoliberal multiculturalism is distinguished from its predecessor in that its sphere is increasingly global, it affirms the role of the market in organizing society, it is largely abstracted from specific groups and their struggles, and invokes a deracialized language. One of the hallmarks of neoliberal multiculturalism is the election of President Barack Obama. His election illustrates the degree to which anti-racism has been deracialized and disconnected from the larger mass of Black people. Mainstream discourse never mentions "racism", rather people talk about "difference", "bias" and "diversity". Those who claim polluters are criminals and racists are not allowed to participate in state initiatives. For instance, the EPA's Environmental Justice Collaborative Problem-Solving (CPS) programme is intended only for communities willing to engage in consensus building and dispute resolution.

Different forms of EJ activism can be discerned under neoliberal multiculturalism. One thread focuses on the cultural and racial diversification of environmentalism. Included are organizations that promote people of colour in mainstream environmentalism, as well as efforts to broaden the contours of environmentalism itself (Carter 2016). This includes highlighting the voices and experiences of those typically cast outside the environmental fold (Angulano et al. 2012). Although this concern has existed from the beginning of EJ, it appears to have grown relative to other sectors of the movement. Such initiatives are entirely in keeping with neoliberal multiculturalism: an emphasis on diversity, racial identities, inclusion and recognition. They are also largely devoid of significant material demands, including improving the actual physical environment.

A different example of neoliberal multicultural EJ can be seen in the American Legislative Exchange Council (ALEC). ALEC is a kind of lobbyist formation that actually brings industry and congressional leaders together to develop neoliberal legislation. It has a standing committee, "Energy, Environment and Agriculture", that has written over 70 pieces of legislation, and has even crafted its own "Environmental Justice Principles" and passed a "Resolution on Environmental Justice". The Principles state that everyone should be treated equally, while also rejecting any additional regulations. "Existing federal, state, and local regulation, properly implemented and enforced, are sufficient to assure protection" (Center for Media and Democracy nd). Of course, this is contrary to the essence of EJ – the idea that some communities are disproportionately impacted. But it exemplifies neoliberal multiculturalism by affirming equality, assuming the market is the best solution, and evincing completely deracinated language. Race is no longer a power relation, indeed, it is not even mentioned. By adopting EJ language, while simultaneously blocking any meaningful change, ALEC is ensuring that vulnerable communities will continue to be polluted, sickened, and die. When a group like ALEC adopts EJ in order to normalize violence, we have to ask, "what does EJ really mean?" Such a question can only be answered by developing critical EJ studies (Pellow and Brulle 2005), which, given its emphasis on power, would enable us to explore the dynamic nature of racial formations.

Conclusion

In this chapter I have argued for the need to analyse changes in the US racial formation. While many researchers have detailed the impact of neoliberalism on EJ, few have considered concomitant shifts in the racial formation. Yet, the EJ movement transcends at least two distinct eras of state anti-racisms, each embodying a distinct set of anti-racist possibilities. The EJ movement arose during liberal multiculturalism in the 1980s, as a last expression of the CRM, and has continued into the neoliberal multiculturalism of the twenty-first century. A close examination of both eras suggests that EJ activists have been labouring in an extremely hostile environment. This may explain why the EJ movement has achieved only limited gains in improving the physical environments of vulnerable communities.

While Melamed offers us a useful framework for understanding recent racial history, I suggest that the US racial formation is shifting once again. There is a resurgence of anti-racist activism as seen in Black Lives Matter, a deepening political consciousness on the part of everyone, including progressives, the left and the many rights. Thanks to Donald Trump, political correctness is a thing of the past, which enables a more honest picture of US racism to emerge. These are significant changes, especially when coupled with growing economic inequality and the fact that global warming can no longer be ignored. A radicalized EJ movement is precisely what is needed to provide critical grassroots leadership to the crisis at hand.

References

Angulano, C., Milstein, T., De Larkin, I., Chen, Y., and Sandoval, J. (2012). "Connecting Community Voices: Using a Latino/a Critical Race Theory Lens on Environmental Justice Advocacy." *Journal of International and Intercultural Communication*, vol. 5, pp. 124–43.

Atlas, M. (2001). "Rush to Judgment: An Empirical Analysis of Environmental Equity in U.S. Environmental Protection Agency Enforcement Actions." *Law & Society Review*, vol. 35, pp. 633–82.

Bjork-James, S. (2015). "The Charleston Shooter Has Plenty of Company on the Internet." *Racism Review*, June 20. www.racismreview.com/blog/author/sophie/

Carter, E. (2016). "Environmental Justice 2.0: New Latino Environmentalism in Los Angeles." *Local Environment: The International Journal of Justice and Sustainability* 21 (1): 3–23.

Center for Media and Democracy (nd) ALEC Exposed. Available at: www.alecexposed.org/wiki/ALEC_Exposed (accessed July 17, 2015).

da Silva, D. (2007). *Toward a Global Idea of Race*. Minneapolis: University of Minnesota.

De Genova, N. (2012). "The 'War on Terror' as Racial Crisis." In Martinez HoSang, D., O. LaBennett, and L. Pulido (eds.), *Racial Formation in the Twenty-First Century*. Berkeley: University of California Press, pp. 246–75.

Deloitte Consulting. (2011). Evaluation of the EPA Office of Civil Rights. Final Report, March 21. Washington, DC. Available at: http://epa.gov/epahome/ocr-statement/epa-ocr_20110321_final report.pdf (accessed July 13, 2015).

Duggan, L. (2003). *The Twilight of Equality: Neoliberalism, Cultural Politics, and the Attack on Democracy*. Boston: Beacon Press.

Dunbar-Ortiz, R. (2014). *An Indigenous Peoples' History of the United States*. New York: Basic Books.

Faber, D. (1998). "The Political Ecology of American Capitalism: New Challenges for the Environmental Justice Movement." In Faber, D. (ed.), *The Struggle for Ecological Democracy: Environmental Justice Movements in the United States*. New York: Guilford Press, pp. 27–59.

Faber, D. (2008). *Capitalizing on Environmental Injustice: The Polluter-Industrial Complex in the Age of Globalization*. Lanham, MD: Rowman & Littlefield Publishers.

Foster, S. & Cole, L. (2001). *From the Ground Up: Environmental Racism and the Rise of the Environmental Justice Movement*. New York: New York University Press.

Gerstle, G. (2002). *American Crucible: Race and Nation in the Twentieth Century*. Princeton: Princeton University Press.

Gilmore, R.W. (2007). *Golden Gulag: Prisons, Surplus, Crisis, and Opposition in Globalizing California*. Berkeley: University of California Press.

Gordon, H. & Harley, K. (2005). "Environmental Justice and the Legal System." In Pellow, D. and Brulle, R. (eds.), *Power, Justice and the Environment: A Critical Appraisal of the Environmental Justice Movement*. Cambridge: MIT Press, pp. 153–70.

Guana, E. (2015). "Federal Environmental Justice in Permitting." In Konisky, D. (ed.), *Failed Promises: Evaluating the Federal Government's Response to Environmental Justice*. Cambridge: The MIT Press, pp. 57–83.

Harrison, J. (2015). "Coopted Environmental Justice? Activists' Roles in Shaping EJ Policy Implementation." *Environmental Sociology* 1 (4): 241–55.

Heynen, N., McCarthy, J., and Robbins, P. (2007). *Neoliberal Environments: False Promises and Unnatural Consequences*. New York: Routledge.

Holifield, R. (2007). "Neoliberalism and Environmental Policy." In Heynen, N., J. McCarthy, and P. Robbins (eds.), *Neoliberal Environments: False Promises and Unnatural Consequences*. New York: Routledge, pp. 202–16.

Horsman, R. (1981). *Race and Manifest Destiny: The Origins of American Racial Anglo-Saxonism.* Cambridge: Harvard University Press.

Huntington, S. (2004). *Who Are We? The Challenges to America's National Identity.* New York: Simon & Schuster.

Inwood, J. (2015). "Neoliberal Racism: The 'Southern Strategy' and the Expanding Geographies of White Supremacy." *Social & Cultural Geography* 16 (4): 407–23.

Johnson, H. (1991). *Sleepwalking Through History: America in the Reagan Years.* New York: Anchor Books.

Kohl, E. (2014). "Surviving Unregulatable Places: Race, Gender, and Environmental Justice in Gainesville, GA." PhD thesis. Athens, GA: University of Georgia.

Konisky, D. (2009). "Inequities in Enforcement? Environmental Justice and Government Performance." *Journal of Policy Analysis and Management,* vol. 28, pp. 102–21.

Konisky, D. & Reenock, C. (2013). "Compliance Bias and Environmental (In)Justice." *The Journal of Politics,* vol. 75, pp. 506–19.

Konisky, D. & Reenock, C. (2015). "Evaluating Fairness in Environmental Regulatory Enforcement." In Konisky, D. (ed), *Failed Promises: Evaluating the Federal Government's Response to Environmental Justice.* Cambridge: The MIT Press, pp. 173–203.

Kun, J. & Pulido, L. (2013). *Black and Brown in Los Angeles: Beyond Conflict and Cooperation.* Berkeley: University of California Press.

Lavelle, M. & Coyle, M. (1992). "Unequal Protection: The Racial Divide in Environmental Law: A Special Investigation." *The National Law Journal,* vol. 15, pp. S1–S12.

Lievanos, R. (2012). "Certainty, Fairness, and Balance: State Resonance and Environmental Justice Policy Implementation." *Sociological Forum,* 27 (2): 481–503.

London, J., Karner, A., Sze, J., Rowan, D., Gambirrazio, G. and Niemeier, D. (2103). "Racing Climate Change: Collaboration and Conflict in California's Global Climate Change Policy Arena." *Global Environmental Change,* 23 (4): 791–99.

López, I.H. (2014). *Dog Whistle Politics: How Coded Racial Appeals Have Reinvented Racism and Wrecked the Middle Class.* New York: Oxford University Press.

López, I.H. (1996). *White by Law: The Legal Construction of Race.* New York: New York University Press.

Lynch, M., Stretesky, P. and Burns, R. (2004). "Determinants of Environmental Law Violation Fines Against Petroleum Refineries: Race, Ethnicity, Income and Aggregation Effects." *Society & Natural Resources,* vol. 17, 333–47.

Mank, B. (2008). "Title VI." In Gerrard, M. & S. Foster (eds.), *The Law of Environmental Justice: Theories and Procedures.* Chicago: American Bar Association, pp. 22–65.

Mann, E. (1991). *LA's Lethal Air.* Los Angeles: Labour/Community Strategy Center.

McGurty, E. (2007). *Transforming Environmentalism: Warren County, PCBs, and the Origins of Environmental Justice.* New Brunswick: Rutgers University Press.

Melamed, J. (2011). *Represent and Destroy: Rationalizing Violence in the New Racial Capitalism.* Minneapolis: University of Minnesota.

Mennis, J. (2005). "The Distribution and Enforcement of Air Polluting Facilities in New Jersey." *Professional Geographer,* vol. 57, 411–22.

Noonan, D. (2015). "Assessing the EPA's Experience with Equity in Standard Setting." In Konisky, D. (ed.), *Failed Promises: Evaluating the Federal Government's Response to Environmental Justice.* Cambridge: The MIT Press, pp. 85–116.

Pellow, D. & Brulle, R. (2005). *Power, Justice and the Environment: A Critical Appraisal of the Environmental Justice Movement.* Cambridge: The MIT Press.

Peña, D. (1999). *Chicano Culture, Ecology, Politics: Subversive Kin.* Tucson: University of Arizona Press.

Perkins, T. (2015). "From Conflict to Collaboration: The Evolving Relationship between California Environmental Justice Activists and the State, 1980s–2010s." Paper presented at the Association of American Geographers. Chicago.

Pulido, L. (1996). *Environmentalism and Economic Justice: Two Chicano Struggles in the Southwest.* Tucson: University of Arizona Press.

Pulido, L. (1997). "Community, Place and Identity." In Jones, J.P., H. Nast and S. Roberts (eds.), *Thresholds in Feminist Geography.* Lanham, MD: Rowman & Littlefield, pp. 11–28.

Pulido, L. (2000). "Rethinking Environmental Racism: White Privilege and Urban Development in Southern California." *Annals of the Association of American Geographers,* vol. 90 (1): 12–40.

Pulido, L. (2006). *Black, Brown, Yellow and Left: Radical Activism in Los Angeles.* Berkeley: University of California Press.

Pulido, L. (2008). "FAQs: Frequently Asked Questions on Being a Scholar/Activist." In Hale, C. (ed.), *Engaging Contradictions: Theory, Politics and Methods of Activist Scholarship.* Berkeley: University of California Press, pp. 341–66.

Pulido, L. (2015). "Geographies of Race and Ethnicity I: White Supremacy vs White Privilege in Environmental Racism Research." *Progress in Human Geography*, vol. 39 (6): 1–9.

Pulido, Laura (2016) "Flint Michigan, Environmental Racism and Racial Capitalism." *Capitalism Nature Socialism* 27 (3): 1–16.

Pulido, L., Kohl, E. and Cotton, N. (2016). "State Regulation and Environmental Justice: The Need for Strategy Reassessment." *Capitalism Nature Socialism* 27 (2): 12–31.

Ringquist, E. (1998). "A Question of Justice: Equity in Environmental Litigation, 1974–1991." *Journal of Politics*, vol. 60, pp. 1148–65.

Robinson, C. (2000). *Black Marxism: The Making of the Black Radical Tradition.* Chapel Hill: University of North Carolina Press (orig. pub. 1983).

Silbergeld, E., Mandrioli, D. and Cranor, C. (2015). "Regulating Chemicals: Law, Science, and the Unbearable Burdens of Regulation." *Annual Review of Public Health*, vol. 36, pp. 175–91.

Smith, A. (2012). "Indigeneity, Settler Colonialism, White Supremacy." In Martinez HoSang, D., O. LaBennett and L. Pulido (eds.), *Racial Formation in the Twenty-First Century.* Berkeley: University of California Press, pp. 66–93.

Winant, H. (2002). *The World is a Ghetto: Race and Democracy Since World War Two.* New York: Basic Books.

Woods, C. (2017). *Development Drowned and Reborn: The Blues and Bourbon Restorations in Post-Katrina New Orleans.* Athens: University of Georgia Press.

3

SOCIAL MOVEMENTS FOR ENVIRONMENTAL JUSTICE THROUGH THE LENS OF SOCIAL MOVEMENT THEORY

Diane M. Sicotte and Robert J. Brulle

Environmental justice movements (or EJMs) are environmental social movements confronting a diverse array of issues, including conflicts over resource extraction, pollution and contamination, and climate justice. EJMs are embedded in a web of other social movements including movements for racial equality and the rights of Indigenous people and the poor, farmers, workers and many others. While environmental social movement groups range widely in size from small informal groups to large organizations fully institutionalized into political systems, EJM groups tend to be grassroots groups at the smaller end of the scale (Brulle 2000). Eschewing the "Cult of Wilderness" that privileges pristine environments over the needs of people, EJM groups everywhere champion the "environmentalism of the poor" (Anguelovski and Martinez-Alier 2014). In this chapter, we focus on EJMs through the lens of social movement theory.

Social movement studies in sociology

Although a relatively new area of inquiry for sociologists, the study of social movements has grown steadily since its inception in the mid-1970s to become one of the largest subfields in US sociology, and, to a lesser degree, in others around the world. It emerged out of a critique of the dominant structuralist-functionalist view of "collective behaviour" as irrational and an aberration that disturbed the smooth functioning of society. By the late 1960s, this view had come to seem an outdated artefact of the 1950s as the rise of Civil Rights, anti-Vietnam war, feminist, gay liberation and reform-oriented environmentalism in the US challenged the distribution of rights and resources. Instead, sociologists began to view social movements as rational, self-interested and instrumental social action (McAdam and Boudet 2012).

Political economy (which emphasized a Marxist view of the relationship between social movements and systems of economic production) was the first theoretical perspective to challenge existing structuralist-functionalist theory (Tilly et al. 1975). It was soon followed by Resource Mobilization Theory, which held that activists must form organizations and gather resources such as adherents and money in order to effectively challenge existing arrangements (McCarthy and Zald 1977). Some Marxist-inspired researchers used historical work to emphasize the continuing importance of structural strains giving rise to collective behaviour, which they argued was not irrational but instead a disruptive tactic used in the social movement actions of poor people who lacked access to "resources" (Piven and Cloward 1977).

Political Process Theory (also called Political Opportunity Theory) soon supplied another perspective that challenged (or supplemented) Resource Mobilization Theory. Political Process Theory focused on elements of the political scene such as opportunity and threat, state repression, and elite alignments as factors either sparking or constraining mobilization (Tilly 1978; McAdam 1982). At present, the three perspectives above are the dominant theoretical perspectives in the social movement subfield of US sociology (McAdam and Boudet 2012).

Social movement theorists have criticized the methodology of social movement studies, claiming that researchers tend to focus exclusively on large, successful social movements rather than small or failed movements. They argue that this has obscured the true frequency of both mobilization and movement success, thus keeping researchers from understanding how external conditions affect the likelihood of mobilization (McAdam and Boudet 2012).

Emergence and nature of environmental justice movements

In the United States, EJMs emerged in the late 1970s from a critique of bureaucratic "reform" environmentalism (Brulle 2000), and of the wilderness conservation focus of environmental organizations, which ignored the environmental concerns of urban, poor, and racial/ethnic minority communities (Taylor 2000). Since that time, the discourse and issue focus of environmental justice has "gone global," and EJM groups are found throughout the world, although sometimes taking on a form different from that of the grassroots groups common to US EJMs (Walker 2012; see also Chapter 39).

Although EJMs first emerged in a wealthy Global North nation, EJM activism coalesced around environmental threats to human rights, health and safety. The focus of EJMs on these issues of survival, together with the rapid emergence of EJM groups in the Global South, contradict the "New Social Movements" theory of environmentalism, which holds that the emergence of new environmental movements was generated by the transition from scarcity to relative affluence and security (Inglehart 1977). While US environmental activists tend to be highly educated middle-class people, EJM activists tend to be drawn from a wide spectrum of society including the poor and marginalized (Rootes and Brulle 2013). Instead of regarding social inequality as a problem separate from environmental degradation, EJM activists hold that social inequality is closely related to (and even a cause of) degraded environments (Taylor 2000).

Although environmentalism provided an overarching framework for EJMs, they also originated from many other social movements. In urban areas, EJM groups originated from movements for urban hygiene and occupational and public health, which is reflected in their current inclusion as issues of environmental justice (Faber and McCarthy 2003; Gottlieb 2005; Taylor 1997). With the recent emphasis on energy justice and climate justice among EJM groups and actors, EJMs have expanded their focus to gas drilling, oil trains and pipelines transporting oil and gas, international treaties restraining greenhouse gas loading, and justice in mitigation and adaptation to climate change at all geographic scales (Roberts and Parks 2009).

This multi-issue focus has meant that, in every part of the world, EJM groups tend to be "hybrids," in which activist groups work on more than one issue, and are part of a network of activist groups. But social movement theorists disagree about the efficacy of hybrid activism. Some theorize that it enhances the strength of social movement groups because it stimulates recruitment of activists from other movements, and increases connections with other social movements (Heaney and Rojas 2014). Others argue that the focus and power of EJMs have been weakened because they attempt to address too many issues (Benford 2005).

Political process/political opportunity

Political Process (also called Political Opportunity) theorists argue that, prior to mobilization, there must first be some type of political opportunity that gives EJM actors hope that they can succeed (Tarrow 1994). Although it adds a vital contextual dimension to understandings of social change efforts, the concept of political opportunity has been critiqued for its elasticity, and the under-specification of the conditions under which it might give rise to or constrain mobilization (Meyer and Minkoff 2004). But in general, social movement theorists define political opportunity as increased or increasing access to the existing political system, divisions within the elite, the availability of elite allies, and decreases in violent repression by the state (McAdam 1996).

The timing of the emergence of EJM groups (in Argentina after the end of dictatorship in 1984, in Chile during the return to civilian rule in 1989, and in South Africa in the post-apartheid era of the early 1990s), seems to confirm the importance of reduced repression to EJM mobilization (Khan 2002; Urkidi and Walter 2011). In the US, the passage of Executive Order 12898 directing federal agencies to consider the disparate environmental impacts of their activities was signed by President Clinton in 1994. Although its passage was the result of sustained EJM activism, it also signalled that US EJMs had an elite ally, which seemed to coincide with increased EJM activity. This illustrates the "intrastate and dynamic" nature of political opportunity, in which social movement activists modify the actions of the state, creating their own political opportunities (Tarrow 1996).

However, David Pellow has critiqued this view of political opportunity as too focused on the state, and thus lacking an understanding of the entire context that confronts EJM groups (also see Chapter 4). The political and economic changes taking place from the 1970s to the mid-1990s have increased the power and influence of multinational corporations, while decreasing the relative power of nation-states (Pellow 2001). In the US, the oversized influence of corporate interests on the political system and close alliances between politicians and corporate campaign funders constitutes the major obstacle in the political opportunity structures confronting EJMs. Legislation limiting US greenhouse gas emissions has failed to pass, due to Republicans funded by fossil fuel corporations, factions in the Democratic Party dependent upon fossil fuels, and the political polarization of the public between climate change believers and climate "sceptics" (McCright and Dunlap 2011). But severe storms such as Hurricane Katrina and Superstorm Sandy constitute political opportunities to link costly and devastating disasters together with human-driven climate change, and to underscore the unequal vulnerability of poor people of colour when facing disaster (Caniglia et al. 2015; Bullard and Wright 2009).

Mobilization

For poor people, who are used to hardships, examples throughout history show that collective action as a response to even the worst provocations is relatively uncommon; thus, something beyond miserable conditions must take place in order for people to mobilize. For structural strain theorists, this "something" is a "sequence or combination of severe social dislocations" (Piven and Cloward 1977). This strand of social movement theory posits that social movements result from ruptures or strains in the sociopolitical order, and seeks to specify under what conditions these will result in activism (Snow et al. 1998).

The rupture or strain can take the form of an accident or disaster, the intrusion of an unwanted facility or process into the sacred territory of people's land, home, neighbourhood or family, or a disruption in people's subsistence routines (Snow et al. 1998). One or more of these are frequently the conditions around which EJMs form. Many groups recently brought into the EJM social movement family by the new focus on climate change (e.g., La Via Campesina) address threats to local survival that go beyond climate change, such as low prices for and limited access to the food grown by farmers (McKeon 2015).

McAdam and Boudet, who studied communities "at risk for mobilization" against energy-generating facilities in the US, theorize that the causes of mobilization for rights-based movements (such as Civil Rights, Anti-Apartheid or feminist movements) differ from those of EJMs (which they refer to as "NIMBY movements"). They argue that rights-based movements are sparked by cracks or ruptures in existing power structures, which constitute political opportunities for marginalized groups; EJMs are instead sparked by threats to local health or quality of life (McAdam and Boudet 2012).

Other theorists take issue with structural strain theorists, and instead focus on "resource mobilization," or the availability of resources (moral and cultural resources, as well as time, effort and money) as the strongest factor predicting collective action. These may be resources that come from within the movement, or they may come from sources external to the movement. Generally, resource mobilization theorists hold that larger, more formalized organizations allow social movement actors to aggregate resources, which increases the effectiveness of the social movement (Edwards and McCarthy 2004). Given this perspective, private funding from philanthropic foundations created to distribute monies from wealthy donors should be an important resource fuelling environmental justice activism; yet, in 2000, US environmental justice groups received only 1.5 per cent of all foundation grants given to environmental groups. The bulk of the money went to preservation and conservation groups, which lack EJMs' radical critique of social inequality (Brulle and Jenkins 2005). However, a group of activist funders reacted to this neglect of EJM groups by coordinating efforts to support EJM groups (Faber and McCarthy 2005).

McAdam and Boudet (2012) argue that all social movements begin as local struggles, but not all are successful enough to undergo a "scale shift" from local to national. The issue of scale is of much interest to scholars of EJMs, for both empirical and theoretical reasons: aside from the spatial ambiguities that are inherent in each effort to determine whether environmentally unjust conditions exist, scale also shapes how EJMs present themselves and their grievances (Kurtz 2003). But regardless of scale, McAdam and Boudet found that EJMs mattered greatly: the *absence* of EJM mobilization was highly correlated with the building of new energy-generating facilities (McAdam and Boudet 2012).

Collective identity and framing

Within the field of social movement studies, collective identity can be defined as a person's broader cognitive, moral and emotional connection with a broader community. Collective identity is theorized to be an important factor in many aspects of movements, including recruitment, retention, collective claims of injustice, tactical decision making, and the outcomes of social movement efforts (Polletta and Jasper 2001). Collective identities connected with histories of struggle against racism are very important aspects of EJM groups and movements in the US and South Africa (Bullard 1993; Capek 1993; Walker 2012).

But in order to be maintained, collective identities require activists to draw and police boundaries between their identity and that of others; thus, they can also have a divisive effect on social movements (Gamson 1995). In the US, the strong focus on racial/ethnic identities, and on EJMs as struggles against environmental racism, are powerful sources of solidarity between people of colour, and between people of colour and anti-racist whites. But this solidarity carries with it risks, as the focus on racism has factionalized US EJMs and over-shadowed other sources of environmental injustice, such as social class inequality and poverty (Agyeman 2005; Brulle and Pellow 2005). The mistaken impression that poor and working-class whites are safe from environmental hazards can tactically weaken US EJMs by excluding these groups (Brulle and Pellow 2005).

Outside the US and South Africa, collective identities in EJMs tend to cohere around issues other than race. In Latin America, collective identities in EJMs tend to be associated with struggles against colonialism (Urkidi and Walter 2011), while in the UK, collective identities based on social class inequality are more salient (Walker 2012). In India, EJM activists' distrust of the government stems from its post-colonial brand of environmentalism, in which elite voices dominate and a true public sphere does not exist (Williams and Mawdsley 2006).

Almost everywhere in the world that EJMs cohere, they seem to begin as local struggles rooted in intense attachment to place and to local identities (Anguelovski and Martinez-Alier 2014; Fresque-Baxter and Armitage 2012; Pena 2005; also see Chapter 47). Although this aspect of EJMs has been examined more by geographers than by social movement theorists, place-based collective identity determines much about EJMs, including their mobilization and their choices of strategies and tactics. This aspect of EJMs seems to align with the idea of environmental hazards as a "threat:" people with more intense attachments to a place tend to fight for "their" place when it is threatened (Kaltenborn 1998; Stedman 2002).

Framing is the process through which EJM activists present both collective identities and grievances to the world. Through collective action frames, social movement actors character-ize situations as unjust, attribute blame, propose solutions, and try to mobilize onlookers into their struggle (Snow et al. 1986). EJMs have spoken to the public through the process of "frame bridging," through which they have incorporated two powerful ideas ("environ-ment" and "justice"), thus creating a more broad and inclusive frame (Snow et al. 1986; Taylor 2000). It is this linking of local troubles to wider social inequalities (such as racism) that has allowed EJMs to address social inequalities on the national (and even global) scales (Kurtz 2003).

These "scale frames" have allowed local EJM groups to link up with national-level groups and diffuse collective action frames from the local to the national level and vice-versa (Capek 1993; Kurtz 2003). Yet, as with collective identities focused around race, some aspects of the emphasis on local identities and injustices can keep EJMs from succeeding. The plurality of local, place-based perspectives embedded in the collective identities and collective action frames of EJMs can keep them from recognizing common struggles and becoming part of a much broader movement (Harvey 1996). When this occurs, EJM protest can sometimes take the form of many isolated local struggles that do not coalesce around a common definition of the problem (Sicotte 2012). This is not a trivial manner, as researchers have found that when local EJMs fail to attract attention and resources from outside the local area, they are less likely to succeed in blocking the building of an unwanted polluting facility (Toffolon-Weiss and Roberts 2005).

Others argue that including the concerns of these varied local struggles has made the collective action frames of EJMs too broad and inclusive, which prevents them from speaking in one coherent voice and dilutes their radical critique of existing social and economic arrangements (Benford 2005).

Strategies and tactics of EJMs

Although EJMs are mostly local movements, they do best when embedded in large networks of activists (Pellow 2001; Toffolon-Weiss and Roberts 2005). In this respect, the spread of the Internet and other electronic forms of communication has enabled collaboration across borders, and facilitated ties between activists (Castells 2012).

Both strong and weak ties between activists have diffused protest strategies and tactics used by EJM groups rapidly throughout the world (Strang and Soule 1998). Strong ties between EJM groups networked through the same NGO, Global Community Monitor, resulted in the diffusion of bucket brigades from Louisiana in the US to Durban in South Africa. Bucket brigades are used by EJM activists on the fence line of large, polluting industrial sites. Activists use a simple plastic bucket to sample ambient air, which is then analysed to measure concentrations of toxic air pollutants (Scott and Barnett 2008; also see Chapters 8 and 23).

Weak ties can also diffuse protest tactics when activists see EJM protest activities covered in the media, and adopt particularly memorable tactics for their own protests. This was done by survivors of the 1984 release of deadly MIC gas from the Union Carbide plant in Bhopal, India, when the Dow Chemical Corporation (which had purchased Union Carbide) announced that it would be a corporate sponsor of the 2012 Olympics in London. On opening day, the survivors held a "die-in" at the games, dramatizing the suffering and death of Bhopal survivors for international media (Botelho and Zavistoski 2015). The disaster at Love Canal in the late 1970s had been an international story, and newspapers had disseminated images of die-ins mounted by EJM activists (Shabecoff 2003).

But this does not mean that the tactical choices of EJM groups are totally their own. As EJMs become increasingly embedded within both domestic and transnational networks of other NGOs, their tactical choices are constrained by the group: each EJM group must craft a message and an arsenal of tactics that is in harmony with their NGO partners, but different enough to stand out from them (Hadden 2015).

Climate justice and transnational EJM networks

Within the last twenty years, EJM organizations have of necessity become increasingly integrated within transnational networks as they work on environmental injustices generated by climate change (see Chapter 29). These injustices include deaths from climate-related disasters, sea level rise that threatens the existence of low-lying countries, and inequality in the financial capacity to adapt to climate change (Ciplet et al. 2015). The sector of EJM organizations addressing these conditions is the climate justice movement, a loosely linked international network of organizations and individuals focused on the earth's changing climate and the extraction and burning of fossil fuels (Caniglia et al. 2015; Ciplet et al. 2015; Roberts and Parks 2009). Some researchers mark the beginning of the climate justice movement at the COP-6 climate change negotiations in the Netherlands in 2000, when US-based EJM leaders such as sociologist Robert Bullard and UK-based grassroots groups such as The Rising Tide began to focus on the victimization of the poor due to climate-linked conditions (Roberts and Parks 2009). The Rising Tide called for repayment of the "ecological debt" owed by wealthy nations in the north to the south, the control of land by Indigenous peoples, and a "just transition" to a low-carbon society (The Rising Tide UK 2012).

According to some observers, climate justice in the US is no longer understood as an "environmental" issue: environmental damage is linked together with (and caused by) social and economic inequality (Ciplet et al. 2015). Such frame-bridging is a typical feature of EJM discourse. But the turn toward climate change among EJMs may have made the collective action frames of EJMs more coherent by uniting harms from climate change with harms from fossil fuel extraction and burning (such as threats to health and safety from mining, oil and gas drilling, fuel transport and refining, and air pollution). The climate justice collective action frame attributes blame for both disproportionate harms, and the profligate and continuing use of fossil fuels, to societal power imbalances (Harlan et al. 2015).

The discourse of EJMs working toward climate justice illustrates the multi-scale social processes of the capitalist world economy (Williams 1999) and makes conceptual links between local "sacrifice zones" where the dirtiest and most destructive extractive activities occur, the policy decisions of national leaders, and the actions of multinational fossil fuel corporations (O'Rourke and Connolly 2003). Unconventional gas drilling (or "fracking"), a risky and environmentally damaging new technique for extracting natural gas from rock formations, has created a fossil fuel feeding frenzy, drawing more people into EJMs (Food and Water Watch 2015). Fracking also gives rise to new hazards such as the transportation of flammable oil and gas through pipelines and on railroads (Burton and Stretesky 2014), and uses enormous quantities of freshwater, which pulls in more EJM activists.

Climate justice is the latest in an increasing number of EJM issues that are transnational in nature (such as the export of banned pesticides, e-waste and plastic waste to the Global South), which makes it necessary for EJM activists to work together across borders. Accordingly, EJMs embedded in international networks do not limit themselves to targeting only environmental regulators, but also target corporations in campaigns such as the Computer TakeBack Campaign, which demanded that computer manufacturers deal with the e-waste they produce instead of dumping it in the Global South (Pellow 2007). While activists must work to overcome the tensions inherent in working together across a gulf of economic inequality (Pellow 2007), the number of EJM groups involved in transnational EJM networks is steadily growing.

Some argue that transnational social movements present the best chance for global solidarity generating a political force powerful enough to ensure human survival (Ciplet et al. 2015). However, Global South EJMs that tend to embrace more radical views of social justice are being marginalized and excluded from meetings on climate change (Caniglia et al. 2015).

Responses to EJMs

According to social movement theorists, marginalizing and excluding activists is only one of many possible ways that societal power players can respond to pressures from EJMs. Other responses include providing some concessions while denying activists access to true power; or "buying off" activists with money or positions of pseudo-power in place of any real change, a tactic known as co-optation (Gamson 1990).

Co-optation can be discerned when challenging groups alter their claims and tactics to ones that can be pursued without disrupting politics as usual (Meyer and Tarrow 1998). At climate treaty negotiations, the co-optation of some of the largest US-based environmental organizations has legitimized the drafting of treaties that are ineffective in limiting deadly greenhouse gas emissions, impeding the work of EJM activists (Ciplet et al. 2015). Within EJMs, it can occur through community partnerships with academic researchers, particularly when there are pressures to divert energies away from their communities and into activities that benefit universities (Wing 2005). Co-optation has also occurred through disagreements between EJM actors, which in the US have led to federal environmental justice grants that fund non-controversial projects rather than confronting environmental situations that threaten health and violate human rights (Harrison 2015). Social movement theory suggests that EJMs should not seek to be at the table with power players, as their only power lies in disrupting the smooth functioning of the society that oppresses them (Piven and Cloward 1977).

Co-optation is useful to those in power because it preserves the illusion of democratic functioning and the appearance of legitimacy. But when these factors are not of concern, power players may simply respond to EJMs with violent repression. Repression of social movements has been defined by social movement theorists as violence or the threat of violence by government authorities aimed at increasing the cost of activism (Davenport 2007). These types of responses to social movements have always menaced activists in Global South nations with authoritarian histories or military governments; in such nations, EJM activists have been killed, imprisoned or executed for coming into opposition with ruling elites (Osaghae 1995; Wolford 2008). But violence has also been used against US social movements, particularly labour movements (Goldberg 1991).

In confronting issues of fossil fuel extraction and climate justice, EJMs are inevitably brought into conflict with multinational corporations and their allies (see Chapters 38 and 45). While these dynamics are not new to EJMs (who have long confronted mining, chemical and waste disposal corporations), oil and gas corporations have almost unlimited economic resources, and can call on corporate-supported government officials in every nation; and it is naïve to suppose that these officials would not respond with violent repression if EJMs mount militant attacks on corporate interests.

Social movement theorists have identified two conditions that seem to predict violent repression of social movements: radical goals which threaten established power structures; or social movements that are (or are perceived to be) weak, such as those of poor people (Earl 2003). Unfortunately, EJMs are at risk for repression due to both characteristics.

Social movement theory on the future of EJMs

Since the early 1980s, EJMs have forced the broader environmental movement and society to redefine "the environment" as including not just wilderness areas but also urban communities and rural sacrifice zones, and to recognize the connections between social inequality and environmental damage. While they have enjoyed some important successes (such as stopping the proliferation of incinerators in the US), it is not certain whether they will be able to generate enough power to prevail in the daunting struggles that confront them at present (Pellow and Brulle 2005).

EJMs are theorized to arise from disruptions in people's daily lives caused by environmental threats perceived to invade their bodies and homes, placing their survival at risk. These threats are perceived as deeply inequitable assaults on people who are poor, or are of racial or ethnic minority groups. The linking together of environmentally degraded conditions with social injustice has given rise to powerful collective action frames, but also shaped EJMs into localized, place-based movements. While some are linked into transnational networks, others remain small and localized.

But social movement theory suggests that localized EJMs are much less likely to succeed; in order to grow, gather resources, and effectively resist environmental threats, EJMs must recognize their local struggle as part of a much larger one, and must reach across the borders of their collective identity to recruit others dealing with environmental threats (Harvey 1996).

This will involve the establishment of new, broader collective identities, which will allow them to shift scale and become more unified.

Climate injustice may be a type of environmental injustice so all-encompassing that only the most privileged remain untouched by fossil fuel extraction and transport, heat waves, droughts, storms or flooding disasters. The climate justice movement thus has potential as the basis of a broader identity through which to mobilize for extensive, systemic change. However, realizing this potential will mean the creation of a truly global movement, in which activists all over the world work closely and effectively together; this will not be easy.

EJMs present a radical critique of the rampant and extreme social inequalities that create environmental injustice. In their analysis, it is not possible to bring about environmental justice without fundamental changes in the distribution of money and power. If EJMs succeed in becoming institutionalized into society, this puts them at risk for co-optation, which would completely undermine their effectiveness. Instead of striving for acceptance or a seat at the table, social movement theory would direct EJM activists to try to create political opportunity by targeting and breaking the smooth alignment of corporations with the political order. This implies the use of disruptive (rather than conciliatory) tactics, which may be greeted with repression.

But EJM activists argue that their lives are already threatened due to climate change, chemical contamination, mineral or fuel extraction, water privatization or some other environmentally unjust condition; regardless of the results, they *must* fight (Perreault 2008). Social movement studies can contribute theories about the characteristics of EJMs that may lead to their success or failure, but ultimately cannot predict either outcome.

References

Agyeman, J. 2005. *Sustainable Communities and the Challenge of Environmental Justice* (New York: NYU Press), p. 17.

Anguelovski, I. and J. Martinez-Alier. 2014. "The 'Environmentalism of the Poor' Revisited: Territory and Place in Disconnected Global Struggles." *Ecological Economics*, 102, 167–176.

Benford, R. 2005. "The Half-Life of the Environmental Justice Frame: Innovation, Diffusion and Stagnation," in D.N. Pellow and R.J. Brulle (eds.) *Power, Justice and the Environment: A Critical Appraisal of the Environmental Justice Movement* (MIT Press), pp. 37–54.

Botelho, B. and S. Zavistoski. 2015. "'All the World's a Stage:' The Bhopal Movement's Transnational Organizing Strategies at the 2012 Olympic Games." *Social Justice*, 41, 1–2, 169–185.

Brulle, R.J. 2000. *Agency, Democracy and Nature: The US Environmental Movement from a Critical Theory Perspective* (MIT Press).

Brulle, R.J. and J.C. Jenkins. 2005. "Foundations and the Environmental Movement: Priorities, Strategies and Impacts," in D.R. Faber and D. McCarthy (eds.) *Foundations for Social Change: Critical Perspectives on Philanthropy and Popular Movements* (Rowman & Littlefield), pp. 151–174.

Brulle, R.J. and D.N. Pellow. 2005. "The Future of EJ Movements," in D.N. Pellow and R.J. Brulle (eds.) *Power, Justice and the Environment: A Critical Appraisal of the Environmental Justice Movement* (MIT Press), pp. 293–300.

Bullard, R.D. 1993. "Anatomy of Environmental Racism and the Environmental Justice Movement," in R.D. Bullard (ed.) *Confronting Environmental Racism: Voices from the Grassroots* (South End Press), pp. 15–40.

Bullard, R.D. and B. Wright. 2009. "Race, Place and the Environment in Post-Katrina New Orleans," in R.D. Bullard and B. Wright (eds.) *Race, Place and Environmental Justice after Katrina* (Boulder, CO: Westview Press), pp. 19–48.

Burton, L. and P. Stretesky. 2014. "Wrong Side of the Tracks: The Neglected Human Costs of Transporting Oil and Gas." *Health and Human Rights*, 16, 1, 82–92.

Caniglia, B.S., R.J. Brulle and A. Szasz. 2015. "Civil Society, Social Movements and Climate Change," in R.E. Dunlap and R.J. Brulle (eds.) *Climate Change and Society: Sociological Perspectives* (Oxford University Press), pp. 235–268.

Capek, S. 1993. "The 'Environmental Justice' Frame: A Conceptual Discussion and an Application." *Social Problems*, 40, 1: 5–24.

Castells, M. 2012. *Networks of Outrage and Hope* (Polity).

Ciplet, D., J.T. Roberts and M.R. Khan. 2015. *Power in a Warming World: The New Global Politics of Climate Change and the Remaking of Environmental Inequality* (Cambridge, MA: MIT Press).

Davenport, C. 2007. "State Repression and Political Order." *Annual Reviews of Political Science*, 10: 1–23.

Earl, J. 2003. "Tanks, Tear Gas, and Taxes: Toward a Theory of Movement Repression." *Sociological Theory*, 21, 1: 44–68.

Edwards, B. and J.D. McCarthy. 2004. "Resources and Social Movement Mobilization," in D.A. Snow, S.A. Soule and H. Kriesi (eds.) *The Blackwell Companion to Social Movements* (Blackwell), pp. 116–152.

Faber, D. and D. McCarthy. 2005. "Breaking the Funding Barriers: Philanthropic Activism in Support of the Environmental Justice Movement," in D.R. Faber and D. McCarthy (eds.) *Foundations for Social Change: Critical Perspectives on Philanthropy and Popular Movements* (Rowman & Littlefield), pp. 175–210.

Faber, D. and M. McCarthy. 2003. "Neo-Liberalism, Globalization and the Struggle for Ecological Democracy: Linking Sustainability and Environmental Justice," in J. Agyeman, R.D. Bullard and B. Evans (eds.) *Just Sustainabilities: Development in an Unequal World* (MIT Press), pp. 38–63.

Food and Water Watch. 2015. "Global Frackdown." www.globalfrackdown.org/about/

Fresque-Baxter, J.A. and D. Armitage. 2012. "Place Identity and Climate Change Adaptation: A Synthesis and Framework for Understanding." *WIREs Climate Change*, 3, 251–266.

Gamson, W.A. 1990. *The Strategy of Social Protest* (New York: Wadsworth).

Gamson, J. 1995. "Must Identity Movements Self-Destruct? A Queer Dilemma." *Social Problems*, 42, 3: 390–407.

Goldberg, R.A. 1991. *Grassroots Resistance: Social Movements in Twentieth-Century America* (Long Grove, IL: Waveland Press).

Gottlieb, R. 2005. *Forcing the Spring: The Transformation of the Environmental Movement* (Island Press), pp. 83–96.

Hadden, J. 2015. *Networks in Contention* (University of Maryland Press).

Harlan, SL., D.N. Pellow, J.T. Roberts, S.E. Bell, W.G. Holt and J. Nagel. 2015. "Climate Justice and Inequality," in R.E. Dunlap and R.J. Brulle (eds.) *Climate Change and Society: Sociological Perspectives* (Oxford University Press), pp. 127–163.

Harrison, J.L. 2015. "Coopted Environmental Justice? Activists' Roles in Shaping EJ Policy Implementation." *Environmental Sociology*, 1, 3, 1–15.

Harvey, D. 1996. *Justice, Nature and the Geography of Difference* (Oxford, UK: Blackwell).

Heaney, M.T. and F. Rojas. 2014. "Hybrid Activism: Social Movement Mobilization in a Multimovement Environment." *American Journal of Sociology*, 119, 4, 1047–1103.

Inglehart, R. 1977. *The Silent Revolution: Changing Values and Political Styles among Western Publics* (Princeton University Press).

Kaltenborn, B.B. 1998. "Effects of Sense of Place on Responses to Environmental Impacts: A Study among Residents in Svalbard in the Norwegian High Arctic." *Applied Geography*, 18, 169–189.

Khan, F. 2002. "The Roots of Environmental Racism and the Rise of Environmental Justice in the 1990s," in D.A. McDonald (ed.), *Environmental Justice in South Africa* (Ohio University Press), pp. 15–48.

Kurtz, H.E. 2003. "Scale Frames and Counter-Scale Frames: Constructing the Problem of Environmental Injustice." *Political Geography*, 22, 8: 887–916.

McAdam, D. 1982. *Political Process and the Development of Black Insurgency, 1930–1970* (Chicago: University of Chicago Press).

McAdam, D. 1996. "Conceptual Origins, Current Problems, Future Directions," in D. McAdam, J.D. McCarthy and M.N. Zald (eds.), *Comparative Perspectives on Social Movements* (Cambridge University Press), pp. 23–40.

McAdam, D. and H.S. Boudet. 2012. *Putting Social Movements in Their Place: Explaining Opposition to Energy Projects in the United States, 2000–2005.* (Cambridge, UK: Cambridge University Press), pp. 1–19.

McCarthy, J.D. and M.N. Zald. 1977. "Resource Mobilization and Social Movements: A Partial Theory." *American Journal of Sociology*, 82, 6: 1212–1241.

McCright, A. and R.E. Dunlap. 2011. "The Politicization of Climate Change and Polarization in the American Public's View of Global Warming, 2001–2010." *The Sociological Quarterly*, 52, 2, 155–194.

McKeon, N. 2015. "La Via Campesina: The 'Peasants' Way' to Changing the System, Not the Climate." *Journal of World-Systems Research*, 21, 2, 241–249.

Meyer, D.S. and D.C. Minkoff. 2004. "Conceptualizing Political Opportunity." *Social Forces*, 82, 4, 1457–1492.

Meyer, D.S. and S.G. Tarrow. 1998. "A Movement Society: Contentious Politics for a New Century," in D.S. Meyer and S. Tarrow (eds.) *The Social Movement Society* (Rowman & Littlefield), pp. 1–28.

O'Rourke, D. and S. Connolly. 2003. "Just Oil? The Distribution of Environmental and Social Impacts of Oil Production and Consumption." *Annual Review of Environmental Resources*, 28, 587–617.

Osaghae, E.E. 1995. "The Ogoni Uprising: Oil, Politics, Minority Agitation and the Future of the Nigerian State." *African Affairs*, 94, 376, 325–344.

Pellow, D.N. 2001. "Environmental Justice and the Political Process: Movements, Corporations and the State." *The Sociological Quarterly*, 42, 1, 47–67.

Pellow, D.N. 2007. *Resisting Global Toxics: Transnational Movements for Environmental Justice* (MIT Press).

Pellow, D.N. and R.J. Brulle. 2005. "Power, Justice and the Environment: Toward Critical Environmental Justice Studies," in D.N. Pellow and R.J. Brulle (eds.) *Power, Justice and the Environment: A Critical Appraisal of the Environmental Justice Movement* (MIT Press), pp. 1–22.

Pena, D. G. 2005. "Autonomy, Equity and Environmental Justice," in D.N. Pellow and R.J. Brulle (eds.) *Power, Justice and the Environment: A Critical Appraisal of the Environmental Justice Movement* (MIT Press), pp. 131–152.

Perreault, T. 2008. "Popular Protest and Unpopular Policies: State Restructuring, Resource Conflict, and Social Justice in Bolivia," in D.V. Carruthers (ed.) *Environmental Justice in Latin America: Problems, Promise and Practice* (Cambridge, MA: MIT Press).

Piven, F.F. and R. Cloward. 1977. *Poor Peoples' Movements: Why They Succeed, How They Fail* (Vintage).

Polletta, F. and J.M. Jasper. 2001. "Collective Identity and Social Movements." *Annual Review of Sociology*, 27, 283–305.

Roberts, J.T. and B.C. Parks. 2009. "Ecologically Unequal Exchange, Ecological Debt, and Climate Justice." *International Journal of Comparative Sociology*, 50, 3–4: 385–409.

Rootes, C. and R. Brulle. 2013. "Environmental Movements," pp. 1–7 in *The Wiley-Blackwell Encyclopedia of Social and Political Movements*, edited by D.A. Snow, D. della Porta, B. Klandermans, and D. McAdam (Blackwell).

Scott, D. and C. Barnett. 2008. "Something in the Air: Civic Science and Contentious Environmental Politics in Post-Apartheid South Africa." *Geoforum*, 40, 3, 373–382.

Shabecoff, P. 2003. *A Fierce Green Fire: The US Environmental Movement* (Island Press).

Sicotte, D. 2012. "Saving Ourselves by Acting Locally: The Historical Progression of Grassroots Environmental Justice Activism in the Philadelphia Area, 1981–2001," in B. Black and M. Chiarappa (eds.) *Nature's Entrepot: Philadelphia's Urban Sphere and its Environmental Thresholds* (University of Pittsburgh Press), pp. 231–249.

Snow, D.A., E.B. Rochfort, S.K. Worden and R.D. Benford. 1986. "Frame Alignment Processes, Micromobilization, and Movement Participation," *American Sociological Review*, 51, 4: 464–481.

Snow, D., D. Cress, L. Downey and A. Jones. 1998. "Disrupting 'the quotidian:' Reconceptualizing the Relationship between Breakdown and the Emergence of Collective Action." *Mobilization: An International Quarterly*, 3, 1: 1–22.

Stedman, R.C. 2002. "Toward a Social Psychology of Place: Predicting Behavior from Place-Based Cognitions, Attitude and Identity." *Environmental Behavior*, 34, 561–581.

Strang, D. and S.A. Soule. 1998. "Diffusion in Organizations and Social Movements: From Hybrid Corn to Poison Pills." *Annual Review of Sociology*, 24, 265–290.

Tarrow, S. 1994. *Power in Movement* (Cambridge University Press).

Tarrow, S. 1996. "States and Opportunities: The Political Structuring of Social Movements," in D. McAdam, J. D. McCarthy and M. N. Zald (eds.) *Comparative Perspectives on Social Movements: Political Opportunities, Mobilizing Structures and Cultural Framings* (Cambridge, UK), pp. 41–61.

Taylor, D.E. 1997. "American Environmentalism: The Role of Race, Class and Gender in Shaping Activism 1820–1995," *Race, Gender & Class*, 5, 1: 16–62.

Taylor, D.E. 2000. "The Rise of the Environmental Justice Paradigm." *American Behavioral Scientist*, 43, 4: 508–580.

The Rising Tide UK. 2012. "The Rising Tide Coalition for Climate Justice Political Statement." (Accessed June 2, 2015). Available at: http://risingtide.org.uk/about/political

Tilly, C. 1978. *From Mobilization to Revolution.* (New York: McGraw-Hill).

Tilly, C., L. Tilly and R. Tilly. 1975. *The Rebellious Century, 1830–1930.* (Cambridge, MA: Harvard University Press).

Toffolon-Weiss, M. and T. Roberts. 2005. "Who Wins, Who Loses? Understanding Outcomes of Environmental Justice Struggles," in D.N. Pellow and R.J. Brulle (eds.) *Power, Justice and the Environment: A Critical Appraisal of the Environmental Justice Movement* (MIT Press), pp. 77–90.

Urkidi, L. and M. Walter. 2011. "Dimensions of Environmental Justice in Anti-Gold Mining Movements in Latin America." *Geoforum,* 42, 683–695.

Walker, G. 2012. *Environmental Justice: Concepts, Evidence and Politics* (Routledge Press).

Williams, G. and E. Mawdsley. 2006. "Postcolonial Environmental Justice: Government and Governance in India." *Geoforum,* 37, 660–670.

Williams, R.W. 1999. "Environmental Injustice in America and its Politics of Scale." *Political Geography,* 18, 1: 49–73.

Wing, S. 2005. "Environmental Justice, Science, and Public Health," in T.J. Goehl (ed.) *Essays on the Future of Environmental Health: A Tribute to Dr. Kenneth Olden* (Washington, DC: National Institute of Environmental Health Sciences), pp. 54–63.

Wolford, W. 2008. "Environmental Justice and the Construction of Scale in Brazilian Agriculture." *Society and Natural Resources,* 21, 7: 641–655.

4

ENVIRONMENTAL JUSTICE MOVEMENTS AND POLITICAL OPPORTUNITY STRUCTURES

David N. Pellow

Environmental justice movements and social movement theory

Environmental justice (EJ) movements have led action campaigns for many years around a range of issues, including mining, the hazardous waste trade, air pollution, climate change, hydroelectric power, incineration, pesticides and agrochemical production, forest protection, and the defence of communities on the frontlines of environmental injustice, including Indigenous and First Nations, people of colour, farmworkers, immigrants, women, and gender-oppressed peoples (Clapp 2001; Pellow 2007). These groups play critical roles in (re)defining and (re)framing issues that are often narrowly cast by nation-states, corporations, and mainstream NGOs as "environmental" challenges that must be addressed through a western technoscientific expertise, into questions of social justice, democracy, and power sharing. For example, while the global climate change treaty process convened by the United Nations has been deeply disappointing to these movements, EJ activists along with many others have succeeded at introducing an understanding within the UN of how historical and contemporary inequalities among nation-states reveal that those populations contributing the least to the problem of climate change are bearing the brunt of the socioecological consequences (Caniglia et al. 2015).

EJ movements and activist networks have also authored a number of important documents that offer a vision and a path toward a future marked by social justice, ecological sustainability, and democratic governance, including the Principles of Environmental Justice, the Universal Declaration of the Rights of Mother Earth, the Cochabamba Principles, and the Bali Principles of Climate Justice (Ciplet et al. 2015). Taken together, these documents have served as guiding frameworks for EJ movements as they support social change efforts at the local, regional, national, and global scales from small community fora to global treaty negotiations at the United Nations (Keck and Sikkink 1998).

Within the field of social movement theory, the political opportunity structure (POS) model has been useful for explaining the formation, growth, decline, and suppression of social movements (McAdam 1982; Tarrow 1998; see also Chapter 3). Theorizing collective action as the confluence of elements within and outside of social movements, the POS model improved upon earlier theories of collective behaviour and resource mobilization. For example, the success of the Civil Rights and Farm Workers' movements in the US was, to a large degree, the result of institution building and organizing within African American and Chicana/o communities, in combination with the opening of opportunities within larger political structures (Jenkins and Perrow 1977). Four key components of the POS model include: 1) the relative openness or closure of the institutionalized political system; 2) the stability of elite networks that typically underlie a political system. If elites are divided, this may be an opportunity for social movements to exploit; 3) the presence of elite allies who sympathize with or support social movements; and 4) the state's capacity or willingness to repress movements.

There are three contributions I make in this chapter with respect to the POS model. The first is to articulate the need for greater attention to the role of race and gender in producing and shaping political opportunities vis-à-vis social movements. The second is to demonstrate the need for more emphasis on the role of nonhuman (or more-than-human) forces in contributing to the terrain of political opportunities. And the third is to outline a path for redefining political opportunities beyond the state with respect to EJ movements and EJ studies.

Social movement theory, Critical Race Theory, and Women of Colour Feminism

While the literature on political opportunity structures (POS) is attentive to the question of how state-based politics functions to allow openings for and exclusions of the voices of social movement actors, there is surprisingly little indication as to how those structures are shaped by racialized and gendered meanings and politics (for an important exception on gender politics, see Shriver et al. 2013). This gap provides an opportunity to integrate key insights from Critical Race Theory and Women of Colour Feminism with social movement theory (see also Chapters 2 and 7).

At the core of Critical Race Theory is the view that race, racism, and racialization are central, driving forces in society. This includes, but is not limited to, the ways that race is implicated in shaping the state, legal systems, educational institutions, labour markets, the economy, language, and national identity (Delgado and Stefancic 2012; Goldberg 2002; Kurtz 2010). Because race is so central to the way human societies operate (particularly in the West), it is also deeply involved in how environmental injustices unfold and impact human populations. Women of Colour Feminist scholars demonstrate that race always intersects with gender, sexuality, class, ability, citizenship and other social categories of difference, to reflect the ways that social structures function in society (Melamed 2011). They have also illuminated the ways that various institutions, policies, practices, behaviours, and discourses *produce* race, gender, and sexuality and myriad uneven hierarchies of privilege and subordination (Hong 2006). This scholarship's relevance for POS research lies in its capacity to reveal how political opportunity structures that social movements face are also reflective of existing racial and gender structures, and it can offer insights into how social movements draw on racialized and gendered meanings, ideas, tactics, and strategies to challenge, change, and produce new political opportunities. Race and gender are what Peter Newell terms "mediating structures in global environmental politics." With respect to environmental justice struggles, these dimensions of social inequality are most pertinent ". . . around the question of who has rights to environmental protection and who bears the burden of waste and pollution" (Newell 2005: 74), but they also reflect the political system and structures that create the conditions of those inequalities and the privilege and suffering that follow from them. In other words, as Newell writes, "[t]hey are relevant to understanding causation (the distribution of benefit from environmental destruction), process (which social groups make these key decisions and through what decision making structures) and distribution (of hazard and harm)" (Newell 2005: 71).

Given that race, class, and gender are highly correlated with who enjoys environmental privileges and who does not, one finds that the immediate reasons for these unequal outcomes are often located in the operations of state decision-making bodies, agencies, and corporations that tend to represent the interests of wealthy, white, and male executives and officials, and reflect their worldviews. Thus it is imperative that social movement scholars more closely examine the myriad ways in which political opportunity structures are deeply racialized and gendered, and how that shapes access and possibilities for change for social movements. Accordingly, I propose the terms *racialized political opportunity structure* and *gendered political opportunity structure* in hopes that they may become a generative part of the social movement theory lexicon.

Political ecological opportunity structures and the role of nonhuman natures

Here I draw on urban political ecology, a body of scholarship known for developing the concept of "socionatures" – the idea that human and nonhuman bodies, spaces, structures, and environments are inseparable and involved in co-production (Heynen et al. 2006; Holifield et al. 2010). Political theorist Jane Bennett offers a perspective that emerges from this field in which she connects human and nonhuman agents in a framework of politics and action. Bennett presents what she terms a "vital materialist" theory of politics and democracy that places the interactions among human and nonhuman agents at the centre. She draws from Bruno Latour and his rejection of the culture/nature divide and its relevance for a new kind of politics. Latour preferred the concepts "collective" and "parliament of things," by which he meant an ecology of human and nonhuman elements that might cohere into some sort of polity. The point is that we must question a definition of politics based primarily on granting and exercising rights for humans and expand our awareness of, recognition of, and respect for the agency of nonhuman natures (Bennett 2009).

Therefore political ecology extends beyond the category of the human to include the more-than-human world (from nonhuman animals to the built environment) as subjects of oppression and as agents of social change. Worms, viruses, ants, water, rocks, mountains, fish, elephants, krill, air/wind, roads, buildings, cell phones, and trees, are just some of the myriad nonhuman beings and things that are impacted by environmental injustice but that also exert their own influence on the character of those conflicts in particular and on the course and trajectory of human society and ecology more generally. Urban political ecology's relevance for EJ studies is that it offers a framework "that helps to untangle the interconnected economic, political, social, and ecological processes that together go to form highly uneven landscapes" (Heynen 2013) and offers the possibility for deepening EJ studies' capacity for considering the ways that environmental inequalities are structured within and across species and space and how those processes are interconnected.

I propose the term *political ecological opportunity structure* to recognize that both human and nonhuman forces shape and constitute the myriad political structures and opportunities in which social movements function. Consider the effects of heavy rain or a snowstorm on voter turnout on a national election day (something represented in Jose Saramago's satire *Seeing*) or the effect of opinions and discourses surrounding climate change on how voters and social movements mobilize support for political and social change in their communities. Nonhuman natures figure in politics and can play a significant role in shaping our political structures.

Therefore it is imperative that social movement scholars more closely examine the various modes through which political opportunity structures are shaped by more-than-human natures, through the interactions and flows between humans and nonhuman forces, and how that shapes access and possibilities for change for social movements. I believe attention to *political ecological opportunity structures* can have a productive impact on social movement theory.

Political opportunity structures beyond the state

Elsewhere I have argued that a major limitation of the POS model is that it is rooted in a state-centric perspective (Pellow 2001). That is, it is generally assumed that social movements seek access to state institutions and decision-making power. POS research tends to present the state as the primary movement target or vehicle of reform. Many movements indeed share this state-centric orientation. However, numerous movements increasingly view the nation-state as either weakened by corporate power or as a site of inherent and violent exclusions and therefore a hindrance to social change efforts. Furthermore, an increasing number of movements embrace an anti-statist or anarchist orientation, seeking change through grassroots, consensus-based decision-making that views autonomy from the state as both a means and goal for social change (Amster et al. 2009). Peter Newell (2005: 71) argues that

> we have to take a more critical look at the role of the state ... in the production and reproduction of environmental justice .... This work takes as given the limits of traditional remedies provided by states and international institutions on environmental and social issues, acknowledging the complicity of states in acts of environmental degradation which impact most severely on the poor and the marginalized.

In other words, drawing on Women of Colour Feminist, Critical Race Theory, and Anarchist perspectives, one can conclude that states are institutional forces that, by definition, practice exclusion, control, and violence (in addition to their other functions), and so statecraft involves the management and manipulation of our everyday existence and mobility, regardless of our consent (Hong 2006). Scholars from these interdisciplinary fields increasingly view the modern state as inherently wedded to racist, classist, heteropatriarchal, speciesist, and authoritarian forms and practices, suggesting an opportunity for EJ scholars to consider thinking beyond the state, because to do otherwise risks limiting the reach of EJ politics to a framework that demands participation and inclusion within, and recognition by, institutions that are funda-mentally committed to anti-EJ principles (Amster et al. 2009; Smith 2011). One might call this an *anarchist political opportunity structure* framework, but the idea is that there are opportunity structures that social movements produce themselves that are (or seek to remain) outside the ambit of the state and exist in a political imaginary that is deliberately outside of the state's control. *Anarchist political opportunity structures* that EJ activists produce enable social change through anti-hierarchical and anti-capitalist tactics and strategies that reduce or refuse a reliance on the state.

In the following sections, I present a discussion of two examples of how EJ movements frame their struggles in the context of dynamic and repressive political opportunity structures, revealing how new ways of thinking about the POS are unfolding from their efforts.

Idle No More

Idle No More is a transnational network of Indigenous or First Nations peoples in Canada who, in late 2012, began advocating for a renewed commitment to sovereignty, treaty rights, human rights, racial and gender justice, and environmental justice. The movement was sparked by a number of parliamentary bills and proposals launched by the Canadian national government that threatened Indigenous sovereignty and environmental protection. Activities by Indigenous protesters and their allies took place in public spaces like shopping malls and street intersections across North America and in other places such as New Zealand and Palestine. The language from Idle No More links the quest for Indigenous sovereignty to the fight against environmental racism, patriarchy, and neoliberalism, through a vision of decolonization of the land and Native territories. The Idle No More manifesto states:

> The taking of resources has left many lands and waters poisoned – the animals and plants are dying in many areas in Canada. We cannot live without the land and water. We believe in healthy, just, equitable and sustainable communities and have a vision and plan of how to build them.
>
> *(Idle No More nd)*

Confronting racialized and gendered political opportunity structures that reflect a devaluation of First Nations communities in general and First Nations women in particular, Idle No More determined that, as a result of government cuts in funding for housing support, women's shelters and related programmes, it would launch a campaign to address the housing crisis within First Nations communities. The INM media advisory on this topic announced:

> Canada is experiencing a growing housing crisis that encompasses all people; it's particularly affecting Indigenous women, two-spirit people and their families. Neeve Nutarariaq, an Inuit woman, is now living in a tent with her family in Igloolik, Nunavut, because the housing shortage is at crisis levels. This is only one example of this emergency situation. Idle No More is not waiting for the federal government to fulfil Treaty terms and promises; Indigenous children and their parents are in need of immediate housing repairs and houses. We will start by building or repairing one house. In time, we hope to grow so that we can reach all Nations.
>
> *(Idle No More 2015)*

Three relevant points unfold from the above statement. First is the attention that INM pays to questions of intersectionality, with a particular focus on the ways that Canadian governmental policies are gendered in their negative impacts on women and two-spirit people. This intersectional and feminist focus is consistent with INM's efforts across a number of campaigns and is reflected in its major demands, which include a call to "actively resist violence against women and hold a national inquiry into missing and murdered Indigenous women and girls" – stemming from numerous reports that hydraulic fracturing (fracking) and other forms of mineral extraction have been associated with the influx of company personnel who have perpetrated rape, sex trafficking, sexual slavery, kidnapping, murder, and other forms of violence directed primarily at Indigenous women and girls (McGill 2014). This language and focus signal INM's recognition of the ways in which the federal Canadian government constitutes a *gendered political opportunity structure* that reinforces patriarchal and masculinist approaches to policy and that this orientation has profound impacts on First Nations communities. Second, within an environmental justice framework that (re)defines "the environment" as those spaces and places where we live, work, play, learn, and pray, the housing environment is one of the most critical components of this perspective. Thus, INM's focus on housing fits well within the EJ movement's efforts to recast what activists mean when they use the term "environment," revealing the view that housing is an EJ issue. Third is INM's decision not to wait on the federal government to fulfil its treaty obligations – in other words, this is a progressive orientation that insists that people and social movement networks can organize and exist beyond the state, suggesting possibilities for an *anarchist political opportunity structure*. Indigenous activists have been frustrated that, in general, neither the Canadian government nor the tribal governments have taken seriously the need to address concerns over environmental justice and gender justice that the fracking industry has produced. This has served as a catalyst for organizers to build strength within their communities that is not reliant on either form of government. In a statement that reflects all three of these observations – the importance of home, intersectionality, and activism beyond the state – Kandi Mossett, an energy and climate justice organizer for the Indigenous Environmental Network, declared:

> It's so hard to do this work when so many people are against you. But we do. We get together in our communities and we organize. And we take back the power because nobody is going to do it for us.
>
> *(McGill 2014)*

With respect to the concept of *political ecological opportunity structures*, Idle No More activist and poet Leanne Simpson sums up the interactive character of human and more-than-human natures in First Nations environmental justice politics when she writes:

> I stand up anytime our nation's land base is threatened because *everything we have of meaning comes from the land*—our political systems, our intellectual systems, our health care, food security, language and our spiritual sustenance and our moral fortitude.
>
> *(Simpson 2012; italics added)*

Simpson further argues that the language of "natural resources" that corporations and the Canadian government employ extends beyond the fossil fuels, minerals, fish, and timber and other more-than-human actors to engulf First Nations *human* communities as well. She states:

> Indigenous communities, particularly in places where there is significant pressure to develop natural resources, face tremendous imposed economic poverty. Billions of dollars of natural resources have been extracted from their territories, without their permission and without compensation. That's the reality. *We have not had the right to say no to development, because ultimately those communities are not seen as people, they are seen as resources.*
>
> *(Klein 2013; italics added)*

That view of human beings – in particular, Indigenous peoples – as "resources" rather than as people reflects centuries of racist policies and practices that the Canadian government and dominant white population have imposed on First Nations communities. It also indicates the way that humans and more-than-human natures are inextricably linked in EJ struggles. Simpson's declaration that First Nations "have not had the right to say no to development" signals the fact that political opportunity structures in Canada are heavily shaped by racist, anti-Indigenous, and patriarchal cultural frameworks that are also rooted in speciesist ideologies that devalue both people and more-than-human natures. As Simpson states:

> Extraction and assimilation go together. Colonialism and capitalism are based on extracting and assimilating. My land is seen as a resource. My relatives in the plant and animal worlds are seen as resources. My culture and knowledge is a resource. My body

is a resource and my children are a resource because they are the potential to grow, maintain, and uphold the extraction-assimilation system. The act of extraction removes all of the relationships that give whatever is being extracted meaning. Extracting is taking. Actually, extracting is stealing – it is taking without consent, without thought, care or even knowledge of the impacts that extraction has on the other living things in that environment. That's always been a part of colonialism and conquest. Colonialism has always extracted the indigenous – extraction of indigenous knowledge, indigenous women, indigenous peoples.

(Klein 2013)

The story Leanne Simpson tells reflects the reality and relevance of *political ecological* and *racialized and gendered opportunity structures* quite powerfully. Ecosystems and nonhuman animals in the agricultural sector and waterways are not only extracted, managed, and produced by corporations and the Canadian state; they play an active role in making racist and patriarchal forms of colonial power and conquest possible. They serve key functions in supporting state and market forces and are often the motivating factors in the emergence of closed (or often undemocratic) political opportunity structures that have negative consequences for First Nations communities in general and for First Nations women in particular. For example, in recent years, neoliberal reforms in Canada have largely consisted of sweeping policy changes that have been directed specifically at rolling back environmental protection and reducing the budgets of government agencies charged with managing and regulating "environmental and natural resources." One of the consequences of these changes has been the growth of large industrial livestock feedlot operations, which produce large volumes of hazardous run off and water pollution that many First Nations communities are downwind and downstream from. Another change (resulting from Bills C-38 and C-45, passed in 2012) was the removal of fish habitat protection from thousands of lakes and rivers, and the failure to recognize First Nations commercial fisheries, without proper Parliamentary debate or consulting First Nations (Idle No More and Defenders of the Land 2013). The political opportunity structure of the federal Canadian government has been largely closed to the participation of First Nations and allied communities who seek to halt or reverse recent neoliberal reforms that are aimed at increasing state and corporate access to more-than-human natures, ecosystems, and nonhuman animals for profit and control. This POS is driven by policy making that occurs outside the realm of public debate, passing significant legal changes as budgetary measures in a system that reflects "a concerted effort by those in power to disconnect, and in some cases remove, indigenous peoples from their resources and land" (Mascarenhas 2015: 173).

On the other hand, nonhuman natures play an active role in the cultural imaginary and the material economy of First Nations, constituting an indispensable collective of actors in any future marked by Indigenous sovereignty and environmental justice.

DECOIN, Global Response, and the Global Environmental Justice Project

Ron Eyerman and Andrew Jamison (1991: 55) argue, "It is precisely in the creation, articulation, and formulation of new thoughts and ideas—new knowledge—that a social movement defines itself in society." If new knowledge is at the centre of what constitutes a social movement, then EJ movements and their scholar allies are in an ideal position to contribute to social change, since we make a living creating and sharing ideas with our students, colleagues, and the public. Importantly, knowledge production can also shape and create political opportunity structures.

In 2008 I joined a collaborative effort with two non-governmental organizations dedicated to environmental justice with the goal of writing a guide for community leaders confronting threats from mining companies operating primarily in Global South communities. The two NGOs were Global Response (based in the US, but now defunct) and DECOIN (Defensa y Conservacion Ecologica de Intag/Intag Defense and Ecological Conservation group, based in Ecuador). Beginning in 1990, Global Response worked in solidarity with Indigenous and other Global South communities whose ecological base and cultural practices are at risk from the activities of transnational corporations and the international financial institutions that support them. This frequently involved threats from oil, gold, silver, and coal mining, timber harvesting, and hydroelectric dam construction. Global Response had members in more than 100 nations who wrote letters urging the key decision makers in these cases to dramatically transform their behaviour to ensure basic protections and sustainable practices, or to cease operations altogether. DECOIN is an organization based in the Intag cloud forest region of Ecuador. Its members have successfully expelled some of the world's largest mining companies from the area, after having experienced enormous environmental harm to the land and human health. DECOIN has also launched sustainable businesses to provide a path toward dignified and non-exploitative living for community members.

The extraordinary damage to ecosystems and human health associated with large-scale mining operations has been documented in the US, Latin America, Europe, Africa, Asia and the Arctic (see, e.g., Chapters 30, 34, 40, 41, 46). Some of the problems include high rates of morbidity and mortality within the surrounding communities and the labour force, fish kills, coral reef destruction, deforestation, soil erosion, and the poisoning of farmland (Gedicks 2001; LaDuke 2005). When communities have fought back against large mining companies and governments, the response often involves intimidation, imprisonment, torture, murder, rape, and other human rights violations. Few written resources exist to assist community activists in their efforts to combat these practices around the world, so Global Response, DECOIN, and the Global Environmental Justice Project (GEJP, formerly the Minnesota Global Justice Project, which I direct) teamed up to produce such a tool.

In March of 2009 we completed our report, titled *Protecting Your Community against Mining Companies and Other Extractive Industries: A Guide for Community Organizers* (Zorrilla et al. 2009). The guide contains several cases that detail how activists and residents can work to secure their communities in the face of predatory actions by transnational corporations seeking to siphon off local minerals. The guide's impact was immediate. Activists in Europe, Canada, East and West Africa, Latin America, Asia, and the United States have used it in their campaigns and in their broader efforts to educate and enlist support from residents in affected areas. An activist from Sierra Leone told us, "my community is fighting a mine right now and this guide is a great help to us." Briana, an activist in a mountainous area of Mexico, provided a testimonial about the guide's impacts in her community. She began with a description of the political context in which the local people find themselves:

> When we began the project, the movement against mining was small and uncoordinated. For years people would come to our office individually or in small groups to ask, always in whispers, for information about the plans for mines in their communities. Specific information about timelines for initiating mining was, and for the most part continues to be, nearly impossible to obtain. Generally, people in the most remote areas tend to be fearful of confronting the authorities.
>
> *(Communication from "Briana," community*
> *activist in Mexico, January 4, 2010)*

The above description of the absence of a space for dissent in the context of authoritarian power structures is quite common in communities where mining companies operate around the world. Fortunately, local activists working with international advocates are often able to produce some level of change, and consequently they are able to shape local political opportunity structures through the dissemination of knowledge and tactics from allies elsewhere. Briana continues:

> But this is fast changing, especially since one mine has already been operating with increasingly strong objections by the local people. The guide, which Global Response generously provided, proved to be very important for this effort. We distributed over 100 copies. Copies of copies are being made by those who received our copies. The guide provides a cool-headed orientation on how to proceed with resistance via national and international legal systems. We were glad to be able to feed information to the key people who are the appropriate ones to organize further. Now it is up to the people to stand firm for the better world we all work for.
>
> *(Ibid.)*

Martin, an activist who works with an EJ organization that confronts extractive companies operating in Indigenous communities in Southeast Asia, also was pleased to use the guide. He went a step further and offered to translate it into a local language, thus making it more accessible to residents:

> Our work is focused on providing indigenous communities with information that they can use to organize, to see through company tactics, to think through alternatives, and generally to support them in making decisions about their own futures. We were excited to read your guide; the analysis, approach and language are very close to our own. Over the last 3 years we've been working on the issue of oil palm plantations in Indonesia. We are developing more written modules to accompany the community film we made about the plantations. We would like to translate your guide on company tactics into Bahasa Indonesia. A lot of the guide is relevant to oil palm and to Indonesia and it would be immediately useful in the dissemination of information to communities.
>
> *(Communication from "Martin," community activist working in United Kingdom and Indonesia, April, 2009)*

While much of the feedback on the guide has been positive, the authors expected that sooner or later the mining industry would come into possession of the document and respond. It did not take long. In early 2010 a documentary film that was financed in part by North American mining corporations active in Ecuador contained a segment that was critical of the anti-mining movement in that nation, specifically targeting Carlos Zorrilla. The documentary argues that the guide distorts facts and encourages extreme tactics to oppose mining operations. Moreover, Ecuadoran President Rafael Correa repeatedly denounced the guide and its authors during press conferences and interviews in the national media, creating a chilling effect among activists inside and outside the region. One thing became clear: we had touched a nerve and produced a document that has threatened at least one aspect of power relationships between Global South communities and transnational corporations.

With respect to the question of *racialized political opportunity structures*, the work of DECOIN and Global Response to challenge the Ecuadoran government and its mining corporation partners exists in tension with that state's long history of racism against its

Indigenous communities that are hit hardest by mining and other forms of industrial extraction. The new Ecuadoran Constitution of 2008 was celebrated by many on the political left around the world, but that document and President Correa both make clear that Ecuador's Indigenous communities will be viewed as partners only as long as they submit to the unilateral management of "strategic sectors" like mining. In fact, Chapter 4, article 57(7) of the Constitution reveals that the state contends that "consultation" not "consent" is the only requirement for development projects on Indigenous lands. This is particularly troublesome since the doctrine of free, prior, and informed consent is a cornerstone of the United Nations Declaration on the Rights of Indigenous Peoples (UNDRIP), and Ecuador is a signatory to that agreement. This is certainly not unique to Ecuador. It reflects the general character of racialized political opportunity structures facing Indigenous peoples around the globe, in which these populations are generally viewed as obstacles to "progress" and the routine functioning of the political system rather than as full participants and valued stakeholders.

When Indigenous activists and their allies from DECOIN confront the Ecuadoran government and its corporate partners, they are frequently jailed, intimidated, and placed under surveillance. These practices are not an aberration. After all, repression of dissent is not something that states engage in on an occasional basis; it is at the core of what states do and why they exist. As Max Weber famously wrote, modern nation states claim to hold a "monopoly on the legitimate use of physical force." In other words, a state is an inherently authoritarian and exclusionary force. In response, Ecuadoran activists from DECOIN and other EJ groups throughout Latin America have promoted local forms of economic development that seek greater autonomy from large corporations and the state, via ecotourism, the production and sale of sustainably grown coffee, of soap, and other goods. These practices necessarily occur within a context overshadowed by the presence and power of the Ecuadoran state, but activists work to reduce their reliance on the state in order to maximize local control over their activities and minimize the risk of repression. These actions suggest possibilities for the creation of practices that gesture toward *anarchist political opportunity structures*, or spaces in which political activities can occur that are not rooted primarily in a state-centric institutional or legal framework and that are driven by grassroots democratic decision-making.

Conclusion

The EJ movements discussed in this chapter articulate a vision of social change that entails building and supporting sustainable and equitable communities in the face of continued onslaughts by states and industry. These activists view the global political economy as shifting risks and hazards from North to South, and from socially privileged communities to socially despised and marginalized communities. They challenge the sources of power in the global political economy – governments, transnational corporations, international financial institutions, environmental groups, and the ideologies that undergird the power imbalances that make environmental injustice possible (see Chapter 38). This is the political economic opportunity structure in which they create points of access to interject their voices, perspectives, information, and bodies, in hopes of creating a more just political space. I argue that these EJ groups and networks also underscore the importance of paying attention to: 1) the myriad ways in which political opportunity structures are deeply racialized and gendered, and how that shapes the access and possibilities for change for social movements; 2) the various modes through which political opportunity structures are shaped by more-than-human natures, through the interactions and flows between humans and nonhuman forces, and how that shapes the access and possibilities for social change movements; and 3) the ways that social movements articulate a politics that seeks independence and autonomy from the nation-state and the capitalist system.

Future scholarship on environmental justice movements and social movement theory might consider paying closer attention to the role of race and indigeneity in the shaping of political opportunity structures at multiple scales of governance – from local and national governments to the United Nations and other global treaty making organizations. This is of particular relevance and concern when considering how policy making unfolds in settler colonial states like Canada and the US and at the global scale with organizations such as the UN seeking to implement and strengthen its Declaration on the Rights of Indigenous Peoples, where the future of Indigenous peoples' existence is at stake. Scholars might also place greater emphasis on the means through which political opportunity structures are influenced by gender ideologies and frameworks that indicate the degree to which hetero-patriarchy and masculinity affect the content and trajectory of policy making. Of course, scholars embracing the concept of intersectionality will note that it is difficult if not impossible to separate race from indigeneity, gender and class, and that political opportunity structures in settler colonial states will necessarily indicate the presence and power of each of these categories simultaneously (see Chapter 7). But if we are taking intersectionality seriously in an environmental justice studies context, then I would argue that we must extend and deepen our understanding of that concept to include the category of species and/or more-than-human natures. I make this claim because 1) the POS of many nations (or what I call *political ecological opportunity structures*) is largely fuelled by and organized around a quest for access to nonhuman natures for the maintenance of societies, states, and economic systems; 2) more-than-human natures and species are therefore integral to understanding how social inequality and privilege are produced and maintained (i.e., through differential access by race, indigeneity, gender, sexuality, and class to ecological materials, energy, nonhuman animals, etc.); and 3) the ideas of race, class, gender, sexuality, indigeneity are what Anne McClintock calls "articulated categories" – that is, they help define each other, so that, for example, we come to understand and think about race through our understanding of gender, and vice versa, and so forth. The same logic applies to more-than-human natures and species and has a long history that philosopher David Theo Goldberg (2002) argues produced "naturalism" – those theories of humanity that place non-Europeans, women, and others in an inherent state of inferiority because of their supposed subhuman (read "animal" or "natural") qualities. In other words, we cannot fully grasp the significance of race, indigeneity, gender, sexuality, class, etc. without considering the role that nonhumans (or our ideas about nonhumans) play in giving life and meaning to those seemingly exclusively human categories.

Finally, scholars from the fields of EJ studies and social movement theory might advance their work appreciably by engaging the possibility that we could define the goals of EJ activism and the very idea of politics itself beyond the formal, institutional, and juridical boundaries of the state. The idea that environmental justice and other social movements, the vast majority of political behaviour, and the concept of justice itself must ultimately remain centred on and tethered to the state – an institutional and cultural framework that has only existed for a small percentage of human history and that is responsible for contributing to if not producing our socioecological crises – seems unnecessarily restrictive and ill-advised. There are innumerable examples of environmental justice and other social movements that place some or even most of their energies on building a future without a primary emphasis on reinforcing state power. I believe it is high time that we spend some of our own energy on listening to and learning from them.

References

Amster, R., A. DeLeon, L.A. Fernandez, A.J. Nocella II, and D. Shannon (Eds.). 2009. *Contemporary Anarchist Studies*. Routledge.

Bennett, J. 2009. *Vibrant Matter: A Political Ecology of Things*. Durham, NC: Duke University Press.

Caniglia, B. Schaefer, R.J. Brulle, and A. Szasz. 2015. "Civil Society, Social Movements, and Climate Change." pp. 2235–2268 in R. Dunlap and R.J. Brulle (Eds.), *Climate Change and Society: Sociological Perspectives*. Oxford University Press.

Ciplet, D.J. T. Roberts, and M.R. Khan. 2015. *Power in a Warming World: The New Global Politics of Climate Change and the Remaking of Environmental Inequality*. Cambridge, MA: The MIT Press.

Clapp, J. 2001. *Toxic Exports: The Transfer of Hazardous Wastes from Rich to Poor Countries*. Ithaca, New York: Cornell University Press.

Delgado, R. and J. Stefancic. 2012. *Critical Race Theory: An Introduction*. 2nd Edition. New York University Press.

Eyerman, R. and A. Jamison. 1991. *Social Movements: A Cognitive Approach*. Pennsylvania State University Press.

Gedicks, A. 2001. *Resource Rebels: Native Challenges to Mining and Oil Corporations*. South End Press.

Goldberg, D.T. 2002. *The Racial State*. New York: Blackwell Publishers.

Heynen, N. 2013. "Urban political ecology I: The urban century." *Progress in Human Geography* 1–7.

Heynen, N., M. Kaika, and E. Swyngedouw (Eds.). 2006. *In the Nature of Cities: Urban Political Ecology and the Politics of Urban Metabolism*. New York: Routledge.

Holifield, R., M. Porter, and G. Walker. 2010. "Introduction—Spaces of Environmental Justice—Frameworks for Critical Engagement." In R. Holifield, M. Porter, and G. Walker (Eds.), *Spaces of Environmental Justice*. Wiley-Blackwell.

Hong, G.K. 2006. *Ruptures of American Capital: Women of Colour Feminism and the Culture of Immigrant Labor*. University of Minnesota Press.

Idle No More. nd. The Manifesto. www.idlenomore.ca

Idle No More. 2015. "Idle No More Launches the One House, Many Nations Campaign." October 7, 2015. www.idlenomore.ca

Idle No More and Defenders of the Land. 2013. "The Termination Plan for Indigenous Peoples." January 28. Powerpoint presentation created for #J-28 Teach-ins.

Jenkins, J.C. and C. Perrow. 1977. "Insurgency of the Powerless: Farm Worker Movements (1946–1972)." *American Sociological Review* 42(2): 249–268.

Keck, M. and K. Sikkink. 1998. *Activists beyond Borders: Advocacy Networks in International Politics*. Ithaca, NY: Cornell University Press.

Klein, N. 2013. "Dancing the World into Being: A Conversation with Idle No More's Leanne Simpson." *Yes! Magazine*. March 8.

Kurtz, H. 2010. "Acknowledging the Racial State: An Agenda for Environmental Justice Research." In R. Holifield, M. Porter, and G. Walker (Eds), *Spaces of Environmental Justice*. Wiley-Blackwell.

LaDuke, W. 2005. *Recovering the Sacred*. South End Press.

McAdam, D. 1982. *Political Process and the Development of Black Insurgency*. Chicago: University of Chicago Press.

McClintock, A. 1995. *Imperial Leather: Race, Gender, and Sexuality in the Colonial Contest*. New York: Routledge.

McGill, M. 2014. "#Frack Off: Indigenous Women Lead Effort against Fracking." *Cultural Survival Quarterly*. December. 38(4).

Mascarenhas, M. 2015. "Environmental Inequality and Environmental Justice." Chapter 10 in K. Gould and T. Lewis (Eds.), *Twenty Lessons in Environmental Sociology*. Oxford University Press. Second edition.

Melamed, J. 2011. *Represent and Destroy: Rationalizing Violence in the New Racial Capitalism*. University of Minnesota Press.

Newell, P. 2005. "Race, class and the global politics of environmental inequality." *Global Environmental Politics* 5(3): 70–94.

Pellow, D.N. 2001. "Environmental Justice and the Political Process: Movements, Corporations, and the State." *The Sociological Quarterly* 42: 47–67.

Pellow, D.N. 2007. *Resisting Global Toxics: Transnational Movements for Environmental Justice*. The MIT Press.

Saramago, J. 2007. *Seeing*. Harvest Books.

Shriver, T., A. Adams, and R. Einwohner. 2013. "Motherhood and opportunities for activism before and after the Czech Velvet Revolution." *Mobilization* 18(3): 267–288.

Simpson, L. 2012. "Aambe! Maajaadaa! (What #IdleNoMore Means to Me)." Decolonization.org. December 21.

Smith, M. 2011. *Against Ecological Sovereignty: Ethics, Biopolitics, and Saving the Natural World*. Minneapolis, MN: University of Minnesota Press.

Tarrow, S. 1998. *Power in Movement: Social Movements and Contentious Politics*. New York: Cambridge University Press. Second edition.

Zorrilla, C., A. Buck, P. Palmer, and D. Pellow. 2009. *Protecting Your Community against Mining Companies and Other Extractive Industries: A Guide for Community Organizers*. Cultural Survival.

5

ENVIRONMENTAL JUSTICE AND RATIONAL CHOICE THEORY

William M. Bowen

Rational choice theory is a framework for understanding social and economic behaviour. It can be applied in making policy or taking action for environmental justice. To either make policy or take action for environmental justice requires decisions, some of which are about what to believe, some about what to prefer, and some about what to say or do. Rational choice theory claims to describe these decisions and prescribe ways to improve the process of making them.

Every day, countless individuals and groups, such as firms, throughout society, face decision situations in which they must choose one out of a range of alternative feasible courses of action in choice sets they face. A choice set is the collection of options which are available to the decision-maker at a given time. Each time a decision is made, the decision-maker can specifically choose to act in a way that is consistent with the creation of conditions of environmental justice in the future, or not. Decision-makers in firms for example that manage characteristic hazardous waste products might have a choice set from which they must select one out of a range of alternative feasible locations for a landfill site. Some sites in the choice set might create inequities in the spatial distribution of hazards and others not. The environmental justice research literature contains plenty of documented instances in which the decision-maker selected a site location that created or exacerbated inequities. For instance, in the 1980s, a Texas study performed by Robert D. Bullard concluded that despite black people only making up 28 per cent of Houston's population, five out of six municipal landfill sites were located in predominantly black neighbourhoods (Been 1993). Similarly, standards of fairness and equity were quite evidently disregarded in the decision in 1984, in Warren County, North Carolina, in which a large amount of soil contaminated by polychlorinated biphenyls (PCBs) was stored in a community that was 84 per cent African American. In the Warren County decision, it was later concluded that the site was chosen on no scientific basis, as it lay 8–10 feet above a water table and would undoubtedly contaminate the drinking water of this area (McGurty 2007).

Choices create or exacerbate environmental injustices when they disregard standards of fairness and equity and lead to outcomes that negatively affect people and/or the environment (Been 1993). Mennis (2005), for instance, argued that state-permits allowing air polluting facilities to operate in New Jersey were concentrated in minority neighbourhoods, and that facilities in areas with a relatively high percentage of minority population tended to have a significantly weaker record of environmental enforcement as compared to other facilities. This outcome could, it would seem, only have been produced by the aggregate of many state-level choices that disregarded fairness and equity made by officials and administrators within state government. Similarly, Lavelle and Coyle (1992) found that the United States Environmental Protection Agency (EPA) more frequently decided to use "containment" tactics rather than permanent remediation of hazardous contamination in minority areas, and this despite the fact that the Congress had explicitly directed permanent remediation of hazardous contamination for everyone. Likewise, Kuehn (1994) found that the decisions to levy penalties against violators of environmental laws in predominantly white communities were higher than those in minority communities, by sometimes as much as 500 per cent. These are only a few of many situations in which individuals and groups made decisions which led to outcomes that negatively affected other people, including minorities.

Individual or group-level choices contribute to the creation of conditions of environmental justice, or not, every day. Some such choices involve the dumping of garbage, construction and demolition debris, toxic and hazardous materials, household appliances, abandoned automobiles, and other waste products, which are often discarded at night to avoid the cost and inconvenience of proper waste disposal. This "midnight dumping" can affect public health. Moreover, areas occupied by minority or low income communities can provide relatively easy targets. Trucking companies, for instance, routinely transport and dispose of toxic liquids. Unfortunately, going as far back as the late summer of 1979, in Warren County, when truckers deliberately leaked 31 000 gallons of PCB-contaminated oil within a 3-foot swath along about 240 miles of rural highway, some have chosen to midnight dump their materials, including in minority or low income areas. The outcomes of choices to midnight dump negatively affect communities and the environment, and are inconsistent with environmental justice. Government officials, too, can make choices that undermine environmental justice: administrators in state-level Departments of Environmental Protection routinely face the choice of whether or not to grant applications by firms that emit toxic chemicals into the environment. If and when pollution permits are disproportionately granted to industrial facilities near minority communities, this creates discriminatory patterns in the locations of toxic chemical emissions (Bowen et al. 2009). In each such situation, the course of action taken by the individual or group may or may not result in environmental justice. When all such choices throughout society are made each day, they aggregate into outcomes in which environmental justice is to a certain degree present or absent. Conditions of improved environmental justice thus cannot and will not be created unless the decisions of individuals and groups lead to aggregate outcomes that are consistent with the standards and principles of equity that characterize environmental justice.

This chapter is guided by the following questions about these decisions. To what extent is rational choice theory useful as a theoretical basis from which to understand and prescribe policies and actions for improving conditions of environmental justice? What are the implications of reliance upon rational choice theory as a basis from which to formulate public policy for environmental justice? What challenges does rational choice theory pose to improved levels of environmental justice? What are the alternatives to rational choice theory as a framework with which to understand environmental justice and prescribe remedies when problems arise? What insight or understanding does rational choice theory suggest for improving environmental justice decisions? The answers to these and similar questions imply that it is important for those individuals interested in achieving improvements in environmental justice to develop grounded, behavioural models of moral decision-making that go beyond formal rational choice theory.

Rational choice theory

Rational choice theory is based upon an idealized model of human decision-making. In its formal form, this model is predicated upon the assumption that choice-making behaviour may be understood on the basis of a set of normative axioms (von Neumann and Morgenstern 1944). These axioms stipulate specifically that decision-makers can make fully rational choices if and only if they have complete knowledge of all the alternatives in their choice sets, they know the entire range of consequences that will follow from the selection of each of the alternatives, they possess a perfectly known and consistent preference ordering across the alternatives, and they use a consistent decision rule for combining their knowledge and preferences. According to rational choice theory, when these axioms are fulfilled decision-makers make rational decisions by selecting from within their choice set of alternative feasible courses of action the one that brings the greatest utility, which is to say the greatest satisfaction, personal value, or pleasure. In this sense, rational choice theory is explicitly utilitarian in nature.

In application, rational choice theory stipulates a decision-making process in which decision-makers follow five steps. The process begins with identifying the problem, then specifying the goals, specifying all of the alternative courses of action available to attain these goals along with the range of outcomes that would be likely to follow from making a commitment to each, evaluating the alternatives in light of these outcomes, and selecting the optimal alternative, which is to say the one that maximizes the decision-makers' utility. In selecting between the alternatives, the golden rule of rational choice theory is to *maximize expected utility*. That is, the theory prescribes that decision-makers should select whichever of the alternative feasible courses of action is expected to bring them the greatest utility. Of course, given that the full range of outcomes of any given decision made today will not occur until sometime in the future, and that the future is always to some degree uncertain, there is always some degree of uncertainty in making rational decisions. Rational choice theory factors this uncertainty in to its model of the decision process by weighting the probability of each alternative with uncertain outcomes by the probability of its occurrence. Thus, the expected utility of each alternative with uncertain outcomes is the sum of its possible outcomes, each weighted by its probability of occurrence. In formal, mathematical terms, the expected utility (EU) of any given alternative feasible course of action is expressed by:

$$EU = P_1 U_1 + P_2 U_2 + \ldots + P_k U_K$$

where P_1 is the probability of a given outcome and U_1 is the utility of that outcome. It is often suggested that the advantages of applying rational choice theory in this way include greater clarity and precision in thinking through decision situations. The theory is also hugely versatile, making it applicable across an unfathomably wide range of decision situations faced by individuals and groups throughout society.

The idea behind applying rational choice theory in making policy or taking action specifically for environmental justice would usually be to help to determine which out of a set of alternative feasible courses of action is most likely with time to bring about environmental justice. This chapter will consider this idea in some detail. In such situations the environment typically contains a hazard or hazards that threaten the well-being of individuals or communities. The environment in this sense is the totality of circumstances in which a threatened individual or community is situated, including the human-made surroundings as well as the natural ecosystem. The hazard might be physical, chemical, biological, or nuclear, but in any case environmental justice considerations arise only when this hazard has been introduced into the environment by human activity (Been 1993). Environmental justice problems arise specifically when the health and well-being of individuals or communities who have not volunteered to expose themselves to the hazard are put unfairly at risk by virtue of their presence in an affected environment.

The term "environmental justice" has evolved over time from giving reference almost exclusively to hazardous materials in low income and minority communities to a much broader social justice movement today. It is now often used in reference to a more general set of issues of equity, often divided into their procedural, geographic and social aspects. The term is now often used to recognize disparities in the burden of the cost on human beings of conducting industrial society as well as an active effort to prevent the injustice. But no matter how the term is used, choices are always required to bring about conditions of environmental justice in the world, and lie at the core of rational choice theory.

Rational choice theory of environmental justice decisions

Although expected utility maximization is the defining characteristic of the concept of rational choice which is the predominant one within the decision science, public policy, risk, and economics literatures, like any model, there are limits to what it can do and to when it should be applied (Jaeger et al. 2001). Descriptions of the actual behaviours of decisions-makers, for instance, indicate that the axioms upon which expected utility maximization is based are seldom fulfilled in practice. The assumption that decision-makers have complete knowledge of the alternatives and the range of outcomes likely to follow from committing themselves to any given alternative must, on the basis of extensive evidence, be rejected in all but the most simplistic models of choice situations. The assumptions that decision-makers possess a perfectly known and consistent preference ordering, and that they use a consistent decision rule for combining their knowledge and preferences, are at odds with the body of research on preference reversals (Tversky et al. 1990). Instead, a substantial body of research evidence shows that decision-makers tend to use simplifying rules, mental shortcuts, or cognitive heuristics which allow them to focus on some aspects of complex decisions, and to ignore others, thereby simplifying their choices, but inducing systematic error. Social scientists have also recognized that decisions are strongly affected by other factors not considered by formal rational choice theory, such as the context within which a choice presents itself, the levels of trust amongst decision-makers and those affected by the decisions, the quantity and sorts of communication channels available, and many others (Ostrom 2010; Ostrom 1998; Kiser and Ostrom 1982; Zeleny 1982).

The recognition, documentation, and understanding that humans have a more complex motivational structure and a greater capability to solve social dilemmas than is posited in formal rational choice theory has led scholars and researchers to create a range of alternative theories (Zey 1992). One prominent alternative theory stipulates that in actual decisions, rational choice is a limiting condition; decision-makers do not make fully rational decisions, but instead make "boundedly rational" ones (Simon 1957). Proponents of boundedly rational decision theory maintain that the formal theory is basically correct, but that decision-makers usually have somewhat less than full knowledge of the alternatives and their consequences, an imperfectly known and somewhat inconsistent preference ordering, and/or a somewhat inconsistent decision rule for combining their knowledge with their preferences. So when making choices they gather information up to the point at which the cost of additional information acquisition begins to outweigh its expected benefits, at which point they "satisfice."

Another alternative (Zeleny 1982) views rational choice not in terms of a search for mathematical maximum utility but rather as the art of balancing multiple objectives. Yet another view stipulates that formal rational choice theory is deeply incorrect, that environmental decisions are not at all rational, but rather that they are characteristically about power, politics, persistent solutions looking for problems, and other such things that have nothing much at all to do with rationality (Cohen et al. 1972). A difficulty with this last view is that it offers very little hope of success for individuals and groups who seek through the use of intelligent reason and argumentation to improve conditions of environmental justice.

One of the most empirically researched and definitive views of rational choice theory is based upon the research of Daniel Kahneman and Amos Tversky. The many carefully observed and documented inconsistencies between the normative axioms of rational choice theory and the actual behaviours of decision-makers led Tversky and Kahneman to conclude that:

> the logic of choice does not provide an adequate foundation for a descriptive theory of decision making. We argue that the deviations of actual behaviour from the norm-ative model are too widespread to be ignored, too systematic to be dismissed as random error, and too fundamental to be accommodated by relaxing the normative system.... We conclude from these findings that the normative and descriptive analyses cannot be reconciled.
>
> *Tversky and Kahneman (1986: S252)*

Thus, insofar as formal rational choice theory is based upon a model that fails on the grounds of consistency with the body of empirical evidence about how individuals make decisions, its applicability as a theoretical basis from which to understand environmental justice and prescribe remedies when problems arise is empirically limited.

The prescriptions of rational choice theory for environmental justice

One of the practical difficulties of relying upon rational choice as a theoretical basis for making decisions related to environmental justice stems from its predictions about individual-level behaviours. Specifically, rational choice theory predicts behaviours that are inconsistent with those that the research on social dilemmas concludes are necessary for purposes of creating collective outcomes consistent with the normative standards of environmental justice. This difficulty arises in large measure due to the fact that rational choice theory models human beings as if individuals are independent of groups, and are self-interested, short-term utility maximizers. Such models have been highly successful at predicting marginal behaviour under conditions of competition, and especially those typified by individuals and firms acting in perfectly competitive markets, in which selective pressures screen out those competitors who do not maximize expected utility. But when standard rational choice models are applied within the context of averting disasters or working out social dilemmas, they are likely to prescribe choices that lead to the very worst possible outcomes, and to do so without stipulating any way that individuals might achieve more productive and sustainable outcomes.

A stark example of how the application of rational choice theory can lead to the very worst possible outcomes may be gleaned from a leaked memo dated December 12, 1991, signed by Lawrence H. Summers, who was then the Chief Economist at the World Bank, and on his way to the Presidency of Harvard University and the Chairmanship of the

US President's Council of Economic Advisors (Summers 1991). The memo, which is presented as Exhibit 1 at the end of this chapter, was about the management and spatial disposition of health impairing pollutants, and specifically their transshipment out of the United States and eventual final location in poor countries. In the memo, Summers assumed his readers to be rational decision-makers who maximize their expected utility, and from this assumption deduced a set of prescriptions for where the pollutants should be finally located. He showed using a deep, long, logical chain of reasoning that rational choice theory logically entails the morally repugnant, ruthless and arrogant conclusion that health impairing pollutants should be shipped out of affluent countries such as the United States areas and in to poorer countries, such as those in Africa, where they will lead to a smaller aggregate loss of utility. Using a very similar chain of reasoning, albeit at a different geographical scale, the assumptions of rational choice theory would logically entail the conclusion that the hazards at issue in environmental justice within the United States should, for purposes of maximizing utility, be located in minority and low income areas.

Facts, standards, and a behavioral theory of choice in environmental justice

While considerations about environmental justice require reference to standards of fairness or equity, formal rational choice theory does not. Indeed, a wide range of conceptual definitions have been proposed for use in stipulating appropriate fairness and equity standards for environmental justice (Wenz 1988; Phillips and Sexton 1999; Schlosberg 2007; Walker 2012). Been (1993), for instance, identified, characterized, and classified these to include 1) equal division of hazards on a per-capita or per-neighbourhood basis; 2) progressive siting, in which more affluent neighbourhoods compensate poorer neighbourhoods by receiving a disproportionately large percentage of the hazards; 3) siting of hazards on the basis of competitive bidding; 4) governmentally-forced internalization of hazards by producers; and 5) fairness in the procedures used to decide a location for the hazard. She also recognized the point made later by Phillips and Sexton (1999) that the selection of any one of these concepts has its own distinct implications for policy, as well as for prescribing the appropriate remedy in situations when wrongs have occurred, and problems have arisen, some of which may be incompatible with others. Schlosberg (2007) argued that in the end none of them can be singled out and used alone in all situations in which environmental justice considerations occur. Walker (2012) recognized the range of such concepts, observed that they are all explicitly prescriptive or normative in nature, distinguished them from empirical evidence about the facts of situations, on one hand, and from the explanations of why things are the way they are on the other, and pointed out that all three (normative concepts, facts, and explanations) are basic elements of environmental justice claim-making.

The distinction between normative concepts and facts is important in environmental justice for a number of reasons. One is that it goes a long way toward avoiding a major thought-stultifying error known as the "is–ought" confusion, which is a pernicious source of bewilderment in that it implies that the way in which decisions are currently being made cannot be improved (Kadane and Larkey 1983). The "is–ought" confusion, as described by Simon (1965: 137), results from failure to distinguish between normative statements about standards used to evaluate facts, on one hand, and empirical statements about the facts thus evaluated on the other. To be clear, empirically-oriented "is" statements are of a different order from normatively-oriented "ought" statements. On the "is" side are declarative, empirical statements related to descriptive or factual claims. On the "ought" side are imperative statements related to judgments, evaluations, norms or principles of conduct. "Ought" statements propose standards and contain the word "should." Both "is" statements and "ought" statements can be debated, but the two are nevertheless decisively asymmetrical in at least one fundamental way. That is, while the decision to accept any given set of facts does not and cannot in and of itself create those facts, the decision to create a standard does create that standard, even if only tentatively (Popper 1961). Failure to recognize this distinction leads logically to the conclusion that a given standard can be reduced to the facts to which it is applied, and thus to the implication that the way in which decisions are currently being made cannot be improved. The many practical implications of the failure to recognize this distinction when considering rational choice and action are among the main reasons why Rescher (1988) argued that only a normative theory of rational choice can be adequate to the complexities of the subject.

This line of reasoning also leads to a view in which only a normative theory of decision and action can be adequate to the task of guiding decisions for the purpose of creating a more environmentally just future. But while explicit recognition of the asymmetry between facts and standards stands to greatly improve the clarity and quality of thought about the nature and limitations of rational choice in environmental justice, it does not go very far in terms of prescribing how to go about actually making decisions that lead to the necessary and desired improvements. It is only to take a first step toward the development of a coherent, useful, and widely agreed upon behavioural approach to a rational choice theory of collective action applicable in environmental justice. Such a theory must also identify and describe the ways in which the decisions of individuals and groups need to improve to be consistent with what is known about rational choice and collective action, one on hand, and with equitable burdens of environmental hazards, judged according to whatever standard is used to define equity, on the other. In application, the purpose of such a theory would be to enable decisions intended to create or improve conditions of environmental justice to be guided by an explicitly behaviourally-oriented framework that enables coherent description and explanation of how myriad individuals and groups acting within the context of their own sets of local circumstances, facts, values, and institutions make decisions that aggregate into the outcomes we observe and associate with environmental justice, or the lack thereof. Such a theory would ideally be well grounded in empirical research, consistent with what is known about behaviour, and useful in terms of enhancing decision-making. Although Ostrom's (1998: 1) work on "social dilemmas" (occurrences in which "individuals in inter-dependent situations face choices in which the maximization of short-term self-interest yields outcomes that leave all participants worse off than feasible alternatives") goes a long way toward developing such a theory, that work remains far from complete. Currently, and unfortunately, such a theory does not yet exist.

Implications

Environmental justice is most clearly understood as a proposal to adopt a particular set of principles or standards, not a fact or set of facts but rather a standard against which to judge such facts, or sets of facts. Understood this way, environmental justice appears to be something that can be created by the aggregated outcomes of myriad decisions made by individuals and groups acting within a framework of social institutions over time. Every day, countless decision-makers throughout society face decision situations in which they must choose one out of a range of alternative feasible courses of action in a choice set. Insofar as these decisions are based upon the set of normative axioms that underlie formal rational choice theory, such theory is clearly applicable to environmental justice. But insofar as these axioms are not met, rational choice theory is of limited applicability. Accordingly, this chapter has contended that the decision-making processes necessary for making improvements in conditions of environmental justice are inadequately described in terms of formal rational choice theory, and that only a normative theory of rational choice would suffice for helping to make such improvements.

Inasmuch as social scientists take interest in this proposal, one important contribution that can be made is to work toward the development of a coherent, useful, and widely-agreed-upon behavioural approach to a rational choice theory of collective action for environmental justice. Formal models of rational decision-making based upon expected utility theory will not, in the end, lead to improved conditions of environmental justice. Not only are the actual choices made by people and groups far too complex to be fully described by formal rational choice models, but the assumption implicit in advocating the use of rational choice models for this purpose – that somehow greater levels of utility maximization will bring about increases in the rates of adoption – is both logically and empirically unjustified. Formal rational choice theory does not include any standards of fairness or equity in its model of decision-making, and these are the core values of interest in environmental justice. Rather, rational choice theory treats fairness and equity as if they can be captured and represented completely in terms of the decision-makers' utility. Rational choice theory thus does not distinguish what is from what ought to be, a fatal oversight in any decision theory useful for purposes of guiding decisions toward improvements in conditions of environmental justice.

For a behavioural approach to a rational choice theory of collective action to be coherent and applicable in environmental justice, it must recognize that the only way to create pervasive conditions of environmental justice is through widespread consideration, discussion and advocacy. Such theory must therefore be far-reaching, multi-faceted and inclusive enough to encompass the domain of collective human action when faced with social dilemmas. In particular, it must include normative elements related to the spread of morality, justice, and equity in highly complex ecological, economic, and socio-technical systems, at large.

In this regard, the research agenda described by Ostrom (1998) is exceedingly promising. She described the "second generation models" of choice that she and her colleagues worked on for years, ones based upon reciprocity, reputation and trust rather than any sort of utility maximization or other overly narrow conception of what rationality is all about. These models assume that there "is a general theory of human behaviour that views all humans as complex, fallible learners who seek to do as well as they can under the constraints that they face and who are able to learn heuristics, norms, rules, and how to craft rules to improve achieved outcomes" (Ostrom 1998: 9). This leads to a research focus in such areas as how people learn and employ heuristics, norms, rules in social dilemmas; the strategies they learn and develop for reciprocity; the roles of trustworthiness and face-to-face communication in the attainment of desirable collective choice outcomes; and the importance of building a reputation for keeping promises and performing actions with short-term costs but long-term net benefits. Considerations such as these, rather than ones designed to cultivate the limited concerns of any particular academic discipline, are far more conducive to the possibility of keeping a perspective on how complex and many-sided rationality is.

The implications of the "is–ought" distinction in application to environmental justice, and particularly the failure of formal rational choice to make this distinction, are substantial. Especially when theorizing about the nature and use of rationality specifically within the context of proposals to improve the levels of justice and equity in the spatial distribution of hazards throughout society, only a fundamentally normative theory will be adequate to the task. The creation of improvements in environmental justice requires improvements in moral decision-making, not utility maximization. The application of formal rational choice theory will not only not suffice to enhance the likelihood of such improvements, but it may even lead to dysfunctional outcomes, as illustrated by Lawrence Summers' leaked memo. Social scientists and theorists who want on one hand to present a theory of rational choice, such as expected utility theory, while at the same time avoiding the vexing empirical and behavioural complexities of collective decisions and actions taken for normative purposes on the other, are engaged in a futile venture condemned from the start to an overly narrow and misguided conception of what rational human choice is all about.

References

Been, V. (1993). "What's fairness got to do with it? environmental justice and the siting of locally unwanted land uses." *Cornell Law Review*, vol. 78, pp. 1001–1085.

Bowen, W.M., Atlas, M., & Lee, S. (2009). "Industrial agglomeration and the regional scientific explanation of perceived environmental injustice." *Annals of Regional Science*, vol. 43(9), pp. 1013–1031.

Cohen, M.D., March, J.G., & Olsen, J.P. (1972). "A garbage can model of organizational choice." *Administrative Science Quarterly*, vol. 3(1), pp. 1–25.

Fabio C., Ashok, A., Waitz, J.A., Yim, S. H.L., Barrett, S. R.H., & Caiazzo, F. (2013). "Air pollution and early deaths in the United States. Part I: Quantifying the impact of major sectors in 2005." *Atmospheric Environment*, vol. 79, pp. 198–208.

Jaeger, C.C., Ortwin, R., Rosa, E.A., & Webler, T. (2001). *Risk, uncertainty, and rational action*. London: Earthscan Publications, Ltd.

Kadane, J.B. & Larkey, P.D. (1983). "The confusion of 'is' and 'ought' in game theoretic contexts." *Management Science*, vol. 29(12), pp. 1365–1379.

Kiser, L.L. & Ostrom, E. (1982). "The three worlds of action: a metatheoretical synthesis of institutional approaches." In Haynes, K.E., A. Kuklinski, & O. Kultalahti (eds), *Pathologies of urban processes*. Jyvasyla, Finland: Finnpublishers.

Kuehn, R.R. (1994). "Remedying the unequal enforcement of environmental laws." *Journal of Civil Rights and Economic Development*, vol. 9(2), pp. 625–668.

Lavelle, M. & Coyle, M. (1992). "Unequal protection: the racial divide in environmental law, a special investigation." *The National Law Journal*, vol. 15(3), pp. S1–S12.

McGurty, E. (2007) *Transforming environmentalism: Warren County, PCBs, and the origins of environmental justice*. New Brunswick, NJ: Rutgers University Press.

Mennis, J. (2005). "The distribution and enforcement of air polluting facilities in New Jersey." *Professional Geographer*, vol. 57(3), pp. 411–422.

Neumann, J. von & Morgenstern, O. (1953). *Theory of games and economic behavior*. Princeton, NJ, Princeton University Press, 1944, second ed. 1947, third ed. 1953.

Ostrom, E. (1998). "A behavioral approach to the rational choice theory of collective action: Presidential Address, American Political Science Association, 1997." *The American Political Science Review*, vol. 92(1), pp. 1–22.

Ostrom, E. (2010). "Beyond markets and states: polycentric governance of complex economic systems." *American Economic Review*, vol. 100, pp. 1–33.

Phillips, C.V. & Sexton, K. (1999). "Science and policy implications of defining environmental justice." *Journal of Exposure Analysis and Environmental Epidemiology*, vol. 91(1), pp. 9–17.

Popper, K.R. (1961). "Facts, standards, and truth: a further criticism of relativism." *The open society and its enemies*, pp. 369–396.

Rescher, N. (1988). *Rationality: a philosophical inquiry into the nature and rationale of reason*. Oxford: Clarendon Press.

Schlosberg, D. (2007). *Defining environmental justice: theories, movements and nature*. Oxford: Oxford University Press.

Simon, H.A. (1957). *Models of man*. New York: Wiley.

Simon, H. (1965). *Models of discovery*. New York: D. Reidel Publishing Company.

Summers, L. (1991). Hearing before the Committee on Finance United States Senate One Hundred Third Congress First Session on the Nomination of Lawrence H. Summers to be Undersecretary of the Treasury for International Affairs. pp. 117–147. www.finance.senate.gov/library/hearings/download/?id=7735172f-99ef-4424-a608-866d965d0e61

Tversky, A. & Kahneman, D. (1974). "Judgment under uncertainty: heuristics and biases." *Science*, vol. 185(4157), pp. 1124–1131.

Tversky, A. & Kahneman, D. (1986). "Rational choice and the framing of decisions." *The Journal of Business*, vol. 59(4: Part 2). *The Behavioral Foundations of Economic Theory*. pp. S251–S278.

Tversky, A., Slovic, P., & Kahneman, D. (1990). "The causes of preference reversal." *The American Economic Review*, 80(1): pp. 204–217.

Walker, G. (2012). *Environmental justice: concepts, evidence and politics*. London: Routledge.

Wenz, P.S. (1988). *Environmental justice*. Albany, NY: Suny Press.

Zeleny, M. (1982). *Multiple criteria decision making*. New York: McGraw-Hill Book Company.

Zey, M. (1992) (ed.). *Decision-making: alternatives to rational choice models*. Newbury Park, California: Sage Publishers.

Exhibit 1: Memorandum from Lawrence H. Sommers

TO: Distribution
FR: Lawrence H. Summers
Subject: GEP [*Global Economic Prospects*]

'Dirty' Industries: Just between you and me, shouldn't the World Bank be encouraging MORE migration of the dirty industries to the LDCs [Least Developed Countries]? I can think of three reasons:

1) The measurements of the costs of health impairing pollution depends on the foregone earnings from increased morbidity and mortality. From this point of view a given amount of health impairing pollution should be done in the country with the lowest cost, which will be the country with the lowest wages. I think the economic logic behind dumping a load of toxic waste in the lowest wage country is impeccable and we should face up to that.

2) The costs of pollution are likely to be non-linear as the initial increments of pollution probably have very low cost. I've always thought that under-populated countries in Africa are vastly UNDER-polluted, their air quality is probably vastly inefficiently low compared to Los Angeles or Mexico City. Only the lamentable facts that so much pollution is generated by non-tradable industries (transport, electrical generation) and that the unit transport costs of solid waste are so high prevent world welfare enhancing trade in air pollution and waste.

3) The demand for a clean environment for aesthetic and health reasons is likely to have very high income elasticity. The concern over an agent that causes a one in a million change in the odds of prostate cancer is obviously going to be much higher in a country where people survive to get prostate cancer than in a country where under 5 mortality is 200 per thousand. Also, much of the concern over industrial atmosphere discharge is about visibility impairing particulates. These discharges may have very little direct health impact. Clearly trade in goods that embody aesthetic pollution concerns could be welfare enhancing. While production is mobile the consumption of pretty air is a non-tradable.

The problem with the arguments against all of these proposals for more pollution in LDCs (intrinsic rights to certain goods, moral reasons, social concerns, lack of adequate markets, etc.) could be turned around and used more or less effectively against every Bank proposal for liberalization.

6

THE POLITICAL ECONOMY OF ENVIRONMENTAL JUSTICE

Daniel Faber

The environmental injustices of capitalism

Capitalism is a for-profit economic system in which private businesses compete for survival in the marketplace. Without an adequate rate of profit, and hence rate of capital accumulation, these businesses (and the larger economy) would lapse into economic crisis. As a general rule, capital will always strive to maximize profits by increasing productivity while minimizing the costs of production and distribution. However, pollution abatement devices and restoration mandated by environmental policies in the advanced capitalist countries usually increase costs and reduce productivity, and are considered to be an added expense that business is resistant to absorb (O'Connor 1998).

Without prohibitions and the threat of punitive actions by state regulatory agencies or the courts, it is simply more profitable for corporations to pollute and leave ravaged landscapes unrestored. So, instead of "internalizing" $10 million in costs for the installation of a "scrubber" to clean the air of pollutants coming out of the smokestack, a corporation will displace or "externalize" this expense onto society in the form of air pollution and other environmental health problems. In addition to the over 7 million people worldwide estimated by the World Health Organization (WHO) to be killed by air pollution each year, these social losses (or "negative externalities") also take the form of long-term damage to human health; the destruction or deterioration of property values and the premature depletion of natural wealth; and the impairment of less "tangible" values associated with environment quality and the loss of community (Kapp 1950: 13).

But not all people are equally impacted by the social and ecological costs of capitalist production. In order to bolster profits and competitiveness, corporations and state institutions embrace various strategies for displacing negative environmental externalities that are simultaneously the most *economically efficient* and *politically expedient* (Faber 2008). Most citizens see the act of releasing dangerous toxins into the air and water as a form of anti-social behaviour, a violation of their fundamental rights as citizens to a clean and healthy environment. Residents will seldom "choose" to see their family members or neighbours poisoned by industrial pollution, especially if they are aware of the dangers. In fact, the successful imposition of such public health dangers is symptomatic of a lack of democracy (Kapp 1950). Once aware of the dangers, affected residents are likely to oppose the offending facility. As a result, capital often

61

adopts more cost-effective practices for exploiting natural resources and disposing of pollutants that offer the path of least political resistance.

Following the path of least resistance often means targeting the most disempowered communities in society for the most ecologically hazardous industrial facilities, toxic waste sites, and natural resource extraction and energy development schemes. The less political power a community of people possesses, the fewer resources (time, money, education) that people within have to defend themselves from potential threats; the lower the level of community awareness and mobilization against potential ecological threats, the more likely they are to experience arduous environmental and human health problems. The weight of the ecological burden upon a community is dependent upon the balance of power between capital, the state, and social movements responding to the needs and demands of the populace (Faber 1998: 27–59). And in capitalist countries such as the United States, it is working class neighbourhoods and poor communities of colour that most often experience the worst problems. In fact, greater inequality in the distribution of power and wealth is likely to engender greater environmental inequality. Today, environmental inequality is now increasing faster than income inequality in the US (Boyce et al. 2014).

Susceptibility to experiencing negative externalities is deeply related to social positionality – where a person or group of people are situated in multiple power structures centred on class, gender, race, ethnicity, citizenship, and more (Walker 2012). The various social positions or "identities" held in these power structures intersect to create different social "axes" of advantage and disadvantage. A poor working class African-American woman encounters multiple disadvantages in comparison to the control capacity exercised by a white, middle class woman (or male). In the US, communities that lack *control capacity* are typically made up of racial and ethnic minorities, as well as the white working class (Schnaiberg 1994). For those members of the socially and spatially segregated "underclass," powerlessness is even more pervasive. America's undocumented immigrants, Chicano farmers, Indigenous peoples, and other dispossessed peoples of colour are the ones being *selectively victimized* to the greatest extent by environmental health abuses (Johnston 1994). As part of the country's *subaltern* experiencing multiple forms of political domination, economic exploitation, and cultural oppression, they are effectively devalued in American society (Pulido 1996). The resulting environmental injustices take the form of noxious industrial pollutants, and hazardous waste sites being situated in poor African-American communities in the rural South (Bullard 1994), or undocumented Mexican workers labouring in the pesticide-soaked agricultural fields of California, Texas, and Florida. In short, the concentration of environmental and health hazards among such disempowered peoples is creating ecological sacrifice zones – areas where it is simply dangerous to breathe the air or take a drink of water (Lerner 2010). As such, ecological sacrifice zones serve as locations where polluting corporations can substantially lower the costs of compliance with environmental regulations.

In this light, environmental injustices are rooted in processes of capital accumulation and power structures that confer social class advantages and white privilege (Sicotte 2016: 13). And when analysing environmental inequality, we should be aware that there are multiple political-economic forces at work that give the injustice a particular context and form (Holifield 2001). The working class in general, and poorer people of colour in particular, face a greater "quadruple exposure effect" to environmental health hazards. This first takes the form of higher rates of "on the job" exposure to dangerous substances used in the production process; and the second as greater neighbourhood exposure to toxic pollutants (Morello-Frosch 1997). Faulty cleanup efforts implemented by the government or the waste treatment industry often magnify these problems (Lavelle and Coyle 1992; O'Neil 2005). Poorer communities, women, and people of colour face greater dislocation, health problems, and loss of livelihood as a result of energy and natural resource extraction (Martinez-Alier 2002; Bell 2013). The final piece to the quadruple exposure effect comes in the form of greater exposure to toxic chemicals in the household, commercial foods and a variety of consumer products. As demonstrated in the case of Flint, Michigan, lead poisoning continues to be a leading health threat to children, particularly poor children and children of colour living in older, dilapidated housing with lead pipes. Black children are now five times more likely than white children to have lead poisoning. Taken together, it is clear that people of colour experience a disparate exposure to environmental hazards where they "work, live, and play" (Alston 1990).

In this chapter I will analyse the political-economic processes that serve to: 1) promote the mobility of ecologically hazardous industries into disempowered communities of colour and white working class neighbourhoods; 2) restrict the ability of the disempowered to move out of dangerous areas for safer neighbourhoods; 3) facilitate the dislocation of the disempowered from ecologically revitalized communities; and 4) limit the ability of the disempowered to leave dangerous jobs for safer occupations. Finally, I will outline the globalization of environmental injustices. In the same manner that environmental problems are being displaced onto disempowered communities inside the US, transnational corporations are also exporting harm onto disempowered communities and countries outside of the US.

Promoting the mobility of ecologically hazardous capital into disempowered communities

Following the path of least resistance often leads capital and the state to target white working class communities and communities of colour for the siting of hazardous industrial facilities and waste sites. That "disempowered" communities are to serve as such a pollution haven is often blatantly advertised. A 1984 report by Cerrell Associates for the California Waste Management Board, for instance, openly recommended that industry and the state locate waste incinerators (or "waste-to-energy facilities") in neighbourhoods of "lower socio-economic" status:

> Members of middle or higher-socioeconomic strata (a composite index of level of education, occupational prestige, and income) are more likely to organize into effective groups to express their political interests and views. All socioeconomic groupings tend to resent the nearby siting of major [polluting] facilities, but the middle and upper-socioeconomic strata possess better resources to affectuate their opposition. Middle and higher-socioeconomic strata neighbourhoods should not fall at least within the one-mile and five-mile radii of the proposed site.
>
> *(California Waste Management Board 1984: 42–43)*

The Cerrell Associates report also makes note of research indicating that communities made up of residents that are low-income, Catholic, Republican and/or conservative in political affiliation, of a low educational level, mostly senior citizens, and/or located in the South and Midwest of the United States, tend to exercise less control capacity over the siting of major polluting facilities.

California now has the highest concentration of racial/ethnic minorities living near incinerators and other commercial hazardous waste treatment, storage and disposal facilities (TSDFs). In Greater Los Angeles, for instance, some 1.2 million people live in close proximity (less than two miles) to seventeen such facilities, and 91 per cent of them (1.1 million) are people of colour (Bullard et al. 2007: 58–60). Of course, the question remains: which came first, the city's [Los Angeles] most polluted neighbourhoods or minority residents? Studies sponsored by the California Policy Research Center looked at the character of an area before a TSDF siting and the demographic and other shifts that occurred in the years after a siting. The findings indicate that since the 1970s the neighbourhoods targeted to house toxic storage and disposal facilities have more minority, poor, and blue-collar populations than areas that did not receive TSDFs (Pastor et al. 2001).

California is not alone when it comes to concentrating environmental problems in racially segregated communities. Environmental laws require capital to *contain* pollution sources for more proper treatment and disposal. Once the pollution is "trapped," the manufacturing industry pays for its treatment and disposal. The waste, now commodified, becomes mobile, crossing local, state, and even national borders in search of low-cost areas for treatment, incineration, and/or disposal. All across the US, disempowered communities are being targeted by corporate executives and state officials for the siting of hazardous facilities. Under the leadership of former EPA Administrator William Ruckelshaus, the waste management company Browning-Ferris Industries earned enormous profits (more than $1.6 billion alone in 1986) through an industry-wide *modus operandi* that kept costs down and profits high by locating the more dangerous facilities in neighbourhoods of colour within such cities as Birmingham, San Antonio, and Houston. Practices of "environmental racism" by BFI, Chemical Waste Management, and other "titans of waste" became rampant in the 1980s–90s, and fuelled the growth of the environmental justice (EJ) movement. For the first time in history, people of colour now comprise the majority of the population (56 per cent) living near the nation's commercial hazardous waste facilities (Bullard et al. 2007). African-Americans are also 79 per cent more likely than whites to live in neighbourhoods where industrial pollution is suspected of posing the greatest health danger.

Neighbourhoods undergoing rapid ethnic, racial, and class-based transitions (or "churning") are often the most vulnerable. Towns experiencing "white flight" to the suburbs and a corresponding demographic shift toward newly arrived Latino or Asian immigrants, for instance, often lack the tight community networks, political connections, and social capital necessary to mobilize residents to oppose ecologically hazardous facilities. Communities highly fragmented by peoples of different racial, ethnic, religious, and national-origin identities, class backgrounds, and languages can also be more vulnerable to the "divide and conquer" strategies of capital. Nevertheless, many poor but homogenous communities of colour have strong civic institutions (such as the Church) that build social solidarity and support long histories of struggle on behalf of civil rights. As such, and in contrast to the assumptions of the Cerrell Report, they can pose formidable opposition to corporate polluters (Ringquist 1997). Only those economically depressed communities burdened by poverty, high unemployment, and a marginal tax base will "choose" to accept hazardous facilities. Such a trade-off is sometimes made due to the potential for job creation, enhanced tax revenues and the provision of social services, and other economic benefits. In contrast, communities with a strong economic base and high degree of *control capacity* over the decision-making processes of local government officials and business leaders are better able to block the introduction of environmental hazards (Gould 1991).

It is important to focus on the systemic political-economic drivers of capital mobility by dirty industry into disempowered communities over the more narrow legalistic definitions that emphasize corporate "intentions" to discriminate. The primary goal of capital when siting hazardous operations in communities of colour is to seek out cheap land, favourable zoning laws, less regulation, good infrastructure, and a community less likely to offer opposition.

It is not necessarily to intentionally inflict harm upon working class whites or people of colour (Wolverton 2009, 2012). However, it is the legacy of systemic racism and class exploitation that creates such self-reinforcing social conditions in any given community: residents in the community may have been redlined by banks, reducing home ownership; educational segregation has undereducated the community's residents, producing not just "white flight" but "class flight" as well and leaving those remaining behind much less equipped to challenge such a siting; economic discrimination has forced residents into low-wage service jobs, requiring many to work multiple jobs and thus reducing overall civic involvement that forms the basis of any public efforts to resist a toxic hazard; economic deprivation further makes the promise of a handful of (dangerous) new jobs in the community difficult to resist; racist voter redistricting and gerrymandering has robbed the community of political power; and racist law enforcement and prosecution has incarcerated a whole generation who would otherwise have the energy and drive to oppose such predations on their homes and families. These structural and systemic political-economic conditions and more make it logical for polluting industry to locate operations in poor communities of colour and to do so without being overtly racist (Faber 2008).

Restricting the mobility of the disempowered out of ecologically hazardous communities

A second political-economic process creating environmental injustice involves preventing the mobility of the disempowered out of environmentally contaminated communities.

Wealthy white citizens tend to exercise greater control over community planning processes, including the "exclusionary zoning" of dirty industries and other locally unwanted land uses (LULUs). Wealthier citizens can also better afford to out-migrate (white flight) and purchase access to nicer neighbourhoods, better schools and housing, ecological amenities and a cleaner environment (South and Crowder 1997), suggesting that lower income households are more likely to be "left behind" in areas with hazardous facilities (Banzhaf and Walsh 2008). As stated by Sicotte (2016: 39), the

> affluent can mobilize their economic power to secure environmental privilege for themselves, which can include tasking other communities with disposing of their waste; restricting the access of others to "natural" or beautiful spaces; buffering such spaces from development or industrialization; and enjoying the convenience of living in centrally located places from which the less affluent have been priced out.

Wealthy whites can also employ various tactics for excluding people of colour and poorer whites from escaping hazardous communities by moving into their neighbourhoods. The ability of poor people of colour and ethnic minorities to migrate to "greener" pastures is limited by their lower incomes, zoning and urban planning policies, regressive taxation and discriminatory housing and mortgage lending (redlining) practices (Sicotte 2016: 39; Taylor 2014). Dating back to the National Housing Act of 1924, there is a disturbing historical pattern of mortgage lending in the US that serves to reproduce highly segregated patterns of residential location by race/ethnicity (Oliver and Shapiro 1995; Taylor 2014). Just a handful of towns in Massachusetts, for instance, account for the majority of loans given to African-Americans and Latinos. Just four communities typically receive more than half of all home-purchase loans to African-Americans, while five other communities receive more than half of all home-purchase loans for Latinos (Campen 2004). Each of these communities except one is ranked as among the thirty most environmentally overburdened communities in Massachusetts (Faber and Krieg 2005). In addition, African-Americans and Latinos *at all income levels* are more than twice as likely to be rejected for a home-purchase mortgage loan than are white applicants *at the same income levels*.

In summary, racial and ethnic segregation in the US is a product of the manner in which real estate developers, bankers, industrialists, and other sectors of capital work in coalition with government officials (at all levels) to form policy and planning structures which promote community development conducive to these business interests, i.e., local growth machines (Logan and Molotch 1987). It pays these interests to displace environmental health problems onto these communities where most residents lack health care insurance, possess lower incomes and property values, and are more easily replaced in the labour market if they become sick or die. Such externalities displaced upon the working poor are less likely to impose costs on capital and the larger economic system than if such harm was inflicted on the professional classes in whom there are significant investments of social capital. The siting of ecologically hazardous industrial facilities in communities of colour, as well as "minority move-ins" to already heavily polluted areas, are both governed by the same systemic logic of capitalist accumulation (Been and Gupta 1997). Such acts of environmental racism are perfectly rational from the perspective of capital and the political-economic logic of local growth machines.

Dislocating the disempowered from ecologically revitalized communities

A third political-economic process producing profound environmental injustices involves the dislocation of people of colour and working class whites from economically and ecologically revitalized communities (Faber and McDonough Kimelberg 2014). Urban sustainability initiatives and community redevelopment projects that create open space or otherwise aim to improve the environmental profile of a neighbourhood can trigger increases in real estate prices, rents and property taxes, leading to the economic displacement of the existing residents who had endured the deleterious effects of pollution and ecological degradation (Banzhaf and McCormick 2012). At the same time, the cleanup and revitalization of neighbourhoods can result in what Marcuse (1986) termed "exclusionary displacement," rendering the sustainable city inaccessible for future residents of limited economic means. Housing price increases hit the poor – the vast majority of whom are renters – especially hard. It is typically only landlords and homeowners who stand to capture the property value gains associated with gentrification (Banzhaf and McCormick 2012).

Environmental gentrification (Sieg et al. 2004) is becoming a major issue impacting low-income residents and people of colour in cities across the United States. The elimination of environmental disamenities (such as toxic waste sites) and the creation of environmental amenities (such as parks) can exact a large economic toll on vulnerable residents (Curran and Hamilton 2012; Gamper-Rabindran and Timmins 2011). Black and Latino populations often decrease significantly after the revitalization of land contaminated with toxic chemicals (Essoka 2010). Similarly, as Gould and Lewis (2012) revealed, the restoration of Brooklyn's Prospect Park in the 1990s led to a significant increase in new construction around the park and a corresponding decrease in the racial and socioeconomic diversity of those areas. In many cases, the displaced residents migrate to more environmentally distressed parts of the city where housing costs are lower (Banzhaf and McCormick 2012: 39–41).

It is clear that a failure to adequately address the social justice dimensions of urban sustainability initiatives can contribute to environmental gentrification (Dooling 2009; Gibbs and Krueger 2007; Pearsall 2012; Quastel 2009). Such a failure is typically grounded in "asymmetrical power relations . . . [that] continually influence how and what kinds of 'environmental' issues are addressed" (Tretter 2013: 308). The threats posed by gentrification are now presenting the EJ movement with a "pernicious paradox" (Checker 2011). Are successful EJ struggles by the working class and people of colour to make their urban environments "greener" likely to yield unintended consequences in the form of the eventual displacement and relocation of these same residents into other polluted communities, where rents and housing prices are cheaper? Must lower income residents "reject environmental amenities in their neighbourhood in order to resist gentrification that tends to follow . . ." (Checker 2011: 211)?

Banzhaf and McCormick (2012: 39–41) conclude that although there are exceptions, "the evidence seems clear that in most cases improvements in local environmental conditions do trigger increases in property prices." For example, the cleanup of Superfund and other brownfield sites – and in some cases, even the *anticipation* of future remediation – results in rising land values, housing values and/or rents (Gamper-Rabindran and Timmins 2013; Pearsall 2010). Furthermore, the "greening" of urban spaces is often accompanied by new retail stores, restaurants, and amenities explicitly targeted toward middle and upper class residents (Bryson 2012; Dale and Newman 2009). This remaking of the commercial and social aspects of neighbourhoods can serve to alienate or marginalize lower income residents and especially the homeless (Dooling 2009, 2012), not only because participation is cost-prohibitive to those with limited means but also because it serves to reinforce class-based symbolic and social boundaries within the community (Lamont and Molnar 2002).

Restricting the mobility of the disempowered out of ecologically hazardous occupations

Similar to mainstream environmental policy, most worker health and safety programmes add to the costs of capital and restrict or prevent the use of more profitable (and more hazardous) chemical substances, materials, and production processes. Industries under stronger competitive pressures from low-cost operations overseas are especially eager to avoid "internalizing" costs on such "unproductive expenditures" as worker health and safety, and will instead displace (or "externalize") these costs onto their labour force in the form of dangerous working conditions and exposure to health hazards (Morgenson 2005). According to Liberty Mutual, the nation's largest workers' compensation insurance company, the direct cost of occupational injury and illness is $48.6 billion (nearly $1 billion per week), with another $145 billion to $290 billion in indirect costs. Some 135 workers die *every day* from diseases caused by exposure to toxins in the workplace.

Not all workers face the same level of health threats on the job. Highly skilled workers are more essential to many businesses and often not easily replaced if they become injured or sick, and are therefore provided greater protection by industry and unions. As a result, unskilled and semi-skilled blue-collar workers involved in manufacturing, construction, logging, and agriculture face greater occupational hazards on the job. Workers in these industries are more "expendable," as they are more easily replaced by other people if an injury or death occurs. In fact, economic damages awarded in tort law are in large part based on wage loss. A restaurant worker earning a low hourly wage is simply 'worth' far less than a highly experienced and well-paid company manager. Since the usual penalty for inflicting environmental and/or occupational disease is the 'restitution' of the injured through the payment of compensatory fines rather than criminal penalties or confiscatory fines, the costs almost never approach the economic advantages that accrue to companies that perpetrate injury and death upon American workers. In other words, it can be cheaper to use unsafe technology, poisonous chemicals, and dangerous production processes that kill or maim unskilled workers and pay the fine than to make the workplace safe (Eligman and Rankin Bohme 2005).

As with pollution and other ecological hazards, the imposition of occupational health and safety dangers upon workers is anti-social action on behalf of capital, and is often met with resistance. The most cost-effective path of least resistance on behalf of capital is to select the most disempowered members of society for the most dangerous occupations. The evidence points to the fact that poorer people of colour and immigrant workers are once again being tracked into the most hazardous types of jobs, and that these jobs are becoming ever more hazardous.

Occupational exposures to cancer-causing substances, pesticides and toxic chemicals, and dangerous working conditions are especially prevalent for people of colour. In California, for instance, Hispanic men have a two-and-a-half times greater risk of occupational disease and injury than white men (Robinson 1989).

Despite the implementation of affirmative action programmes and other accomplishments by the Civil Rights Movement over the past three decades, the racial segmentation of labour persists. The continued implementation of informal "job closure" practices by business (and some unions) restricts occupational mobility for racial and ethnic minorities into safer and better paying jobs (Wright 1992). Business owners and managers regularly rank people of differing racial and ethnic backgrounds for specific job categories. White workers are typically placed at the head of the line (or queue) for the most desirable jobs, especially those offering better working conditions, higher pay, and opportunities for advancement. People of colour and ethnic minorities are typically placed at the end of the queue (Kerbo 2011).

The racial segmentation of labour by this method is functional for capital in that the "racialization" of certain occupations depresses wages/benefits, divides labour against itself and inhibits unionization, and provides a large pool of unemployed/underemployed workers that industry can draw from in periods of rapid economic expansion. Just as importantly, the racial segmentation of labour inhibits the ability of workers of colour to escape dangerous jobs for safer occupations, just as racial segregation of communities inhibits the ability of residents of colour to escape ecologically dangerous neighbourhoods for safer areas. To provide a labour force for dirty industry, semi-skilled blue collar workers may be channelled to live in row homes near industrial zones; while unskilled, underemployed members of the "underclass" are pushed into distressed inner-city neighbourhoods, and serve as a reserve army of cheap labour for nearby dirty industries. Not coincidentally, African-Americans, Latinos, and the working poor face much greater risks for living in much closer proximity (location risk) to the most accident-prone facilities (operations risk) in the US (Elliott et al. 2004).

Knowing that occupational mobility is limited, capital can place greater demands upon workers of colour, and also slash costs relating to occupational health and safety programmes. Occupational dangers are even more profound for those unskilled or semi-skilled immigrants and undocumented workers that lack the formal legal protections afforded by US citizenship (Anderson et al. 2000). Mexicans are about 80 per cent more likely to die on the job than native-born workers (compared to 30 per cent in the mid-1990s). Often in the country "illegally," and reluctant to complain about poor working conditions for fear of deportation or being fired, they are nearly twice as likely as the rest of the immigrant population to die at work. Mexicans also make up the largest segment of migrant farmworkers. Over 313,000 of the 2 million farm workers in the United States – 90 per cent of who are people of colour and mostly undocumented immigrants – suffer from pesticide poisoning each year (Perfecto 1992). Struggling to protect farmworkers and their communities from these poisons, the Farmworker Network for Economic and Environmental Justice is now a key component of the environmental justice movement.

Globalization and the export of environmental injustice

The worsening ecological crisis in the Global South is directly related to a global system of economic and environmental stratification in which the United States and other advanced capitalist nations are able to shift or impose a growing environmental burden on weaker states. Given the weakness of national environmental policy in most developing nations, the lack of rigorous environmental laws and sanctions against corporate polluters, the anti-environmental content of the so-called "free-trade" agreements, and a growing willingness by many desperately poor countries to accept long-term environmental problems in exchange for short-term economic gains, the growing mobility of capital is facilitating the export of environmental problems from the advanced capitalist countries to the Global South and sub-peripheral states.

Similar to the displacement of ecological hazards onto the disempowered *inside* their own countries, American and European businesses are also displacing environmental harm upon marginalized communities *outside* of their own countries (Faber 2008; Pellow 2007). This *export of ecological hazard* from the United States and other Northern countries to the less developed countries takes place: 1) *in the money circuit of global capital*, in the form of foreign direct investment (FDI) in domestically owned hazardous industries, as well as destructive investment schemes to gain access to new oil fields, forests, agricultural lands, mining deposits, and other natural resources; 2) *in the productive circuit of global capital*, with the relocation of polluting and environmentally hazardous production processes and polluting facilities owned by transnational capital to the South; 3) *in the commodity circuit of global capital*, as witnessed in the marketing of more profitable but also more dangerous foods, drugs, pesticides, technologies, and other consumer/capital goods; and 4) *in the waste circuit of global capital*, with the dumping of toxic wastes, pollution, discarded consumer products, trash, and other forms of "anti-wealth" produced by Northern industry.

Under the new ecological imperialism brought about by globalization, the prosperity of the United States and other advanced capitalist countries is becoming increasingly predicated on the appropriation of surplus environmental space from the South. By expanding its *ecological footprint* and other forms of *unequal ecological exchange*, economic growth in the North is dependent upon the confiscation of biomass production from the South (Hornborg 2011). In other words, the expansion of wealth on behalf of the US and the core states under globalization fundamentally involves the use of greater quantities of undervalued natural resources from other territories, as well as the increased displacement of environmental harm (such as pollution) to the developing world, and is creating an unparalleled ecological crisis of global dimensions (Jorgenson and Clark 2012).

Defined in terms of Global North versus Global South, corporate-led globalization is seen as magnifying externally- and internally-based environmental injustices to the advantage of the United States and other core nations. In much of the developing world, access to natural resources is being restricted by the transformation of commonly held lands (the commons) into capitalist private property, that is, the "commodification of nature" (Goldman 1998).

Those peoples in the Global South who draw their livelihood directly from the land, water, forests, coastal mangroves, and other ecosystems are becoming displaced in order to supply cheap raw materials for the dominant classes and foreign capital. Labouring in service of this new global order, but receiving few of its benefits, the popular majorities of the developing world – the poor peasants, workers, ethnic minorities and Indigenous peoples who make up the subsistence sector – struggle to survive by moving onto ecologically fragile lands or by migrating to the shantytowns of the cities by the millions to search for employment. Often left with little means to improve the quality of their lives, the world's poor (especially women) are being forced to over-exploit their own limited natural resource base in order to survive (Shiva 2005). In much of the third world, these survival strategies by the popular classes in response to their growing impoverishment are resulting in the widespread degradation and ecological collapse of the environment. As a result, globalization-inspired development models are becoming increasingly unviable in the global South, giving birth to popularly based movements for social and ecological justice – an *environmentalism of the poor* (Martinez-Alier 2002).

Conclusion

The threats posed to the profits of major corporate polluters by the environmental and EJ movement are invoking a profound political backlash on the Right. Spearheaded by agribusiness, oil and gas, mining, timber, petrochemical, and manufacturing industries, these corporate polluters are channelling enormous sums of money into anti-environmental organizations, public relations firms, foundations, think thanks, research centres, and policy institutes, as well as the election campaigns of "pro-business" candidates in both major political parties. Motivated by the real and potential costs of environmental protection, the goal of this "polluter-industrial complex" is neo-liberal regulatory reform – the rollback of the environmental justice policy, worker health and safety, consumer protection, environmental protection, and other state regulatory "burdens" (Faber 2008). Neo-liberals are also blocking or slowing the introduction of more progressive environmental and EJ legislation, weakening or preventing the enforcement of existing regulations, and delegating programmes to financially strapped local and state governments lacking the capacity to assume the task. As a result, the ability of the environmental justice, climate change, and ecology movements to win truly comprehensive policy has been effectively compromised.

The news is not all bad, however. More than ever, people are fighting for their basic right to a clean and healthy environment. In poor African-American and Latino neighbourhoods of small towns and inner cities, depressed Native American reservations, and Asian-American communities all across the country, people who have traditionally been relegated to the periphery of the environmental movement are increasingly challenging the ruination of their land, water, air, and community health by corporate polluters and indifferent governmental agencies. Acting in coalition with the rise of new forms of community-based, working class environmentalism, anti-toxics activism, climate change advocacy, and the clean production movement, the EJ movement is slowly but surely developing networks and long-term strategies for arresting the ecological crisis. As such, the continued growth and prosperity of these EJ organizations and networks is essential to constructing a more inclusive, democratic, and pro-active environmental politics capable of addressing the political-economic roots of environmental injustice.

References

Alston, D. (1990). *We speak for ourselves: social justice, race, and environment.* Washington DC: The Panos Institute.

Anderson, J.T.L., Hunting, K., and Welch, L. (2000). "Injury and employment patterns among Hispanic construction workers." *Journal of Occupational and Environmental Medicine*, vol. 42, no. 2, pp. 176–186.

Atkinson, R. (2003). "Introduction: misunderstood savior or vengeful wrecker? The many meanings and problems of gentrification." *Urban Studies*, vol. 40, pp. 2343–2350.

Banzhaf, H.S. (2008). "Environmental justice: opportunities through markets." *Property and Environment Research Center Policy Series*, vol. 42, pp. 1–25.

Banzhaf, H.S. & McCormick, E. (2012). "Moving beyond cleanup: identifying the crucibles of environmental gentrification." In H.S. Banzhaf (ed.), *The political economy of environmental justice*. Stanford, CA: Stanford University Press, pp. 23–51.

Banzhaf, H.S. & Walsh, R.R. (2008). "Do people vote with their feet? An empirical test of Tiebout's mechanism." *American Economic Review*, vol. 98, no. 3, pp. 843–863.

Been, V. & Gupta, F. (1997). "Coming to the nuisance or going to the barrios? A longitudinal analysis of environmental justice claims." *Ecology Law Quarterly*, vol. 24, no. 1, pp. 1–56.

Bell, S. (2013). *Our roots run as deep as ironweed: Appalachian women and the fight for environmental justice.* Champaign, IL: University of Illinois Press.

Boyce, J., Zwickl, K., and Ash, M. (2014). "Three measures of environmental inequality." Institute for New Economic Thinking Working Group on the Political Economy of Distribution, Working Paper No. 12.

Boyle, M.A. & Kiel, K.A. (2001). "A survey of house price hedonic studies of the impact of environmental externalities." *Journal of Real Estate Literature*, vol. 9, no. 2, pp. 117–144.

Bryson, J. (2012). "Brownfields gentrification: redevelopment planning and environmental justice in Spokane, Washington." *Environmental Justice*, vol. 5, no. 1, pp. 26–31.

Bullard, R.D. (1994). *Dumping in Dixie: race, class, and environmental quality.* Boulder, CO: Westview Press.

Bullard, R.D., Mohai, P., Saha, R., and Wright, B. (2007). *Toxic wastes and race at twenty: 1987–2007 – Grassroots struggles to dismantle environmental racism in the United States.* United Church of Christ Justice and Witness Ministries, pp. 1–68.

Bunce, S. (2009). "Developing sustainability: sustainability policy and gentrification on Toronto's waterfront." *Local Environment*, vol. 14, no. 7, pp. 651–667.

California Waste Management Board. (1984). *Political Difficulties Facing Waste-to-Energy Conversion Plant Siting.* Los Angeles: Cerrell Associates.

Campen, J. (2004). "The colour of money in greater Boston: patterns of mortgage lending and residential segregation at the beginning of the new century." Metro Boston Equity Initiative of the Harvard University Civil Rights Project.

Checker, M. (2011). "Wiped out by the "greenwave": environmental gentrification and the paradoxical politics of urban sustainability." *City & Society*, vol. 23, no. 2, pp. 210–229.

Collins, M., Munoz, I., and JaJa, J. (2016). "Linking 'toxic outliers' to environmental justice communities," *Environmental Research Letters*, vol. 11, no. 1, pp. 1–22.

Curran, W. & Hamilton, T. (2012). "Just green enough: contesting environmental gentrification in Greenpoint, Brooklyn." *Local Environment*, vol. 17, no. 9, pp. 1027–1042.

Dale, A. & Newman, L. L. (2009). "Sustainable development for some: green urban development and affordability." *Local Environment*, vol. 14, no. 7, pp. 669–681.

Dooling, S. (2009). "Ecological gentrification: a research agenda exploring justice in the city." *International Journal of Urban and Regional Research*, vol. 33, no. 3, pp. 621–639.

Dooling, S. (2012). "Sustainability planning, ecological gentrification and the production of urban vulnerabilities." In Dooling, S. & G. Simon (eds.), *Cities, nature and development: The politics and production of urban vulnerabilities*. Farnham, UK: Ashgate, pp. 101–119.

Eckerd, A. (2011). "Cleaning up without clearing out?: A spatial assessment of environmental gentrification." *Urban Affairs Review*, vol. 47, no. 1, pp. 31–59.

Eligman, D.S. & Rankin Bohme, S. (2005). "Over a barrel: political corruption of science and its effects on workers and the environment." *International Journal of Occupational and Environmental Health*, vol. 11, no. 4, pp. 331–337.

Elliott, M.R., Wang, Y., Lowe, R.A., and Kleindorfer, P.R. (2004). "Environmental justice: frequency and severity of chemical industrial accidents and the socioeconomic status of surrounding communities," *Journal of Epidemiology and Environmental Health*, vol. 58, pp. 24–30.

Essoka, J.D. (2010). "The gentrifying effects of brownfields redevelopment." *Western Journal of Black Studies*, vol. 34, no. 3, pp. 299–315.

Faber, D. (1998). "The political ecology of American capitalism: new challenges for the environmental justice movement." In Faber, D. (ed.), *The struggle for ecological democracy: environmental justice movements in the United States*. New York: Guilford Press, pp. 27–59.

Faber, D. (2008). *Capitalizing on environmental injustice: the polluter-industrial complex in the age of globalization*. New York, NY: Rowman & Littlefield.

Faber, D. & McDonough Kimelberg, S. (2014). "Sustainable urban development and environmental gentrification: the paradox confronting the U.S. environmental justice movement," in Hall, H.R., C.C. Robinson, and A. Kohli (eds.), *Uprooting urban America: multidisciplinary perspectives on race, class and gentrification*. New York: Peter Lang Publishers, pp. 77–92.

Faber, D. & Krieg, E.J. (2005). *Unequal exposure to ecological hazards 2005: environmental injustices in the commonwealth of Massachusetts*. Philanthropy and Environmental Justice Research Project, Northeastern University.

Gamper-Rabindran, S. & Timmins, C. (2011). "Hazardous waste cleanup, neighborhood gentrification, and environmental justice: evidence from restricted access census block data." *American Economic Review Papers and Proceedings*, vol. 10, no. 3, pp. 620–624.

Gamper-Rabindran, S. & Timmins, C. (2013). "Does cleanup of hazardous waste sites raise housing values? Evidence of spatially localized benefits." *Journal of Environmental Economics and Management*, vol. 65, no. 3, pp. 345–360.

Gibbs, D. C. & Krueger, R. (2007). "Containing the contradictions of rapid development? New economy spaces and sustainable urban development." In Krueger, R. & D.C. Gibbs (eds.), *The sustainable development paradox: urban political economy in the United States and Europe*. New York, NY: Guilford Press, pp. 95–122.

Gin, J. & Taylor, D.E. (2010). "Movements, neighbourhood change, and the media—newspaper coverage of anti-gentrification activity in the San Francisco Bay area: 1995–2005." In Taylor, D. (ed.), *Environment and social justice: an international perspective* (Research in Social Problems and Public Policy), vol. 18. Bingley, UK: Emerald Group, pp. 75–114.

Goldman, M. (1998). *Privatizing nature: political struggles for the global commons*. New Brunswick: Rutgers University Press.

Gould, K.A. (1991). "The sweet smell of money: economic dependency and local environmental political motivation." *Society and Natural Resources*, vol. 4, no. 2, pp. 133–150.

Gould, K.A. & Lewis, T.L. (2012). "The environmental injustice of green gentrification: the case of Brooklyn's Prospect Park." In DeSena, J. & T. Shortell (eds.), *The world in Brooklyn: gentrification, immigration, and ethnic politics in a global city*. Lanham, MD: Lexington Books, pp. 113–146.

Holifield, R. (2001). "Defining environmental justice and environmental racism." *Urban Geography*, vol. 22, no. 1, pp. 78–90.

Hornborg, A. (2011). *Global ecology and unequal exchange: fetishism in a zero-sum world*. London: Routledge.

Johnston, B.R. (1994). *Who pays the price?: The sociocultural context of environmental crisis*. Washington, DC: Island Press.

Jorgenson, A.K. & B. Clark (2012). "Footprints: the division of nations and nature," In Hornborg, A., B. Clark and K. Hermele (eds.), *Ecology and power: struggles over land and material resources in the past, present, and future*. London: Routledge, pp. 155–167.

Kapp, W.K. (1950). *The social cost of private enterprise*. New York: Schocken Books.

Kerbo, H. (2011). *Social stratification and inequality: class conflict in history, comparative, and global perspective*. 8th Edition. New York: McGraw-Hill.

Lamont, M. & Molnar, V. (2002). "The study of boundaries in the social sciences." *Annual Review of Sociology*, vol. 28, pp. 167–195.

Lavelle, M. & Coyle, M. (1992). "Unequal protection: the racial divide in environmental law," *National Law Journal*, vol. 21, pp. 2–12.

Lerner, S. (2010). *Sacrifice zones: the front lines of toxic chemical exposure in the United States*. Cambridge: MIT Press.

Logan, J. & Molotch, H. (1987). *Urban fortunes: the political economy of place*. Berkeley: University of California Press.

Marcuse, P. (1986). Abandonment, gentrification, and displacement: the linkages in New York City. In Smith, N. and P. Williams, (eds.), *Gentrification of the city*. London, UK: Unwin Hyman, pp. 153–177.

Martinez-Alier, J. (2002). *The environmentalism of the poor: a study of ecological conflicts and valuation*. Northampton, MA: Edward Elgar.

Morello-Frosch, R. (1997). "Environmental justice and California's 'riskscape': the distribution of air toxics and associated cancer and non-cancer health risks among diverse community." PhD dissertation, School of Public Health, Environmental Health Sciences Division, University of California at Berkeley.

Morgenson, V. (2005). *Worker safety under siege: labor, capital, and the politics of workplace safety in a deregulated world.* Armonk, NY: M.E. Sharpe.

O'Connor, J. (1998). *Natural causes: essays in ecological Marxism.* New York: Guilford, 1998.

O'Neil, S.G. (2005). *Environmental justice in the Superfund clean-up process,* PhD Dissertation, Department of Sociology, Boston College.

Oliver, M. & Shapiro, T. (1995). *Black wealth/white wealth: a new perspective on racial inequality.* New York: Routledge.

Pastor, M., Sadd, J., and Hipp, J. (2001). "Which came first? Toxic facilities, minority move-in, and environmental justice." *Journal of Urban Affairs,* vol. 23, no. 1, pp. 1–21.

Pearsall, H. (2010). "From brown to green? Assessing social vulnerability to environmental gentrification in New York City." *Environment and Planning C: Government and Policy,* vol. 28, pp. 872–886.

Pearsall, H. (2012). "Moving out or moving in? Resilience to environmental gentrification in New York City." *Local Environment,* vol. 17, no. 9, pp. 1013–1026.

Pellow, D.N. (2007). *Resisting global toxics: transnational movements for environmental justice.* Cambridge: The MIT Press.

Perfecto, I. (1992). "Farm workers, pesticides, and the international connection." In Mohai, P. and B. Bryant (eds.), *Race and the incidence of environmental hazards: a time for discourse.* Boulder, CO: Westview Press, pp. 177–203.

Pulido, L. (1996). *Environmentalism and economic justice: two Chicano struggles in the southwest.* Tucson: The University of Arizona Press.

Quastel, N. (2009). "Political ecologies of gentrification." *Urban Geography,* vol. 30, no. 7, pp. 694–725.

Ringquist, E.J. (1997). "Equity and the distribution of environmental risk: the case of TRI facilities." *Social Science Quarterly,* vol. 78, pp. 811–818.

Robinson, J.C. (1989). "Exposure of occupational hazards among hispanics, blacks, and non-hispanic whites in California." *American Journal of Public Health,* vol. 79, pp. 629–630.

Schnaiberg, A. (1994). "The political economy of environmental problems and policies: consciousness, coordination, and control capacity." In Freese, L. (ed.), *Advances in human ecology.* Greenwich, CT: JAI Press, pp. 23–64.

Shiva, V. (2005). *Earth democracy: justice, sustainability, and peace.* Boston: South End Press.

Sicotte, D. (2016). *From workshop to waste magnet: environmental inequality in the Philadelphia region.* New Brunswick, NJ: Rutgers University Press.

Sieg, H.V., Smith, K., Banzhaf, H.S., and Walsh, R. (2004). "Estimating the general equilibrium benefits of large changes in spatially delineated public goods." *International Economic Review,* vol. 45, no. 4, pp. 1047–1077.

South, S.J. & Crowder, K.D. (1997). "Escaping distressed neighborhoods: individual, community, and metropolitan influences." *The American Journal of Sociology,* vol. 102, no. 4, pp. 1040–1084.

Taylor, D. (2014). *Toxic communities: environmental racism, industrial pollution, and residential mobility.* New York: NYU Press.

Tretter, E.M. (2013). "Contesting sustainability: "SMART growth" and the redevelopment of Austin's eastside." *International Journal of Urban and Regional Research,* vol. 37, no. 1, pp. 297–310.

Walker, G. (2012). *Environmental justice: concepts, evidence, and politics.* London: Routledge.

Wolverton, A. (2009). "Effects of socio-economic and input-related factors on polluting plants' location decisions." *The B.E. Journal of Economic Analysis & Policy,* vol. 9, no. 1, pp. 1–30.

Wolverton, A. (2012). "The role of demographic and cost-related factors in determining where plants locate: a tale of two Texas cities." In Banzhaf, H.S. (ed.), *The political economy of environmental justice.* Stanford, CA: Stanford University Press, pp. 199–222.

Wright, B.H. (1992). "The effects of occupational injury, illness and disease on the health status of black Americans: a review." In Bryant, B. and P. Mohai (eds.), *Race and the incidence of environmental hazard: a time for discourse.* Boulder, CO: Westview Press.

7

FEMINISM AND ENVIRONMENTAL JUSTICE

Greta Gaard

What defines a "feminist issue"? From Rachel Carson's (1962) critique of pesticides and their effects on birds, water, and children, to Wangari Maathai's (2004) analysis of the interconnections among deforestation, desertification, and women's subsistence farming that prompted her to launch Kenya's Green Belt Movement, a feminist standpoint places women's lives in social and ecological contexts, augmenting the feminist slogan that "the personal is political" – and ecological too. Not merely an academic endeavour or "way of seeing," both feminism and environmental justice (EJ) emerged from the lives of women who recognized their own experiences of injustice as fundamentally interconnected with the health and well-being of others.

The internal and international diversities of the twenty-first century climate justice movement make it difficult to imagine that the intersectionalities among gender, race, class, sexuality, age, ability, species and environment that have coalesced into a global movement are still being discovered by scholars and activists alike (see also Chapter 29). Women comprise an estimated 60–80 per cent of members in environmental organizations worldwide, and an estimated 90 per cent of members in US environmental justice organizations (Hallum-Montes 2012). Yet, environmental sociologists and environmental justice theorists focus on only one or two social markers (i.e., race and/or class) in studying the impact of toxic environments on individuals and communities (see also Chapter 4). At the same time, transnational feminist and gender studies scholars have largely neglected both environment and species as analytical categories for research, focusing their attentions on human issues of gender, race, class, sexuality, age, and to a lesser extent, ability. While the still-radical concept of *intersectionality* articulated by Black feminists (Collins 1990; Crenshaw 1989) co-occurred with the rise of the environmental justice movement, it simultaneously excluded environments from its analysis. Frameworks that incorporate gender, sexuality, age and ability along with race and class are largely absent from environmental justice discourse, thereby obscuring the ways that gender and gendered labour shapes women, men, and trans★[1] persons' experiences of environments and environmental problems. While the concepts of environmental racism, and to a lesser extent, environmental classism are well known, the analytical categories of environmental sexism, environmental heterosexism, environmental ageism and ableism, not to mention environmental speciesism, are almost unthinkable.

One notable exception to this intersectional silence on gender and environment is ecofeminism, which has addressed the linked exploitation of women and nature since the

1980s, exploring the mutually reinforcing exploitations of race, class, gender, species, and sexualities. Initially an activist-based movement with a perspective articulated through multi-vocal anthologies (Caldecott & Leland 1983; Plant 1989; Diamond & Orenstein 1990), ecofeminist philosophy of the 1990s was critiqued for perpetuating an essentialist equation of women/nature, universalizing "woman" in ways that excluded racial differences (Gaard 2011). Ecofeminist scholars responded, developing intersectional analyses that centred race and gender along with species, sexuality, and colonialism, focusing on the material conditions of women's lives and women's marginalized participation in policy, government, and economic systems that affect their lives (Salleh 1997; Hawley 2015). Global ecofeminist alliances such as the Women's Environment and Development Organization (WEDO) were launched, bringing women of multiply diverse races and nationalities into leadership and visibility, advising the United Nations and foregrounding gender within international environmentalisms.

At the same time that both environmental justice and ecofeminist theories were being developed, women whose activism connects human and environmental health were facing harsh criticism and brutal violence. In Bali, for instance, Mardiana Deren, an Indonesian nurse campaigning against palm oil and mining companies, was run over by motorbikes and barely escaped a stabbing; in Brazil, the body of activist Nilce de Souza Magalhães was found below a hydroelectric dam she had publicly opposed, her hands and feet tied with ropes and attached to large rocks that had kept her submerged for six months. In Honduras, the 2015 Goldman Environmental Award-winner and Indigenous activist Berta Cáceres was shot dead in her home, less than a year after her successful campaign stopped a hydroelectric dam that would have flooded native lands and cut off water supplies. "As women," said Cáceres in a 2014 interview, "we are exposed to violence from businesses, governments and repressive institutions – but also to patriarchal violence. It is three times worse for an indigenous woman" (Win 2015).

From the Euro-American social workers of the settlement house movement to the African- and Asian-American, Chicana/Latina and Indigenous women activists fighting toxic wastes and ecologically damaging industrialism and colonialism, women differently situated by race, ethnicity, or nation have framed their eco-justice activism as an extension of women's gendered role as caregivers, or as a response to the linked devaluation of women, communities of colour, poor people and environments.

Around the world, women are on the frontlines of climate justice crises as well as climate justice solutions. As one of the top two countries most responsible for climate change (Clark 2011; Ravilious 2014), the United States has also been a primary location for the growth of feminism and environmental justice, movements with roots in women's urban activism as municipal housekeepers, as advocates for equal suffrage, and for civil and Indigenous rights.

Movement roots

Through the colonization of Indigenous lands, the capture and importation of African slaves, and the conversion of ecological health into economic wealth, the United States' history has deeply affected the shape of US women's activism.

Women's clubs and "social housekeeping"

After the Civil War officially ended slavery, the African-American population shifted from the south to the north and west, and from rural agriculture to urban industrial centres. In these new urban centres, the Black Women's Club movement was launched and grew to prominence through the efforts of middle-class women such as Ida B. Wells-Barnett and Mary Church Terrill, both of whom spoke and wrote ardently for Black equality, woman suffrage, and the rights of working people (Olson 2001). This movement engaged in home and neighbourhood clean-up campaigns, working to reduce the diseases that arose from unsafe air and tuberculosis-contaminated water, inadequate sewerage, and garbage removal. Black women participated in the American Equal Rights Association, and later in both the National Woman Suffrage Association and the American Woman Suffrage Association.

Euro-American middle-class women also made advances in social reform and education that served the larger community. Chemist Ellen Swallow Richards created the new fields of "oekology" and home economics through her investigations of water pollution and water quality, sewage disposal, and her efforts in launching the first school lunch programmes. In 1889, Jane Addams and her paramour Ellen Gates Starr co-founded Hull House, a settlement house in Chicago, and mobilized the idea of "civic housekeeping," encouraging educated women to extend their gendered caregiving to the larger society. The settlement movement addressed a range of social and environmental issues among immigrant and working class communities, from housing, midwifery, and health care to typhoid and sanitation, education and play, art and intercultural studies (Mann 2011).

Both African-American and Euro-American women worked to improve living and working conditions for women and their families, and both groups used women's networks of neighbourhood, church, family and social circles as the basis for organizing. Yet, these activist women tended to operate separately, divided by racial differences even though the movement for women's suffrage began when middle and upper class White women abolitionists were denied entry to an abolitionist meeting. After Black men were granted the vote in 1870, women's exclusion from suffrage on the basis of gender could have unified these activist groups – but White women's racism kept them divided.

Civil rights

From 42-year-old Rosa Parks, who was the third woman to refuse to give up her seat on the bus; to Jo Ann Robinson, who organized the Montgomery Bus Boycotts in 1955; Fannie Lou Hamer, who organized voter registrations and delivered inspirational speeches; Ella Baker, who organized both the Southern Christian Leadership Conference and later helped found the Student Nonviolent Coordinating Committee, two key movement organizations; and Daisy Bates, who led and mentored the first nine African-American students enrolled in Little Rock Central High School in 1957: these and so many more grassroots Black women powered the Civil Rights movement (Olson 2001).

The visibility and control of Black men in the Civil Rights movement has been attributed to a strategic decision for winning a racial battle within a White patriarchal culture. But there were more compelling reasons as well. Black women had enormous empathy for the ways Black men had been emasculated by White racism and slavery: mocked, belittled, and beaten, Black men experienced racial oppression as did Black women. While Black women had suffered sexual slavery at the hands of White slave-owners, after the Civil War ended and Blacks were no longer valuable as "property," the lynchings of Black men united the Black community against White racism. In the movement toward racial equality, reclaiming Black manhood seemed to include embracing male dominance and enforcing the sexist gender roles of the larger White culture. Like other Black ministers, Martin Luther King was known for his belief that women should be good wives and mothers, staying at home and not participating in the movement, yet his early rise to prominence was a result of these grassroots women's efforts, for the male leaders often failed to capitalize on activist momentum or envision radical strategies like those initiated by women – i.e., the Montgomery bus boycott, or the Children's Crusade (Olson 2001).

On August 28, 1963, the diversity of the Black community was present at the March on Washington for Jobs and Freedom, just as it was during the Harlem Renaissance (Garber 1991) – but only the heterosexual Black men were allowed to speak. Two gay Black men, Bayard Rustin and James Baldwin, were excluded from giving speeches of their own: although Rustin was a co-organizer of the march, he was relegated to reading the list of demands after King's speech, while Baldwin was excluded entirely on the grounds that his words would agitate the crowd. The grassroots women who had powered the movement were also silenced, given a "tribute" and a round of applause.

Woman of colour feminisms

As second-wave feminism evolved through the social movements of the 1960s, queer and women-of-colour feminisms emerged as departures from both White feminism and other 1960s social movements such as Gay Rights, Black Power, Chicano Movimiento, and the American Indian Movement (AIM). Attempting "to bridge the contradictions" of experience, Cherríe Moraga and Gloria Anzaldúa (1981) wrote:

> We are the colored in a white feminist movement.
> We are the feminists among the people of our culture.
> We are often the lesbians among the straight.
> We do this bridging by naming our selves and by telling our stories in our own words.

For Anzaldúa's Chicana feminism, identity was a nexus of indigeneity and Spanish conquest, a location on the borderlands producing the intersections of gender, sexuality, ethnicity, and place. Her work laid the groundwork for Chicano environmental justice.

At Wounded Knee on the Pine Ridge reservation in South Dakota, a similar nexus and emergence through place and history produced the Indigenous women's movement. In July 1968, the American Indian Movement (AIM) formed to address Indigenous sovereignty and treaty rights as well as issues of police harassment and racism, the deplorable conditions of Indigenous people in urban centres, and the corporate exploitation of tribal lands through uranium and coal mining. When events at Wounded Knee made it too dangerous to continue activism as affiliates with AIM, women such as Madonna Thunder Hawk and Lorelei DeCora organized Women of All Red Nations (WARN) in 1974, addressing not only treaty rights and the elimination of Indian mascots from sports teams, but also the health of American Indian women, the frequency of forced sterilizations and non-Indian adoptions of Indian children as forms of eugenics. These and many other environmental health issues inspired the Indigenous Women's Network to continue this work in 1985, with activism that laid the foundation for the Indigenous Environmental Network.

Black feminism took shape through the 1973 formation of the National Black Feminist Organization, the 1977 Combahee River Collective Statement, the founding of Kitchen Table Women of Color Press in 1981, and the first volume of Black Women's Studies, *But Some of Us Are Brave* (Hull et al. 1982). Alice Walker (1983) coined the term "womanist" to describe woman-of-colour-feminism, listing its multiple meanings as "a black feminist or feminist of colour" as well as "a woman who loves other women, sexually and/or nonsexually" and concluding "womanist is to feminist as purple is to lavender." Activist-theorists such as

Audre Lorde (1984) articulated Black feminism in pointed and accessible language, influencing White women's liberal and cultural feminisms to replace the single-focus analysis of sexism with a tripartite analysis of gender, race, and class as indissolubly linked in feminist perspectives. In the 1990s black feminist scholars Kimberlé Crenshaw (1989) and Patricia Hill Collins (1990) advanced the concept of intersectionality, emphasizing not just Lorde's inseparability of race, class, gender and other aspects of human identity, but also their simultaneity.

Feminist anti-toxics and health care movements

The feminist health care movement grew out of early twentieth century women's social work, settlement houses, and the battles for urban sanitation and reproductive rights. Women have often critiqued Western medicine's end-of-the-pipeline "disease" approach to health and wellness for its failure to address disease prevention in ways that women need – as caregivers of children, as workers, and as health consumers. Feminist approaches to health and wellbeing affirmed a belief in women's authority, and the view that women themselves can become their own health experts and serve as catalysts for social change.

In 1978, a young mother of two children in the working-class community of Love Canal, Lois Gibbs, read the newspaper reports describing the toxic waste dump beneath her community, the chemical contaminants found in their air and water, and the dump's suspected links to her community's high rates of miscarriage, birth defects, and childhood illnesses. Her community organizing, research and activism eventually forced a complete buyout of all homes in Love Canal at fair market value, a relocation of these families, and the launch of the Environmental Protection Agency's Superfund Program.

Feminists soon made the connection between reproductive cancers and environmental health, and by the mid-1990s a raft of research was published to document this connection. In 1994, activists from the Massachusetts Breast Cancer Coalition who had noted elevated breast cancer rates throughout Cape Cod called for an investigation of their causes, and, inspired by Rachel Carson's work, founded the Silent Spring Institute (SSI). Breast cancer activists uncovered a list of environmental toxins linked with breast cancer, and launched campaigns such as "think before you pink" to challenge the privatization of breast cancer that blamed women for their cancers, rather than the corporations polluting their environments. The widely-read volume *Our Stolen Future* (Colburn et al. 1996) pointed to the role of pesticides, endocrine-disruptors, phthalates, PCBs, dioxins, and other toxic chemicals in affecting cancers and reproductive health for humans and animals alike. Sandra Steingraber's *Living Downstream* (1997) offered the first study to bring together data on toxic releases with newly released data from US cancer registries, presenting these environmental links to cancer as a human rights issue.

Childhood asthma also became a feminist health concern when mothers initiated asthma activism. As with breast cancer, the asthma–environment connection was difficult to prove, despite the fact that asthma is statistically documented as being most severe among urban, lower-income, and minority children (Sze 2004). In 2011, the asthma attack prevalence rate in Blacks was 36.9 per cent higher than the rate in Whites, as 13.4 per cent of Black children experience asthma at a rate almost double that of White children's rate of 7.6 per cent (CDC 2016). While environmental justice organizations like West Harlem Environmental Action, South Bronx Clean Air Coalition, or Mothers of East Los Angeles all formed to evict polluting industries from their neighbourhoods, they still had to battle the inexact science of risk assessment, used by corporations and government officials to blame the victims and protect the polluters. Responding to these activists' concerns, feminist and environmental justice activists, lawyers and scientists developed the Precautionary Principle, which reverses the onus of proof, requiring that "those who seek to introduce chemicals into our environment first show that what they propose to do is almost certainly not going to hurt anyone" (Steingraber 1997: 270).

Together, the Civil Rights activism of the 1950s and 1960s and the feminist environmental health movements of the 1980s laid the foundation for the environmental justice standpoint that healthy environments are integral to civil rights.

Environmental justice and ecofeminism

The 1990s was a decade of renewed environmental activism, making the environmental racism of mainstream environmentalism a matter of public discussion (see, e.g., Chapters 2 and 3).

Two organizations responsible for bringing environmental justice to the attention of the largely White environmental movement – the Southwest Organizing Project (SWOP) and the Southwest Network for Economic and Environmental Justice (SNEEJ) – both had Mestiza women in critical leadership roles (Peña 1998). In a coordinated effort, SWOP sent a letter to the "Group of Ten"[2] mainstream environmental organizations, calling them out for continuing to "support and promote policies which emphasize the clean-up and preservation of the environment on the backs of working people in general and people of colour in particular." Dated March 15, 1990, the letter was signed by a wide range of environmental justice scholars, activists and organizations, notable among them two Mestizo LGBT organizations, and the Citizens Clearinghouse for Hazardous Waste representative, Penny Newman. Two months later, SNEEJ sent out a similar letter to environmental justice allies, enclosing the letter to the Group of Ten, and inviting these ally organizations to confront their own institutionalized racism and to bring people of colour into their Board of Directors, programme structure, and activities. Lois Gibbs, founder of the Citizens Clearinghouse for Hazardous Waste, may have been surprised to be among the nine recipients of this letter: but within the year, her organization was renamed the Center for Health, Environment, and Justice, its staff diversified, and its resources helped build the base for the First National People of Colour Environmental Leadership Summit in 1991, where the Principles of Environmental Justice were first articulated.

In 2002, the Second National People of Colour Environmental Leadership Summit offered another four-day event that attracted over 1400 participants, expanding and extending the environmental and economic justice paradigm to address globalization and international issues (Stein 2004). As Robert Bullard (2012) acknowledged, the summit wouldn't have happened if not for the efforts of women EJ leaders:

> the women solidified their leadership role in Summit II by default—when many of the men who had been in leadership roles walked away. The women probably would not have been able to achieve this level of visibility and power in an otherwise sexist, male-dominated society—which also extends to the environmental justice movement and Summit II planning—had the men stayed.

At Beverly Wright's insistence, there was a "Crowning Women" awards dinner honouring thirteen women leaders.

Although African-American and Latinx activism may seem most prominent, environmental justice organizations have been developed in all communities of colour. The Indigenous

Environmental Movement has fought the targeting of reservation lands for hazardous waste sites and for nuclear, coal, oil, and hydropower generation (LaDuke 1999). Mexican farm workers organized by Cesar Chavez and Dolores Huerta formed the United Farm Workers to reform hazardous working conditions – being sprayed and breathing in hazardous pesticides while working, denied toilets and water in the fields, receiving substandard housing and sanitation, and working alongside their own children and infants, who are at increased vulnerability to toxic chemicals. Asian American activists formed the Asian Pacific Environmental Network (APEN) to organize workplace health activism in the garment and semiconductor manufacturing industries, to address urban redevelopment (i.e., housing and gentrification), and to promote food safety in Asian diets, particularly in the consumption of contaminated fish (Sze 2004).

Like the environmental justice movement, ecofeminism's roots come from earlier movements – 19th and 20th century feminisms, the anti-nuclear movement of the 1970s, the women's spirituality movement, the animal rights movement, the women's anti-toxics and environmental health movements – and ecofeminism's first manifestations also occurred in the 1980s. In England, the women's peace camp at Greenham Common (1981–2000) developed as a form of nonviolent direct action intended to embody the resistance to the presence of cruise and Pershing II missiles, and the possibility of a "limited nuclear war" in Europe. In the US, the links between feminism and ecology were forged at the "Women and Life on Earth: Ecofeminism in the 1980s" conference, and were soon followed by the Women's Pentagon Actions (WPA) of 1980 and 1981, where large numbers of women demonstrated against militarism and its violence against women, children, people of colour, poor people and the earth. Launched in 1985, WomanEarth Feminist Peace Institute was a direct outgrowth of these anti-militarist actions, founded with the intention of creating an ecofeminist educational centre producing theory, conducting research, and supporting political activities that would confront racism head-on (Sturgeon 1997).

Feminist environmentalists were active in other areas exposing intersections of gender, race and species. In 1982, Marti Kheel and Tina Frisco founded Feminists for Animal Rights (FAR), inspired by Aviva Cantor's work linking racism and speciesism, and by Constantia Salamone's activism foregrounding the linked oppressions of women with animals. Recognizing that domestic violence affects not only women and children but also their animal companions, but that women often stayed in battering relationships to protect their animals, FAR worked to establish coordination between battered women's shelters and animal rescue organizations (Adams 1995).

Activists from FAR and from WomanEarth, along with ecofeminists active in the women's spirituality movement, participated as founders of the US Green Movement during the 1980s, and many remained active through the mid-1990s. In the German Greens, ecofeminist Petra Kelly became well known for her work advancing a women's political agenda of peace and gender equality, and in the US Greens, Charlene Spretnak was a founding member and co-author of the Ten Key Values. In India, ecofeminist physicist Vandana Shiva addressed biotechnology, water, forestry, and oil for the ways these issues affect women's livelihoods and environments. Bringing together women from around the world to take action in the United Nations and other international policymaking forums, the Women's Environment and Development Organization (WEDO) was formed in 1991 by former US Congresswoman Bella Abzug and feminist activist and journalist Mim Kelber. WEDO's primary events in the 1990s included the World Women's Congress for a Healthy Planet in 1991, and a number of Women's Caucuses at key UN conferences.

What defines a feminist issue?

When women and those they care about are disproportionately affected, or are the majority of those affected – whether through breast cancer, toxic exposures during pregnancy and lactation, toxic and gendered workplaces, or the siting of polluting industries – these matters become feminist issues. Women's gendered work as caregivers, housekeepers, sanitation engineers, food-preparers and food providers have made matters of environmental health, toxic waste, and healthy food into feminist issues. The predominance of women in an activity, location, or occupation can also help define a feminist issue: i.e., the "pink collar ghettos" of child care centres and clerical work, or the carcinogenic chemicals in menstrual products (Spector 2013).

Water is a feminist issue. Women's gendered associations with water have lent authority to women's activism on behalf of water preservation, water purity, and the integrity of wetlands: Indigenous women have engaged in multiple "water walks" to call attention to Alberta's tar sands or the degradation of water quality in the Great Lakes. Asian women have been active in the fights to stop hydroelectric dams, from Dai Qing's (1998) critiques of the Three Gorges Dam on China's Yangtze River, to Arundhati Roy's (2001) resistance against the dams on the Narmada River. These are all environmental justice struggles because large hydropower dams destroy the subsistence livelihoods of Indigenous and peasant communities and harm fish, wildlife, and ecosystems – all "sacrificed" to divert water and energy to elite urban communities.

In the US, the nail salon industry is a feminist issue, illustrating the intersection of environmental justice, economic justice, immigrant rights and reproductive justice, as Vietnamese and Korean women comprise over 40 per cent of the workforce (NAPAWF 2011), and women are the majority of its patrons. Wages are low, hours are long, and exposure to toxic nail salon products linked to respiratory, cognitive, and reproductive illnesses poses serious health hazards to workers. Workers experience numerous job-related ailments – from skin rashes and nose bleeds to cancer and miscarriages, as a majority of workers are of reproductive age. A large proportion of nail salon workers are undocumented and have limited English proficiency, which increases their vulnerability to exploitation and makes organizing for better conditions especially challenging. In response to these issues, the National Asian Pacific American Women's Forum has created a National Healthy Nail Salon Alliance, and produced a report documenting the workplace hazards and safety precautions needed to protect workers and consumers alike.

Identified from the ways that women investigate matters of concern, feminist methodologies abound in women's ecofeminist and environmental justice. The 1960s feminist slogan that "the personal is the political" can be seen in women's activism for children's and community health: reasoning from individual experience, women investigate their communities to determine if theirs is also a collective experience, and on that basis they move forward to political action. Women's relational skills at networking and community-building enhance their effectiveness as environmental justice researchers and organizers. While dominant male gender socialization leads men (and others seeking recognition as "authorities") to value working independently, women tend to value working collaboratively, recognizing that the strongest position on an issue can be developed when the widest diversity of those affected are brought together.

Whereas traditional Euro-American conservation and preservation both defined "nature" as a place without humans (i.e., "out in nature," "wilderness"), ecofeminism and environmental justice have contested this definition (see also Chapter 39). Just as feminism redefined the personal in terms of the political, environmental justice redefines "environment" as the places where we "live, work, play, and pray." Ecofeminists and Indigenous women define women's own bodies as environments during pregnancy, and define breast milk as both bodily matter and food whose quality can be diminished via the accumulation and concentration of environmental toxins (LaDuke 1999; Steingraber 2001). Both ecofeminists and environmental justice activists have defined social culture as an environment that can be toxic in its expression of sexism, racism, classism and other oppressions (Adamson et al. 2002; Stein 2004). The environments where elders or disabled persons live can also be contexts of environmental injustice, as many elders, largely women, as well as disabled people may live in substandard housing, receive poor or limited nursing care and health care, and even face physical, sexual, or economic abuse (Day 2010; Ray 2013). In all these ways, redefining "environment" utilizes the feminist rejection of culture/nature dualisms that place humans outside of nature: repositioning humans within nature reframes "saving nature" from traditional environmentalism's altruism to the environmental justice recognition that environmental concerns are simultaneously concerns about human well-being.

Convergence: feminist climate justice

By the end of the 1990s, women eco-activists recognized that a wider movement for environmental justice needs alliances between ecofeminists and environmental justice activists (Kirk 1997), making explicit the feminism within environmental justice and the intersectionality articulated within ecofeminism. Collaborations across these woman-powered movements emerged as activists and scholars within each movement listened to one another, utilizing one another's insights and critiques while responding to the escalating problems of climate change. International organizations bringing a feminist environmental justice perspective to climate change now include not only the Women's Environment and Development Organization (WEDO) but the Global Gender and Climate Alliance; GenderCC: Women for Climate Justice; WoMin: an African Ecofeminist Organization; the Indigenous Environmental Network; and the Women's Earth and Climate Action Network (WECAN).

Across the spectrum of race, class, nation, and sexuality, women have founded organizations and been active in the struggle for climate justice. Four Indigenous women founded the Idle No More movement in 2012, and Indigenous women such as Crystal Lameman (Beaver Lake Cree) and Eriel Tchekwie Deranger (Diné) have been prominent and vocal leaders in the resistance against Alberta's tar sands mining. Other women such as Winona LaDuke from White Earth Reservation and Casey Camp-Horinek of the Ponca Nation have served as bridge-builders, working with both the Indigenous Environmental Network and broader climate justice organizations such as Nebraska's Cowboy and Indian Alliance (CIA), 350.Org, WECAN, and the United Nations.

Queer and trans★ climate activists, while participating in many ecofeminist and environmental justice organizations, have also launched grassroots movements such as the Trans and Women's Action Camp (TWAC), Out4Sustainability, VINE Sanctuary for factory-farmed animals, the Lesbian Rangers, queer food movements in urban spaces, and queer farmer communities across the United States. Reclaiming the erotic – an imperative famously articulated by Audre Lorde (1984) – influenced social ecofeminist Chaia Heller in her formation of the "socio-erotic" (1999) and later activists' articulations of ecosexuality (Stephens & Sprinkle 2016). Founded by an Arab-American and a White working class lesbian couple in 2000, VINE Sanctuary embraces an ecofeminist perspective and is run by five queer or trans★-identified activists, providing a haven for animals who have escaped or been rescued from the meat, dairy and egg industries or other abusive circumstances, such as cockfights or pigeon-shoots. VINE activists recognize that today's racial and economic injustices perpetuate both environmental racism and the continued exploitation of animals, exacerbate climate change (Springmann et al. 2016), and rely on exploiting the reproductive capacities of female animals (jones 2011). Dangerous and environmentally destructive factory farms and processing plants are often located in communities of colour, and local citizens must live with the pollution while working at dangerous and degrading jobs. Black feminist vegans have also recognized these connections (Harper 2010), demonstrating the convergences of diverse eco-justice movements arriving at similar insights through their activism, research, conversations and collaborations (Pellow 2014).

Feminist climate intersectionalities

At the Seattle meeting of the World Trade Organization in November 1999, the anti-globalization movement that had been building for years finally took shape, as activists from around the world marched together to oppose the corporate conquest of workers, the environment, food production, women's bodies, animals, Indigenous and Third World people alike. Their activism built unprecedented alliances among labour, environmental, and social justice movements, and changed the shape of environmental justice organizing. Today, the climate justice movement affords similar opportunities for activism and coalition-building, as seen in the activist analyses of severe climate-change weather events such as Hurricane Katrina.

In its approach, landing and aftermath in August 2005 and beyond, Hurricane Katrina galvanized national discourse on the intersections of race, class, and climate justice. Juxtaposed against news footage of freeways jammed with SUVs and other outwardly mobile families fleeing the city as directed, images of people stranded on rooftop islands or wading through floodwaters with children in tow pervaded the public media. While all aspects of climate justice were present in this environmental tragedy, only some were captured on the news media. To create a more inclusive praxis, one that shapes the work of first responders as well as that of activists and scholars, these climate justice intersectionalities still need to be named and recognized.

Environmental sexism

Women and children are more likely to die during and immediately after ecological disasters than men (Seager 2006). In the 1991 cyclone and flood in Bangladesh, 90 per cent of victims were women, for reasons that characterize the gendered effects of disaster at large: warning information is broadcast in urban centres and workplaces, while women are often confined at home. When disaster struck, women waited for male relatives to return for them, as women are culturally prohibited from travelling alone due to risk of sexual assault. Women are not often taught to swim, and could not navigate the rising waters; moreover, while men often escaped alone, women's caregiving responsibilities meant women were trying to flee while pregnant or post-partum, carrying infants and children, with elderly parents in tow. In the 2004 tsunami in Aceh, Sumatra, more than 75 per cent of those who died were women (Gaard 2015). The death of so many mothers had a cascading effect on increased rates of infant mortality, the early marriage of girls, the neglect of girls' education, increased sexual assaults on unprotected girls, and increased trafficking of women and girls in prostitution.

In New Orleans, the fact of gendered wage inequities combined with race had already placed 41 per cent of female-headed households with children below the poverty line, and economic resources affect one's chances for escape and survival (Henrici et al. 2010); thus, the majority of those left behind were women with children, the poor and the elderly. In the immediate aftermath of Katrina, domestic violence and sexual assaults spiked, though as usual both were under-reported and ignored by law enforcement (Burnett 2005).

Environmental ageism

The very elderly and the very young are more at risk in climate crises and in areas of ongoing toxicity, and the majority of elderly populations are women (IWPR 2010). During and immediately after Hurricane Katrina, the majority of deaths occurred disproportionately among the elderly. In the European heat waves of 2003, 2006, and 2015, more elderly persons went without air conditioning but also without food and even water, as extreme heat takes a heavy toll on mobility. Children of all ages, but particularly those under the age of five are at greater risk from exposure to toxics via air, water, food and environments, and climate change exacerbates these risks by releasing and heating these toxics. Children who lose the protection of mothers in climate disasters then face additional assaults in the aftermath.

Environmental ableism

After Hurricane Katrina, horror stories emerged of people in hospitals and nursing homes being left to drown (Ross 2005). While age and ability often co-occur, impairments of hearing, vision, cognition, speech, and mobility can affect people of all ages, making it difficult for them to seek protection in climate crises; for elderly people, these impairments are more likely and more challenging. Young people are at greater risk as well: given the disproportionate racial impact of asthma among urban and lower-income children of colour, the ability of children to breathe while fleeing or surviving climate disasters is a feminist environmental justice issue. Over two decades ago, environmental justice research confirmed that "toxic time bombs" are concentrated in communities with a high percentage of "poor, elderly, young, and minority residents" (Bullard 1994), and while both age and ability are feminist concerns, these factors have yet to shape climate justice research and response.

Environmental heterosexism

Climate change homophobia is evident in the media blackout of LGBTQ people in the wake of Hurricane Katrina, which occurred just days before the annual queer festival in New Orleans, "Southern Decadence," a celebration that drew 125,000 revellers in 2003. The religious right quickly declared Hurricane Katrina an example of God's wrath against homosexuals, waving signs with "Thank God for Katrina" and publishing detailed connections between the sin of homosexuality and the destruction of New Orleans. It is hard to imagine LGBTQ people – mostly people of colour – not facing harassment, discrimination, and violence during and after the events of Katrina, given the fact that Louisiana, Alabama, and Mississippi lack any legal protections for LGBTQ persons and would have been unsympathetic to such reports. Environmental heterosexism also masks the presence of LGBTQ activists already present in the climate justice movement, yet their vulnerability as queers remains. Responding to the shooting at Pulse nightclub on Latinx night in June 2016, eleven queer and trans* activists of colour in the internationally known climate justice organization, 350.Org, sent out a collectively authored message of grief and hope, affirming "our fights are connected," and disclosing that "many of us who are shoulder to shoulder with you in the streets are LGBTQ+," and "as LGBTQ+ climate activists, we need to bring our whole selves to this work." As Suzanne Pharr (1988) has argued, homophobia is rooted in cultural misogyny, and the liberation of women, people of colour, and queers are inextricably interconnected.

Environmental speciesism

When marginalized communities are situated in toxic environments or are hit with climate crises, animals suffer too. Environmental and climate justice analyses are still largely humanist in the ways they address environmental inequities across human diversity and habitats while failing to consider other species. Occasional news footage showed humane rescue operations saving companion animals (Katrina's effects on wildlife were only estimated), but these efforts earned a mixed reception, with some arguing that rescuers were prioritizing animals over Black people. Speciesism obscures the ways that people's lives are lived in relationship with other animals as well as environments, and limits the scope of feminist analysis: nearly half of those who stayed behind during Katrina refused rescue helicopters and boats that offered safety only to humans, and stayed because of their companion animals, a commitment mirrored by battered women, who refuse the safety of domestic violence shelters, knowing that their absence would leave their animals unprotected at home with the batterer. In disasters of both domestic violence and Hurricane Katrina, many died. But the efforts of animal rescue groups and the intensity of the bond between Katrina's multispecies climate refugees prompted the US Congress to pass the Pets Evacuation and Transportation Standards Act (PETS) in 2006, requiring rescue agencies to save both people and their companion animals in disasters (Grimm 2015).

Conclusion

In summer 2016, another revolutionary moment appeared, as the multiple assaults on the queer and trans★ Latinx community at Pulse nightclub in Orlando, the almost weekly shootings of Black men and Black women (#SayHerName) by US police, the killings of Indigenous eco-activists around the world, and the increasingly frequent terrorist assaults ongoing in the Middle East and throughout Europe bring the structural oppression of women, people of colour, animals, queers, environments and economies to the fore. These global events demonstrate the feminist intersections of climate, economic, and environmental justice.

Will feminist climate justice activists be able to seize this moment and mobilize across our diverse communities? If reports in the popular press are any indication (Acha 2016; Awadalla et al. 2015), movement is already underway.

Notes

1 I use "trans★" as Hayward and Weinstein (2015) suggest: trans★ marks "the *with, through, of, in* and *across* that make life possible" (196); trans★ is "'always already' relational" (198). I am aware of the debate on this usage, best articulated via Trans Student Educational Resources (see www.transstudent.org/asterisk). I remain committed to using the terms that people use to describe their own identities. When a term is contested, writers must reference the diverse viewpoints, with the understanding that whatever usage we choose may later be replaced.
2 The ten organizations include the National Wildlife Federation, Sierra Club, National Audubon Society, Environmental Defense Fund, Environmental Policy Institute, Friends of the Earth, Izaak Walton League, The Wilderness Society, National Parks and Conservation Association, and the Natural Resources Defense Council.

References

Acha, M.A.R. (2016). "How young feminists are tackling climate justice in 2016." *The Huffington Post*, March 7. Accessed at www.huffingtonpost.com/maria-alejandra-rodriguez-acha/how-young-feminists-climate-justice_b_9369338.html on 23 July 2016.

Adams, C. J. (1995). "Woman-Battering and Harm to Animals." In C. J. Adams and J. Donovan, eds. *Animals & Women: Feminist Theoretical Explorations*. Durham, N.C.: Duke University Press, 55–84.

Adamson, J., Evans, M., & Stein, R., eds. (2002). *The environmental justice reader*. Tucson, AZ: University of Arizona Press.

"An African ecofeminist perspective on the Paris climate negotiations." (2015). *WoMin: African Women United Against Destructive Resource Extraction*. Accessed at http://womin.org.za/an-african-ecofeminist-perspective.html on 18 July 2016.

Awadalla, C., Coutinho-Sledge, P., Criscitiello, A., Gorecki, J., & Sapra, S. (2015). "Climate change and feminist environmentalisms: closing remarks." *The Feminist Wire*, May 1. Accessed at www.thefeministwire.com/2015/05/climate-change-and-feminist-environmentalisms-closing-remarks/ on 23 July 2016.

Bullard, R.D. (2012). "Crowning women of colour and the real story behind the 2002 EJ summit." Accessed at http://archive-edu-2012.com/open-archive/922695/2012-12-11 on 7/1/2016.

Bullard, R.D., ed. (1994). *Unequal protection: environmental justice and communities of color*. San Francisco, CA: Sierra Club Books.

Burnett, J. (2005). "More stories emerge of rapes in Post-Katrina chaos." *NPR: National Public Radio*, December 21. Accessed at www.npr.org/templates/story/story.php?storyId=5063796 on 20 July 2016.

Caldecott, L. and Leland, S., eds. (1983). *Reclaim the earth: women speak out for life on earth*. London: The Women's Press.

Carson, R. (1962). *Silent spring*. New York: Houghton Mifflin.

Centers for Disease Control (CDC). (2016). *Most recent asthma data*. Accessed at www.cdc.gov/asthma/most_recent_data.htm on 21 July 2016.

Clark, D. (2011). "Which nations are most responsible for climate change?" *The Guardian*. Accessed at www.theguardian.com/environment/2011/apr/21/countries-responsible-climate-change on 23 July 2016.

Colburn, T., Dumanoski, D., and Myers, J.P. (1996). *Our stolen future*. New York: Penguin/ Plume.

Collins, P.H. (1990). *Black feminist thought: knowledge, consciousness, and the politics of empowerment*. New York: Routledge.

Crenshaw, K. (1989). "Demarginalizing the intersection of race and sex: a Black feminist critique of antidiscrimination doctrine, feminist theory and antiracist politics." *University of Chicago Legal Forum*, 140, 130–167.

Day, R. (2010). "Environmental justice and older age: consideration of a qualitative neighbourhood-based study." *Environment and Planning 4*, 2658–2673.

Diamond, I. & Orenstein, G. (1990). *Reweaving the world: the emergence of ecofeminism*. San Francisco, CA: Sierra Club Books.

Friedan, B. (1963). *The feminine mystique*. New York: W. W. Norton & Co.

Gaard, G. (2011). "Ecofeminism revisited: rejecting essentialism and re-placing species in a material feminist environmentalism." *Feminist Formations*, 23:2, 26–53.

Gaard, G. (2015). "Ecofeminism and climate justice." *Women's Studies International Forum*, 49, 20–33.

Garber, E. (1991). "A spectacle in colour: the lesbian and gay subculture of Jazz Age Harlem." *Hidden from history: reclaiming the gay and lesbian past*, ed. M. Duberman, M. Vicinus, and G. Chauncey Jr. London: Penguin Books. 321–323.

Grimm, D. (2015). "How Hurricane Katrina turned pets into people." *BuzzFeed News*, 31 July. Accessed at www.buzzfeed.com/davidhgrimm/how-hurricane-katrina-turned-pets-into-people?utm_term=.ijgkklQdoW#.rmlrrqEpy8 on 20 July 2016.

Hallum-Montes, R. (2012). "'Para el bien común': indigenous women's environmental activism and community care work in Guatemala." *Race, Gender, & Class*, 19:1, 104–30.

Harper, A.B., ed. (2010). *Sistah Vegan: Black female vegans speak on food, identity, health, and society*. New York: Lantern Books.

Hawley, J., ed. (2015). *Why women will save the planet: a collection of articles for friends of the Earth*. London: Zed Books.

Hayward, E. and Weinstein, J. (2015) "Introduction: tranimalities in the age of Trans★ life." *Transgender Studies Quarterly*, 2:2, 195–208.

Heller, C. (1999). *Ecology of everyday life: rethinking the desire for nature.* Montréal: Black Rose Books.

Henrici, J.M., Helmuth, A.S., & Braun, J. (2010) "Women, disasters, and Hurricane Katrina." Institute for Women's Policy Research, August 28. D492. www.iwpr.org

Hull, G.T., Scott, P.B., & Smith, G., eds. (1982). *All the Women Are White, All the Blacks Are Men, But Some Of Us Are Brave: Black women's studies.* New York: The Feminist Press.

Institute for Women's Policy Research (IWPR). (2010). "Women, Disasters, and Hurricane Katrina." Fact Sheet #D492.

Jones, p. (2011). "Fighting cocks: ecofeminism versus sexualized violence." In L. Kemmerer, ed. *Sister species: women, animals, and social justice.* Urbana, IL: University of Illinois Press, 45–56.

Kirk, G. (1997). "Ecofeminism and environmental justice: bridges across gender, race, and class." *Frontiers*, 18:2, 2–20.

LaDuke, W. (1999) *All our relations: Native struggles for land and life.* Boston: South End Press.

Lorde, A. (1984). *Sister Outsider: essays & speeches.* Trumansburg, NY: The Crossing Press.

Maathai, W. (2004). *The Green Belt Movement.* New York: Lantern Books.

Mann, S.A. (2011). "Pioneers of US ecofeminism and environmental justice." *Feminist Formations*, 23:2, 1–25.

Moraga, C. and Anzaldúa, G., eds. (1981). *This bridge called my back: writings by radical women of color.* Watertown, MA: Persephone Press.

"NAPAWF on nail salons." (2011). *National Asian Pacific American Women's Forum.* Accessed at https://napawf.org/programs/reproductive-justice-2/national-healthy-nail-salon-alliance3/ on 18 July 2016.

Olson, L. (2001). *Freedom's daughters: the unsung heroines of the Civil Rights Movement from 1830 to 1970.* New York: Simon & Schuster.

Pellow, D.N. (2014). *Total liberation: the power and promise of animal rights and the radical earth movement.* Minneapolis: University of Minnesota Press.

Peña, D., ed. (1998). *Chicano culture, ecology, politics: subversive kin.* Tucson, AZ: University of Arizona Press.

Pharr, S. (1988). *Homophobia: a weapon of sexism.* New York: Chardon Press.

Plant, J., ed. (1989). *Healing the wounds: the promise of ecofeminism.* Vancouver, BC: New Society Publishers.

Qing, D. (1998). *The river dragon has come! The Three Gorges dam and the fate of China's Yangtze River and its people.* London: M.E. Sharpe.

Ravilious, K. (2014). "The seven deadly sinners driving global warming." *New Scientist*, January 15. Accessed at www.newscientist.com/article/mg22129523-100-the-seven-deadly-sinners-driving-global-warming/ on 23 July 2016.

Ray, S. J. (2013). *The ecological other: environmental exclusion in American culture.* Tucson: University of Arizona Press.

Ross, L. (2005). "A feminist perspective on Katrina." October 10. *ZNet.* Accessed at https://zcomm.org/znetarticle/a-feminist-perspective-on-katrina-by-loretta-ross/ on 20 July 2016.

Roy, A. (2001). *Power politics.* Cambridge, MA: South End Press.

Salleh, A. (1997). *Ecofeminism as politics: nature, Marx and the postmodern.* London: Zed Books.

Seager, J. (2006). "Noticing gender (or not) in disasters." *Geoforum* 37, 2–3.

Spector, K. (2013). "Groundbreaking report exposes chemicals linked to cancer in feminine care products." *EcoWatch*, November 5. Accessed at www.ecowatch.com/groundbreaking-report-exposes-chemicals-linked-to-cancer-in-feminine-c-1881810973.html on 21 July 2016.

Springmann, M., Godfray, H.C.J., Rayner, M., & Scarborough, P. (2016). "Analysis and valuation of the health and climate change cobenefits of dietary change." *Proceedings of the National Academy of Sciences of the USA.* 113:15, 4146–4151.

Stein, R., ed. (2004). *New perspectives on environmental justice: gender, sexuality, activism.* New Brunswick, NJ: Rutgers University Press.

Steingraber, S. (1997). *Living downstream: a scientist's personal investigation of cancer and the environment.* New York: Vintage Books.

Steingraber, S. (2001). *Having faith: an ecologist's journey to motherhood.* New York: Berkley Publishing Group.

Stephens, B. & Sprinkle, A. (2016). "Ecosexuality." In R.C. Hoogland, ed. *Gender: Nature*. London: Macmillan Interdisciplinary Handbook.

Sturgeon, N. (1997). *Ecofeminist natures: race, gender, feminist theory and political action*. New York: Routledge.

Sze, J. (2004). "Asian American activism for environmental justice." *Peace Review* 16:2, 149–156.

Walker, A. (1983). *In search of our mothers' gardens: womanist prose*. Orlando, FL: Harcourt Brace Jovanovich.

Win, T.L. "Women activists defy danger to protect the environment." *Thomson Reuters Foundation News*, 5 August 2015. Accessed at http://news.trust.org//item/20140805084308-4q693/ on 7/7/2016.

8

OPENING BLACK BOXES

Environmental justice and injustice through the lens of science and technology studies

Gwen Ottinger

Social science research on environmental justice (EJ) demonstrates that environmental racism and injustice are structural problems: zoning laws, mortgage lending practices, public hearing procedures, and other institutional arrangements create inequity both in the distribution of environmental harms and in communities' ability to advocate for themselves (Cole & Foster 2001; Walker 2012). Research in the field of Science and Technology Studies (STS) adds to our understanding of the structural nature of environmental injustice by showing how scientific knowledge production practices, technological infrastructures, and the authority afforded to experts all help create an uneven terrain for diverse communities seeking a healthy, safe environment. At the same time, STS scholarship identifies alternatives to mainstream practices – including community-based science and scientist-activism – that stand to further environmental justice (see also Chapters 23 and 24).

This chapter introduces fundamental concepts in STS and shows how this theoretical lens furthers our understanding of both the structures of environmental injustice and the potential for their transformation. The next section focuses on critical studies of scientific knowledge in the environmental justice arena. Against popular narratives that portray science as a neutral arbiter of the facts, having the potential to cut through environmental controversy, STS begins from the premise that knowledge is constructed in particular social contexts and necessarily reflects the implicit and explicit values of its creators (e.g. Traweek 1988; Latour 1987; Douglas 2009). This theoretical stance, the chapter shows, has fostered critical examination of the science invoked in environmental justice controversies and enabled scholars to demonstrate how existing scientific and regulatory practices systematically marginalize the perspectives of communities with environmental health concerns (e.g. Aitken 2009; Allen 2003; Tesh 2000; Frickel et al. 2010). At the same time, scholars taking an STS approach have argued for the legitimacy of community-based knowledge and identified conditions under which collaborations between grassroots groups and sympathetic scientists can support community environmental goals (e.g. Brown 1992; Corburn 2005; Chapters 23 and 24).

Section 3 focuses on technology. Parallel to its understanding of science as inherently social and value-laden, STS argues that technology and society are made in the same process, and that material objects are intertwined with and implicated in particular political orders (Hughes 1986; Latour 1987; Winner 1981). Drawing on this theoretical framework, researchers have shown that technologies – specifically, sociotechnical infrastructures – contribute to large-scale structures of inequity, as design choices determine how environmental hazards are created, distributed, and managed (e.g. Hecht 2012; Jones 2014). Yet the STS literature also suggests that alternative technologies and justice-oriented approaches to innovation could help create more environmentally just sociotechnical infrastructures (e.g. Ottinger 2011).

In combination with normative political theories about the multi-faceted nature of environmental justice (Schlosberg 2007), STS-informed empirical research points to a set of additional science- and technology-related requirements for environmental justice. Section 4 discusses the importance of epistemic justice (Fricker 2009) as an aspect of EJ, the implications for procedural justice of taking community-based knowledge seriously (Ottinger 2013a), and the professional obligations that a value for environmental justice would impose on scientists and engineers. The chapter concludes by summarizing the contributions of STS to understanding environmental justice and suggesting promising ways forward for EJ researchers and advocates alike.

Scientific knowledge and the epistemic dimensions of environmental injustice

A central tenet of STS is that scientific facts are not discovered; rather, they are constructed, in processes in which epistemic concerns, social norms, political values, and pragmatic issues bleed into one another, becoming separable only in retrospect (Latour 1987). Grounded in laboratory ethnographies of the 1980s, this view of science undergirds research in three directions salient to understanding the structures of environmental injustice. First, STS scholars examine the everyday work done by scientists as they design experiments, determine what counts for evidence, and establish standards that enable diverse communities of researchers to work together and have confidence in each other's results (e.g. Traweek 1988; Star & Griesemer 1989). This close examination of scientific practice shows how value judgments of various sorts are not a corruption of science but an unavoidable part of research (Douglas 2009). It also shows the various ways in which scientific knowledge reflects the cultural contexts of its production, including gender stereotypes and culturally specific assumptions about race (e.g. Martin 1991; Haraway 1989).

Second, STS asks how and why certain knowledge claims (and not others) are accepted as fact, especially in the case of controversy. Research in this area adopts a principle of symmetry: rather than explaining a fact's success by its inherent right-ness and blaming the failure of unsuccessful claims on the taint of social and political influences, STS investigations look at the myriad social and epistemic factors that contribute to the success *or* failure of any given knowledge claim. They show the importance of such things as scientists' identities, interpersonal trust, institutional location, and alignment with powerful political discourses in making a knowledge claim into a fact (e.g. Gieryn 1999; Ottinger 2013b), and how national political cultures shape what claims are taken up in policy processes (Jasanoff 2012). Because of their commitment to a symmetrical approach, STS researchers have been willing to take seriously knowledge claims that challenge the scientific mainstream (e.g. Martin 1996; Zavestocki et al. 2004). STS also recognizes knowledge systems and ways of knowing that are quite different from that of Western science, endeavouring to understand them on their own terms instead of using Western science as a measure of their validity (e.g. Adams 2002).

Finally, STS examines how political-economic factors shape the directions and products of scientific research. Increasingly market-oriented public policies are reshaping scientific research (e.g. Moore et al. 2011); as a consequence, financial and intellectual resources flow overwhelmingly to research in the interests of industry or the military (Hess 2007), while research oriented to the needs of less powerful groups is too often left undone (Frickel et al. 2010). At the same time, STS researchers have shown the power of social movements composed of scientists to establish new fields and reorient existing ones, in some cases to more socially and environmentally progressive ends (Frickel & Gross 2005).

Applied to the study of environmental health and justice issues – most often taking the form of grassroots mobilization against local sources of pollution, whether existing or proposed – STS shows how the structures of science put community and activist groups at a disadvantage. Value judgments are inherent in the sciences of environmental quality and environmental health, just as in other sciences, and the values represented in mainstream science seldom reflect those of grassroots groups seeking environmental justice. For example, epidemiologists will in most circumstances not conclude that a disease cluster exists until they are 95 per cent confident that the pattern could not have happened by random chance. This demand for a high level of statistical significance makes them more likely to say disease rates are not elevated when they are, than to claim that disease rates are elevated when they are not – that is, to favour false negatives over false positives. The 95 per cent standard in turn makes it difficult for concerned communities to establish the significance of apparent clusters of illness, especially for small population sizes (Allen 2003; Brown 1992).

Scientists' values and expectations inform their health risk assessments, as well. Assessments of the hazards associated with particular chemicals rely overwhelmingly on laboratory experiments that correspond poorly to the conditions under which people are exposed (Tesh 2000). More holistic community-level health risk assessments, for their part, make assumptions about routes and frequencies of exposure that are apt for white suburbanites but badly misrepresent the lifestyles of immigrant and Indigenous communities (Kuehn 1996; Powell & Powell 2011). More fundamentally, quantitative, scientific approaches can crowd out or silence alternative ways of conceptualizing the environmental issues, including wellness-based frameworks that revolve around things like access to nature and community functioning (Johnson & Ranco 2011), and environmental impact assessments that recognize the history and cultural significance of Indigenous land (Holifield 2012; Tsosie 2012).

An STS approach to environmental and environmental health science, then, shows the partiality of scientists' approaches to understanding issues that affect communities burdened by pollution. In addition, research on the political-economic organization of science points out that many of the questions of most concern to communities are neglected by scientists, creating a pattern of "undone science" around issues such as air quality in the most polluted areas around industrial facilities, and the synergistic effects of exposures to multiple chemicals at once (Frickel et al. 2010). The pattern is especially pronounced in developing nations, which may have few resources for science and/or rely on support from international NGOs, who in turn set research agendas that may better reflect the priorities of global elites than those of the world's poorest, most polluted communities (Carruthers 2008). Although there is clear evidence that environmental health concerns tend to be un- or under-investigated, it is worth noting countervailing trends. After Hurricane Katrina, for example, resources for environmental contamination were not allocated disproportionately to areas with the most social capital, as one might have expected, but flowed to less well-off areas, as well (Frickel 2014). More nuanced models of the conditions under which science is left undone in different national and cultural contexts are called for.

Taking a symmetrical approach to environmental knowledge claims has led STS scholars to acknowledge, and find value in, so-called laypeople's claims to knowledge about health and environmental issues in EJ communities. Community members possess "local knowledge" or "Indigenous knowledge" – awareness and understanding of the local environment based in observation, historical knowledge, and embodied experience – that would be inaccessible to scientists from outside the community or cultural context (Wynne 1992; Tesh 2000; Philip 2004). In many places, they have also moved to render their experiences in more scientific terms through participatory mapping (Allen 2000; Corburn 2005), popular or lay epidemiology (Brown 1992; Lora-Wainwright 2013), and community-based environmental monitoring (Ottinger 2010; Hemmi & Graham 2014) (see also Chapters 23 and 24). Research on these practices points out that they represent values and, in some cases, ways of knowing different from those favoured by scientists. For example, popular epidemiology favours a more pre-cautionary approach to determining whether elevated levels of disease are "significant" (Brown 1992; Allen 2003), and communities using "bucket" air samplers focus on measuring levels of pollution during episodic peaks in the neighbourhood of a polluting facility, in contrast to environmental regulators' goal of generating a "representative" picture of air quality throughout a region (Ottinger 2010). Far from being poor facsimiles of *real* science, these alternative approaches represent legitimate critiques of the values and assumptions embedded in conventional scientific approaches. Further, incorporating community-based knowledge into official environmental assessments and decision-making, STS scholars argue, will lead to more robust environmental policy (Irwin 1995).

Conceptually, STS approaches put the knowledge claims of community and activist groups on an equal footing with the knowledge claims of scientists. Yet research on community battles with polluters and governments also shows how laypeople's knowledge is systematically devalued in decision-making processes (see also Chapter 20). Technical experts are typically given a central role in public hearings, for example, making it difficult for community members who have not mastered technical jargon to participate (see e.g. Cole & Foster 2001; Barandiaran 2015). Further, when non-scientist participants in public forums, including deliberative ones, question, criticize, or try to offer input into scientists' assessments, their comments are routinely deemed irrelevant to the technical issues at hand and reinterpreted as (at best) insight into public sentiment or social factors that scientists must keep in mind (Aitken 2009; Ottinger 2013b). This process of marginalizing community knowledge is facilitated by formal standards for information inputs to the policy-making process, such as Environmental Impact Assessments (Richardson & Cashmore 2011) and "reference methods" for environmental monitoring. These standards circumscribe the information that is actionable, allowable, and/or relevant, and they help to limit participation in decision-making to those with the resources to meet the standards (Barandiaran 2015; Li 2015).

Recognizing that local knowledge and alternative ways of knowing are persistently marginalized in environmental institutions, communities and EJ activists may choose to adopt mainstream scientific standards in their own knowledge-making projects, and in many cases, grassroots groups get support from sympathetic scientists. STS-informed research explores the power dynamics of these alliances (Cable et al. 2005), and offers models for equitable collaboration (Heaney et al. 2007; Corburn 2005). It also considers EJ-oriented collaborations in the context of scientists' careers, showing how scientist-activists manage to maintain their scientific credibility, and how interactions with EJ activists change the way they do their work (e.g. Frickel et al. 2015; Ottinger & Cohen 2013). At the same time, STS scholars point out the trade-offs involved in adopting a scientific logic for community-based studies: while it may increase the likelihood that results will be accepted by regulators and other decision-makers, it also sidelines more fundamental critiques of the values and worldview embedded in standard scientific practices (Kinchy 2012; Kinchy et al. 2014; Ottinger 2009).

STS theories thus help show how science – its implicit values, taken-for-granted worldview, and authority in political processes – contributes to environmental injustice by potentially obscuring the full range of effects that environmental hazards have on disadvantaged groups, and by limiting the ability of non-scientists to participate meaningfully in environmental decision-making. Initially developed through case studies from the United States and Western Europe, STS ideas are now beginning to inform research on environmental justice cases in developing nations (e.g. Barandiaran 2015; Li 2015; Lora-Wainwright 2013). Cross-cultural comparison will be valuable to developing a richer understanding of the dynamics of undone science, local knowledge, technocratic policy processes, and scientist-activism internationally (Ottinger et al. 2016).

Yet the lens provided by STS also has blind spots where further theoretical development is sorely needed. In defending local knowledge and alternative ways of knowing against charges that they are mere misunderstandings of "real" science, STS theory has not grappled with the possibility of true errors in community knowledge claims – even within the context of their culturally specific ways of knowing. Not doing so suggests an extreme relativist position which few within the STS community would espouse (Latour 1999), and having no way to adjudicate the quality of scientific claims potentially leaves EJ scholars and activists without the ability to advocate for including the most relevant and appropriate expertise in political processes (cf. Collins & Evans 2002).

On the other hand, theoretical development in this direction would bolster STS's tendency to put science at the centre of its analysis, which has tended to come at the expense of considering the relative merits of epistemic arguments versus the myriad other tools employed by the EJ movement in making change. Scholars in the field would do well to consider more thoroughly questions such as: under what conditions are epistemic challenges effective? When would communities be better served by focusing on community organizing, direct action, media outreach, and so forth? Research that touches on this topic tends to show that communities seldom "win" epistemic battles, even though their campaigns might succeed on other grounds (Ottinger 2013b; Kinchy 2012). It suggests that activities like participatory monitoring may in some cases be better understood as a form of organizing or a mechanism for political accountability, rather than as epistemic interventions *per se* (Overdevest & Mayer 2008, Kinchy et al. 2014).

Co-constructing injustice: sociotechnical systems and the politics of technology

STS's theoretical lens on technology parallels its approach to science. Artefacts – the devices, machines, and industrial processes colloquially referred to as "technology" – are understood as just one kind of element in complex, heterogeneous, sociotechnical systems or networks that include users, builders, institutions, standards, material properties, and weather patterns, to name a few (Hughes 1986; Law 1987). Research in the field tends to take the system, rather than the artefact, as the primary unit of analysis; even studies that begin with objects frequently use them as points of leverage for understanding how larger systems are configured (deLaet & Mol 2000; Winner 1981). As described by STS, sociotechnical systems are remarkable in their spatial and temporal scope. Not only do, for example, energy systems extend around the globe (see Chapter 31), enrolling users and workers and materials from dissimilar places in a common project (Hecht 2012); the shape of today's systems reflects decisions made in earlier iterations of the system, and today's decisions will constrain what is possible in the future (Jones 2014).

STS scholars working from a sociotechnical systems perspective argue that "society" and "technology," to the extent that they can be thought of as separate, are co-constructed. For an invention to succeed, the inventor must build simultaneously the machine, the markets for distributing it, the social situations in which its use is desirable, the processes for producing it, and so forth (Latour 1987; Hughes 1986; Law 1987). Artefacts are so thoroughly entangled in social relations that they can never be considered neutral in the sense of being mere tools that can be used to any human purpose. Rather, technologies are inevitably coupled with political systems – systems that may favour hierarchy over collaboration, or distribution over centralization, and so on (Winner 1981). While most of the research in this vein examines how systems are co-constructed and what sociotechnical configurations result, this lens has also enabled scholars to ask to what extent particular technological systems are compatible with desired political orders, such as democracy or justice (Sclove 1995; Ottinger 2011).

In the EJ context, taking a sociotechnical systems approach helps to illuminate technology's role in structuring environmental injustice. Many scholars and EJ advocates have made the point that the pollution and hazard issues that communities grapple with are a by-product of industrial processes, an insight highly compatible with an analysis of sociotechnical systems. However, the implications of understanding society and technology as co-produced go further. They prompt us to examine how technological developments foster or reinforce institutions that distribute resources inequitably or limit people's ability to participate in political processes, as well as ways that existing, unjust sociotechnical arrangements make it more difficult for institutions and technology to evolve in more environmentally just directions.

EJ-focused research on technology and sociotechnical systems is less well developed than that on the role of science and expertise in EJ struggles, yet there are several areas where the environmental justice implications of sociotechnical systems emerge strongly from case studies. Burgeoning research on energy systems, for example, makes clear how existing infrastructures of extraction create "sacrifice zones" in the Global South and rural parts of industrialized nations that not only bear the burdens of energy production, but are also disenfranchised within the larger system (Hecht 2012; Li 2015). Some of this work turns our attention to the ways that injustice is hardened into the technological infrastructure itself: Jones (2014) argues that the shift from railroads to pipelines as the primary mode of conveyance for fossil fuels deprived small communities of the commerce that came along with rail shipments and created a one-way flow of resources from rural sites of extraction to population centres; Winner (1986) suggests that nuclear energy, because of the security concerns associated with fissile materials, requires authoritarian management structures largely incompatible with democratic governance. Other sociotechnical systems that help constitute structures of environmental injustice include waste disposal, especially the global traffic in hazardous waste (Iles 2004); industrial agriculture, with its heavy reliance on pesticides and tendency to crowd out small landowners (Harrison 2008; Wolford 2008); and even information systems to ensure the public's "right to know" about local hazards (see Chapter 9), whose designs reflect problematic assumptions about the relationships between marginalized communities, technology, and social change (Galusky 2003; Delborne & Galusky 2011).

STS analysis of sociotechnical systems can also enrich our understanding of disasters as an environmental justice concern (see e.g. Bullard & Wright 2009; Hecht 2013; see also Chapter 28). Where EJ researchers have made the point that the disproportionate suffering of already vulnerable communities in the wake of disaster is a direct result of already-unjust social structures, the emerging area of "Disaster-STS" (Knowles 2014a; Fortun & Frickel 2012) brings into focus the technical aspects of those structures and their role in intensifying the uneven effects of disaster. The STS lens points out that the potential for disaster is built in to large, sociotechnical systems (Matthewman 2014), and that "disaster" should be understood broadly, to include "slow disasters" like toxic chemical contamination, whose scale and complexity outstrip society's ability to respond adequately (Gray-Cosgrove et al. 2015). Responses to disaster are also encoded in sociotechnical systems through normalizing techniques such as risk assessment and cost-benefit analysis, procedures for post-disaster inquiry, and the training and expertise of engineers (Knowles 2014b). Environmental justice issues are hardly neglected in STS research: a robust body of work shows how knowledge about contamination and other effects of disaster is made and contested, often in ways that disadvantage the populations most affected (see Fortun & Frickel 2012 for a review). Yet more remains to be done to situate these dynamics in larger sociotechnical systems, and to develop the sociotechnical systems view in the environmental justice context more generally.

STS researchers have also examined new and emerging "green" and "community-friendly" technologies, arguing that a sociotechnical systems view is necessary to understanding the EJ implications of new developments in technology. For example, researchers have raised questions about the compatibility of solar and wind energy with environmental justice, showing that its ability to ease the environmental burden of power production on marginalized communities will depend very much on how resource extraction, manufacturing, and decision-making are institutionalized (Mulvaney 2013; Raman 2013; Aitken 2009). New, community-oriented information technologies, including crowd-sourcing platforms for reporting pollution and Internet-enabled aerial mapping, are also regarded hopefully (McCormick 2012; Wylie et al. 2014), but how they are institutionalized and how they ultimately intersect with expert systems will be consequential for their ultimate effectiveness (Wylie et al. 2014; cf. Ottinger 2010). As more EJ-oriented technologies emerge, systematic, comparative study of their impacts is clearly called for, in large part because of its potential to inform future innovation.

Toward a new ethic: environmentally just science and technology

Empirical research in STS makes two things clear. First, claims to scientific knowledge are always value laden, and whether they are accepted as "fact" depends very much on how they are situated in the larger social and political milieu. Calls to "just get the facts straight" – by, for example, weeding out studies biased by industry involvement or insisting on public disclosures of information (cf. Shrader-Frechette 2007) – might be rhetorically powerful but are unrealistic in that they rely on an idealized view of science that does not hold up on close examination. A more nuanced understanding of how knowledge is made both underscores the value of organizing to challenge scientific and regulatory standards or best practices (cf. Epstein 1996) and could help communities assess whether it is worthwhile in particular instances to try to engage experts on scientific territory at all. Second, STS analyses show that environmental justice and injustice take material form in both artefacts and the sociotechnical systems of which they are a part. In the quest for environmental justice, then, technology shouldn't be taken as just a backdrop for organizing; the design of sociotechnical systems also needs to be a target of the EJ movement, and activists should consider making common cause with technology-oriented social movements (Hess 2005) such as that around "Inherently Safer Design."

Beyond these pragmatic points, STS findings have implications for theories of environmental justice, especially what it includes, what it requires, and what ethical obligations surround it. Schlosberg (2007) identifies four aspects of environmental justice: distributive justice, procedural justice, recognition, and capabilities. What his analysis does not fully capture, however, is EJ communities' insistence that their local, experiential knowledge must count in environmental decision-making. Local knowledge is mentioned as an example of the recognition that is due communities, and participatory research is suggested as an aspect of robust procedural justice. However, those frameworks are limited in their ability to distinguish between instances where "laypeople" are seen as offering substantive contributions to the collective understanding of environmental problems and, in contrast, instances where their views are welcomed and recognized but framed as something other than insight into scientific or technical issues (Aitken 2009; Ottinger 2013b).

To make this distinction, "epistemic justice" should be considered a fifth aspect of environmental justice. Epistemic justice asks, "are people respected in their capacity *as knowers*?" Epistemic injustices occur where individuals are not respected as knowers because they belong to a structurally disadvantaged group (e.g., they are persons of colour) and are therefore assumed not to be able to offer information of value to collective understandings (Fricker 2009) – as, for example, in cases where women's testimony that their families are ill from chemical exposures is dismissed by regulators as the ranting of "hysterical housewives" (see also Chapter 7). Epistemic injustices take more subtle, systemic forms as well, occurring where members of disadvantaged groups lack the ability to articulate aspects of their experience in a way that renders it legible to people whose experience is taken as the norm, and when their contributions – including not only assertions but also questions – are not acknowledged as relevant (Fricker 2009; Hookway 2010). EJ examples of this kind of "hermeneutic injustice" documented in the STS literature include the routine reinterpretation of community concerns as social rather than technical questions and fenceline communities' struggles to communicate the impact of periodic "blasts" of pollution that are of little significance in experts' framework of chronic and acute exposures (Ottinger 2009).

Working for epistemic justice as well as other aspects of justice would prompt advocates to think about how to offer communities more resources for articulating their experiences as technical questions or scientific contributions. While popular epidemiology and participatory mapping move in that direction, additional resources for making meaning – of, for example, burgeoning environmental monitoring and pollution data – are still necessary. But much of the remedy to epistemic injustice lies in the attitude of the listener: in this case, government regulators and other scientists and engineers who encounter community concerns about environmental issues. Medina (2012) calls for the cultivation of "epistemic virtues," especially curiosity, humility, and open-mindedness. From an EJ standpoint, these should be considered integral to the professional ethics and training of scientists and engineers.

STS research on the nature of science and technology also has implications for the pursuit of procedural and distributive justice. Information tends to be thought of as a fundamental requirement of procedural justice (Chapter 9), in that people cannot participate effectively in decisions without full knowledge of what they are being asked to decide on; suppression or withholding of information thus stands in the way of procedural justice (Shrader-Frechette 2007). This view, however, relies on the assumption that the information that exists at the time of a decision is complete and adequate to making that decision, whereas STS research on knowledge gaps, local knowledge, and the contested and evolving nature of science makes clear that this is seldom the case. Taking those findings seriously would necessitate rethinking standards for procedural justice to include, at minimum, opportunities for on-going consent as knowledge developed (Ottinger 2013a).

From a distributive justice standpoint, the understanding of technology as embedded in and co-producing social systems is consequential: it implies that we cannot talk about designing more environmentally just *social* orders without imagining the technologies that would support them, or pursue "green" innovation without simultaneously inventing institutions that would ensure that the new technologies were compatible with social justice. To this end, EJ activists should be calling for participatory design of everything from industrial facilities to power grids to fuel-efficient vehicles. For their part, engineers should be learning how to collaborate with EJ advocates and others who think differently than they do (Downey et al. 2006), and developing ways to operationalize "justice" as a design criterion (Ottinger 2011).

Conclusion

The theoretical lens of STS situates science and technology among the structures of environmental injustice. Research in the field shows how community-based knowledge and ways of knowing are marginalized by scientific standards and the authority of science in decision-making, as well as how the questions and concerns most important to EJ communities fall within systemic "knowledge gaps" and get obscured by scientists' often-unacknowledged values and assumptions. STS's understanding of technologies as elements of sociotechnical systems further suggests the role that technological design processes play in hard-wiring environmental injustice not only into the built environment but also into the social institutions that manage and govern technology.

For EJ scholars, STS theory raises additional questions, such as: when is epistemic engagement an effective strategy for EJ activists? How do cross-national variations in the authority, status, and resources for science shape science's role in exacerbating (or ameliorating) environmental injustices? How are unjust institutions and social structures perpetuated by technological choices, and how can socially political orders be produced in conjunction with "green" technologies? For EJ advocates, STS research points to additional places where strategic intervention could help dismantle structures of injustice, including standards of scientific proof, monitoring and assessment procedures, and technological design. For scientists, engineers, and other technical practitioners, findings in this area point to new professional obligations, especially embracing community participation in making knowledge and designing technology, and cultivating the humility, curiosity, and open-mindedness needed for those collaborations to succeed.

References

Adams, V. (2002). "Randomized controlled crime: postcolonial sciences in alternative medicine research." *Social Studies of Science*, vol. 32, no. 5–6, pp. 659–690.

Aitken, M. (2009). "Wind power planning controversies and the construction of 'expert' and 'lay' knowledges." *Science as Culture*, vol. 18, no. 1, pp. 47–64.

Allen, B.L. (2000). "The popular geography of illness in the industrial corridor." in Colten, C.E. (Ed.). *Transforming New Orleans and its environs: centuries of change*. Pittsburgh: University of Pittsburgh Press, pp. 178–201.

Allen, B.L. (2003). *Uneasy alchemy: citizens and experts in Louisiana's chemical corridor disputes*. Cambridge, MA: MIT Press.

Barandiaran, J. (2015). "Chile's environmental assessments: contested knowledge in an emerging democracy." *Science as Culture*, vol. 24, no. 3, pp. 251–275.

Brown, P. (1992). "Popular epidemiology and toxic waste contamination: lay and professional ways of knowing." *Journal of Health and Social Behavior*, vol. 33, no. 3, pp. 267–281.

Bullard, R.D. and Wright, B. (2009). *Race, place, and environmental justice after Hurricane Katrina: struggles to reclaim, rebuild, and revitalize New Orleans and the Gulf Coast*. Boulder, CO: Westview Press.

Cable, S., Mix, T. and Hastings, D. (2005). "Mission impossible? Environmental justice activists' collaborations with professional environmentalists and with academics." in Pellow, D.N. and Brulle, R.J. (Eds.). *Power, justice, and the environment: a critical appraisal of the environmental justice movement.* Cambridge, MA: MIT Press, pp. 55–76.

Carruthers, D.V. (2008). *Environmental justice in Latin America: problems, promise, and practice.* Cambridge, MA: MIT Press.

Cole, L.W. & Foster, S.R. (2001). *From the ground up: environmental racism and the rise of the environmental justice movement.* New York: NYU Press.

Collins, H.M. & Evans, R. (2002). "The third wave of science studies: studies of expertise and experience." *Social Studies of Science,* vol. 32, no. 2, pp. 235–296.

Corburn, J. (2005). *Street science: community knowledge and environmental health justice.* Cambridge, MA: MIT Press.

deLaet, M. and Mol, A. (2000). "The Zimbabwe bush pump mechanics of a fluid technology." *Social Studies of Science,* vol. 30, no. 2, pp. 225–263.

Delborne, J. and Galusky, W. (2011). "Toxic transformations: constructing online audiences for environmental justice." in Ottinger, G. and Cohen, B.R. (Eds.). *Technoscience and environmental justice: expert cultures in a grassroots movement.* Cambridge, MA: MIT Press, pp. 63–92.

Douglas, H. (2009). *Science, policy, and the value-free ideal.* Pittsburgh: University of Pittsburgh Press.

Downey, G.L., Lucena, J.C., Moskal, B.M., Parkhurst, R., Bigley, T., Hays, C., Jesiek, B.K., et al. (2006). "The globally competent engineer: working effectively with people who define problems differently." *Journal of Engineering Education,* vol. 95, no. 2, pp. 107–122.

Epstein, S. (1996). *Impure science: AIDS, activism, and the politics of knowledge.* Berkeley, CA: University of California Press.

Fortun, K. & Frickel, S. (2012). "Making a case for disaster science and technology studies." *An STS Forum on the East Japan Disaster,* 8 March, available at: https://fukushimaforum.wordpress.com/online-forum-2/online-forum/making-a-case-for-disaster-science-and-technology-studies/ (accessed 12 February 2016).

Frickel, S. (2014). "Absences: methodological note about nothing, in particular." *Social Epistemology,* vol. 28, no. 1, pp. 86–95.

Frickel, S., Gibbon, S., Howard, J., Ottinger, G. & Hess, D. (2010). "Undone science: charting social movement and civil society challenges to research agenda setting." *Science, Technology & Human Values,* vol. 35, no. 4, pp. 444–473.

Frickel, S. & Gross, N. (2005). "A general theory of scientific/intellectual movements." *American Sociological Review,* vol. 70, no. 2, pp. 204–232.

Frickel, S., Torcasso, R. and Anderson, A. (2015). "The organization of expert activism: shadow mobilization in two social movements." *Mobilization: An International Quarterly,* vol. 20, no. 3, pp. 305–323.

Fricker, M. (2009). *Epistemic injustice: power and the ethics of knowing.* Oxford: Oxford University Press.

Galusky, W. (2003). "Identifying with Information: citizen empowerment, the internet, and environmental anti-toxins movement." In McCaughey, M. & Ayers, M. (Eds.). *Cyberactivism: online activism in theory and practice.* London: Routledge.

Gieryn, T.F. (1999). *Cultural boundaries of science: credibility on the line.* Chicago: University of Chicago Press.

Gray-Cosgrove, C., Liboiron, M. & Lepawsky, J. (2015). "The challenges of temporality to depollution & remediation." *S.A.P.I.EN.S. Surveys and Perspectives Integrating Environment and Society,* vol. 8, no. 1, available at: https://sapiens.revues.org/1740 (accessed 12 February 2016).

Haraway, D.J. (1989). *Primate visions: gender, race, and nature in the world of modern science.* New York: Routledge.

Harrison, J. (2008). "Abandoned bodies and spaces of sacrifice: pesticide drift activism and the contestation of neoliberal environmental politics in California." *Geoforum,* vol. 39, no. 3, pp. 1197–1214.

Heaney, C.D., Wilson, S.M. and Wilson, O.R. (2007). "The West End Revitalization Association's community-owned and -managed research model: development, implementation, and action." *Progress in Community Health Partnerships: Research, Education, and Action,* vol. 1, no. 4, pp. 339–349.

Hecht, G. (2012). *Being nuclear: Africans and the global uranium trade.* Cambridge, MA: MIT Press.

Hecht, G. (2013). "Nuclear janitors: contract workers at the Fukushima reactors and beyond." *The Asia-Pacific Journal,* vol. 11, no. 1.

Hemmi, A. & Graham, I. (2014). "Hacker science versus closed science: building environmental monitoring infrastructure." *Information, Communication & Society,* vol. 17, no. 7, pp. 830–842.

Hess, D.J. (2005). "Technology- and product-oriented movements: approximating social movement studies and science and technology studies." *Science, Technology & Human Values*, vol. 30, no. 4, pp. 515–535.

Hess, D.J. (2007). *Alternative pathways in science and industry*, Cambridge, MA: MIT Press.

Holifield, R. (2012). "Environmental justice as recognition and participation in risk assessment: negotiating and translating health risk at a Superfund site in Indian Country." *Annals of the Association of American Geographers*, vol. 102, no. 3, pp. 591–613.

Hookway, C. (2010). "Some varieties of epistemic injustice: reflections on Fricker." *Episteme*, vol. 7, no. 2, pp. 151–163.

Hughes, T.P. (1986). "The seamless web: technology, science, etcetera, etcetera." *Social Studies of Science*, vol. 16, no. 2, pp. 281–292.

Iles, A. (2004). "Patching local and global knowledge together: citizens inside the US chemical industry." in Jasanoff, S. & Long, M. (Eds.). *Earthly politics: local and global in environmental governance.* Cambridge, MA: MIT Press, pp. 285–308.

Irwin, A. (1995). *Citizen science: a study of people, expertise and sustainable development.* London: Routledge.

Jasanoff, S. (2012). *Science and Public Reason.* London: Routledge.

Johnson, J.R. & Ranco, D.J. (2011). "Risk assessment and Native Americans at the cultural crossroads: making better science or redefining health." in Ottinger, G. and Cohen, B.R. (Eds.). *Technoscience and environmental justice: expert cultures in a grassroots movement.* Cambridge, MA: MIT Press, pp. 179–200.

Jones, C.F. (2014). *Routes of power.* Cambridge, MA: Harvard University Press.

Kinchy, A.J. (2012). *Seeds, science, and struggle: the global politics of transgenic crops.* Cambridge, MA: MIT Press.

Kinchy, A., Jalbert, K. & Lyons, J. (2014). "What is volunteer water monitoring good for? Fracking and the plural logics of participatory science." *Fields of knowledge: science, politics and publics in the neoliberal age*, vol. 27, no. 2, pp. 259–289.

Knowles, S. (2014a). "Engineering risk and disaster: a special issue of *Engineering Studies*, Introduction." *Engineering Studies*, vol. 6, no. 3, pp. 131–133.

Knowles, S. (2014b). "Engineering risk and disaster: disaster-STS and the American history of technology." *Engineering Studies*, vol. 6, no. 3, pp. 227–248.

Kuehn, R.R. (1996). "The environmental justice implications of quantitative risk assessment." *University of Illinois Law Review*, vol. 1996, p. 172.

Latour, B. (1987). *Science in action: how to follow scientists and engineers through society.* Cambridge, MA: Harvard University Press.

Latour, B. (1999). *Pandora's hope: essays on the reality of science studies.* Cambridge, MA: Harvard University Press.

Latour, B. (2009). *Politics of nature: how to bring the sciences into democracy.* Cambridge, MA: Harvard University Press.

Law, J. (1987). "Technology and heterogeneous engineering: the case of Portuguese expansion." in Bijker, W.E., Hughes, T.P. and Pinch, T.J. (Eds.), *The social construction of technological systems: new directions in the sociology and history of technology*, Cambidge, MA: MIT Press, pp. 111–134.

Li, F. (2015). *Unearthing conflict: corporate mining, activism, and expertise in Peru.* Durham, NC: Duke University Press.

Lora-Wainwright, A. (2013). "The inadequate life: rural industrial pollution and lay epidemiology in China." *The China Quarterly*, vol. 214, pp. 302–320.

Martin, B. (1989). "The sociology of the fluoridation controversy." *Sociological Quarterly*, vol. 30, no. 1, pp. 59–76.

Martin, B. (1996). "Sticking a needle into science: the case of polio vaccines and the origins of AIDS." *Social Studies of Science*, vol. 26, no. 2, pp. 245–276.

Martin, E. (1991). "The egg and the sperm: how science has constructed a romance based on stereotypical male-female roles." *Signs*, vol. 16, no. 3, pp. 485–501.

Matthewman, S. (2014). "Dealing with disasters: some warnings from Science and Technology Studies (STS)." *IDRiM Journal*, vol. 4, no. 1, pp. 1–11.

McCormick, S. (2012). "After the cap: risk assessment, citizen science and disaster recovery." *Ecology and Society*, vol. 17, no. 4.

Medina, J. (2012). *The epistemology of resistance: gender and racial oppression, epistemic injustice, and the social imagination.* Oxford: Oxford University Press.

Moore, K., Kleinman, D.L., Hess, D. and Frickel, S. (2011). "Science and neoliberal globalization: a political sociological approach." *Theory and Society*, vol. 40, no. 5, pp. 505–532.

Mulvaney, D. (2013). "Opening the black box of solar energy technologies: exploring tensions between innovation and environmental justice." *Science as Culture*, vol. 22, no. 2, pp. 230–237.

Ottinger, G. (2009). "Epistemic fencelines: air monitoring instruments and expert-resident boundaries." *Spontaneous Generations: A Journal for the History and Philosophy of Science*, vol. 3, no. 1.

Ottinger, G. (2010). "Buckets of resistance: standards and the effectiveness of citizen science." *Science, Technology & Human Values*, vol. 35, no. 2, pp. 244–270.

Ottinger, G. (2011). "Environmentally just technology." *Environmental Justice*, vol. 4, no. 1, pp. 81–85.

Ottinger, G. (2013a). "Changing knowledge, local knowledge, and knowledge gaps STS Insights into procedural justice." *Science, Technology & Human Values*, vol. 38, no. 2, pp. 250–270.

Ottinger, G. (2013b). *Refining expertise: how responsible engineers subvert environmental justice challenges.* New York: NYU Press.

Ottinger, G., Barandiaran, J. and Kimura, A.H. (2016). "Environmental justice: knowledge, technology, and expertise." in Miller, C., Smith-Doerr, L., Felt, U. and Fouche, R. (Eds.). *The handbook of science and technology studies*, 4th ed. Cambridge, MA: MIT Press, pp. 1029-1057.

Ottinger, G. & Cohen, B. (2013). "Environmentally just transformations of expert cultures: toward the theory and practice of a renewed science and engineering." *Environmental Justice*, vol. 5, no. 3.

Overdevest, C. & Mayer, B. (2008). "Harnessing the power of information through community monitoring: insights from social science." *Texas Law Review*, vol. 86, p. 1526.

Philip, K. (2004). *Civilizing natures: race, resources, and modernity in colonial South India.* New Brunswick, NJ: Rutgers University Press.

Powell, M. & Powell, J. (2011). "Invisible people, invisible risks: how scientific assessments of environmental health risks overlook minorities – and how community participation can make them visible." in Ottinger, G. and Cohen, B.R. (Eds.). *Technoscience and environmental justice: expert cultures in a grassroots movement.* Cambridge, MA: MIT Press, pp. 179–200.

Raman, S. (2013). "Fossilizing renewable energies." *Science as Culture*, vol. 22, no. 2, pp. 172–180.

Richardson, T. & Cashmore, M. (2011). "Power, knowledge and environmental assessment: the World Bank's pursuit of 'good governance'." *Journal of Political Power*, vol. 4, no. 1, pp. 105–125.

Schlosberg, D. (2007). *Defining environmental justice: theories, movements, and nature.* Oxford: Oxford University Press.

Sclove, R. (1995). *Democracy and technology.* New York: Guilford Press.

Shrader-Frechette, K. (2007). *Taking action, saving lives: our duties to protect environmental and public health.* Oxford: Oxford University Press.

Star, S.L., & Griesemer, J.R. (1989). "Institutional ecology, 'translations,' and boundary objects: amateurs and professionals in Berkeley's Museum of Vertebrate Zoology, 1907–39." *Social Studies of Science*, vol. 19, no. 3, pp. 387–420.

Tesh, S.N. (2000). *Uncertain hazards: environmental activists and scientific proof.* Ithaca, NY: Cornell University Press.

Traweek, S. (1988). *Beamtimes and lifetimes: the world of high energy physicists.* Cambridge, MA: Harvard University Press.

Tsosie, R. (2012). "Indigenous peoples and epistemic injustice: science, ethics, and human rights." *Washington Law Review*, vol. 87, no. 4, pp. 1133–1201.

Walker, G. (2012). *Environmental justice: concepts, evidence and politics.* London: Routledge.

Winner, L. (1981). "Do artifacts have politics?" *Daedalus.* vol. 109, no. 1.

Winner, L. (1986). *The whale and the reactor: a search for limits in an age of high technology.* Chicago: University of Chicago Press.

Wolford, W. (2008). "Environmental justice and the construction of scale in Brazilian agriculture." *Society & Natural Resources*, vol. 21, no. 7, pp. 641–655.

Wylie, S.A., Jalbert, K., Dosemagen, S. and Ratto, M. (2014). "Institutions for civic technoscience: how critical making is transforming environmental research." *The Information Society*, vol. 30, no. 2, pp. 116–126.

Wynne, B. (1992). "Misunderstood misunderstandings: social identities and public uptake of science." *Public Understanding of Science*, vol. 1, no. 3, pp. 281–304.

Zavestocki, S., Brown, P., McCormick, S., Mayer, B., D'Ottavi, M. & Lucove, J.C. (2004). "Patient activism and the struggle for diagnosis: Gulf War illnesses and other medically unexplained physical symptoms in the US." *Social Science & Medicine*, vol. 58, no. 1, pp. 161–175.

9

PROCEDURAL ENVIRONMENTAL JUSTICE

Derek Bell and Jayne Carrick

Introduction

Most advocates of environmental justice are not only concerned about the distribution of environmental goods and bads; they are also concerned about the fairness of the process of environmental policy- and decision-making. One of the reasons for the unfair distribution of environmental benefits and burdens is that the decisions that transform the environment are usually made by people who enjoy the benefits rather than the burdens (Schlosberg 2013; Walker 2012). Historically, most people have been excluded or marginalized by the institutions – at all scales from the local to the global – that make policies and decisions that change the environmental conditions in which we live. Today, gross inequalities of political authority, power and influence remain the norm in environmental decision-making. In environmental justice studies, this is known as *procedural environmental injustice*.

In this chapter, we introduce the idea of procedural environmental justice. First, we consider how procedural environmental justice is related to other aspects of environmental justice, especially distribution and recognition. Second, we examine the relationship between procedural environmental justice and related concepts in environmental political theory. Third, we distinguish three accounts of the central principle of procedural environmental justice: equality; proportionality; and plurality. Fourth, we outline a more detailed account of procedural environmental justice, developed by Hunold and Young in their discussion of hazardous waste facility siting, to illustrate the demandingness of ideal conceptions of procedural environmental justice. Fifth, we contrast the ideal with one of the most notable attempts to institutionalize procedural environmental justice, the Aarhus Convention. Sixth, we suggest that the 'gap' between ideals of procedural environmental justice and even the best attempts to institutionalize procedural environmental justice in practice is large enough to make it difficult to use ideal principles to evaluate actual institutions. We conclude by suggesting that developing a robust framework for assessing the relative injustice of actual environmental institutions might be an important step forward in the study – and, subsequently, the reduction – of procedural environmental injustice.

Procedures, distribution and recognition

In the last fifty years, distributive justice has been the central focus of liberal political theory (Rawls 1972; Dworkin 1981, 2002; Cohen 1989). However, there have been many dissenting voices within and beyond liberalism that have emphasized the importance of procedural justice and 'justice as recognition' (Young 1990; Honneth 1995; Fraser 1997; Jones 2006; see Chapter 10). Environmental justice theorists, inspired by environmental justice advocates and environmental justice policy in the US (Schlosberg 2007; Sze et al. 2009; US EPA 2015), and drawing on work of Iris Marion Young (1990) and other critics of the distributive paradigm in liberal political theory (Lake 1996; Walker 2012), have emphasized procedural justice alongside distributive justice.

At the very least, an unfair institutional framework for making environmental decisions is unlikely to promote a fair distribution of environmental benefits and burdens (Morello-Frosch 2002; Stephens et al. 2001; Sze et al. 2009). As Walker (2012: 47) suggests, procedural injustice is an 'explanation or cause of [distributive] injustice'. So, procedural environmental justice is a means for promoting distributive environmental justice. Moreover, environmental justice theorists have noted the 'centrality of procedure in producing inequitable distribution'; procedural environmental justice is a key condition for distributive environmental justice (Schlosberg 2007: 51).

However, procedural environmental justice is not only instrumentally valuable. It is also important for its own sake. We should be concerned about an unfair decision-making process even if, hypothetically, it produced fair outcomes. Political equality requires that institutional structures that have profound effects on our lives are fair. Shrader-Frechette's (2002: 24) account of the Principle of Prima Facie Political Equality captures the importance of both distributive and procedural (or, as she calls it, 'participative') environmental justice:

> To correct problems of environmental justice, it will be necessary to improve the principles and practices of distributive justice – equal apportionment of social benefits and burdens . . . [and] reform the principles and practices of participative justice – equal rights to self-determination in societal decision making.

We might interpret this account in two ways. First, political equality is constituted by both distributive and procedural justice. We fulfil the principle – or realize the value – of political equality if and only if our institutions are procedurally fair and their outcomes are distributively fair. Second, political equality underpins or justifies – and is something different from – both distributive and procedural justice.

The second interpretation resonates with the emphasis that leading theorists place on 'justice as recognition' (Schlosberg 2007; Walker 2012). The recognition of others as 'equals' irrespective of race, ethnicity, class, gender or ideology (or any other 'group' identity) requires institutions that are procedurally just and outcomes that are distributively just. Underpinning these requirements is an attitude or disposition of 'equal respect' (Young 1990). As Schlosberg (2007: 51) suggests, 'environmental justice activists often see their identities devalued and make direct connection between the defence of their communities and demand for respect'. The distributive and procedural environmental injustice that their communities are suffering is experienced as *mis*recognition or lack of respect.

Different forms of injustice tend to maintain and reinforce each other. A group that does not enjoy equal respect is likely to be excluded or marginalized in decision-making and, therefore, also likely to suffer distributive disadvantage. Similarly, a group that suffers distributive disadvantage may not have the resources to participate effectively in decision-making even when it is not formally excluded and, therefore, its members' 'voice' will not be present in the public realm and they may not be recognized as equals by other citizens (Schlosberg 2007: 51). The inter-connectedness of procedures, distribution and recognition is important because it suggests that we are unlikely to tackle one form of injustice successfully without tackling the others. However, this does not prevent us from analysing the three types of justice separately. In this chapter, we focus on procedural environmental justice referring to distribution and recognition only where the inter-connections are important for thinking about the idea or practice of procedural environmental justice.

Democracy, citizenship and public participation

The idea of *procedural* environmental justice is closely related to other key ideas in environmental political theory that have received more attention in the literature, including public participation in environmental decision-making, 'environmental democracy' and 'environmental citizenship'. Sherry Arnstein's (1969) classic study of public participation in planning still serves as one reference point for those thinking about the relative merits of different forms of public participation in environmental decision-making. There have been many subsequent attempts to develop criteria for assessing the quality and fairness of public participation processes in environmental decision-making as well as many attempts to design toolkits for practitioners (and activists) seeking to promote or facilitate public participation in environmental decision-making (Renn et al. 1995; Hunold & Young 1998). These discussions are only rarely presented as attempts to promote procedural environmental justice but many accounts of procedural environmental justice define it or operationalize it in terms of fair public participation in environmental decision-making (Brisman 2013). Therefore, the literature on public participation is an important resource for theorists and practitioners trying to understand and promote procedural environmental justice.

There is also an extensive literature on 'environmental democracy', which is relevant for discussions of procedural environmental justice (Dryzek 1995; Doherty & de Geus 1996; Meadowcroft 1997; Lidskog & Elander 2010). The idea of environmental democracy was especially prominent in environmental political theory in the mid-1990s when it was presented as a response to the eco-authoritarianism of William Ophuls' (1977) *Ecology and the politics of scarcity*. Advocates of environmental democracy argue that the problem is liberal representative democracy – in a capitalist economy – not democracy *per se* (Biegelbauer & Hansen 2011). Some propose more radical forms of participatory – and often local – democracy while others argue that the solution is more deliberative democracy (Dryzek & Schlosberg 2000; Lidskog & Elander 2010). The 'deliberative turn' in contemporary political theory has been especially prominent in environmental political theory and environmental politics has been a favourite testing ground for deliberative experiments (Schlosberg 1999, 2007, 2013; Smith 2003; Niemeyer 2013; Gunderson 2014). The environmental democracy literature has not often talked explicitly about procedural environmental justice but it has focused on imagining and developing institutions that ensure a fair voice for all groups and promote fair outcomes by requiring participants to justify their proposals to other participants. Most accounts of procedural environmental justice draw on discussions of participatory and deliberative democracy (Schlosberg 2007; Walker 2012).

The links to work on environmental and ecological citizenship are less clear but also important (Dobson 2003; Bell & Dobson 2006; Hayward 2006; Latta 2007). Ecological citizenship is often presented as a complex set of virtues or dispositions that underpin or enable environmentally-aware behaviours (Dobson 2003; Connelly 2006; Jagers et al. 2014).

The ecological citizen is concerned about the effects of their actions – and usually the actions of others – on the environment and people now and in the future. Ecological citizens are motivated by a concern for distributive environmental justice, and will engage in environmental decision-making with the aim of promoting distributive environmental justice (as they understand it) (Dobson 2003; Bell 2005). In this sense, ecological citizenship tightens the connections between procedural and distributive environmental justice; fair institutions and citizens with a sense of (distributive) environmental justice are likely to produce an environmentally just outcome.

Conceptions of procedural environmental justice

We have seen that there are close connections between procedural environmental justice and environmental democracy. At the heart of most democratic theories (and theories of procedural environmental justice) is the notion of political equality. However, political egalitarianism, like egalitarianism more generally, can be understood in a variety of ways. At a minimum, it may require only that all citizens have the formal legal right to vote in elections and to stand for political office. On most accounts, especially in environmental political theory, it requires rather more: a genuine opportunity for all to participate directly in environmental decision-making; an equal 'voice' in environmental decision-making; or an equal 'say' or 'vote' in environmental decision-making (Shrader-Frechette 2002: 28). The underlying thought is that a fair procedure for environmental decision-making is one in which power is shared equally among the (potential) participants in the decision-making process. Different accounts interpret the notion of 'equal power' in different ways because they conceptualize power in environmental decision-making in different ways. However, they take it for granted that a fair procedure is one in which there is equality of something: 'opportunity to participate' or 'voice' or 'say' or 'vote'.

The identification of justice with some form of equality is familiar from discussions of distributive justice. However, theorists have also proposed alternative distributive principles, while maintaining a commitment to the fundamental equality of persons (Parfit 1997; Miller 2001). Similarly, alternatives to the principle of equality have been proposed in recent discussions of procedural environmental justice. Two notable alternatives are the principle of proportionality and the principle of plurality (Bell and Rowe 2012; Schlosberg 1999).

The principle of proportionality suggests that power in environmental decision-making should be distributed among persons in proportion to their relative stakes in the outcome of the decision (Bell and Rowe 2012). So, if one person has a much greater stake in the outcome of a decision than another person, they should also have much more power in the decision-making process. Bell and Rowe (2012), drawing on Brighouse and Fleurbaey (2010), suggest that a person's stake should be determined by how much they are affected by the potential outcomes of the decision and we should 'measure' how much they are affected by considering the potential impact on their capabilities (Bell 2011). In the context of environmental decision-making, the principle of proportionality builds on the challenge to conventional political boundaries set out by advocates of the 'all-affected principle' (Goodin 2007). Many environmental policies and decisions have profound effects beyond conventional political boundaries, such as states and cities, but people outside of those boundaries are rarely formally (or practically) included in the decision-making process. The all-affected principle suggests that everyone (potentially) affected should be included – as an equal – in the decision-making process. The principle of proportionality endorses the inclusivity of the all-affected principle but rejects the idea of equal power because simple majoritarianism is not a fair decision-making procedure when some people will be much more significantly affected than others by a decision (Brighouse and Fleurbaey 2010).

The principle of plurality also challenges the egalitarian interpretation of procedural justice but this time the concern is that emphasizing equality 'suppresses differences' (Young 1990: 11) and limits the 'diversity of the stories of injustice' (Schlosberg 2007: 179). The focus on designing institutions that attempt to equalize power in environmental decision-making processes is grounded in a liberal conception of persons as autonomous individuals who are capable of identifying and articulating their own individual interests as well as acting from a shared conception of (distributive) environmental justice. The reality of environmental injustice is that those who suffer it belong to specific social groups. Moreover, they usually suffer it because their group is *mis*recognized and excluded from or marginalized in environmental decision-making; and their experience of it is shaped and coloured by their collective as well as their individual identities and histories. Therefore, we cannot hope to tackle environmental injustice without taking seriously the 'politics of difference', where 'group differences should be acknowledged in public policy and in the policies and procedures of . . . institutions, in order to reduce actual or potential oppression' (Young 1990: 11). Procedural environmental justice is only likely to be achieved as a result of the development of a pluralistic environmental justice movement, which transforms civil society and, ultimately, our formal political institutions in ways that ensure the recognition and inclusion of a plurality of voices in environmental decision-making (Schlosberg 2007).

Institutionalizing procedural environmental justice: ideal principles

Central to all three conceptions of procedural environmental justice outlined in the previous section are commitments to more inclusive participation and a fairer distribution of power in environmental decision-making. From these core ideas, theorists and practitioners have developed accounts of the detailed principles that should govern environmental decision-making processes. Hunold and Young's (1998) account of fair procedures for siting hazardous waste facilities is a good example. It is underpinned by the principle of plurality and the idea of a communicative democracy but the five principles for a fair procedure that they propose could be endorsed by advocates of equality and proportionality.

Their first principle, 'inclusiveness', requires equal recognition for all and a concerted effort to reach out to minority communities that often 'lack effective organization' to represent them (Hunold & Young 1998: 88). Second, a fair procedure will provide for 'consultation over time'; in contrast to 'sporadic' consultations (Hunold & Young 1998: 89). This is essential to maximize 'social knowledge': it makes it more likely that the understanding of the issue that develops through the decision-making procedure and determines the decision that is made will be shaped at all stages in the process by the contributions of all participants (Hunold & Young 1998: 89). Their third principle requires the design of decision-making procedures to eliminate 'gross power disparities' by providing 'weaker parties' with 'informational or economic support to compensate for the imbalance' (Hunold & Young 1998: 89). This might, for example, be achieved by public funding of 'counter studies' commissioned by communities or groups that would otherwise lack the time, expertise or economic resources to challenge well-resourced planners or developers from the public or private sector (Hunold & Young 1998: 89).

Fourth, a fair procedure is one with 'shared decision-making authority': the decision is made jointly by the participants in the process (Hunold & Young 1998: 90). In a fair process, none of the stakeholders, including the 'public officials', have the power to 'unilaterally make decisions', therefore, inclusive participation extends beyond the discussion to the decision and all participants have reason to pay attention to the interests, needs and views of other parties (Hunold & Young 1998: 90). Finally, the fifth principle is 'authoritative decision making' (Hunold & Young 1998: 90). The decision arrived at by the participants in the process should be the final decision; it should not be a recommendation that can be overturned by politicians or public officials.

Together, these five principles set very stringent standards for a fair procedure that are very rarely – perhaps never – met by environmental decision-making processes at the local, national or international scale. Accounts like Hunold and Young's may be considered as located on a continuum that ranges from the more 'ideal' to the less 'ideal' (Swift 2008). Towards the less ideal end, more attention is paid to whether it is politically feasible to implement the proposed principles in environmental decision-making processes without radically transforming the existing political institutions, political culture or political economy. In contrast, more ideal accounts may require very significant changes in political, cultural and economic structures. Hunold and Young's five principles, taken together, are at the more ideal end of the continuum. In contemporary representative democracies with (neo)liberal political cultures and capitalist political economies, we are not likely to see authoritative environmental decision-making processes that are fully inclusive from agenda setting to decision-making and in which resources are provided to 'weaker parties' to eliminate 'gross power disparities' (Bell 2015).

Institutionalizing procedural environmental justice: principles in practice

We may be a long way from institutional arrangements that would realize ideal procedural environmental justice but there have been many attempts to develop and implement less than ideal but fairer than usual decision-making procedures. These are often local or small-scale experiments with different forms of consultation or different participatory or deliberative forums, such as citizens' juries, consensus conferences or deliberative polls (Smith 2003; Elstub 2010). However, there have also been some national and international attempts to create fairer processes for environmental decision-making (Tomlinson 2015). The most notable example with international reach is probably the United Nations Economic Commission for Europe (UNECE) Convention on Access to Information, Public Participation in Decision-Making and Access to Justice in Environmental Matters (commonly referred to as the Aarhus Convention) (European Commission 2015).

The Aarhus Convention, established in 1998 and ratified by 47 parties, including 46 countries and the European Union (UNECE 2016), comprises three 'pillars': access to information; public participation in decision-making; and access to justice in environmental matters. The three pillars establish procedural rights and place correlative duties on states. To fulfil their duties, states should establish appropriate institutional arrangements to secure the procedural environmental rights of their citizens (and others affected by their decisions) (Mason 2010). The procedural rights established by the Aarhus Convention fall well short of the kind of idealized account of procedural environmental justice proposed by Hunold and Young. However, the Aarhus rights do reflect central concerns of most accounts of procedural environmental justice, including Hunold and Young's, for example, the rights of access to information and public participation in decision-making.

The right of access to information, which is set out in Articles 4 and 5 of the Convention, requires that public authorities provide environmental information on request. Article 4 defines 'environmental information' quite broadly, imposes time limits for dealing with requests for information and introduces a 'public interest' test to limit the number and range of cases where public authorities can legitimately refuse to provide information that has been requested. Article 5 requires that public authorities proactively collect environmental information and make that information 'effectively accessible' to the public. So, the Convention supports effective access to information, which is essential to reduce the 'gross power disparities' between different actors, especially well-resourced public authorities and corporations on the one hand and ordinary citizens on the other hand (Hunold & Young 1998: 89). However, it does not require public funding for counter-studies, which might allow less powerful actors to re-define problems and challenge the dominant understanding of what counts as relevant environmental information.

The right of public participation in environmental decision-making is similarly positive but modest in its ambition. There is a commitment to inclusive participation but no requirement to proactively facilitate participation by disadvantaged or minority groups. In this sense, the Convention establishes a formal commitment to an equal right to participate in environmental decision-making without addressing the pluralist's concerns about how difficult it is for some groups to participate under the prevailing institutional norms. The Convention asserts the right to participate in local environmental decisions that are likely to have direct effects on those living in the area but it is much more circumspect in its commitment to public participation in the development of plans and programmes (which may be at the regional or the national scale) and policies and executive regulations (at the national scale). As a result, it fails to take seriously the multi-scalar character of environmental politics and tends to limit public involvement to defensive action designed to protect a 'place' from an already planned 'threat'. The required form of public involvement is 'sporadic' and restricted to particular developments; it is not strategic, broad-ranging or extended over time. Finally, the Convention requires that when public authorities make decisions 'due account is taken of the outcome of the public participation' (Article 6). However, this falls well short of shared decision-making authority. The public authorities make the decision; comments from the public are just one source of information to be considered (Stec et al. 2000).

The Aarhus Convention has influenced the development of European Union and national government environmental legislation. It has improved the availability of environmental information and has provided a starting point for the institutionalization of more public participation in environmental decision-making (Pedersen 2011). However, case law suggests that the Convention sometimes requires less than we might expect. For example, there are some aspects of climate policy where the Court of Justice of the European Union has not upheld the requirement to make information available (Peeters & Nóbrega 2014). There is also some way to go before all of those who have ratified the Convention are consistently fulfilling its requirements (Pedersen 2014; Peeters & Nóbrega 2014).

In sum, the Aarhus Convention does not institutionalize ideal principles of procedural environmental justice. However, it does seek to address *in practice* some of the injustices that are common features of environmental decision-making procedures. Putting its three principles into practice would be a step in the right direction even if it does nothing to tackle the more profound limitations of environmental decision-making in a liberal capitalist world order.

Assessing procedural environmental injustice in environmental decision-making

The Aarhus Convention is just one attempt to create or promote a fairer institutional framework for environmental decision-making. It is notable because of its international reach. Many attempts to promote fairer environmental decision-making are more local and time-limited. For example, a local or national government may go beyond the standard processes required by planning policy or environmental regulation to promote a 'fairer' process for a particular environmental decision. As with the Aarhus Convention, this kind of approach is limited because it fails to properly acknowledge the multi-scalar character of environmental politics and tends to promote 'defensive' participation rather than involvement in the co-authorship of strategic environmental planning. However, in a non-ideal world, we should still be concerned about degrees of injustice in environmental decision-making. Good examples of local (and temporary) institutional innovations will fall short of ideal procedural environmental justice, but might be less unjust than other existing institutions.

Intuitively, we make comparative judgements about institutional innovations and new procedures for environmental decision-making. However, it seems more difficult to develop a robust framework for assessing the relative merits of non-ideal procedures. This difficulty is partly the result of competing accounts of ideal procedural environmental justice. If we have different ideals, they may lead us to different conclusions about the relative merits of the institutions that we are comparing. However, the more significant problem is that ideals of procedural environmental justice may not help us to assess the comparative merits of actual procedures. As we saw, environmental decision-making procedures that complied fully with the Aarhus Convention would fail to satisfy Hunold and Young's five principles. The 'gap' between ideal principles and actual institutions is very large. To compare the merits of actual institutions we need to be able to 'measure' or 'rank order' the degree or extent to which procedures satisfy (and fail to satisfy) ideal principles of procedural environmental justice. However, ideal principles may provide little guidance here.

Consider, for example, Hunold and Young's five principles. The first principle, inclusiveness, does appear to be a matter of degree: a decision-making process may include a lesser or greater number of societal groups. However, the application of the principle may be more difficult: How do we identify distinct groups? Is inclusion all-or-nothing or is it a matter of degree? What are the sufficient conditions that need to be met for us to judge that a group is included? The second principle, consultation over time, also appears to be a matter of degree: a decision-making process may have consultation for longer or shorter periods of time. Again, using the principle to compare actual processes requires more detailed specification: Do all methods of consultation count equally when we compare procedures? Should we think about time in a decision-making process in terms of an account of the stages of decision-making? Is consultation equally important at all stages or does it matter more at the beginning of the decision-making process? The third principle, eliminate gross power disparities, appears to be a matter of degree: a decision-making process can allow larger or smaller power disparities. However, applying the principle may be theoretically and empirically very challenging: How should we conceptualize 'power' in environmental decision-making processes? What are the sources (or resources) of power and how are they related to each other? How should we 'measure' relative power or describe relations of power in environmental decision-making?

So, the first three principles do appear to be a matter of degree; in contrast, the fourth and fifth principles are all-or-nothing conditions and are rarely satisfied in environmental decision-making processes. The fourth principle, shared decision-making authority, requires that the decision is made jointly by the participants in the process; however, it is more common for public authorities to allow participants to provide 'input' into a decision that is made by public officials. Similarly, the fifth principle, authoritative decision-making, requires that the decision arrived at by the participants in the process should be the final decision; however, elected politicians or public officials usually have the final decision. If the fourth and fifth principles are all-or-nothing and are very rarely satisfied, they can do very little work in the comparative assessment of actual institutions.

One way of responding to this problem is to use weaker versions of the fourth and fifth principles when we are assessing non-ideal procedures. So, we might shift from thinking about decision-making *authority* to thinking about decision-making *power* or *influence*. The participants in the decision-making process may not have the authority to make the decision jointly, but the public officials with decision-making authority may *de facto* devolve more or less power or influence over the decision to the participants. They may, for example, be more or less responsive to the ideas and arguments of the participants in the process. Of course, it will be much more difficult to determine whether power or influence has been devolved to the participants (and whether it has been devolved fairly among them) than to determine whether the participants have decision-making authority that is both shared and final.

The difficulty of developing a robust framework for the comparative assessment of actual institutions has not prevented scholars from undertaking case studies of environmental decision-making. These studies are usually based on observation of the process or interviews with participants in the process. In many cases, the strengths and weaknesses of the process are intuitively assessed against a list of principles, which are often drawn from deliberative democratic theory (like Hunold and Young's five principles) (Hartley & Wood, 2005; Hindmarsh & Matthews 2008). Alternatively, some studies focus more directly on trying to identify the relations of power that exist within the decision-making process (Bickerstaff & Walker 2005; Berardo 2013). Many provide insights into environmental decision-making processes but the absence of a widely accepted and theoretically robust framework for thinking about 'degrees' of fairness in practice makes it difficult to compare the results. We might hope that growing interest in non-ideal theory might combine with the 'empirical turn' in deliberative democracy, systematic studies of relations of power in political science and more engagement with practitioners to produce improved theoretical frameworks for comparing the procedural injustice(s) of actual environmental decision-making institutions. One promising direction might be to examine how work on the formal and informal 'rules' of environmental institutions might be combined with assessment of the capabilities and resources of affected parties to develop a framework for understanding relations of power within decision-making processes (Bell 2011; Bell & Rowe 2012).

Conclusion

To summarize: We have located the idea of procedural environmental justice in relation to other aspects of environmental justice and related concepts in environmental political theory. We distinguished three interpretations of the idea of political equality and outlined some ideal principles of procedural environmental justice. We argued that the 'gap' between ideals of procedural environmental justice and even the best attempts to institutionalize procedural environmental justice in practice is large enough to make it difficult to use ideal principles to evaluate actual institutions. There is still important work for ideal theorists to do, specifically, on the proportionality principle and the development of principles of procedural environmental justice that are capable of dealing with environmental policies and decisions that have quite different effects in different places (and over different time periods). However, the development of non-ideal theory that bridges the gap between ideals of procedural environmental justice and empirical studies of actual institutions is particularly important. Developing a robust framework for assessing the relative injustice of actual environmental institutions could be an important step forward in the study – and, subsequently, the reduction – of procedural environmental injustice.

References

Arnstein, S. (1969). "A ladder of citizen participation." *Journal of the American Institute of Planners*, vol. 35, no. 4, pp. 216–224.

Bell, D. (2005). "Liberal environmental citizenship." *Environmental Politics*, vol. 14, no. 2, pp. 179–194.

Bell, D. (2011). *Procedural justice and local climate policy in the UK*. York: Joseph Rowntree Foundation.

Bell, D. & Dobson, A. (eds.) (2006). *Environmental citizenship*. London: The MIT Press.

Bell, D. & Rowe, F. (2012). "Are climate policies fairly made?" viewed 15 July 2015, www.jrf.org.uk/sites/files/jrf/climate-change-policies-summary.pdf.

Bell, K. (2015). "Can the capitalist economic system deliver environmental justice?" *Environmental Research Letters*, vol. 10, no. 12.

Berardo, R. (2013). "The coevolution of perceptions of procedural fairness and link formation in self-organizing policy networks." *The Journal of Politics*, vol. 75, no. 3, pp. 686–700.

Bickerstaff, K. & Walker, G. (2005). "Shared visions, unholy alliances: power, governance and deliberative processes in local transport planning." *Urban Studies*, vol. 42, no. 12, pp. 2123–2144.

Biegelbauer, P. & Hansen, J. (2011). "Democratic theory and citizen participation: democracy models in the evaluation of public participation in science and technology." *Science and Public Policy*, vol. 38, no. 8, pp. 589–597.

Brighouse, H. & Fleurbaey, M. (2010). "Democracy and proportionality." *Journal of Political Philosophy*, vol. 18, no. 2, pp. 137–155.

Brisman, A. (2013). "The violence of silence: some reflections on access to information, public participation in decision-making, and access to justice in matters concerning the environment." *Crime Law and Social Change*, vol. 59, no. 3, pp. 291–303.

Cohen, G. (1989). "On the currency of egalitarian justice." *Ethics*, vol. 99, pp. 906–944.

Connelly, J. (2006). "The virtues of environmental citizenship." In Bell, D. and A. Dobson (eds.), *Environmental citizenship*. London: The MIT Press.

Davoudi, S. & Brooks, E. (2014). "When does unequal become unfair? Judging claims of environmental injustice." *Environment and Planning A*, vol. 46, pp. 2686–2702.

Dobson, A. (2003). *Citizenship and the environment*. Oxford: Oxford University Press.

Doherty, B. & de Geus, M. (1996). *Democracy and green political thought: sustainability, rights, and citizenship*. London: Routledge.

Dryzek, J.S. (1995). "Political and ecological communication." *Environmental Politics*, vol. 4, no. 4, pp. 13–30.

Dryzek, J.S. (1997). *The politics of the earth: environmental discourses*. Oxford: Oxford University Press.

Dryzek, J.S. & Schlosberg, D. (2000). *Deliberative democracy and beyond: liberals, critics, contestations*. Oxford: Oxford University Press.

Dworkin, R. (1981). "What is equality? Part 2: equality of resources', *Philosophy and Public Affairs*, vol. 10, pp. 283–345.

Dworkin, R. (2002). *Sovereign virtue: the theory and practice of equality*. Cambridge, MA: Harvard University Press.

Elstub, S. (2010). "Linking micro deliberative democracy and decision-making: trade-offs between theory and practice in a partisan citizen forum." *Representation*, vol. 46, no. 3, pp. 309–324.

European Commission. (2015). *The Aarhus Convention*, viewed 7 October 2015, http://ec.europa.eu/environment/aarhus/

Fraser, N. (1997). *Justice interruptus: critical reflections on the "postsocialist" condition*. London: Routledge.

Goodin, R.E. (2007). "Enfranchising all affected interests, and its alternatives." *Philosophy and Public Affairs*, vol. 35, no. 1, pp. 40–68.

Gunderson, R. (2014). "Habermas in environmental thought: anthropocentric Kantian or forefather of ecological democracy?" *Sociological Inquiry*, vol. 84, no. 4, pp. 626–653.

Hartley, N. & Wood, C. (2005). "Public participation in environmental impact assessment – implementing the Aarhus Convention." *Environmental Impact Assessment Review*, vol. 25, no. 4, pp. 319–340.

Hayward, T. (2006). "Ecological citizenship: justice, rights and the virtue of resourcefulness." *Environmental Politics*, vol. 15, pp. 435–446.

Hindmarsh, R. & Matthews, C. (2008). "Deliberative speak at the turbine face: community engagement, wind farms, and renewable energy transitions in Australia." *Journal of Environmental Policy & Planning*, vol. 10, no. 3, pp. 217–232.

Honneth, A. (1995). *The struggle for recognition: the moral grammar of social conflicts*. Cambridge, MA: Polity Press.

Hunold, C. & Young, I.M. (1998). "Justice, democracy, and hazardous siting." *Political Studies*, vol. 46, no. 1, pp. 82–95.

Jagers, S.C., Martinsson, J. and Matti, S. (2014). "Ecological citizenship: a driver of pro-environmental behaviour?" *Environmental Politics*, vol. 23, no. 3, pp. 434–453.

Jones, P. (2006). "Equality, recognition and difference." *Critical Review of International Social and Political Philosophy*, vol. 9, no. 1, pp. 23–46.

Lake, R.W. (1996). "Volunteers, NIMBYs, and environmental justice: dilemmas of democratic practice." *Antipode*, vol. 28, no. 2, pp. 160–174.

Latta, A. (2007). "Locating democratic politics in ecological citizenship." *Environmental Politics*, vol. 16, pp. 377–393.

Lidskog, R. & Elander, I. (2010). "Addressing climate change democratically, multi-level governance, transnational networks and governmental structures." *Sustainable Development*, vol. 18, no. 1, pp. 32–41.

Mason, M. (2010). "Information disclosure and environmental rights: the Aarhus Convention." *Global Environmental Politics*, vol. 10, no. 3, pp. 10–31.

Meadowcroft, J. (1997). "Planning, democracy and the challenge of sustainable development." *International Political Science Review*, vol. 18, no. 2, pp. 167–189.

Miller, D. (2001). *Principles of social justice*. Cambridge, MA: Harvard University Press.

Morello-Frosch, R.A. (2002). "Discrimination and the political economy of environmental inequality." *Environment and Planning C-Government and Policy*, vol. 20, no. 4, pp. 477–496.

Niemeyer, S. (2013). "Democracy and climate change: what can deliberative democracy contribute?" *Australian Journal of Politics and History*, vol. 59, no. 3, pp. 429–448.

Ophuls, W. (1977). *Ecology and the politics of scarcity*. San Francisco: W.H. Freeman and Company.

Parfit, D. (1997). "Equality and priority." *Ratio*, vol. 10, pp. 202–221.

Pedersen, O.W. (2011). "Price and participation: the UK before the Aarhus Convention's Compliance Committee." *Environmental Law Review*, vol. 13, no. 2, pp. 115–123.

Pedersen, O.W. (2014). "The price is right: Aarhus and access to justice." *Civil Justice Quarterly*, vol. 33, no. 1, pp. 13–17.

Peeters, M. & Nóbrega, S. (2014). "Climate change-related Aarhus conflicts: how successful are procedural rights in EU climate law?" *Review of European, Comparative and International Environmental Law*, vol. 23, no. 3, pp. 354–366.

Rawls, J. (1972). *A theory of justice*. Oxford: Oxford University Press.

Renn, O., Webler, T. & Wiedemann, P.M. (1995). *Fairness and competence in citizen participation: evaluating models for environmental discourse*. Dordrecht: Kluwer Academic.

Schlosberg, D. (1999). *Environmental justice and the new pluralism: the challenge of difference for environmentalism*. Oxford: Oxford University Press.

Schlosberg, D. (2007). *"Defining environmental justice: theories, movements, and nature."* Oxford: Oxford University Press.

Schlosberg, D. (2013). "Theorising environmental justice: the expanding sphere of a discourse." *Environmental Politics*, vol. 22, no. 1, pp. 37–55.

Shrader-Frechette, K.S. (2002). *Environmental justice: creating equality, reclaiming democracy*. Oxford: Oxford University Press.

Smith, G. (2003). *Deliberative democracy and the environment*. London: Routledge.

Stec, S., Casey-Lefkowitz, S. & Jendroska, J. (2000). *The Aarhus Convention: an implementation guide*. New York: United Nations.

Stephens, C., Bullock, S. & Scott, A. (2001). *Environmental justice: rights and means to a healthy environment for all*, viewed 22 January 2016, http://hdl.handle.net/10068/548056

Swift, A. (2008). "The value of philosophy in nonideal circumstances." *Social Theory and Practice*, vol. 34, pp. 363–87.

Sze, J., London, J., Shilling, F., Gambirazzio, G., Filan, T. & Cadenasso, M. (2009). "Defining and contesting environmental justice: socio-natures and the politics of scale in the Delta." *Antipode*, vol. 41, no. 4, pp. 807–843.

Tomlinson, L. (2015). *Procedural justice in the United Nations Framework Convention on Climate Change: negotiating fairness*. Switzerland: Springer International Publishing.

UNECE. (2016). *Aarhus Convention: map of parties*, viewed 31 January 2016, www.unece.org/env/pp/aarhus/map.html

US Environmental Protection Agency. (2015). *Environmental justice, basic information*, viewed 22 January 2016, www3.epa.gov/environmentaljustice/basics/

Walker, G. (2012). *Environmental justice: concepts, evidence and politics*. Abingdon, Oxon: Routledge.

Young, I.M. (1990). *Justice and the politics of difference*. Princeton: Princeton University Press.

10

THE RECOGNITION PARADIGM OF ENVIRONMENTAL INJUSTICE

Kyle Whyte

Introduction

In this chapter I seek to identify some of the theoretical features of the *recognition paradigm* of environmental injustice – where a paradigm simply refers to a criterion whose presence in someone's social circumstances is used by activists, scholars, politicians, and others to identify that injustice is occurring. People invoke paradigms of environmental injustice, either directly or implicitly, to reveal or prove the occurrence of injustices happening as part of their efforts to raise awareness of wrongdoing and motivate legal and policy solutions that will make for improved futures for people experiencing injustices. Some of the most common paradigms seek to show that societal institutions, such as government regulatory agencies, fail to ensure that some people, such as Indigenous peoples or people of colour in the US, have levels of protection from environmental hazards that are the *same* as or *equal* to those of other populations. These paradigms suggest the widely accepted notion that people deserve equal treatment.

However, the recognition paradigm diverges from some of the common paradigms because its criteria of environmental injustice concern the failure of some human groups to respect or acknowledge *difference*, as opposed to sameness or strict equality. I have two goals in this chapter. First, I seek to introduce readers to some of the major ideas of the recognition paradigm, though I will not explore any of the more detailed debates about whether recognition is a distinct paradigm or merely an extension of other paradigms. I will show, starting with a brief analysis of two cases, why some people see recognition as a significant and distinct paradigm for understanding environmental injustice. Second, I will then suggest to readers that the recognition paradigm is uniquely positioned to understand the ecological dimensions of environmental injustice, which has important implications, for some communities, about what remedies are necessary for achieving justice.

Two cases of environmental injustice

Detroit 48217 (a zip or postal code reference for a particular area of the city) has some of the greatest pollution of all zip codes in the state of Michigan and the United States. Of the nearly 10,000 people living there, more than 88 per cent are African-American and 25 per cent live below the poverty level. The majority of residents have an income level of less than $15,000 a year (Atkin 2014). Residents also live among almost 30 different industries, including Marathon Petroleum, Severstal Steel, U.S. Steel, and Detroit Edison's Rouge Power Plant, giving Detroit 48217 a toxic burden level 46 times the state average (Yu et al. 2013: 14–15). The elevated air toxin levels possibly explain why many residents of Detroit suffer health problems that are worse on average in comparison to other Michiganders. Residents suffer from cancer at a rate of 624.1 cases per 100,000 people, almost 100 cases per 100,000 higher than the state average. Adults suffer from asthma 50 per cent more often than the average Michigan adult, while asthma hospitalization rates are three times that of the state average, and asthma related deaths are twice as high as the state average. Rhonda Anderson, a Detroit environmental justice (EJ) activist, says:

> The community is just inundated with all these polluting sources, and their primary impact is on people's health and their quality of life . . . . Even if people are able to live life to 70, 80, what kind of life is that if the last 20 years you have cancer, cardiovascular disease, if you have asthma?
>
> *(Atkin: 2014)*

Moreover, residents of 48217 face challenges when trying to take meaningful action to change their situation. They do not have the financial resources to support legal remedies even though some of the industries are out of compliance. For example, Marathon Oil has violated its permitted limits for particulate matter and carbon monoxide emissions from various parts of its refinery; yet raising awareness of these violations has not translated into improving the lives of residents (Michigan Environmental Justice Coalition). The US Environmental Protection Agency (EPA) denies most civil rights claims for reasons including the lack of claimants' legal knowledge or a lack of relevance if private companies do not receive EPA funding (Lombardi 2015). EPA and Michigan's Department of Environmental Quality have not developed an assessment tool that can capture the types of combined, cumulative impacts that residents such as those in 48217 can use (Schlanger 2016). Finally, when industries apply for permits or permit renewal, EPA and the state may hold public meetings or hearings, but residents rarely participate in the scheduling or planning of these events and there is no guarantee that their views or experiences get taken seriously.

Consider another case of environmental injustice on the other side of the Great Lakes region. The Mohawk nation of Akwesasne, with 12,000 Tribal members, has homelands that are now interfered with by the US and Canada through the instantiation of the border between the two countries. Both the St Regis Indian Reservation (US side) and the Akwesasne Mohawk Reserve (Canadian side) are among the most polluted areas in both the US and Canada. The causes of pollution are US corporations, such as General Motors (GM), The Aluminum Company of America (ALCOA), and Reynolds Metals, who established industrial plants along the St Lawrence river during the 20th century, releasing toxicants such as polychlorinated biphenyls (PCBs), dibenzofurans, dioxins, polyaromatic hydrocarbons, fluorides, cyanide, aluminium, arsenic, chromium, and styrene (Arquette et al. 2002). Pollution has destroyed the Mohawks' fishery, producing problems such as build-ups in toxicants in mothers' milk. There are also cultural impacts when longstanding practices such as fishing end. Henry Lickers, a Tribal member, claims that ". . . a whole section of your language and culture is lost because no one is tying those nets anymore . . . that whole social infrastructure that was around the fabrication of that net disappeared" (Hoover 2013: 5).

Members of Akwesasne have mobilized to remedy these pollution problems. Yet US and Canadian governments and non-Indigenous scientists often propose solutions that will not work because they require Mohawks to simply stop their traditional practices to avoid exposure to contaminants. Arquette et al. (2002) describe how people do not understand that

> in Akwesasne, as in many other communities, potentially serious adverse health effects can result when people stop traditional cultural practices in order to protect their health from the effects of toxic substances. When traditional foods such as fish are no longer eaten, alternative diets are consumed that are often high in fat and calories and low in vitamins and nutrients.

Eating these diets is linked to "type II diabetes, heart disease, stroke, high blood pressure, cancer, and obesity" (Arquette et al. 2002: 261). In Akwesasne, the peoples' being unable to practise their own cultural traditions produces adverse health outcomes – even when the US and Canada try to create solutions that will lessen exposure. This is why, for example, fish advisories posted by US agencies that tell people not to eat fish do not necessarily help the situation. For the advisories are telling people to stop practising their cultures and ways of life for the sake of not being exposed to contaminants that Tribal members are not responsible for introducing in the first place (Hoover 2013).

Paradigms of environmental injustice

Detroit 48217 and Akwesasne feature criteria of different *paradigms* of environmental injustice. Again, a paradigm refers to some criterion whose fulfilment in a social situation can be used to identify that an injustice, that is, a systematic wrongdoing, is occurring. In terms of what injustice means, moral wrongdoing is systematic when some members or human groups of a society suffer simply because of some morally arbitrary feature that is voluntarily or involuntarily attached to them, such as their age, race, disability, class, gender, indigeneity, or ethnicity. A *morally arbitrary feature* is an attribute whose presence should not be determinative of whether a person or human group suffers more than others. In these cases, just because the residents of Detroit 48217 are predominantly African-American and poor, they do not have the same access to clean air and other environmental conditions as other residents of the city, state and the US. Moreover, they do not have the same opportunities to influence the policies, processes, and decisions that affect their well-being. In Akwesasne, US and Canadian settler societies do not respect the cultural integrity and unique health needs of the Mohawk community.

Across the two cases, different criteria can be identified for why systematic wrongdoing is occurring. In Detroit 48217, two paradigms are certainly discernible. Race and class determine whether some US citizens face greater environmental risks and worse health outcomes than other US citizens (paradigm 1). US citizens who bear environmental and health burdens are also those who lack opportunities to change their situations in US law, policy and democratic processes (paradigm 2). Both paradigms involve criteria that identify how people who are otherwise the same as US citizens are treated differently. But in the case of Akwesasne, peoples' differences are overlooked, which raises the possibility of a third paradigm. For fish advisories and strategies for risk-avoidance require Akwesasne Tribal members to give up their cultural integrity, economic vitality and political self-determination as peoples distinct from other settler, Indigenous, arrivant and migrant peoples in North America. In the recognition paradigm, injustice occurs when differences are not acknowledged or respected (paradigm 3).

Before moving on to what recognition means, I will first set the stage by discussing the meaning of the first two paradigms referenced in the previous paragraph, which I will refer to as the *distributive* and *procedural* paradigms (see also Chapters 5, 6 and 9).

Distributive and procedural paradigms of environmental justice

In the US in the 1970s and 1980s, numerous studies and reports began to show suspicious patterns of systematic wrongdoing. A 1987 study by the United Church of Christ Commission for Racial Justice, *Toxic Wastes and Race*, revealed that commercial hazardous waste facilities in five southern states are sited in communities with the highest demographic compositions of racial and ethnic residents, including African-Americans. The study was motivated, in part, by the direct actions of members of the African-American community in Warren County, North Carolina, who opposed the establishment of a PCB disposal landfill in their community on the basis that the siting process was racist. In 1994, a US congressional study showed that there were 600 open (noncompliant) dumps of municipal solid waste on Indian and Alaska Native lands, many of which were used by federal agencies such as the Bureau of Indian Affairs (Indian Lands Open Dump Cleanup Act) (Grijalva 2008). As people in the US became more aware of these injustices, it was also becoming apparent elsewhere that there has been a longstanding global problem affecting people of colour, Indigenous peoples and people of the Global South. In the 1990s, for example, the Nam Theun-Hinboun hydropower project 210MW dam in Laos displaced around 6000 people in 25 villages, Laos being one of the most indebted poor countries of the Gobal South (Shoemaker 1998).

In light of these patterns, leaders of communities of colour, faith-based organizations, anti-toxics advocates and scholars began pointing out case after case in which citizens of the US and other countries were more likely to face environmental risks owing to their race, ethnicity, gender and class (Bullard 1990; Hofrichter 1993). These findings showed that conversations about civil rights had not yet considered the connections between discrimination and people's capacity to live, work and play in clean, healthy and safe environments. The US-based theory of EJ grew alongside critical legal studies on intersectionality, or the idea that different forms of oppression, such as racism, sexism, and so on, are interlocking and produce suffering that is often invisible to members of privileged populations (Crenshaw 1991). The first two paradigms of injustice often arise from an understanding of the nature of wrongdoing in which social and economic institutions (societal institutions) in a particular nation state are failing to protect the citizens for whom they are responsible. Societal institutions include environmental agencies, corporations, and nongovernmental organizations, among others. That is, the vicious pattern of systematic wrongdoing concerns the breakdown of societal institutions' responsibilities to protect citizens equally. We can refer to these failures of societal institutions as failures of distributive justice, or the idea that some populations face unequal burdens of environmental hazards for morally arbitrary reasons, such as their race, ethnicity, gender, and so on.

The solutions then concerned creating laws and policies and building public awareness about systematic wrongdoing in relation to the distribution of environmental hazards. In 1994, US President Clinton issued Executive Order 12898, which requires all federal agencies to consider the impacts on communities of colour and Tribal communities in their decisions regarding actions that may affect these communities. The previously cited Indian Open Dumps Act is another example. Many lawyers, such as Luke Cole and Sheila Foster, began to explore and use law to show that environmental burdens violated civil rights (Cole and Foster 2001). When activists and scholars examined who shaped these societal institutions, they began to see a more vicious type of wrongdoing. It was not only the fact that societal institutions were misallocating environmental hazards, but also that the people most harmed did not participate actively in these societal institutions. Black, brown and Indigenous people were not prominent as leaders in the agencies responsible for allocating toxic waste repositories. Tribes were not written into the statute governing municipal solid waste as they were supposed to be – that is, as sovereign nations capable of managing their own environmental programmes. Communities of colour lacked the resources to pay lawyers to fight against environmental injustice. Most environmental organizations were primarily white in the makeup of their leadership and staff (Taylor: 2014).

These issues related to influence refer to procedural justice (see Chapter 9). Procedural injustice occurs when persons have no voice or capacity to exercise self-determination in decision-making processes that affect their lives – and there are no morally relevant reasons why they lack voice or information. People working in environmental ethics, especially Kristin Schrader-Frechette, addressed these problems as tied to deep moral issues of free, prior and informed consent and political equality. Kristin Schrader-Frechette refers to procedural justice as "a principle of participative justice" that works to "ensure that there are institutional and procedural norms that guarantee all people equal opportunity for consideration in decision-making" (Shrader-Frechette 2002: 28). When these institutional and procedural norms do not exist in social institutions related to the environment, procedural environmental injustice occurs.

Procedural justice has been pursued through a number of solutions for reform. US agencies, such as the Environmental Protection Agency (EPA), have created committees such as EPA's National Environmental Justice Advisory Committee, that are made up of representatives of communities of colour. Nongovernmental organizations have tried to recruit more people of colour and create EJ programmes in cities that were not typically considered part of "wilderness" conservation, such as the Sierra Club's Detroit Environmental Justice Office. The US federal government requires that Tribes be consulted on federal actions that concern them. Though some would argue that the following are not really attempts to establish justice, it is also true that corporations, for example, have created specific positions for negotiating environmental justice *issues* and often have people of colour or Indigenous persons serving in these roles.

Distributive/procedural versions of environmental injustice defend the equality of all citizens before the societal institutions of their nation states. Environmental injustice is a systematic failure of societal institutions to ensure all citizens who are subject to the *same* laws and policies should be treated equally in the decision-making processes and distributions of outcomes of the societal institutions of nation states to which they belong. The wrongdoing of environmental injustice refers to a failure of societal institutions to protect all citizens who are owed similar treatment.

The recognition paradigm of environmental justice

The recognition paradigm of environmental injustice suggests other criteria for identifying systematic wrongdoing. Instead of looking for violations of sameness, such as equal treatment before the law, the recognition paradigm identifies failure in societal institutions to acknowledge or respect *difference* as a source of systematic wrongdoing (see also Chapters 2, 7, 40 and 41). The idea of the injustice as failure to respect or acknowledge difference goes back to ancient traditions of justice. The Haudenosaunee *Kaswentha* (treaty belt with the Dutch from the 1700s) features two rows of purple wampum beads representing the societies as different types of vessels, a birch bark canoe and a ship, each having its own laws and cultures, yet who must share the same river (Ransom, 1999). Among other things, the *Kaswentha* suggests that differences should be respected and acknowledged if the societies are to avoid wrongdoing against each other. Consider some cases where disrespect of difference is at stake regarding environmental injustice.

El Pueblo Para El Aire y El Agua Limpio, a community organization formed in predominantly Latino Kettleman City, California, resisted a proposed waste incinerator in the 1980s. The rural, mostly Hispanic community members found out that there was a toxic waste dump in their backyards of which they were unaware and that a waste incinerator was being proposed nearby. The community members were unaware of all that was going on because the legal procedures for notifying stakeholders sidestepped the fact that many community members spoke only Spanish and did not have access to one of the newspapers where information about the siting was featured (Figueroa 2001; Cole and Foster 2001). In another case, Kennecott Mining, a subsidiary of Rio Tinto, sought to place a mine in the territory of the Keweenaw Bay Indian Community that would disrupt caretaking practices associated with a sacred rock, Eagle Rock. Yet the state mining permit programme, while it seeks to protect cultural heritage, only counted archeological findings that were structured like Western buildings (e.g. churches) – hence protection of the sacred rock did not factor into the permitting process (Bienkowski 2012).

The United Nations' programme, Reducing Emissions from Deforestation and Degradation (REDD+) seeks to reduce greenhouse gases in the atmosphere by conserving forests. The programme involves creating a market by which payments are made to developing countries to avoid chopping down or degrading forests. Emerging literature is showing that countries such as Tanzania or Panama, that implement REDD+ programmes, displace Indigenous peoples who are living in and using forests. In Tanzania, the Warufiji peoples use the mangrove forests to cultivate rice that they eat for subsistence. The state and environmental organizations portray ricing practices as bad environmental stewardship, hence justifying the removal of these peoples from their sources of subsistence (Beymer-Farris and Bassett 2012).

Or consider tourism, such as in Uluṟu-Kata Tjuṯa National Park (Australia). The Aṉangu people (the legally recognized owners) jointly manage the area with the Australian government as a tourist location, one of the major attractions being Uluṟu, the world famous monolithic rock. For the Aṉangu, the rock has historical and contemporary cultural significance as, among other things, the basis for how they envision their futures as peoples. On a sign for tourists, the Aṉangu post a prohibition against climbing the rock. Yet, non-Aboriginal Australians and visitors largely ignore the prohibition, and engage in climbing the rock *en masse*. The Aṉangu do not police these activities because they see policing and punitive measures as colonial and do not want to repeat them on others (Figueroa and Waitt 2008).

The recognition paradigm explains each of these cases as examples of systematic wrongdoing because the societal institutions of a dominant society fail to acknowledge or respect differences across human groups. Lack of acknowledgement and disrespect produce negative consequences for the nondominant groups, such as the Aṉangu, the Keewenaw Bay Indian Community or the Latino residents of Kettleman City. Figueroa (2006), the first philosopher to consider recognition as an aspect of environmental justice, refers to misrecognition as failure to acknowledge or respect different peoples' environmental identities and heritages. For Figueroa, environmental identities and heritages refer to

> the amalgamation of cultural identities, ways of life, and self-perceptions that are connected to a given group's physical environment .... Environmental identity is

closely related to environmental heritage, where the meanings and symbols of the past frame values, practices, and places we wish to preserve for ourselves as *members of a community.*

(Figueroa 2006).

Injustice, then, occurs when societal institutions are organized in ways that fail to acknowledge or respect the environmental identities and heritages of certain populations. That is, societal institutions fail to recognize human social difference.

Differences can be cultural, including language, or can involve land- or water-based practices associated with people's subsistence and flourishing, such as their economies (e.g. ricing) or political structures (e.g. Anangu lack of policing). In Kettleman City, the use of Spanish by members of an isolated community was not considered in light of the dominance of the English language in the state of California, though there is no morally relevant reason why English must be dominant. Their linguistic difference, then, was not respected, which set them up to be exposed to greater risk. In the case of the Kennecott mine, the Tribe's cultural practices derive from a different cosmology than US settler understandings of cultural heritage. Human cultural difference extends beyond the language one speaks or one's customs. The Anangu seek to exercise their own structures of political self-determination on lands of which they are legal owners through protecting landscapes that are important to their visioning practices through prohibition customs that are consistent with their morality.

McGregor (2009) argues that difference includes both cultural difference, as just discussed, but also non-human difference given that plants, animals, insects and ecosystems are also implicated in EJ situations. For her, EJ is

about justice for all beings of Creation, not only because threats to their existence threaten ours but because from an Aboriginal perspective justice among beings of Creation is life-affirming .... In the Anishinaabe world view, all beings of Creation have spirit, with duties and responsibilities to each other to ensure the continuation of Creation. Environmental justice in this context is much broader than 'impacts' on people. There are responsibilities beyond those of people that also must be fulfilled to ensure the processes of Creation will continue.

(McGregor 2009: 27)

For McGregor, then, environmental injustices occur against non-humans when their habitats are destroyed, for example, through water pollution.

Schlosberg has also argued for "participatory parity" for the natural world: "The goal is more broadly the recognition of the consideration of the natural world in human decision-making" (Schlosberg 2009: 158). He adds that "in order to attain both environmental and ecological justice, we must be sure that views from the margins, the remote, and the natural world are recognized and represented, either directly or through proxies" (2009: 187). This involves finding ways to recognize natural agency; for Dryzek (2000: 149, quoted in Schlosberg 2009: 191) the "recognition of agency in nature therefore means that we should listen to signals emanating from the natural world with the same sort of respect we accord communication emanating from human subjects, and as requiring equally careful interpretation." Plumwood (2002: 178) offers another recognition-based conception of justice for non-humans with her notion of "intentional recognition," or the idea that "recognition of the other's agency is in turn central to any kind of negotiation or mutual adjustment process, it is important to cultivate the ability and the conceptual basis for such recognition." The recognition paradigm of environmental injustice, then, looks closely at whether societal institutions respect and acknowledge not only human cultural and moral difference, but also non-human difference – what Figueroa (2013) has called *interspecies justice*.

Yet respect for and acknowledgement of difference can be coopted by dominant parties who seek to hide rather than actually rectify the injustices they are perpetrating against others. In Canada, for example, scholars have challenged Canada's process of "recognizing" Indigenous peoples' sovereign and cultural differences and Indigenous identities (Lawrence 2003; Coulthard 2014). Coulthard, for example, analyses how the Canadian settler state continued to dispossess Dene peoples of their land for the sake of resource extraction even while the state used rhetorics of honouring Indigenous cultural rights and political autonomy. Respect for cultural integrity served to satisfy settler Canadians that they were not doing anything wrong even when Dene peoples were still being dispossessed of their lands. Nadasdy (2005) discusses how in Canadian joint environmental management the norm of respect for Indigenous peoples' traditional ecological knowledge masks the reality that Aboriginal leaders do not participate in key management decisions about the jointly managed lands and waters in the first place. Acknowledgement and respect for difference is often a smokescreen that obscures the continuance of oppression against nondominant groups such as Indigenous peoples. One of the most obvious cases of this is political reconciliation between settler states and Indigenous peoples, where what often occurs, as it did in Canada, is that the apologies and documentation of colonial oppression do not translate into returning significant amounts of lands and waters to Indigenous peoples.

Conclusion: the ecological dimensions of the recognition paradigm

Recognition justice, as described earlier, presents criteria of acknowledgement and respect for difference as part of identifying environmental injustices. One of the reasons why cooptation is possible is that genuine recognition is deeply ecological. That is, certain forms or expressions of differences – from cultural practices to structures of political self-determination – require a particular range of ecological processes and access to large territories of land and water eco-systems. Cultures in which participants depend on particular medicinal plants require access and availability to those plants. Economies that require control over food sources require adequate baselines of natural resources and microbiomes that sustain the necessary habitats. Structures of political self-determination, then, must have the capacity to protect sufficient areas of land for the sake of maintaining cultural integrity and economic vitality.

Some scholars offer accounts of injustice that focus on systematic wrongdoing (environmental injustice) as disruptions of the ecological processes required for economic vitality, cultural integrity and political self-determination (Maldonado et al. 2013). Maldonado and colleagues work with the Isle de Jean Charles Band of Biloxi-Chitimacha-Choctaw Indians, a Tribe currently facing multiple environmental stressors. Loss of wetlands and barrier islands destroyed natural protection against extreme weather events – the island is now the main storm buffer for other communities in Louisiana to its north. The island now experiences flooding from hurricanes, which was previously unknown. Most of the isle's

> [t]rees, medicinal plants, gardens and trapping grounds are gone. There are ongoing health problems due to industrial contamination and the continuing development of toxic industries, chemicals from dispersants, oil spills, including the 2010 BP oil disaster, and post-storm debris contaminating the air, soil and water.
>
> *(Maldonado et al. 2013: 606)*

Now, climate induced sea-level rise will destroy the remaining capacity of residents to live on the isle, which will be gone in another 30–40 years. Finally, as a result of their having to find refuge away from European and US settlers (to avoid death from settler colonialism), their isolation resulted in the US not recognizing the Band as a sovereign government, which denies the Band specific benefits and protections associated with US federal recognition. They have been excluded from multiple coastal restoration efforts, including the Army Corps' Morganza-to-the-Gulf-of-Mexico Hurricane Protection Project and Louisiana's 50-year Master Plan for Coastal Restoration (Maldonado et al. 2013).

According to Maldonado et al., the Tribe is "looking not just for community and cultural restoration but also for traditional livelihood development to once again be a self-sustaining community." Yet the reason why they face these injustices goes back to how they have been repeatedly denied the ecological processes required for them to express their cultural, economic and political differences. Historically, the Tribe subsisted from a seasonal lifestyle of fishing, trapping and hunting in what is now the Terrebonne Parish, Louisiana. The region was large, with multiple opportunities to adjust to seasonal changes. In response to settler colonialism, they adapted their culture to surviving off of the island terrestrial and aquatic ecosystems as a place of refuge. Today, the Tribe's culture and water-based settlements and livelihoods are threatened by both the causes of climate change and climate change impacts. Since the 1950s, the Isle has shrunk from 5 miles by 12 miles to 1/4 mile by 2 miles from coastal erosion and saltwater intrusion. The shrinking occurred initially due to oil and gas companies dredging canals and cutting pipelines. Dike and levee construction, the damming of the Mississippi River, diverse flood control measures, and industrial agriculture development further contributed to the subsidence of parts of the island (Maldonado et al. 2013).

In the case of the Isle de Jean Charles, the Tribe face environmental injustices due to hundreds of years of settler colonialism that destroyed the ecological processes required for the Tribe to flourish with its own culture, economy and structure of political self-determination. It is hard to imagine solutions for the Tribe that do not entail transformation of the ecosystems in the region. The recognition paradigm of injustice, then, would suggest that failure to respect or acknowledge difference ultimately is also about the undermining of the ecological conditions required for any society to express difference in the first place. But it is not just Indigenous peoples that seek ecological remedies to environmental injustice. White's work on the Detroit Black Community Food Security Network, for example, discusses how some African-American Detroiters have sought to reestablish the ecological processes required for them to express their unique food-based cultures, economics and structures of political decision-making (White 2011a, 2011b). It is also possible to make the case that the recognition-based arguments concerning non-humans, such as McGregor's or Schlosberg's cited earlier, could also be understood in relation to ecological processes.

The ecological dimensions of the recognition paradigm help to understand why certain solutions proposed by dominant parties to recognizing difference are inadequate. A dominant party, such as the US federal government, or a multinational corporation, cannot possibly respect Indigenous rights to practise their cultures, for example, if Indigenous peoples cannot gather the plants required for certain ceremonies. Dominant parties cannot claim to acknowledge rights to food sovereignty for diverse groups, from African-Americans to Indigenous peoples to Latino communities, if the ecological processes are not present for these groups to produce foods that are culturally meaningful and profitable for them in regional and global marketplaces. While the first two paradigms, distribution and procedural paradigms, are certainly capable of expressing the importance of ecological processes, the recognition paradigm can be understood as doing so more emphatically because the very expression of cultural, economic and political differences is often rooted in particular ecosystems. For this reason, important research areas, from political ecology to resilience, are showing in many cases how the possibility of expressing difference is tied to the maintenance of certain ecosystems or the capacity to live with ecological processes that facilitate a society's adaptation to change. The future of work in the recognition paradigm of environmental injustice will perhaps be more and more connected to environmental science fields.

References

Adger, W.N., J.M. Pulhin, J. Barnett, G.D. Dabelko, G.K. Hovelsrud, M. Levy, et al. (2014). "Human Security." In C.B. Field, V.R. Barros, D.J. Dokken, K.J. Mach, M.D. Mastrandrea, T.E. Bilir, et al. (eds.), *Climate Change 2014: Impacts, Adaptation, and Vulnerability. Part A: Global and Sectoral Aspects. Contribution of Working Group Ii to the Fifth Assessment Report of the Intergovernmental Panel on Climate Change*, Cambridge, United Kingdom: Cambridge University Press.

Arquette, M., M. Cole, K. Cook, B. LaFrance, M. Peters, J. Ransom, et al. (2002). "Holistic Risk-Based Environmental Decision-making: A Native Perspective." *Environmental Health Perspectives*, vol. 110(2), pp. 259–264.

Atkin, E. (2014). "Meet the People Fighting Pollution in Michigan's Most Toxic ZIP Code." *Think Progress*. https://thinkprogress.org/meet-the-people-fighting-pollution-in-michigans-most-toxic-zip-code-666588d612a4#.c6lviybxd

Bennett, T.M., N.G. Maynard, P. Cochran, R. Gough, K. Lynn, J. Maldonado, et al. (2014). "Indigenous Peoples, Lands, and Resources." In J.M. Melillo, T.T.C. Richmond and G.W. Yohe (eds.), *Climate Change Impacts in the United States: The Third National Climate Assessment*. Washington, DC: US Global Change Research Program.

Beymer-Farris, B.A. & T.J. Bassett. (2012). "The Redd Menace: Resurgent Protectionism in Tanzania's Mangrove Forests." *Global Environmental Change*, vol. 22(2), pp. 332–341.

Bienkowski, B. (2012). "Sacred Water, New Mine: A Michigan Tribe Battles a Global Corporation." *Environmental Health News*, June 12.

Bullard, R.D. (1990). *Dumping in Dixie: Race, Class, and Environmental Quality*. Boulder: Westview Press.

Cole, L.W. & S.R. Foster. (2001). *From the Ground Up: Environmental Racism and the Rise of the Environmental Justice Movement, Critical America*. New York: New York University Press.

Coulthard, G.S. (2014). *Red Skin, White Masks: Rejecting the Colonial Politics of Recognition*. Minneapolis: University of Minnesota Press.

Crenshaw, K. (1991). "Mapping the Margins: Intersectionality, Identity Politics, and Violence against Women of Color." *Stanford Law Review*, vol. 43, pp. 1241–1299.

Figueroa, R.M. (2001). "Other Faces: Latinos and Environmental Justice." In B.E. Lawson and L. Westra (eds.), *Faces of Environmental Racism: Confronting Issues of Global Justice*. Boston: Rowman and Littlefield Publishers.

Figueroa, R.M. (2006). "Evaluating Environmental Justice Claims." In J. Bauer (ed.), *Forging Environmentalism: Justice, Livelihood, and Contested Environments*.New York: M.E. Sharpe, pp. 360–76.

Figueroa, R.M. (2013). "Risking Recognition: New Assessment Strategies for Environmental Justice and American Indian Communities." *American Philosophical Association Newsletter on Indigenous Philosophy*, vol. 12(2), pp. 4–10.

Figueroa, R.M. and G. Waitt (2008). "Cracks in the Mirror: (Un)covering the Moral Terrains of Environmental Justice at Uluru-Kata Tjuṯa National Park." *Ethics, Place and Environment*, vol. 11(3), pp. 327–349.

Grijalva, J.M. (2008). *Closing the Circle: Environmental Justice in Indian Country*. Durham, NC: Carolina Academic Press.

Hofrichter, R. (1993). *Toxic Struggles: The Theory and Practice of Environmental Justice*. Philadelphia, PA: New Society Publishers.

Hoover, E. (2013). "Cultural and Health Implications of Fish Advisories in a Native American Community." *Ecological Processes*, vol. 2(1), pp. 1–12.

Lawrence, B. (2003). "Gender, Race, and the Regulation of Native Identity in Canada and the United States: An Overview." *Hypatia*, vol. 18(2), pp. 3–31.

Lombardi, K. (2015) "Environmental Racism Persists, and the EPA is One Reason Why." The Center for Public Integrity. www.publicintegrity.org/2015/08/03/17668/environmental-racism-persists-and-epa-one-reason-why.

Maldonado, J.K., C. Shearer, R. Bronen, K. Peterson & H. Lazrus. (2013). "The Impact of Climate Change on Tribal Communities in the US: Displacement, Relocation, and Human Rights." *Climatic Change*, vol. 120(3), pp. 601–614.

McGregor, D. (2009). "Honouring Our Relations: An Anishnaabe Perspective on Environmental Justice." In J. Agyeman, P. Cole and R. Haluza-Delay (eds.), *Speaking for Ourselves: Environmental Justice in Canada*. Vancouver, BC: University of British Columbia Press.

Michigan Environmental Justice Coalition. "Environmental Justice in Michigan." https://michigan environmentaljusticecoalition.wordpress.com/environmental-justice-in-michigan/

Nadasdy, P. (2005). "The Anti-Politics of TEK: The Institutionalization of Co-Management Discourse and Practice." *Anthropologica*, Vol. 47(2), pp. 215–231.

O'Neill, C. (2000). "Variable Justice: Environmental Standards, Contaminated Fish, and 'Acceptable' Risk to Native Peoples." *Stanford Environmental Law Journal*, vol. 19(1), pp. 3–119.

Plumwood, V. (2002). "Environmental Culture: The Ecological Crisis of Reason." London: Routledge.

Ransom, J.W. (1999). *The Waters. Words that Come Before All Else*. Akewsasne Territory/Cornwall, ON, Canada: Native North American Travelling College, pp. 25–33.

Schlanger, Z. (2016). "Choking to Death in Detroit: Flint isn't Michigan's only Disaster." Newsweek. www.newsweek.com/2016/04/08/michigan-air-pollution-poison-southwest-detroit-441914.html

Schlosberg, D. (2009). *Defining Environmental Justice: Theories, Movement, and Nature*. Oxford: Oxford University Press.

Shoemaker, B. (1998). *Trouble on the Theun-Hinboun: A Field Report on the Socio-Economic and Environmental Effects of the Nam Theun-Hinboun Hydropower Project in Laos*. Berkeley, CA: International Rivers Network.

Shrader-Frechette, K.S. (2002). *Environmental Justice: Creating Equality, Reclaiming Democracy, Environmental Ethics and Science Policy Series*. Oxford: Oxford University Press.

Taylor, D.E. (2014). "The State of Diversity in Environmental Organizations: Mainstream NGOs, Foundations and Government Agencies." Green 2.0 Working Group. Retrieved from http://diversegreen.org/report/ (3/21/17)

White, M. (2011a). "D-Town Farm: African American Resistance to Food Insecurity and the Transformation of Detroit." *Environmental Practice*, vol. 13(4), pp. 406–417.

White, M. (2011b). "Sisters of the Soil: Urban Gardening as Resistance in Detroit." *Race/Ethnicity: Multidisciplinary Global Contexts*, vol. 5(1), pp. 13–28.

Yu, S., G. Rivera, S. Turner-Handy, S. Todd, M. Naimi & S. Sagovac. (2013). Detroit, MI: The Detroit Environmental Agenda.

11

A CAPABILITIES APPROACH TO ENVIRONMENTAL JUSTICE

Rosie Day

Introduction

The capabilities approach was developed as a way of conceptualizing and assessing social and economic development. Its well-known founders are Indian economist Amartya Sen, and American philosopher Martha Nussbaum. Their main aim was to change the terms in which we think about human wellbeing and in which we assess deprivation, equality, and ultimately claims of justice and injustice. Although the capability approach has been hugely influential in development thinking and wider social policy for nearly three decades, its impact on environmental justice work has been fairly modest to date. As I shall argue in this chapter though, it provides a potentially constructive and helpful framework that lends itself well to thinking about how environmental matters become matters of justice and injustice. It has resonances with other distributional and procedural formulations of justice as well as justice as recognition, but is distinct in its core focus. Before elaborating further on how the approach can be taken up in environmental justice, as well as some challenges in doing so, I start with some brief explanation of its key ideas and how Sen and Nussbaum respectively argue that it should be used to think about development and social justice.

What are capabilities?

Capabilities are what we are focusing on if we ask the question, 'what is each person able to do and to be?' (Nussbaum 2011: 18). They are valued 'beings and doings'. They can range from the relatively trivial – e.g. I am able to go to the cinema – to the profound – e.g. I am able to vote in national governmental elections; I am able to be healthy. The totality of the things a person is able to do and to be is termed their 'capability set'.

Sen and Nussbaum use the term 'functionings' to indicate the pursuits and states of being that a person is actively engaged in, such as attending school, undertaking paid work, having meaningful relationships. Capabilities are the *opportunity* to engage in valued functionings, so for example the opportunity to work, to have relationships with others. The capability approach takes capabilities rather than functionings as the object of concern because it is argued that individuals should have the freedom to choose which functionings they engage in at any given time, otherwise life would be intolerably prescribed. Sen calls a capability therefore 'an aspect of freedom' (2010: 287): it includes the ability to freely choose to do something (or not) as well as the ability to actually do it.

Capabilities in the way that the term is used in the capabilities approach are not just internal to the individual, i.e. the term does not refer to innate abilities and skills (although these may also be involved). In this sense, the meaning does not completely correspond with the more general use of the word. The capabilities approach recognizes explicitly that in order for people to be able to engage in valued functionings, certain social and material arrangements need to be in place, and likewise that the social and material environment at any time will affect people's ability to engage in valued functionings. Capabilities therefore are not just a matter for the individual to work on, but for society to consider, assess, and ideally to promote for its citizens, through its programmes, policies and ways of working.

Since the 1990s and before, Sen and Nussbaum have argued that capabilities are a better focus for measuring the degree of development of any given society, and for making comparisons between societies, communities and individuals, than the usual alternative approaches of measuring either income/wealth/GDP on the one hand, or subjective satisfaction (utility) on the other (see e.g. Sen 1992, 1999, 2010; Nussbaum 2000, 2011; Nussbaum and Sen 1993). Their critique of GDP or income per capita approaches to evaluating development highlighted a number of flaws, including the hiding of differences between individuals and groups by aggregate measures, and the obliviousness of monetary based measures to many things that matter to people's real quality of life, such as health and political participation. Alternative approaches that measure perceived satisfaction or the fulfilment of subjective preferences they argue to be unsatisfactory for the reason that some people may be more easily satisfied than others and those suffering from some form of deprivation may not be able easily to realize their lack, or conceive of radically alternative lives. Repeatedly they argue that because it is what people can be and do that actually matters, this is what should be assessed.

A major influence of the capability approach was its underpinning of the United Nations Development Programme's Human Development Index, used in annual Human Development Reports from 1990 onwards to compare countries on their development levels (UNDP, nd). The HDI is not a full realization of the capabilities approach as it measures development in terms of per capita income, life expectancy and education only, aggregated at national level, but nevertheless this represented a paradigm shift from the previous dominant conceptualizations which were confined to economic measures. Since the early reports, more measures such as the gender development index have been added to the headline measure to improve its scope, to the approval of capability theorists (Nussbaum 2011).

Capabilities, inequality and justice

It is the contention of the capabilities approach that inequalities should be evaluated in terms of capabilities, measured at the individual level, i.e., the things that people are able to do and be. As such, we might see it as a distributional approach to justice, but with capabilities as the object of concern, rather than primary goods (such as income), which a more Rawlsian approach for example might take (Rawls 1971). However, it is both more and less than a distributional theory of justice: less because it does not offer, or intend to offer, a full account of what a fair distribution would be, either of capabilities or anything else; and more because it is a thicker notion of justice than one concerned only with distributional patterns (as is Rawls' theorization, but in a different way).

The approach does have important implications for thinking about distributions. The shift of focus to outcomes (what people can be and do) rather than inputs (i.e. resources people have) is because, as they explain, individuals are not all able to convert resources to outcomes at the same rate. The ability to do so will vary depending on personal characteristics, circumstances, environment and social context (see Sen 1999: 70). Different individuals therefore are likely to need different levels of resources to reach similar outcomes. Thus, the approach explicitly recognizes diversity and individual and group difference. Neither Nussbaum nor Sen stipulate that equality in capability sets must be the goal, but if that was the chosen aim in any instance, the approach would point towards potentially uneven allocation of resources as necessary in order to achieve broadly equal outcomes (i.e. equity rather than equality).

The approach also has strong commitments to notions of procedural justice (see Chapter 9) and human rights (see Chapter 13). To explore this further it is necessary to acknowledge the differences between Sen and Nussbaum and their respective developments of the approach. Their differences become clear in relation to how they tackle the fundamental questions of: What capabilities matter? And how much of any given capability is needed?

Taking Nussbaum first: she argues that in pursuing justice (and in any practical endeavour to improve human development) it is necessary to distinguish between capabilities which are trivial – earlier I used the example of being able to go to the cinema – which we are not really so worried about, and those which are more fundamental, such as being able to be in good health, which we are really concerned with. In response to the question of what capabilities matter, she proposes a list of ten Central Capabilities (see Table 11.1), which she drew up through extensive recourse to a variety of philosophical sources. She contends that the list is open-ended rather than definitive, but nevertheless should meet with 'overlapping

Table 11.1 Nussbaum's ten Central Capabilities (adapted and abridged from Nussbaum 2011: 33–34)

1.	Life	Being able to live a normal length life.
2.	Bodily health	Being able to have good health, to be adequately nourished and have good shelter.
3.	Bodily integrity	Being free from physical assault; having freedom of movement; having opportunities for sexual satisfaction; having reproductive choice.
4.	Senses, imagination and thought	Being able to use the senses, to imagine, think and reason in a way informed and cultivated by adequate education. Having freedom of expression and religious freedom. Being able to have pleasurable experiences and avoid non-beneficial pain.
5.	Emotions	Being able to have attachments to other people and things. Being able to experience and express emotions. Not having emotional development blighted.
6.	Practical reason	Being able to form a conception of the good and engage in critical reflection regarding one's life. Liberty of conscience.
7.	Affiliation	a) Being able to live with and toward others, engage in social interaction, relationships and empathy; b) Having the social bases of self-respect; being treated as of equal worth to others.
8.	Other species	Being able to live with concern for and in relation to animals, plants and nature.
9.	Play	Being able to laugh, play and enjoy recreational activities.
10.	Control over one's environment	a) Political: having the right of political participation and free speech; b) Material: being able to hold property on an equal basis to others; freedom from unwarranted search and seizure; being able to seek employment on an equal basis to others.

consensus' in pluralistic societies given its plural roots and lack of anchoring to any specific belief set. Nussbaum sees this list as akin to a list of human rights that individuals should be entitled to as a matter of justice, and sees it as the state's responsibility, ideally to be constitutionally enshrined, to ensure the social conditions for individuals to hold this capability set. She terms her approach a 'partial theory of social justice' (2011: 40), in that it lays out a set of minimum entitlements that she believes are a necessary condition of justice, but not a list of all that is needed to live a full life in all contexts, nor any proposal for what a fair distribution of opportunities beyond the list might look like. She acknowledges that it would be necessary to provide further specification on some particulars in local context – indeed, the list is very abstract – and in some cases on thresholds to be achieved as a minimum for justice (for example, how many years of education should someone be entitled to?). This further specification she believes should take place through democratic processes in national contexts. Although a universalist approach to some degree therefore, it allows flexibility for contextual variation in specifics.

Sen, on the other hand, has tended to focus on the informational and comparative use of the capability approach for evaluating and deciding between real alternatives, rather than formulating any notion of ideal arrangements which may not be achievable in practice; in this sense, his is not a transcendental approach. As such, although he also acknowledges that not all capabilities will be of equal import, he resists defining or suggesting any set of essential or priority capabilities in the way that Nussbaum does. He argues that the capabilities that are held to be essential, as well as the relative importance accorded to them, will vary from context to context, and from project to project. Therefore, he argues, the capabilities of concern and their relative weighting should be defined in context, through democratic, deliberative processes. Sen puts great value on deliberative democracy as a good in itself, and sees it as a crucial aspect of the approach, allowing reflection and consensus building, even if that consensus does not stretch to all details. Thus, although democratic specification of capabilities and thresholds comes into Nussbaum's approach, deliberative democracy is conceptually more fundamental and procedurally more central to Sen's, whilst he does not take the essential human rights angle that Nussbaum is committed to.

Capabilities and the environment

Neither Sen nor Nussbaum in their earlier writings on the capability approach make a great deal of reference to environmental concerns, but in later writing they both address them to some extent, although it remains somewhat in passing. In *The Idea of Justice* (2010), Sen addresses sustainable development, and praises the Brundtland Commission's highlighting of the importance of the environment and ecological integrity to human wellbeing. Although he does not dismiss the idea of the environment having a value in its own right, and emphasizes humans' responsibility to other, less powerful species, it is the services to humans both present and future that he is most interested in. From the idea of sustainable development he embraces the concern for future, as well as current, generations and proposes that not compromising their *capabilities* (rather than the Brundtland language of needs) ought to be a concern in a concept of 'sustainable freedom' (2010: 251; see also Sen 2013). Nussbaum in her 2011 book similarly highlights the challenge of how to incorporate consideration of the capabilities of future generations, with reference to environmental conditions, as something important for future capability-focused work to address. At the same time, she departs a little from the anthropocentric focus of sustainable development to spend some time advocating concern for the welfare of non-human animals, suggesting that an 'expanded notion of dignity' to include that of other species might underpin a concept of the basic entitlements of non-human species, though acknowledging that we are not yet at a point of broad social consensus on such matters. Others have taken up discussion of sustainability in capabilities terms (e.g. Ballet et al. 2013), generally finding the framework useful but incomplete, for example Pelenc et al. (2013) suggest that sustainable freedom would need to be integrated with notions of responsibility and voluntary self-restraint (see also Rauschmayer and Lessmann (2013) and other papers in that special issue).

A notable paper by Holland (2008) builds on Nussbaum's approach and proposes that the list of Central Capabilities should be extended to include 'being able to live one's life in the context of ecological conditions that can provide environmental resources and services that enable the current generation's range of capabilities; to have these conditions now and in the future' (2008: 324). She terms this additional capability a 'meta-capability' in that it underpins many of the others. While Sen, Nussbaum and others have tended to position environmental conditions as instrumental to capabilities, Holland's move intentionally positions environmental integrity as something that is essential, irreducible and non-substitutable, rather than a circumstantial condition that might conceivably be substituted with something else. She argues that its inclusion in this way is commensurate with Nussbaum's arguments that the central capabilities on her list have complex interdependencies and some are prefigurative of others, viz. the capabilities of practical reason and of affiliation. My own view is that Holland's proposed capability is rather incongruent with the others in Nussbaum's set in that it is more circumstantial, and therefore more intuitively positioned as part of the social and material pre-requisites of human capabilities, but Holland's argument is defensible given that arguably the ontological division between capabilities and the arrangements that enable them is fuzzy (see Smith and Seward 2009; also Schlosberg's (2007: 30–31) reading of what capabilities are).

Capabilities and environmental justice

If justice is about capabilities, and capabilities involve environmental conditions, then it is a short step to conclude that environmental conditions can be a matter of justice, expressed in capability terms – which brings us into the realm of environmental justice.

If we take Holland's position, which builds on Nussbaum's human rights style capability approach, then we are already at the point of arguing that all individuals have a right – that they can expect the state to protect – to live in an environment that provides the necessary resources and services to enable their other essential capabilities. Any contravention of this would constitute an injustice.

If on the other hand we don't position the freedom to live in such environmental conditions as a capability in itself, then justice and injustice claims can be made in terms of environmental conditions being pre-requisites of other essential capabilities: thus they are derivative rights. Many capabilities are likely to draw on environmental resources of one form or another whilst others will be compromised by environmental hazards and environmental deprivations. Facets of the environment in which people live then, will be essential to the realization and maintenance of a broad capability set. Where capabilities are a matter of justice, the implicated environmental aspects therefore become a matter of justice, unless the capability can be supported in some other way. With colleagues, I have made this argument in the related area of energy justice (Day et al. 2016), as have Sovacool et al. (2014) (see also Chapter 31). In that case, it is actually important that energy is seen as a derived, circumstantial need, rather than 'access to energy' being designated an essential capability in itself, because in the context of climate change and energy shortage, it may well be preferable if capabilities can be supported by alternative means, without drawing on energy resources.

Positioning environmental conditions as a necessary precondition of many important capabilities also works better when considering environmental justice through taking Sen's more open approach, where the capabilities that are considered important or essential to wellbeing are not defined *a priori*, but rather through deliberation among those affected and concerned. Such deliberation may not lead to living in a healthy environment or one with ecological integrity being defined as a priority capability – either in absolute or comparative terms – but environmental conditions and quality would still be implicated as a matter of justice if a necessary pre-condition to the realization of capabilities that are held to be high priority.

Often though in the environmental justice arena we are not starting with a concern about capabilities in a general sense, and what environmental conditions they need, but rather with a specific environmental issue, hazard, or amenity, and a need to make judgements regarding how it can be managed fairly. To illustrate how a capabilities perspective can be applied in such cases, I will briefly discuss the example of ambient air pollution (see also Chapter 26).

Outdoor air pollution is an environmental hazard that affects millions of urban dwelling people worldwide (WHO 2014a). It is a large and increasing problem in rapidly developing economies such as China and India (Wang and Hao 2012; Guttikunda and Calori 2013), caused mainly by growing numbers of motor vehicles and by the burning of coal, wood and animal dung; but cities in Europe, North America and Australasia also regularly exceed safe limits, largely due to vehicle emissions (WHO 2014b). Poor air quality is thought to contribute to the premature deaths of well over 3.5 million people annually worldwide (WHO 2014c). The effects of such pollution are not equal, however: older people, children, and those with pre-existing lung and heart conditions have a higher risk of illness and death from high air pollution episodes (Annesi-Maesano et al. 2003; Anderson et al. 2003). It is also the case that air pollution is not evenly distributed, and it is often, though not always, linked with other indicators of disadvantage such as higher poverty or social deprivation, or minority racial groups (e.g. Jerret et al. 2001; Pearce et al. 2006; Brainard et al. 2002).

If we consider Nussbaum's list of Central Capabilities, exposure to high levels of air pollution is clearly likely to compromise some of them: the capability of life in some cases, and bodily health in many others, thus giving the basis for claims of injustice. The compromising (and, it follows, the injustice) will be greater for people with higher exposure of course, and for people with higher sensitivity. Poor air quality may have indirect effects on other capabilities such as the ability to engage in recreation or to maintain social relationships (play and affiliation) – if, for example, a sensitive person's ability to leave the house is significantly curtailed in high air pollution episodes. Following Nussbaum's reasoning then, because it is the state's duty to provide the conditions for individuals to hold the full set of essential capabilities, it becomes the state's duty to ensure that people do not suffer from such poor air quality that their capabilities are affected in these ways. This implies that legislation, policies and planning regarding permissible emissions, and industrial, transport and urban planning will be required, as a matter of justice.

Furthermore, this protection applies to all individuals and it is not a specific level of air quality that is prescribed, but a level of safety, or freedom from harm. Some more sensitive people may need more attention than others, in order to ensure their health. Thus, taking a capabilities approach would justify, even require, paying attention to specific populations such as children or older people, and taking particular measures to protect them. For example, work in California has examined air quality in the vicinity of schools, on the basis that children's air quality is a particular concern due to their developing bodies and greater sensitivity (Pastor et al. 2006). Capabilities reasoning would endorse this approach and given the finding of high respiratory risk from air pollution in the vicinity of many schools, especially those with more poorer and minority children, it would also endorse the researchers' recommended remedial measures in and around schools, such as improved ventilation, traffic restrictions and building new schools away from busy roads. In general, as it may be hard to reach all sensitive individuals with specific measures, ideally air quality should be high enough that even fairly sensitive people are not unduly affected. However, a capabilities approach might also justify other decisions, such as state subsidy for necessary medication, or provision of recuperative trips, or any other selective intervention deemed to be effective. The point is that unequal resources may need to be spent and differential treatment might be needed in order to support a sufficiency of important capabilities for all individuals.

If we go with Sen's approach, there are no specifically defined priority capabilities that must be supported, but we can think about applying the approach to evaluate alternatives: for example, if an industrial facility might be located in an area where jobs were needed, but there were also concerns over the effects of the added pollution it might bring. The alternatives of the community hosting or not hosting the facility would need to be evaluated in terms of which had the most positive effect on the valued functionings and capabilities of individuals in the community, with community members being free to decide on what those valued functionings and capabilities are, and on how to weight them (for example, health might be considered more important than jobs). Sen's approach calls for *collective* deliberation and decision-making on the important capabilities and on the weighing up of alternatives; it is not meant to be a matter of individual consideration of personal priorities followed by voting. All the same, outcomes for individuals remain the central concern, and it is conceivable therefore that within a place-based 'community', groups with differing claims and views would emerge and need to engage – herein lies the value of the deliberation.

Using capabilities to evaluate alternatives in this way, although pragmatic, could run the risk of lacking vision and failing to imagine possibilities not immediately on the table. A slightly different way to use Sen's more open approach would be to use community deliberation to define a set of priority capabilities, with generalized weightings, that could be used as a basis for making decisions about proposals but also for strategic planning. This would then be applicable in a similar way to Nussbaum's list as considered above, but it would have a greater degree of contextualization. Plans would need to be considered in terms of their effect on the prioritized capability sets of individuals. Taking a specific hazard such as air pollution, it would be an injustice to the extent that it affected individuals' abilities to achieve a decent level of any capabilities agreed to be essential.

Environmental justice scholarship and the capabilities approach

Edwards et al. (2015: 5) claim that 'capabilities is rapidly becoming the core theoretical edifice within which to understand and theorize (environmental) justice'. It is certainly gathering interest, but more concrete applications are to date rarer in environmental justice work. The first substantial discussion of its relevance in theoretical terms came in Schlosberg's (2007) book *Defining Environmental Justice*, where he discussed it alongside distributional, procedural and recognition based approaches, arguing that it usefully integrated the concerns of those. Walker (2009) soon after made a similar argument, as well as seeing the appeal of its reflexive approach. Other theoretical discussion has been offered by Holland (2008), as discussed, and other scholars interested in sustainability and intergenerational justice in capabilities terms. More recently Edwards et al. (2015) have sought to develop theoretical work on capabilities in the context of environmental justice by adjoining it with the theorization of wellbeing, although their discussion of wellbeing largely in psychological terms perhaps risks imposing a rather narrower (albeit more theorized) conceptualization of capabilities than the original approach.

In terms of its application to specific issues, the capabilities concept has been used lightly as part of a multi-dimensional justice framework in discussion of air pollution in Newcastle, UK by Davoudi and Brooks (2014) and touched on by Whitehead (2009) in an article about 'ordinary' environmental justice and urban forestry, but neither really develops its application. Although not using a justice discourse so explicitly, interesting critical work on Payments for Ecosystem Services (PES) has attempted to define more sophisticated PES options by conceptualizing payments in terms of increasing capabilities for communities that are stewards of services provided to other communities (Polishchuk and Rauschmayer 2012 theoretically; Kolinjivadi et al. 2015 in a Nepal study of water management). As noted earlier, recent work on energy poverty and energy justice has also turned to the capabilities approach, but again to date not yet fully mobilized it in application (Sovacool et al. 2014; Day et al. 2016).

Schlosberg and Carruthers (2010) make a more sustained case for the application of a capabilities framework to theorizing environmental injustices to Indigenous communities, and illustrate this with discussion of cases from the US and Chile. Here they argue that the approach needs to be developed to incorporate a notion of 'community capabilities' and 'community functioning', which they argue are especially pertinent in Indigenous communities where environmental injustices threaten social reproduction and the practice and survival of culture, not only individual wellbeing (more on this below). Schlosberg (2012) has gone on to make similar arguments about community capabilities with regard to climate justice (see also Chapter 29). Schlosberg's and Schlosberg and Carruthers' understanding of capabilities and functionings, though, is a little different from my reading of Sen and Nussbaum in that they see capabilities as arrangements that enable healthy individual and community functioning ('the capabilities necessary for functioning' 2010: 16), with the functioning being therefore the implied normative goal, rather than capabilities being the opportunity to engage in specific functionings and therefore capability sets being the object of concern, which is my interpretation. However, as noted earlier, the ontological distinction between capabilities and the arrangements that enable them is difficult to make and so these positions may not be entirely distinct.

Advantages and challenges

The emphasis that the capability approach puts on focusing on what people can actually be and do as a basis for thinking about equality and justice is appealing. When brought into environmental justice arenas, the approach provides a framework that invites recognition of the links between the environment and many important outcomes in people's lives; indeed, formulating claims of environmental justice and injustice in capabilities terms means that these links need to be explicitly argued for. A great range of such links and arguments might be made, and objective and subjective dimensions with plural forms of evidence can be embraced.

The approach provides space for addressing different needs, and in doing so is more sophisticated than a simplistic equality approach. In taking the individual as the basic unit of analysis, it is more open to the emergence of important differences within communities and collectives – unlike approaches that keep their focus at the level of communities, neighbourhoods, or other spatial or collective units.

The approach's focus on the individual, though, is sometimes criticized, or discussed as a shortcoming (e.g. Schlosberg and Carruthers 2010; Schlosberg 2012; Dean 2009) for failing to see how capabilities, and claims for justice and injustice in those terms, might apply to groups or communities, and/or because it prioritizes individual freedom over solidarity. Nussbaum (2000) explains that the reason for assessing capabilities at individual level is to avoid overlooking systematic inequalities that occur within social units such as the family, for example according to gender, a particular concern of hers. Hence, the capabilities of all individuals must be given attention. This doesn't have to mean though that different individuals' capabilities are not understood as intertwined. Still, some capabilities only make sense at a collective level, as Schlosberg and Carruthers have been keen to point out with respect to cultural identity and cultural reproduction. However, as Sen 2010 (p. 246) argues, there is nothing in the approach that excludes the notion of group capabilities, but they are only ascribed value within the approach to the extent that individuals ascribe value to them. The approach does not preclude collective action or collective interest, but individuals must support that and feel that it has value to them in some sense. The individual focus of the capability approach has also been interpreted as Western-biased (Gaspar 1997; Dean 2009) but this is something that both Sen and Nussbaum vigorously refute, citing numerous non-western, especially Asian, political, cultural and philosophical reference points.

Apart from the approach not providing a complete concept of justice, or a roadmap for how to achieve just outcomes, as already discussed, the most significant set of criticism is that relating to the perceived inoperability of the approach. This is in part connected to the amount of deliberation that it calls for, especially in Sen's version – without any guidance on how to organize a good deliberative process, no mean feat in itself as the large literature on that topic attests – and in part connected to the difficulty of measuring capabilities, which are things that people *could* do, not necessarily what they *are* doing. With regard to the latter, in practice applications of the approach have sometimes worked with functionings (what people can be observed as doing) as more practical to gather data on, or a combination of functionings and capabilities (see e.g. Burchardt and Vizard 2011; Anand et al. 2009). With regard to the former, there is no doubt that defining the capability sets of interest, with attendant thresholds and weights, is a significant challenge in research and practice. Nevertheless, several examples of guidance and good practice exist for building on (e.g. Alkire 2007; Robeyns 2003; Burchardt and Vizard 2011). The scale of analysis has a bearing of course – comparisons at large scales, such as the HDI, may find it practical only to work with aggregated, quantitative measures, whereas community level projects are likely to find deliberative processes easier.

Mobilizing the approach across scales is a further potential challenge. For Nussbaum, the important scale is the state, as that is where she locates the responsibility for delivering the Central Capabilities. Sen's approach is flexible enough to be applied at different scales but its reliance on deliberation probably inclines it more to smaller scale applications. Schlosberg (2012) nevertheless ambitiously proposes the use of a form of capabilities framework at international level for constructing (collective) claims relating to climate justice and injustice. He does not provide a full answer to the challenges of finding an appropriate arena for international negotiation or of how to deliver and enforce international justice in the absence of a global legislative or judicial body, but nevertheless the idea of introducing capabilities thinking to international negotiations is intriguing.

Conclusions

The capabilities approach is persuasive in its argument that we should focus on real outcomes for people as the basis for assessments of inequality and for claims of justice and injustice. In environmental justice work, this means it can be applied very effectively based on the explicit making of links between environmental issues and multi-dimensional human wellbeing. It pushes us to recognize the many ways in which environmental resources underpin our essential functioning, and the ways in which quality of life and human dignity are undermined by poor quality, hazardous environments. It is fundamentally anthropocentric, but no different in that from most formulations of environmental justice, and more recently some authors have been interested in the possibilities of extending its scope to non-human species.

Sen's and Nussbaum's developments of the approach have some important differences, with Nussbaum's being more normative and based around an idea of state responsibility to its citizens, whilst Sen's is more in the vein of a conceptual framework to be adapted to varied contexts. Neither, however, provide, or aim to provide, a full account of justice. They do though provoke examination of the distribution of resources and conditions that underpin what people are able to do and to be, including environmental resources; they also really take seriously people's different characteristics and circumstances, and how they affect what people need in order to flourish, which gives significant room to a kind of justice as recognition (see Chapter 10). Both also take democratic decision-making very seriously, with Sen's approach being especially built around deliberative processes. The approach leads therefore to a quite comprehensive understanding of environmental justice.

There are undoubtedly challenges in applying the approach in practice, which goes some way to explaining why capabilities thinking in environmental justice to date has stayed rather conceptual. Nevertheless, because of its flexibility, emphasis on contextual specification and practical rather than idealist aims, it is in many ways an ideal approach for scholars, activists and practitioners to try out in various contexts and more work in this vein would certainly enrich the environmental justice field.

References

Alkire, S. (2007). "Choosing dimensions: the capability approach and multi-dimensional poverty". *Chronic Poverty Research Centre Working Paper 88*. Oxford: Oxford Poverty and Human Development Initiative.

Anand, P., Hunger, G., Carter, I., Dowding, K., & van Hees, M. (2009). "The development of capability indicators". *Journal of Human Development and Capabilities*, vol. 10, pp. 125–152.

Anderson, H.R., Atkinson, R.W., Bremner, S.A., & Marston, L. (2003). "Particulate air pollution and hospital admissions for cardiorespiratory diseases: are the elderly at greater risk?" *European Respiratory Journal*, vol. 21, Suppl. 40, pp. 39–46.

Annesi-Maesano, I., Agabiti, N., Pistelli, R., Couilliot, M.-F., & Forastiere, F. (2003). "Subpopulations at increased risk of adverse health outcomes from air pollution". *European Respiratory Journal*, vol. 21, Suppl. 40, pp. 57–63.

Ballet, J., Koffi, J.-M., & Pelenc, J. (2013). "Environment, justice and the capability approach". *Ecological Economics*, vol. 85, pp. 28–34.

Brainard, J., Jones, A., Bateman, I., & Lovett, A. (2002). "Modelling environmental equity: access to air quality in Birmingham, England". *Environment and Planning A*, vol. 34(4), pp. 695–716.

Burchardt, T., & Vizard, P. (2011). "'Operationalizing' the capability approach as a basis for equality and human rights monitoring in twenty-first century Britain". *Journal of Human Development and Capabilities*, vol. 12, pp. 91–119.

Davoudi, S. & Brooks, E. (2014). "When does unequal become unfair? Judging claims of environmental injustice". *Environment and Planning A*, vol. 46, pp. 2686–2702.

Day, R., Walker, G., & Simcock, N. (2016). "Conceptualising energy use and energy poverty using a capabilities framework". *Energy Policy*, vol. 93, pp. 255–264.

Dean, H. (2009). "Critiquing capabilities: the distractions of a beguiling concept". *Critical Social Policy*, vol. 29, pp. 261–278.

Edwards, G., Reid, L., & Hunter, C. (2015). "Environmental justice, capabilities and the theorization of well-being". *Progress in Human Geography*, 2015, pp. 1–16.

Gaspar, D. (1997). "Sen's capability approach and Nussbaum's capabilities ethic". *Journal of International Development*, vol. 9(2), pp. 281–302.

Guttikunda, S. & Calori, G. (2013). "A GIS based emissions inventory at 1 km × 1 km spatial resolution for air pollution analysis in Delhi, India". *Atmospheric Environment*, vol. 67, pp. 101–111.

Holland, B. (2008). "Justice and the environment in Nussbaum's 'capabilities approach': why sustainable ecological capacity is a meta-capability". *Political Research Quarterly*, vol. 61(2), pp. 319–332.

Jerrett, M., Burnett, R., Kanarolglou, P., Eyles, J., Finkelstein, N., Giovis, C., & Brook, J. (2001). "A GIS-environmental justice analysis of particulate air pollution in Hamilton, Canada". *Environment and Planning A*, vol. 33, pp. 955–973.

Kolinjivadi, V., Gamboa, G., Adamowski, J., & Kosoy, N. (2015). "Capabilities as justice: analysing the acceptability of payments for ecosystem services (PES) through 'social multi-criteria evaluation'". *Ecological Economics*, vol. 118, pp. 99–113.

Nussbaum, M. (2000). *Women and human development: the capabilities approach*. Cambridge UK, New York NY: Cambridge University Press.

Nussbaum, M. (2011). *Creating capabilities: the human development approach*. Cambridge MA: Harvard University Press.

Nussbaum, M. & Sen, A. (eds.), (1993). *The quality of life*. Oxford: Oxford University Press.

Pastor, M., Morello-Frosch, R., & Sadd, J. (2006). "Breathless: schools, air toxics, and environmental justice in California". *The Policy Studies Journal*, vol. 34(3), pp. 337–362.

Pearce, J., Kingham, S., & Zawar-Reza, P. (2006). "Every breath you take? Environmental justice and air pollution in Christchurch, New Zealand". *Environment and Planning A*, vol. 38, pp. 919–938.

Pelenc, J., Lompo, M.K., Ballet, J., & Dubois, J.-L. (2013). "Sustainable human development and the Capability Approach: integrating environment, responsibility and collective agency". *Journal of Human Development and Capabilities*, vol. 14(1), pp. 77–94.

Polishchuk, Y. & Rauschmayer, F. (2012). "Beyond 'benefits'? Looking at ecosystem services through the capability approach". *Ecological Economics*, vol. 81, pp. 103–111.

Rauschmayer, F. & Lessmann, O. (2013). "The capability approach and sustainability". *Journal of Human Development and Capabilities*, vol. 14(1), pp. 1–5.

Rawls, J. (1971). *A theory of justice*. Oxford: Oxford University Press.

Robeyns, I. (2003). "Sen's capability approach and gender inequality: selecting relevant capabilities". *Feminist Economics*, vol. 9(2–3), pp. 61–92.

Schlosberg, D. (2007). *Defining environmental justice: theories, movements and nature*. Oxford: Oxford University Press

Schlosberg, D. (2012). "Climate justice and capabilities: a framework for adaptation policy". *Ethics and International Affairs*, vol. 26(4), pp. 445–461.

Schlosberg, D. & Carruthers, D. (2010). "Indigenous struggles, environmental justice, and community capabilities". *Global Environmental Politics*, vol. 10(4), pp. 12–35.

Sen, A. (1992). *Inequality reexamined*. Cambridge MA: Harvard University Press.

Sen, A. (1999). *Development as freedom*. Oxford: Oxford University Press.

Sen, A. (2010). *The idea of justice*. London: Penguin Books.

Sen, A. (2013). "The ends and means of sustainability". *Journal of Human Development and Capabilities*, vol. 14(1), pp. 6–20.

Smith, M.L. & Seward, C. (2009). "The relational ontology of Amartya Sen's capability approach: incorporating social and individual causes". *Journal of Human Development and Capabilities*, vol. 10, pp. 213–235.

Sovacool, B.K., Sidortsov, R.V., & Jones, B.R. (2014). "Deciphering energy justice and injustice". In Sovacool, B.K., Sidortsov, R.V., & Jones, B.R., *Energy security, equality and justice*. Abingdon and New York: Routledge.

United Nations Development Program, not dated. *Human development reports*. http://hdr.undp.org/en accessed 30/01/16.

Walker, G. (2009). "Environmental justice and normative thinking". *Antipode*, vol. 41, pp. 203–205.

Wang, S. & Hao, J. (2012). "Air quality management in China: issues, challenges, and options". *Journal of Environmental Sciences*, vol. 24(1), pp. 2–13.

Whitehead, M. (2009). "The wood for the trees: ordinary environmental injustice and the everyday right to urban nature". *International Journal of Urban and Regional Research*, vol. 33(3), pp. 662–81.

World Health Organization. (2014a). "Air quality deteriorating in many of the world's cities". News release 7th May 2014, available at www.who.int/mediacentre/news/releases/2014/air-quality/en/ accessed 31/01/16.

World Health Organization. (2014b). *Ambient (outdoor) air pollution in cities database 2014*. Available at www.who.int/phe/health_topics/outdoorair/databases/cities/en/ accessed 30/01/16.

World Health Organization. (2014c). "Ambient (outdoor) air quality and health, factsheet no. 313". Available at www.who.int/mediacentre/factsheets/fs313/en/ accessed 30/01/16.

12

VULNERABILITY, EQUALITY AND ENVIRONMENTAL JUSTICE

The potential and limits of law

Sheila R. Foster

Introduction

Like housing and employment and other material goods, the environment is an area where fairly stark racial and ethnic disparities continue to persist, even 50 years after the enactment of historic civil rights laws in the United States. In particular, empirical evidence from across the United States, and increasingly from other parts of the world, demonstrates that there is a very high correlation between where people live, their socioeconomic status (race, ethnicity, Indigenous status, and income) and their rates of exposure to environmental pollution and toxins (Chakraborty and Green 2014). In the United States, poor ethnic minority communities are significantly more likely to live in communities with, or live nearby, multiple polluting facilities and contaminated land (UCC 2007).

The racial and ethnic disparities in the distribution of polluting facilities and other environmental hazards result in part from historic zoning practices in the early twentieth century that separated immigrant and African American communities by building tenements and other low-income housing in industrial districts (Rabin 1999). These practices continued later on in the twentieth century as African American communities, in particular, were zoned for mixed residential, industrial and commercial uses as contrasted with many non-minority communities which were zoned predominantly residential (Arnold 1998). Zoning and other land use practices, which allow residential and industrial uses to be situated close to one another, have made ethnic and minority populations quite vulnerable to the placement in their neighbourhoods of additional polluting facilities which over-expose them to environmental hazards (Taylor 2014).

Today, the processes that result in the concentration of polluting sources in poor, minority communities are neutral on their face. Modern zoning and land use laws, although embedded in the history of racially biased zoning, are largely governed by local administrative processes. Local officials are attentive to special interests like developers and homeowners who push hard to bend the rules to suit their particular interests. Some legal scholars contend that the politically oriented and market-influenced determinations which allow sitings of new facilities in minority neighbourhoods are evidence that racial factors are not at play in such decisions

(Blais 1996). Others posit that because housing markets are "dynamic," the disproportionate location of polluting facilities results from whites and higher income individuals leaving polluted neighbourhoods, while continued racial discrimination in housing constrains the mobility of low-income people of colour, leaving them trapped in these neighbourhoods (Been 1994; Been and Gupta 1997). The most recent empirical evidence testing of this hypothesis concludes that, even given housing "market dynamics," there is still disproportionate siting of polluting facilities in neighbourhoods that are predominantly ethnic minority at the time of the siting (Pastor et al. 2001; Mohai and Saha 2015a)

While appreciating the complex causality behind racial disparities in environmental hazard disposal, advocates have turned to courts and environmental regulators to better protect and shield impacted minority communities from additional environmental hazard exposure (Cole and Foster 2000). However, as this chapter explains, US courts and regulators enforcing anti-discrimination laws, and those enforcing environmental laws, have not been able to provide relief to these communities for reasons endemic to the shortcomings of both legal regimes. Because courts and regulators in the United States were the first to grapple with environmental justice as a legal issue, this chapter focuses exclusively on American law (particularly federal law as state and local law are beyond the scope of this chapter). Although many other legal regimes around the world do not share the limitations of US antidiscrimination law set out below (Oppenheimer et al. 2012), and environmental justice claims might fare better under those regimes, the lessons of this chapter are nonetheless instructive for the kinds of tensions that might exist when applying equality norms to environmental law and regulation.

Given the limitations of both antidiscrimination law and environmental regulation, and the difficulty of integrating equality norms into environmental decision making, this chapter suggests that one way to align these two frameworks is through the lens of vulnerability analysis. Vulnerability analysis has been embraced by equality theorists as an alternative to the limitations of antidiscrimination law and by social scientists to analyse and measure the ways that some subpopulations are more susceptible to the harms from climate change and environmental hazard events such as hurricanes and floods. Thus, a fertile area of research for environmental justice scholars and policymakers is figuring out how to utilize vulnerability metrics in regulatory and legal analysis to better protect populations and communities most susceptible to disproportional environmental pollution exposure.

The limits of antidiscrimination law

The US Supreme Court has increasingly interpreted constitutional and statutory language to embrace a very narrow conception of discrimination (Siegel 1997). Particularly as regards issues of race and gender discrimination, the Court has over the last 30 years imposed difficult requirements of proof where individual or group based discrimination is alleged. The result is that it is almost impossible to prove that discrimination against racial and ethnic minorities even exists any more (Oppenheimer 2003; Clermont and Schwab 2004).

The narrowing of discrimination in constitutional and civil rights law

As a matter of constitutional interpretation, for claims brought under the equal protection clause, federal courts regularly dismiss discrimination claims for failure to prove explicit bias, or discriminatory intent, and failure to show a specific causal link between that bias and alleged discriminatory treatment. As the Supreme Court ruled in *Washington v. Davis*, 426 U.S. 229 (1976), simply pointing to a pattern or history of harm towards a particular group, or complete exclusion of a group – from the jury, housing or employment opportunities, and the like – does not suffice to infer that discrimination is at work, even if there is no credible alternative explanation for these disparities. Under equal protection jurisprudence, as the Court later clarified in *Personnel Adm'r of Massachusetts v. Feeney* 442 U.S. 256 (1979), a finding of discrimination can exist only upon proof that the decision maker selected a course of action specifically "because of" and not just "in spite of" its discriminatory consequences (even if those consequences were foreseeable and predictable).

Statutory claims based on a theory of disparate impact (or indirect) discrimination have not fared much better in US federal courts, despite the powerful language in the iconic Supreme Court case, *Griggs v. Duke Power Co.*, 401 U.S. 424 (1971). There, the Court ruled that the 1964 Civil Rights Act reached "not only overt discrimination but also practices that are fair in form but discriminatory in operation." Evidence of stark disparities between historically underrepresented minority groups and majority groups was for many years deemed by courts to be powerful evidence of practices and procedures that, although neutral on their face, operated to "freeze the status quo" of prior and historical discriminatory practice.

In more recent cases, such as *Texas Department of Housing and Community Affairs v. The Inclusive Communities Project* 135 S.Ct. 2507 (2015), the Court has made clear that disparate impact claims remain an important legal tool to invalidate practices such as zoning and land use laws that function unfairly to disproportionally harm or burden minority groups, even if due to unconscious bias. Nevertheless, disparate impact claims have been limited by the requirement that a challenger must prove a specific causal link between a discrete policy of the decision maker (an employer, public official, agency, etc.) and the statistical disparity that harms a minority group (Texas Department v. the Inclusive Communities Project, 2015: 19–20). As the Court explained in *Wards Cove Packing v. Antonio*, 490 U.S. 642 (1989: 653), "a robust causality requirement ensures that [r]acial imbalance . . . does not, without more, establish a prima facie case of disparate impact and thus protects defendants from being held liable for racial disparities they did not create." A strict causation requirement is an arguable retreat on the Supreme Court's language in *Griggs* suggesting that it was the combination of historical and continuing discrimination, along with certain practices and procedures of the decision maker, which created what it called "built-in headwinds" (or obstacles) for minority groups and constituted a rebuttable presumption that discrimination was a likely explanation for the disparities (Foster 1997).

Specific intent and causation requirements did not always characterize the Court's jurisprudence. The Court's antidiscrimination jurisprudence, developed in the 1970s and 1980s, was quite openly based on the assumption that racial and ethnic bias, and hence discrimination, provided an explanation for otherwise unexplained deviations from what would be expected in a race-neutral world (Selmi 1997). The presumption was that discrimination not only existed in American society but that its effects persisted and were detectable in significant or stark racial and ethnic disparities in voting, housing, jury selection and employment. Absent good and convincing justifications for these disparities, the Court created a legal presumption that discrimination was the cause, or at least "a" cause, of these disparities. This was true even when there was no allegation or showing of explicit or conscious bias at work in producing these disparities.

Today, however, the judicial assumption about discrimination is the exact opposite; bias and discrimination are no longer presumed to explain otherwise unexplained racial or gender disparities (Foster 1997). Racial disparities, such as disproportionate environmental pollution exposure, are often presumed to be a natural or neutral part of the functioning of the market. This turns out to be a very difficult presumption to overcome and is almost never overcome in cases challenging the disproportionate siting of polluting facilities in minority neighbourhoods.

Environmental justice in the courts

Minority communities facing the prospect of additional polluting facilities have turned, not surprisingly, to federal courts to remedy what they see as discrimination (either conscious or unconscious) in the application of environmental and land use laws. Whether invoking constitutional or statutory claims, challengers have run up firmly against the limits of anti-discrimination law, including insistence on strict requirements of intent and causation. No challenger in an environmental justice case has been able to make even a circumstantial showing of discriminatory intent.

In cases such as *R.I.S.E. v. Kay*, 768 F. Supp. 1141 (E.D. Virginia, 1991), affirmed 977 F.2d 573 (4th Circuit, 1992), there is rarely, if ever, direct evidence that decision makers exhibited racial bias in the selection of the host site or neighbourhood. For this reason, challengers rely upon a circumstantial showing of discriminatory intent established by the Supreme Court in *Village of Arlington Heights v. Metropolitan Housing Development Corp.*, 429 U.S. 252 (1997). Moreover, courts routinely accept neutral, nondiscriminatory reasons offered by state and local officials for why facilities are clustered or in a targeted area. As in *R.I.S.E. v. Kay* and in *Bean v. Southwestern Waste Management Company*, 482 F.Supp. 673 (1979), these reasons range from the "suitability" of the site for hosting a new polluting facility – usually because the site is zoned for industrial uses – to proximity to transportation routes for vehicles carrying waste (or other materials) to and from the site.

Notwithstanding intent and causation requirements, the rare court has on occasion taken seriously the problematic land use history that often underlies the ease with which officials place polluting facilities in already heavily contaminated poor, minority communities. In *Miller v. City of Dallas*, 2002 U.S. Dist. LEXIS 2341 (N.D. Tex. 2002), a federal district (lower) court found that the environmental quality of the neighbourhood was undoubtedly shaped by intentional race discrimination (even if there was no direct evidence that the decision makers were themselves biased) given the history of the area. There, residents of a predominantly African American and Hispanic neighbourhood in Dallas, Texas, alleged that the city maintained a pattern of inferior zoning, flood protection, protection from industrial nuisances, landfill practices, streets and drainage in minority neighbourhoods. The court upheld the equal protection claim (as against a motion for summary dismissal by the City) and found that the effect of the City's practices along with its "sordid history of . . . racially-segregated zoning and related policies . . . offers substantial circumstantial evidence of discriminatory intent." The court found the following facts compelling: zoning for the neigh-bourhood is residential, but the area lies immediately adjacent to heavy industrial uses; the city considered overt racial segregation as a legitimate policy goal for land use decisions through the 1940s; and the city knew that Cadillac Heights would be an industrial area when it designated the area a "Negro development."

Although a clear outlier, this case nevertheless illustrates the appeal and the potential that civil rights and antidiscrimination law has had for the environmental justice movement. It has been through the Civil Rights movement, its activists, and its legal legacy that environ-mental justice advocates recognize that the disproportionate impact of environmental hazards is not random nor the result of "neutral" decisions but rather a product of the same social and economic forces underlying other forms of racial segregation and inequality (Cole and Foster 2000). The impetus to want to frame the environmental degradation of communities of colour within this larger context is in part what drove environmental justice advocates to try to import the civil rights framework into environmental law.

The limits of environmental law and regulation

Much of pollution control law is fundamentally utilitarian in its orientation and largely unconcerned with distributional inequities in pollution exposure. A primary goal of environmental regulation is to provide the maximum amount of environmental protection efficiently to the most people; in other words, to reduce aggregate societal or community exposure levels to harmful pollutants and toxins (Hornstein 1992). As such, when deciding whether to issue, for example, an air emissions permit to a new industrial or commercial facility, the main inquiry is whether the polluting source is in compliance with the maximum level set by regulators of emitting a particular pollutant or chemical (Kuehn 1996).

Environmental compliance, risk assessment and regulatory blindspots

To comply with many environmental statutes – such as the Clear Air Act, the Clear Water Act, and the Resource Conservation and Recovery Act (covering the disposal of municipal and hazardous waste) – polluting facilities are required to install the right pollution control technology capable of reducing exposure to a level that is both safe and practicably feasible. Because pollution limits are aimed at areawide, or ambient, pollution reduction they typically are not concerned with the ways in which risk or exposure may be distributed within or among subpopulations and thus tend to miss geographical "hot spots" that may not have significant areawide impacts (Abel 2008). These kind of highly localized, concentrated (and cumulative) impacts have not traditionally been a concern of US environmental law (Lazarus 1992). Moreover, the more recent market-based regulatory mechanisms, such as emissions trading, while providing industry with regulatory flexibility and cost-savings, are similarly indifferent to where facilities are located and may even encourage pollution "hot spots" (Kaswan 2013: 161).

Further, environmental statutes that contain health-based pollution control standards – used to determine the amount of ambient air pollution exposure that is safe, the amount of pesticide exposure that is safe, or the amount of fish that can be safely consumed – are often based on faulty assumptions that ignore variability in exposure between different social and economic groups (O'Neill 2000). For example, pollution limits traditionally have been based on the consumption and exposure patterns of white, male bodies even as poor and minority groups (such as subsistence fishermen and farmworkers) suffer from highly elevated toxic exposure based on their consumption and/or work patterns (O'Neill 2000). As such, environmental regulatory standards have too often failed to adequately protect the most vulnerable ethnic and low-income populations (US GAO 2000).

There is another, but related, reason why environmental regulators have not been able to adequately address the situation of communities with multiple polluting sources – all governed under different environmental statutes or regulations and which emit a mix or cocktail of pollutants in these communities. In the case of multiple, synergistic interactions between pollutants, the inadequacy and uncertainty that is characteristic of environmental science renders risk assessment processes ineffective at capturing and addressing the aggregation of risks. Thus, while regulators can assess the health effects of one pollutant or one source, they are less able to assess risks from the synergy of multiple pollutants or the risks from multiple exposures through different pathways to the same pollutant (Kuehn 1996). So rules that limit mercury, for example, coming from the smokestack of a polluting facility may not account for other pathways to mercury exposure – e.g. contaminated water or mercury-contaminated fish that subsistence populations tend to consume more. Nor will they account for the impacts of mercury exposure and the exposure to other potentially hazardous or toxic substances in a community.

Environmental regulators' response to these regulatory shortcomings has been mixed. On the one hand, when presented with quantitative data that demonstrate how regulatory standards leave specific populations over-exposed to environmental toxins, environmental regulators have responded by altering those standards. A key example is the US Environmental Protection Agency's (EPA's) decision to revise its methodology for setting water quality standards to incorporate a higher default fish consumption rate based on evidence that Native American subsistence populations tend to consume far greater quantities of self-caught fish than the general population and, thus, were disproportionately harmed by the existing standard (72 Fed. Reg. 18504: 1998).

On the other hand, when advocates for environmental justice have pushed regulators to integrate equality norms into decisions about where to place new polluting facilities, as a way to address some of these deficiencies in regulatory standard setting and the science of risk assessment, the fit has been uncomfortable at best. Regulators often default to technical environmental pollution control standards and are quick to conclude that there is no harm to surrounding communities by concentrated polluting sources if individual sources are in compliance with existing environmental emissions limits. However, even if these standards reflect the best of the science of risk assessment, they still contain numerous uncertainties and are not well set up to capture the social complexity and spatial inequalities in the distribution of pollution (Yang 2002).

Enforcing distributive justice and equality norms at the EPA

One way advocates have sought to work around the shortcomings of the uncertainties inherent in environmental standard setting and risk assessment is to try to meld antidiscrimination concepts such as disproportionate impact into regulatory analysis as a way to capture the types and range of harms suffered by overburdened minority and low-income communities (see also Chapter 22). They have done so by enforcing civil rights norms, such as disparate impact, at the administrative level to challenge permits for new polluting facilities in communities of colour under the authority of both President Clinton's Executive Order on Environmental Justice and federal agency regulations issued under Title VI of the Civil Rights Act of 1964. Both the Executive Order and the Title VI administrative regulations direct federal agencies to avoid actions which would result in disparate impacts on minority (and in the case of the Executive Order, low-income) populations (Executive Order No. 12898; U.S. EPA Draft Title VI Guidance, 2000).

Despite the agency's willingness to import the disparate or disproportionate impact standard into its review of decisions to permit new polluting facilities in minority and/or low-income communities, not one environmental justice challenge has been successful to date. EPA's Environmental Appeals Board (EAB), for instance, has ruled consistently in a number of administrative challenges to permits that the agency has discretionary authority, pursuant to Executive Order 12898, to assess whether a permitted facility will result in a disproportionate, adverse impact on the surrounding minority and/or low-income community (Foster 2000). In each case brought before it, however, the EPA has consistently defaulted to health-based pollution control standards to conclude that the host community will not suffer "adversely" from additional polluting sources even if it would result in a "disparate impact" on that community as measured by increased exposure, odour, noise, increased vehicular traffic, degraded infrastructure and decreased property values (Foster 2000).

The agency has reasoned in this way not only when analysing its authority under pollution control statutes like the Clean Air Act and the Safe Drinking Water Act to grant relief from additional permitted facilities in minority communities, but also when analysing its obligation under its *Guidance for Investigating Title VI Administrative Complaints Challenging Permits* (US EPA 2000) (Foster 1999; Kaswan 2013). Pursuant to Title VI of the Civil Rights Act of 1964, EPA officials have the authority to overturn permitting decisions, reform agency procedures, withhold funding from state agencies, or refer cases to the Justice Department for prosecution based on the finding of a racially disproportionate impact in the enforcement of its laws. Despite the filing of over hundreds of complaints alleging racial discrimination in the permitting of new polluting facilities, the EPA's Office of Civil Rights has not made a formal finding of disparate impact discrimination in 22 years (Lombardi et al. 2015).

The experience at the EPA demonstrates that efforts to meld or integrate civil rights and equality norms – namely, the concept of disparate impact discrimination – into environmental decision making have been quite unsuccessful. We can view this failure as a fundamental "mismatch" between antidiscrimination and environmental law paradigms, which appear to largely be operating in very different normative universes (Yang 2002). If so, then the question is: How should we address this tension?

Environmental justice and vulnerability

One way that environmental justice advocates and environmental regulators might begin to reconcile this tension is through the discursive, conceptual, and scientific realm of vulnerability. Vulnerability analysis, as this section explains, is being utilized both by legal scholars concerned with questions of equality and by social scientists concerned with the relationship between social inequality and exposure to environmental hazards. As such, vulnerability analysis has the potential to provide the missing conceptual and practical link between equality norms and environmental regulation. In other words, it might speak to regulators used to thinking in quantitative and risk-based ways and also be employed as a tool by environmental justice advocates to quantitatively capture the factors that render certain communities so vulnerable to disproportionate exposure to pollution.

There is already some attention to vulnerability in environmental statutory law. For instance, the EPA may consider especially sensitive populations – people with asthma, emphysema or other conditions rendering them particularly vulnerable to pollution exposure – in determining what pollutant levels would be adequately protective of public health (Lazarus and Tai 1999: 631–632). Moreover, climate scientists understand the concept of ecological vulnerability as concerning itself with the structural characteristics of a place, population sensitivity, and/or capacity to adapt. The literature on climate change recognizes that vulnerability assessment is a distinct metric from risk assessment in that while the latter involves an assessment of the risks inherent in the physical phenomenon itself (e.g. exposure to chemical X, hurricane, etc.), the former concerns itself with the interaction of physical risks with social and economic systems (e.g. impoverished communities living on land already contaminated with chemical X, island nations surrounded by warming oceans, etc.) (Sarewitz et al. 2003). Similarly, the emerging field of "disaster justice" recognizes that demographic characteristics such as class and race can influence a community's hazard-risk index as much as its location (Verchick 2012).

Vulnerability and equality theory

From the perspective of equality law and theory, a vulnerability lens similarly requires a broader, more structural assessment of disadvantage and inequality. This approach to equality has been wonderfully developed by legal theorist Martha Fineman, who argues in favour of vulnerability analysis to replace equal protection jurisprudence's focus on discrimination against certain individuals or groups resulting from racial and gender bias (whether conscious or unconscious) (Fineman 2008, 2012). Under her vulnerability analysis, equal protection would focus on the structural advantages and disadvantages associated with "social roles, positioning, and functioning" (Fineman 2008).

Fineman offers up a vulnerable subject that is *both* universal and relational. On the one hand, all of us are vulnerable to harm, whether intentional or not, at different points in our lives depending on our own abilities and disabilities as well as on social and institutional contexts (Fineman 2012). On the other hand, there are real variations in human "social location" that render different individuals or social groups particularly vulnerable. As she argues, we are all "differently situated within webs of economic and institutional relationships that structure our options and create opportunities" (Fineman 2012: 1754–55). These variations in human social location, Fineman argues, are the most significant focus for vulnerability analysis. The opposite of vulnerability for her is "resiliency" – the ability to obtain and utilize the assets necessary to succeed in various institutional settings (Fineman 2008).

Three aspects of Fineman's vulnerability framework are helpful for analysing the issue of environmental justice. First, Fineman's vulnerability approach represents an inquiry that is not focused just on targeted discrimination or bias against defined groups, but rather is concerned with the ways in which privilege and favour are conferred on limited segments of the population by the state directly and through the institutions it brings into existence (and subsequently regulates and maintains). Second, this inquiry does not force equality analysis to reduce itself to one individual trait or characteristic – race, or gender, or class – but rather forces us to articulate the way that disadvantage is structurally produced, and the ways that this production may shift over time and across identity characteristics. Finally, Fineman re-orients the state away from a restrained state under the reigning classical liberal model towards a "responsive state" which takes responsibility for the institutions and private actors that structure vulnerability – specifically the ways in which the state has "responded to, shaped, enabled, or curtailed its institutions" and allocated resources that "privilege and protect some while tolerating the disadvantage of others" (Fineman 2008: 19–21).

Although Fineman's theory of vulnerability is useful for antidiscrimination and equality analysis generally, it has practical limitations. Notably, this theory has no basis in the methodological approach of courts in the United States when analysing issues of discrimination and inequality which, as we have seen, is focused on requirements of intent and causation. However, as recently noted by scholars in Europe, the concept of vulnerability has gained a bit of traction (though not un-problematically so) in the European Court of Human Rights (Peroni and Timmer 2013). A second limitation of Fineman's theory is the risk, at least in the social context of the United States, of diluting the salience of race as a potent social category around which so much economic and social vulnerability is structured. Race is what one might call a 'sticky' identity, which it is difficult to manoeuvre around and avoid, even for individuals well-endowed with the kind of assets that enable resiliency.

Nevertheless, the ultimate utility of Fineman's framework, at least for equality analysis, is that it offers tools to more finely articulate and conceptualize the complex ways in which inequality gets produced in a given context, pushing us past a single frame of reference for this vulnerability – even one as powerful as racial bias. In other words, a vulnerability analysis can help to uncover the persistence of environmental racism "in a world without racists" (Bonilla-Silva 2006). Race operates very differently across social contexts (and social locations), and constrains in very different ways depending on the other social and economic characteristics with which it intersects (see Chapter 2). The empirical evidence on the disproportionate distribution of environmental hazards certainly suggests the importance of accounting for what equality theorists refer to as intersectionality (Crenshaw 1989) – the intertwining effects of race and class – as well as the need to account for the host of other factors that create very distinct ecologically and socially vulnerable communities (see Chapters 2, 4, and 7). A focus on vulnerability, as opposed to bias-based discrimination, acknowledges the contextual, intersecting and structural mechanisms through which much of inequality and disadvantage is produced in our society.

Researchers are still seeking a better understanding of the processes and root causes of environmental injustices and, specifically, the role that racial discrimination, market dynamics, and sociopolitical factors play (Mohai and Saha 2015b). While race is a significant predictor, for example, of where hazardous waste facilities are located (and likely to be located), so too is income, as studies have shown (Bullard et al. 2007). Other factors such as the strength of a community's social capital can determine how vulnerable it is to the siting of noxious land uses. As one study has shown, a lot of demographic change – e.g. "ethnic churning" – taking place in a community weakens the social ties between residents and thus their political power, increasing their susceptibility to siting toxic facilities (Pastor et al. 2001). On the other hand, there is very little attention in these empirical studies to other factors such as gender and age (among others) when assessing the increased susceptibility, vulnerability, and ability to cope with environmental hazard exposure and threats (Cutter 1995).

Social vulnerability analysis and environmental justice

Social scientists such as Susan Cutter have long used vulnerability as both a conceptual lens and a practical tool to assess and predict the risks and impacts of a single environmental hazard – chemical release, a flood, or a hurricane – as well as the ability of populations to respond to these events. Through the creation of empirical metrics to measure 'social vulnerability' – e.g., the social vulnerability index (SOVI) – these researchers have been able to capture the array of factors which shape the susceptibility of certain populations and communities to harms from environmental hazard events (Cutter et al. 2003). For example, there is a general consensus in the literature that social vulnerability determinants include demographic characteristics (such as race, ethnicity, income, etc.), resource access, political access, social capital, physical disability, infrastructure, and housing stock, among other factors (Cutter et al. 2000). Social Vulnerability Analysis (SVA), which has been extensively developed in the literature, describes the relationship between these social characteristics and biophysical vulnerability to hazards (better documenting who is at risk), as well as the distribution of tangible and intangible hazard impacts on particular subpopulations or communities (Cutter et al. 2003; Cutter and Finch 2008). A similar literature is developing these factors for assessing community disaster resiliency (Cutter et al. 2014).

It is important to note that, despite the widespread use in the social sciences of SVA and the SOVI, there are methodological challenges to measuring social vulnerability in the hazards context (Preston et al. 2011), including the need for researchers to pay closer attention to context in understanding social vulnerability drivers (Rufat et al. 2015). Moreover, researchers are continuing to find ways to improve social vulnerability methodology, through uncertainty analysis (Tate 2013) and moving away from narrative assessments of vulnerability to quantitative metrics (Mustafa et al. 2011). Nevertheless, despite its methodological shortcomings, SVA and SOVI remains quite useful in capturing the geography of vulnerability to climate and other ecological hazards.

The social vulnerability model is geographically oriented, or place-based, and incorporates a range of factors – biophysical and social indicators – as they affect particular places and the people who live there (Cutter 1996). As such, when social vulnerability researchers analysed the confluence of biophysical and social vulnerabilities manifested in the Hurricane Katrina disaster (in New Orleans, particularly), they identified the factors that would have predicted the distribution of some of the worst impacts on the part of New Orleans that suffered the most from the impact of the storm. These factors included the geography of the city, which contained higher and lower lying areas, the location of its public housing stock in the most undesirable lower lying areas of the city, along with the history of racial segregation, the natural risks of flooding in lower lying areas, and the inaccessibility of these populations to the goods, services and emergency response personnel that would allow them to escape the worst part of the storm (Cutter 2006).

A similar set of metrics is being developed by researchers to determine environmental justice areas based on more than the demographics of race and class. The Environmental Justice Screening Method (EJSM), for instance, was developed as an approach to assess cumulative impacts from environmental and social stressors across neighbourhoods (Sadd et al. 2011). The EJSM uses roughly 30 health, environmental, climate and social vulnerability measures to map neighbourhoods on three different dimensions: 1) proximity to hazards; 2) exposure to air pollution; and 3) social and health vulnerability (Pastor et al. 2013). A very similar set of metrics, the Cumulative Environmental Vulnerability Assessment (CEVA), also has been devised to capture a composite of a health index, a cumulative environmental health index and a social vulnerability index (Huang and London 2012).

Although not explicitly using either of these new vulnerability-based environmental justice metrics, the EPA in its recent internal Environmental Justice Guidance documents is notably moving towards a more contextual analysis of "environmental justice concerns" that goes beyond a simple metric such as "disproportionate impact" (US EPA Guidance 2015: 7–10, 13–15). Noting that its statutory regulatory authorities provide a "broader basis for protecting" minority and low income populations without demonstrating that the impacts of regulatory action will be disproportionate, the agency then goes on to note that environmental justice concerns can result from a combination of several factors (US EPA Guidance 2015: 7, 13). These factors include proximity and exposure to emission sources stemming from "evolving mixed land use patterns," unique and "non-traditional" exposure pathways linked to cultural background or socioeconomic status, physical infrastructure such as poor housing or the presence of legacy pollutants such as lead, multiple stressors and cumulative impacts, the inability or difficulty to fully participate in environmental decision making (as a result of a host of social, political, and economic barriers), and higher risk profiles based on the inability to resist or tolerate particular environmental stressors (due to chronic illness, age, disability, reduced access to health care, etc.) (US EPA Guidance 2015: 13–14). Once an environmental justice concern has been identified from this analysis, regulators are required to formulate a response, which could result in a stricter regulatory standard or more protective measures put in place (US EPA Guidance 2015: 17).

Conclusion

Vulnerability analysis is not a panacea for the failure of courts or environmental agencies to enforce civil rights norms and guarantees. However, particularly from the standpoint of equality, it can help to overcome the tendency of public officials (and scholars) to dismiss environmental justice challenges for failure to link disproportionate exposures to explicit racial bias or as the result of a single bad actor or policy. As Fineman has argued, vulnerability analysis forces the state to be responsive to the conditions that it helped to bring into existence – here, the history of local land use and regulatory decisions that has created vulnerable communities and populations. Moreover, the "science" of social vulnerability – identifying and mapping the varied and complex factors giving rise to vulnerable communities – can be effectively used to prod environmental regulators and local officials to be more responsive to our most vulnerable communities. In other words, the places that score highly on environmental justice vulnerability indexes should be given special consideration in monitoring, permitting, and enforcement as well as public involvement and economic development (Huang and London 2012).

As both state and federal environmental regulators begin to use and incorporate social vulnerability metrics into their harm assessment and rule-making, the question will be how to create legal and administrative structures that hold agencies responsible and accountable for instantiating antidiscrimination and equality norms into their decisions. This is admittedly more difficult without the force of civil rights law, which historically has been used to call attention to, and remedy, inequality across various spheres of social and economic life. Environmental law and regulation, on the other hand, leaves much to the discretion of enforcing agencies whose priorities can shift across presidential administrations. Moreover, the focus on cost-benefit analysis to justify environmental regulation is in deep tension with equality norms, particularly distributive justice (Driesen 1997). As such, while vulnerability holds much promise in bridging the gap between environmental and civil rights legal regimes, there will need to be more research into whether social vulnerability analysis can guide environmental regulators to decisions that are not simply based on utilitarian impulses or which employ imperfect risk assessment methodology to crowd out a more holistic assessment of vulnerability to environmental hazards.

References

Abel, T.D. (2008). "Skewed Riskscapes and Environmental Injustice: A Case Study of Metropolitan St. Louis." *Environmental Management*, vol. 42, pp. 232–48.

Arnold, C.A. (1998). "Planning Milagros: Environmental Justice and Land Use Regulation." *Denver University Law Review*, vol. 76, pp. 1–149.

Been, V. (1993). "What's Fairness Got to Do with It: Environmental Justice and the Siting of Locally Undesirable Land Uses." *Cornell Law Review*, vol. 78, pp. 1001–1085.

Been, V. (1994). "Locally Undesirable Land Uses in Minority Neighbourhoods: Disproportionate Siting or Market Dynamics?" *Yale Law Journal*, vol. 103, pp. 1383–1422.

Been, V. and Gupta, F. (1997). "Coming to the Nuisance or Going to the Barrios? A Longitudinal Analysis of Environmental Justice Claims." *Ecology Law Quarterly*, vol. 24, pp. 1–56.

Blais, L. (1996). "Environmental Racism Reconsidered." *North Carolina Law Review*, vol. 75, pp. 75–151.

Bonilla-Silva, E. (2006). *Racism Without Racists: Colourblind Racism and the Persistence of Racial Inequality in the United States*. Lanham, Maryland: Rowman & Littlefield Publishers.

Bullard, R.D., Mohai, P., Saha, R., and Wright, B. (2007). *Toxic Wastes and Race at Twenty, 1997– 2007*. Cleveland, OH: United Church of Christ.

Chakraborty, J. and Green, D. (2014). "Australia's First National Level Quantitative Environmental Justice Assessment of Industrial Air Pollution." *Environmental Research Letters*, vol. 9, pp 1–10.

Clermont, K.M. and Schwab, S.J. (2004). "How Employment Discrimination Plaintiffs Fare in Federal Court." *Journal of Empirical Legal Studies*, vol. 1, pp. 429–458.

Cole, L. and Foster, S. (2000). *From the Ground Up: Environmental Racism and the Rise of the Environmental Justice Movement*. Albany, NY: New York University Press.

Crenshaw, K. (1989). "Demarginalizing the Intersection of Race and Sex: A Black Feminist Critique of Antidiscrimination Doctrine, Feminist Theory and Antiracist Politics." *The University of Chicago Legal Forum*, vol. 140, pp. 139–167.

Cutter, S.L. (1995). "The Forgotten Casualities: Women, Children and Environmental Change." *Global Environmental Change: Human and Policy Dimensions*, vol. 5, pp. 181–194.

Cutter, S.L. (1996). "Vulnerability to Environmental Hazards." *Progress in Human Geography*, vol. 20, pp. 529–539.

Cutter, S. (2006). "The Geography of Social Vulnerability: Race, Class and Catastrophe." *Understanding Katrina: Perspectives from the Social Sciences*, http://understandingkatrina.ssrc.org/Cutter/ Social Science Research Council.

Cutter, S., Mitchell, J., and Scott, M. (2000). "Revealing the Vulnerability of People and Places: A Case Study of Georgetown County, South Carolina." *Annals of the Association of American Geographers*, vol. 90, pp. 713–737.

Cutter, S.L., Boruff, B.J., and Shirley, L.W. (2003). "Social Vulnerability to Environmental Hazards." *Social Science Quarterly*, vol. 84, pp. 242–261.

Cutter, S.L. and Finch, C. (2008). "Temporal and Spatial Changes in Social Vulnerability to Natural Hazards." *Proceedings of the National Academy of Sciences*, vol. 105(7), pp. 2301–2306.

Cutter, S.L, Ash, K.D., and Emrich, C.T. (2014). "The Geographies of Community Disaster Resilience," *Global Environmental Change*, vol. 29, pp. 65–77.

Driesen, D. (1997). "The Societal Cost of Environmental Regulation: Beyond Administrative Cost-Benefit Analysis." *Ecology Law Quarterly*, vol. 24, pp. 546–617.

Exec. Order No. 12898, 3 C.F.R. 856 (1995.) Federal Actions to Address Environmental Justice in Minority Populations and Low-Income Populations.

Fineman, M.A. (2008). "The Vulnerable Subject: Anchoring Equality in the Human Condition." *Yale Journal of Law & Feminism*, vol. 20, pp. 1–23.

Fineman, M.A. (2010). "The Vulnerable Subject and the Responsive State." *Emory Law Journal*, vol. 60, pp. 251–276.

Fineman, M.A. (2012). "Beyond Identities: The Limits of an Antidiscrimination Approach to Equality." *Boston University Law Review*, vol. 92, pp. 1713–1769.

Foster, S. (1997). "Intent and Incoherence." *Tulane Law Review*, vol. 72, pp. 1065–1175.

Foster, S. (1999). "Piercing the Veil of Economic Arguments against Title VI Enforcement." *Fordham Environmental Law Review*, vol. 10, pp. 331–346.

Foster, S. (2000). "Meeting the Environmental Justice Challenge: Evolving Norms in Environmental Decision Making." *Environmental Law Reporter*, vol. 30, pp. 10992–11005.

Hornstein, D.T. (1992). "Reclaiming Environmental Law: A Normative Critique of Comparative Risk Analysis." *Columbia Law Review*, vol. 92, pp. 562–633.

Huang, G., and London, J.K. (2012). "Cumulative Environmental Vulnerability and Environmental Justice in California's San Joaquin Valley." *International Journal of Environmental Research and Public Health*, vol. 9, pp. 1593–1608.

Kaswan, A. (2013). "Environmental Justice and Environmental Law." *Fordham Environmental Law Review*, vol. 24, pp. 149–179.

Kuehn, R.R. (1996). "The Environmental Justice Implications of Quantitative Risk Assessment." *University of Illinois Law Review*, vol. 1996, pp. 103–172.

Lazarus, R. (1992). "Pursuing 'Environmental Justice': The Distributional Effects of Environmental Protection." *Northwestern University Law Review*, vol. 87, p. 787.

Lazarus, R. and Tai, S. (1999). "Integrating Environmental Justice into EPA Permitting Authority." *Ecology Law Quarterly*, vol. 26, pp. 618–678.

Lombardi, K., Buford T., Green R., et al. (2015) "Environmental Justice Denied: Environmental Racism Persists, and the EPA Is One Reason Why." Center for Public Integrity, available at www.public integrity.org/2015/08/03/17668/environmental-racism-persists-and-epa-one-reason-why

Mohai, P. and Saha, R. (2015a). "Which Came First, People or Pollution? Assessing the Disparate Siting and Post-Siting Demographic Change Hypotheses of Environmental Injustice." *Environmental Research Letters*, vol. 10, p. 115008.

Mohai, P. and Saha, R. (2015b). "Which Came First, People or Pollution? A Review of Theory and Evidence from Longitudinal Environmental Justice Studies." *Environmental Research Letters*, vol. 10, p. 125001.

Mustafa, D., Ahmed, S., Saroch, E., and Bell, H. (2011). "Pinning Down Vulnerability: From Narratives to Numbers," *Disasters* 35(1): pp. 62–86.

O'Neill, C.A. (2000). "Variable Justice: Environmental Standards, Contaminated Fish, and 'Acceptable' Risk to Native Peoples," *Stanford Environmental Law Journal*, vol. 19, pp. 1–118.

Oppenheimer, D.B. (2003). "Verdicts Matter: An Empirical Study of California Employment Discrimination and Wrongful Discharge Jury Verdicts Reveals Low Success Rates for Women and Minorities." *U.C. Davis Law Review*, vol. 37, pp. 511–566.

Oppenheimer, D.B., Foster, S., and Han, S. (2012). *Comparative Equality and Antidiscrimination Law: Cases, Codes, Constitutions and Commentary*. New York: Foundation Press Thomson/West.

Pastor, M., Sadd, J., and Hipp, J. (2001). "Which Came First? Toxic Facilities, Minority Move-in, and Environmental Justice." *Journal of Urban Affairs*, vol. 23, pp. 1–21.

Pastor, M., Morello-Frosch, R., and Sadd, J. (2013). "Screening for Justice: Proactive Spatial Approaches to Environmental Disparities," *EM*, http://community-wealth.org/sites/clone.community-wealth.org/files/downloads/article-pastor-et-al_1.pdf

Peroni, L. and Timmer, A. (2013). "Vulnerable Groups: The Promise of an Emerging Concept in European Human Rights Convention Law, *International Journal of Constitutional Law*, vol. 11, pp. 1056–1085.

Preston, B.L., Yuen, E.J., and Westaway, R.M. (2011). "Putting Vulnerability to Climate Change on the Map: A Review of Approaches, Benefits, and Risks." *Sustain Sci*, vol. 6(2), pp. 177–202.

Rabin, Y. (1999). "Expulsive Zoning: The Inequitable Legacy of Euclid." in Haar, C.M. and Kayden, J.S. (eds.), *Zoning and the American Dream: Promises Still to Keep*, Chicago: APA Press, pp. 106–108.

Rufat, S., Tate, E., Burton, C., and Maroof, A. (2015). "Social Vulnerability to Floods: Review of Case Studies and Implications for Measurement." *International Journal of Disaster Risk Reduction*, vol. 14, pp. 470–86.

Sadd, J., Pastor, M., Morello-Frosch, R., Scoggins, J., and Jesdale, B.M. (2011). "Playing It Safe: Assessing Cumulative Impact and Social Vulnerability through an Environmental Justice Screening Method in the South Coast Air Basin, California." *International Journal of Environmental Research and Public Health*, vol. 8, pp. 1441–1459.

Sarewitz, D., Pielke, R.J., and Keykhah, M. (2003). "Vulnerability and Risk: Some Thoughts from a Political and Policy Perspective." *Risk Analysis*, vol. 23, pp. 805–810.

Selmi, M. (1997). "Proving Intentional Discrimination: The Reality of Supreme Court Rhetoric." *Georgetown Law Journal*, vol. 86, pp. 279–350.

Siegel, R. (1997). "Why Equal Protection No Longer Protects: The Evolving Forms of Status-enforcing State Action," *Stanford Law Review*, vol. 49, pp. 1111–1148.

Tate, E. (2013). "Uncertainty Analysis for a Social Vulnerability Index." *Annals of the Association of American Geographers*, vol. 103(3), pp. 526–543.

Taylor, D.E. (2014). *Toxic Communities: Environmental Racism, Industrial Pollution and Residential Mobility*. New York: New York University Press, pp. 1–352.

United Church of Christ Justice and Witness Ministries (UCC). (2007). *Toxic Wastes and Race at Twenty: 1987–2007—Grassroots Struggles to Dismantle Environmental Racism in the United States*, available at www.precaution.org/lib/toxic_wastes_and_race_at_20.090601.pdf

US Environmental Protection Agency (US EPA). (2000). "Draft Revised Guidance for Investigating Title VI Administrative Complaints Challenging Permits." 65 *Federal Register* 39649.

US Environmental Protection Agency (US EPA). (2015). www3.epa.gov/environmentaljustice/resources/policy/considering-ej-in-rulemaking-guide-final.pdf

US General Accounting Office (GAO). (2000). *Pesticides: Improvements Needed to Ensure the Safety of Farmworkers and Their Children*. Washington, DC.

Verchick, R. (2012). "Disaster Justice: The Geography of Human Capability." *Duke Environmental Law & Policy Forum*, vol. 23, pp. 23–71.

Yang, T. (2002). "Melding Civil Rights and Environmentalism: Finding Environmental Justice's Place in Environmental Regulation." *Harvard Environmental Law Review*, vol. 26, pp. 1–32.

13

ENVIRONMENTAL HUMAN RIGHTS

Kerri Woods

Introduction

Environmental harms, such as pollution, biodiversity loss, deforestation, soil erosion, and climate change, can have sustained and severe impacts on both human and nonhuman life. Insofar as this is true, it seems plausible to claim that environmental harms constitute rights violations, at least where the beings harmed can be said to be bearers of rights. Certainly, we can observe the increasing use of human rights law as a mechanism to protect against or seek redress for environmental harms. Alongside this, we find growing interest within environmental political theory in the idea of environmental (human) rights, but there is much work to be done before we understand when and how environmental harms give rise to legitimate claims of rights, held by what sort of agents or beings, and why. Indeed, we need to know whether nonhuman beings and future human beings can properly be held to have environmental rights. We also need to understand the relationship between harms that are diffuse, mediated and aggregated, and rights claims which are paradigmatically individualistic.

My aim here is to defend the view that the growing inclusion of environmental issues in both human rights theory and practice points to a significant shift in our collective under-standing of the relationship between human beings and nonhuman nature, which I hold to be significant for the achievement of environmental justice, even as debate persists about both the content and the normative standing of environmental human rights. I begin with a discussion of the relationship between the environment and human rights. I then turn to Henry Shue's influential notion of 'standard threats' to explore how environmental harms give rise to rights and duties. I also reflect on the ways in which concepts of harm and rights are divergently used in relation to environmental justice by political theorists and those involved in human rights practice. Finally, I reflect briefly on whether future humans and nonhumans can be bearers of human rights, and what this means for the project of achieving environmental justice. I conclude that recognition of environmental issues as human rights issues prompts reflection on the human as an ecologically embedded being.

The environment and human rights

There are two senses in which we might expect human rights, as a legal, political and ethical framework, to provide a way of addressing environmental problems. Firstly, human rights are understood to be the most fundamental rights recognized in law. Legally binding human rights treaties, such as the International Covenant on Civil and Political Rights, and the International Covenant on Economic, Social and Cultural Rights (both 1966), are endorsed by the overwhelming majority of governments worldwide. To say that a human right has been violated is to say that a fundamental wrong has been committed, and redress is urgently called for. Given both the gravity and the urgency of many environmental problems, one might expect the human rights framework to be an appropriate one through which to address environmental harms.

The second reason why we might expect this relates to the transnational character of many environmental harms and of many human rights bodies. Human rights are, in principle, rights of all humans everywhere. In legal practice we have rights against our own governments, but these can be pursued, in some jurisdictions at least, in transnational courts. Indeed, the Inter-American Court of Human Rights and the European Court of Human Rights have delivered a number of pro-environmental decisions in recent decades (Gearty 2010). The United Nations has recognized the interrelationship between environmental justice and human rights in a number of initiatives, notably in the Sustainable Development Goals which succeeded the Millennium Development Goals. There is currently no global treaty of specifically environmental rights, but some have called for such a treaty in relation to climate change (including three separate international proposals presented around the Paris Climate talks in December 2015), and some national constitutions recognize specific environmental rights as human rights. As John Knox, the United Nations High Commissioner for Human Rights Special Rapporteur puts it:

> Human rights are grounded in respect for fundamental human attributes such as dignity, equality and liberty. The realization of these attributes depends on an environment that allows them to flourish. At the same time, effective environmental protection often depends on the exercise of human rights that are vital to informed, transparent and responsive policymaking. Human rights and environmental protection are inherently interdependent.
>
> *(2012: 5)*

Clearly, then, in both practical and conceptual terms, there is a close relationship between environmental justice and environmental human rights. Where patterns of environmental harm are caused by the actions of humans – either individuals or collective agents such as states, companies, or international organizations – we can at least in principle think in terms of a right having been violated. Given both the scale of environmental harms that we are currently witnessing, and the fact that we know and understand that human choices and actions are causing these harms, it is intuitively plausible to appeal to environmental human rights as a framework for achieving environmental justice.

What is less obvious is whether the human rights framework, with its inevitable focus on *the human*, can deliver justice for nonhumans suffering anthropogenic environmental harms (Woods 2010; Gearty 2010). Indeed, it is striking that early theorists of environmental justice, and environmental ethics more broadly, tended not to cast their arguments in terms of human rights. We might think of there being (at least) two different senses of environmental justice: 1) Environmental justice as concerned with environmentally produced or exacerbated social inequalities, both locally (as in Louisiana after Hurricane Katrina) and globally (as in the inequalities arising from unequal economic relationships between the Global North and the Global South, and the environmental externalities that fall disproportionately on the latter, perpetuating wider injustices). 2) Environmental justice as concerned with justice *to* the environment or between humans and nonhuman entities (sometimes referred to as ecological justice to differentiate it from sense 1)). Advocates of both these approaches to environmental justice might be disposed to question some things that environmental human rights, particularly as practised in current legal systems, take for granted. For example, environmental human rights, in virtue of being *human* rights, can hardly avoid charges of anthropocentrism. Making a claim for environmental justice through the idiom of human rights may plausibly be said not to disrupt dominant ways of thinking about the relationship between human and nonhuman nature, since the human rights framework implicitly accepts the human as the locus of moral value.

Advocates of environmental human rights have debated whether nonhumans can be held to be legitimate claimants of human rights, and if they cannot, whether human rights as a legal and political tool can be radical enough to challenge structures of consumption and production that serve to perpetuate environmental injustice (Hancock 2003; Grear 2011). It seems to me that environmental human rights can largely be defended against the charge that insofar as they are inherently anthropocentric, they cannot provide an adequate response to the harms that environmental injustice produces and sustains from the point of view of humans. I say more about this below, but the case for environmental human rights delivering justice to the nonhuman environment is less certain (I leave open here what justice for the nonhuman environment might be).

Another familiar criticism relates to the thought that the legal rights framework is state-centric, and thus entrenches an institutional order that (a) at least contributes to the maintenance of unequal North-South power relations, and (b) often frustrates effective action to tackle transboundary environmental harms (Gearty 2010; Woods 2010). Insofar as human rights are typically rights against the citizen's own state, this inevitably privileges the claims of some whilst excluding others who may well be affected by environmental decisions made in neighbouring, or indeed, distant states. To this extent, environmental human rights will be in conflict with versions of the all-affected principle that would grant a political voice to persons affected by environmental decisions in states other than their own and which has many environmentalist supporters (e.g. Eckersley 2004). Moreover, the notion of sovereign authority over natural resources, which is jealously guarded by states, is held to be both a practical obstacle to the development of environmental cooperation across state borders and indicative of an orientation towards nonhuman nature that is premised on notions of ownership rather than stewardship.

But we should note, too, that human rights practice has led to the development of more reflexive norms of sovereignty, and that human rights function both as legal mechanisms for holding powerful agents, mostly governments, to account, and as political norms (such as in the SDGs and in UNHCR Reports) that serve as markers of good practice and provide us with a framework for assessing the ethical fitness of policy initiatives. Moreover, procedural environmental human rights, such as those recognized in the landmark Aarhus Convention (1998), have proved effective in helping citizen groups to exercise rights to be informed of environmental decisions that may have harmful impacts within and beyond the borders of jurisdictions of the relevant courts, and citizens have been able to use such rights to leverage media coverage and political pressure that does precisely (albeit weakly) uphold the all-affected principle (see also Chapters 9 and 50 on the Aarhus Convention).

I do not pretend that there is nothing in the criticisms of those who note the dominance of Northern voices in setting the terms of such 'good' practice (see, e.g., Martinez-Alier 2002), and the problematic history of the concept of human rights itself, having emerged in significant part from the idea of natural rights, which were not human rights at all, but rather rights held by propertied, white, Christian men only (Woods 2014) (see also Chapter 7). But the very protean character of human rights that some sceptics of rights lament is in fact evidence of the remarkable potential for human rights to adapt and to evolve: at the same time as our understanding of the most fundamental harms affecting human beings has developed and given rise to recognition of new rights claims, so too ideas about who and what belong within the community of rights-bearers called human beings have been contested and expanded. For example, it was once the case that women were the property of men; we now have a global human rights treaty affirming not only the legal equality of women but committing governments to eliminating discrimination against women (the 1979 Convention on the Elimination of All Forms of Discrimination Against Women). The history of human rights struggles enacted by and on behalf of marginalized groups offers hope to the environmental movement in terms of the conceptual and legal changes that recognition of rights can bring. Insofar as environmental issues come to be accepted amongst the pantheon of human rights, we might argue that this in fact represents a significant shift in our thinking about the relationship between humans and nonhumans, and a substantial step towards environmental justice.

Harm, rights and duties

Henry Shue (1992) held that human rights, or in his terms 'basic rights', protect against what he called 'standard threats', that is, threats, or potential harms, which are universal in their character. A standard threat is one that every human being everywhere ought to be protected against. Although ways of life are hugely diverse across the globe, the thought here is that there are some protections and resources that are needed for any human in any community to have a chance of a decent life. Shue's account of standard threats yields rights to basic security and basic subsistence. From his more recent work on climate change we can expect he would be sympathetic to the claim that environmental harms now constitute a standard threat.

Shue's model is a useful way of thinking about human rights from an environmental point of view. If we were to ask, starting from a blank slate, what kinds of threats might emerge that would undermine the well-being of any person, regardless of their political, social, cultural or other context, we might immediately think of environmental issues. Seen in this way, environmental rights might be among the most truly universal of human rights, despite their notable absence from the Universal Declaration of Human Rights and the International Covenants.

Let us consider some of the ways in which environmental issues might affect human rights. Toxic pollution harms human health, and undermines food chains and agricultural systems. Climate change directly impacts upon rights to life, health and well-being (Caney 2010). Deforestation and coastal erosion threaten people's homes. Resource scarcity, which is increasing as the human population increases, is a major factor in many armed conflicts, which in turn have significant human rights impacts. Environmental goods – healthy soil, breathable air, the resilience that biodiversity provides and the carbon sinks and ecosystem services that forests provide – underpin human rights associated with work, health, shelter, security and family life. And the role of rainforests and other habitats in the cultural rights of Indigenous peoples is now well recognized, including in specific human rights treaties (Knox 2012). There are many obvious and direct ways in which environmental problems can harm humans.

That said, environmental harms do not necessarily map well onto the normative architecture of rights. There are (at least) two sides to this: The first relates to the agents that the rights framework identifies. Specific recognition has been agreed at international level for the group rights – that is the collective rights – of Indigenous peoples in relation to the important role that their lands play in their collective well-being (see also Chapter 11). But for the most part, human rights continue to track an individualist logic – rights are held by individual human beings against (first and foremost) their own governments. This raises questions both about the conceptions of value underpinning human rights, and about the role and status of sovereignty as an organizing principle for political communities. I will say more about this below.

The second side to consider in relation to the question of how environmental harms map on to the normative architecture of rights relates to the content of rights. Consider, for instance, the problem of toxic pollution. One way of responding to this threat in terms of rights might be to say that people have the right to live in an unpolluted environment, but that is implausible. That right, if upheld, would bring the global economy to an immediate halt, which would have severe human rights impacts. Production processes for all kinds of goods, on which lives and livelihoods depend, entail the production of pollution, much of it harmful. Indeed, what counts as 'harmful' is not a simple matter. For example, the atmospheric concentration at which people experience health impacts will vary from person to person, related to characteristics such as age, body size and underlying health conditions (see Chapter 26). So, should the condition for a right to an unpolluted environment be set at toxin concentrations that are acceptable to those most sensitive to toxins (in many cases this will be young infants), or should it be set at a level acceptable for the average person, or some other level? These sorts of questions have of course been much discussed within the literature on environmental justice as well as environmental policy (Shrader-Frechette 2007). The key point here is that it is not obvious that framing the question in terms of rights clarifies anything in these debates, given that environmental human rights claims will be weighed against other human rights claims, including rights to work and rights to property.

Except perhaps the right to life, most human rights are not absolute. When trying to understand what exactly a right is a right to, environmental rights can be most clearly understood in terms of thresholds, as James Nickel (1993) argued in his seminal defence of a right to a safe environment. There are other rights that work in this way – the right to education, for example, entitles individuals to access a minimum level of education; it sets a threshold below which individuals should not be allowed to fall. In practice, this threshold varies enormously from country to country, so while in some jurisdictions the right to education may entail state support from primary school to university, in others the right will be satisfied if all children are able to attend primary school, with or without paying fees. Similarly, Nickel tells us that

> the environment, or the level of safety from environmental risks, should be satisfactory or adequate for health. [...] The fact that terms such as 'satisfactory' and 'adequate' are vague is not a significant problem in this context. [...] International human rights typically set broad normative standards that can be interpreted and applied by appropriate legislative, judicial, or administrative bodies at the national level.
>
> *(Nickel 1993: 285)*

Note that in the case of education, this 'threshold' right attaches to a distinct individual. Environmental rights can be claimed by individuals, but they are properly collective rights.

Thus, in the 1993 Philippine case *Minors Oposa v. The Secretary of the Department of the Environment and Natural Resources* a group of individual children acted as claimants on behalf of future generations. This celebrated case specifically addressed an issue of timber licensing, but turned on the question of whether the rights of future generations could impose duties on present agents, specifically the Philippine government. This case illustrates an important point for the threshold model of environmental rights: The threshold at which the content of rights is set will be a political decision as well as a scientific one, but the conditions for meeting a given context will be influenced – potentially substantially so – by intergenerational environmental practices. In other words, while we might wish today to have an environment free from toxic pollution, our ability to uphold that right is constrained by the environmental choices of past generations, and we in turn will have huge impacts on the environmental rights thresholds available to generations that will come after us. So while Nickel is right that the 'threshold' model, allowing for variation, need not be a problem from the point of view of there being such a thing as an environmental human right, it is nevertheless the case that the environmental right is importantly different from other rights, and its realization is more vulnerable to impacts beyond the control of any single government than most other human rights. The harms that environmental human rights might be invoked to protect against can, of course, have a single immediate cause, such as an oil spill. But in many cases the sorts of harms that we are concerned with here will be of a different character from the sorts of harms that human rights practice is historically adapted to protecting against, where patterns of harm and redress, cause and effect, are much less diffuse and mediated.

There is interesting divergence between human rights practice and the ways in which environmental political theorists have developed accounts of environmental human rights. Much human rights practice relates to so-called procedural rights – e.g. rights to be informed of prospective developments and their environmental impacts, rights to assemble and to freedom of expression to protest and raise awareness about environmental issues (see Chapter 9). Another prominent strain of human rights practice relates to activists' appeals to environmental rights already recognized in national constitutions (over 90 countries worldwide currently contain such rights to some degree) and/or campaigns to embed or strengthen such rights. There are a few environmental theorists who have focused on this aspect of human rights practice (e.g. Hayward 2005). The potential value of constitutional rights is illustrated in the *Minors Oposa* case already mentioned: Here a group of citizens was able to use a constitutional right to argue that the right of future generations to a healthy environment was being compromised by the extensive licensing of the timber industry in the country. In countries where democracy and the rule of law is less secure the most significant human rights for environmentalists may well be the right to life, the right against torture, and rights to fair legal processes.

In contrast, the most sustained work by political theorists on environmental human rights has been in relation to climate change, and has focused on using human rights as a justificatory device to support the claim that present generations owe a rights-based duty to future generations to tackle climate change now. This has not typically related to specific actions, as in the *Minors Oposa* case, but more to a just distribution of benefits and burdens across generations. Given that climate change is widely seen as being amongst the most urgent moral and political problems facing us today (see Chapter 29), and given that human rights have become the authoritative medium in which to advance moral and political claims, it is not surprising that theorists of climate justice have drawn on human rights to explicate what justice demands here.

A lot of work on climate justice invokes analogies to illustrate the wrongfulness of climate change. One much-discussed example is Robert Elliot's (1988) time capsule, booby trapped by someone at time *t*. Whether the capsule is opened at *t*+1 or *t*+80 makes no difference to the wrongfulness of the action, and thus whether the person injured by the booby trap is alive at the point at which the booby trapper set it, or is yet to be born (i.e. belongs to a future generation) is irrelevant to the moral issue at stake. The person injured has a right not to be harmed, and the booby trapper has a duty correlative to that right not to set the trap. By analogy, it matters not whether the people who have rights against anthropogenic climate change are or were alive at the point at which harmful concentrations of greenhouse gases (GHGs) were released into the atmosphere; human beings, including future humans, have a right to life, health and a decent standard of living. Insofar as it is clear that climate change represents a threat to all three of these widely recognized human rights, we, the current generation, have a duty to take action on climate change (Caney 2010).

But what precisely does that mean we have a duty to do? GHGs are part of our economy, as well as part of our biophysical processes – we literally breathe out a GHG. Simon Caney writes that, 'people are entitled to those emissions necessary for them to attain a decent standard of living' (2010: 544), but not to those above that level. The distinction between 'subsistence' and 'luxury' emissions is a familiar one in this literature. On this account it seems that actions that generate GHG emissions up to a decent standard of living are not harmful – indeed, we are entitled to them, not least because we have a *human* right to the conditions for a minimally decent life – whereas *the same* actions that generate GHG emissions beyond this level *are* harmful, and are thus a violation of future people's rights. As I have argued elsewhere (Woods 2015), this represents a significant shift in the ways in which harm, rights and duties are connected. In an authoritative analysis of the concept of harm that predates much of the discussion of climate change, Joel Feinberg discusses cases of environmental harms where the impact is cumulative and the causal responsibility diffuse:

> In these contexts, no prior standard of wrongfulness exists. There is nothing inherently wrongful or right-violating in the activity of driving an automobile, generating electricity, or refining copper. These activities can be meaningfully condemned only as violations of an authoritative scheme of allocative priorities.
>
> *(Feinberg 1974: 230)*

So, the content of an environmental human right that tries to address problems like climate change – that is, problems where the causes are not direct and singular as in the booby-trapped time capsule, but rather multiple, diffuse, spread across space, time and a huge number of uncoordinated agents – will depend both on thresholds that are going to be determined by a number of variables, and on the construction of some kind of allocative scheme that will allow agents to say that some everyday instances of polluting activity violate a right, while others do not. As noted above, the threshold will be determined both by political factors and by biophysical ones: it will be a factor of many variables including the lifestyle decisions that the community collectively and individually decides constitutes 'a decent standard of living', the size of the population, the level of technology, as well as, crucially, the availability and maintenance of ecological resources.

Because, as already discussed, the actions of the current generation will play a role in setting what sorts of rights-thresholds will be available to future generations, what climate justice theorists implicitly require of human rights is a re-thinking of the time-horizons that rights stretch across, and with this, the bounds of the political community that rights encompass. While the legal and sociological paradigm from which human rights developed was primarily backward-looking, in that it was a means to protest against and seek redress for wrongs committed (generally, by the state and against specific individuals), contemporary work on climate justice reorients human rights to look forwards, to articulate and justify the duties of the present generation (individuals, corporations and governments) towards collective future generations, whose members are unspecified and largely unspecifiable from the point of view of the duty-bearer. In doing so, this work implicitly extends the evolution of human rights, both in terms of the content of the rights and in terms of the membership of the community of rights-bearers.

Rights bearers and environmental justice

At this point it is worth pausing to ask, who can be a bearer of environmental human rights? How far can the seemingly elastic notion of human rights be stretched, and what can it bring to environmental justice? Clearly, present generation human beings can be said to be bearers of human rights. I have already indicated that there are examples from both theory and practice that support the claim that future generations of human beings are also bearers of rights that have legal and moral standing now, but this remains controversial. The *Minors Oposa* judgment, for example, though widely celebrated, in fact only recognized the rights of future persons in a very circumscribed way, and though environmental rights are recognized in over 90 national constitutions, the rights of future generations are mentioned in only a handful (Knox 2012).

The status of the rights of future generations is even more contested at a theoretical level (see also Chapters 31 and 50). Much has been made of Derek Parfit's 'non-identity problem', which holds that a person cannot be said to have been harmed relative to what their life would otherwise have been, if it is the case that the allegedly harmful action was among the factors that determined their having existed. So, for example, if an environmentally harmful policy was among the factors that led to a particular couple having a child, it is implausible for the child to claim that her life was harmed by the environmental policy. In this case, her life *would not otherwise have been at all*, and so she cannot have a rights-based claim against the purportedly harmful act, according to Parfit (1987). Many sophisticated discussions of Parfit's work have explored its salience for climate justice in recent years (see, inter alia, Gosseries 2008; Tremmel 2009; Brännmark 2016). The development of analogies like Elliot's booby-trapped time capsule was intended to defeat the problems Parfit raises, and some commentators on this debate take a common sense view that it is 'obvious' that climate change does harm future generations, and insofar as that is the case then we can meaningfully speak of their rights being violated (see, e.g., Bell 2011). The debate persists, in philosophy seminars at least.

In practice human rights have an intriguing dual character. They both reinforce a state-centric vision of the political community, and undermine it. International human rights law is firmly entrenched in the idea that the sovereign state is the basic unit of international politics. As already indicated, our human rights do not entail universal human duties; on the contrary, we have rights first and foremost against our own sovereign government. Moreover, governments are committed to upholding human rights by international treaties, though there are, of course, widespread failures on this. Because of the sovereign authority over natural resources, among other things, only through cooperation at national government level can global environmental problems be addressed. In many ways, then, recognition of human rights defines a political community. If human rights extend into the future, then the political community defined by those rights is also extended, which can only be a positive thing from the point of view of environmental justice if it is true that human short-sightedness is among the factors that have led us to our current state of environmental destruction. However, both the legal and the most sustained theoretical elaborations of the rights of future generations have focused on the rights held by future generations of a particular country against their particular government, not the rights of all future generations against all those currently living (Brännmark 2016; Hiskes 2009). Paradoxically, perhaps, parochialism seems to be a future of intergenerational human rights.

But the other face of human rights – the one located in protest movements and activism – is determinedly internationalist, globalist, cosmopolitan in outlook. Human rights protesters look to their friends and collaborators in other countries for support, and support is often forthcoming, as in cases where organizations like Amnesty International mobilize public opinion to apply pressure on governments halfway across the world to uphold the human rights of their citizens. As a social movement (see Chapters 3 and 4), the diverse groups that might loosely be called the human rights movement share much in terms of aims, strategies and values with environmental justice movements (Stammers 1999; Gearty 2010). It is for these reasons that human rights are simultaneously associated with both the entrenchment of state sovereignty and the erosion of state sovereignty.

From the point of view of environmental activists, however, a much more pressing question might well be whether the human rights framework is or can be hospitable to justice for the nonhuman environment. In this chapter I have thus far said almost nothing about the rights of animals, let alone trees, plants or ecosystems. Many will feel that an approach to environmental justice that does not have space for the claims and moral standing of such beings is hardly addressed to *environmental* justice at all. Of course it is the case that a broad understanding of human interests gives us very good reasons to promote and protect the interests of nonhumans to a significant degree, and thus one can argue that a robust account of environmental human rights must depend upon a conception of the human as an inherently ecologically embedded being (Woods 2010). Understood in these terms, the integrity of ecosystems matters for us independently of any direct connection to the human economy or to rights-related needs. But even on this model, the justification for any action to protect nonhuman entities from anthropogenic environmental harms is firmly rooted in the human as the locus of moral value, rather than in the moral value of the nonhuman for its own sake (see also Chapter 37). Insofar as this is true, it is hard to see how the human rights framework could yield a full account of environmental justice *qua* justice for or to the environment.

Nevertheless, it seems valid to me to argue that the very inclusion of environmental concerns within the human rights framework is significant for our understanding of our relations to nonhuman nature. Recognizing environmental rights as human rights prompts us to take seriously the idea that we are ecologically embedded beings. If environmental threats are among the standard threats that any human has a right against, then they are among the most important and fundamental things that support any meaningful and decent human life.

Conclusion

Human rights are paradigmatically individualistic, humanistic, and rooted in the sovereign state. And yet, human rights are also paradigmatically tools of resistance and change. The growing recognition of environmental human rights, in both theory and practice, may represent a powerful tool in the struggle for environmental justice. This can be the case in terms of how communities resist particular environmental injustices – through claiming their rights to be informed about development and to participate in decision-making, through exercising their rights to assembly and free speech, through invoking constitutional rights promoting a healthy environment for future generations or arguing that such rights should be embedded in national constitutions.

But it may be even more important for the struggle for environmental justice that recognition of environmental human rights prompts us to reflect on the human in human rights: Environmental human rights raise questions about the bounds of the political community that rights demarcate – who is a rights bearer? Does a human being have to be currently living to have their rights protected? Can a future person have rights, can a great ape, can a rainforest or a river? What sorts of things can be 'standard threats' to the well-being of human beings, and what do these threats tell us about the nature of human beings? If environmental rights are human rights, then the human is an ecologically embedded being. That we have always been so may be obvious, but it has not always been part of our social, political, economic and cultural consciousness. By raising these questions, the practice of environmental human rights opens up space for the re-shaping of the idea of human rights itself, and given the fundamental role that human rights play in our conception of a decent society and a legitimate government, this re-shaping is potentially significant for any project of environmental justice.

Future research will pick up and develop these questions, about the ways in which emerging notions of environmental rights, in both theory and practice, re-shape our understanding of human rights, and about the complicated question of who is and should be recognized as a bearer of 'human' rights. Alongside these more abstract debates, there is also important work to be done in continuing to reflect critically on the extent to which the deep interrelation of human rights and environmental issues is fully understood in our legal, political and social responses at all levels of governance.

References

Bell, D. (2011). 'Does Anthropogenic Climate Change Violate Human Rights?' *Critical Review of International Social and Political Philosophy*, 14, 2, 99–124.

Brännmark, J. (2016). 'Future Generations as Rights Holders'. *Critical Review of International Social and Political Philosophy*, 19, 6, 680–698.

Caney, S. (2010). 'Climate Change, Human Rights, and Moral Thresholds'. In S. Humphreys (ed.), *Human Rights and Climate Change*, Cambridge: Cambridge University Press.

Eckersley, R. (2004). *The Green State: Rethinking Democracy and Sovereignty*, London: MIT Press.

Elliot, R. (1988). 'The rights of future people'. *Journal of Applied Philosophy*, 6, 2, 159–69.

Feinberg, J. (1974). 'The Rights of Animals and Future Generations'. In W. Blackstone (ed.), *Philosophy and Environmental Crisis*, Athens, Georgia: University of Georgia Press.

Gearty, C. (2010). 'Do Human Rights Help or Hinder Environmental Protection?' *Journal of Human Rights and the Environment*, 1, 1, 7–22.

Gosseries, A. (2008). 'On Future Generations' Future Rights'. *Journal of Political Philosophy*, 16, 4, 446–474.

Grear, A. (2011). 'The Vulnerable Living Order: Human Rights and the Environment in a Critical and Philosophical Perspective'. *Journal of Human Rights and the Environment*, 2, 1, 23–44.

Hancock, J. (2003). *Environmental Human Rights: Power, Ethics and Law*. London: Ashgate.

Hayward, T. (2005). *Constitutional Environmental Rights*. Oxford: Oxford University Press.

Hiskes, R. (2009). *The Human Right to a Green Future*. Cambridge: Cambridge University Press.

Knox, J. (2012). 'Report to the United Nations Human Rights Council of the Independent Expert on the issue of human rights obligations relating to the enjoyment of a safe, clean, healthy and sustainable environment.' United Nations, A/HRC/22/43.

Martinez-Alier, J. (2002). *The Environmentalism of the Poor: A Study of Ecological Conflicts and Valuation*. Cheltenham: Edward Elgar.

Nickel, J. (1993). 'The Human Right to a Safe Environment: Philosophical Perspectives on its Scope and Justification'. *Yale Journal of International Law*, 18, 281–96.

Parfit, D. (1987). *Reasons and Persons*. Oxford: Oxford University Press.

Shrader-Frechette, K. (2007). 'Human Rights and Duties to Alleviate Environmental Injustice: The Domestic Case'. *The Journal of Human Rights*, 6, 107–30.

Shue, H. (1992). *Basic Rights: Subsistence, Affluence and US Foreign Policy*. Princeton, NJ: Princeton.

Stammers, N. (1999). 'Social Movements and the Social Construction of Human Rights'. *Human Rights Quarterly*, 21, 4, 980–1008.

Tremmel, J.C. (2009). *A Theory of Intergenerational Justice*. London: Earthscan/Routledge.

Woods, K. (2010). *Human Rights and Environmental Sustainability*. Cheltenham: Edward Elgar.

Woods, K. (2014). *Human Rights*. Basingstoke: Palgrave Macmillan.

Woods, K. (2015). 'On Climate Justice, Motivation and Harm'. In M. Thorseth and D. Birnbacher (eds.), *The Politics of Sustainability: Philosophical Perspectives*, London: Routledge.

14

SUSTAINABILITY DISCOURSES AND JUSTICE

Towards social-ecological justice

Ulrika Gunnarsson-Östling and Åsa Svenfelt

Introduction

In the process of facing up to environmental problems like climate change and biodiversity loss, but also issues like poverty and health, the sustainability concept has gained importance in decision-making at different levels of society. In spite of its significance, or maybe because of it, sustainability is sometimes vaguely defined and sometimes defined in different and antagonistic ways (Redclift 2005; Gunnarsson-Östling et al. 2013). Several parallel sustainability discourses exist that mirror different views of human relations to nature. Nature in some sustainability discourses is regarded as something distant that needs to be preserved and protected from humans, and rights to nature and non-human beings are emphasized (e.g. Humphrey 2002). In contrast, there are very anthropocentric discourses that do not value pristine nature but instead focus on how to utilize natural resources to make profit. Human-made capital is thus seen as substitutable for natural capital. These discourses differ in their view of nature, in what are defined as problems and what are seen as solutions, but also in their view on justice.

Since different interpretations of sustainability generate different descriptions of important contemporary problems and also what their solutions might be, we argue that the concept needs to be carefully explained when used in different contexts. This is particularly the case in relation to how justice and patterns of access to ecosystem goods and the distribution of environmental bads figure in sustainability discourses, because such issues are often neglected in sustainability policy and planning (e.g. Movik 2014; Bradley et al. 2008).

In this chapter we consider how the relationship between sustainability and justice has been problematized and present an overview of sustainability discourses that are present in current debate and policy formulations. Through this analysis we identify overlaps and gaps as well as some key tensions within the related concept of sustainable development. Our aim is to highlight how different views on nature and justice within sustainability discourses suggest different solutions and transformations, but also to discuss how elements of different discourses can be merged to avoid silo-based policy where environmental quality and human equality are separated. We then consider how this framework can be applied and operationalized in policy, planning and decision-making.

Just sustainability

In the existing literature, a critical debate has been opened up about how sustainability discourses typically focus more on environmental stewardship and biodiversity than cultural diversity, and on inter-generational equity rather than intra-generational justice (Agyeman et al. 2003). These arguments emerge from engagement with environmental justice research, which, as detailed across various chapters in this Handbook, shows that disenfranchised, low-income, and/or minority populations are generally more at risk of being exposed to environmental hazards than other groups. Such research has been undertaken initially in the US, but increasingly also in other parts of the world. The necessary linkage between environmental justice and sustainable development is stressed by Agyeman et al. (2003: 78), who write:

> Sustainability, we argue, cannot be simply a 'green' or 'environmental' concern, important though 'environmental' aspects of sustainability are. A truly sustainable society is one where wider questions of social needs and welfare, and economic opportunity, are integrally related to environmental limits imposed by supporting ecosystems. This emphasis upon greater equity as a desirable and just social goal, is intimately linked to a recognition that, unless society strives for a greater level of social and economic equity, both within and between nations, the long-term objective of a more sustainable world is unlikely to be secured.

In order to bring together the two discourses of environmental justice and sustainability, the authors coined the term 'Just Sustainability'. Agyeman (2005) sees this as a bridge between the New Environmental Paradigm and the Environmental Justice Paradigm. While the New Environmental Paradigm, according to Agyeman (2005), focuses on environmental stewardship, the Environmental Justice Paradigm integrates social justice and environmental concerns. Merging these perspectives, according to Agyeman (2005) means focusing on the interdependence of social justice, economic well-being and environmental stewardship in terms of quality of life both here and now and for future generations. There is also a strong focus on deliberation given that '*process* is as important as *product*' in the 'Just Sustainability Paradigm' (Agyeman 2005: 89).

There are also examples of those who have attempted to address both social justice and planetary boundaries simultaneously. Kate Raworth (2012), for example, stresses the need for considering not only environmental issues when striving for sustainable development, but also social issues. However, the extent of the literature on the relation between sustainability and justice is limited. In order to develop this more systematically in what follows we look closely at a selection of sustainability discourses and especially their understanding of justice and nature.

Three interpretations of sustainability

In order to select which discourses to focus on we started out from other classifications of sustainability discourses such as Dryzek and Schlosberg (2005), Harvey (1996) and Martínez-Alier (2015), who together identify a spread of discourses, from those that see nature as setting limits to human activity to those that see no limits which human ingenuity cannot overcome. Our categorization of discourses presented below draws on these sources but also reflects our own scientific traditions, background and experience. The key differences we identify between the discourses are summarized in Table 14.1.

Table 14.1 Key differences between the discourses

	Eco-localism	*Green growth*	*Limits to growth*
Nature	Within local carrying capacity, local supply	Resource input	Basis for the economy, finite. Social-ecological systems
Justice	Direct participation	Individual choice	Rights to participate, distributive justice, diverse justice approaches
Solutions	Local systems, currencies, food supply and production	Technological development, end of pipe, market based	Define boundaries, co-management, local bottom up solutions, global entitlements
Change	Localize, enhance embeddedness	Adjust to current trends, internalize environmental effects	Stay away from limits. Limit/ decrease resource throughput and economic growth

Eco-localism

Eco-localism is a discourse that centres on bioregional principles, the localization of production and consumption and alternative solutions to the globalized trading systems as a means of enhancing sustainability (e.g. O'Hara and Stagl 2001; Gray 2007; Carr 2004). Trade can induce economic growth but is also sometimes seen as an accelerator of the destruction of natural ecosystems. Daly (1996) argues that international trade disrupts economic life in communities, and that it can lead to decreased power for people in communities, as decisions that affect them are taken on the other side of the world. Furthermore, trade leads to a situation in which we loosen communities from local capacities and constraints and instead import capacities from elsewhere, which makes it difficult to keep within ecological limits (Daly 1996). Bioregionalism, on the other hand, strives for place-based economies that produce primarily for the local community and only secondarily for trade, and can be seen as a reaction and opposition to the globalized system (Carr 2004). Here we use the term eco-localism in line with Curtis (2003), as the term can encompass a diversity of research and approaches related to increased self-sufficiency and localized markets. What they have in common is that they see the creation of local, self-reliant, community economies as a road to environmental sustainability.

The kinds of change and solutions that are discussed in the realms of eco-localism are those that enhance local self-sufficiency and that strengthen local social relations and trust. Face-to-face democratic decision-making and other direct democracy solutions, rather than centralized ones, are advocated (Hahnel 2007). Local energy generation (Späth and Rohracher 2014), local currencies, such as time banks and convertible local currencies (e.g. Dittmer 2013), local farmers' markets, community supported agriculture (SCA) and urban farming are all recurrent examples (e.g. Hinrichs 2000). Local food systems, it is argued, re-embed food production within a community and thereby can develop an authentic relationship between producer and consumer (Hendrickson and Heffernan 2002).

The view of nature in this kind of discourse implies a strong focus on place and the connection between local societies and their biophysical environments (Gray 2007). There are also arguments (e.g. Church 2014) that bioregional ideals implemented in urban areas could contribute to a shift in the human relationship to the environment and natural resources. The carrying capacity of a region, i.e. the population size that can be sustained by the natural resource base in the long term (Norse 1992), is an important concept and sets a boundary for consumption.

The justice perspective is not particularly explicit in the eco-localism discourse. Some argue that there are 'just' characteristics, for example Leach (2013) claims that localized economies are more inclusive, that there is better opportunity for civic engagement, that economic power can be redistributed and thereby inequalities may be reduced. Also, although it does not seem to be an explicit assumption, it is possible that a localized economy could reverse the injustices in unequal ecological exchange, i.e. decreasing international trade that allows for a net transfer of resources from developing to industrialized countries (e.g. Hornborg 2009; Jorgenson 2009).

On the other hand there is criticism of eco-localism that it is unreflective about justice consequences and that the implications might be problematic, leading to undemocratic, unrepresentative, elitist and reactionary solutions and strategies (e.g. Hinrichs 2003; DuPuis and Goodman 2005; Born and Purcell 2006). Hinrichs (2000) reflects, for example, upon the social inequalities of many farmers' markets in that they mostly serve well-educated middle-class consumers. Hahnel (2007) points out that there are more visions than models of rules and procedures for decision-making in community-based economies, hence important issues of community-based decision-making have not been addressed. Hahnel (2007) also argues that severe inequalities could arise if communities were completely self-sufficient, because some communities would be better off than others, on the basis of their access to natural, physical, and human capital.

Limits to growth

A second key sustainability discourse focuses on the more macro scale question of limits to growth. An emphasis on economic growth has for a long time dominated the economic system and the political agenda of many countries. Some argue that economics lost the connection to physical reality and to the idea that the economy is limited by nature at the start of the twentieth century (Norgaard 2000). Increased production and consumption has certainly produced an enhanced standard of living for many, but also negative consequences for both humans and the environment. However, neither the negative effects nor the gains of growth and acceleration in the global economy are evenly distributed (e.g. UNDP 2013). Increasingly, the question of limits and physical boundaries for the economy has been addressed and placed on the agenda. Meadows et al. (1972), for example, modelled continued exponential growth and concluded that the result would be that the resource base would collapse. Steffen et al. (2011) argue that the enhancement of human material wealth has led to scarcity of critical resources and degradation of ecosystem function to such an extent that human activity now outweighs natural geophysical processes as main driving forces. In this new epoch, 'the Anthropocene', society needs to govern the systems away from the hostile states that we are heading towards.

Several economists have addressed the physical limits to economic activities and proposed alternatives. Georgescu-Roegen (1971, 1977) introduced bioeconomics, which stresses the biological basis for economic activities and limitations in accessible resources that are also unevenly distributed and unequally appropriated (see also Chapter 39). Daly (1996) argued for a radical shift from the growth economy into the steady state economy. Growth, i.e. increase in the physical scale of material and energy throughput for production and consumption, should be replaced by a constant throughput, which means that a steady state economy can develop without growing. Jackson (2009) argues that any vision of prosperity has to address the question of limits in order to be credible.

Several approaches exist for dealing with sustainability in an economy that is limited by and acknowledges environmental thresholds and boundaries for resource use. In ecological economics, for example, environmental, economic and social problems are studied together, and not separately, as had often been the case before (e.g. Jansson and Zucchetto 1978; Costanza and Daly 1987). The environment is then seen as a prerequisite for our ability to create well-being (McMichael et al. 2005) and the view of nature in this line of thinking, in contrast to neoclassical economics, is that the economic system is a basis for and an integral part of society. Berkes and Folke (1998: 4) used the term 'social-ecological systems' (SES) to emphasize that social and ecological systems are interlinked and interdependent. In order to support development of sustainable and resilient social ecological systems that acknowledges limits, researchers have defined boundaries for what can be emitted into natural systems and taken out from them in order to stay away from unsustainable trajectories, so called 'planetary boundaries' (Rockström et al. 2009; Steffen et al. 2011). This, it is argued, can serve to encourage growth in sectors of the economy that use fewer resources (Victor 2010).

Another option would be to limit growth itself (Victor 2010) and to scale down economic activity and instead opt for degrowth (e.g. Schneider et al. 2010). Even if the economy stagnates, it is argued, resources cannot circulate within the economy forever. Degrowth can, according to Schneider et al. (2010), mean an equitable downscaling of production and consumption that increases human well-being and reduces the societal throughput of energy and raw materials. Degrowth approaches can range from, for example, criticizing the narrow focus on GDP growth in society, to a strategy for reaching environmental goals and to a strategy to transform and replace capitalistic modes of production and distribution (Ott 2012). Sharing, simplicity, conviviality, care and the commons can indicate what the degrowth society might be like, according to Kallis et al. (2015). In the degrowth discourse justice is a key issue, because it points at the injustices in the current system, and structural changes and alternatives for 'just degrowth' are being explored (Muraca 2012). The strategies and solutions related to environmental justice in the degrowth literature are diverse and can range from community supported agriculture and transition towns to solidarity economy, ethical purchasing groups, equal ecological footprints and CO_2/capita emissions (Muraca 2012; Martínez-Alier 2012), which mirror in some respects those advocated by eco-localism approaches.

Collaborative solutions and governing the commons are common denominators of both degrowth and ecological economics approaches. Hardin's (1968) 'tragedy of the commons' explains the overuse of resources as a function of the difficulties involved in properly managing open access or common pool resources. In order to manage such resources, Ostrom (1990) proposes cooperation, with institutional arrangements fitted to local ecosystems and rules that need to be understood as legitimate by the users, hence the scale and interaction between scales are important. Solutions and approaches should therefore include and learn from local users and communities, and are participatory in that sense. In co-management, for example, the governance of social-ecological systems should acknowledge rights of local users and power should be shared between community, regional and national levels rather than be top-down (e.g. Armitage et al. 2008; Dietz et al. 2003). In this sense, participative justice aspects are addressed.

Green growth

Green growth is currently one of the most dominant sustainability discourses. It is exemplified by the Green Growth Knowledge Platform (GGKP) – a global network established in 2012 that brings together international organizations and experts to support a transition to a 'green economy' (GGKP 2014). As Ferguson (2015: 17) notes, 'the concept of a "green economy" has been part of environmental discourse for more than two decades' although in the aftermath of the global financial crisis in 2008 it was drawn into the foreground. Green growth's core idea – having economic growth together with significant environmental protection – might seem rather straightforward, but it differs from standard economics, which views environmental degradation mainly as a market failure. Both green growth and standard economics are liberal discourses where the market economy and its view of people as rational individuals is seen as a good ground for creating a just society. However, the green growth discourse suggests alternatives to GDP that also include social and environmental indicators.

Thus, green growth can be seen as a continuation of the discourse of *ecological modernization*, which emphasizes technological solutions as a way to decouple economic growth from environmental degradation. Ecological modernization makes ecological deficiency a driving force to modernize society out from its crisis (Hajer 1996). Thus, environmental degradation is not an impediment to economic growth, but a catalyst for growth through the quest for clean technologies, energy efficient products and management systems such as road pricing and smart production systems. Ecological modernization was recognized as a promising policy alternative in Western countries in the early 1980s (Hajer 1996) but has been viewed by its critics as an attempt to green capitalism, sustain business-as-usual and de-radicalize sustainable development (Barry 2003). Others see ecological modernization as strategic environmental management, with emphasis on environmental improvements in the private sector which mean that eco-efficiency gains can be achieved without transforming society (Buttel 2000).

The Green Growth discourse acknowledges natural resources as 'fundamental for the economy and for well-being' (OECD 2015: 21). As in ecological modernization, environmental harm as well as degradation and depletion of natural resources caused by human activities is acknowledged and therefore an efficient use of natural resources is seen as important. Natural resources are, however, seen as exchangeable with other resources so that nature is mainly viewed as a resource input and as something that can be effectively replaced with human-made substitutes.

The green growth discourse acknowledges that the consequences caused by extraction, production, consumption and recycling of natural resources are often displaced, because of cheaper labour in poorer countries, and are sometimes global in nature (like climate change) (see e.g. OECD 2015: 40). However, policy recommendations focus not on these environmental injustices but instead on resources productivity and an effective flow of materials through the economic system as a move towards green growth (OECD 2015: 46). The assumption is that a green economy will generate resources that enhance human well-being and social justice while environmental risks and the use of finite resources are minimized.

Sustainability discourses and just sustainability

As discussed earlier, the idea of just sustainability is to bring justice and sustainability discourses and principles together in productive ways. To what extent do the main sustainability discourses we have discussed achieve this? Eco-localism could potentially lead to more people having a stake in how local resources are used. Thus, people could become active decision-makers, users and care-takers of their local commons. However, depending on who owns the land and resources, eco-localism could also mean that a few individuals are in control and become powerful. There may also be justice implications in reversing today's global unequal ecological exchange between countries. Environmental degradation resulting from consumption would become more local. For example, e-waste (discarded electronic appliances) would not be transferred from rich to poor countries and cause environmental and health problems (e.g. Robinson 2009), but would be taken care of in the producing and consuming country. However, eco-localism would not necessarily safeguard from poverty and new injustices could turn up at the local level. Also, even if production is localized, many environmental threats are still global – for example climate change – and the impacts are not necessarily evident in the local region.

The limits to growth discourse could, since it strongly emphasizes maintained ecological function, be interpreted as a kind of 'justice to nature' or eco-centric approach (e.g. Humphrey 2002), although it is mainly framed in the literature as being about benefits to humans and society in the long term. Explicit justice perspectives featuring in the literature on social-ecological systems (e.g. Ostrom 1990; Dietz et al., 2003) are mainly about inclusion of users in management and management decisions. Ecological economist Costanza (1989: 5), however, argues that sustainability is inherently about limits, and that issues of equity and distribution are also issues of limits, because 'we do not have to worry so much about how an expanding pie is divided, but a constant or shrinking pie presents real problems'. Justice and distribution of ecosystem goods and services is an underexplored field of research that only recently has come into focus (e.g. Movik 2014; Sievers-Glotzbach 2013). The Anthropocene (e.g. Steffen et al. 2011) and related concepts such as resilience have been criticized for lacking a perspective on power (Hornborg 2009) and for hiding the fact that it is a minority of people (not the entire species – the Anthropos as such) that determine natural conditions in the Anthropocene. In this perspective, the theoretical framework of the Anthropocene is merely an attempt by natural scientists to extend their world-views to society (Malm and Hornborg 2014).

In the degrowth version of limits to growth, justice is strongly emphasized. However, whether justice is enhanced by degrowth approaches still has to be established. Many approaches depend on active citizens and since different individuals have a different ability to take advantage of opportunities, injustices can prevail also in a degrowth economy. Muraca (2012: 544) brings up the example of the equal individual share of CO_2 emissions, and that it might lead to 'significant injustice in the absence of coordinating institutions that through politics of redistribution counteract discrimination and exclusion and provide formal and substantial conditions for participatory parity and the good life'. One of the core questions when discussing nature and justice remains: What is going to be distributed and according to what principles?

In the green growth paradigm justice is about creating a well-functioning global market together with environmental protection so that people here and now, as well as future generations, should be able to make their own individual choices within a global market economy. Even though the discourse acknowledges that environmental negative consequences are often displaced today, the focus is on resource productivity and effective material flows assuming that green growth will sort the displacement out. Thus, green growth gives preference to the market and technical solutions over a coherent concept of justice. The implicit argument is that greater justice will come with greener and more economic growth. Ferguson (2015: 21–22) points out that the main criticism of the green economy and green growth is that it is dependent on decoupling economic growth from environmental degradation and resource consumption – something that has so far not been feasible (even if this does not prove it cannot happen in the future).

Also, green growth builds on the liberal-economic idea that a just distribution results from the aggregation of self-interested individuals' preferences. However, individual liberties may conflict. To take Campbell's (2006: 95) example: 'In relation to planning matters, the nature of interests is often complex and problematic; for example, individuals generally both desire clean air and to be able to drive their car(s) freely. Our preferences are therefore often inconsistent and overlapping'. Even if an individual knows that driving causes greenhouse gas emissions, this is not something that affects the individual immediately. Perhaps the effects are not even visible where the car driver lives, but rather elsewhere. Thus, the green growth discourse alone does not seem to offer a just sustainability.

While eco-localism, limits to growth and the green growth paradigm all focus on resources, the environmental justice discourse focuses on everyone's right to a clean environment and also a just distribution of environmental burdens and the right to participate in environmental decision-making. On the other hand the environmental justice discourse has not traditionally focused on today's distribution of environmental goods at a global scale, even though more global perspectives are now emerging. The clean environment that people should have access to in the environmental justice perspective has primarily been the local environment, where people 'work, live and play'. Also the local living environment is urban, and not primarily a site for local use of natural resources as in eco-localism. Indeed, we would argue that environmental justice has not traditionally acknowledged the environmental support systems, both local and distant, that are a basis for the survival of communities.

The explicit views on justice put forward by the discourses presented in this chapter concentrate on humans. In more academic discourses, such as political ecology, humans' power over nature is also highlighted. For example, feminist political ecology, in its effort to tell stories other than the dominant ones, describes humans as part of a Nature (with a capital N) that orders the way we live, including the practical everyday life of women and men and the fluidity between masculine and feminine forces (Walsh 2015; see also Chapter 7). In this view, Nature is a living being (of which humans are part) with rights and therefore we should not endanger natural systems. Therefore, discussions about the allocation of resources not only between people, but also between species, can be relevant.

With this discussion, we hope we have highlighted the importance of dealing with justice issues in an *explicit* way in decision-making concerning the environment. In sustainability policies and debates justice is often unspoken, which means that the tensions we have identified are hidden. We will summarize these ideas below as a need to explicitly work with a set of principles of social-ecological justice.

Towards social-ecological justice

According to Allen et al. (2003), sustainability should be considered in terms of sustainability of what, for whom, for how long, and at what cost. One obvious conclusion from our review is that the answers to these questions are divergent, sometimes overlapping but also contradictory between different discourses and interpretations of sustainability. They entail different proposals for solutions and focus on different spatial and temporal scales. We therefore see the need for expanding the scope of perspective, in terms of spatial and temporal scales, and also types of solutions considered when discussing environmental justice and sustainability. Without such a move there is a risk of sub-optimization in planning, policy and decision-making. This can, for example, mean that too narrow a focus in spatial scale can mask consequences on other scales. If, for example, the environment is seen as where we live, work and play, supporting ecosystems at other scales may not be considered and human well-being may consequently be harmed (e.g. Heynen 2003; Sundkvist et al. 2005). It can also mean that a focus on future generations fails to address properly just distribution within present populations, and vice versa. And regarding solutions, several approaches neglect distributional issues and power perspectives (e.g. limits to growth and green growth) and others are more reactive to environmental impact and change (e.g. environmental justice) rather than being proactive (as for example limits to growth). Also, we are concerned that a purely critical perspective is insufficient and therefore call for more normative suggestions on approaches to finding solutions.

As we see it, there is a need both for understanding the ecological basis for sustaining societies and highlighting the distribution of environmental resources and environmental impacts between different groups in society. However, those two perspectives are rarely explicitly handled or discussed in sustainability discourses. The ecological support systems and their services are often disregarded in urban and industrialized societies, since globalization has dispersed our dependence on ecosystems over the globe and made us seemingly independent of the local resource base. Similarly, environmental justice perspectives have been neglected because human suffering from environmental degradation can often occur at a large distance from the end-consumer and affected groups may therefore have little influence on policy and planning.

In order to take forward planning, policy and decision-making that addresses power and influence, both inter and intra-generational justice, both local and distant support systems, both visible and non-apparent dependence on ecosystems, both environmental benefits and burdens, we argue that social-ecological justice should include the following characteristics:

- Decisions are based on how the system/case/society depends upon and influences local ecosystems and social-ecological systems in other regions.
- Conflicts, complexity, dynamics, and uncertainty are acknowledged.
- There is a just distribution of environmental goods and services, between and within both communities and generations.
- There is a just distribution of environmental bads (environmental burdens), between and within both communities and generations.
- Principles for just distribution are discussed and defined in a deliberative process.
- Justice permeates planning, policies and production and consumption.
- There is awareness and identification of who is included, who decides, and where power is located.
- Environmental decisions are based on and shaped by affected groups/peoples/communities.

These characteristics are anthropocentric in the sense that distribution and power is attributed to humans. It could also be argued that non-human entities should be considered in a justice perspective. This perspective has not been sufficiently explored in a planning and policy context, but would be interesting to develop in future research. These principles of social-ecological justice can be used as an analytical framework and as a checklist in planning, policy and decision-making. We are not proposing to create yet another discourse, but we aim to stimulate and create a discussion and a practice for a more comprehensive approach to dealing with both the ecological basis for sustaining societies and the distribution of environmental resources and environmental impacts between different groups in society. We have taught masters students on this subject for eight years, and they have used these principles to assess real planning/environmental cases and to develop suggestions for how to implement more socio-ecologically just planning/environmental engineering (e.g. Ackebo et al. 2013). The students have drawn up proposals on how, for example, not only municipalities but also the mining industry can promote social-ecological justice. The specific proposals – plans or policy recommendations – are dependent on the specific study object, but the analytical framework could in the future be developed into an assessment framework, perhaps as part of environmental assessment tools and/or social impact assessments. It is a holistic approach that combines a view on nature as a system that humans are part of with a view on justice that means striving towards a just distribution of the access to environmental goods on different scales. We argue that this joint perspective is vital in view of the sustainability challenges we are facing, or else sustainable transitions may not be just, and may cause conflicts and power struggles between groups and scales.

References

Ackebo, J., et al. (2013). *What is the potential to create a just social-ecological system in Fisksätra/Saltsjöbaden?* Report from the Ecosystem support and Environmental Justice course (AG2803), TRITA-INFRA-FMS ISSN: 1652-5442; 2013:02, KTH Royal Institute of Technology.

Agyeman, J. (2005). *Sustainable communities and the challenge of environmental justice.* New York: New York University Press.

Agyeman, J., Bullard, R.D. & Evans, B. (2003). *Just sustainabilities: development in an unequal world.* Cambridge, MA: The MIT Press.

Allen, T.F.H., Tainter, J.A. & Hoekstra, T.W. (eds) (2003). *Supply-side sustainability.* New York: Columbia University Press.

Armitage, D., Marschke, M. & Plummer, R. (2008). "Adaptive co-management and the paradox of learning." *Global Environmental Change*, vol. 18, pp. 86–98.

Barry, J. (2003). "Ecological modernisation." In Page, E. & Proops, J. (eds.) *Environmental thought.* Cheltenham: Edward Elgar Publishers, pp. 191–213.

Berkes, F. & Folke, C. (1998). "Linking social and ecological systems for resilience and sustainability." In Berkes, F. & Folke, C. (eds.) *Linking social and ecological systems: management practices and social mechanisms for building resilience.* Cambridge University Press.

Born, B. & Purcell, M. (2006). "Avoiding the local trap: scale and food systems in planning research." *Journal of Planning Education and Research*, vol. 26, pp. 195–207.

Bradley, K., Gunnarsson-Östling, U. & Isaksson, K. (2008). "Exploring environmental justice in Sweden: how to improve planning for environmental sustainability and social equity in an 'eco-friendly' context." *Projections, MIT Journal of Planning*, vol. 8, pp. 68–81.

Bullard, R.D. (1990). *Dumping in Dixie: race, class, and environmental quality.* Boulder, CO: Westview.

Bullard, R.D. (ed.). (1993). *Confronting environmental racism: voices from the grassroots.* Boston MA: South End Press.

Bullard, R.D. (2001). "Environmental justice in the 21st century: race still matters." *Phylon*, vol. 49, pp. 151–171.

Buttel, F.H. (2000). "Ecological modernization as social theory." *Geoforum*, vol. 31, pp. 57–65.

Campbell, H. (2006). "Just planning: the art of situated ethical judgement." *Journal of Planning Education and Research*, vol. 26, pp. 92–106.

Carr, M. (2004). *Bioregionalism and civil society: democratic challenges to corporate globalism.* Vancouver: UBC Press.

Church, S.P. (2014). "Exploring urban bioregionalism: a synthesis of literature on urban nature and sustainable patterns of urban living." *S.A.P.I.EN.S*, vol. 7(1).

Costanza, R. (1989). "What is ecological economics?" *Ecological Economics*, vol. 1, pp. 1–7.

Costanza, R. & H.E. Daly (eds.). (1987). "Ecological economics." *Ecological Modeling* [Special Issues] 38(1) and (2).

Curtis, F. (2003). "Eco-localism and sustainability." *Ecological Economics*, vol. 46, pp. 83–102.

Daly, H.E. (1996). *Beyond growth: the economics of sustainable development.* Boston: Beacon Press.

Dietz, T., Ostrom, E. & Stern, P.C. (2003). "The struggle to govern the commons." *Science* 302(5652), pp. 1907–12.

Dittmer, K. (2013). "Local currencies for purposive degrowth? A quality check of some proposals for changing money-as-usual." *Journal of Cleaner Production*, vol. 54, pp. 3–13.

Dryzek, J.S. & Schlosberg, D. (2005). *Debating the Earth: the environmental politics reader. 2nd Edition.* Oxford and New York: Oxford University Press.

DuPuis, E.M. & Goodman, D. (2005). "Should we go 'home' to eat? toward a reflexive politics of localism." *Journal of Rural Studies*, vol. 21, pp. 359–371.

Ferguson, P. (2015). "The green economy agenda: business as usual or transformational discourse?" *Environmental Politics*, vol. 24, pp. 17–37.

Georgescu-Roegen, N. (1971). *The entropy law and the economic process.* Cambridge, MA: Harvard University Press.

Georgescu-Roegen, N. (1977). "Inequality, limits and growth from a bioeconomic viewpoint." *Review of Social Economy*, vol. XXXV, pp. 361–375.

Gray, R. (2007). "Practical bioregionalism: a philosophy for a sustainable future and a hypothetical transition strategy for Armidale, New South Wales, Australia." *Futures*, vol. 39, pp. 790–806.

Green Growth Knowledge Platform. (2014). www.greengrowthknowledge.org. Accessed 2015/10/06.

Gunnarsson-Östling, U., Edvardsson Björnberg, K., & Finnveden, G. (2013). "Using the concept of sustainability to work: interpretations in academia, policy, and planning." In Metzger, J. & Rader Olsson, A. (eds.) *Sustainable Stockholm: exploring urban sustainability in Europe's greenest city.* London: Taylor & Francis, pp. 51–70.

Hahnel, R. (2007). "Eco-localism: a constructive critique." *Capitalism Nature Socialism*, vol. 18, pp. 62–78.

Hajer, M.A. (1996). "Ecological modernisation as cultural politics". In Lash, S., Szerszynski, B., & Wynne, B. (eds.) *Risk, environment and modernity: towards a new ecology.* London, Thousand Oaks and New Delhi: SAGE Publications, pp. 246–268.

Hardin, G. (1968). "The tragedy of the commons". *Science*, vol. 162, pp. 1243–1248.

Harvey, D. (1996). *Justice, nature and the geography of difference.* Oxford: Wiley-Blackwell.

Hayward, T. (2007). "Human rights versus emissions rights: climate justice and the equitable distribution of ecological space". *Ethics & International Affairs*, vol. 21, pp. 431–450.

Hendrickson, M.K. & Heffernan, W.D. (2002). "Opening spaces through relocalization: locating potential resistance in the weaknesses of the global food system." *Sociologia Ruralis*, vol. 42, pp. 347–369.

Heynen, N.C. (2003). "The scalar production of injustice within the urban forest." *Antipode*, vol. 35, pp. 980–998.

Hinrichs, C.C. (2000). "Embeddedness and local food systems: notes on two types of direct agricultural market." *Journal of Rural Studies*, vol. 16, pp. 295–303.

Hinrichs, C. (2003). "The practice and politics of food system localization." *Journal of Rural Studies*, vol. 19, pp. 33–45.

Hornborg, A. (2009). "Zero-sum world: challenges in conceptualizing environmental load displacement and ecologically unequal exchange in the world-system." *International Journal of Comparative Sociology*, vol. 50(3–4), 237–262.

Humphrey, M. (2002). "The foundations of ecocentrism." In *Preservation versus the people?: nature, humanity, and political philosophy.* Oxford Scholarship Online, pp. 1–6.

Jackson, T. (2009). *Prosperity without growth: economics for a finite planet.* London: Earthscan.

Jansson, A.-M., & Zucchetto, J. (1978). "Man, nature and energy flow on the island of Gotland." *Ambio*, vol. 7, pp. 140–149.

Jorgenson, A.K. (2009). "The sociology of unequal exchange in ecological context: a panel study of lower-income countries, 1975–2000." *Sociological Forum*, vol. 24, pp. 22–46.

Kallis, G., Demaria, F., & D'Alisa, G. (2015). "Degrowth." In D'Alisa, G., Demaria, F., and Kallis, G. (eds.) *Degrowth: a vocabulary for a new era.* New York and London: Routledge, pp. 1–17.

Leach, K. (2013). "Community economic development: localisation, the key to a resilient and inclusive local economy?" *Local Economy*, vol. 28, pp. 927–931.

Malm, A. & Hornborg, A. (2014). "The geology of mankind? A critique of the Anthropocene narrative." *The Anthropocene Review*, vol. 1, pp. 62–69.

Margules, C.R. & Pressey, R.L. (2000). "Systematic conservation planning." *Nature*, vol. 405, pp. 243–253.

Martínez-Alier, J. (2012). "Environmental justice and economic degrowth: an alliance between two movements." *Capitalism, Nature, Socialism*, vol. 23, pp. 51–73.

Martínez-Alier, J. (2015). "Environmentalism, currents of." In D'Alisa, G., Demaria, F., & Kallis, G. (eds.) *Degrowth: a vocabulary for a new era.* New York and London: Routledge, pp. 37–40.

McMichael, A., Scholes, R., Hefny, M., Pereira, E., Palm, C., & Foale, S. (2005). "Linking ecosystem services and human wellbeing", Chapter 3 in Capistrano, D., Samper K.C., Lee, M.J., & Raudsepp-Hearne, C. (eds.) *Ecosystems and human well-being: multiscale assessments, volume 4.* Findings of the Sub-global Assessments Working 54 Group of the Millennium Ecosystem Assessment. Washington, Covelo, London: Island Press, pp. 43–60.

Meadows, D.H., Meadows, D.L., Randers, J., & Behrens, W.W. (1972). *The limits to growth: a report for the club of Rome's project on the predicament of mankind.* New York: Universe Books.

Movik, S. (2014). "A fair share? Perceptions of justice in South Africa's water allocation reform policy." *Geoforum*, vol. 54, pp. 187–195.

Muraca, B. (2012). "Towards a fair degrowth-society: justice and the right to a 'good life' beyond growth." *Futures*, vol. 44, pp. 535–45.

Norgaard, R.B. (2000). "Ecological economics". *BioScience*, vol. 50, p. 291.

Norse, D. (1992). "A new strategy for feeding a crowded planet." *Environment*, vol. 34, pp. 6–11, 32.

OECD (2015). *Material resources, productivity and the environment.* OECD Green Growth Studies, Paris: OECD Publishing.

O'Hara, S.U. & Stagl, S. (2001). "Global food markets and their local alternatives: a socio-ecological economic perspective." *Population and Environment*, vol. 22, pp. 533–553.

Ostrom, E. (1990). *Governing the commons: the evolution of institutions for collective action.* The political economy of institutions and decisions. Cambridge: Cambridge University Press.

Ott, K. (2012). "Variants of de-growth and deliberative democracy: a Habermasian proposal." *Futures*, vol. 44, pp. 571–81.

Raworth, K. (2012). "A safe and just space for humanity: can we live within the doughnut?" *Nature*, vol. 461, pp. 1–26.

Redclift, M. (2005). "Sustainable development (1987–2005): an oxymoron comes of age." *Sustainable Development*, vol. 13, pp. 212–227.

Robinson, B.H. (2009). "E-waste: an assessment of global production and environmental impacts." *Science of the Total Environment*, vol. 408(2), pp. 183–191.

Rockström, J., Steffen, W., Noone, K., Persson, Å., Chapin, F.S., Lambin, et al. (2009). "A safe operating space for humanity." *Nature*, vol. 461, pp. 472–475.

Schlosberg, D. (2004). "Reconceiving environmental justice: global movements and political theories." *Environmental Politics*, vol. 13, pp. 517–540.

Schneider, F., Kallis, G., & Martínez-Alier, J. (2010). "Crisis or opportunity? Economic degrowth for social equity and ecological sustainability." Introduction to this special issue. *Journal of Clean Production*, vol. 18, pp. 511–518.

Sievers-Glotzbach, S. (2013). "Ecosystem services and distributive justice: considering access rights to ecosystem services in theories of distributive justice." *Ethics, Policy & Environment*, vol. 16, pp. 162–176.

Späth, P. & Rohracher, H. (2014). "Beyond localism: the spatial scale and scaling in energy transitions." In Padt, F., Opdam, P., & Polman, N. (eds.) *Scale-sensitive governance of the environment.* Somerset, NJ: John Wiley & Sons.

Steffen, W., Persson, Å., Deutsch, L., Zalasiewicz, J., Williams, M., Richardson, K., et al. (2011). "The anthropocene: from global change to planetary stewardship." *Ambio*, vol. 40(7), pp. 739–761.

Sundkvist, Å., Milestad, R., & Jansson, A.-M. (2005). "On the importance of tightening feedback loops for sustainable development of food systems." *Food Policy*, vol. 30, pp. 224–239.

UNDP. (2013). "Humanity divided: confronting inequality in developing countries." United Nations Development Program. www.undp.org/content/dam/undp/library/Poverty Reduction/Inclusive development/Humanity Divided/HumanityDivided_Full-Report.pdf.

Victor, P. (2010). "Questioning economic growth." *Nature*, vol. 468(7322), pp. 370–371.

Walsh, C. (2015). "Life, nature and gender otherwise: feminist reflections and provocations from the Andes." In Harcourt, W. & Nelson, I.L. (eds) *Practising feminist political ecologies: moving beyond the 'green economy'.* London: Zed books, pp. 101–128

PART II

Methods in environmental justice research

15

SPATIAL REPRESENTATION AND ESTIMATION OF ENVIRONMENTAL RISK

A review of analytic approaches

Jayajit Chakraborty

Introduction

The academic literature on environmental justice (EJ) analysis encompasses a wide range of quantitative case studies that seek to examine "the environmental equity hypothesis – whether environmental risk burdens are distributed evenly across people and places, or if racial/ethnic minority and economically disadvantaged individuals are disproportionately exposed to pollution and related health risks" (Chakraborty 2009: 675). While various kinds of environmental hazard sources and many different study areas have been examined, this equity hypothesis has been typically tested by: 1) comparing the socio-demographic characteristics of areas exposed to specific environmental risks to the characteristics of areas that are not exposed; or 2) analysing the magnitude or significance of the association between indicators of environmental exposure/risk and variables describing the socio-demographic characteristics of the residential population. While the first approach requires researchers to delineate the spatial extent and size of the area that is exposed to environmental hazard sources, the second approach is based on quantifying the magnitude of the risk burden resulting from human exposure (known or potential) to hazard sources.

Since the 1980s, a variety of analytic approaches have been used in EJ research to estimate the geographic boundaries of areas exposed to environmental hazard sources and measure exposure to environmental risks. This chapter seeks to provide an overview and critical assessment of these approaches. The goal here is to examine the commonly used methods and related assumptions, discuss their strengths and limitations, and trace how methodological approaches for assessing environmental exposure for EJ research have evolved over time. For the purpose of this review, the different methodologies that have been employed in prior EJ studies to derive the geographic definition and measurement of risk exposure are classified into four broad categories: (a) spatial coincidence analysis; (b) distance-based analysis; (c) plume-based analysis; and (d) cumulative exposure analysis. The chapter concludes by highlighting key limitations of current approaches and data sources, and identifying future research needs associated with spatial representation and quantification of environmental risk for EJ analysis.

Spatial coincidence analysis

In the context of EJ analysis, spatial coincidence refers to an approach that assumes environmental exposure to be confined only to the boundaries of pre-defined geographic entities or census enumeration units that contain hazard sources. The most widely used technique, referred to as *unit-hazard coincidence* (Mohai and Saha 2006), utilizes the location of a hazard source within each spatial unit as a proxy for environmental exposure. The socio-demographic characteristics of units containing a hazard (host units) are statistically compared to all others (non-host units) in the study area to evaluate disproportionate exposure/risk. Several early, influential, and widely cited EJ studies conducted at the national level in the US have used the presence or absence of hazardous facilities within ZIP codes (United Church of Christ 1987; Goldman and Fitton 1994) or census tracts (Anderton et al. 1994; Been 1995) to estimate risk burdens. National and state level studies have also used the county as a spatial unit for coincidence analysis (Hird 1993; Daniels and Friedman 1999; Elliott et al. 2004).

The choice of geographic entity to represent the host spatial unit has been the subject of considerable debate (McMaster et al. 1997; Mennis 2002) and researchers have examined how EJ results from the unit-hazard coincidence method vary across multiple spatial scales (Cutter et al. 1996; Taquino et al. 2002; Baden et al. 2007; Noonan et al. 2009). Although these studies are not comparable because of dissimilarities in study areas and hazards examined, their findings suggest that different units of analysis potentially lead to different EJ conclusions regarding variables such as race/ethnicity or income. Data aggregated at higher levels such as a county or metropolitan area (coarse spatial resolution), however, have been documented to be less reliable as indicators of disproportionate burdens than data aggregated to smaller units (finer spatial resolution) such as census block groups (McMaster et al. 1997; Sheppard et al. 1999; Maantay 2002). It is generally acknowledged that using the smallest practicable unit of analysis yields the most accurate results (Maantay 2007; Chakraborty et al. 2011), while the use of larger areal units often increases the strength and significance of statistical relationships between environmental risk indicators and socio-demographic variables (Cutter et al. 1996; Taquino et al. 2002).

Regardless of the geographic entity selected, there are several limitations associated with the unit-hazard coincidence method (Mohai and Saha 2006; Chakraborty et al. 2011). First, most applications do not usually distinguish between spatial units that host one hazard source and those in which multiple hazard sources are located. Second, this approach does not account for boundary or edge effects. These effects deal with the possibility that an environmental hazard source could be so close to the boundary of its host spatial unit that a neighbouring non-host unit could be equally or more exposed to the same hazard source. Unless the hazard is located near the centre of the spatial unit, the representativeness of the socio-demographic data used to analyse EJ becomes questionable. Third, the unit-hazard coincidence method assumes that the environmental exposure is restricted only to the boundaries of the host spatial units. However, pre-defined geographic entities such as census units are unlikely to represent the shape or size of the area potentially exposed to the entire range of environmental risks associated with a hazard source.

An illustration of the limitations of the unit-hazard coincidence method is provided in Figure 15.1, which depicts the locations of 20 industrial pollution sources across census tracts in a hypothetical county. Most of these point sources are located near the boundaries of their host tracts and close to adjacent non-host tracts. Given the spatial distribution of these industrial hazards, it appears unlikely that their adverse effects (e.g. air pollution) are confined solely to the host unit boundaries. This problem becomes more pronounced when the size

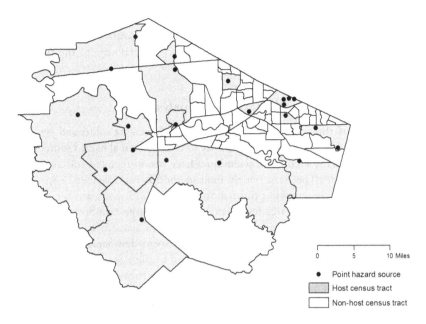

Figure 15.1 Selection of host and non-host units (census tracts) based on spatial coincidence

of census units varies substantially within a study area. An additional limitation is that all host tracts are treated equally, although the number of hazard sources within each host tract is often unequal, as shown in Figure 15.1.

The inability to distinguish between host spatial units based on the number or magnitude of hazards can be addressed by summing the number of facilities or the quantity of pollutants released within each unit. Several EJ studies have extended the basic spatial coincidence approach by estimating the frequency of toxic facilities within census tracts (Burke 1993; Fricker and Hengartner 2001) and ZIP codes (Ringquist 1997), as well as the number of airborne toxic releases within counties (Cutter and Solecki 1996). Since certain databases such as the US Environmental Protection Agency (EPA)'s Toxic Release Inventory (TRI) provide detailed data on annual quantities of toxic chemicals released at each facility, a more refined assessment of the magnitude of pollution released within each host unit is possible. While several EJ studies have relied on the pounds of emitted pollutants from industrial facilities (Boer et al. 1997; Daniels and Friedman 1999), others have used toxicity or risk indicators to weight annual releases within each spatial unit (Bowen et al. 1995; Perlin et al. 1995; McMaster et al. 1997; Brooks and Sethi 1997; Kershaw et al. 2013; Chakraborty and Green 2014). Since public databases such as the TRI (US), National Pollutant Release Inventory (Canada), and National Pollutant Inventory (Australia) do not include toxicity data for released chemicals, researchers have utilized surrogate measures such as threshold limit values (TLVs) or toxic equivalency potential (TEP) scores to weight the quantity of emissions for each pollutant. Although TLVs or TEPs are available for many industrial chemicals, it remains a problematic measure for risk assessment because these are intended to only assess occupational safety among a healthy worker population (Maantay 2002).

The incorporation of data on the quality and quantity of pollution emitted from each hazard source has allowed researchers to distinguish between host spatial units on the basis of the magnitude of potential environmental risk and improve upon the basic unit-hazard coincidence method that examines the mere presence of hazards. Applications of spatial coincidence analysis that utilize emissions or toxicity data, however, are still limited by their inability to consider the exact geographic location of the hazard within the host spatial unit, and determine the geographic boundaries of potential exposure to the risks posed by the hazard.

Distance-based analysis

Several limitations of the spatial coincidence approach can be addressed by measuring risk exposure on the basis of distance or proximity to environmental hazard sources. A variety of simple and advanced distance-based techniques have been suggested and implemented in the EJ research literature. The most widely used method is *buffer analysis*, a spatial analytic technique provided by geographic information systems (GIS) software for creating new polygons around point, line, or area features on a map. A large number of EJ studies have used GIS-based circular buffers around point sources of hazards to identify areas and populations exposed to their environmental risks. The socio-demographic characteristics of areas lying inside buffer zones are statistically compared to the rest of the study area (outside the buffers) to determine disproportionate exposure to the hazards of concern. Figure 15.2 provides a typical example of buffer analysis, based on circles of radii two miles centered at each pollution source in the county. The underlying census tracts can be used to estimate the social characteristics of the population residing within these buffer zones.

The radii of circular buffers in EJ studies have ranged from 100 yards (Sheppard et al. 1999) to 3 miles (Mohai and Saha 2006). Distances of 0.5 and 1.0 mile from facilities of concern have been used most frequently (US GAO 1995; Chakraborty and Armstrong 1997; Neumann et al. 1998; Bolin et al. 2000, 2002; Baden and Coursey 2002; Maantay 2007; Mohai and Saha 2007; Kearney and Kiros 2009; Konisky and Schario 2010). Although most applications of buffer analysis in EJ research have focused on point sources, linear or network

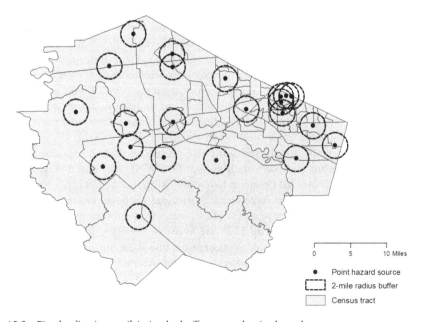

Figure 15.2 Fixed radius (two-mile) circular buffers around point hazard sources

buffers have also been utilized to estimate areas and people exposed to pollution from roadways and railways (Jacobson et al. 2005; Chakraborty and Zandbergen 2007; Maantay 2007; Margai 2010).

Buffer analysis for EJ research has undergone a number of enhancements in the last two decades. Instead of limiting their analysis to a single radius or buffer, some EJ studies have constructed several circular rings at increasing distances from point sources of hazards (Neumann et al. 1998; Perlin et al. 1999; Sheppard et al. 1999; Atlas et al. 2002; Pastor et al. 2004; Walker et al. 2005; Sadd et al. 2011; Wilson et al. 2012). The use of multiple concentric buffers has allowed researchers to investigate how the choice of the distance metric influences the results of EJ analysis. To explicitly account for spatial overlap of buffers generated from different sources, some EJ studies have estimated a hazard density index which gives each spatial unit in the study area a quantitative score based on aggregating the areas of circular buffer zones that are intersected by the spatial unit (Bolin et al. 2002; Grineski and Collins 2010). Measures of proximity to environmental hazard sources have also been enhanced by using GIS software to generate continuous statistical surfaces of distance to the nearest hazardous facility and calculate aggregated statistics of the surface within each spatial unit (Mennis 2002; Mennis and Jordan 2005; Downey 2006a; Raddatz and Mennis 2013; Chakraborty et al. 2014). To distinguish between buffers on the basis of the magnitude of environmental risk, some EJ studies have summed pollutant concentrations, release volumes, or toxicity-weighted emissions within a fixed radius of each pollution source in the study area (Neumann et al. 1998; Harner et al. 2002; Downey and Crowder 2011; Kershaw et al. 2013).

Circular buffer analysis remains a widely used approach because it can be easily implemented using GIS software, provides a simple visual representation (circles centered at point sources), and makes statistical comparisons between potentially exposed (inside circle) and non-exposed (outside circle) areas and populations convenient. Additionally, it provides a more accurate spatial representation of environmental exposure than the spatial coincidence because it does not assume that the adverse effects are limited to the boundaries of host spatial units. There are, however, several limitations associated with the application of circular buffers in EJ analysis. First, the facility or hazard representing the centre of the circle is assumed to be small enough to be treated as a point. For undesirable land uses such as Superfund sites that are large in size, a circular buffer may not accurately depict the area surrounding the site if the radius is too small. Some hazardous sites need to be delineated as a polygon instead of a point and the buffer should be constructed around the polygon (Liu 2001). Although previous EJ studies have not considered this issue, the shape and size of the hazard source needs to be first examined before deciding which type of buffer is appropriate.

A second limitation is that the radius of the circular buffer is usually chosen arbitrarily and buffers around all hazard sources in a study area have the exact same radius. The nature and quantity of hazardous substances stored or released at each individual facility have rarely been incorporated in the determination of buffer radii to reflect the spatial extent of potential exposure. The operational parameters of emission releases (e.g. release height, exit velocity, exit temperature) are also not typically considered in the determination of the buffer size.

A third problem is the underlying assumption that the adverse effects of a hazard are restricted only to the specified circular area or distance, while areas outside the buffer remain unaffected. While this binary or dichotomous assumption makes comparisons convenient, the results are highly sensitive to the choice of buffer radius. A discrete measurement (e.g. within 1 mile of a facility) is also unlikely to reflect a more continuous or gradual reduction in environmental exposure with distance from the hazard. Using multiple concentric buffers can overcome this limitation to a certain degree, but the determination of the number of buffers to use and choice of radii remain ambiguous and do not necessarily result in a more accurate representation of potential exposure.

Continuous distances, based on the calculation of the exact distance between locations of the potentially exposed population and hazard sources, represent an alternative to the use of discrete buffer analysis. Several EJ studies have utilized the distance from the centroid of each census unit to their nearest hazard source as a continuous measure of potential exposure (Pollock and Vittas 1995; Gragg et al. 1996; Stretesky and Lynch 1999; Mennis 2002; Wilson et al. 2012; Raddatz and Mennis 2013). The analysis of continuous distances has also been improved by generating a cumulative distribution function (CDF) to compare specific population subgroups (e.g. White vs. non-White residents). A CDF is typically depicted as a line graph that provides the number or percentage of observations falling below every threshold value. Applied to any set of hazard sources, a CDF can be plotted to indicate how the size of a potentially exposed subgroup (as a percentage of the total population in the study area) increases with proximity from the hazard source. Several EJ studies have demonstrated that the CDFs are particularly well suited for assessing disproportionate exposure because they overcome the limitations of choosing arbitrary and discrete buffer distances (Waller et al. 1997, 1999; Chakraborty and Zandbergen 2007; Fitos and Chakraborty 2010).

Instead of assuming that the adverse impacts of a hazard decline with distance in a linear fashion, a few studies have utilized curvilinear distance decay functions to model residential proximity. Pollock and Vittas (1995) hypothesized three functional forms of exposure (linear, square root, and natural logarithm) with respect to distance from TRI facilities in Florida, and selected the natural logarithm of the distance to the nearest facility as a proxy for exposure. A GIS-based distance decay modelling technique was developed by Downey (2006b) and applied to examine proximity to TRI facilities in Detroit. While this technique was found to be flexible enough to incorporate any appropriate distance decay function, several different curvilinear and reverse curvilinear functions were used to estimate neighbourhood proximity to TRI activity. More recently, Downey and Crowder (2011) used a 1.5-mile curvilinear distance decay function to calculate proximity indicators that were summed to assess tract level exposure to TRI emissions.

An inherent limitation of this distance decay approach is that researchers are unaware of the actual and precise rate at which the negative impacts of an environmental hazard decline with increasing distance (distance decay rate). The mathematical functions used to calculate distance decay are typically based on assumptions about the distance decay process rather than on precise knowledge of the process (Downey 2006b). Regardless of the distance decay rate or function utilized, it is important to consider that distance serves only as a surrogate for environmental exposure and the actual extent of exposure may not be a simple function of distance. In reality, distance decay functions are likely to vary substantially based on circumstances of release, types and quantities of substances released, and local meteorological conditions (e.g. wind, temperature, topography).

Although distance-based approaches for EJ analysis have evolved from the use of discrete circular buffers to continuous functions, they are still limited by the fact that proximity may not always provide a valid proxy for exposure to risk. Additionally, distance-based methods fail to consider directional biases in the distribution of environmental risks by assuming that their adverse effects are identical and uniform in all directions. Although physical processes do not always operate in a symmetrical or isotropic manner, distance-based analyses assume that environmental exposure is not dependent on local wind direction and other factors influencing the movement and dispersal of pollutants.

Plume-based analysis

To provide a more accurate spatial representation of the area potentially exposed to environmental risks, several EJ studies have used detailed information on toxic chemical emissions and local meteorological conditions to model the environmental fate and dispersal of pollutants released from the hazard source. *Geographic plume analysis* is a methodology that integrates environmental dispersion modelling with GIS to estimate areas and populations exposed to airborne releases of toxic substances (Chakraborty and Armstrong 1997). Air dispersion models typically combine data on the volume and physical properties of a released chemical with site-specific information and atmospheric conditions to estimate pollutant concentrations downwind from the emission source. This information is used to identify the spatial extent or boundary of the area potentially exposed to the chemical's spreading plume, or the plume footprint. The footprint represents the area where ground-level concentrations of the pollutant are predicted to exceed a user-specified limit or threshold value (Figure 15.3).

EJ studies using geographic plume analysis have often relied on ALOHA (Areal Locations of Hazardous Atmospheres), a dispersion model developed by the US National Oceanic and Atmospheric Administration (NOAA) and the EPA to support emergency responses to hazardous chemical accidents. The ALOHA model has been applied to generate, at each facility of concern, a single plume footprint (Chakraborty and Armstrong 1996), a composite footprint based on historical weather patterns (Chakraborty and Armstrong 1997, 2004), or plume-based circular buffers whose radii are based on worst-case chemical release scenarios (Chakraborty 2001; Margai 2001; Chakraborty and Armstrong 2001; Margai and Barry 2011). EJ studies have also utilized the Industrial Source Complex Short Term (ISCST) air dispersion model (Dolinoy and Miranda 2004; Fisher et al. 2006; Maantay et al. 2009), ash

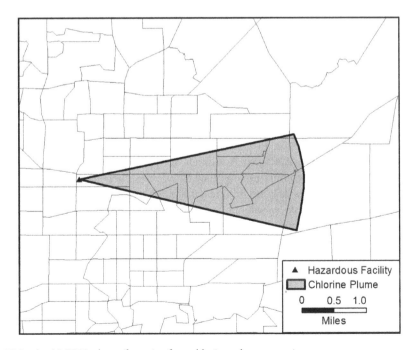

Figure 15.3 An ALOHA plume footprint for a chlorine release scenario

deposition models (Bevc et al. 2007), noise pollution models (Most et al. 2004), and the Community Multiscale Air Quality model (Gray et al. 2013, 2014).

Air dispersion modelling allows the concentration of toxic pollutants released from a hazard source and their estimated risks to decrease continuously with increasing distance from the release source, as well as vary according to compass direction. Plume-based analysis thus addresses the problems of previous approaches which assume that residing in either a census unit containing a hazard (spatial coincidence analysis) or within a specific distance from a hazard (distance-based analysis) leads to environmental exposure. There are, however, certain limitations associated with this approach. First, dispersion models typically require large volumes of site-specific and facility-specific information, such as the facility's stack height and diameter, gas exit velocity and exit temperature, detailed emissions data on each chemical released (e.g. average hourly quantities and rates), and meteorological data (e.g. average monthly or hourly wind speed and direction). The input information necessary for air dispersion modelling is rarely available for all pollution sources in a study area. Second, some dispersion models such as ALOHA assume that topography is always flat and are unable to provide accurate concentration estimates when the atmosphere is stable or wind speeds are low. Third, the creation of a comprehensive data set using plume-based analysis that includes all pollution sources and chemical emissions in a study area can be difficult, time-consuming, and expensive. Consequently, few regional or metropolitan scale plume modelling-based datasets have been constructed and those that exist focus only on specific types of hazards, as discussed in the next section.

Cumulative exposure analysis

In recent years, EJ researchers have argued that conventional chemical-specific and source-specific assessments of environmental exposure are unlikely to reflect the multiple environmental stressors faced by vulnerable communities that can act additively or synergistically to increase adverse health risks (Sexton and Linder 2010). Consequently, several EJ studies have developed numerical indicators for assessing cumulative impacts and environmental disparities based on aggregating variables that represent hazard proximity, air pollution exposure, health risk, and social vulnerability (Sadd et al. 2011; Huang and London 2012; Meehan et al. 2012). Quantitative EJ research in the last decade has increasingly relied on public datasets that provide modelled estimates of cumulative exposure to specific pollutants and related public health risks. Two national scale databases that have been used extensively for EJ analysis in the US include the EPA's Risk-Screening Environmental Indicators (RSEI) model and the National-Scale Air Toxic Assessment (NATA). These datasets are particularly appropriate for EJ research not only because they allow researchers to estimate the potential health risks associated with specific environmental hazards and spatial units, but also because the exposure analysis and risk assessment techniques used to derive these data consider factors such as wind speed, wind direction, air turbulence, smokestack height, and the rate of chemical decay and deposition (Chakraborty et al. 2011).

The RSEI model can be used to estimate potential human health risks from air pollutants based on toxicity and atmospheric dispersion of chemicals emitted by facilities in the EPA's TRI database. For each individual TRI site and pollutant, the RSEI integrates information on the facility location, the quantity and toxicity of the chemical, fate and transport through the environment, the route and extent of human exposure, and the number of people affected for up to 44 miles (101 km) from the source of release. The ambient concentration of each TRI pollutant is determined for each square kilometer of the 101 km by 101 km grid in which the facility is centered. EJ studies have merged risk scores from the RSEI grids with census socio-demographic data to analyse disproportionate exposure to TRI pollutants in the entire US (Ash and Fetter 2004; Downey 2007; Downey and Hawkins 2008; Zwickl et al. 2014) and in urban areas such as Milwaukee (Collins 2011), Philadelphia (Sicotte and Swanson 2007), St. Louis (Abel 2008), and Seattle (Abel and White 2011). The RSEI public release data, however, only provide facility-specific hazard and risk scores for each TRI facility. Disaggregated information on individual grid cells impacted by the toxic releases from TRI facilities is available in the RSEI Geographic Microdata (RSEI-GM), which provides exposure and risk estimates for each square grid cell in the entire US, with each cell value incorporating information from multiple TRI facilities. Although the disaggregated data are not released publicly due to their large size and complexity, the EPA has made the RSEI-GM data available to the research community.

Since the pollution plumes used to obtain the risk estimates can extend in any direction for up to 101 km from a TRI facility, the RSEI modelling technique had the advantage of allowing hazards and emissions in a particular spatial unit to affect people living in other units. Additionally, the spatial resolution of the latest version of the RSEI-GM database (810 m by 810 m grids) exceeds that of other US databases. However, several simplifying assumptions have to be made to model pollutant concentrations for thousands of TRI facilities and hundreds of thousands of releases across the entire US. Each facility in the RSEI database, for example, is given a single smokestack height estimate that is often based on the median smokestack height for an entire industry. Additionally, the RSEI model assumes constant emissions rates and uses chemical decay estimates that are not necessarily accurate (Downey 2006b, 2007).

The EPA's NATA, designed to guide air pollution reduction and related prioritization efforts, has emerged as a reliable data source for estimating exposure concentrations and public health risks associated with inhalation of hazardous air pollutants (HAPs) from different types of emission sources. While criteria air pollutants include common contaminants such as particulate matter, sulphur dioxide, nitrogen oxides, ozone, carbon monoxide, and lead, HAPs (also known as air toxics) include 188 specific substances identified in the Clean Air Act Amendments of 1990 that are known to cause, or suspected of causing, cancer and other serious health problems, including respiratory, neurological, immune, or reproductive effects (EPA 2008). Census tract level estimates of lifetime cancer risk from the first NATA (1996) have been utilized for EJ analysis in Maryland (Apelberg et al. 2005), California (Pastor et al. 2005), and major metropolitan areas of the US (Morello-Frosch and Jesdale 2006). Recent EJ studies have used the 1999 or 2005 versions of the NATA database to examine the social distribution of cancer or respiratory health risks in several US metropolitan areas such as El Paso (Collins et al. 2011; Grineski et al. 2013a), Houston (Linder et al. 2008; Chakraborty et al. 2014), Miami (Grineski et al. 2013b; Collins et al. 2015) and Tampa Bay (Chakraborty 2009, 2012).

An important advantage of the NATA is its spatial compatibility with census socio-demographic data: the modelled risk estimates are available at the level of spatial units (tracts) for which socio-demographic data are published by the US Census. Additionally, it provides health risk estimates of public health risks (cancer, respiratory, and neurological) for ambient exposure to multiple categories of emission sources (point, non-point, on-road and non-road mobile). The NATA thus allows EJ analysis to extend beyond major stationary sources such as TRI facilities and include smaller emitters and transportation-related pollution. However, the NATA is somewhat limited by its specific focus on only one category of air pollutants (HAPs) and one avenue of human exposure (inhalation) to HAPs. The adverse health risks of exposure to criteria air pollutants and through other pathways such as ingestion or skin contact are not included. Although the use of RSEI or NATA data represents a significant improvement over the spatial coincidence and distance-based approaches, these are often based on necessary and problematic assumptions and may not be as accurate as many researchers think.

Concluding discussion

This review has explored how the assessment of environmental exposure and risk in quantitative EJ research has evolved from comparing the prevalence of socially disadvantaged populations in geographic entities hosting hazard sources and discrete buffer zones to more sophisticated techniques that utilize continuous distance metrics, plume-based analysis, and modelled estimates of health risk from cumulative environmental exposure. Although a majority of quantitative EJ studies have focused on locational or distributional inequities, the emergence of national scale exposure databases such as the RSEI-GM and NATA have extended the range and complexity of EJ research questions that can be addressed. For example, recent studies have used the RSEI-GM data to identify the worst corporate air polluters in the US, based on the disproportionate social impacts of exposure to air releases of industrial toxic chemicals (Ash and Boyce 2011; Collins et al. 2016). Similarly, the NATA has been used to examine relationships between exposure to ambient air pollution, socioeconomic status, and adverse health outcomes (Grineski et al. 2013a; Brink et al. 2014), as well as the EJ implications of hazard awareness and residential decision-making priorities at the household level (Collins et al. 2015).

In spite of significant improvements in data availability and methodology, the spatial representation and quantification of environmental risk for EJ research still remain constrained by several limitations that could be addressed in future work. While the analytic focus has been on the assessment of exposure to pollution and related health risks, it is important to consider that these may not be the only set of negative impacts imposed by environmental hazards. The presence of a hazard can also result in psychological stress, social stigmatization, property value decline, and loss of employment, in addition to adversely affecting sense of community and local economic activity (Downey 2007). Since these effects cannot be analysed on the basis of plume-based techniques or environmental datasets discussed in this chapter, there is a growing need to explore alternative data sources and methodologies for addressing community vulnerability to environmental hazards.

Another shortcoming of current approaches and databases is that these are designed primarily to assess exposure to air pollution. More research is necessary to develop methods and models for analysing the EJ consequences of exposure to water pollution, which follows more complex patterns of dispersion that are more difficult to model than air emissions. When a pollutant is carried away downstream by a running body of water, dilution can be complicated by factors such as the presence of other entering stream segments, biological and chemical interactions, deposition to sediments, and the treatment and removal of a water pollutant sent to a publicly owned treatment plant. Issues of potential human exposure via drinking water, contact with contaminated soils, and indoor air pollutants have also received limited attention in quantitative EJ research. Additional research into these areas could provide analysts with better data, methods, and tools to investigate adverse and disproportionate exposure to their impacts. To address current gaps in the assessment of exposure and risk, future EJ research should continue to explore the use of both primary and secondary data sources, geographically detailed datasets, and emerging pollution modelling techniques, in conjunction with geospatial technologies such as GIS, global positioning systems (GPS), and remote sensing. More individual and household level analysis is also required to better understand and document environmentally unjust outcomes, as well as formulate equitable environmental policies.

References

Abel, T.D. (2008). "Skewed riskscapes and environmental injustice: A case study of metropolitan St. Louis." *Environmental Management*, vol. 42, no. 2, pp. 232–248.

Abel, T.D. and White, J. (2011). "Skewed riskscapes and gentrified inequities: Environmental exposure disparities in Seattle, Washington." *American Journal of Public Health*, vol. 101, no. S1, pp. S246–S254.

Anderton, D.L., Anderson, A.B., Oakes, J.M. and Fraser, M.R. (1994). "Environmental equity: The demographics of dumping." *Demography*, vol. 31, no. 2, pp. 229–248.

Apelberg, B.J., Buckley, T.J. and White, R.H. (2005). "Socioeconomic and racial disparities in cancer risk from air toxics in Maryland." *Environmental Health Perspectives*, vol. 113, no. 6, pp. 693–699.

Ash, M. and Boyce, J.K. (2011). "Measuring corporate environmental justice performance." *Corporate Social Responsibility and Environmental Management*, vol. 18, no. 2, pp. 61–79.

Ash, M. and Fetter, T.R. (2004). "Who lives on the wrong side of the environmental tracks? Evidence from the EPA's risk-screening environmental indicators model." *Social Science Quarterly*, vol. 78, pp. 793–810.

Atlas, M. (2002). "Few and far between? An environmental equity analysis of the geographic distribution of hazardous waste generation." *Social Science Quarterly*, vol. 83, no. 1, pp. 365–378.

Baden, B.M. and Coursey, D.L. (2002). "The locality of waste sites within the city of Chicago: A demographic, social, and economic analysis." *Resource and Energy Economics*, vol. 24, no. 1, pp. 53–93.

Baden, B.M., Noonan, D.S. and Turaga, R.M.R. (2007). "Scales of justice: Is there a geographic bias in environmental equity analysis?" *Journal of Environmental Planning and Management*, vol. 50, no. 2, pp. 163–185.

Been, V. (1995). "Analyzing evidence of environmental justice." *Journal of Land Use & Environmental Law*, vol. 11, no. 1, pp. 1–6.

Bevc, C.E., Marshall, B.K. and Picou, J.S. (2007). "Environmental justice and toxic exposure: Toward a spatial model of physical health and psychological well-being." *Social Science Research*, vol. 36, no. 1, pp. 48–7.

Boer, J.T., Pastor, M., Sadd, J.L. and Snyder, L.D. (1997). "Is there environmental racism? The demographics of hazardous waste in Los Angeles County." *Social Science Quarterly*, vol. 78, no. 4, pp. 793–810.

Bolin, B., Matranga, E., Hackett, E.J., Sadalla, E.K., Pijawka, K.D., Brewer, D. and Sicotte, D. (2000). "Environmental equity in a Sunbelt city: The spatial distribution of toxic hazards in Phoenix, Arizona." *Global Environmental Change Part B: Environmental Hazards*, vol. 2, no. 1, pp. 11–24.

Bolin, B., Nelson, A., Hackett, E.J., Pijawka, K.D., Smith, C.S., Sicotte, D., et al. (2002). "The ecology of technological risk in a sunbelt city." *Environment and Planning A*, vol. 34, no. 2, pp. 317–339.

Bowen, W.M., Salling, M.J., Haynes, K.E. and Cyran, E.J. (1995). "Toward environmental justice: Spatial equity in Ohio and Cleveland." *Annals of the Association of American Geographers*, vol. 85, no. 4, pp. 641–663.

Brink, L.L, Benson, S.M., Marshall, L.P. and Talbott, E.O. (2014). "Environmental inequality, adverse birth outcomes, and exposure to ambient air pollution in Allegheny County, PA, USA." *Journal of Racial and Ethnic Health Disparities*, vol. 1, no. 3, pp. 157–162.

Brooks, N. and Sethi, R. (1997). "The distribution of pollution: Community characteristics and exposure to air toxics." *Journal of Environmental Economics and Management*, vol. 32, no. 2, pp. 233–250.

Burke, L.M. (1993). "Environmental equity in Los Angeles (93-6)." National Center for Geographic Information and Analysis (NCGIA): Technical Report 93-6, Santa Barbara, CA.

Chakraborty, J. (2001). "Acute exposure to extremely hazardous substances: An analysis of environmental equity." *Risk Analysis*, vol. 21, no. 5, pp. 883–894.

Chakraborty, J. (2009). "Automobiles, air toxics, and adverse health risks: Environmental inequities in Tampa Bay, Florida." *Annals of the Association of American Geographers*, vol. 99, no. 4, pp. 674–697.

Chakraborty, J. (2012). "Cancer risk from exposure to hazardous air pollutants: Spatial and social inequities in Tampa Bay, Florida." *International Journal of Environmental Health Research*, vol. 22, no. 2, pp. 165–183.

Chakraborty, J. and Armstrong, M.P. (1996). "Using geographic plume analysis to assess community vulnerability to hazardous accidents." *Computers, Environment and Urban Systems*, vol. 19, no. 5, pp. 341–356.

Chakraborty, J. and Armstrong, M.P. (1997). "Exploring the use of buffer analysis for the identification of impacted areas in environmental equity assessment." *Cartography and Geographic Information Systems*, vol. 24, no. 3, pp. 145–157.

Chakraborty, J. and Armstrong, M.P. (2001). "Assessing the impact of airborne toxic releases on populations with special needs." *The Professional Geographer*, vol. 53, no. 1, pp. 119–131.

Chakraborty, J. and Armstrong, M.P. (2004). "Thinking outside the circle: Using geographical knowledge to focus environmental risk assessment investigations." In Janelle, D., Warf, B. and Hansen, K. (ed.) *World Minds: Geographical Perspectives on 100 Problems*. Dordrecht, The Netherlands: Kluwer Academic Publications.

Chakraborty, J. and Zandbergen, P.A. (2007). "Children at risk: Measuring racial/ethnic disparities in potential exposure to air pollution at school and home." *Journal of Epidemiology and Community Health*, vol. 61, no. 12, pp. 1074–1079.

Chakraborty, J., Maantay, J.A. and Brender, J.D. (2011). "Disproportionate proximity to environmental health hazards: Methods, models, and measurement." *American Journal of Public Health*, vol. 101, no. (S1), pp. S27–S36.

Chakraborty, J., Collins, T.W., Grineski, S.E., Montgomery, M.C. and Hernandez, M. (2014). "Comparing disproportionate exposure to acute and chronic pollution risks: A case study in Houston, Texas." *Risk Analysis*, vol. 34, no. 11, pp. 2005–2020.

Chakraborty, J. and Green, D. (2014). "Australia's first national level quantitative environmental justice assessment of industrial air pollution." *Environmental Research Letters*, vol. 9, no. 4, 044010. DOI: 10.1088/1748-9326/9/4/044010.

Collins, M.B. (2011). "Risk-based targeting: Identifying disproportionalities in the sources and effects of industrial pollution." *American Journal of Public Health*, vol. 101, no. S1, pp. S231–S237.

Collins, T.W., Grineski, S.E., Chakraborty, J. and McDonald, Y.J. (2011). "Understanding environmental health inequalities through comparative intracategorical analysis: Racial/ethnic disparities in cancer risks from air toxics in El Paso County, Texas." *Health & Place*, vol. 17, no. 1, pp. 335–344.

Collins, T.W., Grineski, S.E. and Chakraborty, J. (2015). "Household-level disparities in cancer risks from vehicular air pollution in Miami." *Environmental Research Letters*, vol. 11, 015004 DOI: 10.1088/1748-9326/10/9/095008.

Collins, M.B., Munoz, I. and JaJa, J. (2016). "Linking 'toxic outliers' to environmental justice communities." *Environmental Research Letters*, vol. 11, 015004. DOI: 10.1088/1748-9326/11/1/015004.

Cutter, S.L. and Solecki, W.D. (1996). "Setting environmental justice in space and place: Acute and chronic airborne toxic releases in the southeastern United States." *Urban Geography*, vol. 17, no. 5, pp. 380–399.

Cutter, S.L., Holm, D. and Clark, L. (1996). "The role of geographic scale in monitoring environmental justice." *Risk Analysis*, vol. 16, no. 4, pp. 517–526.

Daniels, G. and Friedman, S. (1999). "Spatial inequality and the distribution of industrial toxic releases: Evidence from the 1990 TRI." *Social Science Quarterly*, vol. 80, pp. 244–262.

Dolinoy, D.C. and Miranda, M.L. (2004). "GIS modelling of air toxics releases from TRI-reporting and non-TRI-reporting facilities: Impacts for environmental justice." *Environmental Health Perspectives*, vol. 112, no. 17, pp. 1717–1724.

Downey, L. (2006a). "Using geographic information systems to reconceptualize spatial relationships and ecological context." *American Journal of Sociology*, vol. 112, no. 2, pp. 567–612.

Downey, L. (2006b). "Environmental racial inequality in Detroit". *Social Forces*, vol. 85, no. 2, pp. 771–796.

Downey, L. (2007). "US metropolitan-area variation in environmental inequality outcomes." *Urban Studies*, vol. 44, no. 5, pp. 953–977.

Downey, L. and Crowder, K. (2011). "Using distance decay techniques and household-level data to explore regional variation in environmental inequality." In Maantay, J. and McLafferty, S. (eds.) *Geospatial Analysis of Environmental Health*. New York: Springer.

Downey, L. and Hawkins, B. (2008). "Race, income, and environmental inequality in the United States". *Sociological Perspectives*, vol. 51, no. 4, pp. 759–781.

Elliott, M.R., Wang, Y., Lowe, R.A. and Kleindorfer, P.R. (2004). "Environmental justice: frequency and severity of US chemical industry accidents and the socioeconomic status of surrounding communities." *Journal of Epidemiology and Community Health*, vol. 58, no. 1, pp. 24–30.

Environmental Protection Agency (EPA). (2008). *Health Effects Notebook for Hazardous Air Pollutants.* Retrieved from: www.epa.gov/ttn/atw/hlthef/hapindex.html.

Fisher, J.B., Kelly, M. and Romm, J. (2006). "Scales of environmental justice: Combining GIS and spatial analysis for air toxics in West Oakland, California." *Health & Place*, vol. 12, no. 4, pp. 701–714.

Fitos, E. and Chakraborty, J. (2010). "Race, class, and wastewater pollution: Spatial and environmental injustice in an American metropolis: A study of Tampa Bay, Florida." In Chakraborty, J. and Bosman, M.M. (ed). *An American Metropolis: A Study of Tampa Bay, Florida.* Amherst, NY: Cambria Press.

Fricker, R.D. and Hengartner, N.W. (2001). "Environmental equity and the distribution of toxic release inventory and other environmentally undesirable sites in metropolitan New York City." *Environmental and Ecological Statistics*, vol. 8, no. 1, pp. 33–52.

Glickman, T.S. (1994). "Measuring environmental equity with Geographical Information Systems." *Renewable Resources Journal*, vol. 12, pp. 17–21.

Goldman, B.A. and Fitton, L.J. (1994). *Toxic Wastes and Race Revisited: An Update of the 1987 Report on the Racial and Socioeconomic Characteristics of Communities with Hazardous Waste Sites.* Washington DC, Center for Policy Alternatives.

Gragg, R.D., Christaldi, R.A., Leong, S. and Cooper, M. (1996). "The location and community demographics of targeted environmental hazardous sites in Florida." *Journal of Land Use & Environmental Law*, vol. 12, no. 1, pp. 1–24.

Gray, S.C., Edwards, S.E. and Miranda, M.L. (2013). "Race, socioeconomic status, and air pollution exposure in North Carolina." *Environmental Research*, no. 126, pp. 152–158.

Gray S.C., Edwards, S.E., Schultz, B.D. and Miranda, M.L. (2014). "Assessing the impact of race, social factors and air pollution on birth outcomes: A population-based study." *Environmental Health* 2014, vol. 13, p. 4.

Grineski, S.E. and Collins, T.W. (2010). Environmental injustices in transnational context: Urbanization and industrial hazards in El Paso/Ciudad Juárez. *Environment and Planning A*, vol. 42, no. 6, pp. 1308–1327.

Grineski, S.E., Collins, T.W., Chakraborty, J. and McDonald, Y.J. (2013a). "Environmental health injustice: Exposure to air toxics and children's respiratory hospital admissions in El Paso, Texas." *The Professional Geographer*, vol. 65, no. 1, pp. 31–46.

Grineski, S.E., Collins, T.W. and Chakraborty, J. (2013b). "Hispanic heterogeneity and environmental injustice: Intra-ethnic patterns of exposure to cancer risks from traffic-related air pollution in Miami." *Population and Environment*, vol. 35, no. 1, pp. 26–44.

Harner, J., Warner, K., Pierce, J. and Huber, T. (2002). "Urban environmental justice indices." *The Professional Geographer*, vol. 54, no. 3, pp. 318–331.

Hird, J.A. (1993). "Environmental policy and equity: The case of Superfund." *Journal of Policy Analysis and Management*, vol. 12, no. 2, pp. 323–343.

Huang, G. and London, J.K. (2012). "Cumulative environmental vulnerability and environmental justice in California's San Joaquin Valley." *International Journal of Environmental Research and Public Health*, vol. 9, no. 5, pp. 1593–1608.

Jacobson, J.O., Hengartner, N.W. and Louis, T.A. (2005). "Inequity measures for evaluations of environmental justice: a case study of close proximity to highways in New York City." *Environment and Planning A*, vol. 37, no. 1, pp. 21–43.

Kearney, G. and Kiros, G.E. (2009). "A spatial evaluation of socio demographics surrounding National Priorities List sites in Florida using a distance-based approach." *International Journal of Health Geographics*, vol. 8, no. 1, p. 33.

Kershaw, S., Gower, S., Rinner, C. and Campbell, M. (2013). "Identifying inequitable exposure to toxic air pollution in racialized and low-income neighbourhoods to support pollution prevention." *Geospatial Health*, vol. 7, no. 2, pp. 265–278.

Konisky, D.M. and Schario, T.S. (2010). "Examining environmental justice in facility-level regulatory enforcement." *Social Science Quarterly*, vol. 91, no. 3, pp. 835–855.

Linder, S.H., Marko, D. and Sexton, K. (2008). "Cumulative cancer risk from air pollution in Houston: Disparities in risk burden and social disadvantage." *Environmental Science & Technology*, vol. 42, no. 12, pp. 4312–4322.

Liu, F. (2001). *Environmental Justice Analysis: Theories, Methods, and Practice*. New York: CRC Press.

Maantay, J. (2002). "Mapping environmental injustices: Pitfalls and potential of geographic information systems in assessing environmental health and equity." *Environmental Health Perspectives*, vol. 110, Suppl 2, pp. 161–171.

Maantay, J.A. (2007). "Asthma and air pollution in the Bronx: Methodological and data considerations in using GIS for environmental justice and health research." *Health and Place*, vol. 13, no. 1, pp. 32–56.

Maantay, J.A., Tu, J. and Maroko, A.R. (2009). "Loose-coupling an air dispersion model and a geographic information system (GIS) for studying air pollution and asthma in the Bronx, New York City." *International Journal of Environmental Health Research*, vol. 19, no. 1, pp. 59–79.

Margai, F.L. (2001). "Health risks and environmental inequity: A geographical analysis of accidental releases of hazardous materials." *The Professional Geographer*, vol. 53, no. 3, pp. 422–434.

Margai, F.M. (2010). "Toxic chemicals: Disparate patterns of exposure and health outcomes." In *Environmental Health Hazards and Social Justice: Geographical Perspectives on Race and Class Disparities*. Washington DC: Earthscan, pp. 114–138.

Margai, F.M. and Barry, F.B. (2011). "Global geographies of environmental injustice and health: A case study of illegal hazardous waste dumping in Cote d'Ivoire." In Maantay, J. and McLafferty, S. (eds.) *Geospatial Analysis of Environmental Health*. New York: Springer.

McMaster, R.B., Leitner, H. and Sheppard, E. (1997). "GIS-based environmental equity and risk assessment: Methodological problems and prospects." *Cartography and Geographic Information Systems*, vol. 24, no. 3, pp. 172–189.

Meehan, A.L., Faust, J.B., Cushing, L., Zeise, L. and Alexeeff, G.V. (2012). "Methodological considerations in screening for cumulative environmental health impacts: Lessons learned from a pilot study in California." *International Journal of Environmental Research and Public Health*, vol. 9, no. 9, pp. 3069–3084.

Mennis, J. (2002). "Using geographic information systems to create and analyse statistical surfaces of population and risk for environmental justice analysis." *Social Science Quarterly*, vol. 83, no. 1, pp. 281–297.

Mennis, J.L. and Jordan, L. (2005). "The distribution of environmental equity: Exploring spatial nonstationarity in multivariate models of air toxic releases." *Annals of the Association of American Geographers*, vol. 95, no. 2, pp. 249–268.

Mohai, P. and Saha, R. (2006). "Reassessing racial and socioeconomic disparities in environmental justice research." *Demography*, vol. 43, no. 2, pp. 383–399.

Mohai, P. and Saha, R. (2007). "Racial inequality in the distribution of hazardous waste: A national-level reassessment." *Social Problems*, vol. 54, no. 3, pp. 343–370.

Morello-Frosch, R. and Jesdale, B.M. (2006). "Separate and unequal: Residential segregation and estimated cancer risks associated with ambient air toxics in US metropolitan areas." *Environmental Health Perspectives*, vol. 114, no. 3, pp. 386–393.

Most, M.T., Sengupta, R. and Burgener, M.A. (2004). "Spatial scale and population assignment choices in environmental justice analyses." *The Professional Geographer*, vol. 56, no. 4, pp. 574–586.

Neumann, C.M., Forman, D.L. and Rothlein, J.E. (1998). "Hazard screening of chemical releases and environmental equity analysis of populations proximate to toxic release inventory facilities in Oregon." *Environmental Health Perspectives*, vol. 106, no. 4, pp. 217–226.

Noonan, D.S., Turaga, R.M.R. and Baden, B.M. (2009). "Superfund, hedonics, and the scales of environmental justice." *Environmental Management*, vol. 44, no. 5, pp. 909–920.

Pastor, M., Morello-Frosch, R. and Sadd, J.L. (2005). "The air is always cleaner on the other side: Race, space, and ambient air toxics exposures in California." *Journal of Urban Affairs*, vol. 27, no. 2, pp. 127–148.

Pastor, M., Sadd, J.L. and Morello-Frosch, R. (2004). "Waiting to inhale: The demographics of toxic air release facilities in 21st-century California." *Social Science Quarterly*, vol. 85, no. 2, pp. 420–440.

Perlin, S.A., Setzer, R.W., Creason, J., and Sexton, K. (1995). "Distribution of industrial air emissions by income and race in the United States: An approach using the toxic release inventory." *Environmental Science and Technology*, vol. 29, no. 1, pp. 69–80.

Perlin, S.A., Sexton, K. and Wong, D.W. (1999). "An examination of race and poverty for populations living near industrial sources of air pollution." *Journal of Exposure Analysis & Environmental Epidemiology*, vol. 9, pp. 29–48.

Perlin, S.A., Wong, D. and Sexton, K. (2001). "Residential proximity to industrial sources of air pollution: Interrelationships among race, poverty, and age." *Journal of the Air & Waste Management Association*, vol. 51, no. 3, pp. 406–421.

Pollock, P.H. and Vittas, M.E. (1995). "Who bears the burdens of environmental pollution? Race, ethnicity, and environmental equity in Florida." *Social Science Quarterly*, vol. 76, no. 2, pp. 294–310.

Raddatz, L. and Mennis, J. (2013). "Environmental justice in Hamburg, Germany." *The Professional Geographer*, vol. 65, no. 3, pp. 495–511.

Ringquist, E.J. (1997). "Equity and the distribution of environmental risk: The case of TRI facilities." *Social Science Quarterly*, vol. 78, no. 4, pp. 811–829.

Sadd, J.L., Pastor, M., Morello-Frosch, R., Scoggins, J. and Jesdale, B. (2011). "Playing it safe: Assessing cumulative impact and social vulnerability through an environmental justice screening method in the south coast air basin, California." *International Journal of Environmental Research and Public Health*, vol. 8, no. 5, pp. 1441–1459.

Sexton, K. and Linder, S.H. (2010). "The role of cumulative risk assessment in decisions about environmental justice." *International Journal of Environmental Research and Public Health*, vol. 7, no. 11, pp. 4037–4049.

Sheppard, E., Leitner, H., McMaster, R.B. and Tian, H. (1999). "GIS-based measures of environmental equity: Exploring their sensitivity and significance." *Journal of Exposure Analysis and Environmental Epidemiology*, vol. 9, pp. 18–28.

Sicotte, D. and Swanson, S. (2007). "Whose risk in Philadelphia? Proximity to unequally hazardous industrial facilities." *Social Science Quarterly*, vol. 88, no. 2, pp. 515–534.

Stretesky, P. and Lynch, M.J. (1999). "Environmental justice and the predictions of distance to accidental chemical releases in Hillsborough County, Florida." *Social Science Quarterly*, vol. 80, no. 4, pp. 830–846.

Taquino, M., Parisi, D. and Gill, D.A. (2002). "Units of analysis and the environmental justice hypothesis: The case of industrial hog farms." *Social Science Quarterly*, vol. 83, no. 1, pp. 298–316.

United Church of Christ Commission for Racial Justice. (1987). *Toxic Wastes and Race in the United States: A National Report on the Racial and Socio-economic Characteristics of Communities with Hazardous Waste Sites.* New York: United Church of Christ.

US Government Accountability Office (GAO). (1995). *Demographics of People Living Near Waste Facilities.* Washington DC: Government Printing Office.

Walker, G., Mitchell, G., Fairburn, J. and Smith, G. (2005). "Industrial pollution and social deprivation: Evidence and complexity in evaluating and responding to environmental inequality." *Local Environment*, vol. 10, no. 4, pp. 361–377.

Waller, L.A., Louis, T.A. and Carlin, B.P. (1997). "Bayes methods for combining disease and exposure data in assessing environmental justice." *Environmental and Ecological Statistics*, vol. 4, no. 4, pp. 267–281.

Waller, L.A., Louis, T.A. and Carlin, B.P. (1999). "Environmental justice and statistical summaries of differences in exposure distributions." *Journal of Exposure Analysis and Environmental Epidemiology*, vol. 9, pp. 56–65.

Wilson, S.M., Fraser-Rahim, H., Williams, E., Zhang, H., Rice, L., Svendsen, E. and Abara, W. (2012). "Assessment of the distribution of toxic release inventory facilities in metropolitan Charleston: An environmental justice case study." *American Journal of Public Health*, vol. 102, no. 10, pp. 1974–1980.

Zwickl, K., Ash, M., and Boyce, J.K. (2014). "Regional variation in environmental inequality: Industrial air toxics exposure in U.S. cities." *Ecological Economics*, vol. 107, pp. 494–509.

16

ASSESSING POPULATION AT RISK

Areal interpolation and dasymetric mapping

Juliana Maantay and Andrew Maroko

Introduction

Today, the need for visualization of population data is increasingly crucial, not just for descriptive purposes – to show the geographic extent and density of populations – but also for spatial analytical and predictive modelling purposes, in order to inform risk assessments and public policy formation on many urban issues, including environmental justice concerns (Lwin and Murayama 2011; Maantay et al. 2009; Moon and Farmer 2001).

This chapter provides a description and comparison of techniques used to estimate population counts and socio-demographic characteristics in areas potentially impacted by environmental and other hazards, or to ascertain which populations do not have adequate access to health-promoting land uses and facilities. These techniques can be used to facilitate environmental justice and health equity analyses. Environmental impacts are often spatially represented by discrete bounded areas (e.g. circular, plume, or network buffers) or continuous surfaces (e.g. outputs from dispersion modelling or land use regression). Because these impact exposure areas will normally not conflate geographically with population data and their units of aggregation, it is necessary to devise a method whereby the disparate units of analysis can be harmonized and reconciled. The challenge is to determine the optimal method to estimate the affected population, either within discrete boundaries or within a certain pollutant concentration, as derived from a continuous surface, for example.

To that end, we examine and compare the relative merits of methods including spatial coincidence/selection (such as centroid containment), and disaggregation techniques (such as areal interpolation, filtered areal weighting, and cadastral-based dasymetric mapping), using an example of a potentially impacted population living near a controversially sited waste water treatment plant in Harlem, New York.

Population estimation for environmental health and justice analysis

Geographic Information Science (GISc) methods have been used in environmental justice research primarily to analyse the spatial relationships between sources of pollution (and other environmental burdens and hazards) and the socio-demographic characteristics of potentially affected populations. More recently, they have also been used to estimate and locate populations affected by lack of geographic access to health-promoting activities and facilities, such as open spaces for physical recreation, establishments selling healthy food options, and medical providers. With GISc, we try to map instances of environmental injustice, usually by plotting the locations of facilities or land uses suspected of posing an environmental and human health hazard or risk, and then determining the racial, ethnic, and socioeconomic characteristics of the potentially affected populations, as compared to a reference population. Many of the techniques that are described below would either not be feasible to do without utilizing GISc or would be extremely cumbersome and time-consuming. Quantitative environmental justice and health equity analyses have essentially been made possible because of GISc methods.

In order to be able to estimate the counts of vulnerable sub-populations who may be adversely affected by noxious land uses or events or who suffer from lack of access to health-promoting facilities, we need to have an accurate count of people potentially impacted by environmental hazards, within the geographic extent of the hazardous conditions or at-risk locations. This can be challenging to calculate because population information is most frequently based on data aggregated by census tracts, postal ZIP codes, or counties, in the US, and equivalent administrative or government jurisdictional boundaries in other countries. It is rare that environmental and health conditions coincide with these arbitrary and usually artificial boundaries. Hence, there is a need to figure out how to parse the data, and arrive at the best representation of population distribution within non-coincidental areas.

Spatial representation of impact areas

Areas affected by environmental impacts are generally depicted in one of two ways: either as discrete areas, bounded by some distance-based buffer zone (Chakraborty and Maantay 2011); or as a continuous surface, based on pollutant concentration or dispersion. These main types of spatial representation are listed and illustrated below, and discussed in more detail in Chapter 15.

Buffers

Constant-distance buffers (fixed-distance buffers)
Example: A buffer of 2500 feet around a polluting facility, representing the likely distance that specific smokestack emissions will travel through the air (Figure 16.1a).

Concentric buffers (multiple-ring buffers)
Example: Concentric buffers around points, representing the smokestacks of industrial plants, with several different potential impact distances, which allows for a sensitivity analysis to be performed (Figure 16.1b).

Variable buffers
Example of variable point buffer: The smokestacks of polluting facilities are shown as points, and different sized impact buffers might be based on the different toxicity levels of the pollution mix emitted from the stacks of each of these noxious facilities, with the more toxic ones having larger impact buffers. The buffers in this case would not necessarily reflect the actual geographic extent of the impact, but rather be an indication of the magnitude/hazardousness of the impact (Figure 16.1c).

Example of variable line buffer: In developing a realistic buffer to indicate potential impacts from roadway vibration or noise, variability could be based on vehicular count.

Wider, busier roads would have a larger impact buffer than secondary or local roads, and the buffer around each type of road would vary accordingly (Figure 16.1d).

Plume buffers
Example: A plume buffer based on modelled outputs would show an approximation of the flow of air pollutants from a smokestack, using information about pollutant type, prevailing wind direction and wind speed at the site, exit velocity of the emissions from the stack, and other features of the pollutant and landscape morphology that would affect the geographic distribution of the pollutant (Figure 16.1e).

Network buffers
Example: A network buffer of streets would show the catchment area (or "walking-shed") of population having access to a healthy food store (or a park entrance), if we assume a normal optimal walking distance of no more than a quarter of a mile. By contrast, a fixed-distance circular buffer of a quarter of a mile around the store or park entrance would not as realistically capture the population having good access, since it would not accurately reflect actual walking distance along sidewalks or streets (Figure 16.1f).

Continuous surfaces

Environmental conditions also can be shown as a continuous surface, such as can be obtained from dispersion modelling, land use regression, or interpolated ambient environmental indicators. Example: a continuous surface of air pollution values is aggregated to the units that will be analysed, such as by taking an average of the pollutant concentrations across a census tract, which can then be linked with socio-demographic data for each tract (Figure 16.1g).

Methods of population estimation

Spatial coincidence

In this simplest method of determining the affected or impacted population (Figure 16.2a), the geographic unit either contains an environmental "good" or "bad," or it does not. If it does, then all the population within that unit is considered "affected" by the facility or land use. This is a binary analysis, and of course is quite inexact, since population that is close to the facility or land use, but on the other side of the containing boundary line would be considered not impacted, or not exposed, or without access, whereas a population within the geographic unit but far away from the facility or land use would be considered to be impacted. This method is adequate for broad-brush analyses, for instance a nation-wide study at the county level, but is not sufficiently nuanced for a more fine-grained city- or community-scale analysis.

Selection by proximity buffers (distance-based analysis)

In utilizing proximity buffers (Figure 16.2b–d), there are a few important decisions to make in order to estimate affected population within a buffer zone. The underlying population data is generally obtained from a census or other demographic database, and the data is typically aggregated by census enumeration district (tract, block group, etc.), or by postal ZIP code, county, or state, depending upon the geographic extent of the study and its level

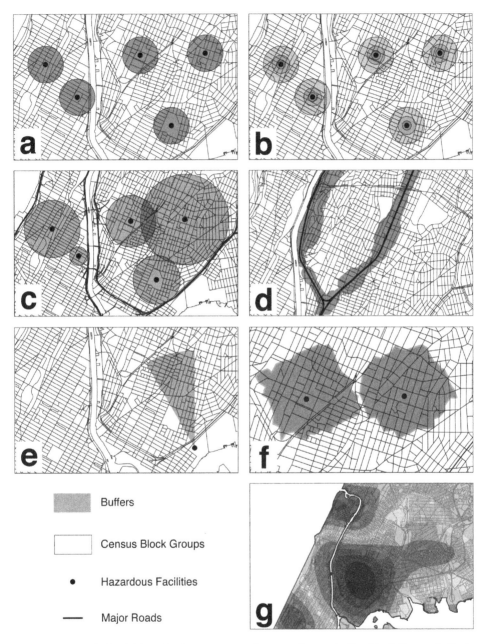

Legend:

Buffers

Census Block Groups

● Hazardous Facilities

— Major Roads

Figure 16.1   Methods of delineating impact areas: a. Constant–distance buffers (fixed–distance buffers); b. Concentric buffers (multiple–ring buffers); c. Variable point buffer; d. Variable line buffer; e. Plume buffer; f. Network buffers; g. Continuous surface

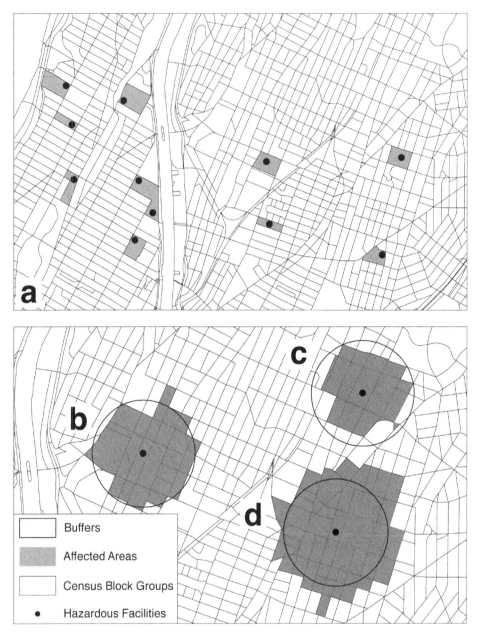

Figure 16.2 Selection of "impacted" population: a. Spatial coincidence; b. Centroid containment; c. Complete polygon containment; d. Polygon intersection

of detail. When the proximity buffers are overlaid on the unit of aggregation, they will not conflate – they will not be spatially co-incident. Some units will be partially within the buffers and only a few may lie entirely within the buffer. In order to mitigate this spatial incongruity, it will be necessary to decide which of the units (census tracts, for instance) to include in calculating the affected population. The typical way of doing this in a GIS is to select the tracts that meet certain criteria. We could select 1) Complete polygon containment: all the tracts that fall completely within the buffers; 2) Centroid containment: all the tracts whose centroid (the geometric centre of the unit) falls within the buffers; or 3) Polygon intersection: all the tracts that intersect with or are contained within the buffer. The 50 per cent areal containment method is also frequently used in EJ analyses, which entails including a polygon in the selection set if at least 50 per cent of its area is within the buffer. However, this method generally yields a very similar result to that obtained by centroid containment. Selection choice #1 is generally the most exclusive (containing the fewest tracts and therefore the least population), while #3 is generally the most inclusive (containing the most tracts, and the maximum population). The buffers can be fixed distance, concentric (multiple-ring) buffers, variable distance buffers, or plume buffers, or based on a continuous surface, but the problem of calculating the population within the affected areas remains the same.

Areal interpolation

The problem of non-coincident spatial boundaries and the frequent need to transfer data from one set of zones to another (e.g. population within census block groups versus population within a flood zone) is a long-standing dilemma, and is not addressed well by choropleth mapping, which simply distributes the population or other data evenly throughout the spatial unit (Maantay and Maroko 2009). Typically, the issue of rectifying attribute data from different sets of spatial units is handled by a procedure called *areal interpolation* (Figure 16.3a). This very simple method redistributes the source data (e.g. population) based solely on area proportions.

In areal interpolation, we again are using buffers or a continuous surface to delineate the potential areas of exposure, access, or impact, and census data or similar for the underlying population data. But, as opposed to the selection method, we will try to apportion the population in each tract according to the amount of the tract that is within the buffer. So, for example, if the buffer intersects the tract in such a way that 20 per cent of the tract is within the buffer and 80 per cent is outside the buffer, then we would make the assumption that 20 per cent of the tract's population was within the buffer, and therefore "affected." One major deficiency of this method is that it is based on the erroneous assumption that population is distributed evenly and equally throughout the unit of aggregation. This is rarely the case, and tracts may have population primarily in one part of the tract and not in other parts.

Dasymetric mapping

The underlying concept of dasymetric mapping involves the process of disaggregating spatial data to a finer unit of analysis, using additional (or "ancillary") data to help refine locations of population or other phenomena being mapped (Holt et al. 2004; Maantay et al. 2007; Mennis 2003). Dasymetric methods range from basic to more complex in order to achieve greater accuracy and a more realistic portrayal of population distribution, and are discussed below in order of ascending complexity.

Filtered areal weighting

Filtered areal weighting (FAW; Figure 16.3b) takes into account that some parts of a census tract or other geographic unit do not have any population, and therefore when apportioning population according to the amount of the tract that is within the buffer, these unpopulated areas should be excluded. Conventionally, the areas that are excluded are parks, open spaces, and water bodies. The buffered areas are "filtered" by an ancillary data set or sets, which are used to mask out the non-populated areas. In the simplest version, a binary redistribution is used: areas are deemed either inhabited or non-inhabited. If, from the land cover data, one can infer that an area is uninhabited, no population from the population source layer would be assigned to that polygon or pixel, leaving all of the population to be distributed to the remaining areas. In essence, the ancillary data set in this example acts to mask the census tract data so that the uninhabited land is left devoid of population (Eicher and Brewer 2001). This type of mapping often utilizes land cover data from satellite images to create the filtering or masking information, in combination with areal interpolation, and is usually considered a simple form of dasymetric mapping.

A further refinement of the filtered areal weighting method is to exclude, in addition to parks, open spaces, and water bodies, those areas which are typically uninhabited or very sparsely inhabited, such as industrial or commercial areas. The locations of these areas can be ascertained by either land cover data from satellite imagery or land use or zoning maps from municipal or county sources, and then used to mask the population, as in basic areal weighting.

Cadastral-based dasymetric mapping

Cadastral-based dasymetric mapping (Figure 16.3c) is a still further refinement of filtered areal weighting. In addition to excluding areas that are typically uninhabited, it uses additional ancillary data, such as property lot (cadastral) data, containing, for instance, the number of residential units or residential square footage per lot, to redistribute spatial data in a more accurate and logical manner. Building volume as derived from building footprint data has also been used to develop a more accurate picture of population distribution (Wu et al. 2005). Details such as housing tenure, ownership, and values can also be used to assess socioeconomic characteristics of proximate households.

This is considered an improvement over the disaggregation methods used in filtered areal weighting, which can be of limited utility in urban areas, due to, among other reasons, the lack of distinction typically made between land cover and land use, and spatial resolution that is too low for the application. Complete information about land use, and more importantly, population density, is not reflected clearly in the land cover data (Forster 1985).

Moreover, in a spatially heterogeneous urban area such as New York City, simply knowing whether or not an area comprises residential land use is insufficient for calculating population, since residential buildings range from containing one household to several hundred households on one lot, and population density can vary widely even within the relatively small area of a census block group, or the even smaller city block. This type of cadastral data is now available for most urban areas in the United States and other developed countries, since its primary purpose is for property tax collection.

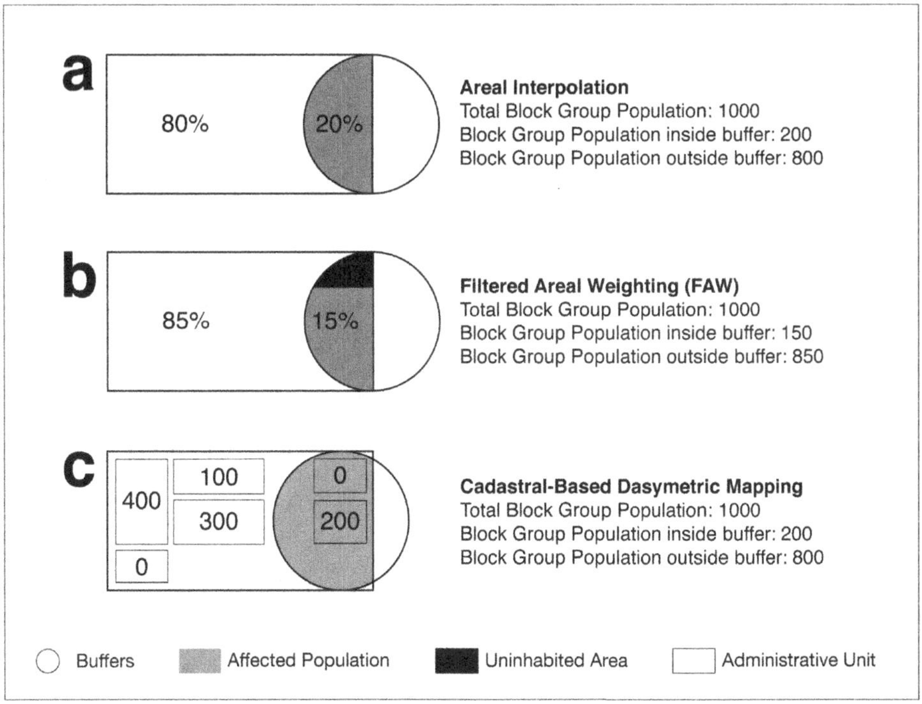
a
80%
20%
Areal Interpolation
Total Block Group Population: 1000
Block Group Population inside buffer: 200
Block Group Population outside buffer: 800
b
85%
15%
Filtered Areal Weighting (FAW)
Total Block Group Population: 1000
Block Group Population inside buffer: 150
Block Group Population outside buffer: 850
c
400
100
300
0
0
200
Cadastral-Based Dasymetric Mapping
Total Block Group Population: 1000
Block Group Population inside buffer: 200
Block Group Population outside buffer: 800
Buffers
Affected Population
Uninhabited Area
Administrative Unit

Figure 16.3 How "impacted" population numbers are estimated by various methods: a. Areal interpolation; b. Filtered areal weighting; c. Cadastral-based dasymetric mappping

Case study of the North River Waste Water Treatment Plant, New York City

In this case study, we use the environmental justice implications of the North River Waste Water Treatment Plant as an illustration of the various methods to calculate populations and sub-populations potentially affected by a specific land use (Figure 16.4). The siting of the North River Waste Water Treatment Plant (WWTP), on the Hudson [North] River in Harlem, NYC, was very contentious, and more than 20 years later is still considered to be an example of environmental injustice to the largely minority community nearby. The plant was completed in 1994, but planning for the facility began nearly a century earlier, in 1914, when the city decided that a WWTP was needed to handle the waste water of the entire west side of Manhattan. At that time, most of Manhattan's waste water was released as raw sewage into the Hudson [North] River or the East River. Site selection for the WWTP began in earnest in the late 1950s, due largely to the burgeoning population of Manhattan, accompanied by a big push in the nascent environmental movement to clean up the polluted Hudson River. A number of possible sites were considered, including one at West 72 Street, in a neighbourhood that was at that time just beginning to gentrify. There was a huge outcry from the middle- and upper-class residents against this facility being located in their neighbourhood, and a politically connected mobilization to prevent the plant from being sited there. Robert Moses, infamous as New York's "Power Broker," also helped to scuttle the plans for the West 72 Street WWTP.

The city then moved their prospects to a 28-acre, 8-block stretch between 137 and 145 Streets in the largely black community of Harlem. Design studies were started, and detailed

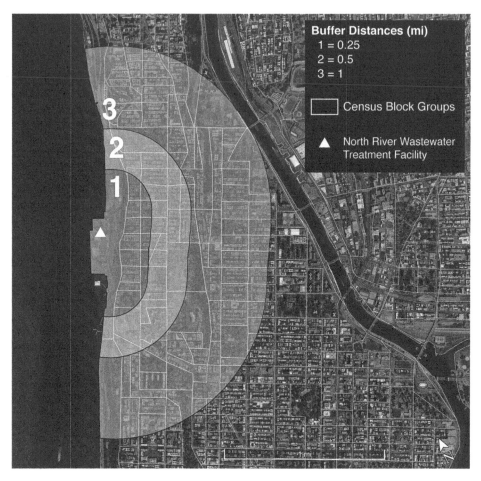

Figure 16.4 Aerial image of case study area, showing the location of the North River WWTP with concentric buffers indicating potential impact areas

plans were completed by 1971. The City Planning Commission approved of the plans long before any public hearings were held, and there was no legal requirement at that time for environmental impact assessments or public input process (Bunch 1994; Gladwell 1993; Mandell 1993; NACCHO 2011a, b). A look at the 1990 census data bears witness to the stark difference in socio-demographic characteristics between the neighbourhoods surrounding the two possible sites: The population around the originally proposed West 72 Street site was 84 per cent white and 8 per cent black, with an average household income of $123 000 per year, and 8.5 per cent of its people below the poverty line, while the population around the eventually built West 145 Street site was 15 per cent white and 60 per cent black, with an average household income of $26 000 per year, and 34 per cent of its people below the poverty line. Although New York City's government and private development interests have been dumping burdensome facilities in poor and minority communities probably since the city's 17th century founding, this was one of the first such decisions in modern times to be such a blatant and highly visible example of discriminatory siting, and one whose controversies received so much attention from the public and the press.

The plant, which is one of 14 WWTPs in NYC, handles 125 million gallons of waste water per day in dry weather, and is designed to handle 340 million gallons in wet weather (storm water is also processed in addition to waste water and sewage from buildings). To help mitigate the adverse impacts on the surrounding community, a new sports and recreation centre was built on top of the WWTP. Since the completion of the plant and the park, complaints from nearby residents about sickening odours and increased respiratory illness have continued without abating. The air quality problems stemming from the plant beneath the park have made using the park less than ideal for passive and active recreation.

> Shortly after construction, community members complained about the overpowering smell of rotten eggs coming from the plant. The smell became more potent during the summer months and travelled as far down as 120th Street and up to 157th Street—almost a two-mile distance. Residents had been forced to keep windows closed and stay inside to avoid the stench. Many have also reported itchy eyes, shortness of breath, and respiratory symptoms.
>
> *(NACCHO, 2011b)*

Comparison of methods in the case study

In order to demonstrate the potential differences in results that can be a product of population estimation techniques, concentric (multi-ring) fixed-distance buffers were created around the North River Waste Water Treatment Plant (at a quarter of a mile, half a mile, and 1 mile) and different estimation techniques were used, which included three selection methods, areal weighting, filtered areal weighting, and cadastral-based dasymetric disaggregation (Figure 16.5). Total population as well as major racial/ethnic breakdowns were estimated as "impacted" based on the methods and each method's output are compared. Demographic, planimetric, and cadastral data were from as near 2000 as possible, in order to approximate the affected populations soon after the plant came online.

Data used

Population demographics were collected from the 2000 US census at the block group level for New York County, NY (borough of Manhattan). *Block Groups* (BGs) are statistical divisions of *census* tracts, are generally defined to contain between 600 and 3000 people, and are used to present data (www.census.gov/geo/. . ./gtc_bg.html). In size, they are roughly intermediate between a census tract and a census block. Table P004 from the summary file 1 (100 per cent sample) was used to create variables for total population, Hispanic/Latino, non-Hispanic white, non-Hispanic black, non-Hispanic Asian, and non-Hispanic other (which represents residents who do not self-identify as any of the previous population groups).

Parks and open space data were downloaded from NYC open data and represent spatial features such as cemeteries, courts, and tracks as well as state and city parks. (https://data.cityofnewyork.us/Recreation/New-York-City-Open-Spaces)

Cadastral data used was from the MapPLUTO dataset, which is a "spatialized" version of Primary Land Use Tax Lot Output (PLUTO) data, provided by the NYC Department of City Planning (www.nyc.gov/html/dcp/html/bytes/archive_pluto_mappluto.shtml). This information was used for defining the number of residential units (e.g. number of apartments) in each tax lot.

The waste water treatment plant was delineated using the MapPLUTO data above and manually modified using aerial photos in order to create as accurate a boundary as possible.

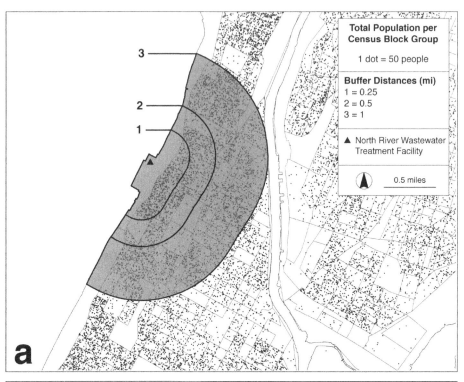

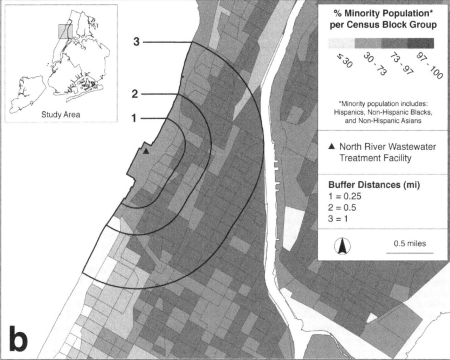

Figure 16.5   a. Case study area dot density map of total population distribution by census block group;
b. Per cent minority population by block group

Selection by proximity buffers (distance-based analysis)

Simple selection is generally the fastest method for estimating populations within specified distances from a location (in this case the water treatment plant polygon). Three selection definitions were used: Complete polygon containment: the census block groups which fall completely within the buffers (most exclusive selection method); Centroid containment: those BGs which have their centroids (geographic centres) within the buffers; and Polygon intersection: those BGs which intersect the buffers (most inclusive selection method). To demonstrate the inclusive and exclusive nature of each of these methods, the half-mile buffer resulted in 23, 29, and 39 block groups being selected using the complete polygon containment, centroid containment, and polygon intersection methods, respectively.

Once the impacted block groups are identified, the demographic groups are simply summed, which results in estimates of the populations of interest which meet the criteria as described above. Not surprisingly, population estimates range with respect to method. Total population within half a mile was estimated as approximately 48 000, 57 000, and 71 000 for the complete polygon containment, centroid containment, and polygon intersection methods, respectively.

Areal interpolation

In order to perform areal interpolation in this case study, a number of steps were taken to prepare the data. First, the areas of each block group were calculated (A_0). Next, the block group polygons were combined with the buffers using the "union" function in ArcGIS. This results in the block group polygons which are intersected by the buffers being split, however the block groups which are either completely within the buffers, or completely outside of the buffers, will be unaffected. Areas were then recalculated to represent the potential new shapes (A_1). Finally, a simple formula is used which redistributes the original population (P_0) to new population estimates (P_1) based on the area ratios. It can be written as

$$P_1 = P_0 * A_1/A_0$$

Using this method, if $A_1 = A_0$, then $P_1 = P_0$. However, if A_1 is, for instance, half as large as A_0, then P_1 will also be half as large as P_0. When compared with the selection methods above, the estimates generated by areal interpolation tend to be most similar to the centroid method, with outputs falling in between the more inclusive and exclusive techniques.

Filtered areal weighting

Filtered areal weighting is similar conceptually to simple areal interpolation; however, selected unpopulated areas are removed in order to get a more accurate estimate of where the population actually resides. In this case study, parks and open spaces were masked (filtered) from the census block group polygons using geoprocessing functions in ArcGIS before the area of the block groups (A_0) is calculated. What follows is functionally identical to simple areal interpolation, with the (filtered) block groups being split by the buffers, new areas being calculated (A_1), and finally the new populations being estimated by multiplying the original block group population by the ratio of the filtered areas. In this case study, the outputs are similar to the simple areal interpolation, with slightly larger estimates at the half-mile buffer. However, depending on the characteristics of the region being studied, the units of aggregation used in the analysis, and the buffer size, the results could show much larger differences between the methods.

The refined areal weighting method was also undertaken for the case study area, but in this instance the results were very similar to the cadastral-based dasymetric disaggregation method (see below). It should be noted that these two methods do not always yield similar results, depending upon the data used in the particular case, and the two methods might show quite different results.

Cadastral-based dasymetric disaggregation

The final estimation technique that will be demonstrated in this case study is also the most labour- and data-intensive. By assuming that population distribution is biased by an ancillary variable other than land area, in this case number of residential units, we should be able to achieve a more robust and nuanced estimate, and one which more accurately reflects actual conditions. In order to achieve this, cadastral data from MapPLUTO aggregated to the tax lot (cadastre) was overlain with the block group polygon layer as well as the buffer layers. Tax lots which fall within the block group and buffer layers were given their attribute information (buffer distances, block group identifiers, demographics, etc.). The total number of residential units in each block group was then calculated and appended to each tax lot. Next, a population estimate was created for each lot by applying a formula similar to that of simple areal interpolation, where the estimated tax lot population (PL_1) is equated to the original block group population (P_0) multiplied by the ratio of residential units in the tax lot (RU_1) over the total number of residential units in the block group (RU_0). These data can now be manipulated to create aggregates based on exposure or impact (e.g. inside a buffer). The results from the cadastral-based dasymetric disaggregation in this case study tend to be slightly more inclusive than those of areal interpolation or filtered areal weighting for the half-mile buffer. This will vary case by case, depending upon the particular way the population is distributed in any given study area and within the units of aggregation.

Case study results

Results from the case study are shown in Table 16.1, with both the absolute number population estimates, as well as comparisons of each method to the cadastral-based dasymetric disaggregation as per cent difference. The cadastral-based method was used as the standard for comparison amongst the methods because it is considered the most robust and is based on more realistic distribution of the population data (Eicher and Brewer 2001; Holt et al. 2004; Maantay et al. 2007; Martin 2006; Mennis 2003; Poulsen and Kennedy 2004). It can be seen that as the buffer distances increase, the relative differences in the estimates decrease. This suggests that if the impact areas of interest are quite large with respect to the units of aggregation that house the socio-demographic data (in this case, census block groups), the potential error introduced by the various methods becomes less marked. For instance, the polygon intersection selection method (most inclusive) has a difference of 82 per cent when compared to the dasymetric method used, whereas the complete polygon containment method has a difference of −42.7 per cent. However, when examining the outputs for the one-mile buffer, those differences drop to 6.5 per cent and −22.1 per cent for the polygon intersection and complete polygon containment methods, respectively. It is also important to note that although many of the demographic groups examined follow the total population trends, there is still a fair amount of variation. For instance, when looking at the centroid

Table 16.1 Comparison of population estimates. Counts of estimated impacted populations as well as per cent difference of each estimate from the cadastral-based dasymetric disaggregation for each buffer size.

Estimation Method	Buffer Dist.	Population Count Estimates						Per cent Difference from Cadastral–based Dasymetric Disaggregation					
		Total Population	Hispanic/ Latino	Non–Hispanic White	Non–Hispanic Black	Non–Hispanic Asian	Non–Hispanic Other	Total Population	Hispanic/ Latino	Non–Hispanic White	Non–Hispanic Black	Non–Hispanic Asian	Non–Hispanic Other
Polygon Containment	1/4 mile	13,010	9,953	538	2,039	240	240	-42.7	-38.6	-30.4	-59.1	-27.1	-44.3
Centroid Containment		20,741	14,803	775	4,444	334	385	-8.7	-8.6	0.2	-10.8	1.5	-10.6
Polygon Intersection		41,515	29,047	1,290	9,801	477	900	82.8	79.3	66.8	96.7	44.9	109.0
Areal Interpolation		23,092	16,593	780	4,952	327	441	1.7	2.4	0.8	-0.6	-0.7	2.4
Filtered Areal Weighting		23,161	16,652	786	4,952	328	442	2.0	2.8	1.7	-0.6	-0.3	2.7
Cadastral-based Dasym. Disag.		22,713	16,198	773	4,983	329	431	N/A	N/A	N/A	N/A	N/A	N/A
Polygon Containment	1/2 mile	47,853	31,599	1,434	13,134	575	1,111	-20.8	-14.1	-37.7	-30.5	-39.7	-24.5
Centroid Containment		56,687	35,188	1,643	17,802	684	1,370	-6.2	-4.4	-28.6	-5.8	-28.3	-6.8
Polygon Intersection		71,630	40,540	3,494	24,364	1,418	1,814	18.6	10.2	51.8	29.0	48.6	23.4
Areal Interpolation		57,491	35,242	2,286	17,621	938	1,404	-4.8	-4.2	-0.7	-6.7	-1.7	-4.5
Filtered Areal Weighting		58,009	35,401	2,410	17,786	988	1,424	-4.0	-3.8	4.7	-5.8	3.5	-3.2
Cadastral-based Dasym. Disag.		60,408	36,792	2,302	18,889	954	1,471	N/A	N/A	N/A	N/A	N/A	N/A
Polygon Containment	1 mile	143,944	69,855	8,361	59,222	2,707	3,799	-22.1	-14.2	-26.9	-29.1	-26.3	-20.9
Centroid Containment		187,486	82,413	11,780	84,687	3,752	4,854	1.5	1.3	3.0	1.4	2.1	1.1
Polygon Intersection		196,714	84,413	13,071	89,925	4,193	5,112	6.5	3.7	14.3	7.7	14.1	6.5
Areal Interpolation		180,889	80,109	11,570	80,782	3,714	4,714	-2.1	-1.6	1.2	-3.2	1.1	-1.8
Filtered Areal Weighting		180,727	80,081	11,573	80,649	3,712	4,712	-2.2	-1.6	1.2	-3.4	1.1	-1.8
Cadastral-based Dasym. Disag.		184,774	81,389	11,433	83,478	3,674	4,801	N/A	N/A	N/A	N/A	N/A	N/A

containment method, there is a tendency to undercount all population when compared to the dasymetric method. However, the underestimation of non-Hispanic black residents is more severe than the undercounting of non-Hispanic white residents which could result in a mischaracterization of the exposed groups and the obfuscation of potential environmental justice implications. As such, it is extremely important when conducting EJ analyses to attempt to produce the most reliable and robust population estimates.

Conclusions

Accurate estimates of impacted populations can have profound influences on research findings, and ultimately on the political process and in decision-making. This case study has demonstrated the differences amongst various methods in an "after-the-fact" siting situation, as a worked example of the relative merits of the methods. Clearly, using these techniques as a pro-active analysis during the environmental impact assessment process, other public review procedures, or as part of a preliminary planning initiative would be much more beneficial.

The comparison of population estimation methods used in the case study finds that although spatial coincidence/selection is typically the quickest and most frequently used method to calculate at-risk population, this tends to result in more general estimates, depending upon the size of the geographic unit of aggregation in relation to the impact exposure area. On the other hand, population disaggregation techniques such as the cadastral-based dasymetric method are more computationally intensive, but often yield a more robust and reliable estimate.

This method will be useful in many disparate fields and serve many purposes, for instance emergency management operations and implementation; police operations; criminal justice; fire and ambulance services; utility providers; and any other crucial public support systems dependent upon population information. Additionally, urban planning applications will benefit from better spatial data on potentially impacted sub-populations, since understanding the locational characteristics of target populations would allow for more equitable resource allocation in spheres such as community infrastructure development, provision of open space and recreational opportunities, transportation access, and necessary environmental facilities (Maantay et al. 2007).

For EJ analyses, more precise estimations of populations within impact extents and better locational information on population and sub-population distribution can be beneficial in "making the case" to decision-makers and policy experts. The more realistic and accurate results obtained through these methods can contribute to more confidence in crafting equitable solutions, and potentially increased political support for EJ policies and remedies.

There are a number of limitations inherent in population estimations and assessing populations at risk. With all of the methods discussed here, even dasymetric techniques, assumptions are made regarding the distribution of population (e.g. the assumption that where there are more residential units, there is greater population).

Secondly, the population numbers rely on census data, which are known to often be inaccurate, with serious problems of undercounting certain populations, especially in large urban areas. Thirdly, census data only take into account the residential location of population, who may not be at home as many hours as they are in school or at work, thereby over-estimating some of the population that might be impacted (Maantay et al. 2008). There have been a number of recent studies exploring the daily spatial mobility of individuals, to demonstrate the heterogeneity of many people's "activity spaces" (Chaix et al. 2012, 2013; Harrison et al. 2014; Matthews 2011). This is an important topic for future study, since only then we will be able to base exposure, access, and impact assessments on the actual locations that people occupy throughout their typical day, rather than relying solely on their residential locations.

Although we believe that the cadastral-based dasymetric disaggregation technique will generally yield the most reliable results for population estimation, it is a much more labour- and data-intensive method, and therefore may not be appropriate for every analysis. The extra effort involved may not provide a concomitant increase in reliability or accuracy in every situation, and the difference in the complexity of the methods might not produce significantly different results. This will also depend upon the degree of exactitude required by the specific analysis being undertaken. If time permits, very likely the best and most thorough approach is to conduct exploratory spatial data analysis (ESDA), whereby several methods are used to view the data from as many different vantage points as possible, and then selecting the one that most faithfully reflects reality (or a combination/averaging of results from various techniques). Often, presenting the reader with the results of several different methods is the most transparent and honest approach. Exploring your data as much as possible in the initial phase of a research project is always the best way to start, where feasible. Data exploration and visualization through mapping and spatial analysis often provides a more robust understanding of the data, as well as improved clarity in viewing the phenomena under study, which will lead to better design of further analyses and additional hypothesis generation, in an iterative fashion (Maroko et al. 2011).

Every "place" will be different: how population is distributed within the study extent and within each unit of aggregation, for instance, can radically affect the outcome of the analysis, as well as influence whether the selection of the various methods described in this chapter makes a marked difference in the results. The researcher or analyst must understand how different methods of population estimation might obscure their findings for the specific place and phenomenon they are investigating, dependent on the data and time available for the analysis.

Acknowledgements

The National Oceanic and Atmospheric Administration's Cooperative Remote Sensing Science and Technology Center (NOAA-CREST) provided critical support for this work under NOAA grant number NA17AE162. The statements contained within this chapter are not the opinions of the funding agency or the US government, but reflect the authors' opinions.

The authors would also like to thank Adam Jessup, cartographer and graphic designer extraordinaire (and a graduate student in Lehman College's MS-GISc Program), for his work on creating the maps, diagrams, and other figures accompanying the text.

References

Bunch, W. (1994). "$52 Million won't flush stink." *New York Newsday*, June 30.

Chaix, B., Kestens, Y., Perchoux, C., Karusisi, N., Merlo, J., and Labadi, K. (2012). "An interactive mapping tool to assess individual mobility patterns in neighbourhood studies." *American Journal of Preventive Medicine*, 43(4): 440–50.

Chaix, B., Meline, J., Duncan, S., Merrien, C., Karusisi, N., and Perchoux, C. (2013). "GPS tracking in neighbourhood and health studies: a step forward for environmental exposure assessment, a step backward for causal inference?" *Health and Place*, 21: 46–51.

Chakraborty, J. and Maantay, J.A. (2011). "Proximity analysis for exposure assessment in environmental health justice research." In Maantay, J.A. and McLafferty, S. (eds.) *Geospatial Analysis of Environmental Health*, Dordrecht, NL: Springer-Verlag, pp. 111–138.

Eicher, C. and Brewer, C. (2001). "Dasymetric mapping and areal interpolation: implementation and evaluation." *Cartography and Geographic Information Science*, 28: 125–138.

Forster, B.C. (1985). "An examination of some problems and solutions in monitoring urban areas from satellite platforms." *International Journal of Remote Sensing*, 6(1): 139–151.

Gladwell, M. (1993). "Letter from the park: play area hailed at first flush now appears ripe for criticism." *Washington Post*, May 28.

Harrison, F., Burgoine, T., Corder, K., van Sluijs, E., and Jones, A. (2014). "How well do modelled routes to school record the environments children are exposed to?: a cross-sectional comparison of GIS-modelled and GPS-measured routes to school." *International Journal of Health Geographics* 13(1): 5.

Holt, J.B., Lo, C.P., and Hodler, T.W. (2004). "Dasymetric estimation of population density and areal interpolation of census data." *Cartography and Geographic Information Science*, 31: 103–121.

Lwin, K.K. and Murayama, Y. (2011). "Accuracy assessment of GIS based building population estimation algorithm." In Murayama, Y. and Thapa, R. (eds.) *Spatial Analysis and Modelling in Geographical Transformation Process: GIS-based Applications*, Dordrecht, NL: Springer-Verlag, pp. 99–112.

Maantay, J.A. and Maroko, A.R. (2009). "Mapping urban risk: flood hazards, race, and environmental justice in New York." *Applied Geography*, 29: 111–124.

Maantay, J.A., Maroko, A., and Herrmann, C. (2007). "Mapping population distribution in the urban environment: the Cadastral-based Expert Dasymetric System (CEDS)." *Cartography and Geographic Information Science*, 34(2): 77–102.

Maantay, J.A., Maroko, A.R., and Porter-Morgan, H. (2008). "A new method for mapping population and understanding the spatial dynamics of disease in an urban area: asthma in the Bronx, New York City." *Urban Geography*, 29(7): 724–738.

Maantay, J.A., Maroko, A.R., and Culp, G. (2009)."Using geographic information science to estimate vulnerable urban populations for flood hazard and risk assessment in New York City." In Showalter, P., and Lu, Y. (eds.) *Geotechnical Contributions to Urban Hazard and Disaster Analysis*, Dordrecht, NL: Springer-Verlag, pp. 71–97.

Mandell, J. (1993). "A state park in Harlem: nature in a surreal setting: a sewage plant by any other name still stinks." *New York Newsday*, May 27.

Maroko, A.R., Maantay, J.A., and Grady, K. (2011). "Using geovisualization and geospatial analysis to explore respiratory disease and environmental health justice in New York City." In Maantay, J.A. and McLafferty, S. (eds.) *Geospatial Analysis of Environmental Health*, Dordrecht, NL: Springer-Verlag, pp. 39–66.

Martin, D. (2006). "An assessment of surface and zonal models of population." *International Journal of Geographical Information Systems*, 10(8): 973–89.

Matthews, S.A. (2011). "Spatial polygamy and the heterogeneity of place: studying people and place via egocentric methods." *Communities, Neighborhoods, and Health*. Dordrecht, NL: Springer-Verlag, pp. 35–55.

Mennis, J. (2003). "Generating surface models of population using dasymetric mapping." *The Professional Geographer*, 55(1): 31–42.

Moon, Z.K. and Farmer, F.L. (2001). "Population density surface: a new approach to an old problem." *Society and Natural Resources*, 14: 39–49.

National Association of City and County Health Officials (NACCHO). (2011a). *A Neighborhood Fights Back: Building the North River Waste Water Treatment Plan in West Harlem*. www.rootsof healthinequity.org

National Association of City and County Health Officials (NACCHO). (2011b). *Polluting Sites in Northern Manhattan*. www.rootsofhealthinequity.org/polluting-sites-in-manhattan.php

Poulsen, E. and Kennedy, L.W. (2004). "Using dasymetric mapping for spatially aggregated crime data." *Journal of Quantitative Criminology*, 20(3): 243–62.

Wu, S., Qiu, X., and Wang, L. (2005). "Population estimation methods in GIS and remote sensing: a review." *GIScience and Remote Sensing*, 42(1): 80–96.

17
APPLICATION OF SPATIAL STATISTICAL TECHNIQUES

Jeremy Mennis and Megan Heckert

Introduction

Environmental justice is an inherently spatial issue, as one of its central concerns revolves around the relationship between the spatial distribution of environmental risk and the spatial distribution of traditionally disenfranchised population groups, such as the poor, the vulnerable (e.g. children and the elderly), and racial and ethnic minorities. Indeed, the driving question behind much environmental justice research addresses environmental equity, i.e. whether the burden of environmental risk falls disproportionately on certain subgroups of population (Downey 2005). Questions of environmental equity have been widely addressed quantitatively through statistical analysis. Typically, a set of location- or area-based observations are used as a basis for encoding measures of environmental risk and socioeconomic character, for which a spatial association is investigated. The significance and magnitude of this association is often considered evidence for socioeconomic inequality in the distribution of environmental risk. An analogous approach can be applied to analyses of health outcomes or access to environmental amenities instead of environmental risk.

It is well known that the statistical analysis of spatial data often fails to meet the assumptions of regression and other conventional inferential statistical techniques, such as the independence of observations, which can bias statistical results and lead to erroneous interpretations of the statistical analysis. Various spatial statistical techniques have thus been developed to address these issues which are unique to spatial data analysis. Since the inception of academic research on environmental justice, issues of spatial data and spatial analysis have been recognized as key to improving environmental justice investigations (Anderton et al. 1994; Maantay 2007; Mennis 2002). Consequently, over the last two decades an increasing number of environmental justice researchers have adopted spatial statistical approaches.

The purpose of this chapter is to review the application of spatial statistics in environmental justice research and the specific spatial statistical techniques that have been employed by environmental justice researchers. The chapter begins with a description of the conventional statistical approach to environmental justice which has been critiqued and which serves as the basis for considerations of spatial statistical issues. The chapter then addresses the issue of spatial scale of analysis as embodied in the modifiable areal unit problem (MAUP). Several specific spatial statistical techniques which have been prominently used in environmental justice research are then presented, including spatial clustering, spatial econometric modelling, multilevel modelling, and other techniques. The chapter concludes with a brief discussion of the future of spatial statistical analysis in environmental justice research. For simplicity we address the analysis of environmental risk in our general discussions of various spatial statistical techniques, as historically this has been the most common focus of environmental justice research. However, we note that the spatial statistical techniques described here can be applied analogously to quantitative environmental justice studies focusing on environmental amenities and health outcomes.

The conventional statistical approach

We use the term 'conventional' to describe an analytical approach used by many environmental justice studies from the early 1980s through the present (e.g. CRJ 1987; Anderton et al. 1994) as a point of departure for our discussion of spatial statistical approaches to environmental justice research, as spatial statistics have often been employed as a response to shortcoming of the conventional approach. Given a hypothesis that there is an association between an indicator of socioeconomic status and environmental risk, several analytical decisions arise: First, how to represent socioeconomic status and environmental risk quantitatively; second, the form of the association; and third, the analytical technique used to test the hypothesis. The conventional approach typically uses socioeconomic data collected by a government census agency and made available spatially aggregated to some set of spatial units, as with, for instance, data on poverty rate or the percentage of the total population identifying as minority within US Census Bureau tracts. Data on environmental risk, such as the locations and amounts of pollutant releases, are often gathered from government agencies, such as the US Environmental Protection Agency.

The association between the environmental risk and the socioeconomic status is typically identified through data processing that integrates the environmental and population data via a spatial relationship. Often, the environmental risk data is aggregated to the spatial units of the population data to encode the presence of an environmental risk, the distance to an environmental risk, or some other measure of the magnitude of environmental risk associated with the population residing within a spatial unit (see Chapters 15 and 16). This data processing typically produces a data set in which a single observation (i.e. a census spatial unit or a postal code) is attributed with values of socioeconomic status and environmental risk to facilitate analysis. A statistical technique is then applied to test whether there is a significant association (i.e. spatial relationship) between socioeconomic status and environmental risk, where such an association may be considered evidence of environmental inequity.

Scale of analysis

Scope and resolution

In quantitative studies of environmental justice, scale of analysis refers to both the spatial scope and the spatial resolution of an analysis. Scope refers to the size of the study area, for instance whether a study encompasses data for an entire nation, for a single state or province, or for a specific city. The scope of analysis is particularly important when interpreting the generalizability of the findings, where studies with a smaller scope (e.g. the breadth of a single city) as compared to a larger scope (e.g. the breadth of an entire nation) may offer evidence of environmental equity or inequity that is limited to the study area.

Spatial resolution refers to the size of the spatial unit which serves as a single observation in the analysis, which is particularly germane to quantitative environmental justice studies because most analyses use spatially aggregated socioeconomic data to represent the distribution of population characteristics. For example, an analyst may use demographic data spatially aggregated to the level of US Census Bureau tracts, counties, or states in the US. While in many cases the resolution is associated with the scope of analysis, for instance one might do an analysis at the census tract resolution for the scope of a single city and an analysis of counties for the scope of a state, spatial resolution and scope are distinct characteristics; one can do a tract level analysis at the scope of a single city or for the entire US. The choice of scope and resolution is often driven by data availability, but should also be driven by the study question; for example, if one is interested in environmental inequity due to state level environmental policies, one may want to choose the nation as the scope of analysis and the state as the resolution so that comparisons among different state level policies can be made.

The Modifiable Areal Unit Problem (MAUP)

The issue of resolution is particularly relevant to environmental justice because of the well-known Modifiable Areal Unit Problem (MAUP), which refers to the fact that statistical results can differ depending on the nature of the spatial units used to encode data (Fotheringham and Wong 1991). There are two permutations of the MAUP. First, analytical results may differ depending on the size of the spatial units, and second, analytical results may differ depending on the partitioning scheme of the data (even when the size and number of the units remains the same). For example, the upper left of Figure 17.1 shows a point process of white and black points. Two different partitioning schemes, each with the same number and resolution of spatial units, are shown in the upper left and lower right of the figure, with text showing the percentage of white points in each spatial unit. Clearly, changing the partitioning scheme of data aggregation changes the depiction of the spatial distribution of percentage white points, even when both schemes are applied to the same underlying point process. The same principle can be seen when employing a different resolution of spatial aggregation, as shown in the lower right of Figure 17.1, where twice as many spatial units, each with half the area of those shown previously, are used to calculate the within-unit percentage of white points.

The impact of the MAUP on environmental justice analyses has been recognized explicitly since at least the mid-1990s (Anderton et al. 1994; Cutter et al. 1995; Zimmerman 1994) and expounded on many times since (Baden et al. 2007; Downey 2006; Mennis 2002). Consequently, researchers have made recommendations on choosing the appropriate spatial unit of analysis in environmental justice research (Williams 1999; Yandle and Burton 1996), although in practice the resolution is often limited by the availability of the environmental and socioeconomic data.

One approach to the MAUP is to repeat the same statistical analysis using data aggregated at different resolutions, when possible, in order to investigate whether the analytical results differ (Baden et al. 2007; Noonan et al. 2009; Taquino et al. 2002). However, it should be noted that a difference in analytical results among different data resolutions should not simply be viewed as a problem, per se, but as possible evidence for the mechanisms driving the spatial relationship that may be observed between population character and a measure of environmental risk. Some researchers have avoided the issues associated with the use of spatially aggregated population data by using household level data, which may be mapped to point locations as opposed to over a spatial unit (Collins et al. 2015; Crowder and Downey 2010; Pais et al. 2014).

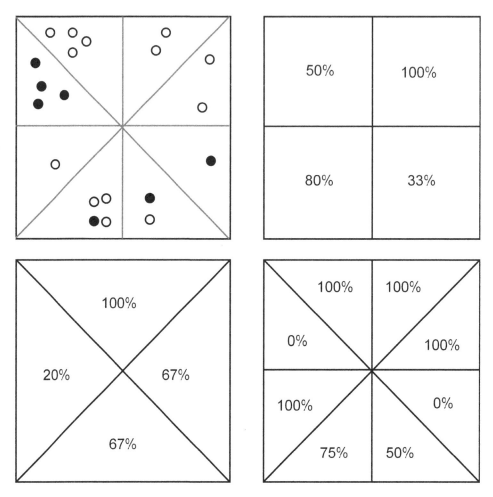

Figure 17.1 An illustration of how data aggregation affects the summary representation of a point process (upper left) using two different partitioning schemes at the same resolution (upper right and lower left) and at a finer resolution (lower right). The percentage of points which are white is shown in each polygon.

Quantifying environmental risk

Researchers have employed a variety of ways to quantify the spatial relationship between the location of an environmental hazard and population character (Chakraborty and Maantay 2011; McMaster et al. 1997; Mennis 2002), which may be categorized according to spatial coincidence-based, distance-based, and plume-based approaches (see Chapter 15). Central to all these approaches is the use of Geographic Information Systems (GIS) software, which supports the representation of spatial data and the analysis of spatial relationships. In the spatial coincidence approach, also referred to as the unit-hazard coincidence approach (Mohai and Saha 2006), one is typically interested in the co-location of a point-referenced hazard location with the spatial unit of population data. In the distance-based approach one can define distance buffers from the hazardous facility and identify those spatial units within the buffer (Mennis 2002). The plume-based approach aims to implement a more sophisticated spatial representation of environmental risk by incorporating the amount and toxicity of any chemical releases, or by modelling the dispersion of toxic release via a plume model (Bevc et al. 2007; Chakraborty and Armstrong 1995; Maantay et al. 2009).

For environmental justice research focusing on environmental amenities, such as green-space, recreational opportunities, and nutritious food, the measurement of accessibility to the amenity is given priority, as opposed to the measurement of exposure to risk. While approaches similar to those described above can also be used to model accessibility, e.g. spatial containment and distance-based approaches, the network modelling capabilities of many GIS packages offer more sophisticated approaches to modelling accessibility assuming travel over road networks, barriers to accessibility, and points of access. These approaches are demon-strated for environmental justice analyses by Comber et al. (2008) and Heckert (2013) in studies of accessibility to parks and greenspace, who show that different parameterizations of accessibility can influence analytical results.

Spatial clustering

Hotspot mapping

Perhaps the most basic question one can ask concerning the spatial distribution of environmental risk is whether the distribution is random or spatially clustered. If, indeed, the distribution is clustered, then one would want to know the form of the clustering, e.g. where are the areas of high versus low risk and the distance threshold over which the clustering occurs? The terms 'hotspot map' or 'heat map' are used colloquially to indicate a map that shows areas of high and low concentrations of a set of points or a variable value. Such maps can take the form of a conventional choropleth map that shows the density of a point process summarized to a set of spatial units (e.g. the number of hazardous facilities per unit area by county) or a continuous variable (percentage of the population identifying as Hispanic in a county). Hotspot maps of point distributions are also often mapped using kernel density estimation (KDE) which, given a circular kernel with a radius (distance) d, generates a density λ for every coordinate location l on the surface, expressed as

$$\lambda(l) = \frac{\#(C(l, d))}{\pi d^2}$$

where $\#(C(l, d))$ represents the count of the number of points (e.g. hazardous facility locations) within a distance d from location l (Lloyd 2010).

Ripley's K function

In cases where environmental risk is represented by a set of point data, as with a set of facili-ties releasing toxic chemicals into the environment, one can carry out a cluster analysis to identify areas of significant point clusters and, thus, high risk. To this end, some environ-mental justice studies have employed Ripley's K function (Ripley 1976), a point clustering technique which has the advantage of testing the significance of clustering over multiple distance thresholds, thus addressing the issue of scale of analysis, noted above. Ripley's K for a specific distance d may be expressed as

$$K(d) = \frac{\#(C(l, d))}{\bar{\lambda}}$$

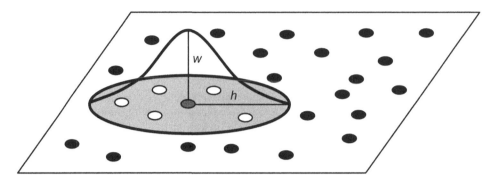

Figure 17.2 A diagram of geographically weighted regression (GWR), where for a distribution of point
observations, a regression equation is fit to a subset of points within a kernel defined by
bandwidth *h* (shown as a shaded circle) according to the weighting function (shown as the
curve over the shaded circle), where the weight *w* of each observation *j* (each shown as a
white point) is assigned as a function of its distance from the focal observation *i* (shown as
the grey point in the centre of the kernel).

where $\bar{\lambda}$ is the density over the entire study region. Examples of the use of Ripley's *K* in
environmental justice research include work by Fisher et al. (2006), who investigate the scale
of hazardous facility clustering in the San Francisco Bay area in California, and Hillier et al.
(2009), who employ a local adaptation of the *K* function to identify locations of significant
clustering of unhealthy billboard advertisements near child-serving institutions.

Spatial autocorrelation

For data encoding the magnitude of an environmental risk variable mapped on to a tessellation of spatial units, as with data processed to calculate environmental risk scores over say, a
set of census tracts, a test of spatial autocorrelation is typically employed to identify spatial
clustering in the variable. Spatial autocorrelation refers to the level of spatial dependency in
spatial data, where one can inquire whether spatial units located near one another tend to
have particularly similar (or dissimilar) values as compared to spatial units located farther apart.
In calculating spatial autocorrelation, it is necessary to encode the weight between every pair
of units, i.e. the relative influence that one would expect one unit would have on another
based on their spatial relationship. These weights are encoded in the spatial weights matrix
that specifies the weight between every pair of units (Lloyd 2010). A simple parameterization
assigns a weight of one to all pairs of units within a certain distance of each other, and a weight
of zero to all other pairs of units. Other common parameterizations include Queen's contiguity, in which neighbouring units are given a weight of one if they are connected by an edge
or a point and a weight of zero if not. Rook's contiguity is defined similarly but weights of
one are restricted to pairs of units that share an edge. Distance-based weights matrices assign
a weight determined as a function of the distance between units, often measured according
to the distance between unit centroids when using a tessellated polygonal data set (Figure
17.2).

A commonly used statistical test of spatial autocorrelation is Moran's *I*, which can provide
a single test statistic that indicates the direction (i.e. whether nearby values are particularly
similar or dissimilar), magnitude, and significance of spatial dependency, expressed as

$$I_i = \frac{n \sum_{i=1}^{n} \sum_{j=1}^{n} w_{ij} (y_i - \bar{y})(y_j - \bar{y})}{\sum_{i=1}^{n} (y_i - \bar{y})^2 \left(\sum_{i=1}^{n} \sum_{j=1}^{n} w_{ij} \right)}$$

where y is a variable, i is a spatial unit, j is a neighbouring spatial unit of i, n is the number of units, w is an element of a spatial weights matrix that specifies the weight between spatial units i and j, and the attribute values y_i have the mean $\bar{y}$ (Lloyd 2010). Various parameterizations of the spatial weights matrix can produce substantially different estimations of I.

Local spatial autocorrelation

For investigations of clustering in environmental risk data, local measures of spatial autocorrelation are typically employed, which can reveal where particularly high or low concentrations of environmental risk occur. The Local Indicators of Spatial Autocorrelation (LISA) statistic (Anselin 1995), as the name implies, adapts the Moran's I statistic to identify local clustering, expressed as

$$I_i = z_i \sum_{j=1}^{n} w_{ij}, j \neq i$$

where z_i are deviations from the mean $(y_i - \bar{y})$. Such an approach facilitates mapping that depicts statistically significant clusters of spatial units with particularly high (low) values which are surrounded by other spatial units with particularly high (low) values. Environmental justice examples include investigations of local clustering of lifetime cancer risk using the US National Air Toxics Assessment (NATA) data set at the census tract level (Lievanos 2015), paediatric lead exposure in Chicago (Oyana and Margai 2010), and vulnerability to climate change (Wilson et al. 2010). Two variables can be simultaneously incorporated in the LISA statistic in order to identify clusters of spatial units with a high value in one variable (e.g. economic disadvantage) surrounded by spatial units with a high value in another variable (e.g. environmental risk). This approach is demonstrated by Ogneva-Himmelberger et al. (2015) in a study of vulnerable populations (e.g. youth and the elderly) and exposure to air pollution from hog farming in North Carolina and by Zou et al. (2014) in a county-level study of air pollution and demographic character in the urban US.

Another statistical technique used for identifying local clustering is the Getis G_i (Getis and Ord 1992), expressed as

$$G_i = \frac{\sum_{j=1}^{n} w_{ij} y_j}{\sum_{j=1}^{n} y_j}, j \neq i$$

The Getis G_i has been used in environmental justice research to identify clusters of animal feeding operations in Ohio (Lenhardt and Ogneva-Himmelberger 2013), clusters of TRI facilities in an examination of lead exposure and crime in Florida (Lersch and Hart 2014), and co-occurring clusters of transportation generated particulate matter with incidences of asthma and lung cancer in Massachusetts (McEntee and Ogneva-Himmelberger 2008).

Incorporating spatial effects into predictive models

Spatial econometric modelling

Ordinary least squares (OLS) regression models have historically been the most common approach in environmental justice research for testing the predictive power of one variable on another variable, or the association between two variables (e.g. environmental risk and socio-economic status), while controlling for the confounding effects of other explanatory variables. Given a continuous outcome variable y, the regression equation may be expressed as

$$y = \beta_0 + \beta_1 x_1 + \beta_2 x_2 + \cdots \beta_n x_n + \varepsilon$$

where x_1, x_2, . . .x_n are a set of independent variables 1 through n, β is a coefficient to be estimated, and ε is the error term. Though widely used, OLS regression can be problematic in two ways. First, the method relies on the assumption of independence of observations, a condition which rarely holds true for the spatial variables used in environmental justice research which often exhibit spatial autocorrelation. A second potential problem of OLS is that the residual error terms may also be spatially autocorrelated, a problem which is not identified with typical OLS diagnostics but which nevertheless may indicate that confidence intervals may be biased and OLS results should not be taken at face value. Many environmental justice studies employ Moran's I to test for spatial autocorrelation in OLS residuals (e.g. Chakraborty 2011; Grineski et al. 2013); the LISA statistic has also been applied for this purpose (Mennis 2011).

Simultaneous autoregressive (SAR) models, also known as spatial econometric models (Anselin 1988), provide an alternative to OLS regression that explicitly incorporates spatial effects in the model, in particular spatial autocorrelation. Two types of spatial econometric models are commonly used in environmental justice research; the first, the spatial lag model, incorporates a term representing spatial autocorrelation in the dependent variable as an explanatory variable in the model, while the latter does the same for the OLS regression model error term. Both types are typically calculated using maximum likelihood estimation.

The spatial lag model estimates the value of the dependent variable y for any observation as a function of the values of y at nearby observations, where the term in the regression equation that captures the nearby values of y is referred to as the spatial lag term. The spatial weights matrix is employed to calculate the value of the spatial lag term individually for each observation, such that for any observation i the spatial lag represents a weighted mean of the values of the dependent variable nearby to i. The spatial lag model thus modifies the OLS model to

$$y = \beta_0 + \beta_1 x_1 + \beta_2 x_2 + \cdots \beta_n x_n + \rho \mathbf{W} \mathbf{y} + u$$

where $\mathbf{W}$ specifies a spatial weights matrix, $\mathbf{y}$ is vector of outcome variable values, and $\mathbf{W}\mathbf{y}$ is thus a spatial lag term with autoregressive parameter ρ, and u is a spatially-independent error term. The spatial error model, on the other hand, incorporates the spatial autocorrelation as a component of the error term rather than the dependent variable, specified as

$$y = \beta_0 + \beta_1 x_1 + \beta_2 x_2 + \cdots \beta_k x_n + \nu \mathbf{W} \varepsilon + u$$

where ε is vector of error terms, and $\mathbf{W}\varepsilon$ is thus a spatial error lag term with autoregressive parameter ν.

The choice between the spatial lag and spatial error models can be made based on theoretical considerations, where the spatial lag model is often used when one hypothesizes some sort of spatial contagion or diffusion process, and the spatial error model is used when one wishes to treat the spatial autocorrelation as a nuisance to be accounted for in the model. Another commonly used approach is to apply both models and then choose one based on a combination of factors, including best model fit, lowest residual spatial autocorrelation, and lowest Akaike Information Criterion (AIC) score (Kissling and Carl 2008). The choice can also be made based on Lagrange Multiplier test statistics (Anselin 2005). It is also not uncommon to try different approaches to parameterizing the spatial weights matrix, using the AIC scores to determine the best approach.

Chakraborty (2011) used spatial econometric modelling to assess the relationship between race, class, and cancer risk from exposure to transportation-related air pollution in the Tampa Bay, Florida region after showing that both the lifetime cancer risk and OLS model residuals exhibit spatial autocorrelation. The use of spatial econometric models significantly reduced spatial autocorrelation in the model residuals and resulted in lower AIC scores, indicating better model fit. The results of the spatial econometric models also differed somewhat from the OLS results, highlighting the importance of using spatial approaches to address the unmet assumptions of OLS regression. Similar studies have used SAR models to explore the extent to which race, class, and other socio-demographic measures are predictive of distributions of exposure to the urban heat island effect (Mitchell and Chakraborty 2014), flooding (Grineski et al. 2015a), air toxics exposure (Chakraborty 2009; Germani et al. 2014; Raddatz and Mennis 2013), vegetation (Pham et al. 2012), and trees planted on public rights of way (Landry and Chakraborty 2009).

Multilevel modelling

Another approach to incorporating spatial effects into regression models employs multilevel modelling, which is appropriate when data are conceptualized as being hierarchically nested. Such an approach is common in environmental health research, and has been used occasionally in environmental justice research. Health and environmental justice data can often adhere to a nested structure, where individual level data (level 1) may be spatially nested within neighbour-hoods (level 2), e.g. households nested within census tracts, or, alternatively, where individual level variables are measured repeatedly over time (Schwartz et al. 2011). Multilevel modelling allows one to test for level 2 effects on level 1 outcomes (i.e. in a random intercept model where the intercept is allowed to vary among level 2 observations), as well as for interactions between level 1 and level 2 explanatory variables (i.e. in models where both intercepts and slopes are allowed to vary among level 2 observations). The basic random intercept multilevel model of a dependent variable y and a single explanatory variable x may be expressed using the following three equations

$$y_{fg} = \beta_{0g} + \beta_{1g} x_{fg} + \varepsilon_{fg}$$

$$\beta_{0g} = \gamma_{00} + \gamma_{01} z_g + u_{0g}$$

$$\beta_{1g} = \gamma_{10} + u_{1g}$$

where f denotes a level 1 observation nested within a level 2 observation g, β is a parameter to be estimated, z is level 2 variable, and u_g is the random effect at level 2 (Snijders and

Bosker 2012). The slope of a level 1 variable may be allowed to vary by inserting additional variables into the equation for β_{1g}.

As an example of multilevel modelling for environmental justice research, Collins et al. (2014) investigate neighbourhood and individual level measures of economic deprivation on respiratory health among Hispanic children in El Paso, Texas. Ard (2015) employs a three level multilevel model incorporating nesting in both space and over time to model the changing relationship between air toxics and socioeconomic status, encoded at the census block group level for the US. McLeod et al. (2000) investigate the relationship between air pollution and various indicators of socioeconomic status using multilevel modelling to show how such relationships differ across regions of England and Wales. As an alternative approach Collins et al. (2015) argue that generalized estimation equation (GEE) techniques may be more appropriate for environmental justice studies that incorporate spatial nesting of explanatory variables because the focus of such research is typically on the influence of population subgroup membership on an individual risk or health outcome, and not necessarily on the effect of some external environmental condition. Such an approach was also used in an environmental justice analysis of access to recreational facilities in Paris, France (Billaudeau et al. 2011).

Simulation and Bayesian techniques

These spatial statistical approaches have also been augmented by alternative parameterization techniques and approaches to significance testing that also deserve mention. Some researchers (Chakraborty and Armstrong 2001; Sheppard et al. 1999) have employed randomization approaches to generate robust confidence intervals for significance testing in order to account for potential bias due to non-independence of observations, an approach implemented by comparing observed patterns of environmental risk in empirical data to a series of spatial simulations.

Bayesian regression has also been used for spatial statistical analysis of environmental justice, which can address issues of model over-fitting by employing Markov chain Monte Carlo (MCMC) simulation techniques for parameter estimation (LeSage and Pace 2009). Chun et al. (2012) employ Bayesian estimation for Poisson regression to account for non-independence of observations and consequent spatial autocorrelation, which they note is applicable to many environmental justice analyses where the outcome is a count variable (e.g. the number of hazardous facilities contained with each spatial unit). Burda and Harding (2014) develop a semiparametric Bayesian proportional hazard model to investigate socio-economic inequity of clean-up durations for Superfund sites, while Rojas (2011) employs a space-time Bayesian model to investigate disease inequity among Indigenous people of Chile.

Geographically weighted regression (GWR)

Environmental justice researchers have also noted that the relationships between socio-economic status and environmental risk may vary from place to place, a statistical property referred to as spatial non-stationarity. Geographically weighted regression (GWR) is a statistical technique that facilitates the exploration and visualization of spatial nonstationarity (Fotheringham et al. 2002) and has been used by environmental justice researchers to investigate how inequity in environmental risk varies over space. Conceptually, GWR can be understood to conduct a separate OLS regression individually for each observation in the data set, where for each regression the remaining observations are weighted according to their spatial relationship to the focal observation. GWR can be expressed as

$$y_i = \beta_0(l_i) + \beta_1(l_i)\, x_{i1} + \beta_2(l_i)\, x_{i2} + \cdots \beta_k(l_n)\, x_{in} + \varepsilon_i$$

The weight of each observation is based on a distance decay function away from *i*, and may take the form of a variety of functions which can be tuned via a bandwidth setting such that the weight approaches zero at some threshold distance (Figure 17.2). For a tessellation of spatial units, the distances between units are determined using the unit geometric centroids. The results of GWR yield a set of parameter estimates, associated *t*-values, and goodness of fit statistics for each observation that can be mapped to illustrate areas of positive and/or negative relationships among the explanatory and outcome variables (Mennis 2006). Environmental justice research using GWR has shown that relationships among indicators of race and class with environmental risk often vary over space. Gilbert and Chakraborty (2011) demonstrate such a study focusing on cancer risks due to air toxics in Florida, while Mennis and Jordan (2005) show how the relationships of race and class with air toxics differ across regions of New Jersey. Similar studies have addressed race and respiratory disease in New York City (Maroko et al. 2011), race and lead exposure from air toxic releases in Tampa Bay, Florida (Lersch and Hart 2014), and socioeconomic status and wheezing due to exposure to air pollution in El Paso, Texas (Grineski et al. 2015b).

Conclusion

The use of spatial statistics has grown substantially in environmental justice research over the past 20 years as environmental justice researchers have recognized the necessity of spatial statistical approaches in a field whose research questions have a substantial spatial component and often analyse spatial data. Over this period spatial statistics have become increasingly accessible to researchers as these techniques have been implemented in general statistics software packages (e.g. SPSS [IBM, Inc.], Stata [StataCorp LP]), stand-alone spatial statistics software (e.g. Geoda [University of Chicago], R [R Foundation for Statistical Computing]), or as analytical components in GIS packages (e.g. GRASS GIS [Open Source Geospatial Foundation], ArcGIS [ESRI, Inc.]). The availability of these spatial statistics tools has allowed environmental justice researchers to expand the scope of their research to address new kinds of research questions, which can form the basis for future research.

While the vast majority of statistical environmental justice research has used spatially aggregated population data, the opportunity to use household level data affords researchers the ability to avoid many of the pitfalls associated with the use of spatially aggregated data elicited above. In addition, incorporating data at the household level allows for modelling individual decision-making regarding residential choice and other relevant behaviours at the household level, which can serve to elucidate not only emergent patterns of environmental inequity but also the mechanisms by which such inequities might have been created. As such, several newer studies seek to 'downscale' to address outcomes at the individual and household level (e.g. Collins et al. 2015; Pais et al. 2014).

The integration of geospatial technologies with mobile computing and wearable sensors also affords a much higher level data granularity to address individual outcomes than was possible until recently. For example, global positioning systems (GPS) technology embedded in mobile phones can be used to generate streaming location data so as to capture the daily activities and travel patterns of individuals. These data can be used to develop person-level measures of environmental risk or accessibility to environmental amenities at a far more granular level of detail as compared to measuring exposure or accessibility based on the location of a residence within a census tract or similar spatial unit (Mennis and Mason 2011), as has often been done in environmental justice research. Wearable sensors can also be used to collect individual level data on exposure to air pollution, ambient noise, and other environmental hazards or stressors, as well as biophysical indicators of health responses to such environmental stressors, such as skin conductance and heart rate (Stahler et al. 2013). Data derived from location tracking devices and/or mobile sensors are typically temporal as well as spatial, and can be massive in volume. Thus, such data sets demand new types of spatial statistical techniques and approaches to summarize and assess inequities in environmental risk among socioeconomic groups.

Finally, while much statistical environmental justice research has been associative in nature, future research has the opportunity to investigate the mechanisms and outcomes of environmental inequity in a richer way. For instance, spatiotemporal statistics can provide a framework for integrating spatial effects and longitudinal models that illustrate causation in linking environmental risk and health outcomes. In addition, disparities in specific health outcomes may be modelled as a function of inequities in exposure to environmental risk via the integration of spatial statistics with path, or mediated, statistical models. Such an approach addresses the potential health consequences of environmental inequity, pointing the way towards addressing health disparities through the lens of environmental justice policy.

References

Anderton, D.L., Anderson, A.B., Oakes, J.M., and Fraser, M.R. (1994). "Environmental equity: The demographics of dumping." *Demography*, vol. 31, pp. 229–248.

Anselin, L. (1988). *Spatial Econometrics: Methods and Models*. Dordrecht, NL: Kluwer Academic Publishers.

Anselin, L. (1995). "Local indicators of spatial association – LISA." *Geographical Analysis*, vol. 27, no. 2, pp. 93–115.

Anselin, L. (2005). *Exploring Spatial Data with GeoDa: A Workbook*. Spatial Analysis Laboratory, Department of Geography, University of Illinois, Urbana-Champaign.

Ard, K. (2015). "Trends in exposure to industrial air toxins for different racial and socioeconomic groups: A spatial and temporal examination of environmental inequality in the U.S. from 1995–2004." *Social Science Research*, vol. 53, pp. 375–390.

Baden, B.M., Noonan, D.S., & Turaga, R.M. (2007). "Scales of justice: Is there a geographic bias in environmental equity analysis?" *Journal of Environmental Planning and Management*, vol. 50, pp. 163–185.

Bevc, C.A., Marshall, B.K, & Picou, J.S. (2007). "Environmental justice and toxic exposure: Toward a spatial model of physical health and psychological well-being." *Social Science Research*, vol. 36, pp. 48–67.

Billaudeau, N., Oppert, J.-M., Simon, C., Charreire, H., Casey, R., Saize, P., et al. (2011). "Investigating disparities in spatial accessibility to and characteristics of sport facilities: Direction, strength, and spatial scale of associations with area income." *Health and Place*, vol. 17, pp. 114–121.

Burda, M. & Harding, M. (2014). "Environmental justice: Evidence from Superfund clean-up durations." *Journal of Economic Behavior & Organization*, vol. 107, Part A, pp. 380–401.

Chakraborty, J. (2009). "Automobiles, air toxics, and adverse health risks: Environmental inequities in Tampa Bay, Florida." *Annals of the Association of American Geographers*, vol. 99, no. 4, pp. 674–697.

Chakraborty, J. (2011). "Revisiting Tobler's first law of geography: Spatial regression models for assessing environmental justice and health risk disparities." In Maantay, J.A. & S. McLafferty (eds.), *Geospatial Analysis of Environmental Health*, Springer Science and Media, pp. 337–356.

Chakraborty, J. & Armstrong, M.P. (1995). "Using geographic plume analysis to assess community vulnerability to hazardous accidents." *Computers, Environment and Urban Systems*, vol. 19, no. 5–6, pp. 341–356.

Chakraborty, J. & Armstrong, M.P. (2001). "Assessing the impact of airborne toxic releases on populations with special needs." *The Professional Geographer*, vol. 53, no. 1, pp. 119–131.

Chakraborty, J. & Maantay, J.A. (2011). "Proximity analysis for exposure assessment in environmental health justice research." In Maantay, J.A. & S. McLafferty (eds.), *Geospatial Analysis of Environmental Health*, Springer Science and Media, pp. 111–138.

Chun, Y., Kim, Y., & Campbell, H. (2012). "Using Bayesian methods to control for spatial autocorrelation in environmental justice research: An illustration using toxics release inventory data for a Sunbelt county." *Journal of Urban Affairs*, vol. 34, no. 4, pp. 419–439.

Collins, T.W., Kim, Y., Grineski, S.E., & Clark-Reyna, S. (2014). "Can economic deprivation protect health? Paradoxical multilevel effects of poverty on Hispanic children's wheezing." *International Journal of Research and Public Health*, vol. 11, no. 8, pp. 7856–7873.

Collins, T.W., Grineski, S.E., Chakraborty, J., Montgomery, M.C., & Hernandez, M. (2015). "Downscaling environmental justice analysis: Determinants of household-level hazardous air pollutant exposure in greater Houston." *Annals of the Association of American Geographers*, vol. 20, no. 4, pp. 684–703.

Comber, A., Brunsdon, C., & Green, E. (2008). "Using a GIS-based network analysis to determine urban greenspace accessibility for different ethnic and religious groups." *Landscape and Urban Planning*, vol. 86, pp. 103–114.

CRJ (Commission for Racial Justice). (1987). *Toxic Wastes and Race in the United States: A National Report on the Racial and Socioeconomic Characteristics of Communities with Hazardous Waste Sites.* New York: United Church of Christ Commission for Racial Justice.

Crowder, K. & Downey, L. (2010). "Inter-neighbourhood migration, race, and environmental hazards: Modelling microlevel processes of environmental inequality." *The American Journal of Sociology*, vol. 115, no. 4, pp. 1110–49.

Cutter, S.L., Holm, D., & Clark, L. (1995). "The role of geographic scale in monitoring environmental justice." *Risk Analysis*, vol. 16, no. 4, pp. 517–26.

Downey, L. (2005). "Assessing environmental inequality: How the conclusions we draw vary according to the definitions we employ." *Sociological Spectrum*, vol. 5, no. 3, pp. 349–369.

Downey, L. (2006). "Using geographic information systems to reconceptualize spatial relationships and ecological context." *The American Journal of Sociology*, vol. 112, no. 2, pp. 567–612.

Fisher, J.B., Kelly, M., & Romm, J. (2006). "Scales of environmental justice: Combining GIS and spatial analysis for air toxics in West Oakland, California." *Health and Place*, vol. 12, pp. 701–714.

Fotheringham, S.A., Brunsdon, C., & Charlton, M.E. (2002). *Geographically Weighted Regression: The Analysis of Spatially Varying Relationships.* Chichester, UK: Wiley.

Fotheringham, A.S. & Wong, D.W.S. (1991). "The modifiable areal unit problem in multivariate statistical analysis." *Environment and Planning A*, vol. 23, pp. 1025–44.

Germani, A.R., Morone, P., & Testa, G. (2014). "Environmental justice and air pollution: A case study on Italian provinces." *Ecological Economics*, vol. 106, pp. 69–82.

Getis, A. & Ord, J.K. (1992). "The analysis of spatial association by use of distance statistics." *Geographical Analysis*, vol. 24, pp. 189–206.

Gilbert, A. & Chakraborty, J. (2011). "Using geographically weighted regression for environmental justice analysis: Cumulative cancer risks from air toxics in Florida." *Social Science Research*, vol. 40, no. 1, pp. 273–286.

Grineski, S.E., Collins, T.W., Chakraborty, J., & McDonald, Y.J. (2013). "Environmental health injustice: Exposure to air toxics and children's respiratory hospital admissions in El Paso, Texas." *The Professional Geographer*, vol. 65, no. 1, pp. 31–46.

Grineski, S., Collins, T.W., Chakraborty, J., & Montgomery, M. (2015a). "Hazardous air pollutants and flooding: A comparative interurban study of environmental injustice." *GeoJournal*, vol. 80, no. 1, pp. 145–158.

Grineski, S.E., Collins, T.W., & Olvera, H.A. (2015b). "Local variability in the impacts of residential particulate matter and pest exposure on children's wheezing severity: A geographically weighted regression analysis of environmental health justice." *Population and Environment*, vol. 37, no. 1, pp. 22–43.

Heckert, M. (2013). "Access and equity in greenspace provision: A comparison of methods to assess the impacts of greening vacant land." *Transactions in GIS*, vol. 17, no. 6, pp. 808–827.

Hillier, A., Cole, B.L., Smith, T.E., Yancey, A.K., Williams, J.D., Grier, S.A., & McCarthy, W.J. (2009). "Clustering of unhealthy outdoor advertisements around child-serving institutions: A comparison of three cities." *Health and Place*, vol. 15, pp. 935–945.

Kissling, W.D. & Carl, G. (2008). "Spatial autocorrelation and the selection of simultaneous autoregressive models." *Global Ecology and Biogeography*, vol. 17, pp. 59–71.

Landry, S. & Chakraborty, J. (2009). "Street trees and equity: Evaluating the spatial distribution of an urban amenity." *Environment and Planning A*, vol. 41, no. 11, pp. 2651–2670.

Lenhardt, J. & Ogneva-Himmelberger, Y. (2013). "Environmental injustice in the spatial distribution of concentrated animal feeding operations in Ohio." *Environmental Justice*, vol. 6, no. 4, pp. 133–139.

Lersch, K.M. & Hart, T.C. (2014). "Environmental justice, lead, and crime: Exploring the spatial distribution and impact of industrial facilities in Hillsborough County, Florida." *Sociological Spectrum*, vol. 34, no. 1, pp. 1–21.

LeSage, J. & Pace, R.K. (2009). *Introduction to Spatial Econometrics*. Boca Raton, FL: CRC Press.

Lievanos, R.S. (2015). "Race, deprivation, and immigrant isolation: The spatial demography of air-toxic clusters in the continental United States." *Social Science Research*, vol. 54, pp. 50–67.

Lloyd, C.D. (2010). *Local Models for Spatial Analysis*. Boca Raton, FL: CRC Press.

Maantay, J. (2007). "Asthma and air pollution in the Bronx: Methodological and data considerations in using GIS for environmental justice and health research." *Health & Place*, vol. 13, no. 1, pp. 32–56.

Maantay, J.A., Tu, J., & Maroko, A.R. (2009). "Loose-coupling an air dispersion model and a geographic information system (GIS) for studying air pollution and asthma in the Bronx, New York City." *International Journal of Environmental Health Research*, vol. 19, no. 1, pp. 59–79.

Maroko, A., Maantay, J.A., & Grady, K. (2011). "Using geovisualization and geospatial analysis to explore respiratory disease and environmental health justice in New York City." In Maantay, J.A. & S. McLafferty (eds.), *Geospatial Analysis of Environmental Health*, Springer Science and Media, pp. 39–66.

McEntee, J.C. & Ogneva-Himmelberger, Y. (2008). "Diesel particulate matter, lung cancer, and asthma incidences along major traffic corridors in MA, USA: A GIS analysis." *Health and Place*, vol. 14, pp. 817–828.

McLeod, H., Langford, H., Jones, A.P., Stedman, J.R., Day, R.J., Lorenzoni, I., & Bateman, I.J. (2000). "The relationship between socio-economic indicators and air pollution in England and Wales: Implications for environmental justice." *Regional Environmental Change*, vol. 1, no. 2, pp. 78–85.

McMaster, R.B., Leitner, H., & Sheppard, E. (1997). "GIS-based environmental equity and risk assessment: Methodological problems and prospects." *Cartography and Geographic Information Science*, vol. 24, no. 3, pp. 72–189.

Mennis, J. (2002). "Using geographic information systems to create and analyse statistical surfaces of population and risk for environmental justice analysis." *Social Science Quarterly*, vol. 83, no. 1, pp. 281–297.

Mennis, J. (2006). "Mapping the results of geographically weighted regression." *The Cartographic Journal*, vol. 43, no. 2, pp. 171–179.

Mennis, J. (2011). "Integrating remote sensing and GIS for environmental justice research." In X. Yang (ed.), *Urban Remote Sensing: Monitoring, Synthesis, and Modelling in the Urban Environment*, Chichester, UK: Wiley, pp. 225–237.

Mennis, J.L. & Jordan, L. (2005). "The distribution of environmental equity: Exploring spatial non-stationarity in multivariate models of air toxic releases." *Annals of the Association of American Geographers*, vol. 95, no. 2, pp. 249–268.

Mennis, J. & Mason, M.J. (2011). "People, places, and adolescent substance use: Integrating activity space and social network data for analyzing health behavior." *Annals of the Association of American Geographers*, vol. 101, no. 2, pp. 272–291.

Mitchell, G. & Chakraborty, J. (2014). "Urban heat and climate justice: A landscape of thermal inequity in Pinellas County, Florida." *Geographical Review*, vol. 104, no. 4, pp. 459–480.

Mohai, P. & Saha, R. (2006). "Reassessing racial and socioeconomic disparities in environmental justice research." *Demography*, vol. 43, no. 2, pp. 383–399.

Noonan, D.S., Turaga, R.M.R., & Baden, B.M. (2009). "Superfund, Hedonics, and the scale of Environmental Justice." *Environmental Management*, vol. 44, pp. 909–920.

Ogneva-Himmelberger, Y., Huang, L., & Xin, H. (2015). "CALPUFF and CAFOs: Air pollution modelling and environmental justice analysis in the North Carolina hog industry." *ISPRS International Journal of Geo-Information*, vol. 4, pp. 150–171.

Oyana, T.J. & Margai, F.M. (2010). "Spatial patterns and health disparities in pediatric lead exposure in Chicago: Characteristics and profiles of high-risk neighbourhoods." *The Professional Geographer*, vol. 62, no. 1, pp. 46–65.

Pais, J., Crowder, K., & Downey, L. (2014). "Unequal trajectories: Racial and class differences in residential exposure to industrial hazard." *Social Forces*, vol. 92, no. 3, pp. 1189–1215.

Pham, T.-T.-H., Apparicio, P., Seguin, A.-M., Landry, S., & Gagnon, M. (2012). "Spatial distribution of vegetation in Montreal: An uneven distribution or environmental inequity?" *Landscape and Urban Planning*, vol. 107, pp. 214–224.

Raddatz, L. & Mennis, J. (2013). "Environmental justice in Hamburg, Germany." *The Professional Geographer*, vol. 65, no. 3, pp. 495–511.

Ripley, B.D. (1976). "The second-order analysis of stationary processes." *Journal of Applied Problems*, vol. 13, pp. 255–266.

Rojas, F. (2011). "Poverty determinants of acute respiratory infections in the Mapuche population of ninth region of Araucania, Chile (2000–2005): A Bayesian approach with time-space modelling." In Maantay, J.A. & S. McLafferty (eds.), *Geospatial Analysis of Environmental Health*, Springer Science and Media, pp. 411–442.

Schwartz, J., Bellinger, D., & Glass, T. (2011). "Expanding the scope of risk assessment: Methods of studying differential vulnerability and susceptibility." *American Journal of Public Health*, vol. 101, no. 1, S1, pp. S102–S109.

Sheppard, E., Leitner, H., McMaster, R.B., & Hongguo, T. (1999). "GIS based measures of environmental equity: Exploring their sensitivity and significance." *Journal of Exposure Analysis and Environmental Epidemiology*, vol. 9, no. 1, pp. 18–28.

Snijders, T.A.B. & Bosker, R.J. (2012). *Multilevel Analysis: An Introduction to Basic and Advanced Multilevel Modelling* (Second edition), London: Sage.

Stahler, G.J., Mennis, J., & Baron, D. (2013). "Geospatial technology and the exposome: New perspectives on addiction." *American Journal of Public Health*, vol. 103, no. 8, pp. 1354–1356.

Taquino, M., Parisi, D., & Gill, D.A. (2002). "Units of analysis and the environmental justice hypothesis: The case of industrial hog farms." *Social Science Quarterly*, vol. 83, no. 1, pp. 298–316.

Williams, R.W. (1999). "The contested terrain of environmental justice research: Community as a unit of analysis." *Social Science Journal*, vol. 36, no. 2, pp. 313–28.

Wilson, S.M., Richard, R., Joseph, L., & Williams, E. (2010). "Climate change, environmental justice, and vulnerability: An exploratory analysis." *Environmental Justice*, vol. 3, no. 1, pp. 13–19.

Yandle, T. & Burton, D. (1996). "Methodological approaches to environmental justice: A rejoinder." *Social Science Quarterly*, vol. 77, no. 3, pp. 520–27.

Zimmerman, R. (1994). "Issues of classification in environmental equity: How we manage is how we measure." *Fordham Urban Law Journal*, vol. 21, pp. 633–669.

Zou, B., Peng, F., Wan, N., Mamady, K., & Wilson, G.J. (2014). "Spatial cluster detection of air pollution exposure inequities across the United States." *PLOS ONE*, vol. 9, no. 3.

18

HISTORICAL APPROACHES TO ENVIRONMENTAL JUSTICE

Christopher G. Boone and Geoffrey L. Buckley

Early work in environmental justice focused on the immediate need of analysing the distribution of environmental burdens, such as toxic facilities or landfills, to see if low-income and ethnic/racial minority communities live with a disproportionately high concentration of environmental hazards. In more cases than not, disproportionate burdens for minority communities turned out to be the case and it provided important evidence for environmental justice activists and scholars (Mohai and Saha 2007). Investigations into the causes of environmental injustices led to a sharper focus on the fairness of process rather than distribution alone. This can include fairness in the application of environmental and other laws, development and maintenance of fair institutions, fairness in decision-making, and recognition of marginalized groups as stakeholders in decisions (Schlosberg 2007).

Historical approaches provide an important and useful mechanism for understanding the origins, causes, and legacies of present day environmental injustices, both for process and outcome. However, working with historical data has its challenges. Gathering data can be time consuming and labour intensive. Written records may not exist and even when they do, gaps and inaccuracies often prevent us from drawing definitive conclusions. Living informants may not be available and when they are their recollection of past people, places, and events may be biased or unreliable. Sometimes we are presented with too much information, making the challenge one of selection and interpretation (Harris 1978; Kipping et al. 2013). Despite these limitations, historical data sets are useful to environmental justice investigators. When used with caution, census data, company records, government documents, meeting minutes, newspapers, oral histories, and photographic archives cast light on the processes that produce the landscapes and patterns we see today (Massard-Guilbaud and Rodger 2011).

In this chapter, we outline some of the historical methods scholars have used to explicate the landscape of environmental injustices. We draw extensively on environmental justice work we have conducted for the Baltimore Ecosystem Study, an urban long-term ecological research project sponsored by the US National Science Foundation (Grove et al. 2015).

Legacies and environmental burdens

One of the limitations of many environmental justice analyses is the use of single snapshots in time, dictated usually by the date or year of the most recent data set (Mohai and Saha 2015).

A typical approach is to compare the demographics of places using decennial census data (e.g. 2010) and the distribution of an environmental hazard such as the US Environmental Protection Agency's Toxics Release Inventory in the same year. Examining the distribution of environmental burdens in relation to where different social groups live is an important step, but this dominant approach provides little understanding of the processes that created unfair burdens in the first place (Taylor 2014). It is important and worthwhile to analyse the fairness of current distributions, but unwise to infer process based on correlations of present day conditions. Cities are landscapes produced by accumulations of past and present decisions (Colten 2002). In our own research we have found that many legacies of past decisions and processes decades old persist down to the present. The redlining of Baltimore by the federal Home Owners Loan Corporation in the 1930s, for example, set in motion decades of disinvestment in primarily African American communities (Boone et al. 2009). Redlined boundaries from the 1930s still define to a large extent where the majority of vacant properties presently exist (Grove et al. 2015). In Phoenix, redlining of Latino and African American neighbourhoods in the 1930s set in motion a pattern of disinvestment that prescribes outcomes to this day (Bolin et al. 2005). Cities are dynamic, ever changing landscapes but some aspects of the built environment and urban life endure.

The city of Commerce, California, located east of downtown Los Angeles, has long been one of the hotspots for toxic facilities in the United States. Using archival evidence, Boone and Modarres (1999) show the importance of zoning decisions made by the Los Angeles County Regional Planning Commission more than eight decades ago in the creation of this highly polluted city. In 1929, the commission designated this area of the county, and no other, as suitable for heavy industry. It had locational advantages, including open farmland ready to be developed and three railroads passing through it, but minutes from the zoning board show that racist attitudes towards non-Anglo populations played a role in specifically designating this area as ready for industrial land use. Noteworthy as well is where the board did not zone for high density commercial and industrial: the wealthier and whiter west end of LA County (Boone 2005). The development of Commerce into a toxic city was related to racist thoughts about the largely Latino and East Asian population in that part of the county and a deflection of unwanted land uses from the well-to-do west end, a material expression of white and social privilege (Pulido 2000).

Elsewhere in Los Angeles County, Pulido et al. (1996) demonstrate using archival evidence how Latinos came to occupy the most polluted districts of Torrance. Designed as an 'industrial garden city' in the early part of the twentieth century, Torrance was pitched as an ideal community for white workers. However, some of the lowest-paying jobs were granted to people of Mexican origin who were permitted to live only in a small residential area zoned for heavy industry. Pulido et al. (1996) draw on letters, magazines, and official documents (such as deeds) to show that in Torrance there was a pronounced racialized division of labour and restrictive housing choices that confined Mexicanos to living in the most polluted neighbourhoods.

In the case of Phoenix, Bolin et al. (2005) explain how the present concentration of polluting industries in low income and Hispanic communities is the result of racial discrimination that extends to the founding of the city. Promoted as an Anglo city, in contrast to the identity of some of the Mexican and colonial Spanish towns in the US Southwest, Phoenix nevertheless always had a Hispanic population (in addition to an African American population). In 1887, when Phoenix was a small town of only 5000 people, a rail line bisected the settlement, creating an Anglo north and Latino and black south. South Phoenix became the dumping ground for the growing city in the twentieth century, receiving sewage from the north side of the tracks, housing the stockyards and dirty industries, leaving its largely Latino and African American residents to live in deplorable conditions. Lack of adequate housing, running water, and sanitation facilities contributed to one of the highest infant mortality rates in the country in the 1920s and 1930s. Restrictive covenants and deeds barring Mexicans and African Americans from living north of the rail line meant crowded conditions and disinvestment, as did the redlining by the Home Owners Loan Corporation in the 1930s. Restrictive deeds also meant that minorities could not take advantage of low interest mortgage loans from the Federal Housing Authority to purchase suburban homes in the 1940s and 1950s. This historical study provides important insights on the role of institutional racism – related to land use zoning, infrastructure investment, and housing – in shaping outcomes past and present, and provides a useful method for explaining environmental injustice formation (Bolin et al. 2005).

For Baltimore, analyses of environmental justice patterns have generated unexpected results, considering the long history of racism endured by African Americans in that city. Here, majority white neighbourhoods are more likely than majority black neighbourhoods to contain toxic facilities (Boone 2002, 2008). Baltimore is and was a very segregated city that historically hemmed black families into dense and relatively expensive housing through deed restrictions, covenants, and for a brief period municipal ordinances (Power 1983; Olson 1997). Following a hostile incident when a black lawyer moved into a majority white neighbourhood, the City passed an ordinance in 1911 that created separate white and 'coloured' blocks. The ordinance had support from influential corners, including the Peabody Heights Improvement and Protection Association, which also advocated for deed restrictions against Jews (Merse et al. 2009; Buckley and Boone 2011). In 1917, the ordinance and others like it were declared illegal by the US Supreme Court, but deed restrictions imposed by property owners persisted well into the 1960s and 1970s, even after the Fair Housing Act of 1964. The net effect was to segregate African Americans into areas of West and East Baltimore and some sections of South Baltimore. It happens that these neighbourhoods were located away from sought-after factory jobs where white workers had preference (Olson 1997). What had been seen as an amenity in early twentieth century Baltimore – living close to well-paying factory jobs – later translated into the location of Toxics Release Inventory sites. Although a present day analysis does not reveal a disproportionate burden of TRI sites in black communities, the current landscape of hazardous facilities is nevertheless the product of a long history of racism and disadvantage for African Americans in Baltimore. The processes that led to these seemingly just outcomes of fewer toxic facilities in black neighbourhoods were patently unfair.

While most environmental justice research focuses on urban areas, a growing body of work examines inequity in rural locations. Recent studies have used newspaper accounts, corporate documents, and interviews to track how public health and environmental crises have unfolded in central Appalachia over the past several decades. While the social and environmental costs of living near mountaintop removal mining sites are well documented, exposure to polluting industries also presents risks (Morrone and Buckley 2011). Along the Ohio River, for instance, C8 (perfluorooctanoic acid) contamination from the DuPont Corporation's Washington Works facility has tainted drinking water supplies and divided communities. Such is the case in Little Hocking, Ohio. While some residents maintain that DuPont has been a good neighbour and steady employer, others disagree, arguing that the company withheld scientific evidence for years and exposed its workers and their families to harmful chemicals. Whether the company's actions constitute an environmental injustice is openly debated today, pitting workers with few alternative employment opportunities against activists and others whose health appears to be compromised by C8 exposure (Kozlowski and Perkins 2016). Downstream, in Cheshire, Ohio, emissions from a large coal-fired power plant led to the purchase and dissolution of the small river town by American Electric Power (AEP). While the sale of the town made headlines in national newspapers, historical research showed that friction between residents and AEP had been building for years (Buckley et al. 2005). What this study reveals is that environmental justice activism can be the result of an accumulation of difficulties over a long period of time rather than a response to a single, episodic event.

Legacies and environmental benefits

Environmental justice scholarship has traditionally focused on environmental burdens, such as toxic sites or hazardous waste facilities, but a growing body of work examines equity in the distribution of and access to environmental amenities such as green space or tree canopy cover (Lawrence 2008; Wolch et al. 2014). While most studies examine present day conditions, some use historical approaches to reveal inequities (e.g. Boone et al., 2009; Wells et al. 2008). In Baltimore today, African American neighbourhoods enjoy greater access to parks within walking distance than do white neighbourhoods, although the latter generally have access to more acres. Rather than conclude that African Americans benefit from a just distribution of parks, however, historical research shows just the opposite. Park Board meeting minutes, court records, and newspaper accounts point to a long history of de jure and de facto segregation, which precluded blacks from participating in a wide variety of recreational activities even when they lived in close proximity to a neighbourhood park. Only after large numbers of whites left the city for the suburbs starting in the 1950s was the African American population able to spread out and gain access to all of the city's parks and recreational facilities, including golf courses and swimming pools (Wells et al. 2008). Not only did African Americans "inherit" this system of parks, they played no role in the decision-making process that led to its design and distribution.

Regarding tree canopy cover, typically seen as an environmental benefit, legacy effects of past decisions regarding tree planting have present day consequences. Management strategies from decades ago favoured tree planting in certain districts over others, sometimes under the influence of neighbourhood improvement and protection associations that proved adept at attracting amenities such as parks and trees while promoting segregation and deflecting unwanted land uses elsewhere (Buckley 2010; Buckley and Boone 2011). In other cases, however, an environmental justice determination is more difficult to make. Today, sections of East Baltimore are virtually treeless despite the best efforts of city foresters since the 1950s. When these areas were inhabited by immigrants from southern and eastern Europe and their descendants, residents opposed planting programmes on the grounds that they preferred "clean, uncluttered concrete" (Buckley 2010). Starting in the 1960s large numbers of African Americans migrated to this part of the city and inherited a landscape devoid of trees. This demographic shift did not signal a change in attitude toward urban trees. While not vehemently opposed to more trees, residents have questioned the ability of the city to manage its street trees and voiced concern about gentrification and displacement. They argue that the city has more pressing problems to solve, including crime and trash collection, before it embarks on an aggressive tree planting campaign (Battaglia et al. 2014).

In Washington, DC, several blocks of the Trinidad neighbourhood, a predominantly African American community, possess extensive canopy coverage while others nearby are bare. Closer inspection reveals that many of the trees are very old and dying. Historical census data and archival records indicate that the trees were planted in the 1940s when the neighbourhood was predominantly white. Residents complain that the city does not adequately cull and prune the trees and that roots and fallen limbs are responsible for property damage. Although residents of Trinidad inherited a leafy landscape, poor maintenance has created a present day condition where trees have lost a lot of their amenity value. Meanwhile, residents of Georgetown, an affluent and largely white enclave, circumvent city maintenance efforts by holding fundraisers and hiring private arborists (Buckley et al. 2013).

Longitudinal studies

Another historical approach used in environmental justice research is longitudinal analysis, the examination of change over time. Widely used in public health research, longitudinal analysis has been employed to explore causality between environmental conditions and health outcomes of individuals and groups. Alcock et al. (2014), for example, tracked individuals over time using the British Household Panel Survey to demonstrate that mental health improved for those who moved closer to green spaces. In early environmental justice work, some researchers examined whether minority groups or polluting industry moved in first as a way of assessing if industry deliberately located in minority communities (Been and Gupta 1997). The reasoning is that if minorities moved into a neighbourhood after a polluting industry was established, then industry could not be blamed and the resulting situation should not be interpreted as an environmental injustice. One of the difficulties of "which came first" studies is that they may not take into account other forms of discrimination (e.g. housing, employment) that limit choices of where minority communities can live or feel they can live. These institutional forms of racism may create environmental injustices regardless of whether polluting industry or minority communities moved in first (Pulido 2000). To understand the complex dynamics of land use change requires engagement with a broader set of social, economic, and biophysical drivers than timing of demographic change.

With these cautions in mind, longitudinal analyses can nevertheless provide some insights into environmental injustice formation, the persistence or change of environmental inequity over time, and temporal coincidence with some larger societal shifts (Mohai and Saha 2015). Based on unexpected findings in Baltimore, Boone et al. (2014) asked the question of whether the current concentration of polluting industries in white neighbourhoods is a recent phenomenon or one with a longer history. The Environmental Protection Agency's Toxics Release Inventory (TRI) dates back only to 1987 but since the screening for the TRI sites begins with specific heavy industry and petrochemical facility codes, the investigators were able to use Dun and Bradstreet business directories to identify likely polluting industries beginning in 1960 and match those locations to historical census data and boundaries. Analysis of the concentration of polluting industry at ten year intervals from 1960 to 2010 showed that from 1960 to 1980, neighbourhoods with lower income and lesser educated families had higher concentrations of polluting industries, while from 1990 to 2010, it was primarily white and less educated neighbourhoods that had the highest concentrations. This longitudinal analysis showed a shift in environmental inequity correlated with income (1960–1980) to one correlated with race (1990–2010), while low education attainment remained significantly correlated with polluted neighbourhoods for the entire half century. On the environmental benefits side, another team working for the Baltimore Ecosystem Study examined historical census data from 1960 to 2000 and showed that the best predictor of current tree canopy cover in the Gwynns Falls Watershed of Greater Baltimore was the percentage of residents in professional occupations in 1960. The demographic characteristics of who lived in the neighbourhoods in 1960, perhaps the individuals who planted and cared for the trees still standing, turned out to be the most important variable for understanding the distribution of a current environmental amenity (Boone et al. 2010).

Longitudinal analysis can be used to study the effects of broad social change on local environmental conditions. Saha and Mohai (2005), for example, test whether growing environmental awareness in the United States in the 1970s influenced environmental inequity patterns in Michigan. They examine the distribution of hazardous facilities (hazardous waste treatment, storage and disposal facilities) from 1950 to 1990 and the demographic and housing characteristics, using the decennial census, of surrounding neighbourhoods. The census data were weighted according to the proportion of the census tracts that fell within a one-mile circular buffer of the hazardous waste facility. The statistical analyses confirmed the authors' hypothesis. Prior to 1970, most hazardous waste facilities were sited near white and higher than average income neighbourhoods. The facilities that were sited after 1970, however, tended to be located in or near neighbourhoods that had disproportionately high non-white populations, lower mean family incomes, and lower housing values. Heightened awareness of the dangers of hazardous waste and toxins after the Love Canal incident in the early 1980s and the Bhopal explosion in 1984 may have intensified the NIMBY (Not In My Back Yard) forces that deflected unwanted land uses away from socially privileged neighbourhoods. The interesting conclusion from this study is that growing environmental awareness in the 1970s and 1980s may be responsible for worsening patterns of environmental inequity.

Since the unit of analysis in environmental justice research tends to be at the community or neighbourhood level, longitudinal analysis remains a challenge. In the United States, census categories and boundaries change over time. By 1960, individuals were able to choose their own race designation on census forms, although the categories were fixed by the Census Bureau. Identifying oneself as Hispanic, a problematic term for many, was not an option until 1970. Census boundaries shift over time to keep contained populations relatively similar while taking into account land use and demographic changes. The result is that neighbourhood boundaries can move over time, making longitudinal analysis challenging. One of the most difficult issues of using community level data provided by the census is the inability to track changes in location of individuals, the gold standard in public health. Panel studies, surveys, and interviews, although time consuming and expensive, are alternative approaches that can explore the dynamics of environmental exposure and environmental justice formation (Mohai and Saha 2015).

Benefits and burdens of historical approaches

Despite the clear value of historical approaches, they have not been employed in environmental justice to a large degree. Similar to any set of methodologies, historical analyses have the benefits we have described but also significant costs (see Table 18.1). Investment of time working in archives, for instance, can be onerous. Success working in archives requires patience and perseverance, as days can go by without any useful information being uncovered. Many researchers enter archives without sufficient training or experience, which can further delay progress (Kipping et al. 2013). Others are overwhelmed by the sheer volume of documents and are unable to discern quickly what is relevant to the research question. Bias in what information is recorded and what is omitted can be difficult to discern, especially in older documents. Some major national newspapers are indexed, which can ease searches, but many of the smaller papers that served communities of colour are not. Interviews with living informants or their descendants can provide insights that official documents may not, but

Table 18.1 Benefits and burdens of selected historical approaches to environmental justice research

Method	Benefits	Burdens	Method examples
Archival (official documents, letters, newspapers)	• detailed information captured in documents • good source for formal institutions • identify key individuals and motivations, as well as political conflicts	• difficult to search • time consuming to search and code • information gaps • publisher bias (esp. newspapers)	Bolin et al. (2005); Boone (2005); Boone et al. (2009); Hurley (1997); Pulido et al. (1996); Taylor (2014); Washington (2005)
Interviews and ethnographies	• lived experience of real people • direct answers to questions	• inaccurate recollections • limited number of informants especially for older periods • time to code and analyse responses	Gandy (2002); Bataglia et al. (2005); Darby (2012); Hernandez et al. (2015)
Longitudinal data	• relatively consistent data points over time (e.g. census data) • ability to measure trends	• some data not digital • must infer process based on observations over time	Boone (2008); Boone et al. (2015); Pastor et al. (2001); Mohai and Saha (2005, 2015); Mitchell and Norman (2012)

recall and bias can be a problem. However, speaking to people rather than "listening" to documents can provide a sense of daily life and experience that is valuable for environmental justice researchers. Interviews and ethnographies allow researchers to ask questions and get direct answers, another advantage over archival methods.

A promising approach for environmental justice inquiry is to use longitudinal data, much of it now available in digital format, which can speed up analysis, and is especially important for national level inquiries. The National Historical Geographic Information System (www.nhgis.org/) at the University of Minnesota Population Center, for example, is an excellent resource for historical census data and boundaries that can be employed in time series analyses. Categorical consistency over time remains an issue so some caution must be employed in the use of historical census and other data sets. Longitudinal analyses can begin to unravel processes, but other qualitative information is necessary to fully understand environmental injustice formation. Combining archival, interview, and longitudinal methods requires a sophisticated team of investigators and talented persons to synthesize the information and results. Although a combination of these methods can provide a rich comprehension of environmental justice dynamics, it is difficult to do so beyond single places, making generalization and theory building challenging.

Lessons from the past for building a sustainable future

The primary argument of this chapter is that historical approaches provide insights into the processes that create environmental injustices in the present as well as the past. Such approaches also elucidate the multiple factors and complexities that generate uneven environmental consequences for different groups of people (Pellow 2002; Washington 2005). These lessons provide pathways for addressing environmental injustices in the present and avoiding them in the future by highlighting some of the key factors in environmental injustice formation. Unfair housing and zoning practices should receive special scrutiny based on findings from many of the historical analyses in the environmental justice literature. Recognition and participation of historically disadvantaged groups in planning and other decision-making institutions is another important corrective based on the research we have summarized. Decision-makers and community groups can use knowledge of the past to build a fairer, more inclusive future by compensating for past wrongs and designing fair processes that actively seek to avoid environmental injustices (Colten 2007). Future work in environmental justice should not forget the past and the rich understanding it can provide of the processes many seek to remedy in order to build more just futures.

References

Alcock, I., White, M.P., Wheeler, B.W., Fleming, L.E., & Depledge, M.H. (2014) "Longitudinal effects on mental health of moving to greener and less green urban areas", *Environmental Science & Technology*, vol. 48, no. 2, pp. 1247–1255.

Battaglia, M., Buckley, G.L., Galvin, M., & Grove, J.M. (2014) "It's not easy going green: Obstacles to tree-planting programs in east Baltimore", *Cities and the Environment*, vol. 7, no. 2, article 6.

Been, V. & Gupta, F. (1997) "Coming to the nuisance or going to the barrios? A longitudinal analysis of environmental justice claims", *Ecology Law Quarterly*, vol. 24, no. 1, pp. 1–56.

Bolin, B., Grineski, S., & Collins, T. (2005) "The geography of despair: Environmental racism and the making of South Phoenix, Arizona, USA", *Human Ecology Review*, vol. 12, no. 2, pp. 156–168.

Boone, C.G. (2002) "An assessment and explanation of environmental inequity in Baltimore", *Urban Geography*, vol. 23, no. 6, pp. 581–595.

Boone, C.G. (2005) "Zoning and environmental inequity in the industrial east side", In Deverell, W. & G. Hise (eds.), *Land of sunshine: An environmental history of metropolitan Los Angeles*. Pittsburgh: University of Pittsburgh Press, pp. 167–178.

Boone, C.G. (2008) "Improving resolution of census data in metropolitan areas using a dasymetric approach: Applications for the Baltimore Ecosystem Study", *Cities and the Environment*, vol. 1, no. 1, article 3.

Boone, C.G. & Modarres, A. (1999) "Creating a toxic neighborhood in Los Angeles County: A historical examination of environmental inequity", *Urban Affairs Review*, vol. 35, no. 2, pp. 163–187.

Boone, C.G., Buckley, G.L., Grove, J.M., & Sister, C. (2009) "Parks and people: An environmental justice inquiry in Baltimore, Maryland", *Annals of the Association of American Geographers*, vol. 99, no. 4, pp. 1–21.

Boone, C.G., Cadenasso, M.L., Grove, J.M., Schwarz, K., & Buckley, G.L. (2010) "Landscape, vegetation characteristics, and group identity in an urban and suburban watershed: Why the 60s matter", *Urban Ecosystems*, vol. 13, no. 3, pp. 255–271.

Boone, C.G., Fragkias, M., Buckley, G.L., & Grove, J.M. (2014) "A long view of polluting industry and environmental justice in Baltimore", *Cities*, vol. 36, pp. 41–49.

Buckley, G.L. (2010) *America's conservation impulse: A century of saving trees in the Old Line State*. Chicago: Columbia College and the Center for American Places.

Buckley, G.L. & Boone, C.G. (2011) "'To promote the material and moral welfare of the community': Neighborhood improvement associations in Baltimore, Maryland, 1900–1945." In Rodger, R. & G. Massard-Guilbaud (eds.), *Environmental and social justice in the city: historical perspectives*. Cambridge: White Horse Press, pp. 43–65.

Buckley, G.L., Bain, N.R., & Swan, D.L. (2005) "When the lights go out in Cheshire", *The Geographical Review*, vol. 95, no. 4, pp. 537–555.

Buckley, G.L., Whitmer, A.C., & Grove, J.M. (2013) "Parks, trees, and environmental justice: Field notes from Washington, DC", *Applied Environmental Education & Communication*, vol. 12, no. 3, pp. 148–162.

Colten, C.E. (2002) "Basin Street blues: Drainage and environmental equity in New Orleans, 1890–1930", *Journal of Historical Geography*, vol. 28, no. 2, pp. 237–257.

Colten, C.E. (2007) "Environmental justice in a landscape of tragedy", *Technology in Society*, vol. 29, no. 2, pp. 173–179.

Darby, K.J. (2012) "Lead astray: Scale, environmental justice and the El Paso smelter", *Local Environment*, vol. 17, no. 8, pp. 797–814.

Gandy, M. (2002) *Concrete and clay: Reworking nature in New York City*. Cambridge: MIT Press.

Grove, J.M., Cadenasso, M.L., Pickett, S.T.A., Machils, G.E., & Burch, W.R. (2015) *The Baltimore School of Urban Ecology: Space, scale, and time for the study of cities*. New Haven: Yale University Press.

Harris, C. (1978) "The historical mind and the practice of geography". In Jey, D. & M. Samuels (eds.), *Humanistic geography: Prospects and problems*. London: Croom Helm, pp. 123–37.

Hernandez, M., Collins, T.W., & Grineski, S.E. (2015) "Immigration, mobility, and environmental injustice: A comparative study of Hispanic people's residential decision-making and exposure to hazardous air pollutants in Greater Houston, Texas", *Geoforum*, vol. 60, pp. 83–94.

Hurley, A. (1997) "Fiasco at Wagner Electric: Environmental justice and urban geography in St. Louis", *Environmental History*, vol. 2, no. 4, pp. 460–81.

Kipping, M., Wadhwani, R.D., & Bucheli, M. (2013) "Analyzing and interpreting historical sources: A basic methodology." In Bucheli, M. & R.D. Wadhwani (eds.), *Organizations in time: History, theory, methods*. Oxford: Oxford University Press, pp. 305–330.

Kozlowski, M. & Perkins, H.A. (2016) "Environmental justice in Appalachia Ohio? An expanded consideration of privilege and the role it plays in defending the contaminated status quo in a white, working-class community", *Local Environment*, vol. 21, no. 10, pp. 1288–1304.

Lawrence, Henry W. (2008) *City trees: A historical geography from the Renaissance through the nineteenth century*. Charlottesville: University of Virginia Press.

Massard-Guilbaud, G. & Rodger, R. (2011). *Environmental and social justice in the city: Historical perspectives*. Cambridge: The White Horse Press.

Merse, C.L., Buckley, G.L., & Boone, C.G. (2009) "Street trees and urban renewal: A Baltimore case study", *The Geographical Bulletin*, vol. 50, no. 2, pp. 65–82.

Mitchell, G. & Norman, P. (2012) "Longitudinal environmental justice analysis: Co-evolution of environmental quality and deprivation in England, 1960–2007", *Geoforum*, vol. 43, no. 1, pp. 44–57.

Mohai, P. & Saha, R. (2007) "Racial inequality in the distribution of hazardous waste: A national-level reassessment", *Social Problems*, vol. 54, no. 3, pp. 343–370.

Mohai, P. & Saha, R. (2015) "Which came first, people or pollution? A review of theory and evidence from longitudinal environmental justice studies", *Environmental Research Letters*, vol. 10, no. 12, p. 125011.

Morrone, M. & Buckley, G.L. (eds.) (2011) *Mountains of injustice: Environmental and social justice in Appalachia*. Athens: Ohio University Press.

Olson, S.H. (1997) *Baltimore: The building of an American city*. Baltimore: The Johns Hopkins University Press.

Pellow, D.N. (2002) *Garbage wars: The struggle for environmental justice in Chicago*. Cambridge: MIT Press.

Power, G. (1983) "Apartheid Baltimore style: The residential segregation ordinances of 1910–1913", *Maryland Law Review*, vol. 42, no. 2, pp. 289–328.

Pulido, L. (2000) "Rethinking environmental racism: White privilege and urban development in Southern California", *Annals of the Association of American Geographers*, vol. 90, no. 1, pp. 12–40.

Pulido, L., Sidawi, S., & Vos, R.O. (1996) "An archaeology of environmental racism in Los Angeles", *Urban Geography*, vol. 17, no. 5, pp. 419–439.

Saha, R. & Mohai, P. (2005) "Historical context and hazardous waste facility siting: Understanding temporal patterns in Michigan", *Social Problems*, vol. 54, no. 2, pp. 618–648.

Schlosberg, D. (2007) *Defining environmental justice: Theories, movements, and nature*. Oxford: Oxford University Press.

Taylor, D.E. (2014) *Toxic communities: Environmental racism, industrial pollution, and residential mobility*. New York: New York University Press.

United States Environmental Protection Agency (2016). Toxics Release Inventory (TRI) Program [Online], Available: www.epa.gov/toxics-release-inventory-tri-program Accessed 5 July 2016.

Washington, S.H. (2005) *Packing them in: An archaeology of environmental racism in Chicago, 1865–1954*. Lanham: Lexington Books.

Wells, J., Buckley, G.L., & Boone, C.G. (2008) "Separate but equal? Desegregating Baltimore's golf courses", *The Geographical Review*, vol. 98, no. 2, pp. 151–170.

Wolch, J.R., Byrne, J., & Newell, J.P. (2014) "Urban green space, public health, and environmental justice: The challenge of making cities 'just green enough.'" *Landscape and Urban Planning*, vol. 125, pp. 234–244.

19

THE ETHICS OF EMBODIED ENGAGEMENT

Ethnographies of environmental justice

Catalina M. de Onís and Phaedra C. Pezzullo

Ethnography ("ethno" = culture and "graphy" = writing) involves firsthand research to interpret practices within a particular context through qualitative methods, including but not limited to interviews and participant observation. As we will argue, ethnography has been vital to the development of environmental justice as a movement and as an academic research topic due, at least in part, to its ethical commitment to fostering dialogue between the researcher and the researched. Of note for environmental justice, *critical ethnography* is defined by Madison (2012: 5) as a practice that "begins with an ethical responsibility to address processes of unfairness or injustice within a particular lived domain." Rather than romanticizing "critical distance" as a criterion of academic research, Conquergood (1982: 5, 9–11) invites critical ethnographic scholars to engage in "genuine conversation" through a "dialogical performance."

Another reason why ethnography is particularly significant to environmental justice research is its attention to embodiment and experiential knowledge. For example, when questioned for evidence of her toxic chemical exposures, Charlotte Keyes, long-time environmental justice activist and founder of Jesus People against Pollution, stated: "the evidence is in my body" (Lerner 2005: 218). Keyes' claim evinces the importance of resisting privileged epistemologies by foregrounding the lived experiences of those most directly and severely impacted by environmental injustices. By attending to embodied experiences, ethnographies document aspects of environmental injustice that are all too often erased or otherwise elided (Pezzullo 2003).

Ethnographies unfold in various ways. Ethnographic studies may be conducted by members of an environmental community or movement, as well as by those who identify as somehow "outside." They may be single- or multi-sited and last for various durations of time, ranging from a few hours to several years. They may involve a range of embodied participation practices, from interviewing to volunteering to working for an organization.

In this chapter, we summarize ethical motivations and choices related not only to embodiment, but also voice, engagement, dialogue, and future trends. We organize these themes by addressing autoethnography, participant observation, interviews, and emergent practices. Overall, we hope to make evident the reasons why environmental justice research benefits from ethnographic practices, as well as the range of ethnographic dilemmas this research raises.

"We speak for ourselves": autoethnography

Autoethnography is not simply autobiography, or the writing of one's life story, but draws on personal experiences to help understand broader cultural values, challenges, and practices that exceed an individual. From an environmental justice perspective, autoethnographies aim to foreground authors' own biases, limitations, and interconnections to better understand environmental injustices and justice. Is my story relevant to how I came to my research project? When can my own experience in the field illuminate or further marginalize the injustices a community is describing? How does the way I respond in an interview shape what the interviewee will say next?

Most importantly, considering one's own position in relation to environmental justice research is important because the environmental justice movement persuasively has argued that it matters who is allowed to speak and when. As late legal advocate Alston (1990) emphasized, "we speak for ourselves." Inclusion has been a significant principle in making space at the decision-making table for those most directly and disproportionately impacted by pollution burdens. It, therefore, is not uncommon to find environmental justice scholarship that includes an autoethnographic narrative.

The foundational Black environmental justice sociologist Bullard regularly recalls how his own research and advocacy turned to environmental justice as a result of a request when he was a first year professor at Texas Southern University in 1978. He was asked to study the city he was living in for waste patterns in relation to racial demographics for a class action lawsuit, *Bean v. Southwestern Waste Management*:

> my graduate students and I mapped the location of every major landfill site in Houston using pushpins on paper. If we noticed a hill in the usually flat landscape, we investigated it because a change in topography often indicated a dump. We found that although at that time Blacks made up just over one-fourth of Houston's population, five out of five city-owned landfills (100 percent) and six of the eight city-owned incinerators (75 percent) were sited in Black neighborhoods. After my study for the Bean case, my career became linked with the environmental justice movement, and I have since then had the opportunity to work with communities all over the world.
>
> *(Bullard 2014: 28)*

Environmental justice movement leaders and scholars such as Bullard often defy rigid categories separating academics from those who are most impacted by environmental injustices.

Some autoethnographic work is performed on stage, in addition to the page. For example, Madison's (2007, 2010) critical ethnographic practices have included *Water Rites*, a multimedia performance based on critical ethnographic research in Ghana about water as a human right. This critical orientation engages the researcher's embodied relations to pollution and other forms of environmental injustice. In another example, Vignes (2008) drew on her own experiences with Hurricane Katrina in her performance of *Hang It Out to Dry*, which presented the storm not as something to be forgotten like "water under the bridge," but as waves of a disaster with ongoing traumas and injustice. *Cry You One* (Cryyouone.com 2015) is a much larger, ongoing project in the southern Louisiana region that exemplifies how performances can be interactive and constructed in conjunction with digital storytelling (see also Chapter 20).

There are also environmental justice scholars who begin as activists and then become academics. One of us was asked by a reviewer to be transparent that her work was "autoethnographic". We quote the response to this request at length as a way of explaining the limits of autoethnographic labels:

> Although I do not dwell heavily on my personal experiences in this book, some still consider this project a kind of "auto-ethnography" insofar as I describe and discuss tours that I both observed as a researcher and participated as an activist. Yet, I should pause to explain what I mean when I claim to be a part of the environmental justice movement. The environmental justice movement, in part, is based on a sense of location. It began in communities that felt they were or still are disproportionately targeted for pollution because of their racial identity and/or economic status. Many residents of these areas argue that they have had no choice about their activism—they have to do something in order to survive ... I, on the other hand, am not poor nor am I a person of color. I do not live in a grassroots community of the movement. I am middle class and live with white privilege. I *choose to visit* places that have been polluted and to struggle for environmental justice. There is a difference. ... [F]or those of us who literally go out of our way to participate in a movement, our "belonging" does mean something different. There is something distinct at stake in the performances of those of us who are able to choose more freely our associations.
>
> *(Pezzullo 2007)*

As illustrated here, what constitutes autoethnography as a specific genre and ethnography more broadly is complex and even labelling a study as "autoethnographic" reflects an ethical decision for academics.

"Being there made a difference": embodied participant observation

Lindlof and Taylor define *participant observation* generally as "the craft of experiencing and recording events in social settings" (2002: 134). This method requires two approaches: 1) embodied participation or the involvement of a researcher in a particular cultural practice; and 2) observation or the study, reflection, and sharing of insights derived from such an intimate epistemology, or way of knowing. The assumption is that travelling matters, as a Bay Area breast cancer and anti-toxic activist said: "It's *not* that I didn't know. I *did* know. And I've heard *many* people from that community talk. But, *being there* made a difference" (Pezzullo, 2007: 19).

Participant observation in environmental justice research can take many forms. Environmental justice researchers have embodied the role(s) of witness, performer, fellow worker, note taker, consumer, protestor, community member, and other positions in a host of situations, including cultural and city expos (Sze 2015), neighbourhood group meetings (Anguelovski 2015), international climate negotiations (Joshi 2014), local rallies and protests (Urkidi & Walter 2011), toxic tours (Pezzullo 2003, 2007), water rights events and rituals (Madison 2007; Mehta et al. 2014), farmers' markets (Alkon 2008), workplaces employing low-income community members (Wilson et al. 2009), after-school urban farm and garden educational initiatives (Travaline & Hunold 2010), forest walking programmes and conservation groups (Pinder et al. 2009), governmental and environmental agency proceedings (de Onís 2016), citizen advisory boards (Pezzullo 2001), public education meetings for environmental projects (Sze et al. 2009), and much more.

To undertake participation, ethnographic researchers are confronted with the necessity of reflecting on their positionalities and relationships with different environments and communities. Some ethnographers study groups with racial and ethnic backgrounds different from their own (Checker 2005). Others engage communities with which they share a common heritage. For example, Peña (2005) and Pulido (1996) have written about numerous environmental brutalities and challenges encountered by Chicanx communities. Informing his fieldwork, Peña heads the Acequia Institute, a non-profit foundation that works on water and agricultural projects with community members in the upper Rio Grande area of southern Colorado. Pulido studied environmental racism and socio-economic barriers in US Mexican-origin communities. From charting the anti-pesticides struggles of the United Farm Workers to efforts by members of Ganados del Valle/Livestock Growers of the Valley in New Mexico, her (1996) scholarship uncovers possibilities for resistive practices opposed to racial violence and how race and place are co-constitutive.

Ethnographers may select single- or multi-sited studies, which may be of short or long duration. Some ethnography is based on long-term familiarity with a place, which holds the potential to offer in-depth insights about environmental justice exigencies and techniques for resisting injustice, as well as their evolution. Wright's (2005) environmental justice research, school curriculum development, and outreach in New Orleans, Louisiana spans multiple decades. Her role as the founding director of the Deep South Center for Environmental Justice evinces how prolonged advocacy for liveable, healthy lives in a specific place can assist in toxic site remediation and effective engagement with intersectional environmental concerns, including displacement and unemployment. Depoe's (2004) environmental justice research and community activism also is motivated by decades of living near the Fernald nuclear site in Ohio. Sze (2007: 23) argues her seven years of participant observation in four low-income New York City neighbourhoods enabled her to chart how environmental justice "groups were able to use a discourse of environmental justice despite their conflicts and contradictions."

No matter how much time is spent with a community, trust is necessary to undertake ethical ethnographic research. Calling her work "activist ethnography," Checker (2005) volunteered with a local community group for 13 months, during which she engaged in grant writing, after-school programme leading, and website creation and maintenance. She notes the importance of building reciprocal relationships – giving something back to the community in exchange for them making her research possible – and nods toward the implications for building community trust. Checker (2005: 194) explicates the expertise and everyday practices that may be reciprocated after trust with a community is established over time: "I eventually found that participating in HAPIC aided my research, and sometimes my research aided my participation in HAPIC."

Multi-sited ethnographic studies of environmental justice are typically shorter term, but they use comparative observations to analyse the successes, challenges, and stakes of community activism circulated and diverged across ethnic, regional, national, and other cultural borders. Marcus (1998: 79–80) defines "multi-sited ethnography" as studying "the circulation of cultural meanings, objects, and identities in diffuse time-space. This mode defines for itself an object of study that cannot be accounted for ethnographically by remaining focused on a single site of intensive investigation." For example, Pezzullo's (2007) study of toxic tours as an advocacy practice drew upon multiple sites in North America with a focus on southern Louisiana, San Francisco, and Matamoros, Mexico. Likewise, Anguelovski's (2015) fieldwork followed tactics employed by activists living in urban neighbourhoods in Barcelona, Boston, and Havana. Alkon's (2008) study of two farmers' markets in the San Francisco Bay area, one in a community that is predominantly affluent and White, the other in one generally low-income and Black, reveals the different environmental commitments and needs shaping diverse discourses in these communities. Alkon's participant observation makes possible the discovery of multiple meanings of the environment that help us to better understand and participate in local food movements cognizant of socio-economic and racial difference that might otherwise be ignored, essentialized, or misread.

In addition to creating possibilities for studying important differences between privileged and underprivileged communities, comparative multi-sited ethnographic research also may shed light on environmental justice principles related to sovereignty, decolonization, and democratization. From studying and participating in the Brazilian environmental justice network, for example, Roberts (2007) reflects on why environmental organizations in the United States are more likely to partner with more radical organizations abroad than at home. Powell's (2015) fieldwork with the Navajo Nation on energy controversies particularly underscores how an environmental topic such as decentralized power may resonate and be negotiated with an Indigenous perspective on sovereignty – the right to community self-determination and local, sustainable energy generation. Casas-Cortés et al. (2008: 71; 2013) also argue that studying social movements in general warrants an ethnographic, place-based approach to explore what they call "knowledge-practices." They draw on comparative fieldwork on a North American Indigenous environmental justice network, Chicago's Direct Action Network collective, and Italy's alter-globalization movement to establish the ways in which embodied research practices over time with a community may provide additional insights to better understand social movements as ongoing processes. These insights include findings about how meaning-making informs collective action and how troubling readings of particular individuals as "irrational" impede environmental justice.

"Aquí es el que duele, el que sabe": interviews

"Here are the human beings that hurt, the human beings that know." This translated remark is derived from one of our personal interviews with a nurse and an environmental justice activist in Vieques, Puerto Rico (de Onís 2016). By interviewing the "human beings that know," researchers enact another dimension shaping ethnographies of environmental (in)justice. As Madison (2012: 28) notes, "The interview is a window to individual subjectivity and collective belonging: *I am because we are and we are because I am.*" These interview relationships often unfold in three different forms that we will define accordingly: oral history, personal narrative, and topical interview.

Oral history involves "a recounting of a social historical moment reflected in the life or lives of individuals who remember them and/or experienced them" (Madison 2012: 28). Ethnographers may draw on already existing oral histories for study, contribute interviews to extant archives, or build their own oral history collections in the absence of other records. Such efforts create openings for sharing these voices with different publics. For instance, Pezzullo and Depoe (2010) draw on a digital oral history archive, in part developed by Depoe, in order to study the everyday impacts of living near and working in a nuclear weapons facility. Endres (2011), who has constructed American Indian oral histories of nuclear waste sites, argues that the archival nature of oral histories, that is, knowing they will be made available for future researchers and publics, is a significant part of their value.

In their study of the impacts of Marine Protected Areas on local fishing communities in Guam and the Northern Marianas Islands, Richmond and Kotowicz (2015) used oral histories to advance environmental justice in at least two ways. First, their oral histories showed how these protected areas risked creating injustice, by limiting local community member access to culturally and economically significant waters. In this case, oral histories uncovered and foregrounded cultural values rooted in place and the "global inequities based on class, ethnicity, and geopolitical position" implicated in different conservation techniques (p. 124). In addition, using interviews to address a dearth of paper records on the topic, the researchers were able to provide "the most accurate and longest running information about travel to and use of the Islands Unit waters" (p. 120).

Oral history also can uncover a hidden transcript (Scott 1990) that reminds us of the importance of carefully considering a range of cultural perspectives and positions that might not be apparent without fieldwork or intimate community knowledge. For example, Wanzer-Serrano (2015: 169) draws on oral history in his study of the New York Young Lords and members' decolonial performative rhetoric striving for "community control" in the face of numerous environmental and other structural injustices. As the group mobilized around various issues, including demanding adequate municipal garbage services and testing for lead poisoning, some of the Young Lords' lesser-known struggles involved the "revolution within the revolution": an effort by women in the group to call attention to machismo by male leaders (p. 104). By sharing "Testimonios de Transgresión"/"Testimonies of Transgression" (p. 94), Wanzer-Serrano chose "to privilege the voices of women in the organization so that their stories of struggle can recalibrate the male-centred tales that are most often told" (p. 93). Discussion of physical violence against women in the group "is rarely retold in public forums—and is missing from all published accounts" (p. 101).

Interviews also enable ethnographers to generate *personal narrative*, defined as "an individual perspective and expression of an event, experience, or point of view" (Madison 2012: 28). In a study of meaning-making in the discourses and practices of environmental justice activists, Allen et al. (2007) present interlocutors' self-identifications and journeys to environmental activism. They show how environmental action emerges from "the significance of identities— durable subjectivities and self-understandings, both collective and individual—that develop through cultural activities within 'figured worlds' of environmental justice and environmentalism and through dialogues with people and groups both inside and outside the movement" (p. 105). They contend that studying these activists' practices of interpretation and meaning-making is key for understanding how movements build commitments for environmentalism and environmental justice (see also Chapter 20).

Finally, the *topical interview* focuses on participants' perspectives on "a particular subject, such as a program, an issue, or a process" (Madison 2012: 28). Ethnographic studies using topical interviews typically seek perspectives from a wide range of individuals, including public officials, community members, and environmental group leaders. Lawhon (2013), for example, chronicles the tensions and possibilities resulting from e-waste discourses in South Africa by attending to different actors' use of the ecological modernization and environmental justice frameworks. Moore (2009) explores garbage strikes as an important resource for political change, guided by interview questions shaped by her interest in the relationship between waste and urban areas. Place and Hanlon (2011: 168) document First Nation responses to a mining site expansion project in British Columbia. Using a topical focus and interview exchanges, they report that Indigenous community members expressed concerns about the project regarding "three major themes: environment, human health, and doubts and mistrust about the EIA [Environmental Impact Assessment] process itself."

Reflecting on positionality and building relationships of trust are crucial to ethnographic approaches to interviewing. Park and Pellow discuss both of these issues in their critique of the "flipside" of environmental injustice: environmental privilege (2011: 3). Interviewing environmentalists and Latinx immigrants, they contend that environmental privilege occurs when "some groups can access spaces and resources, which are protected from the kinds of ecological harm that other groups are forced to contend with everyday [sic]" (p. 4). They assert (p. 26):

> In the course of our research, we found ourselves in a fortunate position in which immigrant workers—some of whom were in quite vulnerable situations—trusted us with their personal stories of immigration, work and labour, and raising families. They gave us candid reflections on how they survive, whom they work for, and the nature of local politics. We were able to gain their trust because of our relationships with well-respected community leaders from social service, religious, media, and advocacy organizations.

Park and Pellow's mention of establishing community trust calls attention to the relationships community members and researchers forge and negotiate and the different risks involved for each, given different (and sometimes shared) power differentials, affiliations, and stakes.

These interviewer-interviewee exchanges are vital for recuperating forgotten, ignored, or silenced stories to ensure that those disciplined by technocratic, patriarchal practices (which too often discipline emotionality and emplaced cultural ways of knowing) are heard and taken seriously. Di Chiro long has amplified the voices of those most impacted by environmental decisions and discourses (1992). She has published interviews testifying to how some environmental movement organizations have marginalized environmental justice communities (2000), how Indigenous voices and rights are oppressed in most genetic research (2007), and how some anti-toxic discourse may marginalize transgender bodies (2010). Similarly, Kaplan (1997: 156) uses ethnographic interviews to document and bring attention to "calls for justice" of women's movements living in sacrifice zones. As demonstrated above, ethnography, achieved via participant observation and interviews, serves an imperative scholarly and activist function to ensure these marginalized movements do not "disappear from view."

Ethnographic futures for a more environmentally just world

We close by gesturing toward ongoing challenges and possibilities that environmental justice ethnographers face today. Like any research method, ethnography is not without its limitations. For example, working across difference, building trust with community members, translating conversations in the field for academic forums, and blending qualitative and quantitative approaches together are just a few areas that can pose challenges for ethnographers and community members. However, we contend that obstacles to ethnography must not deter researchers from the participatory possibilities that this method affords, especially since ethnography resonates with many environmental justice principles (e.g. inclusion, sovereignty, decolonization, and democratization).

As we contemplate the future of ethnographic research to address environmental justice concerns, encouraging and enacting collaboration across disciplines is a fruitful endeavour (see Chapter 21). Holifield et al. (2009: 593) explain: "The [environmental justice] field has also begun to find productive engagements with other academic traditions, each of which has stimulated the asking of new questions and the exploration of new analytical and explanatory strategies." Consideration of how ethnography might be mobilized to explore intersecting environmental, climate, and energy justice concerns – whether (trans)/urban or (trans)/rural – provides an exemplar to consider.

We also find that cross-cultural comparisons might advance ethnographic environmental justice research (Carter et al. 2013: 144). Exemplifying the crossing of not only disciplinary boundaries but also geographic and cultural borders, Ranganathan and Balazs (2015: 417) studied drinking water marginalization in India and California; they used a blend of archival work, interviews, surveys, and mapping to guide their collaborative, comparative approach "in which scholars across the North–South divide talk to and learn from each other." This collaboration may raise challenges caused by cultural and communication differences, as well as varying institutional expectations and timelines. Relatedly, attention to translation – whether it involves navigating disciplinary differences, explaining complex technical language, and/or literally translating from one language to another – is a key consideration for ethnographic research (de Onís 2016).

Universities increasingly seem to be valuing "public engagement" or "engaged scholarship," which can resonate with environmental justice ethnographic research approaches that we have outlined. Frontline environmental justice communities also continue to find value in these interactions, at times, to help them address and resist environmental injustices. There is an ethical burden to sharing these stories in spaces where they might not be heard otherwise. As a 93-year-old domestic worker noted in an interview: "I don't want my good name and what I'm telling you to be tossed around up there at that there University like some ol' rag" (Madison 1998: 276). Listening, therefore, is only part of the ethical choices one faces; how and what to share where continues to pose challenges once the researcher begins to write.

Finally, it is worthwhile to consider what new and emerging media technologies afford for ethnographies of environmental justice. Despite the ongoing environmental injustices perpetuated by e-waste, media technologies can assist in presenting new ethnographic insights about environmental crises and injustices in interactive modes of circulation. For example, environmental justice studies have employed PhotoVoice, which involves training community-member researchers to take photographs that can be used to initiate discussions about environmental justice concerns. Conversations resulting from the PhotoVoice approach also can be shared with various decision-makers who might not otherwise encounter or consider the perspectives of marginalized voices (Harper et al. 2009). Others, such as Vannini and Taggart (2013: 307), include in their footnotes URL hyperlinks to the audio versions of interviews.

We have noted the importance of autoethnography, participant observation, and interviews in the field, in academic journals, on stage, and as digital archives or interactive storytelling projects. We also have outlined challenges and possibilities that environmental justice ethnographers face, including interdisciplinary, cross-cultural, public engagement, and multimodal research. Taylor (2011: 291) emphasizes that environmental justice movement research methods are protean and "will continue to expand." With pressing and colliding ecological, economic, and social crises facing the world today, ethnography must remain a vital approach for those studying the frontlines of environmental injustice.

References

Alkon, A. (2008). "Paradise or pavement: The social constructions of the environment in two urban farmers' markets and their implications for environmental justice and sustainability." *Local Environment*, vol. 13, pp. 271–89.

Allen, K., Daro, V., & Holland, D.C. (2007). "Becoming an environmental justice activist." In Sandler, R. and P.C. Pezzullo (eds.), *Environmental justice and environmentalism*. Cambridge, MA and London: MIT Press.

Alston, D. (1990). *We speak for ourselves: Social justice, race and environment*. Washington, DC: Panos Institute.

Anguelovski, I. (2015). "Tactical developments for achieving just and sustainable neighborhoods: The role of community based coalitions and bottom-to-bottom networks in street, technical, and funder activism." *Environment and Planning C*, vol. 33, pp. 1–23.

Bullard, R. (2014). "Environmental justice and the politics of garbage: The mountains of Houston." *Cite: The Architecture and Design Review of Houston*, vol. 93, pp. 28–33.

Carter, E.D., Silva, B., & Guzmán, G. (2013). "Migration, acculturation, and environmental values": The case of Mexican immigrants in central Iowa." *Annals of the Association of American Geographers*, vol. 103, pp. 129–47.

Casas-Cortés, M.I., Osterweil, M., & Powell, D.E. (2008). "Blurring boundaries: Knowledge-practices in contemporary social movements." *Anthropological Quarterly*, vol. 81, no. 1, pp. 17–58.

Checker, M. (2005). *Polluted promises: Environmental racism and the search for justice in a southern town.* New York: NYU Press.

Conquergood, D. (1982). "Performing as a moral act: Ethical dimensions of the ethnography of performance." *Literature in Performance*, vol. 5, no. 1, pp. 1–13.

Cryyouone.com. (2015). *Cry You One.* [online] Available at: www.cryyouone.com/ Accessed 25 Sep. 2015.

de Onís, K.M. (2016). "'Pa' que tú lo sepas': Experiences with co-presence in Puerto Rico." In McKinnon, S., R. Asen, K. Chávez, and R. Howard (eds.), *Text + field: Innovations in rhetorical method.* State College, PA: Pennsylvania State Press, pp. 101–116.

Depoe, S.P. (2004). "Public involvement, civic discovery, and the formation of environmental policy: A cooperative analysis of the Fernald Citizens Task Force and the Fernald Health Effects Subcommittee." In Depoe, S.P., J.D. Delicath, and M. Aepli (eds.), *Communication and public participation in environmental decision-making.* Albany: State University of New York Press, pp. 157–73.

Di Chiro, G. (1992). "Defining environmental justice: Women's voices and grassroots politics." *Socialist Review*, vol. 22, no. 4, pp. 92–130.

Di Chiro, G. (2000). "Bearing witness or taking action?: Toxic tourism and environmental justice." In Hofrichter, R. (ed.), *Reclaiming the environmental debate: The politics of health in a toxic culture.* Cambridge, MA: MIT Press, pp. 275–300.

Di Chiro, G. (2007). "Indigenous peoples and biocolonialism: Defining the 'science of environmental justice' in the century of the gene." In Sandler, R. and P.C. Pezzullo (eds.), *Environmental justice and environmentalism: The social justice challenge to the environmental movement.* Cambridge, MA: MIT Press, pp. 251–83.

Di Chiro, G. (2010). "Polluted politics? Confronting toxic discourse, sex panic, and eco-normativity." In Mortimer-Sandilands, C. and B. Erickson (eds.), *Queer ecologies: Sex, nature, politics, desire.* Bloomington, IN: Indiana University Press, pp. 199–230.

Endres, D. (2011). "Environmental oral history." *Environmental Communication: A Journal of Nature and Culture*, vol. 5, no. 4, pp. 485–98.

Harper, K., Steger, T., & Filčák, R. (2009). "Environmental justice and Roma communities in Central and Eastern Europe." *Environmental Policy and Governance*, vol. 19, no. 4, pp. 251–68.

Holifield, R., Porter, M., & Walker, G. (2009). "Introduction spaces of environmental justice: Frameworks for critical engagement." *Antipode*, vol. 41, pp. 591–612.

Joshi, S. (2014). "Environmental justice discourses in Indian climate politics." *GeoJournal*, vol. 79, pp. 677–91.

Kaplan, T. (1997). *Crazy for democracy: Women in grassroots movements.* New York: Routledge.

Lawhon, M. (2013). "Dumping ground or country-in-transition? Discourses of e-waste in South Africa." *Environment and Planning*, vol. 31, pp. 700–15.

Lerner, S. (2005). *Diamond: A struggle for environmental justice in Louisiana's chemical corridor.* Cambridge, MA: The MIT Press.

Lindlof, T. & Taylor, B. (2002). *Qualitative communication research methods.* Thousand Oaks, CA: Sage.

Madison, D.S. (1998). "Performance, personal narratives, and the politics of possibility." In Dailey, S.J. (ed.), *The future of performance studies: Visions and revisions.* Annandale, VA: National Communication Association, pp. 276–286.

Madison, D.S. (2007). "Performing ethnography: The political economy of water." *Performance Research*, vol. 12, pp. 16–27.

Madison, D.S. (2010). *Acts of activism: Human rights as radical performance.* Cambridge: Cambridge University Press.

Madison, D.S. (2012). *Critical ethnography: Method, ethics, and performance*, Second edition. Thousand Oaks, CA: Sage.

Marcus, G.E. (1998). *Ethnography through thick and thin.* Princeton, NJ: Princeton University Press.

Mehta, L., Allouche, J., Nicol, A., & Walnycki, A. (2014). "Global environmental justice and the right to water: The case of peri-urban Cochabamba and Delhi." *Geoforum*, vol. 54, pp. 158–66.

Moore, S.A. (2009). "The excess of modernity: Garbage politics in Oaxaca, Mexico." *The Professional Geographer*, vol. 61, no. 4, pp. 426–37.

Peña, D.G. (2005). *Mexican Americans and the environment: Tierra y vida.* Tucson, AZ: University of Arizona Press.

Park, L.S. & Pellow, D.N. (2011). *The slums of Aspen: Immigrants vs. the environment in America's Eden.* New York: NYU Press.

Pezzullo, P.C. (2001). "Performing critical interruptions: Rhetorical invention and narratives of the environmental justice movement." *Western Journal of Communication*, vol. 64, no. 1, pp. 1–25.

Pezzullo, P.C. (2003). "Resisting 'National Breast Cancer Awareness Month': The rhetoric of counter-publics and their cultural performances." *Quarterly Journal of Speech*, vol. 89, pp. 345–65.

Pezzullo, P.C. (2007). *Toxic tourism: Rhetorics of travel, pollution, and environmental justice.* Tuscaloosa, AL: University of Alabama Press.

Pezzullo. P.C. & Depoe, S.P. (2010). "Everyday life and death in a nuclear world: Stories from Fernald." In Brouwer, D.C., and R. Asen (eds.), *Public modalities: rhetoric, culture, media, and the shape of public life.* Tuscaloosa, AL: University of Alabama Press, pp. 85–108.

Pinder, R., Kessel, A., Green, J., & Grundy, C. (2009). "Exploring perceptions of health and the environment: A qualitative study of Thames Chase Community Forest." *Health Place*, vol. 15, pp. 349–56.

Place, J. & Hanlon, N. (2011). "Kill the lake? Kill the proposal: Accommodating First Nations' environmental values as a first step on the road to wellness." *GeoJournal*, vol. 76, pp. 163–75.

Powell, D.E. (2015). "*The rainbow is our sovereignty*: Rethinking the politics of energy on the Navajo Nation." *Journal of Political Ecology*, vol. 22, pp. 27–52.

Pulido, L. (1996). *Environmentalism and economic justice: Two Chicano struggles in the southwest.* Tucson, AZ: University of Arizona Press.

Ranganathan, M. & Balazs, C. (2015). "Water marginalization at the urban fringe: Environmental justice and urban political ecology across the North–South divide." *Urban Geography*, vol. 36, pp. 403–23.

Richmond, L. & Kotowicz, D. (2015). "Equity and access in marine protected areas: The history and future of 'traditional indigenous fishing' in the Marianas Trench Marine National Monument." *Applied Geography*, vol. 59, pp. 117–24.

Roberts, J.T. (2007). "Globalizing environmental justice." In Sandler, R. and P.C. Pezzullo (eds.), *Environmental Justice and Environmentalism: The Social Justice Challenge to the Environmental Movement.* Cambridge, MA: MIT Press, pp. 285–307.

Scott, J.C. (1990). *Domination and the arts of resistance: Hidden transcripts.* New Haven, CT and London: Yale University Press.

Sze, J. (2007). *Noxious New York: The racial politics of urban health and environmental justice.* Cambridge, MA and London: MIT Press.

Sze, J. (2015). *Fantasy islands: Chinese dreams and ecological fears in an age of climate crisis.* Oakland, CA: University of California Press.

Sze, J., London, J., Shilling, F., Gambirazzio, G., Filan, T., & Cadenasso, M. (2009). "Defining and contesting environmental justice: Socio-natures and the politics of scale in the Delta." *Antipode*, vol. 41, pp. 807–43.

Taylor, D.E. (2011). "The evolution of environmental justice activism, research, and scholarship." *Environmental Practice*, vol. 13, no. 4, pp. 280–301.

Travaline, K. & Hunold, C. (2010). "Urban agriculture and ecological citizenship in Philadelphia." *Local Environment*, vol. 15, pp. 581–90.

Urkidi, L. & Walter, M. (2011). "Dimensions of environmental justice in anti-gold mining movements in Latin America." *Geoforum*, vol. 42, pp. 683–95.

Vannini, P. & Taggart, J. (2013). "Voluntary simplicity, involuntary complexities, and the pull of remove: The radical ruralities of off-grid lifestyles." *Environment and Planning A*, vol. 45, pp. 295–311.

Vignes, D.S. (2008). "Hang it out to dry: A performance script." *Text and Performance Quarterly*, vol. 28, pp. 351–65.

Wanzer-Serrano, D. (2015). *The New York Young Lords and the struggle for liberation.* Philadelphia, PA: Temple University Press.

Wilson, D., Beck, D., & Bailey, A. (2009). "Neoliberal-parasitic economies and space building: Chicago's southwest side." *Annals of the Association of American Geographers*, vol. 99, pp. 604–26.

Wright, B. (2005). Living and dying in Louisiana's "Cancer Alley." In R.D. Bullard (ed.), *The Quest for Environmental Justice: Human Rights and the Politics of Pollution.* San Francisco, CA: Sierra Club Books.

20

STORYTELLING ENVIRONMENTAL JUSTICE

Cultural studies approaches

Donna Houston and Pavithra Vasudevan

Introduction

In early 2012, community members in Warrenton, North Carolina, were busy planning the commemoration of an event that profoundly shaped the town's local environmental history. Facilitated by Rev. William Kearney, long-time residents, activists, representatives from local churches, local government, and not-for-profit organizations met to discuss how best to commemorate the thirty-year anniversary of the community protest over the siting of a PCB landfill in Warren County – now widely acknowledged as the spark that ignited the Environmental Justice Movement in the United States (Bullard 2007; McGurty 2009). The community discussions leading up to the thirtieth anniversary were imbued with hopeful socio-ecological imaginings for the future, as well as the residual social impacts of the PCB toxic legacy. For some, mobilizing the "birthplace of environmental justice" story as a cultural resource had transformative possibility for local community development; while for others, the anniversary brought up sedimented feelings of anger and distrust over the siting of the PCB landfill and its later remediation (Vasudevan 2013).

The public commemoration of environmental justice in Warren County was both unsettling and cathartic. It highlights the significant long-term impacts that environmental injustice can have on local communities, long after the contaminated site is cleaned up. Environmental injustice does not just leave material "scars" on the landscape; it leaves cultural scars too, the legacies of which evoke a range of memories, emotions, affects, values and practices (Storm 2014). The commemoration event highlights the complex interplay of culture and environmental justice – where local activists employ heritage practices to narrate the local and global consequences of historical and ongoing injustices, and to mobilize communities in struggling for change (Banerjee and Steinberg 2015).

The legacies of the PCB landfill in Warren County – material and storied – highlight the ways in which culture, history and place are entangled with environmental justice (Houston 2013; Vasudevan 2013). Circulating within and between structural inequities and the politics of legal recognition in environmental justice struggles are meanings, memories, practices and values – cultural phenomena that help individuals, communities, activists, and advocates make sense of and communicate with each other about the social impacts and underlying causes of pollution and unwanted land uses – often in the absence of officially recognized evidence and forms of public accountability. Indeed, one of the early lessons from Warren County, where the local community failed to stop the PCB landfill, was that the *story* of environmental justice that emerged there has enduring material effects (Alston 2010; Pezzullo 2001).

In this chapter, we focus on cultural studies approaches to environmental justice activism and research. We bring together theories and methodologies from cultural studies to frame environmental justice as a mode of "storytelling" that performs and communicates what it means to live with environmental injury. Storytelling, as we define it here, is an enlarged concept encompassing relational, emotional, embodied, place-based knowledge and practices (Brickell and Garrett 2015; Cameron 2012). It reflects varied cultural contexts and realities whereby individuals and communities come to interpret and articulate environmental crises and impacts. It is therefore important to acknowledge that modes of storytelling in environmental justice are "more-than-discursive" (Cameron 2012). As we will discuss in the following sections of the chapter, storytelling encompasses narratives and performances capable of disrupting boundaries between real and imagined, particularly as they are represented in Western scientific knowledge (Blaser 2010). Storytelling does not disregard scientific or policy perspectives about environmental in/justice but it does call into account the types of social realities they produce and for whom.

Storytelling involves the enactment of "cultural methodologies" – which include fictional, non-fictional, autobiographical, testimonial, and scientific accounts of environmental injustice. Environmental justice "stories" are narrated and enacted through different types of cultural media, including: film, documentary film, poetry, plays, novels, graphic novels, social activist media, place-based tours and walks, public commemoration and participatory citizen science projects. While not exhaustive, this list highlights the diversity of cultural methods and also the sense in which environmental justice stories are performed in the world (Houston 2008; Vasudevan 2012). "Cultural methodologies" in environmental justice are enacted by environmental scholars and activists, often in ways that blur boundaries between academic and community knowledge, scientific and political authority, and personal experience. Utilizing methods ranging from tours of toxic sites to citizen science efforts to gather evidence of harm, communities impacted by environmental injustices construct stories relating their lived experiences to theories of socio-environmental change. These cultural practices transcend boundaries between lay people and experts and reveal complex, intimate relationships between people and localized environmental degradation (Di Chiro 2003, 2008). In what follows, we draw out these insights by introducing some key concepts from cultural studies before going on to discuss some of the typical elements of how environmental justice is storied through narratives and other types of cultural practices.

Environmental justice and cultural studies

In environmental justice research, cultural elements are often subsumed or simplified within framings of "rights" and "justice". Cultural politics, practices, representations and values might intersect with claims for rights and justice, but cultural phenomena tend to be treated as part of the background context upon which struggles unfold. Recently, environmental justice scholars have expanded framings of justice to be more critical and attentive to race and cultural difference (Pulido 1996, 2015; Schlosberg 2007; Walker 2012). Influenced by feminism, postcolonialism, critical race theory, critical Indigenous perspectives, and poststructuralism, expanded notions of justice focus on the importance of "recognition" and "misrecognition" for understanding how environmental inequity intersects with other forms of marginalization (Agyeman et al. 2009; Holifield 2012; Pulido 2000; Schlosberg 2007;

Walker 2012; also see Chapter 10). As Walker (2012: 50-51) argues, the cultural politics of misrecognition comprise different states of symbolic violence and social injury – such as the failure to recognize and respect cultural difference, the failure to recognize the invisibility of systematic nonrecognition experienced by marginalized groups, and the failure to recognize forms of cultural domination and oppression in mainstream cultural framings of environmental issues and conflicts.

The growing emphasis on "justice-as-recognition" provides significant opportunities to explore cultural politics and methods in environmental justice research and activism. Such an approach acknowledges that culture is itself an arena of social and political struggle and that the performative "cultural field" of environmental justice can be studied in its own right (see Adamson et al. 2002 for an excellent example). Cultural approaches enrich understandings of how individuals and communities draw on diverse and historically situated experiences and resources to tell their stories about environmental injustice and to intervene politically in dominant discourses shaping cultural contexts of discard, disregard, disposability and sacrifice (Banerjee and Steinberg 2015; Dickinson 2012; Houston 2008; Pezzullo 2001; Vasudevan 2012; Vasudevan and Kearney 2015).

"Culture", Raymond Williams (1983: 87) famously wrote in *Keywords*, "is one of two or three of the most complicated words in the English language." Culture is a complicated term because it has a long history with various past and contemporary meanings, and because it operates across symbolic, ideological and material domains (Williams 1983). The study of culture is similarly complex, drawing together critical analyses of values, experiences, attitudes, practices, materials and perceptions. The academic field of "cultural studies" emerged in mid-twentieth century Britain as an umbrella term for academic inquiry that investigates how cultural forms, identities, institutions and practices are constituted by and imbricated in relations of power and, in particular, how the relationship between culture and power is differentially experienced as a result of a person or group's race, class, gender or sexuality (Nelson et al. 1992). In the latter part of the twentieth century, cultural studies developed a strong profile in the US, Canada and Australia, expanding into an interdisciplinary field addressing cultural forms and politics in a rapidly globalizing world (Du Gay et al. 1997; Nelson et al. 1992).

Since the 1980s, cultural studies have been influential in shaping approaches to researching identity, everyday life, and power in academic disciplines such as geography, anthropology, sociology, critical education and literary studies. Drawing on Marxist, feminist, postcolonial and poststructural theory, scholars studying culture have developed critical insights and approaches that have come to be associated with "the cultural turn" in social theory (for examples, see Grossberg et al. 1992; Anderson et al. 2002). Cultural theorists have challenged "essentialist" (innate) ideas of culture as "a whole way of life", and "elitist" ideas of culture as belonging only to the upper social classes (Williams 1983). Rather, culture is understood as complex, heterogeneous phenomena with a powerful constitutive role in the "production and circulation of meaning" (Du Gay et al. 1997: 13). In other words, culture is something that is collectively made; it forms social identities, meanings, places, practices and relations of power, and in turn, is formed by them.

If culture is something that is made, then the task of cultural studies is to understand the specific social, historical, subjective and material elements of its making. Key concepts associated with this task relate to modes of meaning-making (thinking, talking and acting upon specific meanings) such as *signification* and *representation* (Du Gay et al. 1997). *Signification* refers to how specific communicative elements, such as text, image, and spoken words are articulated. *Representation* refers to the production of the broader "discursive" or "representational" field, where the articulation and circulation of meanings produce cultural identities, knowledge and values. Cultural meanings are not fixed in time or agreed upon by all members of a society. In order for cultural meanings to circulate and structure contexts for individual and collective experiences, they must also be enacted and practised (Du Gay et al. 1997). Material cultural practices highlight the concrete relationality between the production and circulation of cultural meanings and practices of production and consumption in late capitalist societies.

Cultural studies scholars are thus particularly concerned with how representations are enacted and contested through social, political and spatial relations of power (Mitchell 2000). The intersections between culture, representation and power are important sites of contestation, cooperation and political struggle for environmental justice. Cultural politics shape specific local contexts and imaginaries involving contestations over waste, contamination, toxic exposure, and the consequences of social exclusion from mainstream environmental governance, decision-making and protection. For example, environmental justice scholars and activists working at the interface between space, nature, culture and power have focused on the cultural invisibility of environmental injustice. As Julie Sze (2002: 163) asserts, "environmental justice is a political movement concerned with public policy issues of environmental racism, as well as a cultural movement interested in issues of ideology and representation". For Sze, cultural representations of nature in environmental justice critique dominant ideologies of nature as pure, external and devoid of human labour.

The ideological construction of environment as pure and undefiled nature in mainstream environmental culture privileges the concept of wilderness. With its focus on the lived experiences, histories and places of marginalized peoples and communities, environmental justice activism resituates the environment as the place where "people live, work, play and pray" (Agyeman 2008; Sze 2009; Walker 2009). Agyeman (2008) has noted that this intervention has profoundly reshaped environmental narratives, challenging the ways in which nonhuman nature derives its value from its symbolic and material separation from culture. Environmental justice problematizes such nature–culture boundaries. The vast and unknowable chemical impacts of industrialization on the biosphere make it increasingly difficult to conceive of the existence of remote, pristine nature. Environmental justice has played a key role in the critique of what Lawrence Buell (1998) terms "toxic discourse". Toxic discourse reflects dominant cultural narratives and practices that seek to minimalize, contain and downplay cultural anxieties about the risks and impacts of toxic exposure and environmental pollution (Buell 1998). Yet, these very attempts at containment have produced a proliferation of stories, testimonies and cultural actions in the absence of scientific and policy accountability. Toxic discourse contains subversive and subaltern elements; it produces geographies of containment and proliferation where the boundaries between chemical agents, bodies and places are blurred.

Storytelling as method and practice

One way to make sense of the complexity of environmental justice as cultural phenomena is through the framework of storytelling (Dickinson 2012; Houston 2013). Environmental justice storytelling is a particular form of political intervention that envisions socio-ecological transformation and produces more hopeful futures by narrating the environment as intimately connected to human wellbeing. This narrative tradition has a powerful precedent in Rachel Carson's (1962: 1-3) ground-breaking book *Silent Spring*, whose opening fable foreshadows the "grim spectre" of pesticides devastating farmers' families and all life in the surrounding fields and streams. Widely heralded as a catalyst for contemporary environmentalism in the US, *Silent Spring* effectively combined apocalyptic myth and science writing, making emergent knowledge of toxicity relevant and accessible to the general public (Waddell 2000). Carson's dystopian vision inspired a generation of eco-critical storytelling in the 1960s and 1970s, focused on urban and industrial zones, the "vernacular environments" of hazardous waste (Newman 2012: 32). Building on traditions of activist writing, popular theatre and cultural activism, environmental justice scholars and activists employ storytelling to combat the *invisibility* of toxins and the *historical marginalization* associated with environmental injustice. In this section, we define story and storytelling, explain how storytelling plays a role in activism and scholarship, and briefly consider the contributions and limits of environmental justice storytelling.

"Story" is a slippery concept, variously used to describe narratives of everyday experience that circulate as forms or objects of knowledge; culturally situated practices of collective expression; and narrative or literary modes of communicating scholarly and intellectual work (Cameron 2012). Broadly, stories draw attention to how experiences are interpreted and communicated, encompassing both personal expression, and the ways in which these narratives "exceed the personal and particular" (Cameron 2012: 574) to reflect collective meanings and social identities. Pezzullo (2001) suggests that stories serve two main functions in environmental justice struggles: first, to set the stage for the exploration of potential futures; and second, to enact new possibilities. That is, stories are *inventional*: they reassemble the chronology and causality of past events to produce *new meanings* capable of challenging the unacknowledged narratives that sustain oppressive conditions (Pezzullo 2001).

To illustrate the utility of storytelling as a framework for describing the cultural work of environmental justice, we highlight several components of Walter Benjamin's (1969) iconic description of storytelling – a creative process in which the narrator crafts meaning out of lived experiences to provide the audience with moral lessons. First, storytelling is an *intersubjective practice* derived from oral history traditions, where the narrator brings experiences alive through repeated tellings with interested listeners. Whether sharing individual testimonies in state hearings to challenge official narratives that minimize the severity of living in toxic areas, or organizing memorial events to affirm ongoing connection to degraded land, environmental justice storytelling practices produce fora where alternative forms of evidence are transformed into *public* knowledges, investing audiences in the experiences of marginalized communities as witnesses and prompting ethical action (Pezzullo 2001; Pezzullo 2003; Houston 2013).

Second, stories serve a *moral and practical function*, providing "counsel", which Benjamin (1969: 86) suggests is "less an answer to a question than a proposal concerning the continuation of a story which is just unfolding". Unlike positivist traditions of environmentalism, "making sense" of environmental injury through storytelling is not merely about explanation or categorization; rather, stories function by raising questions about observable patterns in ongoing developments that offer the audience space to interpret and wonder. For example, the accumulation of anecdotal evidence regarding patterns of illness and demography collected by residents of toxic landscapes and community activists in the emergent environmental justice struggles of the early 1980s not only revealed hidden networks of hazardous waste disposal, but subsequently inspired new forms of inquiry regarding the underlying socio-spatial patterns that produce unequal environmental phenomena in the first place (see for example Newman's (2012) discussion of toxic autobiography and geocriticism). By bringing together multiple modalities that exceed reductive policy, scientific or technical explanations – including the experiential/affective, biographical, philosophical/spiritual and historical – storytelling functions to illuminate the deeply layered causalities of environmental injustice and interrogate the taken-for-granted assumptions underpinning industrial development and environmental governance.

Third, Benjamin describes storytelling as a craft of chronicling the secular world, whose raw material is lived experience and whose purpose is interpreting the greater meanings of daily life. For Benjamin, *interpretation* by storytellers involves an ability to see the mystical in the living and inanimate world. Environmental justice storytelling practices bring a mythological and literary vision to "seeing" the apocalyptic and dystopic in seemingly mundane, everyday human-environment relations of toxic landscapes. For example, visualizations of workers attempting to protect themselves from lethal pesticides with cloth rags or descriptions of tap water smelling like sewage collected from communities impacted by industrial agriculture serve to remind us of the heightened stakes of everyday life in toxic areas, and the disconnect between discourses of industrial progress and the lived experience of those who carry the burden of waste (Perkins and Sze 2011). Cultural methodologies of environmental justice raise difficult questions about how to make sense of the "chemical regime of living" (Murphy 2008), where the impacts of industrialization and toxicity complicate the boundaries between human/nonhuman and nature/culture.

While geographical scholarship in the 1990s focused on stories as *discursive* forms that reiterated ideology and power structures at intimate scales, Cameron (2012) notes a resurgence of scholarly interest in the *more-than-discursive* capacities of storytelling that transform how people both interpret and relate to their lived environments in material, non-representational and affective ways. In other words, storytelling is a *material* practice, a "performative way of practicing knowledge that constructs both publicness and ontological difference" (Houston 2013: 9). Stories are productive and world-making, shifting collective understandings to allow for new possibilities to be enacted and embodied through cultural practices that make sense of life in toxic landscapes. The task of "making sense of" evokes discursive, relational, performative, embodied, and affective registers. Rather than (spoken or written) texts alone, environmental justice storytelling encompasses a range of cultural practices, from "image events" – visual interventions achieved through the circulation of images of direct action events (DeLuca 2005) to "toxic tours" – an embodied strategy of grassroots action that inverts eco-tourism to influence publics through witnessing the effects of toxicity in place (Pezzullo 2003, 2009; Di Chiro 2003).

Storied interventions

Storytelling is a pivotal site of contestation in environmental justice struggles (Sze et al. 2009); narratives are "the terrain of contest for social movements" (Pezzullo 2001: 5). Official state narratives frequently minimize or erase the ongoing and unequal devastation borne by vulnerable communities. In response, environmental justice activists use rhetorical strategies, memory practices and creative artistic forms to interrupt and critique the presumed necessity of the official line (Pezzullo 2001; Houston 2013; Krupar 2013). For example, counteracting the state narrative of nuclear development as a peak of human achievement, activists at Yucca Mountain make visible the impacts of radioactive pollution through memory practices and stories that present alternatives to nuclear waste disposal (Houston 2013: 15–16). Banerjee and Steinberg (2015) highlight how cultural discourses and practices can be deployed by local communities as a specific form of "cultural justice". Drawing on Swindler's (1986) "cultural toolkit model", Banerjee and Steinberg (2015) explore three dimensions: *symbologies of place* (cultural symbols and heritages); *historiographies of place* (place-based storytelling) and *social ties and community networks* (inter-community ties between different groups to navigate and address claims of environmental injustice).

Communities facing environmental harm employ storytelling practices to interpret and make sense of unfolding events, as well as to develop collective capabilities and capacities (Hofrichter 1993). Storytelling represents a key strategy for gathering evidence and for shaping the local and translocal contexts of cultural activism. For example, veteran community organizer Naeema Muhammad of the North Carolina Environmental Justice Network (NCEJN) routinely employs storytelling, both in developing the organizing capacity of local communities, and in demanding accountability from government officials[1]. When she first visits a community living in a hazardous environment, Ms Muhammad listens while community members share the experiences that have led them to believe that they are being harmed. She then makes connections between localized events and broader structures, drawing from a vast repertoire of community stories collected through decades of organizing (Interview with Ms. Muhammad, 14 August 2014).

The genre of "personal testimony" (Evans 2002) is a powerful tool for mobilizing grassroots participation within a community, as well as a building block for organizing across isolated communities. Returning to the example from the NCEJN, in the rotational quarterly meetings, host communities are invited to share insights on organizing strategies with visiting communities. Ms Muhammad believes that the development of public fora such as these allows those who are directly impacted by environmental hazards to develop the self-confidence and communication skills necessary for advocating on their own behalf. Individual testimonials serve to demonstrate grassroots knowledge of injustices that are then documented and collected, accumulating in support of the development of legal cases, direct action or other organizing strategies. These rehearsal spaces culminate at the annual NCEJN Summit, attended by impacted communities from across the state and environmental justice advocates, where three grassroots leaders are invited to testify at a Community Speak-Out. Regional and state government officials sitting on a panel are tasked with listening and responding to the charges and claims being presented. Through these processes of storytelling, environmental justice networks build expertise and share knowledge about technical aspects such as citizen science research and environmental/civil rights regulations. Equally significant, however, they create a forum and an audience for the kinds of stories that are absent in mainstream media.

Individual and collective narratives of suffering, illness and awareness of toxicity are a crucial component of environmental justice organizing. As a form of social learning, storytelling allows communities to interpret and share knowledge about the impacts of toxicity, often in the absence of or prior to the availability of scientific evidence. Citizen science efforts in environmental justice movements have sparked tremendous shifts in scientific knowledge production, democratizing environmental health communication and expanding the scope of legitimate scientific expertise through blending of local and scientific knowledges (Hill 2003; Brown et al. 2004; Di Chiro 2008; Kinsella 2004; Corburn 2005; also see Chapters 8, 23 and 24). However, the significance of storytelling extends beyond its value to accumulating knowledge about toxicity and health. Benjamin (1969: 87) usefully contrasts *information* – which he defines as factual knowledge whose importance is contained in the transmission of the fact itself, and therefore in the moment – with *stories*, whose value is accumulated through repetitions of telling and hearing, where meaning is produced not through the mere communication of facts, but what he calls "the epic side of truth, wisdom".

In scholarship and the mass media, communication about environmental justice tends to emphasize the "information" side of this equation, carrying always the burden of providing evidence of harm through scientific or sociological data, where the information in itself is presumed to provide explanation. Yet, the storytelling aspect – *making sense* of events and circulating *meanings* through the communication of experiences – is essential to environmental justice struggles over the long haul. Furthermore, storytelling engenders participatory democracy by emphasizing the human and ethical costs of industrial "progress", necessitating public involvement in debates about risk, hazardous waste and environmental health, and challenging the assumption that toxicity is the domain of scientific and legal expertise alone (Newman 2012).

Making "slow violence" visible

Environmental justice activists and scholars also use creative methods to interrupt taken-for-granted narratives and to imagine alternative possibilities for people and places marginalized by environmental injustice. Dickinson (2012: 58) proposes a *critical environmental justice narrative framework* that employs storytelling "as an instrument to help expose and critique the racializing practices that underlie environmental politics and illuminate the battles of environmental justice activists". In Dickinson's (2012) analysis of ongoing conflict between developers and Indigenous communities in the area surrounding the Petroglyph National Monument, New Mexico, she incorporates a fictional tale of impending crisis, projected five decades into the future, to comment on the absurdity and hypocrisy of pro-development arguments that allowed for highway construction through an Indigenous sacred area. Dickinson envisions how a recently approved road-building project might catalyse residential and commercial development in the area for a period of time, until contamination of drinking water by fracking and the siting of a toxic waste landfill eventually lead to white flight and economic collapse. The imagined response is a massive re-development effort displacing Native Americans from the area in order to construct a "Petroglyph Global Village" that offers Pueblo artisanry and culture as a commodified product for tourist consumption. Dickinson's fable seeks to expose the racist narratives underlying development discourses and reveal the manner in which the state's solutions disregard Indigenous belief systems and undermine political agency.

The interwoven contexts of the personal, public and particular that make story such a powerful intervention also reflect the complexity of working with its imaginative contexts and realities. "Story," observes Griffiths (2007, online), "is sometimes underestimated as something that is easy and instinctive". The idea that stories are "made-up" or can readily be discounted in the absence of conclusive scientific evidence is a serious problem for environmental justice activists and scholars. The task of putting together disparate phenomena into publicly discernible knowledge is a rigorous and demanding process (Griffiths 2007; Gibson-Graham 2008). Far from conveying "made up", overly subjective knowledge, story is "a privileged carrier of truth, a way for allowing multiplicity and complexity at the same time as guaranteeing memorability" (Griffiths 2007). As Griffiths argues, "the conventional scientific method separates causes from one another, it isolates each one and tests them individually in turn. Narrative, by contrast, carries multiple causes along together, it enacts connectivity. We need both methods."

Storytelling is not just about narratives, it is about recognizing emergent cultural, temporal and environmental realities shaped by the material presence of toxics and toxic environments. The relational, embodied and material dimensions of story play significant roles in activist, community and academic cultural work on environmental justice issues. The emphasis is less on proving and disproving the presence of toxic substances in communities and localities (though this remains a key strategy) and more on being able to draw connections and *tell stories* about how environmental in/justice assembles experiences, evidence, politics, knowledge, time and practices (Krupar 2013). Environmental justice struggles do not unfold according to incremental and linear time. While emissions from a local factory certainly may lead to an immediate spike in the incidence of childhood asthma, environmental injustice is often accumulative and unfolds across time-spaces that are not synchronized or well understood. Knowledge about industrial contaminants may become lost over time as toxic legacies are buried or built over. In other cases, the time-spaces between toxic exposure to a toxic or radioactive contaminant and the development of a related illness might be in the decades (Houston and Ruming 2014). Practices of storytelling are key to keeping knowledge about toxic contamination and its temporal displacements alive.

The threads of stories and narratives discussed in this chapter may seem disparate and disconnected. Nixon (2011) refers to the discordant temporalities of environmental injustice as "slow violence". Slow violence is the violence of accumulation and attrition that occurs in places that bear the brunt of environmental sacrifice in its multitude of forms. Nixon (2011), echoing Carson, sees the task of confronting the long and invisible catastrophe of environmental destruction as the struggle to articulate and name a formless, amorphous threat made all the more difficult because this threat unfolds over time and in ways that exceed our current capacities to govern and manage them. In Nixon's (2011: 10) words, "to confront slow violence requires, then, that we plot and give figurative shape to formless threats whose fatal repercussions are dispersed across space and time".

We have argued in this chapter that cultural approaches to environmental justice can offer a different perspective on what environmental justice means in a world unevenly transformed by chemicals and contaminants. Indeed, among the tasks for scholars and activists interested in storytelling environmental justice in its myriad cultural forms, is to "plot and give figurative shape" to experiences of injustice and the ways that it manifests in specific bodies and communities and across time and space. The discordant realities of environmental justice, bound up as they are in the unfolding of the slow violence that poisons communities and ecosystems, calls for critical and creative approaches that forge new connections between academic scholarship, scientific practice and cultural activism. Alongside its powerful critique of the normalization of toxic exposure and the invisibility of slow violence, storytelling environmental justice represents, in manifestly practical ways, collective work towards sustainable and just transformation. Environmental justice scholars and activists working at the intersections of environment and culture are also tasked with the challenge to "plot and give figurative shape" to more hopeful and equitable socio-ecological futures.

Note

1 This is drawn from Pavithra Vasudevan's ongoing collaboration and research with Naeema Muhammad, Co-director/Community Organizer, North Carolina Environmental Justice Network.

References

Adamson, J., Evans, M.M., and Stein, R. (Eds.) (2002). *The Environmental Justice Reader: Politics, Poetics, and Pedagogy*. Tucson: University of Arizona Press.

Agyeman, J. (2008). "Toward a 'Just' Sustainability?" *Continuum – Journal of Media and Cultural Studies*, vol. 22, pp. 751–757.

Agyeman, J., Cole, P., Haluza-DeLay, H., and O'Riley, P. (Eds.) (2009). *Speaking for Ourselves: Environmental Justice in Canada*. Vancouver: University of British Columbia Press.

Alston, D. (2010). "The Summit: Transforming a Movement." *Race Poverty and the Environment*, vol. 2, pp. 14–17.

Anderson, K., Domosh, M., Pile, S., and Thrift, N. (Eds.) (2002). *Handbook of Cultural Geography*. London: Sage.

Banerjee, D. and Steinberg, S.L. (2015). "Exploring Spatial andCultural Discourses in Environmental Justice Movements: A Study of Two Communities." *Journal of Rural Studies*, vol. 39, pp. 41–50.

Benjamin, W. (1969). "The Storyteller: Reflections on the Work of Nikolai Leskow." *Illuminations: Essays and Reflections*. New York: Schocken.

Blaser, M. (2010). *Storytelling Globalization from the Chaco and Beyond*. Durham and London: Duke University Press.

Brickell, K. and Garrett, B. (2015). "Storytelling Domestic Violence: Feminist Participatory Video in Cambodia." *ACME: International Journal for Critical Geographies*, vol. 14, pp. 928–953.

Brown, P., Zavestoski, S., McCormick, S., Mayer, B., Morello-Frosch, R. and Gasior Altman, R. (2004). "Embodied Health Movements: New Approaches to Social Movements in Health." *Sociology of Health and Illness*, vol. 26, pp. 50–80.

Buell, L. (1998). "Toxic Discourse." *Critical Inquiry*, vol. 24, pp. 639–665.

Bullard, R.D. (2007). "25th Anniversary of the Warren County PCB Landfill Protests: Communities of Color Still on Frontline of Toxic Assaults." *Dissident Voice* (29 May).

Cameron, E. (2012). "New Geographies of Story and Storytelling." *Progress in Human Geography*, vol. 36, pp. 573–592.

Carson, R. (1962/2002). *Silent Spring*. New York: Houghton Mifflin.

Corburn, J. (2005). *Street Science: Community Knowledge and Environmental Health Justice*. Cambridge, MA: The MIT Press.

DeLuca, K.M. (2005). *Image Politics: The New Rhetoric of Environmental Activism*. New York: Routledge.

Di Chiro, G. (2003). "Beyond Ecoliberal 'Common Futures': Environmental Justice, Toxic Touring and a Transcommunal Politics of Place." In Moore, D.S., Kosek, J., & Pnadian, A. (Eds.), *Race, Nature and the Politics of Difference*. Durham and London: Duke University Press, pp. 204–232.

Di Chiro, G. (2008). "Living Environmentalisms: Coalition Politics, Social Reproduction, and Environmental Justice." *Environmental Politics*, vol. 17, pp. 276–298.

Dickinson, E. (2012). "Addressing Environmental Racism through Storytelling: Toward an Environmental Justice Narrative Framework." *Communication, Culture and Critique*, vol. 5, pp. 57–74.

Du Gay, P., Hall, S., Jones, L., Mackay, H., and Negus, K. (1997). *Doing Cultural Studies: The Story of the Sony Walkman*. London: Sage.

Evans, M.M. (2002). "Testimonies." In Adamson, J., Evans, M.M., & Stein, R. (Eds.), *The Environmental Justice Reader: Politics, Poetics, and Pedagogy*. Tucson: University of Arizona Press, pp. 29–43.

Gibson-Graham, J.-K. (2008). "Diverse Economies: Performative Practices for 'Other Worlds'." *Progress in Human Geography*, vol. 32, pp. 613–632.

Griffiths, T. (2007). "The Humanities and an Environmentally Sustainable Australia." *Eco-humanities Corner*, vol. 43.

Grossberg, L., Nelson, C., and Treichler, P. (Eds.) (1992). *Cultural Studies*. New York and London: Routledge.

Hill, R.J. (2003). "Environmental Justice: Environmental Adult Education at the Confluence of Oppressions." *New Directions for Adult and Continuing Education*, vol. 99, pp. 27–38.

Hofrichter, R. (1993). "Cultural Activism and Environmental Justice." In Hofrichter, R. (Ed.), *Toxic Struggles: The Theory and Practice of Environmental Justice*. Philadelphia, PA: New Society, pp. 85–97.

Holifield, R. (2012). "Environmental Justice as Recognition and Participation in Risk Assessment: Negotiating and Translating Health Risk at a Superfund Site in Indian Country." *Annals of the Association of American Geographers*, vol. 102, pp. 591–613.

Houston, D. (2008). "Crisis and Resilience: Cultural Methodologies for Environmental Sustainability and Justice." *Continuum*, vol. 22, pp. 179–190.

Houston, D. (2013). "Environmental Justice Storytelling: Angels and Isotopes at Yucca Mountain, Nevada." *Antipode*, vol. 45, pp. 417–435.

Houston, D. and Ruming, K. (2014). "Suburban Toxicity: A Political Ecology of Asbestos in Australian Cities." *Geographical Research*, vol. 52, pp. 400–410.

Kinsella, W.J. (2004). "Public Expertise: A Foundation for Citizen Participation in Energy and Environmental Decisions." In Depoe, S.P., Delicath, J.W., & Elsenbeer, M.A. (Eds.), *Communication and Public Participation in Environmental Decision Making*. Albany: State University of New York Press, pp. 83–95.

Krupar, S.R. (2013). *Hot Spotter's Report: Military Fables of Toxic Waste*. Minneapolis: University of Minnesota Press.

McGurty, E. (2009). *Transforming Environmentalism: Warren County, PCBs, and the Origins of Environmental Justice*. New Brunswick: Rutgers University Press.

Mitchell, D. (2000). *Cultural Geography: A Critical Introduction*. Oxford, UK: Blackwell.

Murphy, M. (2008). "Chemical Regimes of Living." *Environmental History*, vol. 13, pp. 695–703.

Nelson, C., Treichler, P., and Grossberg, L. (1992). "Cultural Studies: An Introduction." In Grossberg, L., Nelson, C., & Treichler, P. (Eds.), *Cultural Studies*. New York and London: Routledge, pp. 1–16.

Newman, R. (2012). "Darker Shades of Green: Love Canal, Toxic Autobiography, and American Environmental Writing." In Foote, S. & Mazzolini, E. (Eds.), *Histories of the Dustheap: Waste, Material Cultures, Social Justice*. Cambridge, MA: The MIT Press, pp. 21–48.

Nixon, R. (2011). *Slow Violence and the Environmentalism of the Poor*. Cambridge, MA: Harvard University Press.

Perkins, T. and Sze, J. (2011). "Images from the Central Valley." *Boom: A Journal of California*, vol. 1, pp. 70–80.

Pezzullo, P.C. (2001). "Performing Critical Interruptions: Stories, Rhetorical Invention, and the Environmental Justice Movement." *Western Journal of Communication*, vol. 65, pp. 1–25.

Pezzullo, P.C. (2003). "Touring 'Cancer Alley,' Louisiana: Performances of Community and Memory for Environmental Justice." *Text and Performance Quarterly*, vol. 23, pp. 226–252.

Pezzullo, P.C. (2009). *Toxic Tourism: Rhetorics of Pollution, Travel, and Environmental Justice*. Alabama: University of Alabama Press.

Pulido, L. (1996). "A Critical Review of the Methodology of Environmental Racism Research." *Antipode*, vol. 28, pp. 142–159.

Pulido, L. (2000). "Rethinking Environmental Racism: White Privilege and Urban Development in Southern California." *Annals of the Association of American Geographers*, vol. 90, pp. 12–40.

Pulido, L. (2015). "Geographies of Race and Ethnicity I: White Supremacy vs White Privilege in Environmental Racism Research." *Progress in Human Geography*, vol. 39, pp. 809–817.

Schlosberg, D. (2007). *Defining Environmental Justice: Theories, Movements, and Nature*. New York: Oxford University Press.

Storm, A. (2014). *Post-Industrial Landscape Scars*. New York: Palgrave Macmillan.

Swidler, A. (1986). "Culture in Action: Symbols and Strategies." *American Sociological Review*, vol. 51, pp. 273–286.

Sze, J. (2002). "From Environmental Justice Literature to the Literature of Environmental Justice." In Adamson, J., Evans, M.M., & Stein, R. (Eds.) (2002), *The Environmental Justice Reader: Politics, Poetics, and Pedagogy*. Tucson: University of Arizona Press, pp. 163–180.

Sze, J. (2009). "Boundaries and Border Wars: DES Technology, Environmental Justice." In White, D.F. & Wilbert, C. (Eds.), *Technonatures: Environments, Technologies, Spaces, and Places in the Twenty-first Century*. Waterloo: Wilfrid Laurier University Press, pp. 125–148.

Sze, J., London, J., Shilling, F., Gambirazzio, G., Filan, T., and Cadenasso, M. (2009). "Defining and Contesting Environmental Justice: Socio-natures and the Politics of Scale in the Delta." *Antipode*, vol. 41, pp. 807–843.

Vasudevan, P. (2012). "Performance and Proximity: Revisiting Environmental Justice in Warren County, North Carolina." *Performance Research*, vol. 17, pp. 18–26.

Vasudevan, P. (2013). "Memory and the Re-invention of Place: The Legacies of Environmental Justice in Warren County, North Carolina." Masters Thesis. University of North Carolina, Chapel Hill.

Vasudevan, P. and Kearney, W.A. (2015). Remembering Kearneytown: Race, Place and Collective Memory in Collaborative Filmmaking. *Area*, doi: 10.1111/area.12238.

Waddell, C. (Ed.) (2000). *And No Birds Sing: Rhetorical Analyses of Rachel Carson's Silent Spring*. Carbondale: SIU Press.

Walker, G. (2009). "Beyond Distribution and Proximity: Exploring Multiple Spatialities of Environmental Justice." *Antipode: A Radical Journal of Geography*, vol. 41(4), pp. 614–636.

Walker, G. (2012). *Environmental Justice: Concepts, Evidence and Politics*. New York: Routledge.

Williams, R. (1983). *Keywords*. London: Fontana.

21

FACILITATING TRANSDISCIPLINARY CONVERSATIONS IN ENVIRONMENTAL JUSTICE STUDIES

Jonathan K. London, Julie Sze and Mary L. Cadenasso

Introduction

The rapid expansion of environmental justice (EJ) studies in the past two decades reflects the vibrancy of the field. Research that brings together social sciences, natural sciences, and the humanities has presented powerful opportunities to analyse the intertwined causes and effects of environmental injustice in low-income communities and communities of colour. But such interdisciplinarity also represents a challenge to the field's definition and coherence. Sze and London define environmental justice scholarship as being at a "crossroads" that offers "a convenient meeting point, but which can also indicate crossed signals, sites of liberation and possibility, seduction, and danger" (2008: 1332). It is this liminal quality that represents the field's greatest strength as "a framework that can engage otherwise disparate disciplinary fields into a transdisciplinary conversation, and that can surface the commonalities of struggle on diverse issues in diverse places and among diverse populations" (2008: 1347). However, the crossroads also has its perils, as disciplines that infuse EJ studies have their own intellectual lineage, theoretical frameworks, methodological approaches, and standards of excellence. These, in turn, can complicate, slow, and sometimes stymie effective collaboration. Ultimately, this liminal quality raises an important question: "if environmental justice can mean almost anything, does it risk a dilution and even loss of meaning and purpose?" (Sze and London 2008: 1332).

Thus, we think it timely to revisit what the "transdisciplinary conversation" looks like and to ask how well the field of EJ studies has managed expansive growth while maintaining analytic grounding. To do so, we present two case studies that illustrate the prospects and perils of collaboration among researchers from multiple disciplines. These case studies draw on our own experience. They are: 1) a study of a planning process intended to protect the ecological and economic values of California's Sacramento Bay Delta region (drawing primarily from Sze et al. 2009); and 2) a study of urban gardening and soil lead contamination in low-income communities and communities of colour in Sacramento, California.

In both cases, we found ourselves in territory outside our disciplinary fields. We needed to find new ways to articulate and integrate our conceptual and methodological approaches to provide meaningful analyses. Our progress was slowed by misunderstandings about how to define and operationalize central concepts, including scale and the relationship between the natural and the social world. We also wrestled with the question of what is "good" science. We struggled to create a cohesive theoretical and methodological map that could encompass the unique contours of our respective disciplines. That is, we sought not merely a multidisciplinary or an interdisciplinary, but a *transdisciplinary* approach. Our path was not linear and did not lead to either total success or failure. However, our work offers useful insights for fellow EJ scholars embarking on transdisciplinary journeys.

Transdisciplinary conversations

Recent scholarship has identified several research approaches including single-disciplinary, multidisciplinary, interdisciplinary, and transdisciplinary modalities (MacMynowski 2007; Miller et al. 2008; Stokols 2006). According to MacMynowski (2007), these modalities form a continuum in which disciplinary methodologies and epistemologies interact, resulting in conflict, tolerance, mutual cooperation and identification, and mutual transformation (Apgar et al. 2009; Jahn et al. 2012; MacMynowski 2007).

Other scholars have gone beyond the idea of inter- or multidisciplinarity to explore ideas of transdisciplinary scholarship. Miller et al. (2008: 3) describe how multidisciplinary projects engage researchers in the study of a common issue, but do so from within their own "epistemological siloes." Interdisciplinary projects, they argue, involve "unified problem formulation, sharing of methods, and perhaps the creation of new questions." But transdisciplinary research requires collaborators to "accept an epistemological perspective unique to the effort, redrawing the boundaries between disciplinary knowledges." This transdisciplinary approach is needed because some problems exceed the capacity of single existing disciplines to characterize and/or to inform effective action (Apgar et al. 2009; Hall et al. 2012; Jahn et al. 2012; Stokols 2006).

Miller et al. (2008) and Miller et al. (2011) propose "epistemological pluralism," which allows scholars to embrace multiple ways of knowing and invites negotiation between discipline-based researchers. This pluralism is necessary to understand the complex linkages between social and ecological systems and to inform policy and action. Eigenbrode et al. (2007: 57) go further, using the term "metadiscipline" to describe "an emergent and sustained epistemological framework spawned by persistent transdisciplinary effort." These approaches can disassemble existing forms of "epistemological sovereignty" that are characterized by separating researchers into disciplinary fiefdoms and hierarchies (Miller et al. 2008; Miller et al. 2010), placing one way of knowing above others, and often favouring the "harder" sciences (MacMynowski 2007; Pennington 2008).

While there are many benefits of transdisciplinary research, it is also a process beset with obstacles. There is no clear *lingua franca*: the same word can be used to refer to different phenomena, making effective communication difficult (Bracken and Oughton 2006; Fazey et al. 2014; Hall et al. 2012). Other barriers to collaboration arise when scholars from different disciplines disagree about reductionism vs. holism, about the social construction of knowledge vs. positivism, and tightly vs. loosely bounded realms of scientific and lay knowledge (Strang 2009; Apgar et al. 2009). These problems are exacerbated when scholars lack the time and capacity to discuss and understand each other's methodologies (Eigenbrode et al. 2007; Strang 2009).

Faced with these obstacles, some EJ scholars may be tempted to retreat behind the walls of their disciplinary fiefdoms. Indeed, while EJ studies encompass works grounded in many disciplines across the humanities there are few truly interdisciplinary or transdisciplinary studies to draw on as precedents (but see Clark et al. 2007). And yet, we believe that striding towards transdisciplinarity is crucial for the vitality of the field. We derive this orientation from several observations about the phenomenon of environmental injustice itself. First, environmental injustice is driven by complex and co-produced social, economic, political, and environmental systems (Bickerstaff and Agyeman 2009; Holifield 2009; Holifield and Schuelke 2015; Sze et al. 2009). Second, environmental injustice operates at multiple spatial and temporal scales (Bulkeley 2005; Kurtz 2002, 2003; Nixon 2011). Third, environmental justice social movements develop repertoires that draw on a variety of cultural traditions, and knowledge systems that can blur the boundaries between what is considered "good science" and "lay knowledge" (Balazs and Morello-Frosch 2013; Harrison 2011; London et al. 2011; Minkler et al. 2008; Schlosberg 2003). This critical approach to science makes environmental justice scholarship well positioned to address what Foucault (1980) terms "power/ knowledge", or the ways in which the construction of disciplinary knowledge can be used (often simultaneously) for liberatory and oppressive ends.

Recent work on transdisciplinary research has proposed practices to help EJ scholars identify and, to some extent, overcome these obstacles. Eigenbrode et al. (2007) provide a "philosophical tool kit" with guiding epistemological questions about underlying disciplinary assumptions, including motivation, methodology, confirmation, objectivity, values, reductionism, and emergence. Miller et al. (2008) propose an "iterative science cycle" to ensure open, honest and non-judgmental discussion to resolve disciplinary conflicts and promote mutual learning. Apgar et al. (2009) emphasize the value of developing holistic world views based on their transdisciplinary action research on climate change with indigenous communities in Central America. Fazey et al. (2014) offer principles for transdisciplinary research including identifying multiple end users and selecting and evaluating knowledge exchange approaches based on how well they help inform action to resolve important social and ecological problems. Scholars of social ecological systems (see especially Fazey et al. 2014; Hall et al., 2012; Stokols 2006) focus on the active engagement of multiple social actors inside and outside the academy in the design, implementation, and application of transdisciplinary action research.

The following case studies illustrate the ways in which we, along with several other colleagues from diverse disciplinary backgrounds, drew on these approaches to navigate the continuously branching crossroads of EJ scholarship. While we do not claim to have achieved a state of transdisciplinarity, we believe our experiences can point the way towards a more meaningful integration of diverse ways of understanding the social, political, economic, and ecological dimensions of environmental injustice.

Case studies

Re-scaling environmental justice

Origins

Starting in 2007, the authors were part of a group of scholars with backgrounds in geography, community development, plant sciences, urban ecology, and American studies, brought together at the University of California, Davis, to conduct an interdisciplinary research project based on environmental justice issues in northern California. We had already enjoyed positive experiences working as part of interdisciplinary teams with each other (London et al. 2008; Sze and London 2008) and with others (Cadenasso et al. 2007; Pickett et al. 2005, 2004). Two members had conducted research examining the environmental politics of an ongoing planning process in the Sacramento-San Joaquin Delta (Shilling et al. 2009) and a third member was completing a doctoral dissertation focused on the Delta's social and environmental history (Gambirazzio 2009). Based on earlier partnerships with environmental justice organizations we learned of their significant concerns about environmental injustices in the region. We developed a project focused on the spatial and temporal dimensions of social and ecological transformation in the Delta. Specifically, the project examined the state-chartered planning process to create what was deemed the "Delta Vision." The Delta Vision intended to "identify a strategy for managing the Sacramento–San Joaquin Delta as a sustainable ecosystem that would continue to support crucial environmental and economic functions" (Blue Ribbon Task Force 2007, 2008). Concerned about the long and sordid history of the domination of water resources by California's economic and political elites, often to the detriment of the ecological and social systems, the study team sought to assess how well the Delta Vision reflected the voices and interests of the many low-income people and people of colour living in and dependent on the Delta.

Our initial research revealed that the Delta Vision plan produced, instead of reduced, environmental injustices because of the ways that policy makers and scientists defined and operationalized scale. The question of scale was in turn bound up with questions about how planners and policy makers characterized the social-ecological system of the Delta, and the role of science in underwriting these socio-ecological constructions. Soon, the core questions guiding our research project became: How did planners and policy makers bound the Delta as both a material place and as a locus of state intervention? What were the scientific and policy processes that constructed these boundaries and whose interests did they serve? What (and whom) was included within these boundaries, what lay outside, and what was the relationship between these inside and outside realms? What conflicts had these concepts and boundaries engendered, and how did they affect whose voices and visions were included in the plan? In particular, how did the approach to scale in the Delta Vision planning process and ultimate plan produce and reproduce environmental injustices in the region?

Towards transdisciplinarity

Our goal was thus to understand the scientific frameworks through which the Delta was measured and made legible as a social-ecological subject in need of public policy interventions and how these interventions shaped a landscape of environmental injustices in the region. We needed to develop a multi-methodological and transdisciplinary approach to critically reflect on our own disciplinary-based visions of scale and the role of science in the discursive and material construction of the Delta.

We settled upon scale as the focal point of our exploration, in part because it is a term that has very different meanings and uses in our respective fields. Like "resilience" or "sustainability," the term "scale" has different resonances that can make talking across the natural science/social science/humanities divide virtually impossible (Silver 2008). We spent a great deal of time learning about what scale meant in our respective fields. This was possible because of our explicit goal of building interdisciplinary partnerships and because we had pre-existing professional and personal relationships allowing us to trust and listen to each other.

Based on this dialogue, we developed a transdisciplinary research design that used historical, geographic and discursive tools to analyse the political and social processes shaping the Delta. Over 12 months, our team (with primary field work roles played by two geography graduate students) conducted over 20 semi-structured interviews with key actors (scientists, policy makers, and activists), analysed planning documents and scientific reports, and were participant observers at Blue Ribbon Task Force meetings in Sacramento and throughout the Delta region. Our major finding was that concerns about environmental justice were marginalized within the Delta Vision process and in the resulting Strategic Plan. This marginalization, we argued, was produced through scientific narratives and models of the Delta built on restrictive spatial and temporal constructions of scale. These constructions erased its deeper social-ecological histories, overlooked present-day socio-economic disparities, excluded the implications of water management decisions in the Delta on upstream communities, and cast environmental justice as a "special interest," not a core value equal to protecting ecosystem health and economic vitality. The Blue Ribbon Task Force's techno-scientific orientation also excluded many members of the public from meaningful participation in the planning process.

Reflections

The Delta research project represented an exciting journey for its participants, leading to new disciplinary perspectives from which to view and conduct our research. The project did face significant challenges. First, while the project drew on the disciplinary backgrounds of all its members, the integration of the ecological science approach to scale lagged behind our engagement with social scientific debates on environmental justice and scale. Second, the slow pace of developing the conceptual framework for the project, coupled with the one year of funding, allowed time for only one direct publication (Sze et al. 2009). Third, the relatively short duration of the project limited the time available for engaging with our community partners in the environmental justice advocacy and policy sectors.

Yet we believe that the benefits of this approach far outweighed the costs, especially over the long term. Based on our extensive discussions across our disciplinary domains, we were able to develop a synthesis that brought together the ecological sciences' views of scale based on spatial and temporal factors and the resolution of units of analysis (Sayre 2005) and the social sciences and humanities' views of scale based on ideas about social construction and politics (Kurtz 2002, 2003; Swyngedouw and Heynen 2003). We also linked ecologists' views of the relationship between human and natural systems (Clark et al. 2007; Pickett et al. 2005) with social sciences' and humanities' concepts of socio- or techno-natures (Loftus 2007; Perreault et al. 2012). We coupled ecologists' analyses of how ecosystems function as "model, meaning, and metaphor" in the natural and social sciences (Pickett et al. 2004) with social science and humanities scholars' analysis of the social construction and politics of knowledge. By using a spatially and temporally expansive view of scale, we addressed critiques of the environmental justice movement that take the movement and the scholarly field of EJ studies itself to task for its local place-based and "militant particularism" (Holifield 2009; Swyngedouw and Heynen 2003). Finally, the project facilitated future transdisciplinary collaborations between the authors and other colleagues (Liévanos et al. 2010; London et al. 2013; Shilling et al. 2009; Sze et al. 2009) and informed the approach of one scholar to a new book project (Sze 2014).

Soil Lead and Urban Gardening (Slug) research

Origins

The Soil Lead and Urban Gardening (SLUG) research project was developed by a team of ecologists and environmental social scientists (including two of the authors of this chapter, who also collaborated on the earlier Delta project) interested in studying the environmental justice dimensions of urban agriculture. Despite the fact that they live in an agricultural region, many low-income people and people of colour in Sacramento's underserved neighbourhoods do not have access to fresh and nutritious food. Environmental justice and food justice organizations have launched programmes to install raised-bed or in-ground gardens in residential yards to improve access and build community cohesion and empowerment. Because these neighbourhoods also tend to be older neighbourhoods with homes, businesses, and roadways built before the ban on lead in house paint (1978) and gasoline (1996), exposure to lead in the soil while gardening is a potential concern.

A research team with expertise in plant and soil science, landscape ecology, environmental planning, and human geography secured funding to study the relationships among ecosystem services associated with home gardening for food production, health, and community empowerment on the one hand, and potential exposure to soil lead on the other. Though each member of the collaboration was motivated by different objectives, the overarching goals of SLUG were to produce knowledge that would support residents' food self-provisioning while minimizing the negative health impacts of lead exposure associated with gardening. Community organizations installed raised-bed gardens in residential yards as an intervention to promote healthy neigbourhoods. Social science researchers were interested in residents' sense of place and perceptions of gardening, soil lead, and other environmental hazards that could be used to inform strategies for community self-empowerment, organizing, and policy advocacy. Ecologists wanted to understand the tradeoff between the ecosystem services of food provisioning and potential exposure to soil lead. Ecologists also sought to quantify soil lead concentrations in residential yards to inform the safest placement of gardens and to determine the relationship between soil lead and features of the urban environment such as buildings, roads, tree canopies, and bare soil.

Towards transdisciplinarity

The SLUG project sought to critically engage the emerging concept of ecosystem services with EJ studies. Ecosystem services are defined as the benefits that people derive either directly or indirectly from ecosystems (e.g. food, water, shade). Conversely, ecosystem disservices are defined as burdens people experience either directly or indirectly from ecosystems (e.g. flooding, extreme heat, air pollution) (Berkes et al. 2000; von Döhren and Haase 2015). A key sustainability goal is to maximize ecosystem services while minimizing ecosystem disservices (Bennett et al. 2009; Raudsepp-Hearne et al. 2010).

Recent critiques of the ecosystem services concept have focused on its limited attention to the political, economic, social, and cultural drivers of racial, ethnic, and class disparities in access to ecosystem services, and burdens from disservices (Carpenter et al. 2009). The concept of ecosystem services is also of limited use in understanding the ways in which natural and social systems co-produce each other, and make it hard to disentangle them (Lyytimäki et al. 2008). It is also problematic as ecosystem services and disservices are often defined *a priori* by scientists, not the people confronting these conditions. To respond to these critiques, scholars have put forward new notions of social-ecological systems (Pickett et al. 2005) and socio-natures (Perreault et al. 2012; Swyngedouw 2007). These allow for greater attention to the political dimensions of how natural and human systems are represented through scientific discourse, and how they shape and are shaped by the actions of the state, capital, and civil society.

With this conceptual framework in place, the SLUG team created an action research framework that could span multiple spatial scales, consider different models of social-ecological systems, and take a reflexive approach to the role of disciplinary knowledge in understanding and acting upon urban agricultural landscapes. Unlike the compressed timeline for the one-year Delta research project, the timeline for the SLUG project benefited from a five-year grant period that allowed for a more extensive development, implementation, reflection, and adaptation process.

The research team and community partner organizations met intensively over the first six months of the project in an "iterative science cycle" (Miller et al. 2008) to develop the research design. After this design phase, the university researchers spoke by conference call twice a month (once as a whole team, and once in separate social science and ecological teams). This telephone contact was necessitated by the location of two of the Co-PIs at other universities. The university and community partners interacted regularly through the implementation of the soil lead testing process (described below), somewhat less frequently by telephone, and once a year in a day-long workshop where the results of the prior year were presented, and plans for the coming year were developed.

The research design combined distinct but interlinked ecological science, social science, and community-based participatory action research methods. The core of the SLUG action research project was collaboration with community organizations to provide free soil lead tests for interested residents participating in the home gardening programme. These tests yielded a rich data set on the profile of soil lead levels in and around garden boxes/beds, and throughout the yards of each property. The garden and property-level data helped answer questions about the factors influencing soil lead levels, such as land cover and management techniques. The project provided residents with a valuable service that could be used to inform garden placement and soil mitigation measures to minimize potential lead exposure. Lastly, the soil lead concentration data from the parcels can be used to spatially model predictions of soil lead concentration at the scale of the entire city to inform policy and advocacy on urban gardening and soil mediation activities.

The social science research sub-team used a range of methods to understand perceptions about home gardening and soil lead contamination at the household, neighbourhood, and city scale. Resident-produced videos, interviews, and surveys administrated before and after the garden installation and/or soil lead tests tracked changes in perceptions of mental and physical health, neighbourhood conditions, self-empowerment and politicization and provided rich ethnographic narratives about the cultural politics of urban gardening and environmental justice.

Reflections

While the SLUG project is ongoing as of this writing, the study has yielded several valuable insights. First, by linking ecological and social science concepts and methods we identified ways in which home gardening in yards with soil lead contamination can be recast from posing a health risk to representing a benefit to both urban sustainability goals and healthy gardening efforts (Schwarz et al. 2016). This double benefit was provided through soil testing and soil management that could enhance the role of gardens as the basis for community mobilization to address lead and other neighbourhood hazards.

Second, we have highlighted the value of field research teams composed of both social and ecological scientists conducting the soil lead tests. This allowed the ecological scientists to record and reflect on their experiences interacting with residents receiving gardens and/ or soil lead tests, providing a rich data source for the social scientists to analyse. Likewise, by assisting with the lead testing, the social scientists gained an appreciation for the material processes of ecological research in ways that would later assist in the more ecologically-oriented collaborative writing projects. Third, the collaboration with SLUG's community partners in developing and implementing the multi-method research design yielded insights that would not have been possible without this engaged scholarship approach (London et al. 2017).

Synthesis

What can the Delta and SLUG projects contribute to the possible transdisciplinary conversations at the crossroads of EJ studies? How can both their successes and shortcomings enrich dialogues based on epistemic pluralism that is capable of "redrawing the boundaries between disciplinary knowledges" (Miller et al. 2008)?

Both the Delta and SLUG projects shared many characteristics of interdisciplinarity, and in some cases, transdisciplinarity, as defined by recent scholarship (Apgar et al. 2009; Bracken and Oughton 2006; Hall et al. 2012; Jahn et al. 2012; MacMynowski 2007; Miller et al. 2008; Stokols 2006; Strang 2009). The Delta and SLUG research teams' experiences highlight the value of investing the time in developing positive working relationships among the members based on a genuine interest in collaboration. This can help the EJ research teams avoid what MacMynowski (2007) calls the "conflict" and "tolerance" stages, and facilitate a move into the "mutual cooperation and identification" stage. They can set their sights on achieving the "mutual transformation" of transdisciplinarity.

Movement towards this mutual transformation goal can be facilitated by what can be understood as "disciplinary humility." This is not to be confused with the denigration of one's own or another's discipline. Indeed, transdisciplinary research demands that team members are well grounded in the concepts and methods from their home disciplines. Researchers need to value the contributions of other disciplines to fill in gaps and blind spots in their own approach and enrich their understandings of topics and questions of mutual interest. This collaborative approach is especially important while conducting research on complex social-ecological systems such as regional waterscapes and urban agricultural landscapes that no one discipline can fully encompass.

For example, in acknowledgement of the problematic role of discipline-based science in mischaracterizing the spatial and temporal scale of the Delta, the Delta research team adopted a reflexive view of how its own disciplinary orientations opened and closed analytical pathways. The SLUG team addressed the politics of knowledge by building collaboration between ecological scientists, social scientists, and community partners to frame, conduct, and apply their research.

As recommended by the scholarship on interdisciplinary and transdisciplinary research, both projects devoted significant time to understanding each other's disciplinary terminology, the development of a shared language and conceptual framework, and "epistemological pluralism" (Miller et al. 2008). This provided opportunities for multiple "iterative science cycles" (Miller et al. 2008).

At the same time, the miscommunications and missteps experienced by the Delta and SLUG teams suggest that transdisciplinarity in EJ studies is best understood as an ongoing process of negotiation and navigation towards an aspirational goal rather than a fixed and stable destination. For example, despite concerted efforts, in the Delta project, the ecological sciences were mainly used descriptively, to set the physical stage upon which the policy conflicts over the Delta Vision process would play out. As a result, the Delta study remained primarily a social sciences and humanities-based project. In contrast, the SLUG project did apply diverse concepts and multiple methods from the ecological and social sciences, placing them on roughly equal footing. However, while there has been transformation in the processes of "mutual cooperation and identification" (MacMynowski 2007), it is still too early to judge whether this has produced a framework that does not merely integrate disciplines but transforms them into a distinct transdisciplinary epistemological framework. And yet, the SLUG project does appear to represent a strong case of "epistemological pluralism" (Miller et al. 2008) in which the boundaries of disciplinary fiefdoms have been blurred in a robust, if sometimes contentious, knowledge democracy. This kind of cognitive justice in which diverse ways of knowing are recognized as a valuable part of a larger knowledge democracy is well-aligned with visions of environmental justice (Schlosberg 1999, 2004).

While it would therefore be too ambitious to say that these projects developed entirely new epistemologies or completely redrew our disciplinary boundaries, we hope that both the achievements and shortcomings of these projects can help facilitate new and productive transdisciplinary conversations at the crossroads of EJ studies. In particular, these cases provide three broader contributions to what Pellow and Brulle (2005), Pellow (2016) and others (Fiskio and Christensen 2014; Holifield et al. 2009) have called "critical environmental justice studies".

First, critical EJ scholars must devote sufficient time throughout the research process for dialogue between the researchers and community partners to develop a meta-language that can span divides between disciplines and between academic and community worlds. Indeed, troubling, if not transcending, the boundaries between different realms of knowledge is necessary to avoid reproducing the power/knowledge hierarchies through which low-income communities and communities of colour are disadvantaged in discursive and material ways. Reflexive discussions should occur about what constitutes legitimate knowledge, how this is represented and communicated in different venues, and who has the authority to narrate these accounts in order to facilitate collaborative and (ideally) liberatory EJ scholarship.

Second, creative and expansive notions of spatial and temporal scale are needed to integrate scholarship from the humanities, natural, and social sciences into a transdisciplinary approach to critical EJ studies. In particular, a critical realist orientation to scale recognizes both the material realities of the world "out there" that can be described in scalar language on the one hand, and the political implications of the methods to spatially and temporally bound phenomena in particular ways, on the other. Indeed, critical EJ studies demands holding these material and discursive understandings in both hands, simultaneously.

Third and finally, critical EJ studies draw together insights from the humanities, natural and social sciences to expand the circle of analysis and concern to the more-than-human realm and consider the ways in which those elements (soil, lead, gardens, salt marshes, fish, and so on) exert their own agency within larger actor networks constituted through public policy, advocacy, and science.

While achieving fluency in transdisciplinarity is the work of a lifetime, the conversations between environmental justice scholars and practitioners cultivated in the process can ensure that field continues to grow and thrive.

Acknowledgements

The authors wish to gratefully acknowledge the full research teams in addition to the co-authors and funding resources that contributed to the projects described in the case studies.

Rescaling Environmental Justice: Fraser Schilling (Co-PI), Trina Filan, Gerardo Gambirazzio. Funding from the UC Davis Faculty Senate Interdisciplinary Research Fund.

Soil Lead and Urban Gardens: Bethany Cutts (Co-PI), Kirsten Schwarz (Co-PI), Cindy Engle, Jill Baty, Heather Smith, Bryn Montgomery, Shaina Meiners, Jeff Marquez, Jeanette Lim, Charles Mason Jr., Katie Valenzuela, Rangineh Azimzadeh, Caliph Assagai, and Chanowk and Judith Yisrael. Funding from the University of California Agriculture and Natural Resources competitive grants program (No. 11–958).

References

Apgar, J.M., Argumedo, A. and Allen, W. (2009). "Building transdisciplinarity for managing complexity: lessons from indigenous practice." *International Journal of Interdisciplinary Social Sciences*, vol. 4, no. 5, pp. 255–270.

Balazs, C.L. and Morello-Frosch, R. (2013). "The three Rs: how community-based participatory research strengthens the rigor, relevance, and reach of science." *Environmental Justice*, vol. 6, no. 1, pp. 9–16.

Bennett, E.M., Peterson, G.D. and Gordon, L.J. (2009). "Understanding relationships among multiple ecosystem services." *Ecology letters*, vol. 12, no. 12, pp. 1394–1404.

Berkes, F., Folke, C. and Colding, J. (2000). *Linking social and ecological systems: management practices and social mechanisms for building resilience.* Cambridge UK: Cambridge University Press.

Bickerstaff, K. and Agyeman, J. (2009). "Assembling justice spaces: the scalar politics of environmental justice in north-east England." *Antipode*, vol. 41, no. 4, pp. 781–806.

Blue Ribbon Task Force. (2007). *Delta Vision.* State of California Resources Agency, [Online], Available: http://deltavision.ca.gov/BlueRibbonTaskForce/FinalVision/Delta_Vision_Final.pdf. (4 January 2016).

Blue Ribbon Task Force. (2008). *Delta Vision Strategic Plan.* State of California Resources Agency. [Online], Available: http://deltavision.ca.gov/StrategicPlanningProcess/StaffDraft/Delta_Vision_Strategic_Plan_standard_resolution.pdf. (4 January 2016).

Bracken, L.J. and Oughton, E.A. (2006). "'What do you mean?' The importance of language in developing interdisciplinary research." *Transactions of the Institute of British Geographers*, vol. 31, no. 3, pp. 371–382.

Bulkeley, H. (2005). "Reconfiguring environmental governance: towards a politics of scales and networks." *Political Geography*, vol. 24, no. 8, pp. 875–902.

Cadenasso, M.L., Pickett, S. and Schwarz, K. (2007). "Spatial heterogeneity in urban ecosystems: reconceptualizing land cover and a framework for classification." *Frontiers in Ecology and the Environment*, vol. 5, no. 2, pp. 80–88.

Carpenter, S.R., Mooney, H.A., Agard, J., Capistrano, D., DeFries, R.S., Díaz, S., Dietz, T., et al. (2009). "Science for managing ecosystem services: beyond the Millennium Ecosystem Assessment." *Proceedings of the National Academy of Sciences*, vol. 106, no. 5, pp. 1305–1312.

Clark, W.C., Kates, R.W., Richards, J.F., Mathews, J.T., Meyer, W.B., Turner, B.L., Pickett, S.T., et al. (2007). "Relationships of environmental justice to ecological theory." *Bulletin of the Ecological Society of America*, vol. 88, no. 2, pp. 166–170.

von Döhren, P. and Haase, D. (2015). "Ecosystem disservices research: a review of the state of the art with a focus on cities." *Ecological Indicators*, vol. 52, pp. 490–497.

Eigenbrode, S.D., O'Rourke, M., Wulfhorst, J.D., Althoff, D.M., Goldberg, C.S., Merrill, K., Morse, W., et al. (2007). "Employing philosophical dialogue in collaborative science." *BioScience*, vol. 57, no. 1, pp. 55–64.

Fazey, I., Bunse, L., Msika, J., Pinke, M., Preedy, K., Evely, A.C., Lambert, E., et al. (2014). "Evaluating knowledge exchange in interdisciplinary and multi-stakeholder research." *Global Environmental Change*, vol. 25, pp. 204–220.

Fiskio, J. and Christensen, J. (2014). "Introduction: critical environmental justice studies in the twenty-first century." *Resilience: A Journal of the Environmental Humanities*, vol. 1, no. 1.

Foucault, M. (1980). *Power/knowledge: Selected interviews and other writings, 1972–1977*. New York, NY: Pantheon.

Gambirazzio, G.C. (2009). *The parallax view: race, land and the politics of place-making in Locke, California*. PhD dissertation. Geography Graduate Group. University of California, Davis.

Hall, K.L., Vogel, A.L., Stipelman, B.A., Stokols, D., Morgan, G. and Gehlert, S. (2012), "A four-phase model of transdisciplinary team-based research: goals, team processes, and strategies." *Translational Behavioral Medicine*, vol. 2, no. 4, pp. 415–430.

Harrison, J.L. (2011). "Parsing 'participation' in action research: navigating the challenges of lay involvement in technically complex participatory science projects." *Society & Natural Resources*, vol. 24, no. 7, pp. 702–716.

Holifield, R. (2009). "Actor-network theory as a critical approach to environmental justice: a case against synthesis with urban political ecology." *Antipode*, vol. 41, no. 4, pp. 637–658.

Holifield, R., Porter, M. and Walker, G. (2009). "Introduction spaces of environmental justice: frameworks for critical engagement." *Antipode*, vol. 41, no. 4, pp. 591–612.

Holifield, R. and Schuelke, N. (2015). "The place and time of the political in urban political ecology: contested imaginations of a river's future." *Annals of the Association of American Geographers*, vol. 105, no. 2, pp. 294–303.

Jahn, T., Bergmann, M. and Keil, F. (2012). "Transdisciplinarity: between mainstreaming and marginalization." *Ecological Economics*, vol. 79, pp. 1–10.

Kurtz, H.E. (2002). "The politics of environmental justice as the politics of scale: St. James Parish, Louisiana, and the Shintech siting controversy." In Herod, A. and Wright, M.W. (eds.), *Geographies of power: placing scale*. Oxford, UK: Blackwell Publishers Ltd, pp. 249–273.

Kurtz, H.E. (2003). "Scale frames and counter-scale frames: constructing the problem of environmental injustice." *Political Geography*, vol. 22, no. 8, pp. 887–916.

Liévanos, R., London, J. and Sze, J. (2010). "Uneven transformations and environmental justice: regulatory science, street science, and pesticide regulation in California." In Ottinger, G. and Cohen, B. (eds.), *Engineers, scientists, and environmental justice: transforming expert cultures through grassroots engagement*. Cambridge MA: MIT Press.

Loftus, A. (2007). "Working the socio-natural relations of the urban waterscape in South Africa." *International Journal of Urban and Regional Research*, vol. 31, no. 1, pp. 41–59.

London, J.K., Schwarz, K., Cutts, B., Cadenasso, M.L., Mason, C., Lim, J., Smith, H. and Valenzuela, K. (2017). "Weaving community-university research and action partnerships for environmental justice." *Action Research Journal*. Prepublished March 14, 2017.

London, J.K., Karner, A., Sze, J., Rowan, D., Gambirazzio, G. and Niemeier, D. (2013). "Racing climate change: collaboration and conflict in California's global climate change policy arena." *Global Environmental Change*. vol. 23, no. 4, pp: 791–799.

London, J.K., Sze, J. and Lievanos, R.S. (2008). "Problems, promise, progress, and perils: critical reflections on environmental justice policy implementation in California." *UCLA Journal of Environmental Law and Policy*, vol. 26, p. 255–289.

London, J.K., Zagofsky, T.M., Huang, G. and Saklar, J. (2011). "Collaboration, participation and technology: the San Joaquin Valley cumulative health impacts project." *Gateways: International Journal of Community Research and Engagement*, vol. 4, pp. 12–30.

Lyytimäki, J., Petersen, L.K., Normander, B. and Bezák, P. (2008). "Nature as a nuisance? Ecosystem services and disservices to urban lifestyle." *Environmental Sciences*, vol. 5, no. 3, pp. 161–172.

MacMynowski, D.P. (2007). "Pausing at the brink of interdisciplinarity: power and knowledge at the meeting of social and biophysical science." *Ecology and Society*, vol. 12, no. 1, pp. 1–20.

Miller, T.R., Muñoz-Erickson, T. and Redman, C.L. (2011). "Transforming knowledge for sustainability: towards adaptive academic institutions." *International Journal of Sustainability in Higher Education*, vol. 12, no. 2, pp. 177–192.

Miller, T.R., Baird, T.D., Littlefield, C.M., Kofinas, G., Chapin III, F.S. and Redman, C.L. (2008). "Epistemological pluralism: reorganizing interdisciplinary research." *Ecology and Society*, vol. 13, no. 2, p. 46.

Minkler, M., Vásquez, V.B., Tajik, M. and Petersen, D. (2008). "Promoting environmental justice through community-based participatory research: the role of community and partnership capacity." *Health Education & Behavior*, vol. 35, no. 1, pp. 119–137.

Nixon, R. (2011). *Slow violence and the environmentalism of the poor.* Cambridge MA: Harvard University Press.

Pellow, D.N. (2016). "Towards a critical environmental studies: Black Lives Matter as an environmental justice challenge." *DuBois Review*, vol. 13, no. 2, pp. 221-236.

Pellow, D.N. and Brulle, R.J. (2005). "Power, justice, and the environment: toward critical environmental justice studies." In Pellow, D.N. and Brulle, R.J (eds.) *Power, justice, and the environment: a critical appraisal of the environmental justice movement.* Cambridge MA: MIT Press. pp. 1–19.

Pennington, D.D. (2008). "Cross-disciplinary collaboration and learning." *Ecology and Society*, vol. 13, no. 2, pp. 1–13.

Perreault, T., Wraight, S. and Perreault, M. (2012). "Environmental injustice in the Onondaga lake waterscape, New York State (USA)." *Water Alternatives*, vol. 5, no. 2, pp. 485–506.

Pickett, S.T.A., Cadenasso, M.L. and Grove, J.M. (2004). "Resilient cities: meaning, models, and metaphor for integrating the ecological, socio-economic, and planning realms." *Landscape and Urban Planning*, vol. 69, no. 4, pp. 369–384.

Pickett, S.T.A., Cadenasso, M.L. and Grove, J.M. (2005). "Biocomplexity in coupled natural–human systems: a multidimensional framework." *Ecosystems*, vol. 8, no. 3, pp. 225–232.

Raudsepp-Hearne, C., Peterson, G.D. and Bennett, E.M. (2010). "Ecosystem service bundles for analyzing tradeoffs in diverse landscapes." *Proceedings of the National Academy of Sciences*, vol. 107, no. 11, pp. 5242–5247.

Sayre, N.F. (2005). "Ecological and geographical scale: parallels and potential for integration." *Progress in Human Geography*, vol. 29, no. 3, pp. 276–290.

Schlosberg, D. (1999). *Environmental justice and the new pluralism: the challenge of difference for environmentalism.* Oxford UK: Oxford University Press.

Schlosberg, D. (2003). "The justice of environmental justice: reconciling equity, recognition, and participation in a political movement." In Light, A. and de-Shalit, A. (eds.) *Moral and political reasoning in environmental practice.* Cambridge MA: MIT Press. pp. 77–106.

Schlosberg, D. (2004). "Reconceiving environmental justice: global movements and political theories." *Environmental Politics*, vol. 13, no. 3, pp. 517–540.

Schwarz, K., Cutts, B.B., London, J.K. and Cadenasso, M.L. (2016). "Growing gardens in shrinking cities: a solution to the soil lead problem?" *Sustainability*, vol. 8, no. 2, p. 141.

Shilling, F.M., London, J.K. and Liévanos, R.S. (2009). "Marginalization by collaboration: environmental justice as a third party in and beyond CALFED." *Environmental Science and Policy*, vol. 12, no. 6, pp. 694–709.

Silver, J.J. (2008). "Weighing in on scale: synthesizing disciplinary approaches to scale in the context of building interdisciplinary resource management." *Society and Natural Resources*, vol. 21, no. 10, pp. 921–929.

Stokols, D. (2006). "Toward a science of transdisciplinary action research." *American Journal of Community Psychology*, vol. 38, no. 1–2, pp. 63–77.

Strang, V. (2009). "Integrating the social and natural sciences in environmental research: a discussion paper." *Environment, Development and Sustainability*, vol. 11, no. 1, pp. 1–18.

Swyngedouw, E. (2007). "Technonatural revolutions: the scalar politics of Franco's hydro-social dream for Spain, 1939–1975." *Transactions of the Institute of British Geographers*, vol. 32, no. 1, pp. 9–28.

Swyngedouw, E. and Heynen, N.C. (2003). "Urban political ecology, justice and the politics of scale." *Antipode*, vol. 35, no. 5, pp. 898–918.

Sze, J. (2014). *Fantasy islands: Chinese dreams and ecological fears in an age of climate crisis.* Berkeley CA: University of California Press.

Sze, J., Gambirazzio, G., Karner, A., Rowan, D., London, J. and Niemeier, D. (2009). "Best in show? climate and environmental justice policy in California." *Environmental Justice*, vol. 2, no. 4, pp. 179–184.

Sze, J. and London, J.K. (2008). "Environmental justice at the crossroads." *Sociology Compass*, vol. 2, no. 4, pp. 1331–1354.

Sze, J., London, J., Shilling, F., Gambirazzio, G., Filan, T. and Cadenasso, M. (2009). "Defining and contesting environmental justice: socio-natures and the politics of scale in the delta." *Antipode*, vol. 41, no. 4, pp. 807–843.

22

CUMULATIVE RISK ASSESSMENT

An analytic tool to inform policy choices about environmental justice

Ken Sexton and Stephen H. Linder

Introduction

Calls for environmental justice typically involve allegations that those living in poverty, particularly people of colour, are at higher comparative risk for a variety of adverse health outcomes because they (a) bear a disproportionate cumulative burden of environmental stressors and (b) are more susceptible to environmentally related illness and injury than the general population (Sexton and Anderson 1993; Sexton 1997; IOM 1999; Fiscella and Williams 2004; Hynes and Lopez 2007; Sexton and Linder 2010). Conventional human risk assessments, however, have not typically addressed these concerns, focusing instead on individual chemicals, discrete health endpoints, single sources or source categories, specific environmental media, and distinct exposure pathways and routes (Sexton and Adgate 1999; NRC 2009; Sexton and Linder 2010). Accordingly, there is mounting apprehension among stakeholders that past risk assessments have been so narrowly constrained as to leave dangerous cumulative health risks undetected, unaddressed, and undeterred (IOM 1999; NEJAC 2004; NRC 2009; Sexton and Linder 2010). This chapter focuses on the potential of cumulative risk assessment to improve policy decisions about environmental justice. Although the emphasis is on regulatory conditions in the United States, lessons may also be applicable to other industrialized countries (for an overview of cumulative risk assessment outside the US see Sexton 2012).

Regulatory risk assessment is the science-based tool used to identify, evaluate, and calculate human health risks resulting from exposures to environmental hazards as a key input to policy decisions about which risks are unacceptable and what, if anything, to do about them (Sexton and Adgate 1999; NRC 2009). Implementation of systematic and quantitative risk assessment as an integral part of regulatory decision making developed out of efforts during the 1970s by the US Environmental Protection Agency (EPA) and US Food and Drug Administration to estimate human health risks from exposure to anthropogenic carcinogens in air, water, and food (Callahan and Sexton 2007; NRC 2009). The regulatory philosophy existing at that time, which accentuated chemical-specific hazards, national command-and-control regulations, and technology-based pollution-control approaches, strongly influenced the subsequent evolution of risk assessment processes and procedures. As a result, most risk assessments conducted by the US EPA over the past 40 years have been narrowly concentrated on individual chemical agents, distinct pollution sources and source categories, and single exposure pathways, environmental media, routes of exposure, and health endpoints (Callahan and Sexton 2007; NRC 2009).

Cumulative risk assessment (CRA) is a science-policy instrument for organizing and analysing scientific information to examine, characterize, and possibly quantify combined adverse effects on human health (or ecological resources) from multiple environmental stressors, including both chemical and nonchemical agents (USEPA 2003; Callahan and Sexton 2007; NRC 2009; Sexton 2015; Sexton and Hattis 2007). The application of CRA is substantively different from traditional risk assessment because it: 1) entails evaluation of cumulative health effects of multiple stressors, as opposed to individual effects of a single stressor; 2) expands the range of environmental agents under examination to include not just chemicals, but also psychological (e.g. residential crowding) and sociological (e.g. joblessness) stressors; 3) explicitly incorporates the concept of vulnerability into the assessment (e.g. differential biological susceptibility and exposure, differential preparedness and ability to recover from stressor effects); 4) emphasizes real-world cumulative exposures experienced by actual people, not theoretical exposures of hypothetical people as is often the case in traditional risk assessments; 5) formally acknowledges that the details (e.g. timing, co-exposure to multiple agents) and history (e.g. continuous, intermittent, simultaneous, sequential) of exposure are important for estimating health risk; 6) considers background exposures (i.e. combined exposure to toxicologically relevant stressors, including those not formally targeted for assessment) that contribute to the risk being evaluated; and 7) allows for the possibility of semi-quantitative or even qualitative analyses and results, depending on the circumstances (Sexton 2012).

The collective nature of CRA embodies several contemporary and distinct ideas. First, harmful entities (e.g. environmental stressors, levels of toxicity), events (e.g. duration and frequency of exposure), and outcomes (e.g. discomfort, dysfunction, disability, disease) accumulate, causing aggregate quantities of multiple stressors and stressor-related effects to increase. Second, for risk assessment purposes, these factors along with as-yet-unspecified contributors can be lumped together to estimate cumulative effects. And third, although it is still uncertain whether interactions among stressors are additive, non-additive, or some combination, one can visualize accumulation of adverse effects as occurring vertically within specific risk categories (e.g. inhalation), and lumping them together as happening horizontally across diverse risk categories (e.g. inhalation, ingestion, dermal absorption). As currently constituted, CRA can encompass either accumulation within a risk classification, lumping across risk classifications, or both.

Guidelines for CRA of chemical mixtures have been published by the US EPA (USEPA 1986, 2000, 2006a) and the Agency for Toxic Substances and Disease Registry (ATSDR 2004), and nominal CRAs have been done for certain classes of pesticides, including organophosphates (USEPA 2002), chloroacetanilides (USEPA 2006b), triazines (USEPA 2006c), and n-methyl carbamates (USEPA 2007). Several investigators have looked at racial and ethnic differentials in residential proximity to known sources of pollution (Sexton and Anderson 1993; Perlin et al. 1995, 1999), and numerous studies have explored environmental justice issues related to cumulative cancer risk from exposure to airborne pollutants (Fox et al. 2004; Tam and Neumann 2004; Apelberg et al. 2005; Morello-Frosch and Jesdale 2006; Linder et al. 2008; Gilbert and Chakraborty 2011). Specific metrics and screening methods for estimating cumulative risk burden have been proposed (Su et al. 2009; Sadd et al. 2011; Rider et al. 2012), and three states – California, Minnesota, and New Jersey – have officially made evaluation of cumulative health risks from multiple stressors an explicit requirement as part of risk management decisions (MPCA 2009a,b, 2010; NJDEP 2009; Cal/EPA 2010;

Ellickson et al. 2011; Alexeeff et al. 2012). Nevertheless, true CRA remains in its infancy and vital theoretical constructs, adequate data, suitable mechanistic models, necessary science-policy procedures, and practical application methods are in short supply (Linder and Sexton 2011; Sexton 2012, 2015).

Overview of cumulative risk assessment

The EPA published a framework for cumulative risk assessment in 2003, which provided a conceptual structure for distinguishing the basic elements and fundamental principles of a systematic process for analysing cumulative health risks (USEPA 2003; Callahan and Sexton 2007). The framework was meant to encourage dialogue on theoretical issues, technical matters, important definitions, and implementation challenges "whether or not the methods or data currently exist to adequately analyze or evaluate those aspects of the assessment" (USEPA 2003: xvii–xx). As shown in Figure 22.1, the EPA framework described three interrelated and generally sequential phases of CRA: 1) phase A – planning, scoping, and problem formulation; 2) phase B – information and data analysis; 3) phase C – interpretation and risk characterization (USEPA 2003; Callahan and Sexton 2007).

In the first phase, the goals, breadth, depth, and focus of the assessment are determined jointly by a team of risk assessors, risk managers, and interested stakeholders. Two major products of this step are a conceptual model that identifies important stressors, related health effects, and relationships among stressors and outcomes, and an analysis plan that stipulates data needs, the approach to be taken, and key features of expected results. The second phase entails development of exposure profiles, examination of stressor interactions, appraisal of population risks, and evaluation of uncertainty and variability. Estimation of cumulative risk for the people and places of interest, along with estimates of uncertainty and variability are the major products from phase B. In phase C, the scientific and science-policy justification for risk estimates is explained, results are described in terms of assessor confidence and public health significance, and a determination is made as to whether the assessment met the goals and objectives set forth in phase A (USEPA 2003; Callahan and Sexton 2007).

In 2009, the National Research Council (NRC 2009) opined that stakeholders, including members of affected communities, are increasingly worried that the restricted scope of traditional, single-chemical risk assessments do not capture combined effects from exposure to multiple chemical and nonchemical stressors, nor do they account adequately for factors that might affect vulnerability. The NRC warned that unless cumulative risks are taken into account, when appropriate, that "risk assessment might become irrelevant in many decision contexts, and its application might exacerbate the credibility and communication gaps between risk assessors and stakeholders" (NRC 2009: 213). The NRC went on to say that the EPA framework and attendant broadening definition of CRA represented "a move toward making risk assessments more relevant to decision-making and to the concerns of affected communities" (NRC 2009: 214).

According to the NRC definition (NRC 2009), CRA necessarily entails three key elements: evaluation of the effects of nonchemical stressors; inclusion of all relevant exposure routes and pathways; and consideration, to the extent necessary, of both background exposures and vulnerability. Based on this characterization, no comprehensive and thorough CRAs have been accomplished so far. Among the practical obstacles that continue to impede implementation are: inadequate and insufficient scientific knowledge about cumulative exposures and related health effects; absence of scientific consensus regarding appropriate conceptual models, diagnostic frameworks, and analytical procedures; lack of a robust

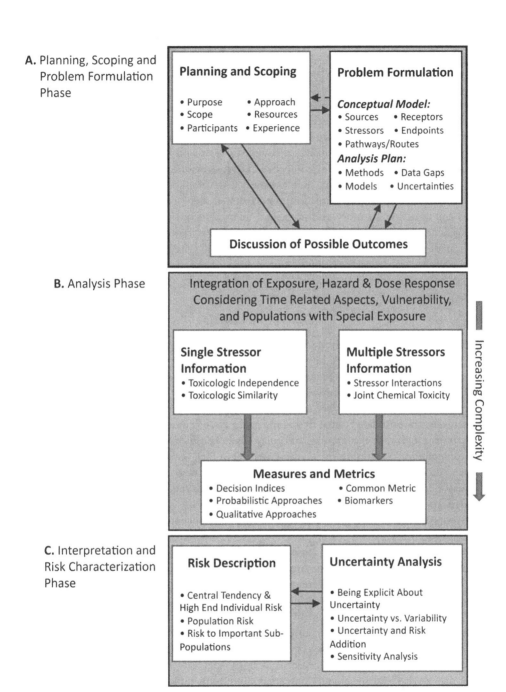

Figure 22.1 Three phases of the cumulative risk assessment process

methodology to constrain the scope of the analysis so that it is manageable and results are relevant and timely; deficiency of real-life experience with actual applications; a concomitant shortage of policies and processes to guide implementation; and unanswered questions about legal authority to use CRA under existing laws and statutes.

Risk assessors and managers thus find themselves confronting two contradictory realities – increasing awareness of the necessity of conducting CRA as part of informed decision making about environmental justice and other environmental health issues, and the stark reality of data deficiencies and intrinsic analytical complexities that hamper evaluation of cumulative health risks. They are, paradoxically, compelled to address broader, more complicated questions about cumulative risk, while striving simultaneously to develop simpler, easier-to-use assessment tools. Further confounding the situation, the expanded scope of CRA is likely to cut across statutes and agencies' statutory authority, and to involve questions and topics beyond the traditional domain of health risk assessment. The fact is that CRA is more theoretically complicated, methodologically complex, computationally challenging, and bureaucratically problematic than classic single-chemical, source-oriented risk assessments (Callahan and Sexton 2007; Sexton and Linder 2010; Sexton 2012, 2015).

Conceptual models

Because there is no empirically derived theory or scientific consensus on an appropriate theoretical framework for CRA, applications and related research have necessarily relied on speculative conceptual models (Linder and Sexton 2011). A theoretical framework provides an intellectual blueprint that serves as a formalized analytic schema to organize critical assumptions, concepts, indicators and propositions into methodical configurations that definitely describe postulated causal factors and pathways, often using box-and-arrow diagrams. It operates as a logic map that grounds and guides model development and evaluation, defining the meaningfulness and utility of indicator variables. This implies that the meaning of key variables, their expected interactions, and their unspecified connections with undefined constructs outside the model are contingent on an assumed theoretical framework.

There are basically three overarching frameworks that have been used to structure the CRA process, and each gives rise to a "family" of conceptual models that share similar theoretical roots (Linder and Sexton 2011). First, there are *multiple stressor models*, which deem health inequality to be caused by exposure to a specified array of environmental stressors. Cumulative stress from multiple environmental factors is assumed to disrupt function and well-being, thereby leading to differential health effects. Second, there are *social determinant models*, which consider health to be primarily a product of social factors, and health inequality to be a product of social inequality. In this conceptualization, both social engagement and psychological coping are adversely affected by worsening material conditions, which ultimately leads to pathophysiological processes and behavioral adaptions that cause impairment and increased risk of death and disease. And third, there are *health disparity models*, which assume that a variety of social and contextual conditions, rather than just social inequality, give rise to unequal health outcomes. The emphasis here is on the impact of determinants like socioeconomic status, ethnicity/race, and gender on individual behaviours and biological responses that contribute to disproportionate health consequences.

All three model families are referred to as "speculative" because rather than functioning as testable analytic plans they serve primarily as conceptual aids and heuristics (Linder and Sexton 2011). The form and function of CRA as well as the direction of related research has been significantly influenced by successive rounds of speculative modelling, more so than by empirical findings. And while these models can simplify abstract theories about causal relationships and make assumptions about hypothesized cause–effect linkages explicit, they need to be tested empirically so that findings are more systematic, robust, and replicable. Ultimately, the choice of a guiding theoretical framework (family of models) is crucial for realistic and reliable assessment of cumulative health risks because it affects virtually every major facet of the evaluation process, including: identification of independent and dependent variables and delineation of interrelationships among confounders and co-factors; specification of appropriate indicators or surrogate measures to characterize the effects of multiple stressors on individuals and communities; determination of the extent to which the connection between cumulative exposure and cumulative health effects can be quantified; enumeration of how vulnerability factors aggravate or modulate adverse effects of chemical and nonchemical stressors; and resolution of whether and how an aggregate index – if effects can be added together – or profile – if they cannot – can be used to represent cumulative risks for a particular community or population.

Practical approaches

Although information deficits, knowledge gaps, and lack of pertinent policies and procedures continue to provide significant implementation challenges, three generic approaches to conduct systematic and comprehensive CRAs have been identified (Sexton 2012, 2015). As shown in Table 22.1, each one aims to answer a different question, starting from a different point of departure, using different analytical strategies. Stressor-based methods use a prospective, bottom-up analytical approach to assess cumulative health risks for a defined set of stressors, while effects-based methods employ a retrospective, top-down analytical approach to explain observed or hypothesized health outcomes in a defined place or population. Vulnerability-based assessment is a hybrid approach that combines elements of both stressor- and effects-based methods. It starts with concern about disproportionate cumulative risks for a particular vulnerable group or population and, depending on the circumstances, can use a prospective bottom-up, retrospective top-down or combined analytical approach. Stressor-based approaches have been driven primarily by regulatory concerns about exposure to chemical mixtures, effects-based approaches by the need for community-based risk assessments of chemical and nonchemical stressors, and vulnerability-based methods by public health concerns about disproportionate cumulative risks for economically and politically disadvantaged populations.

Near-term progress in the application of CRA to issues of environmental justice depends on development of practical implementation aids, such as tiered assessment approaches and phased execution tactics (Sexton 2015). A tiered approach necessitates creation of an ordered framework of categories for data/knowledge-availability related to (a) cumulative exposure, (b) combined health effects, and (c) population vulnerability (see Table 22.2). Because more knowledge and understanding are available for each succeeding tier, associated analyses can be more certain and less cautious. Thus the tiered framework can be used to decide which analytical methods are appropriate for each data/knowledge-availability classification. For example, in the case of cumulative exposure assessment, reasonable worst-case estimates would be used for tier 0, generic exposure scenarios using conservative default assumptions for tier 1, exposure estimates based on limited measurements and models for tier 2, and probabilistic methods for estimating exposure distributions for tier 3.

A phased approach to CRA refers to the explicit identification of key analytical steps and specification of the sequence in which they will occur. As summarized in Table 22.3, the phases will vary according to whether the CRA is stressor-based, effects-based, or vulnerability-based. All, nonetheless, begin with a description of the conceptual model being used, and conclude with an estimate of cumulative health risks and associated uncertainties. Establishing policies and procedures, including guidelines, for tiered-analysis categories and phased-assessment approaches are critical steps in the evolution of CRA from a conceptual ideal to a practical, decision-relevant tool.

Table 22.1 Comparison of stressor-based (bottom-up), effects-based (top-down), and vulnerability-based (hybrid) approaches to cumulative risk assessment (adapted from Sexton 2015)

Attribute	Stressor-based assessment	Effects-based assessment	Vulnerability-based assessment
Analytical strategy	Prospective, bottom-up analysis (evaluate constituent interactions)	Retrospective, top-down analysis (deconstruct and elucidate outcomes)	Hybrid approach: a prospective, bottom-up evaluation; a retrospective, top-down appraisal; or a combined analysis depending on circumstances
Central question	Which potential health effects are associated with a defined set of stressors?	Which stressors explain observed or hypothesized health outcomes?	To what extent are disadvantaged (susceptible) groups affected by cumulative environmental risks?
Starting point	Identification of key stressors and recognition of the populations and health end points potentially influenced by them	Development of a conceptual model incorporating the stressors plausibly associated with critical health outcomes	Recognition of a vulnerable group/population based on differential exposures, differential susceptibility, or a combination of both
Primary emphasis	Analysis of stressor interactions to predict the likelihood and severity of possible future adverse health outcomes	Determination of stressor contributions to observed or hypothesized health outcomes, including consideration of co-exposures and background processes	Establish extent and magnitude of adverse effects from cumulative exposures on vulnerable groups and populations
Typical applications	Chemical mixtures	Combinations of chemical and nonchemical stressors	Either chemical mixtures or combinations of chemical and nonchemical stressors
Driving force(s)	Regulatory decisions about protection of human health from exposure to multiple chemicals (e.g. more true-to-life exposure scenarios)	Analysis of health disparities, concerns about nonchemical stressors, and the need for community-based risk assessments	Public health focus on possible disproportionate cumulative risks experienced by economically and politically disadvantaged groups
Example references	FQPA 1996; SDWA 1996; ATSDR 2004; USEPA 1986, 2000, 2002, 2006a,b,c, 2007; NRC 2008; Meek et al. 2011	Gee and Payne-Sturges 2004; Callahan and Sexton 2007; NRC 2009; Linder and Sexton 2011; Salinas et al. 2012	Fox et al. 2002; deFur et al. 2007; NEJAC 2010; Huang and London 2012; Prochaska et al. 2014; Sadd et al. 2011; USEPA 2015a, b; Sexton and Linder 2010

Table 22.2 Outline for a tiered approach to categorizing data/knowledge sufficiency for cumulative risk assessment (adapted from Sexton 2015)

Scientific information to characterize cumulative exposure	Scientific information to characterize combined health effects	Scientific information to characterize population vulnerabilities
Tier 0 – *insufficient* data on cumulative exposures: use reasonable worst-case estimate of exposures	Tier 0 – *insufficient* data on constituent interactions: dose or effect addition assumed for all mixture constituents	Tier 0 – *insufficient* data on susceptibility factors: assume reasonable worst case for both environmental and receptor vulnerability factors
Tier 1 – *minimal* data on cumulative exposures: use generic exposure scenarios using conservative point estimates	Tier 1 – *minimal* data on constituent interactions: improved potency assessment based on relevant point-of-departure data	Tier 1 – *minimal* data on susceptibility factors: improved vulnerability estimates based on community-level data (e.g. US Census) for the affected population
Tier 2 – *moderate* data on cumulative exposures: use improved exposure assessment based on measurements and models	Tier 2 – *moderate* data on constituent interactions: refined potency assessment based on mode of action for the mixture constituents	Tier 2 – *moderate* data on susceptibility factors: refined vulnerability estimates based on detailed data drawn from the affected population
Tier 3 – *adequate* data on cumulative exposures: use probabilistic methods to estimate exposures, including distributions	Tier 3 – *adequate* data on constituent interactions: PBPK and/or BBDR models used to produce probabilistic estimates of potency	Tier 3 – *adequate* data on susceptibility factors: application of verified formula/method to population-specific data in order to estimate vulnerability quantitatively

Note: PBPK physiologically based pharmacokinetic; BBDR biologically based dose-response

Table 22.3 Phased approach to three forms of cumulative risk assessment (adapted from Sexton 2015)

Phase	Stressor-based (bottom-up) approach	Effects-based (top-down) approach	Vulnerability-based (hybrid) approach
Phase 1	– specify conceptual model describing important stressors and ways they cause effects – identify receptors and endpoints affected by stressors both individually and in combination – establish common denominators for appraisal by identifying common receptors and endpoints	– develop conceptual model describing important stressors and the ways they cause critical effects – establish common denominators for evaluation by identifying common receptors and endpoints	– identify the vulnerable population and/or community – specify conceptual model describing important stressors and the ways they cause critical effects – establish common denominators for evaluation by identifying common receptors and endpoints

(continued)

Table 22.3 Phased approach to three forms of cumulative risk assessment (adapted from Sexton 2015)
(continued)

Phase	Stressor-based (bottom-up) approach	Effects-based (top-down) approach	Vulnerability-based (hybrid) approach
Phase 2	– screen stressors of interest to determine which need to be included in the assessment and which may act in combination	– screen potential stressors to identify an appropriate and manageable number to characterize the problem adequately	– screen stressors affecting the population/ community to establish which should be included in the analysis
Phase 3	– appraise individual effects of stressors and combinations of stressors in the context of the conceptual model – determine how the combined effects of multiple stressors affect critical endpoints – consider relevant psychosocial stressors by characterizing the environmental, cultural, and socioeconomic attributes of those exposed	– appraise the individual effects of individual stressors to determine whether one or a few stressors are predominant	– evaluate individual and combination effects of important stressors to determine whether one or a few stressors predominate
Phase 4	– assess combined effects of stressors taking account of potential interactions among stressors and effects	– assess combined effects of stressors without considering potential interactions	– consider combined effects of important stressors without considering potential for interactions
Phase 5	– not applicable	– gauge combined effects of stressors taking account of potential interactions among stressors and effects	– estimate combined effects of important stressors taking into account potential interactions among stressors and effects

Analysing environmental justice

Despite its growing importance for policy decisions, environmental justice remains for many, including some risk assessors and risk managers, a vague and abstract notion that is difficult to define in concrete, real-world terms. Yet if CRA is to be effective in fostering informed policy choices, it is important to identify the essential elements of environmental justice and define them unequivocally and precisely. Although different definitions are likely to have disparate effects on policy choices, the crucial point is not that any specific definition is "right" or "wrong" but rather that selecting a particular definition has direct implications for the formulation, application, and appraisal of environmental justice policies and related risk assessment activities (Phillips and Sexton 1999). From a risk assessment perspective, a distinct and workable definition is vital so that questions of environmental justice can be examined by stating an unambiguous hypothesis and then testing it empirically in a scientifically rigorous manner. A practical and testable definition for environmental justice must answer five fundamental questions (Phillips and Sexton 1999):

- How are fairness, equity and justice defined? In terms of procedural fairness, outcome fairness, or both?
- How are unfairness, inequity and injustice defined? Which deviations from fair/equitable/just are characterized as unfair/inequitable/unjust?
- What is the scope of concern? Which hazards over what geospatial scale are included in the analysis?
- Which individuals and/or groups are the focus of concern? Which attributes (e.g. race, ethnicity, socioeconomic status, residential location) are shared by those who systematically and involuntarily experience a disproportionate risk from environmental hazards?
- What is the presumed root cause of unfairness, inequity, and injustice? Are disparities caused by political expediency, utilitarian policy decisions, intentional discrimination, or neighbourhood transition and other voluntary choices?

A definitive and explicit operational definition is necessary before society can put lofty and laudable environmental justice principles into practice. If we can't or won't define it, how can we recognize environmental injustice, prioritize problems, and measure progress toward solutions? We need to move away from fuzzy concepts, hazy theories, and ambiguous terminology to establish a clear, concrete, consensus definition for environmental justice as a necessary precondition for estimating and comparing disproportionate health risks. So far, such a definition remains elusive (Phillips and Sexton 1999; Sexton and Linder 2010).

Analytical methodologies

The importance of environmental justice in the context of cumulative risk assessment is widely recognized (Sexton and Anderson 1993; Sexton 1997; IOM 1999; NEJAC 2004, 2010; Hynes and Lopez 2007; Sexton and Linder 2010; USEPA 2013, 2015a,b), and numerous case studies have been published (Fox et al. 2002; Evans and Marcynyszyn 2004; Krieg and Faber 2004; Sadd et al. 2011; Huang and London 2012; Prochaska et al. 2014; Gallagher et al. 2015). Moreover, a variety of practical approaches and methods have been developed to evaluate cumulative exposures, burdens, and risks (USEPA 2004; Barzyk et al. 2010; Medina-Vera et al. 2010; NEJAC 2010; MacDonell et al. 2013). Selected examples of methodologies for analysing environmental justice issues are given in Table 22.4.

The EPA has provided guidance to agency personnel on when (USEPA 2015b) and how (USEPA 2013) to incorporate environmental justice into development of regulations. Environmental justice analysis is triggered, according to the EPA, by "environmental justice concerns," which are defined to mean "disproportionate impacts on minority, low-income, or indigenous populations that exist prior to or that may be created by the proposed action" (USEPA 2013: 2). "Multiple stressors and cumulative impacts" are identified as a "concern"

Table 22.4 Selected examples of a cumulative risk methodology applied to environmental justice (adapted from Sexton 2015)

Method, organization, and reference(s)	Conceptual model (see Linder and Sexton 2011)	General approach
Cumulative Environmental Hazard Index (CEHII) by the University of California-Berkeley (Su et al. 2009)	Multiple stressors	creates an index (from publicly available data) summarizing racial/ethnic and socioeconomic inequalities from the impact of cumulative environmental hazards
Cumulative Risk Screening Methodology by the Cal/EPA (Cal/EPA 2010; Alexeeff et al. 2012)	Multiple stressors	calculates an overall cumulative impact score by multiplying a pollution burden score (sum of exposures, health and environmental effects) by a population characteristics score (sum of sensitivity and socioeconomic factors)
C-FERST by the USEPA (USEPA 2012; Zartarian et al. 2011)	Multiple stressors	provides a web-based GIS tool using publicly available databases (e.g. environmental, socioeconomic, health) to assist communities with identifying and prioritizing cumulative exposures and environmental risks
EJSCREEN by the USEPA (USEPA 2015a)	Multiple stressors	uses 12 environmental indicators, 6 demographic indicators and 12 environmental justice indices to identify areas with potentially elevated cumulative risk burdens from environmental hazards
EJSM by University of California scientists (Sadd et al. 2011)	Multiple stressors	organizes 23 indicator metrics into three categories (hazard proximity and land use; air pollution exposure and risk; social and health vulnerability) and then calculates a cumulative impact score by Census tract
Urban-HEART by the WHO (WHO 2008, 2010a, 2010b)	Social determinants	analyses data on health outcomes and social determinants to construct an urban health equity matrix to identify highest comparative risks for communities or population groups
Human Security Index (HSI) by the UNDP (Hastings 2011; HSI 2012; Salinas et al. 2012)	Social determinants	constructs an overall index from three sub-indices (indicators of economic, environmental, and social well-being) summarizing the cumulative health (or resiliency) of a particular area or population

Note: C-FERST community-focused exposure and risk screening tool; EJSCREEN environmental justice screening and mapping tool; EJSM environmental justice screening method; UNDP United Nations developmental programme; HEART health equity assessment and response tool; WHO = World Health Organization that necessarily triggers an environmental justice analysis (USEPA 2015b: 14). The agency states that:

> Analyzing cumulative impacts from multiple stressors allows a more complete evaluation of a population's risk from pollutants targeted by the action under consideration, particularly when there may be important interaction effects among these multiple stressors and adequate data and methods are available.
>
> *(USEPA 2015b: 14)*

Specific analytic principles are identified to help guide the environmental justice analysis, including: specification of key objectives; understanding of root causes and contributors; identification of data, methods, and analytical needs; recognition of population groups with environmental justice concerns; and selection of an appropriate comparison group (USEPA 2013: 5–6).

When planning a CRA, agency analysts are directed to evaluate exposures, relevant health and environmental outcomes, and other germane factors, both within and across population groups, so as to determine (a) baseline levels, (b) differentials among groups, and (c) how differentials decrease or increase as a result of different regulatory options (USEPA 2013: 19–20). Within the structure of the "scoping" step in phase A of CRA (see Figure 22.1), several key questions are to be addressed (USEPA 2013: 26). Which population groups, as defined by attributes such as geographic location, ethnicity or race, gender or baseline health status, should be part of the assessment? Which health endpoints are to be addressed in the assessment? Which exposure routes and pathways are relevant? Are there specific exposure pathways that may lead to specific effects? What exposure scenarios should be modelled? By answering these questions, analysts establish a set of boundaries for CRA that can be incorporated into step 2 of phase A: "problem formulation," which focuses on determining whether minority, low-income or Indigenous populations experience disproportionate risks (see Figure 22.1). As part of this process, sources and characteristics of stressors contributing to the disproportionate risk are clarified, factors that influence relevant exposures are identified, and population susceptibilities and/or vulnerabilities that cause exposure or risk differentials are characterized. The major products resulting from the problem formulation step are a list of assessment health endpoints, a description of the overarching conceptual model, and a plan for analysing environmental justice in the context of cumulative risks (USEPA 2013: 31).

As part of the "planning, scoping, and problem formulation phase," and to ensure integrity and utility of the process, risk assessors and managers should ask themselves a series of pivotal questions at the outset (Sexton 2015). Is a cumulative assessment really necessary and, if so, why? How will results be used in decision making? How can the scope of the assessment be constrained so that it is both analytically manageable and decision relevant? Which technical challenges and scientific uncertainties are likely to be most influential in risk estimation? Which conceptual framework and analytical methodology is best suited to address the situation/problem at hand? By answering these and related questions, goals and objectives can be agreed on upfront, data gaps identified, computational deficiencies ascertained, and the form of expected results characterized; all of which contributes to making CRAs more methodologically rigorous, analytically tractable, and germane to risk-based decision making (Sexton 2015).

Geospatial scope and data realities

Important risk-related variables, like exposures, health effects, and exposure-effect linkages fluctuate geospatially, which means that they are conditioned by factors such as geographic location and neighbourhood boundaries, the scale of spatial analysis, and the spatial resolution of the aggregated data (Sexton and Linder 2011). Findings from statistical analysis of these kinds of variables depend, consequently, on the geospatial scale of analysis (e.g. county, city, census tract); a truism scientists have termed the "modifiable areal unit problem" or MAUP. For example, the true effect of exposure on disease incidence is not a biological constant but rather is modified by spatial scale. A disease that appears random at one geospatial scale may appear non-random at another, and a statistically significant regression analysis at one spatial scale may be statistically insignificant at a different one. Thus statistical analytic results for key risk-relevant variables change depending on spatial boundaries, scale, and resolution; which is to say, the resolution of exposure and health effects data defines the spatial scale of detectable statistical association and establishes the context within which results must necessarily be interpreted (Sexton and Linder 2011: S84). Examples of the kinds of data sources and related geospatial scales available for cumulative risk assessment are provided in Table 22.5 (Soobader et al. 2006; Sexton and Linder 2011).

As a practical matter, the availability of data is usually the determining factor in decisions about geospatial issues. For example, data on sociodemographic characteristics (e.g. age, race/ethnicity, income, occupation, education), social disadvantage (e.g. uninsured, unemployed, single mother, poverty level), and biologic susceptibility (e.g. pregnant women, infants, elderly,

Table 22.5 Selected examples of data sources and applicable geospatial levels available for assessment of cumulative health risks (adapted from Soobader et al. 2006 and Sexton and Linder 2011)

Data source	Geospatial levels available
Demographic, economic and social variables	
US Census	block group, census tract, MCD, µSA, MSA
Environmental exposure	
AIRS (Aerometric Information Retrieval System)	county, MSA, state
NATA (National-scale Air Toxics Assessment)	census tract, county, MSA, state
TRI (Toxic Release Inventory)	individual facilities, county, state
SDWIS (Safe Drinking Water Info System)	water system, county, MSA, state
SNAP (Superfund NPL Assessment Program)	site locations, county, MSA, sate
Body burden	
NHANES (National Health & Nutrition Exam Survey)	block group, census tract, county, MSA, state
National Report on Human Exposure	national reference ranges by pollutant
Health status	
NHANES (National Health & Nutrition Exam Survey)	block group, census tract, county, MSA, state
NHIS (National Health Interview Survey)	block group, census tract, county, MSA, state
SEER (Surveillance Epidemiology & End Results)	block group, census tract (CA only), MSA, state

Note: MCD Minor Civil Division; µSA micropolitan statistical area; MSA metropolitan statistical area infirm) are typically obtained from the US Census. Accordingly, data are available only at certain spatial scales used by the census, including block groups, census tracts, minor civil divisions, micropolitan statistical areas, and metropolitan statistical areas. Environmental health data, in contrast, are usually obtained from public or private databases (e.g. National Health and Nutrition Examination Survey – NHANES, Toxic Release Inventory – TRI, National Health Interview Survey – NHIS, EPA's Aerometric Information Retrieval System – AIRS, Surveillance Epidemiology and End Results – SEER) at a variety of spatial scales, including block group, census tract, metropolitan statistical area, county, and state. In the end, decisions about geospatial scale for a specific CRA ordinarily depend on the geospatial scale of relevant risk-related data (Sexton and Linder 2011).

Summary and conclusions

Although there is no consensus regarding an operational definition, the concept of *environmental justice* is generally taken to embrace the fundamental principles of fairness and equity that must be brought to bear as part of society's goal to attain adequate protection from harmful effects of environmental hazards for everyone, regardless of race, ethnicity, age, culture, creed, gender or socioeconomic status. The environmental justice paradigm directs our attention to crucial questions of whether economically and politically disadvantaged communities bear a disproportionate burden of environmental health risks, and whether past environmental practices, programmes and policies have been fair, equitable and just. How we answer these questions and decide what to do in response is a function of our values, beliefs, and experiences. Thus, while there is virtual unanimity across various sectors of society that environmental justice is a worthwhile goal, there is a divergent array of opinions about what it means for our daily lives and how to go about achieving it.

Assuming an operational definition is explicitly adopted, cumulative risk assessment offers the promise of a science-based analytic tool to estimate and compare aggregate health risks among various sociodemographic groups and geospatial locations. Using available scientific data and extant knowledge in combination with science-policy assumptions and expert judgment, CRA can be used to appraise combined health risks from exposure to a mixture of both chemical and nonchemical stressors in the context of environmental justice analysis. If adequate data exist, a CRA can be used to answer some key environmental justice questions, like "Are cumulative risks higher for one socioeconomic group than another?" "Do inter-group risk differentials exceed some predetermined threshold for regulatory action?" "What stressors are the main contributors to observed risk differentials and where and how do they originate?" "Which intervention strategies are most likely to address environmental justice concerns in an effective and efficient manner?" But CRA results can only provide limited input into other, more complicated societal questions regarding environmental justice, such as "Is it fair and/or equitable to add another industrial plant or roadway to an already highly polluted community?" "Are mitigation actions warranted in a poor, minority neighbourhood that bears a significantly greater pollution burden than an adjacent neighbourhood that is affluent and white?" These latter questions are really about what kind of society we want to live in, and require policymakers to consider a diversity of factors in addition to health risks, such as economic issues, societal values, legal precedents, political realities, ethical and moral principles, cultural beliefs, and bureaucratic realities.

Application of CRA is meant to broaden the extent of risk analysis to incorporate a more encompassing and real-world set of chemical and nonchemical stressors that are useful to decision making about environmental justice. The goal is to make risk assessment results more realistic by simulating real-life conditions, and thereby make them more relevant to the issues facing decision makers, more responsive to affected communities and other stakeholders, and more reliable as input to risk management decisions. But while many scientists and risk assessors see CRA as an essential decision making tool for identifying, evaluating, and resolving issues of environmental justice, others, including many environmental justice advocates, see it as a discriminatory, expensive, time-consuming, and ethically suspect process that is used to maintain the status quo by those in power. Currently, there is an ongoing debate about whether risk assessment is part of the problem or part of the solution (Sexton and Linder 2010, 2016).

This chapter has briefly examined the historical development of CRA as an aid to decisions about environmental justice, and reviewed the conceptual frameworks and analytical methodologies presently available. Although some have argued that systematic evaluation of health risks experienced by vulnerable populations and communities is both unnecessary and counterproductive – that instead of analysis, resources are better devoted to fixing obvious problems and rectifying observable wrongs – we maintain that CRA is a valuable tool to estimate health risk and associated uncertainties so that policy makers will be able to understand the magnitude of risk inequalities and make informed choices about which risk management options are most cost-effective. We agree that risk assessment must not be used as an excuse for doing nothing, nor should it impair decision makers' discretion. In our view, CRA is in the midst of a transition from theoretical paradigm to a practical and beneficial decision making tool that provides an objective methodology for systematic analysis of environmental justice issues (Sexton and Linder 2010).

Meaningful progress in the continued development and application of CRA requires execution of a coordinated and integrated research strategy that systematically works through a series of interrelated steps (Sexton and Linder 2011):

- Distinguish high-priority communities and populations at increased cumulative risk from exposure to both chemical and nonchemical hazards.
- Stipulate an overarching analytical framework that specifies postulated causal factors and pathways, and use it to guide collection, analysis, and interpretation of empirical data.
- Carry out targeted research on at-risk communities and populations to determine the magnitude, duration, frequency, and timing of relevant exposures, ascertain whether related cumulative health effects are additive, multiplicative, or antagonistic, and elucidate key interaction mechanisms that affect toxicity among mixture components.
- Execute a systematic assessment of cumulative health risks in selected environmental justice communities and/or populations relying on the overarching analytical framework as a guide for methodological evaluation.
- Use empirical data and analytic results to modify, revise, or discard the framework as a practical guide for assessing cumulative risks.
- Repeat the process (steps 1 to 5) as necessary.

The challenge, in addition to obtaining adequate funding, is to structure the research enterprise appropriately so as to generate timely and relevant research results, spur advances in conceptual approaches and theoretical paradigms, and produce diagnostic frameworks that provide pertinent information for better decision making about environmental justice.

References

Alexeeff, G.V., Faust, J.B., August, L.M., Milanes, C., Randles, K., Zeise, L., Denton, J. (2012). "A screening method for assessing cumulative impacts." *International Journal of Environmental Research and Public Health*, vol. 9, pp. 648–659.

Apelberg, B.J., Buckley, T.J., White, R.H. (2005). "Socioeconomic and racial disparities in cancer risk from air toxics in Maryland." *Environmental Health Perspectives*, vol. 113, pp. 693–699.

ATSDR (Agency for Toxic Substances and Disease Registry). (2004). *Guidance Manual for the Assessment of Joint Toxic Action of Chemical Mixtures*. Division of Toxicology, Atlanta, GA.

Barzyk, T.M., Conlon, K.C., Chahine, T., Hammond, D.M., Zartarian, V.G., Schultz, B.D. (2010). "Tools available to communities for conducting cumulative exposure and risk assessments." *Journal of Exposure Science and Environmental Epidemiology*, vol. 20(4), pp. 371–384.

Cal/EPA (California Environmental Protection Agency). (2010). Cumulative Impacts: Building a Scientific Foundation. Office of Environmental Health Hazard Assessment, Sacramento, CA, USA. Available at www.cumulativeimpact.org/document/CIReport123110.pdf [Accessed on 23 June 2014].

Callahan, M.A. & Sexton, K. (2007). "If cumulative risk assessment is the answer, what is the question?" *Environmental Health Perspectives*, vol. 115, pp. 799–806.

deFur, P.L., Evans, G.W., Hubal, E.A.C., Kyle, A.D., Morello-Frosch R.A. (2007). "Vulnerability as a function of individual and group resources in cumulative risk assessment." *Environmental Health Perspectives,* vol. 115, pp. 817–824.

Ellickson, K.M., Sevcik, S.M., Burman, S., Pak, S., Kohlasch, F., Pratt, G.C. (2011). "Cumulative risk assessment and environmental equity in air permitting: Interpretation, methods, community participation and implementation of a unique statue." *International Journal of Environmental Research and Public Health*, vol. 8, pp. 4140–4159.

Evans, G.W. & Marcynyszyn, L.A. (2004). "Environmental justice, cumulative environmental risk, and health among low- and middle-income children in upstate New York." *American Journal of Public Health*, vol. 94(11), pp. 1942–1944.

Fiscella, K. & Williams, D.R. (2004). "Health disparities based on socioeconomic inequalities: Implications for urban health care." *Academic Medicine*, vol. 79(12), pp. 1139–1147.

Fox, M.A., Groopman, J.D., Burke, T.A. (2002). "Evaluating cumulative risk assessment for environmental justice: A community case study." *Enviromental Health Perspectives*, vol. 110(Suppl 2), pp. 203–209.

Fox, M.A., Tran, N.I., Groopman, J.D., Burke, T.A. (2004). "Toxicological resources for cumulative risk: An example with hazardous air pollutants." *Regulatory Toxicology and Pharmacology*, vol. 40, pp. 305–311.

FQPA (Food Quality Protection Act). 1996. Publication L. No. 104–170, 104th Congress, Washington, DC.

Gallagher, S.S., Rice, G.E., Scarano, L.J., Teuschler, L.K., Bollweg, G., Martin, L. (2015). "Cumulative risk assessment lessons learned: A review of case studies and issue papers." *Chemosphere* vol. 120, pp. 697–705.

Gee, G.C. & Payne-Sturges, D.C. (2004). "Environmental health disparities: A framework integrating psychosocial and environmental concepts." *Environmental Health Perspectives*, vol. 112, pp. 1645–50.

Gilbert, A. & Chakraborty, J. (2011). "Using geographically weighted regression for environmental justice analysis: Cumulative cancer risks from air toxics in Florida." *Social Science Research*, vol. 40, pp. 273–286.

Hastings, D. (2011). "The Human Security Index: Potential roles for the environmental and earth observation communities." *Earthzine* May 4, 2011. Available at www.earthzine.org/2011/05/04 [Accessed on June 23, 2014].

HSI (Human Security Index). (2012). Human Security Index. Available at www.humansecurityindex. org/ [Accessed on June 23, 2014].

Huang, G. & London, J.K. (2012). "Cumulative environmental vulnerability and environmental justice in California's San Joaquin Valley." *International Journal of Environmental Research and Public Health*, vol. 9, pp. 1593–1608.

Hynes, H.P. & Lopez, R. (2007). "Cumulative risk and a call for action in environmental justice communities." *Journal of Health Disparities Research and Practice*, vol. 1(2), pp. 29–57.

IOM (Institute of Medicine). (1999). *Toward Environmental Justice: Research, Education, and Health Policy Needs*. Washington, DC: National Academy Press.

Krieg, E.J. & Faber, D.R. (2004). "Not so black and white: Environmental justice and cumulative impact assessments." *Environmental Impact Assessment Review*, vol. 24, pp. 667–694.

Linder, S.H., Marko, D., Sexton, K. (2008). "Cumulative cancer risk from air pollution in Houston: Disparities in risk burden and social disadvantage." *Environmental Science and Technology*, vol. 42, pp. 4312–22.

Linder, S.H. & Sexton, K. (2011). "Conceptual models for cumulative risk assessment." *American Journal of Public Health*, vol. 101(Suppl 1), pp. S74–S81.

MacDonell, M.M., Haroun, L.A., Teuschler, L.K., Rice, G.E., Hertzberg, R.C., Butler, J.P., et al. (2013). "Cumulative risk assessment toolbox: Methods and Approaches for the practitioner." *Journal of Toxicology*, vol. 2013, article 310904, 36 pgs.

Medina-Vera, M., Van Emon, J.M., Melnyk, L.J., Bradham, K.D., Harper, S.L., Morgan, J.N. (2010). "An overview of measurement method tools available to communities for conducting exposure and cumulative risk assessment." *Journal of Exposure Science and Environmental Epidemiology*, 20(4), pp. 359–370.

Meek, M.E., Boobis, A.R., Crofton, K.N., Heinemeyer, G., Raaj, M.V., Vickers, C. (2011). "Risk assessment of combined exposure to multiple chemicals: A WHO/IPCS framework." *Regulatory Toxicology Pharmacology*, vol. 60, pp. S1–S14.

Morello-Frosch, R. & Jesdale, B.M. (2006). "Separate and unequal: Residential segregation and estimated cancer risks associated with ambient air toxics in U.S. metropolitan areas." *Environmental Health Perspectives*, vol. 114, pp. 386–393.

MPCA (Minnesota Pollution Control Agency). (2009a). Cumulative Air Emissions Risk Analysis at the MPCA – Background Document. St. Paul, MN. Available at www.pca.state.mn.us/index.php/view-document.html?gid=144 [Accessed on 23 June 2014].

MPCA. (2009b). How to Conduct a Cumulative Air Emissions Risk Analysis. St Paul, MN. Available at www.pca.state.mn.us/index.php/view-document.html?html?gid=143 [Accessed on 23 June 2014].

MPCA. (2010). Reference Document for Minnesota Statute 116.07, Subdivision 4a. St Paul, MN. Available at http://cumulativeimpacts.org/documents/Reference%20doc%20for%20Minn%20Stat%2020116%2007%20subd4a%205_10.pdf [Accessed on 23 June 2014].

NEJAC (National Environmental Justice Advisory Council). (2004). Ensuring Risk Reduction in Communities with Multiple Stressors: Environmental Justice and Cumulative Risks/Impacts. Office of Environmental Justice, US Environmental Protection Agency, Washington, DC, USA. Available at www.epa.gov/environmentaljustice/resources/publications/nejac/nejac-cum-riskrpt-122104.pdf [Accessed on 23 June 2014].

NEJAC. (2010). Nationally Consistent Environmental Justice Screening Approaches. Report to the US Environmental Protection Agency, Washington, DC. Available at http://epa.gov/environmental justice/resources/publications/nejac/ej-screeningapproaches-rpt-2010.pdf [Accessed on 23 June 2014].

NJDEP (New Jersey Department of Environmental Protection). (2009). A Preliminary Screening Method to Estimate Cumulative Environmental Impacts. Trenton, NJ. Available at www.state.nj.gov/dep/ej/docs/ejc_screeningmethods20091222.pdf [Accessed on 23 June 2014].

NRC (National Research Council). (2008). *Phthalates and Cumulative Risk Assessment: The Tasks Ahead*. Washington, DC: National Academies Press.

NRC. (2009). *Science and Decisions: Advancing Risk Assessment*. Washington, DC: National Academies Press.

Perlin, S.A., Setzer, R.W., Creason, J., Sexton, K. (1995). "The distribution of industrial air emissions by income and race in the United States: An approach using the Toxics Release Inventory." *Environmental Science and Technology*, vol. 29, pp. 69–80.

Perlin, S.A., Sexton, K., Wong, W.A. (1999). "An examination of race and poverty for people living near industrial sources of air pollution." *Journal of Exposure Analysis and Environmental Epidemiology*, vol. 9, pp. 29–48.

Phillips, C.V. & Sexton, K. (1999). "Science and policy implications of defining environmental justice." *Journal of Exposure Analysis and Environmental Epidemiology* vol. 9(1), pp. 9–17.

Prochaska, J.D., Nolen, A.B., Kelley, H., Sexton, K., Linder, S.H., Sullivan, J. (2014). "Social determinants of health in environmental justice communities: Examining cumulative risk in terms of environmental exposures and social determinants of health." *Human and Ecological Risk Assessment*, vol. 20, pp. 980–994.

Rider, C.V., Dourson, M., Hertzberg, R.C., Mumtaz, M.M., Price, P.S., Simmons, J.E. (2012). "Incorporating nonchemical stressors into cumulative risk assessments." *Toxicological Sciences*, vol. 127, pp. 10–17.

Sadd, J.L., Pastor, M., Morello-Frosch, R., Scoggins, J., Jesdale, B. (2011). "Playing it safe: Assessing cumulative impact and social vulnerability through an environmental justice screening method in the south coast air basin, California." *International Journal of Environmental Research and Public Health*, vol. 8, pp. 1441–1459.

Salinas, J.J., Shah, M., Abelbary, B., Gay, J., Sexton, K. (2012). "Application of a novel method for assessing cumulative risk burden by county." *International Journal of Environmental Research and Public Health*, vol. 9, pp. 1820–1835.

SDWA (Safe Drinking Water Act). (1996). Amendments. Publication L. No. 104–182, 104th Congress, Washington, DC.

Sexton, K. (1997). "Sociodemographic aspects of human susceptibility to toxic chemicals: Do class and race matter for realistic risk assessment?" *Environmental Toxicology and Pharmacology*, vol. 4, pp. 261–269.

Sexton, K. (2012). "Cumulative risk assessment: An overview of methodological approaches for evaluating combined health effects from exposure to multiple environmental stressors." *International Journal of Environmental Research and Public Health*, vol. 9, pp. 370–90.

Sexton, K. (2015). "Cumulative health risk assessment: Finding new ideas and escaping from the old ones." *Human and Ecological Risk Assessment*, vol. 21(4), pp. 934–951.

Sexton, K. & Adgate, J.L. (1999). "Looking at environmental justice from an environmental health perspective." *Journal of Exposure Analysis and Environmental Epidemiology*, vol. 9(1), pp. 3–8.

Sexton, K. & Anderson, Y.B (eds). (1993). *Equity in Environmental Health: Research Issues and Needs*. Special Issue of *Toxicology and Industrial Health* vol. 9(5), pp. 679–967.

Sexton, K. & Hattis, D. (2007). "Assessing cumulative health risks from exposure to environmental mixtures—three fundamental questions." *Environmental Health Perspectives*, vol. 115, pp. 825–832.

Sexton, K. & Linder, S.H. (2010). "The role of cumulative risk assessment in decisions about environmental justice." *International Journal of Environmental Research and Public Health*, 7, pp. 4037–49.

Sexton, K. & Linder, S.H. (2011). "Cumulative risk assessment for combined health effects from chemical and nonchemical stressors." *American Journal of Public Health*, vol. 101(S1), pp. 581–588.

Sexton, K. & Linder, S.H. (2016). "Finding fault with health risk assessment: A typology for risk assessment criticism." *Human and Ecological Risk Assessment*, vol. 22, pp. 203–210.

Soobader, M., Cubbin, C., Gee, G.C., Rosenbaum, A., Laurenson, J. (2006). "Levels of analysis for the study of environmental health disparities." *Environmental Research*, vol. 102, pp. 172–180.

Su, J.G., Morello-Frosch, R., Jesdale, B.M., Kyle, A., Jerrett, M. (2009). "An index for assessing inequalities in cumulative environmental hazards with application to Los Angeles, California." *Environmental Science and Technology*, vol. 43, pp. 7626–7634.

Tam, B.N. & Neumann, C.M. (2004). "A human health assessment of hazardous air pollutants in Portland, OR." *Journal of Environmental Management*, vol. 73, pp. 131–45.

USEPA (US Environmental Protection Agency). (1986). Guidelines for the Health Risk Assessment of Chemical Mixtures. Risk Assessment Forum, Washington, DC.

USEPA. (2000). Supplementary Guidance for Conducting Health Risk Assessment of Chemical Mixtures. Risk Assessment Forum, Washington, DC.

USEPA. (2002). Organophosphate Pesticides: Revised Cumulative Risk Assessment. Office of Pesticide Programs, Washington, DC. Available at www.epa.gov/pesticides/cumulative/rra-op/ [Accessed on 23 June 2014].

USEPA. (2003). Framework for Cumulative Risk Assessment. PA/630/P-02/001A. Risk Assessment Forum, Office of Research and Development. Washington, DC.

USEPA. (2004). Toolkit for Assessing Potential Allegations of Environmental Injustice. EPA 300-R-04-002. Office of Enforcement and Compliance Assurance – Environmental Justice. Washington, DC.

USEPA. (2006a). Considerations for Developing Alternative Health Risk Assessment Approaches for Addressing Multiple Chemicals, Exposures, and Effects. National Center for Environmental Assessment, Washington, DC.

USEPA. (2006b). Cumulative Risk from Chloroacetanilide Pesticides. Office of Pesticide Programs, Washington, DC.

USEPA. (2006c). Cumulative Risk from Triazine Pesticides. Office of Pesticide Programs, Washington, DC.

USEPA. (2007). Revised N-Methyl Carbamate Cumulative Risk Assessment. Office of Pesticide Programs, Washington, DC.

USEPA. (2012). The Community-Focused Exposure and Risk Screening Tool (C-FERST). National Exposure Research Laboratory, Research Triangle Park, NC. Available at www.epa.gov/heasd/c-ferst/ [Accessed on July 6, 2015].

USEPA. (2013). Draft Technical Guidance for Assessing Environmental Justice in Regulatory Analysis. Available at http://yosemite.epa.gov/sab/sabproduct.nsf/0/0F7D1A0D7D15001B8525783000673 AC3/$File/EPA-HQ-QA-2013-0320-0002[1].pdf [Accessed on June 29, 2015].

USEPA. (2015a). EJSCREEN: Environmental Justice Screening and Mapping Tool. Available at www2.epa.gov/ejscreen/what-ejscreen [Accessed on June 18, 2015].

USEPA. (2015b). Guidance on Considering Environmental Justice During the Development of Regulatory Actions. Available at www.epa.gov/environmentaljustice/resourses/considering-ej-in-rulemaking-guide-final.pdf [Accessed on June 29, 2015].

WHO (World Health Organization). (2008). Urban HEART: Health Equity Assessment and Response Tool. Final Report. WHO Publications, Kobe, Japan.

WHO. (2010a). Urban HEART: Health Equity Assessment and Response Tool. WHO Publications, Kobe, Japan. Available at www.who.int/kobe.centre/publications/urban_heart.pdf?ua=1 [Accessed on June 23, 2014].

WHO. (2010b). Urban HEART: User Manual. WHO Publications. Kobe, Japan. Available at www.who.int/kobe_centre/publications/urban_heart_manual.pdf?=1 [Accessed on June 23, 2014].

Zartarian, V.G., Schultz, B.D., Barzik, T.M., Smuts, M., Hammond, D.M., Median-Vera, M., Geller, A.M. (2011). "The EPA's community-focused exposure and risk screening tool (C-FERST) and its potential use for environmental justice efforts." *American Journal of Public Health*, vol. 101(S1), pp. 286–294.

23

A REVIEW OF COMMUNITY-ENGAGED RESEARCH APPROACHES USED TO ACHIEVE ENVIRONMENTAL JUSTICE AND ELIMINATE DISPARITIES

Sacoby Wilson, Aaron Aber, Lindsey Wright and Vivek Ravichandran

Introduction

Community-based environmental justice (EJ) organizations engage in community-driven research to address local health issues, including metropolitan air pollution, low-quality infrastructure, pollution-intensive industries, locally unwanted land uses, transportation, and industrial animal production (Bell et al. 2002; Corburn 2002, 2005; Heaney et al. 2007; Israel et al. 2005a, 2005b; Minkler & Wallerstein 2003; O'Fallon et al. 2000; O'Fallon & Dearry 2002; Srinivasan & Collman 2005; Vasquez et al. 2006; O. Wilson et al. 2008; S. Wilson et al. 2007–2008, 2008; Wing 2002, 2005). Through community-driven research, these groups combine grassroots activism, "expert local knowledge" derived from community members' contextual experiences, university partnerships, and community empowerment through research into a framework that moves science from the realm of neutrality, seen in conventional social research, to the realm of values, which strengthens its utility.

The community-driven approach makes research "action-oriented," which: (a) increases community capacity to address EJ and health issues; (b) drives civic engagement by people of colour and low-income stakeholders; (c) balances relationships between local and academic experts; (d) ensures the local relevancy of research; (e) develops an intergenerational pipeline of community leaders knowledgeable about EJ issues; and (f) provides justice and positive environmental and social change through solutions that reduce hazards (Bacon et al. 2013; Bell et al. 2002; Heaney et al. 2007; Minkler & Wallerstein 2003; O'Fallon & Dearry 2002; O'Fallon et al. 2000; O. Wilson et al. 2008; S. Wilson et al. 2007–2008, 2008). The outcomes of this approach include the empowerment of marginalized groups; increased awareness of EJ issues at the local, state, and national levels; and greater environmental health literacy.

This chapter will present community-driven research as an evolutionary process that relies on partnership building and participatory research to advance the reduction of health disparities. Examples of community-driven research in diverse areas among diverse groups in the US will be provided. However, we note that community-engaged research has been conducted in other areas of the world, including Canada, the UK, and Hungary (Conrad & Hilchey 2011; Harper 2009). We conclude with general best practices that should guide community-driven research.

Defining community-driven research

Community-based participatory research (CBPR)

To define community-driven research, we use the definition of community-based participatory research (CBPR), given by Israel et al. (1998: 177) as:

> a collaborative approach to research that equitably involves, for example, community members, organizational representatives, and researchers in all aspects of the research process. The partners contribute 'unique strengths and shared responsibilities'… to enhance understanding of a given phenomenon and the social and cultural dynamics of the community, and integrate the knowledge gained with action to improve the health and well-being of community members.

This method of research is widely used in public health. It goes by many names and variants, but the basic principles are the same, identified by Israel et al. (1998: 178–180) as:

1. Recognizing the community as a unit of identity
2. Building on strengths and resources within the community
3. Facilitating collaborative partnerships in all phases of the research
4. Integrating knowledge and action for mutual benefit of all partners
5. Promoting a co-learning and empowering process that attends to social inequalities
6. Involving a cyclical and iterative process
7. Addressing health from both positive and ecological perspectives
8. Disseminating findings and knowledge gained to all partners.

Central to CBPR are the establishment of partnerships and the use of participatory research methods. The Community–Campus Partnerships for Health (CCPH), a non-profit devoted to forging community–academic partnerships to solve health issues, has developed elements of an "authentic partnership," that drive successful implementation of CBPR (CCPH 2013a, 2013b):

1. Principles of mutual accountability, open communication, shared goals, and transparency
2. Quality processes to ensure healthy partnerships
3. Meaningful outcomes for communities
4. Transformative processes at the individual, community, institutional, and political levels.

CBPR and the community engagement continuum

CBPR is an evolutionary process because of the time and effort it takes for partnerships to successfully develop. To reflect this, the Clinical and Translational Science Awards

Table 23.1 The community engagement continuum (adapted from ASTDR) (CTSA 2011)

Outreach	Consult	Involve	Collaborate	Shared leadership
Some community involvement. *Communication flows from one to the other, to inform.* Provides community with information. Entities coexist. Outcomes: optimally, establishes communication channels and channels for outreach.	*More community involvement.* *Communication flows to the community and then back, answer seeking.* Gets information or feedback from the community. Entities share information. Outcomes: develops connections.	*Better community involvement.* *Communication flows both ways, participatory form of communication.* Involves more participation with community on issues. Entities cooperate with each other. Outcomes: visibility of partnership established with increasing cooperation.	*Community involvement.* *Communication flow is bidirectional.* Forms partnerships with community on each aspect of the project from development to solution. Entities form bidirectional communication channels. Outcomes: partnership building, trust building.	*Strong bidirectional relationship.* Final decision making is at community level. Entities have formed strong partnership structures. Outcomes: broader health outcomes affecting broader community. Strong bidirectional trust built.

Consortium (2011) places community–university research partnerships on a community engagement continuum of which CBPR is a part. This is presented in Table 23.1.

On this continuum, a fully realized CBPR project would fall under "Collaborate," as it engages communities in the research process and strengthens partnerships. The community engagement continuum shows that partnerships may vary from little interaction among communities and researchers to partnerships characterized by shared power, bidirectional communication, and trust.

The examples in the following sections reflect partnerships along all points of the continuum to show how these partnerships develop. These examples are analysed based on the four principles of partnership from CCPH listed above.

Partnerships in community-driven research

Principles of good partnerships

Several studies have critically examined how well public health partnerships between communities and universities adhere to core CBPR principles. For example, Minkler et al. (2008) use a theoretical framework to analyse the partnerships in four CBPR case studies, finding that good partnerships include strong community leadership; active community participation; diverse and specialized skill sets; good networking; shared values and power; an ability and willingness to fight entrenched powers and focus on larger contexts; dialogue and critical reflection to promote respect; and broader institutional, financial, and network support. Bacon et al. (2013) discuss the need for more action in CBPR partnerships and coined the term "community-based participatory action research" (CBPAR) to reflect a greater emphasis on action than CBPR exhibits. Many CBPR projects focus more on process outcomes (i.e. evaluating adherence to CBPR principles) than on implementing interventions, changing policy, or eliminating disparities. Bacon et al. describe CBPAR as a cycle of partnership building, study design, data collection, data analysis, dissemination/results discussion, actions led by the community, evaluation, and broad dissemination to spark further research. As the examples throughout this chapter show, action is central to EJ research.

Case studies in implementing EJ and health partnerships

Caldwell et al. (2015) examine the University of Michigan's Detroit Urban Research Center as a successful case study of CBPR partnerships established through good relationships and effective resource use. They highlight as best practices for successful partnerships trust, flexibility, good project management, shared expectations, and consistent presence of researchers in the community. Recommendations for improving the implementation of CBPR principles within partnerships included trust-building activities, proper mentoring of researchers in CBPR, working with an interested community, and researcher humility. These findings mirror earlier research by Parker et al. (2003) on using CBPR to address air pollution and asthma, in which best practices identified were trust, evaluation, and cost assessment.

Similarly, Garcia et al. (2013) evaluate a community–university EJ partnership among Occidental College, the University of Southern California, and local community groups known as THE (Trade, Health, Environment) Impact Project focused on addressing the impacts of port-related traffic in Los Angeles and Long Beach, CA. THE Impact Project used collaborative, community-based research to empower the community to speak on policy changes on goods movement and air pollution, such as a clean air plan and a comprehensive assessment of health and economic impacts. THE's success was rooted in strong relationships and a unified focus on policy change. Similar outcomes are identified by Wilson et al. (2014) in their evaluation of a community–university partnership centred on EJ and trade in Charleston, SC called the Low Country Alliance for Model Communities. Researchers additionally noted that this partnership trained residents as citizen scientists through such projects as Environmental Justice Radar (EJRADAR) (described later) that have reduced hazards and empowered residents to be involved in land use and zoning decisions.

Community–university partnerships have also been used in rural communities to address adverse impacts of agriculture, as shown by Wing et al. (2008) in their examination of a partnership focused on the impacts of industrial hog production in North Carolina. The partnership allowed residents to collect data over time to determine the relationship between certain quantifiable variables associated with industrial hog farms and health outcomes. The researchers concluded that this participatory approach increased data validity by providing local data unhindered by confounding variables found in a normal epidemiological study. However, they also found a decline in the data's robustness because the use of such localized data decreased the sample size and the ability to compare across populations.

In the rural communities along the US–Mexico border, partnerships have formed around the issue of public water, as Lemos et al. (2002) analyse through the Border WaterWorks program, a collaborative effort to bring public water systems to *colonias*, or poor, rural communities. Researchers identified community involvement; variation in accountability among stakeholders and partners; and the use of public funds as factors influencing project capacity, flexibility, and efficacy. They also stressed the importance of community-specific interventions as a means of giving power to communities.

Community–university EJ partnerships also form around issues of the built environment and chronic diseases. Krieger et al. (2009) describe efforts to improve the built environment in High Point, Seattle, WA by the High Point Walking for Health Project to combat obesity and diabetes. This was a multidisciplinary group that used CBPR to allow residents to gather baseline data and choose interventions along with their partners that they felt would best improve the community. The group focused on the racial and classist dimensions of health inequities and health-related decision-making. Best practices emphasized drawing on the strengths and expertise of the community to enhance intervention and drawing on diversity to allow for greater collaboration.

When community–university partnerships reach the "Collaborate" stage on the community engagement continuum, their implementation of collaborative practices may vary. Arroyo-Johnson et al. (2015) examine two CBPR partnerships related to reducing cancer risks through the use of surveys in the St. Louis, MS area among African-Americans and in the Twin Cities and central Minnesota area among Latino communities. The similar goals of these partnerships were to reduce cancer disparities among disadvantaged groups, implement CBPR interventions, and train CBPR researchers. How each partnership assessed and implemented its CBPR efforts differed, however; for example, on assessing empowerment and expectations of effectiveness, the Missouri group rated community empowerment as a better measure of partnership effectiveness, while the Minnesota group chose future collaboration. This shows that how CBPR is analysed and implemented by communities varies, reinforcing the need for community-specific solutions.

Citizen science in community-driven research

Citizen science and community-owned and -managed research (COMR)

Citizen science is "the involvement of the public in scientific research – whether community-driven research or global investigations" (US Environmental Protection Agency [EPA] 2016). Community-driven EJ research projects use citizen science to help community members address issues in a hands-on way. Citizen science works in conjunction with the development of partnerships as a means of involving community members in assessing their environments and formulating solutions.

Heaney et al. (2007) describe a citizen science framework that works concomitantly with partnership development to surpass the level of engagement of CBPR and reach "Shared leadership" on the community engagement continuum. The framework, community-owned and -managed research (COMR), was designed by the West End Revitalization Association, a community-based EJ and health organization in Mebane, NC. The partnership formed to address poor relationships between university researchers and the community, who felt that researchers were exploiting Mebane's EJ problems to further their own careers instead of finding solutions. COMR builds on CBPR by promoting collaboration with community-based organizations that demonstrate organizational capacity around defined EJ issues. COMR also captures the goal of citizen science by allowing communities to set research questions and collect the data to answer them, but it moves beyond both CBPR and citizen science by requiring that community organizations be funded directly as the sole principal investigators and project managers of research activities. This allows communities to select university "experts" with whom to consult in their research; manage the research process to prioritize, maximize, and leverage available funding; and ensure implementation of solutions.

While CBPR can build capacity, education, and power with citizen science, it is not always applied in a way that produces data to initiate compliance with civil rights, environmental, planning, and public health regulations. It is this gap that COMR addresses, using "science for compliance" (Wilson & Wilson 2015: 26). COMR recognizes the community rather than the university as the centre of knowledge production, project management, decision-making, data collection, learning, and social change. While difficult to implement with groups that lack organizational capacity, it nevertheless serves as a model for community-driven research.

Citizen science and data collection

Citizen science and air pollution

Northridge et al. (1999) were among the first to train youth as citizen scientists, working through a partnership led by WEACT for Environmental Justice in Harlem, New York City, NY to collect data on pollution levels associated with traffic and diesel exhaust. Best practices identified in this early work were community involvement in all stages of the research process, proper data oversight for quality control, and proper training of youth citizen scientists.

Wier et al. (2009) show how citizen science effected political change in San Francisco, CA through a collaboration among the community group PODER and city and university researchers. Using quantitative and qualitative data from surveys, traffic counts, and GIS mapping, researchers identified high traffic levels and poor air quality as issues that ultimately drove the community to advocate for a city-wide initiative to include EJ principles in the formulation of traffic-related decisions. Furthermore, Cohen et al. (2012) demonstrate citizen science as a facilitator of community change in Richmond, CA. Using a citizen-developed and researcher-implemented survey, the team assessed the association between cumulative impacts of stressors and self-reported poor health among Richmond's vulnerable populations, finding among those surveyed high rates of asthma and a strong belief that this was caused by local pollution sources. This recognition of citizens' contextual knowledge as a means of understanding the role of cumulative stressors on health led to a transformative shift in environmental health literacy and capacity to participate in and drive local environmental decision-making.

Similar transformations are described by Petersen et al. (2006), who detail how citizen participants in the Southern California Environmental Justice Collaborative enhanced their civic engagement by helping to renegotiate a California air quality rule. Participants were actively involved in all steps of the policy creation process, using media; partner resources; and community support and power to build the collective efficacy of the collaborative to achieve EJ goals even beyond the air quality rule.

Citizen science, water quality, and infrastructure

Citizen science has also effectively addressed issues with infrastructure and water quality. Heaney et al. (2011) note that citizens in Mebane, NC were underserved prior to the West End Revitalization Association due to disparities in access to publicly regulated sewer and water infrastructure, paved roads, safe housing, and other basic amenities. Through the partnership, citizens assisted in the development of household infrastructure surveys and in the collection of water samples from private wells and recreational waters. Survey results found widespread poor water and infrastructure quality. Private wells, for example, showed highest levels of fecal indicators, and many samples showed above-average levels of pollution. In a related paper, Wilson et al. (2007–2008) present the development of "action strategies" around this issue of infrastructure as having mitigated and publicized the issue, reinforcing COMR's ties between research and action.

Other communities in North Carolina face EJ issues due to lack of adequate amenities. Heaney et al. (2013) describe a partnership between the Rogers-Eubanks Neighbourhood Association and the University of North Carolina formed to combat poor infrastructure and impacts from a local landfill. Researchers followed COMR principles to train residents as citizen scientists to sample water from private wells near the landfill. Results found that contaminants in this water exceeded EPA water quality standards and that recreational water sources had high levels of fecal coliforms. This citizen-gathered information highlighted residents' EJ concerns, improved their environmental health literacy, and helped them receive improved community amenities like new roads and a new community centre. Furthermore, this partnership allowed residents to file discrimination claims under Title VI of the Civil Rights Act, halting a waste transfer station in 2009 (Eynon 2012).

Citizen science and food justice

Citizen science has also addressed food justice issues (see also Chapter 33). Ramirez-Andreotta et al. (2015) present a paradigm of researching food justice that combines CBPR, which they frame as a method of establishing community–researcher partnerships, and PPSR (public participation in scientific research), another term for citizen science. PPSR can fall under the "Involve" and "Collaborate" categories on the community engagement continuum. The PPSR paradigm was implemented among populations at an Arizona Gardenroots site, where residents who sought to understand how a local hazardous waste site impacted food grown through urban gardening were trained to collect soil, air, and water samples. The study increased residents' knowledge and engagement with local officials. It also recognized difficulties in engaging the public to measure health outcomes; maintaining adequate communication among partners; and allowing citizens to analyse their own samples.

Vasquez et al. (2007) examined a collaborative partnership to address food insecurity in Bayview Hunters Point, San Francisco, CA. The partnership addressed the presence of stores that sold low-quality foods and attracted criminal activity, and arose out of a desire to bring higher quality food to the neighbourhood. Residents were trained to co-develop and administer a store survey, collect data on shelf space, and map demographic and store information to identify food insecurity among different community groups. This led to higher citizen involvement in local decision-making and better food access policy. Similarly, Ndirangu et al. (2008) examined the use of CBPR in the Lower Mississippi Delta to assess food justice problems and perceptions of scientific integrity among residents. Through focus groups, the team found challenges in understanding the roles of residents in CBPR and issues with their engagement in the decision-making process. Benefits of the partnership studied included enhanced representation of diverse stakeholders who had concerns about food justice issues, education of residents, and shared power among partners, including traditionally marginalized stakeholders.

Challenges of citizen science

Despite the advantages of citizen science described above, certain factors continue to impede its acceptance as a trusted method of research (see also Chapters 8 and 24). Ottinger (2010), in an analysis of the Bucket Brigade partnership's efforts to measure air quality in Louisiana near a Shell chemical plant, suggests that researchers are loath to trust data collected in unstandardized ways, and citizen science currently lacks such standardization. Standards, according to Ottinger (2010: 249), also contribute to the "struggle between citizens and experts – or more precisely, among social movement groups, academic scientists, regulators, and industries." As Ottinger further notes, EJ is often built around the rejection of standards that make proving harm difficult and allow for small amounts of acceptable risk.

Brown et al. (2010: 3) identify a second barrier for citizen science. Through interviews with practitioners, they found that university researchers often face from university institutional review boards (IRBs) "opposition to two particular CBPR practices: (1) layperson participation in the research process and (2) the report-back of individual results to study participants." However, both Brown et al. and Minkler (2014) offer potential solutions to overcoming this hesitancy, including training IRBs in CBPR practices, giving communities a voice on IRBs, and recognizing the value of relying on local communities to contextualize data collection methods to suit the needs of that particular group. Balazs and Morello-Frosch (2013) also emphasize the power of citizen-collected data by showing how two collaborative studies in California expanded the "rigour," "relevance," and "reach" of the data by giving it credibility and making it applicable to the communities.

In community-driven research, community–university partnerships and citizen science have been important in addressing environmental injustice and related health disparities. We will next discuss the use of Photovoice and Geographic Information Systems (GIS) as emerging methods in helping communities collect or display data that can be translated to positive actions that help build capacity and "inpower" residents.

Methods of community-driven research

Photovoice

Photovoice taps into the contextual knowledge of residents on various issues, including the built environment, food access, and physical activity opportunities (see also Chapter 19). Wang and Burris (1997: 369) define Photovoice as "a process by which people can identify, represent, and enhance their community through a specific photographic technique." It involves training community members to use cameras; encouraging them to photograph aspects of their community that represent the project's research goals; discussing their photographs in groups; and presenting their photographs to fellow community members and decision-makers to facilitate change (Wang & Burris 1997).

Freedman et al. (2014) used Photovoice to assess the impacts of housing conditions on psychosocial health outcomes, such as lack of attachment to place and a sense of lost community autonomy. Photovoice proved useful in identifying relevant socio-environmental themes that could inform community interventions. Photovoice is also an emerging best practice among immigrant communities. Schwartz et al. (2015) examined the relationship between pesticide spraying and childhood asthma among Latino residents of Mexican ancestry in California's San Joaquin Valley using Photovoice and interviews. This study revealed four themes of concern among residents: proximity to pesticides, school boundaries, lack of environmentally friendly spaces, and poor pesticide containment.

Photovoice can also document problems with the built environment in both urban and rural contexts. Redwood et al. (2010) describe the use of Photovoice to document the perceived causes of negative health outcomes through a partnership among African-Americans in Atlanta, GA. The data revealed four themes residents noted as contributing to poor health: 1) poor housing and neighbourhood conditions; 2) discriminatory disinvestment; 3) speculative development leading to potential gentrification; and 4) community displacement. In rural West Virginia, Bell (2008) used Photovoice to help residents identify low community involvement in local politics, which spurred residents to engage with politicians and advocate for community reforms.

While Photovoice is effective in helping residents capture EJ-related concerns and translate qualitative data into action, it is not without its challenges. Freedman et al. (2014), for example, note that their sample size of 18 may not have been representative of the public housing community they served. While they offer no explicit solution, the implication is that a robust process of random sampling and collecting a large pool of participants could reduce data distortions. Similarly, Stedman-Smith et al. (2012), in their study of Photovoice use to reduce pesticide exposure in Minnesota, expressed disappointment in issues of retaining participants due to the multi-week length of the project. However, they suggest that active follow-up with participants could help ensure that they continue to take part.

Public participatory geographic information systems (PPGIS)

Another promising practice for community-engaged EJ research is public participatory geographic information systems (PPGIS) (see also Chapter 24). PPGIS allows communities disproportionately affected by environmental issues to "tell their exposure story" (Wilson 2015: 440) beyond the scope of public meetings by working with GIS, a tool otherwise regarded as the "'privileged' knowledge of experts and institutions" (Jankowski 2009: 1966). PPGIS allows experts and community members to work together to understand the "spatial consequences of locally important problems" (Jankowski & Stasik 1997: 74). PPGIS has also evolved with technology, going from "paper maps and markers" to "digital mapping with markers using Internet PPGIS applications" (Brown 2012: 11).

Jordan (2012) identified community-based GIS (CBGIS), a variant of PPGIS, as a means of engaging marginalized communities in Galena Park, TX in mapping environmental hazards and assessing cumulative impacts. Environmental hazards were mapped using photographs and the expertise of residents who participated in all stages of the data collection process and the post-mapping stage. Thompson (2012), in a similar effort to use PPGIS methods in New Orleans, LA after Hurricane Katrina, recommended accessible mapping software and representative, participatory data collection.

Wilson et al. (2015) describe the development and use of a PPGIS tool known as EJRADAR (www.ejradar.org) as part of the Low-Country Alliance for Model Communities partnership described previously. Community input guided the development of the tool, which stemmed from the desire of residents to map local hazards and assess the cumulative impacts of those hazards. This input improved the quality and utility of the tool, allowing, for example, multiple layers of information to be mapped at once to create a visual representation of cumulative health risk. With EJRADAR, facility attributes and sociodemographic variables can be plotted as different colours to show spatial variation and pinpoint pollution hot spots. Users can also add point, line, and area pollution source data through the website, their smartphones, and even through active Photovoice. Since 2014, community leaders have trained over 125 individuals, and the tool has increased environmental health literacy and empowered residents to be engaged in local environmental decision-making.

Conclusion: best practices for community-driven research

Community-driven research is essential to EJ practitioners and advocates. As this chapter has shown, making communities part of the solution makes the solution relevant to them. Broadly, the following represent best practices and components of success that the partnerships examined above share:

1. *Open communication*

   Community engagement is not easy. Distrust, preconceived notions, and prejudices from both community members and researchers can hamper collaboration. However, as these studies have shown, there is a general need for communication that is honest, forthright, and constant. Open communication must also include the free flow of ideas, opinions, goals, and values in a bidirectional way.

2. *Flexibility*

   Flexibility must also be bidirectional, but it is researchers who must be especially open to changing their thinking, approach, and methodology when working with communities. Community research works because it tailors solutions, which often forces researchers to reconsider the knowledge and ideas they bring to the process if those mechanisms do not adequately serve the community's needs.

3. *Equity*

   Equity in community-driven research refers to the power wielded by all involved in the process, in terms of their knowledge, standing, and resources. While researchers will almost always have access to more resources than communities, it is necessary for these resources to be shared. Environmental injustice arises out of imbalances in power, which demands that its solutions provide a balance. Whether it is grant funding or background knowledge, researchers should ensure access to community participants.

   Equity can, however, be affected by the approach used. For example, COMR arose out of a perceived lack of ownership over research in CBPR (Heaney et al. 2007). As this chapter has shown, community research exists on a spectrum, ranging from minimal community participation to balanced community–university partnerships. Practitioners of CBPR and its many variants must take into account that these projects exist to serve the community. Therefore, the method they choose in their research should reflect the needs of the community rather than the needs of researchers.

4. *Action*

   Successful community-driven research facilitates change. This requires community–researcher partnerships to emphasize finding and implementing solutions to injustices that communities face. While certain variations of community-engaged research place more emphasis on action than others (Bacon et al. 2013; Israel et al. 1998), all successful studies cited in this chapter required a degree of active participation from communities and often led to tangible changes in policy or in the community itself. Photovoice, PPGIS, and citizen science illustrate the actions communities can take, with the help of researchers, to learn about and articulate the issues they face in a way that empowers them to take a stand.

Community-driven research, through the development of community–university partnerships and the implementation of citizen science methods, has emerged as an effective practice to solve EJ and health issues. Through replication of the best practices highlighted in this chapter, researchers can continue to successfully engage communities in a way that educates and empowers. This will help raise the profile of community-driven research as a credible, impactful, and meaningful method of addressing environmental injustice and related health problems.

References

Arroyo-Johnson, C., Allen, M.L., Colditz, G.A., Hurtado, G.A., Davey, C.S., Sanders Thompson, V.L., et al. (2015). "A tale of two community networks program centers: Operationalizing and assessing CBPR principles and evaluating partnership outcomes." *Progress in Community Health Partnerships: Research, Education, and Action*, vol. 9, pp. 61–69.

Bacon, C., deVuono-Powell, S., Frampton, M.L., LoPresti, T., Pannu, C. (2013). "Introduction to empowered partnerships: Community-based participatory action research for environmental justice." *Environmental Justice*, vol. 6, pp. 1–8.

Balazs, C.L. & Morello-Frosch, R. (2013). "The three Rs: How community-based participatory research strengthens the rigor, relevance, and reach of science." *Environmental Justice*, vol. 6, pp. 9–16.

Bell, J.D., Bell, J., Colmenar, R., Flournoy, R., McGehee, M., Rubin, V., et al. (2002). *Reducing health disparities through a focus on communities.* Oakland, CA: PolicyLink.

Bell, S.E. (2008). "Photovoice as a strategy for community organizing in the central Appalachian coalfields." *Journal of Appalachian Studies*, vol. 14, pp. 34–48.

Brown, G. (2012). "Public Participation GIS (PPGIS) for regional and environmental planning: Reflections on a decade of empirical research." *Journal of Urban and Regional Information Systems Association*, vol. 25, pp. 7–18.

Brown, P., Morello-Frosch, R., Brody, J.G., Altman, R.G., Rudel, R.A., Senier, L., et al. (2010). "Institutional review board challenges related to community-based participatory research on human exposure to environmental toxins: A case study." *Environmental Health*, vol. 9, pp. 1–12.

Caldwell, W.B., Reyes, A.G., Rowe, Z., Weinert, J., Israel, B.A. (2015). "Community partner perspectives on benefits, challenges, facilitating factors, and lessons learned from community-based participatory research partnerships in Detroit." *Progress in Community Health Partnerships: Research, Education, and Action*, vol. 9, pp. 299–311.

CCPH (Community–Campus Partnerships for Health). (2013a). *About Us.* Retrieved from: https://ccph.memberclicks.net/about-us

CCPH. (2013b). *Position statement on authentic partnerships.* Retrieved from https://ccph.memberclicks.net/principles-of-partnership

Cohen, A., Lopez, A., Malloy, N., & Morello-Frosch, R. (2012). "Our environment, our health: A community-based participatory environmental health survey in Richmond, California." *Health Education & Behavior*, vol. 39, pp. 198–209.

Conrad, C.C. & Hilchey, K.G. (2011). "A review of citizen science and community-based environmental monitoring: Issues and opportunities." *Environmental Monitoring and Assessment*, vol. 176, pp. 273–291.

Corburn, J. (2002). "Combining community-based research and local knowledge to confront asthma and subsistence-fishing hazards in Greenpoint/Williamsburg, Brooklyn, New York." *Environmental Health Perspectives*, vol. 110, Supplement 2, pp. 241–248.

Corburn, J. (2005). *Street science: Community knowledge and environmental health justice.* The MIT Press. Retrieved from https://mitpress.mit.edu/books/street-science

CTSA (Clinical and Translational Science Awards Consortium) Community Engagement Key Function Committee Task Force on the Principles of Community Engagement. (2011). *Principles of community engagement: Second edition.* (Primer No. 11–7782). Washington, DC: NIH. Retrieved from ww.atsdr.cdc.gov/communityengagement/pdf/PCE_Report_508_FINAL.pdf

EPA (US Environmental Protection Agency). (2016). *What is citizen science?* Retrieved from www.epa.gov/citizen-science/what-citizen-science

Eynon, B.R. (2012). *Rogers Road remediation: Challenges remain.* Retrieved from http://blogs.law.unc.edu/civilrights/2012/12/03/rogers-road-remediation-challenges-remain/

Freedman, D.A., Pitner, R.O., Powers, M.C.F., & Anderson, T.P. (2014). "Using Photovoice to develop a grounded theory of socio-environmental attributes influencing the health of community environments." *British Journal of Social Work*, vol. 44, pp. 1301–1321.

Garcia, A.P., Wallerstein, N., Hricko, A., Marquez, J.N., Logan, A., Green Nasser, E., & Minkler, M. (2013). "THE (trade, health, environment) impact project: A community-based participatory research environmental justice case study." *Environmental Justice*, vol. 6, pp. 17–26.

Harper, K. (2009). "Using Photovoice to investigate environment and health in a Hungarian Romani (Gypsy) community." *Practicing Anthropology*, vol. 31, pp. 10–14.

Heaney, C.D., Wilson, S., & Wilson, O.R. (2007). "The West End Revitalization Association's community-owned and -managed research model: Development, implementation, and action." *Progress in Community Health Partnerships: Research, Education, and Action,* vol. 1, pp. 339–349.

Heaney, C., Wilson, S., Wilson, O., Cooper, J., Bumpass, N., & Snipes, M. (2011). "Use of community-owned and -managed research to assess the vulnerability of water and sewer services in marginalized and underserved environmental justice communities." *Journal of Environmental Health*, vol. 74, pp. 8–17.

Heaney, C.D., Wing, S., Wilson, S.M., Campbell, R.L., Caldwell, D., Hopkins, B., et al. (2013). "Public infrastructure disparities and the microbiological and chemical safety of drinking and surface water supplies in a community bordering a landfill." *Journal of Environmental Health*, vol. 75, pp. 24–36.

Israel, B.A., Schultz, A.J., Parker, E.A., & Becker, A.B. (1998). "Review of community-based research: Assessing partnership approaches to improve public health." *Annual Review of Public Health*, vol. 19, pp. 173–202.

Israel, B.A., Eng, E., Schulz, A.J., & Parker, E.A. (eds.). (2005a). *Methods in community-based participatory research.* San Francisco, CA: Jossey-Bass.

Israel, B.A., Parker, E.A., Rowe, Z., Salvatore, A., Minkler, M., Lopez, J., et al. (2005b). "Community-based participatory research: Lessons learned from the Centers for Children's Environmental Health and Disease Prevention Research." *Environmental Health Perspectives*, vol. 113, pp. 1463–1471.

Jankowski, P. & Stasik, M. (1997). "Spatial understanding and decision support system: A prototype for public GIS." *Transactions in GIS*, vol. 2, pp. 73–84.

Jankowski, P. (2009). "Towards participatory geographic information systems for community-based environmental decision making." *Journal of Environmental Management*, vol. 90, pp. 1966–1971.

Jordan, D.R. (2012). *Exploring the use of geographic information systems as an environmental and social justice advocacy tool for community-based organizations: A case study of Galena Park, Texas.* Unpublished: Georgia State University, Retrieved from http://scholarworks.gsu.edu/geosciences_theses/43

Krieger, J., Rabkin, J., Sharify, D., & Song, L. (2009). "High point walking for health: Creating built and social environments that support walking in a public housing community." *American Journal of Public Health*, vol. 99, pp. S593–S599.

Lemos, M.C., Austin, D., Merideth, R., & Varady, R.G. (2002). "Public–private partnerships as catalysts for community-based water infrastructure development: The border WaterWorks program in Texas and New Mexico colonias." *Environment and Planning C: Government and Policy*, vol. 20, pp. 281–295.

Minkler, M. (2014). "Enhancing data quality, relevance, and use through community-based participatory research." In *What Counts: Harnessing Data for America's Communities.* Edited by Federal Reserve Bank of San Francisco and Urban Institute. Retrieved from www.whatcountsforamerica.org/wp-content/uploads/2014/12/WhatCounts.pdf.

Minkler, M. & Wallerstein, N. (2003). *Community-based participatory research for health.* San Francisco, CA: Jossey-Bass.

Minkler, M., Vasquez, V.B., Tajik, M., & Petersen, D. (2008). "Promoting environmental justice through community-based participatory research: The role of community and partnership capacity." *Health Education & Behavior*, vol. 35, pp. 119–137.

Ndirangu, M., Yadrick, K., Bogle, M.L., & Graham-Kresge, S. (2008). "Community–Academia partnerships to promote nutrition in the Lower Mississippi Delta: community members' perceptions of effectiveness, barriers, and factors related to success." *Health Promotion Practice*, vol. 9, pp. 237–245.

Northridge, M.E., Yankura, J., Kinney, P.L., Santella, R.M., Shepard, P., Riojas, Y., et al. (1999). "Diesel exhaust exposure among adolescents in Harlem: A community-driven study." *American Journal of Public Health*, vol. 89, pp. 998–1002.

O'Fallon, L.R. and Dearry, A. (2002). "Community-based participatory research as a tool to advance environmental health sciences." *Environmental Health Perspectives*, vol. 110, pp. 155–159.

O'Fallon, L.R., Tyson, F., & Dearry, A. (2000). "Improving public health through community-based participatory research and education." *Environmental Epidemiology and Toxicology*, vol. 2, pp. 201–209.

Ottinger, G. (2010). "Buckets of resistance: Standards and the effectiveness of citizen science." *Science, Technology, & Human Values*, vol. 35, pp. 244–270.

Parker, E.A., Israel, B.A., Williams, M., Brakefield-Caldwell, W., Lewis, T.C., Robins, T., et al. (2003). "Community action against asthma: Examining the partnership process of a community-based participatory research project." *Journal of General Internal Medicine*, vol. 18, pp. 558–567.

Petersen, D., Minkler, M., Breckwich Vasquez, V., & Corage Baden, A. (2006). "Community-based participatory research as a tool for policy change: A case study of the Southern California environmental justice collaborative." *Review of Policy Research*, vol. 23, pp. 339–354.

Ramirez-Andreotta, M.D., Brusseau, M.L., Artiola, J., Maier, R.M., & Gandolfi, A.J. (2015). "Building a co-created citizen science program with gardeners neighboring a Superfund site: The Gardenroots case study." *International Public Health Journal*, vol. 7.

Redwood, Y., Schulz, A.J., Israel, B.A., Yoshihama, M., Wang, C.C., & Kreuter, M. (2010). "Social, economic, and political processes that create built environment inequities: Perspectives from urban African Americans in Atlanta." *Family & Community Health*, vol. 33, pp. 53–67.

Schwartz, N.A., von Glascoe, C.A., Torres, V., Ramos, L., & Soria-Delgado, C. (2015). "Where they (live, work and) spray": Pesticide exposure, childhood asthma and environmental justice among Mexican-American farmworkers." *Health & Place*, vol. 32, pp. 83–92.

Srinivasan, S. & Collman, G.W. (2005). "Evolving partnerships in community." *Environmental Health Perspectives*, vol. 113, pp. 1814–1816.

Stedman-Smith, M., McGovern, P.J., & Peden-McAlpine, C.J. (2012). "Photovoice in the Red River basin of the north: A systematic evaluation of a Community–Academic partnership." *Health Promotion Practice*, vol. 13, pp. 599–607.

Thompson, M.M. (2012). "The city of New Orleans blight fight: Using GIS technology to integrate local knowledge." *Housing Policy Debate*, vol. 22, pp. 101–115.

Vasquez, V.B., Minkler, M., & Shepard, P. (2006). "Promoting environmental health policy through community based participatory research: A case study from Harlem, New York." *Journal of Urban Health*, vol. 83, pp. 101–110.

Vasquez, V.B., Lanza, D., Hennessey-Lavery, S., Facente, S., Halpin, H.A., & Minkler, M. (2007). "Addressing food security through public policy action in a community-based participatory research partnership." *Health Promotion Practice*, vol. 8, pp. 342–349.

Wang, C. & Burris, M.A. (1997). "Photovoice: Concept, methodology, and use for participatory needs assessment." *Health Education & Behavior*, vol. 24, pp. 369–387.

Wier, M., Sciammas, C., Seto, E., Bhatia, R., & Rivard, T. (2009). "Health, traffic, and environmental justice: Collaborative research and community action in San Francisco, California." *American Journal of Public Health*, vol. 99, pp. S499–S504.

Wilson, O. & Wilson, S. (2015). "Assessment of a novel environmental justice community–university partnership." In NIEHS (2015). *Advancing Environmental Justice*. Retrieved from www.niehs.nih.gov/research/supported/assets/docs/a_c/advancing_environmental_justice_508.pdf.

Wilson, S.M., Wilson O.R., Heaney, C.D., Cooper, J. (2007). "Use of EPA collaborative problem-solving model to obtain environmental justice in North Carolina." *Progress in Community Health Partnerships: Research, Education and Action*, vol. 1, pp. 327–338.

Wilson, S.M., Wilson, O.R., Heaney, C.D., & Cooper, J. (2007–2008). "Community-driven environmental protection: Reducing the PAIN of the built environment in low-income African-American communities in North Carolina." *Social Justice in Context*, vol. 3, pp. 41–57.

Wilson, O.R., Bumpass, N.G., Wilson, O.M., & Snipes, M.H. (2008). "The West End Revitalization Association (WERA)'s right to basic amenities movement: Voice and language of ownership and management of public health solutions in Mebane, North Carolina." *Progress in Community Health Partnerships: Research, Education, and Action*, vol. 2, pp. 237–243.

Wilson, S.M., Heaney, C.D., Cooper, J., & Wilson, O.R. (2008). "Built environment issues in unserved and underserved African-American neighbourhoods in North Carolina." *Environmental Justice*, vol. 1, pp. 63–72.

Wilson, S.M., Rice, L., & Fraser-Rahim, H. (2011). "The use of community-driven environmental decision making to address environmental justice and revitalization issues in a port community in South Carolina." *Environmental Justice*, vol. 4, pp. 145–154.

Wilson, S., Campbell, D., Dalemarre, L., Fraser-Rahim, H., & Williams, E. (2014). "A critical review of an authentic and transformative environmental justice and health community–university partnership." *International Journal of Environmental Research and Public Health*, vol. 11, pp. 12817–12834.

Wilson, S.M., Murray, R.T., Jiang, C., Dalemarre, L., Burwell-Naney, K., & Fraser-Rahim, H. (2015). "Environmental justice radar: A tool for community-based mapping to increase environmental awareness and participatory decision making." *Progress in Community Health Partnerships: Research, Education, and Action*, vol. 9, pp. 439–446.

Wing, S. (2002). "Social responsibility and research ethics in community-driven studies of industrialized hog production." *Environmental Health Perspectives*, vol. 110, pp. 437–444.

Wing, S. (2005). "Environmental justice, science and public health." In Goehl, T.J. (ed.), *Essays on the future of environmental health research: A tribute to Dr. Kenneth Olden*, pp. 54–63. Research Triangle Park, NC: Environmental Health Perspectives/National Institute of Environmental Health Sciences. Retrieved from www.brown.edu/research/research-ethics/sites/brown.edu.research.research-ethics/files/uploads/Environmental%20justice%20-%20Steve%20Wing.pdf

Wing, S., Horton, R.A., Muhammad, N., Grant, G.R., Tajik, M., & Thu, K. (2008). "Integrating epidemiology, education, and organizing for environmental justice: Community health effects of industrial hog operations." *American Journal of Public Health*, vol. 98, pp. 1390–1397.

24

PARTICIPATORY GIS AND COMMUNITY-BASED CITIZEN SCIENCE FOR ENVIRONMENTAL JUSTICE ACTION

Muki Haklay and Louise Francis

Introduction

When facing local environmental issues, maps are a very effective tool. They can help in collecting facts about an area, bring issues to the table, allow for comparison between areas, act as a tool of communication with local decision makers, and identify key issues for action. While the maps on their own do not alter the power relationships within which communities live their lives, many studies and examples have shown the purposeful role of mapping as part of community mobilization and action (e.g. Chambers 2006; Sieber 2006). The use of maps and geographical information to address community concerns is integrating different areas, including participatory mapping (Chambers 2006), participatory geographic information systems (PGIS) or Public Participation GIS (PPGIS) (Sieber 2006), and the emerging area of citizen science (Haklay 2013).

The aim of this chapter is to provide an overview of the way in which these areas contributed to local community environmental action with a particular focus on the role of participatory mapping and citizen science. The proliferation of accessible techniques for community-based environmental monitoring, combined with practices that emerged from the environmental justice movement, provides the basis for new applications of mapping and sensing that are now within reach of individuals and communities.

The rest of the chapter is structured as follows. The next section provides a brief introduction to geographical and localized elements of environmental justice, PGIS and citizen science. Following this, a detailed discussion of a methodology that has been developed and deployed by us for the past decade is provided. The methodology is illustrated in a case study of community-led noise monitoring at the Pepys Estate in Deptford, London. Following the case study, we conclude with observations on the integration of citizen science, participatory mapping and environmental justice.

Environmental justice, participatory geographic information systems and citizen science

As noted in other chapters in this volume, the history of environmental justice goes back to the 1980s with a clear focus in the US on health concerns due to the location of waste and toxic facilities (Walker 2012). Over the years, the scope of what is included in environmental justice, as well as the scale of the interactions, has grown dramatically. While concepts of justice in liberal philosophy have a long history in Western thought (Sandel 2009), the manner in which justice is understood within environmental justice discourse is more complex in ways that challenge simple definitions (see Chapter 10). As Schlosberg (2007) demonstrates, in environmental justice there is a direct link between aspects of distribution, participation and recognition. Distribution deals with access to resources and has been central to the political theory of justice – for example, is it just to provide the potential to access resources or is it vital to ensure that every individual is capable of access? In addition, within the environmental justice discourse, there is a need to understand that recognition of communities and individuals by wider society and accepting their right to have a say in decisions that influence them are also important. The definition of access to environmental benefits and the impact of burdens is not a simple one, and while some burdens have a proven link to health and wellbeing (e.g. noise), others are culturally and contextually based. As a result, the list of social dimensions that are included in environmental justice discourse includes aspects such as ethnicity, gender, class and inter-generational distribution, while the environmental topics that are covered range from air pollution and urban dereliction to outdoor recreation, mineral extraction and the wider impact of deforestation and climate change (Walker 2012). Yet, while environmental justice discourse went through an expansive process of themes and issues, many of its everyday manifestations are locally based and experienced in the interaction of communities and the local geographic context in which they live.

Throughout the development of environmental justice discourse, mapping has played an important part in understanding and demonstrating the patterns of exposure and benefits, and the use of Geographic Information Systems (GIS) for the analysis of environmental inequalities is common (Mitchell & Walker 2007, Mennis 2011). Some of this analysis has been carried out without the direct involvement of the communities that are affected by the analysis (e.g. Mitchell 2005), but of particular interest here is the use of mapping techniques within participatory settings, as these are the situations in which community mobilization around the process of mapping and gathering evidence about local environmental conditions occurs. The linkage between participatory processes, applications of GIS and environmental justice is in the area of participatory GIS.

While building on existing practices in participatory planning, participatory GIS is different as a result of the use of information and communication technologies and their influence on the process (Chambers 2006). Participatory GIS emerged in the mid-1990s, out of concerns over the societal impact of increasing use of GIS by local and central government and large corporations, with an increased exchange of digital maps within planning and management processes. While the move to GIS streamlined and accelerated decision making processes, it marginalized communities and individuals who did not have access to the systems or to the information that they contain. Participatory applications of GIS try to rectify this problem by providing access to information and technology so those affected by the authorities' use of GIS can benefit from more equitable access to it. The literature in this area uses two related terms: Participatory GIS (PGIS) and Public Participation GIS (PPGIS). These two related areas of practice and research are similar in their methods and overall aims, although with a stronger emphasis on applications in the Global South within the PGIS literature, while the PPGIS is linked to urban planning practice in the Global North (Verplanke et al. 2016; Sieber 2006). In both, there is an explicit attempt to use digital mapping technologies to give voice, amplify, and represent local needs – especially of marginalized groups. Both areas echo concerns related to environmental justice, with Harris et al. (1995) explicitly calling for applications that are 'broad-based, inclusive, gender-sensitive, and biased towards marginalised people' (p. 218). While Sieber (2006) as well as Verplanke et al. (2106) provides a comprehensive review of the PGIS/PPGIS field, in the area of environmental management, Participatory GIS has an additional legitimacy due to the development of Principle 10 of the Rio Declaration (Haklay 2002). Principle 10 emphasized public access to environmental information, participation in decision making and access to justice in environmental matters. The three pillars of Principle 10 were later enshrined into legislation through the Aarhus convention (UNECE 1998), and the movement to provide open data by governments across the world facilitated the potential of using PGIS in cases of environmental justice, as access to basic environmental information that can be manipulated and visualized is critical to the creation of GIS-based representations.

As a result of the implementation of Principle 10 and Aarhus, official environmental information was increasingly provided over the Internet. With increased awareness of the provision of open governmental data in the past decade, access to information has accelerated as well as access to the underlying data sets. However, in addition, a new source of data has emerged since the late 1990s. This, in part, has been made possible by the proliferation of equipment for sensing the environment which became part of routine, large-scale monitoring programmes at local and national levels and also came within reach of non-governmental organizations and community groups. An example of this is the Global Community Monitor – an organization that, since 1998, has developed a method to allow communities to monitor air quality near polluting factories (Scott & Barnett 2009). The sampling is done by members of the affected community using widely available plastic buckets and bags followed by analysis in an air quality laboratory. Finally, the community is provided with guidance on how to understand the results. This activity is termed 'Bucket Brigade' and is used across the world in environmental justice campaigns, for example in the struggle of local African-American residents in Diamond, Louisiana against a polluting Shell Chemical plant (Ottinger 2010).

This type of data collection, carried out outside professional settings (be it university or environmental authority), is now recognized as citizen science (Silvertown 2009), which is defined as: 'the scientific activities in which non-professional scientists volunteer to participate in data collection, analysis and dissemination of a scientific project' (Haklay 2013: 106). Importantly, citizen science also contains within it the notion that science needs to work together with communities and individuals, in ways that appreciate local knowledge and practices (Irwin 1995). Stilgoe (2009: 11) articulated the obligation on scientists in stating that:

> All scientists are citizens, but not all scientists are Citizen Scientists. Citizen Scientists are the people who intertwine their work and their citizenship, doing science differently, working with different people, drawing new connections and helping to redefine what it means to be a scientist.

Citizen science contributes to the wider pool of environmental information by creating local information that represents the issues that the local community is concerned about, as well as up-to-date information that is, at times, difficult or challenging for the authorities to collect.

Methodology for participatory mapping and citizen science

As noted above, environmental justice is context dependent and, as it is framed within specific social and political understandings of place, it is important to first understand the local conditions before turning to the description of the methodologies and approaches that we have used in our studies. In the rest of this chapter, we focus on a detailed explanation of a methodology that we developed originally in London, and later tested in other locations across the UK, Poland, Italy and Finland. Yet, it is important to explain and contextualize the approach within the local conditions. In the UK, environmental justice discourse has been 'imported' by academics and governmental bodies, who sometimes prefer to use the term 'environmental inequalities' (Agyeman & Evans 2004). A significant proportion of UK work has been characterized by a top-down approach, in which researchers use geographical datasets and information to analyse a locale's environmental conditions, then either act with the local community to change the situation or publish the results together with NGOs to advocate policy changes (e.g. Mitchell 2005). This approach to environmental justice evaluation can be attributed, at least partially, to the richness of the national data available in the UK, including the highly detailed digital maps from the Ordnance Survey that lend themselves easily to large-scale analysis. Importantly, until the last decade and the growth in open data, most of the official mapping data were out of reach for community organizations because of the costs associated with purchasing the data, as well as the technical skills required to analyse them effectively. In addition, as environmental decision making is characterized by strong scientific framing of problems and solutions (Haklay 2009), any community that intends to use environmental justice arguments is required to collect evidence to support its claims. Thus, there is a need for evidence gathering, which can be facilitated through participatory mapping and citizen science.

Our methodology emerged in 2007, through the London 21 Sustainability Network project 'A Fairer, Greener London', which aimed to give six marginalized communities the opportunity to develop their own understanding of local environmental justice issues and support action plans to address them. The project was integrated closely with the project 'Mapping Change for Sustainable Communities', which was funded as part of the UrbanBuzz scheme (UrbanBuzz 2009). Both projects were based on accessible GIS technologies and available environmental information sources.

The methodology evolved into a six-stage process that is inherently flexible and iterative so, while the stages are presented as a serial process (see Figure 24.1), the application of the methodology for a specific case is carried out through a discussion with the local community.

Before the process starts, there is a need to understand the local conditions and to ensure that contact has been made with all interested members of the community. If the process is facilitated by an external actor, this person will need to ensure that, in the initial meeting in which the first and second stages happen, the meeting is inclusive and represents all the groups and people within the study area. The initial meeting should happen near or in the area in which the study takes place, and at a time that ensures a high level of participation.

The first stage of the process is the 'introduction to existing public information'. Here, information that is accessible through the Aarhus process and open data are compiled to provide a baseline for the discussion. This includes information from local and national government sources and from environmental protection agencies, frequently on dedicated websites (e.g. a local authority air quality website). Additional information relevant to the issues at hand might come from official sources such as the population census or other authoritative sources about the socio-economic conditions in the area. The reasons for presenting this information are two-fold. First, most of the information is presented in a

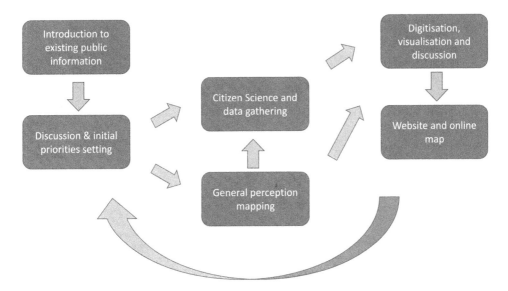

Figure 24.1   Methodology for participatory mapping and citizen science

'supply' rather than 'demand' approach – the information is provided without many outreach attempts and, even if these exist, they are limited in time and scope. Therefore, it is very likely that the participants in the meeting will not be familiar with it – as a succinct response from a participant in one of our early workshops demonstrates: 'this is not community information, expressed in community language that we can understand.' Second, the information used at the beginning of the meeting frames the environmental issues that will be explored, while allowing for flexibility and openness in deciding the exact direction that will be followed by community members. All too often, researchers can have a preconceived idea about the problem that the community faces, and therefore introduce the project as a 'fait accompli' before any detailed discussion takes place about what issue should be addressed. By presenting a portfolio of information and issues, the discussion starts with an emphasis on community control over the process.

The second stage, 'discussion and initial priorities setting', is a facilitated discussion around large-scale maps of the local area (Figure 24.2). The discussion is aimed at identifying issues that the specific project will focus on. An exercise such as 'identify places on the map that you like and dislike' can assist in ensuring that all the participants can use maps effectively. Although digital maps can be used for this purpose, the advantage of the paper map is to ensure that participants who are unfamiliar with technology can contribute to the discussion, without the need to dedicate time for training. The result of the discussion is an agreement on what should be recorded and how. Here, there is a choice between two possible stages (perception mapping or citizen science) or a combination of both.

The discussion needs to establish how the problem that was identified by the community will be addressed through data collection. This can be by establishing the situation on the ground or gathering evidence that can be used in discussions with decision makers and public bodies. What is then required, as in any typical scientific process, is to decide what information will be collected, the data collection protocol, the process of collating the information and, importantly, what analysis or presentation will be needed with the data. Of course, decisions on analysis and visualization for the purpose of communication are taken in stages five and

Figure 24.2 Detailed discussion over local maps

six, but preliminary ideas are useful at this stage, to help consider whether the process will yield the required information. It is also worth carefully considering the protocol for data collection – the instruments that will be used, the information that will be captured, and the level of skills and training that the participants need to carry out the work. The protocol requires special attention – there is a need to balance robust, systematic and consistent data gathering that will be effective and won't be dismissed as 'anecdotal', while taking into account the practices, skills, resources and time availability of those who are involved in the process. For example, assumptions about the availability of smartphones with certain types of sensors might not stand up in a marginalized community, or when working with people of certain age groups. The same is true for expecting people to carry out activities around the common working day when some participants might be shift workers or in other employment with less structured working days. Use of similar instruments and methods to those used by the authorities has the advantage of making it easier to communicate information with them, although that might burden the community with learning the appropriate data collection and interpretation methods.

   Stage three involves 'general perception mapping' – recording qualitative local knowledge that is not about physical features in the area but aspects such as local history, memories, or feelings about places (e.g. dangerous, unpleasant). Perception mapping can be done easily with paper maps that are distributed in the community and completed in a short session by participants – for example, by having a project volunteer reaching out to neighbours and collecting information from them. Volunteers can then digitize the information on a shared online community map (e.g. the community mapping system that was developed in Mapping for Change – see Figure 24.3) (for a detailed discussion of the system see Ellul et al. 2011, 2013).

Frequently, perception mapping is done either together with 'citizen science and data gathering' (stage four), or as a first stage to understand the local environment and pinpoint the specific aspects on which detailed factual data gathering is required. In other cases, early discussions (stage two) lead in a direction that requires citizen science to be the primary source of community information. For the purposes here, we can view any factual data collection (e.g. from levels of pollutants in the air to the location of broken street light) as part of citizen science. As noted above, a detailed discussion about the data collection protocol is required, to ensure that the information will be acceptable to the authorities and might lead to action by them. Citizen science activities always require data collection instruments, even if these are predefined forms with additional maps to locate the observation. The costs of the instruments, their availability and the training that they require should form part of the considerations when setting up such an activity.

Following the data collection process (and often during it), the information should be digitized, analysed and visualized using GIS software (stage five). While the use of online data collection and visualization is attractive, this should be considered carefully. Early discussion might lead to the conclusion that the participants prefer not to share the information openly, either in the early stages of the project or at any stage: for example, when a conflict can occur with the polluting facility owners, or when the information can be used against the impacted community. Thus, consideration of the benefits and the risks of sharing information openly and visibly is necessary. At the same time, the visualization of information has been demonstrated to motivate participants and to provide them with direct feedback of what they have achieved. A solution for this can be a localized GIS, which participants can interrogate, or a password-protected website.

The final stage, 'website and online map', is relevant in cases where the community agrees to share the information online. The online map can be used to share the information with other people in the local community who are not involved directly in data collection and analysis, for example due to lack of time or technical capability. As noted, an online map can also provide a focus for a final feedback discussion with the participants and their wider community on the results of the study, as well as a discussion on what the next steps should be in using the information to lead to a change in local conditions.

As Figure 24.1 shows, there is no simple route through the process. The elements that we have found effective in terms of engagement and empowerment include flexible use of the stages in the process to allow the community to co-design and co-determine the process (see the framework of 'Extreme Citizen Science' in Haklay 2013). The flow of the process emphasizes flexibility, iterative cycles and alternative pathways for different levels of participation so the activity can be inclusive.

Community noise pollution monitoring and action in Pepys Estate, London

To demonstrate the application of the methodology, we describe a case study on noise in the Pepys Estate in Lewisham, London, where residents took part in early activities that led to the consolidation of the methodology. The Pepys Estate is a social housing complex near the banks of the Thames, originally built in 1966. The estate had a complex history of ups and downs, and in the 2000s was classified in the national Index of Multiple Deprivation within the 20 per cent poorer parts of England (see the history of the estate at Municipal

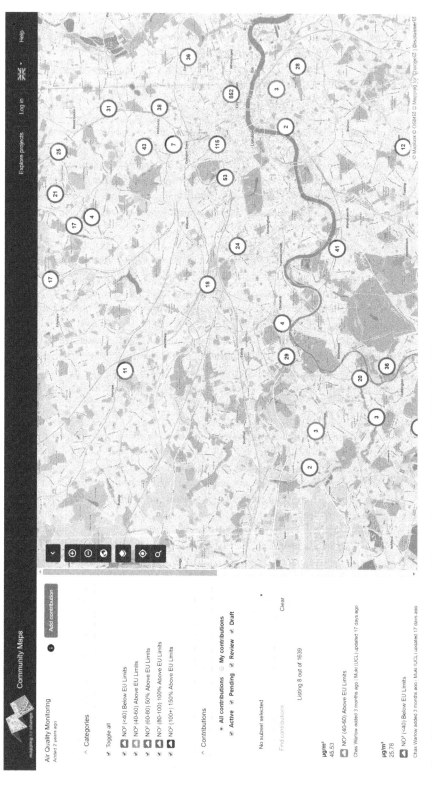

Figure 24.3 Community mapping system (Mapping for change: map tiles © Mapbox ©OpenStreetMap contributors)

Dreams 2015). In stage one of the process, after consultation with the Pepys Community Forum – a local community organization – and dedicated effort to recruit people from across the estate, residents pointed to the daily banging and grinding of a scrapyard near the centre of the estate, in close proximity to both a primary and a nursery school. The residents were frustrated that previous complaints had gone unanswered and felt they had little means to address the situation. Discussions with the local authority revealed that the scrapyard was considered a recycling facility, and was therefore regulated by the environmental authority (The Environment Agency), which limits the power of the local authority to regulate activity there. At the same time, the permission for such a noisy and polluting industrial activity to be located near schools and residential areas was associated with the limited resources and power of the community in the estate. Once the noise problem was identified, a participatory data collection effort ensued, using paper forms and maps, and Class 2 noise meters supplied from electronics shops (where they are available for workplace assessment). Importantly, the participants were interested in expressing their views and perception of the noise that they had measured, thus mixing 'perception mapping' and 'citizen science' in stages three and four. For the Pepys Estate community, the data collection was an opportunity to represent their daily experiences of the nuisance, but to do so experientially and scientifically.

Over seven weeks in early 2008 a group of residents conducted a comprehensive noise pollution mapping survey (stage three). The participants were trained to take noise readings at all hours of the day and night across the whole of the Pepys Estate. In total, over 1000 individual readings were taken across all the sites. In addition to recording dBA sound levels, residents also collected qualitative information expressing how they felt about the noise. They were asked to choose words such as relaxing, annoying or disturbing to describe the sound. In addition, participants were asked to detail the principal noise source and provide any further information on how they were affected by the noise. During the engagement period, completed survey sheets were collected on a weekly basis by the project team and reviewed to make sure a good distribution through time and space was being achieved. Feedback was given to residents about which areas and times needed more attention.

The data was then analysed using a GIS to produce noise pollution maps for the areas (stage four). Qualitative information was summarized using charts and graphs. These results showed that the vast majority of readings were described as either loud, very loud or extremely loud. In addition, results revealed that individuals were quite accurate in their perception of noise levels, suggesting that people's perception of noise is very reliable. Members of the community also found disturbingly high levels of noise, affecting quality of life up to 350 metres from the scrapyard.

At the final stage of the process (stage six), a public meeting was organized with Lewisham Council and the Environment Agency and according to the Community Forum, it showed one of the highest levels of participation in terms of the number of residents who attended it. Over two hours, the results were presented by the residents who had collected the data, and there was a facilitated discussion in which many residents used the opportunity to explain their own frustration and concerns about the impact of noise on them and on their children. The authorities accepted that the effort of the residents provided enough evidence of a problem and carried out their own study, leading to a tightening of the regulations for the operation of the scrapyard.

The study, as well as another study near London City Airport, led to large-scale engagement with communities around London Heathrow Airport, and the use of smartphones as noise sensing devices to create an extensive noise and perception survey of the area (see Becker et al. 2013 for a detailed discussion).

Conclusions

The methodology that we have described above has proved to be effective and useful in assisting communities to address environmental justice concerns. The requirement of the environmental decision making process to rely on scientific information makes the use of citizen science and comprehensive qualitative data collection through perception mapping especially useful for this context. However, we should also notice the limitations and issues that we have identified.

Similar to other PGIS methodologies, such as Participatory 3D Modelling (P3DM), which is used in development plans, and includes building a physical model of the location with the local community (CTA 2016), the methodology relies on high levels of engagement and participation by the local community. In order to be effective, it requires significant effort in ensuring that the process is inclusive and represents all those who would like to have their voice heard through it. There is also a need for constant contact with the participants, to provide them feedback and encouragement throughout the process. This usually results in relatively time limited activity that lasts for four to eight weeks.

Secondly, as a bottom-up approach, there is a risk of misreading or misunderstanding the issues that can make a difference – for example, in terms of local power relationships. The case of the Pepys Estate demonstrates that the local authority was not the body which had the ability to influence the activities of the problematic facility, even though it was the body with which the residents were familiar and comfortable. There is therefore a need to critically evaluate the activity and the purpose of the information that is collected, and plan ahead for the use of the outputs from the project.

Thirdly, there is a territorial issue – in many of the cases in which the methodology was used, we explained to the participants that it would be beneficial to collect data beyond their own neighbourhood, as this could demonstrate the disparity in distribution of environmental nuisances and strengthen their case. We observed that not only were the participants reluctant to use the term 'environmental justice', but the plotting of the data collected usually delineated the boundary of the local neighbourhood. It seemed as if the participants were not interested in or confident about collecting data beyond their area. We can assume that despite participants' internal perceptions about the distribution of environmental burdens, they did not judge the political purchase of using the environmental justice argument to be as strong as a direct complaint about noise, air quality, or (in a case in Katowice, Poland) dog fouling. Understanding the exact reasons for localized data collection and the assessment of political arguments requires further research.

In summary, this chapter explored the intersection between citizen science, participatory GIS and environmental justice, explaining and demonstrating a methodology for carrying out projects. The general methodology allows the participants to understand the official data about their area, and to identify gaps in information and in representation of their local conditions. Following Schlosberg (2007), Walker (2012) notes that the discourse on environmental justice requires a multifaceted approach that not only allows demonstration of the inequity but also addresses aspects of representation. Participation, mapping and local data collection through citizen science provide the means to create such a representation through the sharing of information that carries with it the authority of scientific practice in being reliable, objective and trustworthy. Yet, there is a need to provide space for perceptions, emotions, memories and other qualitative aspects as part of this representation. This qualitative expression should also be seen as part of the participation and representation. The combination of (frequently qualitative) local knowledge and (often quantitative) scientific measurement is never simple or straightforward, and it is a mistake to attempt to simply scale it and treat it quantitatively. The maps and outputs that are produced in the process need to take into account the wishes of the community and what they would like to express through their activities.

Finally, there is also a need to note the dual meaning of citizen science – as Irwin (1995) and Stilgoe (2009) pointed out, citizen science can also be understood to be science in the service of citizens. The methodology proposed here is structured around the approach that the researchers and scientists involved in it accept their role as facilitators, leaving ample space for the participants to shape the project, the data collection methodology and the interpretation of results. Accepting that local knowledge and expertise are on a par with scientific expertise, and developing a more egalitarian mode of knowledge production, is a challenge that frequently arises in the context of using the methodology.

We do not suggest that the methodology is universally applicable in participatory mapping and citizen science in environmental justice cases – the social, environmental, economic and political contexts of each locality and community entail sensitive adaptation that allows for the inclusive and productive collaboration of all those involved.

References

Agyeman, J., & Evans, B. (2004). "'Just sustainability': The emerging discourse of environmental justice in Britain." *The Geographical Journal* 170: 155–164.

Becker, M., Caminiti, S., Fiorella, D., Francis, L., Gravino, P., Haklay, M., et al. (2013). "Awareness and learning in participatory noise sensing." *PLoS One* 8(12): e81638.

Chambers, R. (2006). "Participatory mapping and geographic information systems: Whose map? Who is empowered and who disempowered? Who gains and who loses?" *The Electronic Journal of Information Systems in Developing Countries*. 25.

CTA - Technical Centre for Agricultural and Rural Cooperation ACP-EU (2016). *The Power of Maps: Bringing the Third Dimension to the Negotiation Table*. Netherlands: Proud Press.

Ellul, C., Francis, L., & Haklay, M. (2011). "A Flexible Database-Centric Platform for Citizen Science Data Capture, Computing for Citizen Science Workshop", in *Proceedings of the 2011 Seventh IEEE International Conference on eScience* (eScience 2011) 71–90.

Ellul, C., Gupta. S., Haklay, M., & Bryson, K. (2013). "A platform for location based app development for citizen science and community mapping." In Krisp, J.M. (ed.) *Progress in Location-Based Services*, Lecture Notes in Geoinformation and Cartography, Berlin: Springer, p. 71.

Haklay, M. (2002). "Public environmental information: Understanding requirements and patterns of likely public use." *Area*, 34(1), 17–28.

Haklay, M. (2009). "The contradictions of access to environmental information and public participation in decision making." *Nordic Environmental Social Science 2009*, London, 10–12 June.

Haklay, M. (2013). "Citizen science and volunteered geographic information: Overview and typology of participation." In Sui, D.Z., Elwood, S. & M.F. Goodchild (eds.) *Crowdsourcing Geographic Knowledge*. Berlin: Springer, pp. 105–122.

Harris, T.M., Weiner, D., Warner, T.A., & Levin, R. (1995). "Pursuing social goals through participatory geographic information systems." In Pickles, J (ed.) *Ground Truth: The Social Implications of Geographic Information Systems*. New York: Guilford, pp. 196–222.

Irwin, A. (1995). *Citizen Science: A Study of People, Expertise and Sustainable Development*. London and New York: Routledge.

Mennis, J. (2011). "Integrating remote sensing and GIS for environmental justice research." In Yang, X. (ed.) *Urban Remote Sensing: Monitoring, Synthesis and Modeling in the Urban Environment*, Chichester: John Wiley, pp. 225–237.

Mitchell, G. (2005). "Forecasting environmental equity: Air quality responses to road user charging in Leeds, UK." *Journal of Environmental Management*, 77, 212–226.

Mitchell, G. & Walker, G.P. (2007). "Methodological issues in the assessment of environmental equity and environmental justice." In Deakin, M., Mitchell, G., Nijkamp, P. & R. Vreeker (eds.) *Sustainable Urban Development: The Environmental Assessment Methods*, London: Routledge, pp. 447–472.

Municipal Dreams (2015). *The Pepys Estate, Deptford: for 'the peaceful enjoyment and well-being of Londoners'*, available at https://municipaldreams.wordpress.com/2015/08/04/the-pepys-estate-deptford-for-the-peaceful-enjoyment-and-well-being-of-londoners/

Ottinger, G. (2010). "Buckets of resistance: Standards and the effectiveness of citizen science." *Science, Technology & Human Values*, 35(2), 244–270.

Sandel, M. (2009). *Justice: What's the Right Thing to Do?* New York: Farrar, Straus and Giroux.

Schlosberg, D. (2007). *Defining Environmental Justice: Theories, Movements, and Nature*. Oxford: Oxford University Press.

Scott, D. & Barnett, C. (2009). "Something in the air: Civic science and contentious environmental politics in post-apartheid South Africa." *Geoforum*, 40(3), 373–382.

Sieber, R. (2006). "Public participation geographic information systems: A literature review and framework." *Annals of the Association of American Geographers*, 96(3), 491–507.

Silvertown, J. (2009). "A new dawn for citizen science." *Trends in Ecology & Evolution* 24(9), 467–471.

Stilgoe, J. (2009). *Citizen Scientists: Reconnecting Science with Civil Society*. London: Demos.

United Nations Economic Commission for Europe (UNECE) (1998). *Convention on Access to Information, Public Participation in Decision-making and Access to Justice in Environmental Matters*, ECE Committee on Environmental Policy, Aarhus, Denmark.

UrbanBuzz (2009). *UrbanBuzz: Building Sustainable Communities*. London: UCL.

Verplanke, J., McCall, M.K., Uberhuaga, C., Rambaldi, G. & Haklay, M., (2016). "A shared perspective for PGIS and VGI." *The Cartographic Journal*, 53(4), 308–317.

Walker, G. (2012). *Environmental Justice: Concepts, Evidence and Politics*, London: Routledge.

PART III

Substantive issues in environmental justice research

25

STREAMS OF TOXIC AND HAZARDOUS WASTE DISPARITIES, POLITICS AND POLICY

Troy D. Abel and Mark Stephan

Nearly three decades ago, research on the location of hazardous waste landfills rattled US environmental policy institutions. A United Church of Christ (UCC 1987) study reported that three times as many non-whites resided in zip codes hosting these toxic sites compared to zip codes without landfills. Garnering headlines across the country, the UCC study drew national attention to the distributional scrutiny of race, class and pollution over the next seven years. One reviewer at the time concluded that "declaring an end to 'environmental racism' is an integral part of the national environmental agenda" (Kisch 1994).

Two decades later, another UCC study reported that significant racial and socioeconomic disparities remained in the distribution of commercial hazardous waste landfills (Bullard et al. 2007). In 2015, researchers produced a similar but more comprehensive analyses of the changing demographics near US Transport, Storage, and Disposal Facilities (TSDFs) between 1966 and 1995. While many prior studies found disparities after the siting of a hazardous waste facility and challenged many EJ policy responses, this more recent publication reported evidence of siting disparities in all time periods. These results lead the study's authors to conclude that toxic and hazardous waste disparities are the result of racial discrimination and socio-political vulnerabilities (Mohai and Saha 2015).

Yet no significant policy discussions about environmental justice (EJ) re-emerged and unlike the landmark 1987 UCC report, the 2015 study was unable to reshape the national environmental policy agenda. Likewise, the second UCC report in 2007 and its findings of persistent racial and socioeconomic disparities in the location of toxic and hazardous waste facilities received little attention in the national news cycle. These three studies delineate different waves of empirical studies on the topic of this chapter: the environmental injustice of toxic and hazardous waste disparities. Each one also appeared when the political and policy context was very different. Nonetheless, the problem of where we treat, store, and dispose of toxic and hazardous waste is a uniquely important and central issue in EJ.

The birth of the EJ movement is often traced to 1982 and the resistance of a predominantly African-American community to the disposal of polychlorinated biphenyls (PCBs) in a Warren County North Carolina landfill. Consequently, Congressional representatives asked the General Accounting Office (GAO) to examine the correlation between hazardous waste landfill locations and their surrounding demographics. The GAO completed an analysis of

Warren County and three other communities in the Southeast hosting hazardous waste landfills. African-American populations were found to range from 38 to 90 per cent while between 26 and 42 per cent of the residents had incomes below the poverty level in zip codes with toxic dumps (USGAO 1983).

Subsequently, the UCC study, and many more, focused solely on the location disparities of toxic and hazardous waste sites that animated policy debates in the early nineties (Szasz and Meuser 1997). While the issues involving the intersection of toxics and social justice range widely from lead-paint exposure inequalities, migrant farm worker risks from pesticides, and electronic waste exports, none have captured the research or policy attention like toxic and hazardous waste disposal disparities. We therefore concentrate our review on these disparities and their research and policy intersections in the US. Moreover, comprehending the convergence and then divergence of toxic and hazardous waste disparities research and policy development in the US over the last three decades is fundamental to understanding the complexity of many substantive EJ issues and developing new directions for research and practice.

Our chapter proceeds as follows. First, we describe a policy process framework that helps describe the historical intersections of toxic and hazardous waste disparities and EJ policy developments in the US. We then briefly review a second wave of EJ research emblematic of the politics of policy formation. Third, we discuss the consolidation of policy developments in the Clinton administration, across the states, and a third wave of research on toxic and hazardous waste disparities. Fourth, we describe the paradox of how EJ policy cohesion in the 2000s paradoxically weakens in some parts of the executive branch while other federal programmes revive efforts to address toxic and hazardous waste disparities. Our chapter concludes with a discussion of the latest policy developments and social science advances related to toxic and hazardous waste disparities.

Converging problems, policies, and politics

After the first wave of EJ research in the 1980s that culminated with the UCC study (see Bullard 1990), several policy responses began to develop in the executive branch. First, the US Environmental Protection Agency (EPA) established an environmental equity workgroup in 1990 with representatives from the agency's major programme offices. Second, EPA established an Office of Environmental Equity in 1992 that was later renamed the Office of Environmental Justice (OEJ). This office would organize a national forum, additional studies, and an EJ hotline (USEPA 1994). Third, and most significantly, President Clinton signed Executive Order (EO) 12898 in 1994 requiring all federal agencies to integrate the achievement of EJ into their programmes. A few studies helped advance EJ to "the forefront of the nation's environmental policy agenda" (Sexton and Zimmerman 1999: 428). Yet as we discuss below, getting EJ on the federal policy agenda did not ensure it would remain there or become settled into law.

According to Lester et al. (2001), EJ policy developments of the early 1990s exemplified agenda-setting process theories and in particular, the multiple streams approach (MSA) first developed by Kingdon (1984). This perspective describes three mostly independent streams of problems, policies, and politics that seemingly flow and shift on their own. Then, a convergence of information about public problems, available and tested policy ideas, and political support for action leads to new policies. Sometimes, however, an issue's convergence on the national agenda-setting stage may be followed by divergence and even failure during implementation (Zahariadis and Exadaktylos 2016).

Clinton's executive order could be seen as a coupling of the three streams. EJ was seemingly on its way to join past policy landmarks like the Clean Air, Clean Water, and Endangered Species Acts. However, on eight occasions from 1992 to 1996, Congress failed to enact legislation. EJ initiatives were instead limited to executive branch actions and several subnational efforts. This was hardly the policy transformation suggested by Kingdon's original model. Instead of the original model, we propose that a revised policy streams framework integrating the classic policy stages perspective (Howlett et al. 2015) better describes the *convergence* and then *divergence* of toxic and hazardous wastes disparity research and EJ policy development over the last three decades. While a full treatment of the MSA and stages integration is beyond the scope of this chapter, we highlight several key conceptual elements that help illuminate the ebbs and flows of thirty years of toxic and hazardous waste inequity research and EJ policy development.

First, this more recent iteration of the MSA recognizes the presence of multiple policy windows. For EJ, there never was a strong enough coupling of the problem, policy and political streams to propel a new statute through the Congressional legislative process in the first decade of this issue's history (1983–1994). When Republicans regained control of the national legislative political stream in 1994, national environmental institutions gridlocked and EJ efforts flowed to other channels. For example, states provided an alternative policy window for about a decade. This attention to state EJ policy development faded by the middle of the 2000s however (Abel et al. 2015a).

Second, this MSA and policy stages refinement better describes the nonlinear and turbulent stages of a problem's policy evolution. Instead of just delineating three major stages of a policy's trajectory (agenda setting, policy legitimation, and implementation), the refined perspective adds policy formation and decision-making phases (see Figure 25.1). The former begins with a 'whirlpool' of strategic appraisals where science and politics converge just before policy formation and decision making. Consolidation follows, where a process and programme stream are added to Kingdon's original three (problems, policies, and politics). Policy settlement concludes the complex process of national policy change.

Third, the refined MSA introduces an explicit possibility of a dominant stream influencing the others. "This suggests qualitatively different kinds of policy making at each intersection point depending on exactly which stream guides the current at a particular point in the policy-making process," according to Howlett et al. (2015: 427). For EJ, the problem stream was initially the strongest current pushing the siting of toxic and hazardous waste facilities onto the national environmental policy agenda.

Rising toxic and hazardous waste disparities research

Consistent with the MSA, policy evaluation reports and studies describing the environmental conditions and trends requiring a governmental response were some of the key catalysts in the problem stream. The 1983 GAO study marked the first intersections of civil rights and environmental pollution in a policy analysis setting. This report strongly implied an association between the location of hazardous waste landfills and minority communities that echoed the seminal academic EJ study (Bullard 1983). Moreover, both the GAO and UCC studies were amplified by the 1992 book *Race and the Incidence of Environmental Hazards: A Time for Discourse*. One chapter revisited the UCC study (Lee 1992), and in combination with the GAO report, the first wave of EJ studies shown in the upper left of Figure 25.2 established the empirical standard for a whole series of social science studies in the 1990s. The spatial and statistical coincidence of minorities and hazardous waste sites became the focus of a decade-long debate among social scientists that still echoes today.

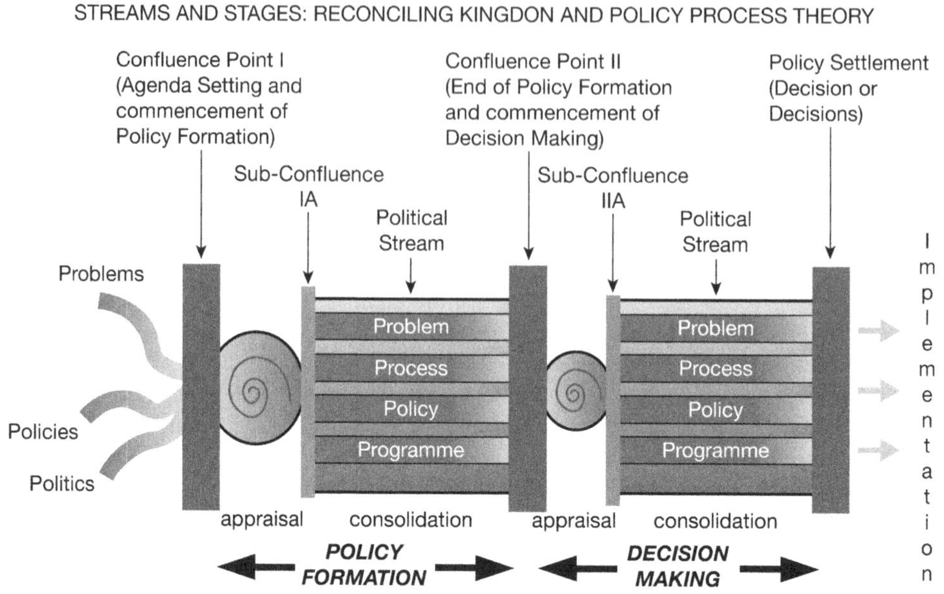

STREAMS AND STAGES: RECONCILING KINGDON AND POLICY PROCESS THEORY

Figure 25.1 Multiple streams and stages approach

Source: *European Journal of Political Research*, Volume 54, Issue 3, pages 419–434, 31 JUL 2014
DOI: 10.1111/1475–6765.12064 http://onlinelibrary.wiley.com/doi/10.1111/1475–6765.12064/
full#ejpr12064–fig–0004

Policy formation and consolidation: 1992–2000

During the 1990s, a second wave of toxic waste disparities research shown in the top left-half problem channel of Figure 25.2 informed and directed a nationalized and more consolidated stream of debate about the environmental injustice problem. Social science publications featured those who declared that minorities were targeted for the disposal of society's toxic and hazardous wastes (Bullard 1994; Goldman and Fitton 1994), studies that critiqued this research's methods, results, and prescriptions (e.g. Anderton et al. 1994), those who pointed to more factors than just discrimination (Been 1994), and those claiming the scholarly debate actually distracted poor and minority communities from more pressing issues (Foreman 1998). Prominent journals also featured academic debates (see Yandle and Burton 1996 and responses), multiple reviews (Brown 1995 among others), and research agendas (e.g. Pellow 2000; Pulido 1996).

Federal appraisals and subnational channels

Between 1992 and 1994, policy appraisals for the toxic and hazardous waste stream churned in the executive and legislative branches. This appraisal process can be turbulent as "political agents . . . consider the matter of it, and how to proceed, as well as issues such as whether their initial assumptions about the 'problem' remain valid" according to Howlett et al. (2015: 426). More studies helped sustain this process. Journalists from the *National Law Journal* reported in 1992 that average civil penalties for environmental law violations were lower in poor and minority neighbourhoods (Lavelle and Coyle 1992). In the same year, the EPA published a report concluding that existing research was "highly suggestive" that minority and low-income populations might experience greater exposures to toxic pollutants than the general population (EPA 1992).

In March of 1993, the Congressional Subcommittee on Civil and Constitutional Rights held two days of EJ hearings that included testimonies from academics, civil rights leaders, Native Americans, and activists (US Congress 1993). The EPA also collaborated with the Agency for Toxic Substances and Disease Registry (ATSDR) and the National Institutes of Environmental Health Sciences (NIEHS) on a workshop addressing "Equity in Environmental Health: Research Issues and Needs" (Sexton and Anderson 1993). Sub-confluence point IA would mark the end of this appraisal stage and, in our view, is best exemplified by Clinton's Executive Order shown in Figure 25.2.

However, the policy settlement prospects for EJ dimmed dramatically in 1994 when Republicans became the majority in Congress. More broadly, Congress gridlocked on all policies environmental after 1994 and led scholars to describe federal policy institutions as a green "labyrinth" that "has channelled tremendous political energies down other policymaking pathways" (Klyza and Sousa 2013: 10). Those other venues included legislative riders, the courts, collaborative processes, executive branch initiatives, and the states. The subnational channel appears in Figure 25.2 as one half of the policy stream. While an EJ programme office solidified at the EPA (the programme stream), the states saw a surge of policy development beyond a gridlocked federal government.

In the mid-1990s, the National Council of State Legislatures (NCSL 1995) identified three levels of state EJ responses and policy intensity: 1) information gathering and provision; 2) prevention through planning and stakeholder involvement; and 3) mitigation with increased environmental enforcement and anti-concentration restrictions on toxic and hazardous waste sites. The latter is seen in the first half of the policy stream in Figure 25.2. For instance, Arkansas achieved the first successful legislative EJ effort one year before Clinton's executive order. In 1993, the state's legislature passed the Environmental Equity Act with the goal of distributing solid waste management sites equitably. This statute required community involvement and prohibited the construction of high impact solid waste management facilities within 12 miles of any existing facility (ABA 2004). Similar policies were adopted in Alabama, California, Georgia, Maryland, and Mississippi. Such substantive and redistributive siting regulations in states would become the exception instead of the rule, however.

"Only six states moved the scales of justice with substantive policies" between 1994 and 2004 according to Abel et al. (2015a: 219). These researchers would also find that state EJ policies were driven primarily by a state's political stream. However, this supportive channel in the 1990s would quickly shift during the first decade of the 2000s. As quickly as the state policy window for EJ was opened by EO 12898, partisan realignments across the states began sliding their windows closed. In fact, twelve states took no action to address EJ (ABA 2004). Nonetheless, the Clinton administration provided an overarching and consolidating political current for EJ's problem, process, policy, and programme streams at the federal level.

A third wave of riskscapes research and community collaborations

A significant body of EJ research after Bush's election also took a subnational turn in a third wave of empirical studies seen in the top right-half of the problem channel. While arguments over scale and temporal dynamics dominated the second wave of toxic and hazardous waste disparity research, new studies moved beyond these debates that sometimes were characterized as trivial. For instance, one researcher argued that because "an environmental justice study done at only one scale or based on one areal unit cannot, by definition, produce a reliable indication of environmental justice or injustice, because one can never tell how the analytical results were affected by the nature of the data aggregation" (Mennis 2002: 285).

POLICY FORMATION

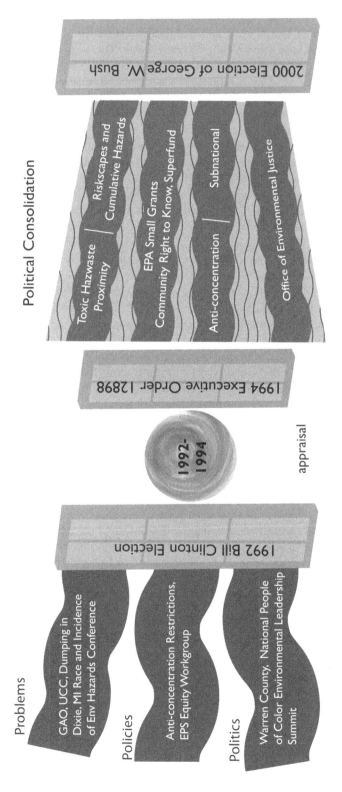

Figure 25.2 Policy formation

The development of Geographic Information Systems (GIS) software also enabled third-wave EJ researchers to apply distance-based methods in regional and local case studies on the one hand (Maantay 2002). On the other hand, the focus on the proximity of poor and minority communities to one kind of toxic or hazard also became recognized as too narrow. Thus, EJ researchers in the mid to late 1990s began expanding toxics disparity research beyond just hazardous waste landfills.

Some of the most important work emerged out of a group studying toxic disparities in Los Angeles (LA), California. In a series of seminal publications from 1996 to 2005, LA became the epicentre of a riskscape focus shown in the top first-half of the problem stream in Figure 25.2. It was defined as the overlapping toxic and hazardous waste pollution created by point and mobile sources posing cumulative and unequal risks for socially vulnerable communities (Morello-Frosch et al. 2001: 572). Like the second wave, cross-sectional analyses (Boer et al. 1997) were juxtaposed with longitudinal studies (e.g. Pastor et al. 2001). However, these studies were not out to demonstrate that one method was superior to another. Instead, scholars in this emerging "LA School" of EJ research demonstrated that progress towards understanding the political economy of toxic and hazardous waste injustices depended on research combining designs, methods, and hazards (Morello-Frosch et al. 2002).

In one of the ground-breaking LA studies, school locations were the centre of the analysis instead of one toxic or environmental hazard. Researchers then examined the surrounding spatial patterns of land use, industrial facilities emitting toxic air emissions, and indices of cancer and noncancer risks. Minority students and Latinos in particular were found to attend schools near hazardous facilities and face higher health risks from air pollution exposures (Pastor et al. 2002). This study's design exemplified what Corburn (2002) would describe as an important alternative to the conventional risk assessment of a single hazard. Such an exposure or cumulative risk assessment instead puts the community at the centre of the analysis and then considers exposures from multiple hazards (see Chapter 22). The "LA School" also led the field in the transformation of EJ research from emphasizing expert and technical processes to community-based and collaborative strategies (Petersen et al. 2006).

EJ activists and a small but growing number of researchers had been calling for just such a democratization of risk assessment. Traceable in part to the emergence of "popular epidemiology" around one of the first Superfund sites in Woburn, Massachusetts, Community-Based Participatory Research (CBPR) emphasizes the participation and influence of non-experts in the process of creating knowledge (see Chapter 23). Embraced by the "LA School" and others in the early 2000s, CBPR emerged as a significant strategy in the EJ policy development process. However, EJ's national political stream began to shift again after the Clinton administration.

Decision making diffused, then reformed: 2000–2010

George W. Bush's 2000 election marked EJ's second significant window for the problem, process policy, and programme streams of toxic and hazardous waste politics. However, instead of progressing towards policy settlement as suggested by the refined MSA, EJ's multiple streams began to drift apart. For example, at the end of Bush's first term, the EPA's Office of Inspector General (OIG) concluded that the agency had not established a clear plan to integrate EJ strategy, goals, or performance benchmarks (USEPA OIG 2004: 16). Two years later, the OIG would again chastise the Bush administration and call for a review of EJ activities and corrective actions (USEPA OIG 2006). The problem was reframed and the prospects of policy settlement faded in the third decade of EJ policy development.

POLICY DRIFT

Political Deconsolidation

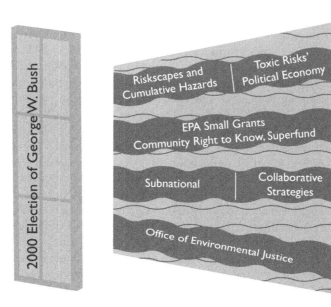

Figure 25.3 Policy drift

Instead of a problem-focused decision making where actors seek solutions and a strategic course of action, a qualitatively different kind of agenda management emerged. Described by policy scholars in earlier work, the diffusion of decision making involves "helping keep a difficult issue down or off a crowded and politicised policy agenda, even at the expense of tackling the problem itself" (Howlett et al. 2013: 18). Early in the first Bush term, for instance, the newly appointed EPA administrator reframed the objective of EJ as a "goal to be achieved for all communities and persons" (Whitman 2001: 1). The EPA's Inspector General documented how this de-emphasized the focus on minority and low-income communities. This was a particularly ironic move given that the title of EO 12898 was "Federal Actions to Address Environmental Justice in Minority Populations and Low-Income Populations" (Clinton 1994). Put differently, redefining the goal as EJ for everyone diffused the decision-making agenda so broadly that it returned EPA to "pre-Executive Order status" (USEPA OIG 2004: 11; also see Holifield 2012). The problem, process, policy and programme streams began to drift apart (see Figure 25.3).

As EJ decision making diffused at the federal level, there were subnational convergences especially in certain locales where community-based and participatory environmental processes were developing independently. These kinds of processes and CBPR in particular harnessed several new currents in environmental policy and politics. These included the democratization of risk assessment (e.g. Stern and Fineberg 1996), participatory and de-centralized environmental policies (e.g. John 1994), and the support of Community Advisory Groups (CAGs) and Technical Assistance Grants (TAGs) in the Superfund programme.

A short case study of Seattle, Washington demonstrates how all three converged for one locale after 2000.

Superfund processes and collaborative programmes

In 2001, the EPA added a five mile stretch of Seattle's Duwamish River to the Superfund's National Priorities List (NPL). A century of heavy industrial use near the river had resulted in the contamination of its sediments with arsenic, dioxins, PCBs, and more than a dozen other toxic and hazardous wastes (USEPA 2015a). Before the river flows into Seattle's Port and the picturesque Elliott Bay, it passes through the Georgetown and South Park neighbourhoods. Fifty per cent of the residents here are non-white and 32 per cent of the households earn less than the poverty level (US Census 2010). These Duwamish Valley communities became the beneficiary of a significant Superfund programme during the political deconsolidation era of EJ: collaborative and community-based strategies.

Sometimes referred to as uncontrolled hazardous waste sites, the cleanup of toxic dumping areas like the Duwamish River became the focus of federal environmental policy in 1980. That year, Congress enacted the Comprehensive Environmental Response, Compensation, and Liability Act (CERCLA). Revised in 1986 with the Superfund Amendments and Reauthorization Act (SARA), Congress expanded the opportunities of citizen participation in cleanup decisions. One mechanism involved administrative support for Community Advisory Groups (CAGs) and a second effort supporting Technical Assistance Grants (TAGs) sought to increase community involvement. TAGs are awarded to eligible community groups to hire their own, independent technical advisors and have been associated with higher cleanup levels (Daley 2007). Communities near Superfund sites are eligible to receive up to $50,000 to get technical assistance and in 2001, a TAG award helped establish the Duwamish River Cleanup Coalition (DRCC).

Representing half a dozen environmental groups and the neighbourhood associations of Georgetown and South Park, the Coalition's mission is to "ensure a cleanup of the Duwamish River that is accepted by and benefits the community and protects fish, wildlife, and human health" (DRCC 2016). In 2004, at one of the worst spots along the Duwamish River's Superfund stretch, a coupling opportunity emerged for the EJ problem, process, policy, and programme streams. A local journalist reported that "a thick, hardened flow of oily asphalt still oozes to the river like lava from a Hawaiian volcano" (Ith 2004: 22). This occurred at the former location of Malarkey Asphalt, where the company had left behind a legacy of toxic and hazardous wastes. Now owned by the Port of Seattle, this location, epitomizing an uncontrolled hazardous waste site, had PCB concentrations up to 4,000 parts per million (PPM) according to one assessment (USEPA and WADOE 2002).

Relabelled Terminal 117 and zoned commercial across the river from the South Park neighbourhood, a second phase of the site remediation by the EPA and the Port aimed for PCB levels allowable for industrial site cleanup levels of 10 PPM at the surface, and 25 PPM below two feet. But with the TAG funded staff of DRCC, Duwamish Valley residents responded in a claim-making (Walker 2012) counter move as the neighbourhood fought for more cleanup and 1 PPM of PCBs; the cleanup standard for residential areas (McClure 2006). One month later, Commissioners for the Port acquiesced and amended the cleanup plan to the more stringent level of 1 PPM (Scott 2006).

DRCC's counterclaim-making has led the Duwamish Valley residents to first reframe, and then literally reclaim a polluted industrial site for a neighbourhood. This victory of the South Park neighbourhood was lauded in a nationally broadcast documentary (Smith and

Young 2009), while others used the Seattle case to demonstrate the emerging movement to redemocratize cities (Purcell 2008). DRCC's success is also consistent with other applications of the MSA, suggesting policy entrepreneurs can be more effective at local scales of government (e.g. Dudley 2013). Conversely, recent political advocacy and more problem research has yet to achieve anything like the policy focus and prominence of the 1990s.

Policies that fail and programmes that survive: 2010–2016

Several scholars have documented EJ's struggle to reach the policy settlement stage. An edited book was titled *Failed Promises* (Konisky 2015) while another report blamed the EPA for the persistence of environmental racism (Lombardi et al. 2015). These all echo one assessment over a decade ago describing how EJ policy's "ambitious goals and initiatives have beached routinely on the shoals of a formidable mix of scientific, political, organizational, and financial obstacles" (Ringquist 2004: 257). While some offer a glimmer of hope for EJ's future (Konisky 2016), the new president's administration is another critical junction for the national political and policy response to the stream of toxic and hazardous waste disparity research.

For example, local EJ efforts in Seattle and elsewhere have been re-energized since 2008 by a supportive political coalition and rejuvenated process, policy and programme streams in the executive branch (see Figure 25.4). Obama's EPA Administrator-designate opened her tenure with the following. "We must take special pains to connect with those who have

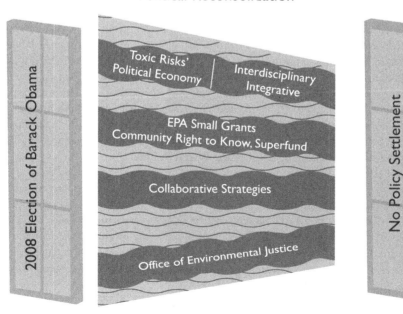

Figure 25.4 Policy reformation

been historically underrepresented in EPA decision-making" (Jackson 2009). A year later, the same administrator identified EJ as one of seven agency priorities (Jackson 2010). Late in each of Obama's two terms, the EPA released *Plan EJ 2014* (USEPA 2011) and *Plan EJ 2020* (USEPA 2016a) that aimed to help the agency integrate EJ into its activities, policies, and programmes. In between, the EPA developed its *Policy on EJ for Working with Federally Recognized Tribes and Indigenous Peoples*; an EJ screening and mapping tool (www.epa.gov/ ejscreen); and two guidance documents for regulatory actions and analysis (USEPA 2015b, 2016b). The EPA also funded 289 community groups with small EJ grants between 2008 and 2016. In short, a supportive political stream reconsolidates the process, policy, and programme channels maintaining EJ's national policy place for now while research in the problem stream continues to evolve.

A fourth wave in the EJ problem stream

The most recent wave of research on toxic air pollution disparities provides a number of new methods to address many of the same old and unsettled questions about proximity to toxic and hazardous waste sites. Several authors have employed interdisciplinary and integrative approaches to more carefully examine causality in the political economy of toxic risks (see Figure 25.4). For example, Crowder and Downey (2010) used multilevel methods that integrate individual and neighbourhood level data with a temporal research design to better make sense of disparities in toxic pollution proximity across different racial and ethnic groups. The authors find evidence of both existing disparities and mobility patterns reinforcing environmental injustices. Likewise, another national multi-level study of residential mobility between 1991 and 2007 found that the odds of starting with low pollution exposure but ending with high levels of exposure were 38 per cent higher for immobile black households than for immobile whites (Pais et al. 2014).

Other researchers employing multiscalar analysis found multiple drivers of inequalities by integrating quantitative data with qualitative methods that can tease out temporal patterns and causality across scales. For example, Hernandez et al. (2015) interviewed US and immigrant Hispanics living in census tracts with high and low risks of exposure to hazardous air pollutants (HAPs) in Houston, Texas. The results revealed that proximity to industrial sources of HAPs was a function of not only economic constraints, but combined with a complex picture of micro-scale factors such as individual aspirations for homeownership and sociocultural security. Such multiscalar methods integrating quantitative and qualitative data have rarely been used by EJ scholars. Mixed-method research designs allow scholars to push beyond the limits of just a structural model of EJ or the presumptions of behavioural individualism.

Like the case study of Baltimore earlier in this volume (see Chapter 18), the combination of outcome and process methodologies is critical to another new direction in EJ research. A small but growing body of integrative research is focusing on environmental gentrification. It occurs when "the bulk of the benefits of a policy that successfully cleans up dirtier neighbourhoods where the poor live may actually be captured by rich households" (Banzhaf and Walsh 2006: 25). In other words, urban polluted neighbourhoods that are cleaned up become attractive to upper-class residents who displace lower-income and non-white households (Checker 2011; Gould and Lewis 2017). Instead of helping achieve EJ, new mixed-methods research reveals how cleanup processes can ironically lead to environmentally unjust outcomes (Abel et al. 2015b; Pearsall 2010).

Policy streams without law: more political uncertainty ahead for EJ

Emboldened by these research advances in the problem stream, the new presidential administration could continue the reconsolidation of EJ towards policy settlement. However, it may instead begin another phase of political deconsolidation that would weaken a programme stream that has provided three decades of support for community-based EJ. On multiple occasions, Seattle's environmental groups, including the DRCC, have received infusions of small grant support from the EPA's EJ programme office. Since 1994, according to one study (Abel and Stephan 2008), the distribution of these kinds of small grants has been EPA's most consistent EJ effort. To date, over 1500 small EJ grants have been awarded with a 1995 peak of 175. Small grants declined to 95 at the end of the Clinton administration and then bottomed-out to a low of ten in both 2006 and 2007 (USEPA 2105c). While there was a rebound during the Obama administration (rising as high as 76 grants in 2010), the annual programme is now bi-annual and it has been described as inconsistent (Daley and Reames 2015) and unfocused (Vajjhala 2010).

These small grants and other collaborative programmes at the EPA aiming to alleviate toxic and hazardous waste disparities are also vulnerable without a statutory foundation. Sousa and Klyza (2007) observed that the broader collaborative turn in national environmental policy was a reconstruction of what Lowi (1979) had decried as "policy without law" in the New Deal era. Unlike the statutory base found in the Clean Air and Water Acts, federal EJ policy could diffuse again or even be rescinded by a future president. Much uncertainty lies ahead for EJ's policy and political fortunes.

Our application of the refined multiple streams approach (MSA) highlights this uncertainty, and the historical developments of both policy responses to, and definitions of the EJ problem. But like the streams of toxic and hazardous waste problems, processes, policies, and programmes, the MSA is at a critical stage of its development as a theory. Policy scholars recently called for MSA research to consider more than agenda setting. New lines of EJ research could benefit from "extending the scope of the MSA to include implementation and perhaps termination, as well as carefully distinguishing between agenda setting and decision making" (Weible and Schlager 2016: 9). Our treatment here responds accordingly and explicates the multiple policy windows EJ has passed through, its turbulent phases between convergent and divergent decision making, and how the shifting political stream first consolidated, then pulled apart, and then reconsolidated the problem, process, policy, and programme channels between 1994 and 2016.

Moreover, the refined MSA framework clearly illuminates the possible termination of, and uncertain future faced by, federal and state EJ policy and programmes. Conversely, added elements of the refined MSA allowed us to identify the importance of a consistent programme stream originating from executive branch offices supported by processes from other policy areas such as Superfund. After nearly 30 years, community-based collaborations confronting toxic and hazardous waste disparities remain the only constant venue of significant EJ activity. These are supported through programmes from EPA's OEJ and its regions but no substantive legislative breakthroughs are on the horizon. There also is far less political interest in the ever-expanding problem stream. New social science research continues to fill this stream with characterizations of America's skewed political economy and how it contributes to toxic and hazardous waste disparities.

References

Abel, T.D., Salazar, D.J. and Robert, P. (2015a) "States of environmental justice: Redistributive politics across the United States, 1993–2004." *Review of Policy Research*, vol. 32, pp. 200–225.

Abel, T.D. and Stephan, M. (2008). "Tools of environmental justice and meaningful involvement." *Environmental Practice*, vol. 10, pp. 152–163.

Abel, T.D., White, J., and Clauson, S. (2015b) "Risky Business: Sustainability and Industrial Land Use across Seattle's Gentrifying Riskscape." *Sustainability*, vol. 7, pp. 15718–15753.

American Bar Association (ABA). (2004). *Environmental Justice for All: A Fifty-State Survey of Legislation, Policies, and Initiatives.* Berkeley, CA: Hastings College of Law, University of California.

Anderton, D., Anderson, A., Oakes, J. and Fraser, M. (1994). "Environmental equity: The demographics of dumping." *Demography*, vol. 31, pp. 229–248.

Banzhaf, H.S. and Walsh, R.P. (2006). *Do People Vote with Their Feet? An Empirical Test of Environmental Gentrification.* Washington, DC: Resources for the Future.

Been, V. (1994). "Locally undesirable land uses in minority neighbourhoods: Disproportionate siting or market dynamics?" *Yale Law Journal*, vol. 103, pp. 1383–1422.

Boer, J.T., Pastor, M., Sadd, J.L. and Snyder, L.D. (1997). "Is there environmental racism? The demographics of hazardous waste in Los Angeles County." *Social Science Quarterly*, vol. 78, pp. 793–810.

Brown, P. (1995). "Race, class, and environmental health: A review and systematization of the literature." *Environmental Research*, vol. 69, pp. 15–30.

Bullard, R.D. (1983). "Solid waste sites and the black Houston community." *Sociological Inquiry*, vol. 53, pp. 273–288.

Bullard, R.D. (1990). *Dumping in Dixie: Race, Class, and Environmental Quality.* 2nd edition. Boulder: Westview Press.

Bullard, R.D. (1994). "Overcoming racism in environmental decision-making." *Environment: Science and Policy for Sustainable Development*, vol. 36, pp. 10–44.

Bullard, R.D., Mohai, P., Saha, R., and Wright, B. (2007). *Toxic Wastes and Race at Twenty: A Report Prepared for the United Church of Christ Justice & Witness Ministries.* Cleveland, Ohio: United Church of Christ.

Checker, M. (2011). "Wiped out by the 'Greenwave': Environmental gentrification and the paradoxical politics of urban sustainability." *City & Society*, vol. 23, pp. 210–229.

Clinton, W. (1994). "Executive Order 12898: Federal actions to address environmental justice in minority populations and low-income populations, February 11, 1994." *Weekly Compilation of Presidential Documents*, vol. 30, pp. 276–279.

Corburn, J. (2002). "Combining community-based research and local knowledge to confront asthma and subsistence-fishing hazards in Greenpoint/Williamsburg, Brooklyn, New York." *Environmental Health Perspectives*, vol. 110, pp. 241–248.

Crowder, K. and Downey, L. (2010). "Inter-neighborhood migration, race, and environmental hazards: Modeling micro-level processes of environmental inequality." *American Journal of Sociology*, vol. 115, pp. 1110–1149.

Daley, D.M. (2007). "Citizen groups and scientific decisionmaking: Does public participation influence environmental outcomes?" *Journal of Policy Analysis and Management*, vol. 26, pp. 349–368.

Daley, D.M. and Reames, T.G. (2015). "Public participation and environmental justice: Access to federal decision making." In Konisky, D.N. (ed.), *Failed Promises: Evaluating the Federal Government's Response to Environmental Justice*, Cambridge, MA: MIT Press, pp. 143–171.

Dudley, G. (2013). "Why do ideas succeed and fail over time? The role of narratives in policy windows and the case of the London congestion charge." *Journal of European Public Policy*, vol. 20, pp. 1139–1156.

Duwamish River Cleanup Coalition (DRCC). "About." [Online], Available: http://duwamish cleanup.org/about/ [22 November 2015].

Foreman, C.H. (1998). *The Promise and Peril of Environmental Justice.* Washington, DC: Brookings Institution.

Goldman, B.A. and Fitton, L. (1994). *Toxic Wastes and Race Revisited: An Update of the 1987 Report on the Racial and Socioeconomic Characteristics of Communities with Hazardous Waste Sites.* Washington, DC: Center for Policy Alternatives.

Gould, K.A. and Lewis, T.L. (2017). *Green Gentrification: Urban Sustainability and the Struggle for Environmental Justice.* Abingdon, Oxon and New York: Routledge.

Hernandez, M., Collins, T.W. and Grineski, S.E. (2015). "Immigration, mobility, and environmental injustice: A comparative study of Hispanic people's residential decision-making and exposure to hazardous air pollutants in Greater Houston, Texas." *Geoforum*, vol. 60, pp. 83–94.

Holifield, R. (2012). "The elusive environmental justice area: Three waves of policy in the US Environmental Protection Agency." *Environmental Justice*, vol. 5, pp. 293–297.

Howlett, M.P., McConnell, A. and Perl, A.D. (2013). "Reconciling Streams and Stages: Avoiding Mixed Metaphors in the Characterization of Policy Processes." *Annual Meeting Paper; American Political Science Association 2013 Annual Meeting*, [Online], Available: https://ssrn.com/abstract =2300442 [23 Oct 2016].

Howlett, M.P., McConnell, A. and Perl, A.D. (2015). "Streams and stages: Reconciling Kingdon and policy process theory." *European Journal of Political Research*, vol. 54, pp. 419–434.

Ith, I. (2004). "The road back from Seattle's Superfund sewer to haven once more." *The Seattle Times* (3 Oct 2004), p. 22.

Jackson, L.P. (2009). *Opening Memo to EPA Employees*, [Online], 23 Jan., Available: http:// blog.epa. gov/administrator/2009/01/26/opening-memo-to-epa-employees/ [15 Oct 2016].

Jackson, L.P. (2010). *Memorandum from Lisa P. Jackson, Administrator to All EPA Employees*, [Online], Available from: http://yosemite.epa.gov/opa/admpress.nsf/0/bb39e443097b5df585257 6a9006a5a86?OpenDocument [23 Feb 2016].

John, D. (1994). "Civic environmentalism." *Issues in Science and Technology*, vol. 10, pp. 30–34.

Kingdon, J.K. (1984). *Agendas, Alternatives, and Public Policies*. Boston: Little, Brown.

Kisch, R.J. (1994). "Putting environmental racism on the national agenda? A review of race and the incidence of environmental hazards: A time for discourse." *Environmental Law*, vol. 24, pp. 1171–1183.

Klyza, C.M. and Sousa, D.J. (2013). *American Environmental Policy: Beyond Gridlock*. 2nd edition. Cambridge, Massachusetts: MIT Press.

Konisky, D.M. (2015). *Failed Promises: Evaluating the Federal Government's Response to Environmental Justice*. Cambridge, MA: MIT Press.

Konisky, D.M. (2016). "Environmental justice delayed: Failed promises, hope for the future." *Environment: Science and Policy for Sustainable Development*, vol. 58, pp. 4–15.

Lavelle, M. and Coyle, M. (1992). "The racial divide in environmental law: Unequal protection." *National Law Journal*, vol. 15, pp. S1–S12.

Lee, C. (1992). "Toxic waste and race in the United States." In Bryant, B.I. and Mohai, P. (eds.), *Race and the Incidence of Environmental Hazards: A Time for Discourse*. Boulder, CO: Westview Press, pp. 28–54.

Lester, J., Allen, D. and Hill, K.M. (2001). *Environmental Injustice in the United States: Myths and Realities*. Boulder, CO: Westview Press.

Lombardi, K., Buford, T. and Greene, R. (2015). *Environmental Racism Persists, and the EPA is One Reason Why*, [Online], Available www. publicintegrity.org/2015/08/03/17668/environmental-racism-persists-and-epa-one-reason-why [3 March 2016].

Lowi, T.J. (1979). *The End of Liberalism: The Second Republic of the United States*. New York, Norton.

Maantay, J. (2002). "Mapping environmental injustices: Pitfalls and potential of geographic information systems in assessing environmental health and equity." *Environmental Health Perspectives*, vol. 110, pp. 161–171.

McClure, R. (2006). "'Malarkey' cleanup pits port, activists: Extent of pollution removal from site is at issue." *Seattle PI*, 24 May 2006, [Online], Available: www.seattlepi.com/local/article /Malarkey-cleanup-pits-port-activists-1204401.php [12 Oct 2016].

Mennis, J. (2002). "Using geographic information systems to create and analyze statistical surfaces of population and risk for environmental justice analysis." *Social Science Quarterly*, vol. 83, pp. 281–297.

Mohai, P. and Saha, R. (2015). "Which came first, people or pollution? Assessing the disparate siting and post-siting demographic change hypotheses of environmental injustice." *Environmental Research Letters*, vol. 10, p. 115008.

Morello-Frosch, R., Pastor, M. and Sadd, J. (2001). "Environmental justice and southern California's 'Riskscape': The distribution of air toxics exposures and health risks among diverse communities." *Urban Affairs Review*, vol. 36, pp. 551–578.

Morello-Frosch, R., Pastor, M., Porras, C. and Sadd, J. (2002). "Environmental justice and regional inequality in southern California: Implications for future research." *Environmental Health Perspectives*, vol. 110, pp. 149–154.

National Conference of State Legislatures (NCSL). (1995). *Environmental Justice: A Matter of Perspective*. Denver, Colorado: National Conference of State Legislatures.

Pais, J., Crowder, K. and Downey, L. (2014). "Unequal trajectories: Racial and class differences in residential exposure to industrial hazard." *Social Forces*, vol. 92, pp. 1189–1215.

Pastor, M., Sadd, J. and Hipp, J. (2001). "Which came first? Toxic facilities, minority move-in, and environmental justice." *Journal of Urban Affairs*, vol. 23, pp. 1–21.

Pastor, J., Sadd, J.L. and Morello-Frosch, R. (2002). "Who's minding the kids? Pollution, public schools, and environmental justice in Los Angeles." *Social Science Quarterly*, vol. 83, pp. 263–280.

Pearsall, H. (2010). "From brown to green? Assessing social vulnerability to environmental gentrification in New York City." *Environment and Planning C: Government and Policy*, vol. 28, pp. 872–886.

Pellow, D.N. (2000). "Environmental inequality formation toward a theory of environmental injustice." *American Behavioral Scientist*, vol. 43, pp. 581–601.

Petersen, D., Minkler, M., Vásquez, V.B. and Baden, A.C. (2006). "Community-based participatory research as a tool for policy change: A case study of the Southern California Environmental Justice Collaborative." *Review of Policy Research*, vol. 23, pp. 339–354.

Pulido, L. (1996). "A critical review of the methodology of environmental racism research." *Antipode*, vol. 28, pp. 142–59.

Purcell, M. (2008). *Recapturing Democracy: Neoliberalization and the Struggle for Alternative Urban Futures.* New York: Routledge.

Ringquist, E.J. (2004). "Environmental justice." In Durant, R.F., Fiorino, D.J. and O'Leary, R. (eds.), *Environmental Governance Reconsidered: Challenges, Choices, and Opportunities*. Cambridge, MA: MIT Press, pp. 255–288.

Scott, A. (2006). "Port OKs $6 million to start cleanup of South Park site; Old asphalt plant; Land tainted with PCBs; EPA wants project to begin before fall." *The Seattle Times*, 28 June, p. E1.

Sexton, K and Anderson, Y.B. (1993). "Equity in environmental health: Research issues and needs." *Toxicology and Industrial Health*, vol. 9, pp. 679–975.

Sexton, K. and Zimmerman, R. (1999). "The emerging role of environmental justice in decision making." In Sexton, K., Marcus, A.A., Easter K.W. and Burkhardt, T. (eds.), *Better Environmental Decisions: Strategies for Governments, Businesses, and Communities*. Washington, DC: Island Press, pp. 419–444.

Smith, H. and Young, R. (2009). *Poisoned Waters,* WGBH Educational Foundation and Public Broadcasting Service. [Online], Available: www.pbs.org/wgbh/pages/frontline/ poisonedwaters/ [15 Oct 2016].

Sousa, D. and Klyza, C.M. (2007). "New directions in environmental policy making: An emerging collaborative regime or reinventing interest group liberalism?" *Natural Resources Journal*, vol. 47, pp. 377–444.

Stern, P.C. and Fineberg, H.V. (1996). *Understanding Risk: Informing Decisions in a Democratic Society*. Washington, DC: National Academies Press.

Szasz, A. and Meuser, M. (1997). "Environmental inequalities: Literature review and proposals for new directions in research and theory." *Current Sociology*, vol. 45, pp. 99–120.

United Church of Christ. Commission (UCC) for Racial Justice (1987). *Toxic Wastes and Race in the United States: A National Report on the Racial and Socio-economic Characteristics of Communities with Hazardous Waste Sites*. New York, N.Y.

US Census. (2010), *American Fact Finder*, [Online], Available: http://factfinder.census.gov [28 Jul 2014].

US Congress. House. Committee on the Judiciary. (1993). *Environmental Justice*. 103rd Congress, 1st session, Serial No. 64, 3 and 4 March. Washington, DC: Government Printing Office, 1994.

US Environmental Protection Agency (USEPA). (1990). *Reducing Risk: Setting Priorities and Strategies for Environmental Protection*. Washington, DC: US EPA.

USEPA. (1992). *Environmental Equity: Reducing Risk for All Communities*. Washington, DC: US EPA: USGPO.

USEPA. (1994). *Environmental Justice Initiatives 1993*. Washington, DC: US EPA.

USEPA. (2011). *Plan EJ 2014*. Washington, DC: US EPA.

USEPA. (2015a). *Environmental Justice Analysis for the Lower Duwamish Waterway Cleanup*. Available from: www3.epa.gov/region10/pdf/sites/ldw/pp/ej_analysis _ldw_feb_2013.pdf [Accessed 13 February 2016].

USEPA. (2015b). *Guidance on Considering Environmental Justice During the Development of Regulatory Actions*. Washington, DC: US EPA.

USEPA. (2015c). *Environmental Justice Small Grants Program*. Available from: www3.epa.gov/ environmentaljustice/grants/ej-smgrants.html [Accessed 22 February 2016].

USEPA. (2016a). *Plan EJ 2020*. Washington, DC: US EPA.

USEPA. (2016b). *Technical Guidance for Assessing Environmental Justice in Regulatory Analysis*. Washington, DC: US EPA.

USEPA Office of the Inspector General (OIG). (2004). *EPA Needs to Consistently Implement the Intent of the Executive Order on Environmental Justice*. Report No. 2004-P-00007. Washington, DC: EPA Inspector General.

USEPA OIG. (2006). *EPA Needs to Conduct Environmental Justice Reviews of Its Programs, Policies, and Activities*. Report No. 2006-P-00034. Washington, DC: EPA Inspector General.

USEPA and Washington State Department of Ecology (WADOE). (2002). *Lower Duwamish Waterway Site Community Involvement Plan*. Seattle, WA: US EPA.

US General Accounting Office (USGAO) (1983). *Siting of Hazardous Waste Landfills and Their Correlation with Racial and Economic Status of Surrounding Communities: Report*. Washington, DC: GAO.

Vajjhala, S. (2010). "Building community capacity? Mapping the scope and impacts of EPA's environmental justice small grants program." In Taylor, D.E. (ed.), *Environment and Social Justice: An International Perspective*, Bingley, UK: Emerald, pp. 353–384.

Walker, G. (2012). *Environmental Justice: Concepts, Evidence and Politics*. New York, NY: Routledge.

Weible, C.M. and Schlager, E. (2016). "The multiple streams approach at the theoretical and empirical crossroads: An introduction to a special issue." *Policy Studies Journal*, vol. 44, pp. 5–12.

Whitman, C.T. (2001). *EPA's Commitment to Environmental Justice*. Memorandum [Online], Available: https://yosemite.epa.gov/opa/admpress.nsf/89745a330d4ef8b9852572a000651fe1/41a2df9798d627a185256aaf0067e435 [15 Oct 2016].

Yandle, T. and Burton, D. (1996). "Reexamining environmental justice: A statistical analysis of historical hazardous waste landfill siting patterns in metropolitan Texas." *Social Science Quarterly*, vol. 77, pp. 477–492.

Zahariadis, N. and Exadaktylos, T. (2016). "Policies that succeed and programs that fail: Ambiguity, conflict, and crisis in Greek higher education." *Policy Studies Journal*, vol. 44, pp. 59–82.

26

AIR POLLUTION AND RESPIRATORY HEALTH

Does better evidence lead to policy paralysis?

Michael Buzzelli

Introduction

This chapter provides a conceptual and empirical overview of air pollution and respiratory health as seen through the lens of environmental justice (EJ). This critical lens could be applied to many contexts in developing and developed countries and many circumstances as well. The focus in this chapter is on the experience of environmental/ambient air pollution and human health in developed countries where there is ample EJ research built up over the last two decades. This chapter begins with a brief discussion of basic concepts: types of pollutants, human exposure and health impacts, hazards, risks and the epidemiologist's 'dose–response' relationship between air pollution exposure and human health outcomes.

While evidence of environmental air pollution and health research is growing more certain all the time, it is now confronted and informed by the EJ movement. The subsequent section therefore concerns itself with the ways in which EJ alters our understanding of relationships between air pollution exposures and health. EJ has not only supplied myriad answers, as we shall see, it has also influenced the kinds of fundamental questions we ask about air pollution exposures and human health impacts.

We conclude this chapter with a discussion of the seemingly intractable policy problems that arise from the growing EJ literature on air pollution and health. While we might otherwise (traditionally) seek to reduce air pollution emissions and exposures for the general population, EJ suggests that we ought to also (or perhaps first) redress disparities between those most and least affected by air pollution. Paradoxically, as the research record mounts with better tools and more refined insights, the nature and extent of the policy challenge appears to grow more complex.

Air pollution and health risks: a primer

When we speak of air pollution and human health, it is important to understand the factors that influence this relationship. The concepts presented here guide our understanding and help build better evidence and, ideally, better policy. This section contains a number of important concepts that are given only brief treatment (e.g. ground level exposure, particle size and respiration) so that the importance and impact of the EJ lens can be put into perspective.

When discussing air pollution and human health, toxicants or pollutants are an obvious starting point. In general air pollution health research often relies upon the air quality management planning of various levels of government, particularly the data captured by fixed-site monitoring stations (Mennell and Bhattacharyya 2002). Depending on the type of station/ network, community priorities and available resources, monitoring measures (daily, continuously) a range of pollutants such as carbon monoxide (CO), oxides of nitrogen (NO_x), sulphur dioxide (SO_2), ozone (O_3), and particulates of various sizes such as inhalable PM_{10} (10 microns in diameter) or fine $PM_{2.5}$. In my community of London, Canada, for example, there exists one monitoring station in the east end of the city located on international airport lands. This arrangement is fairly typical for a medium-sized city of about 400 000: the monitor tracks hourly concentrations of O_3, $PM_{2.5}$ and NO_x. We highlight the air pollution monitoring used in studies discussed below as well. Several international and national agencies, such as the World Health Organization (www.who.int/topics/air_pollution/en/), the US Environmental Protection Agency (www3.epa.gov/airquality/urbanair/) and Health Canada (www.hc-sc.gc.ca/ewh-semt/air/index-eng.php), publish good descriptions of these and other pollutants that are known to affect human health.

Given our interest in a range of air pollutants, what we focus on in health research is the so-called dose–response relationship between pollutants and health effects. While we can think of air pollution sources, emissions and dispersion, our primary interest is in ambient or environmental concentrations at ground level because this is where human exposures occur and may represent a health hazard. We say 'may' because, until we know with reasonable certainty how much we are exposed to and the effect it has on our health, then strictly speaking it is a potential 'hazard'. When a causal link with human health is clearer, we speak of 'risk'. This gives us a reading of what epidemiologists term 'dose'. And once we have a handle on the exposure dosage, we then turn to the main question at hand: the health 'response'. Thus we may be interested in the dose–response relationships for individuals or groups (often target groups such as children or the elderly) over a given time period (immediately or over the life course). The sorts of impacts we typically see in this domain are acute/short-term events (such as asthma, often captured in administrative hospitalization data) and long-term outcomes such as various cancers.

Using a key example, we see the importance of the pollutants in question, their distribution/exposure and the kinds of associated health impacts. Publication of the American 'Reanalysis Project' (Krewski et al. 2000) was an important moment in air pollution health research. This large-scale epidemiologic meta-study helpfully provides a summary of the state of the art of air pollution health research to that time. Reanalysis of the data of the original studies (Harvard Six Cities and American Cancer Society) was aimed at quantifying the mortality risks of air pollution in cities across the United States. Air pollution data were drawn from the same kinds of monitoring stations described above for a large number of urban centres across the country. An important finding was confirmation of a dose–response relationship. For instance, an increase in the daily monitored ambient concentration of inhalable particulate matter (PM_{10}) of between 18 micrograms per cubic metre (ug/m^3) and 24.5ug/m^3 was associated with an elevated risk of daily all-cause mortality of between 1.17 and 1.26. Spread over hundreds of thousands of subjects, this elevated risk represents many additional daily deaths associated with variations in environmental air pollution.

Finally, for the reader to appreciate the complexity of air pollution health research, study design bears brief discussion. Most of our research in this domain is constrained to observational study designs (in which we make observations based on social reality) rather than experimental approaches (in which we apply treatments with use of controls, as we do in a laboratory).

Notwithstanding the nuances of each, air pollution health research is almost always observational because experimental methods are typically viewed as ethically unacceptable (i.e. applying a given pollution exposure to subjects to measure the health response in a given time-frame). In the Reanalysis Project, we have an observational time-series study in which short-term health effects were analysed amongst subjects in city-to-city comparisons (e.g. as many as 154 groups/cities and 552 138 subjects). Cross-sectional studies, by contrast, are easier to undertake and may be suggestive but are regarded as weakest of all observational designs because they are essentially correlative (i.e. not causal) for a single point in time. In general, whatever the design, weight must be placed on the accumulation of consistent evidence rather than the precision of any one study because this kind of research is observational rather than experimental.

Environmental justice: complicating the picture

The epidemiology of air pollution and human health is both complicated and well-established. Similarly, EJ research has also been with us for decades though its message – that individuals and groups of lower socioeconomic position may be at greater risk of pollution exposure and health impacts – has only recently begun to influence our understanding of the impact of air pollution on the population's health. That said, the influence is clear in several important respects.

EJ grows out of a number of movements as discussed elsewhere in this book including the civil rights/libertarian movements in the US. Within academia, political philosophy (Young 1990; Smith 1994) and cognate fields motivate an ethical EJ stance that says the distribution of environmental 'bads' is fundamentally unfair. Regrettable necessities, as Aldo Leopold would say, like air pollution are a necessary by-product of growth and development and yet only select segments of the population benefit from this progress (such as through a rising living standard). Moreover, pollution is often most generated by the same fortunate few. One strand of this reasoning to draw out explicitly is that the focus of EJ is much more on groups (such as racial minorities) rather than the traditional focus on individuals in air pollution epidemiology. Given that the philosophy of these arguments is wonderfully complex, it is no surprise that there is debate about what EJ should mean and how it ought to be applied in research and in social reality (Been 1993; Pulido 1996; Agyeman et al. 2009). So how has the EJ perspective been applied in research on air pollution and human health?

The answer to this question is split into two broad avenues: 1) a body of work within the EJ literature itself, concerned primarily with exposure to air pollution; 2) more mainstream epidemiologic and public health research that has 'heard' the message of EJ in research on the health impacts of exposure. Beginning with the former, we have seen growing evidence of disproportionate air pollution exposure amongst a range of community types and environments. These insights are made possible in part by developments in the 'new science of exposure analysis' (Ott 1995). To be sure, time series and cohort studies continue to supply the bulk of evidence on various measures of air pollution morbidity (e.g. asthma), mortality (e.g. cardiovascular) and related health care utilization (e.g. Pope and Dockery 2006). However, while still working within observational study designs, we now have a richer mix of methods that 'assign' air pollution to study subjects (individuals, communities, areas). In addition to approaches like the Reanalysis Project (Krewski et al. 2000) discussed above, we have also seen the rise of alternatives including use of monitoring networks where available, geostatistical techniques, various kinds of statistical modelling (dispersion, land use regression) and, more recently, mobile monitoring of pollutants (Özkaynak et al. 2013). Each is trying to do the same thing: confidently apply an air pollution exposure to a study subject (often a group or area) at a given time-space moment.

For example, in Vancouver, Canada, Buzzelli et al. (2006) constructed a GIS of neighbourhoods and air pollution values (total suspended particles, TSP) to correspond with the national census at five year intervals from 1976 to 2001. Analysis was undertaken at the neighbourhood (census tract) scale to which air pollution values were assigned using geostatistics. This involved interpolation between a consistent set of 32 monitoring stations to create air pollution 'surfaces' from which annual average TSP values were drawn. This research found both reductions in the region's overall TSP profile as well as significant social geographical change. And yet, in the face of all this flux over 25 years, pollution continued to map systemically and significantly differently across Vancouver's social geography. As shown in Figure 26.1, neighbourhoods marked by low education and single-parent

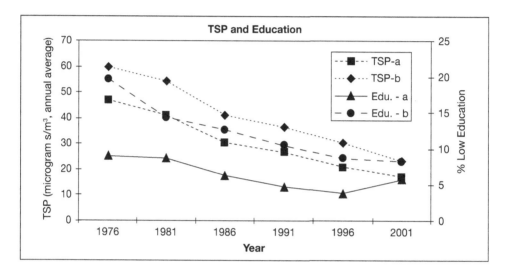

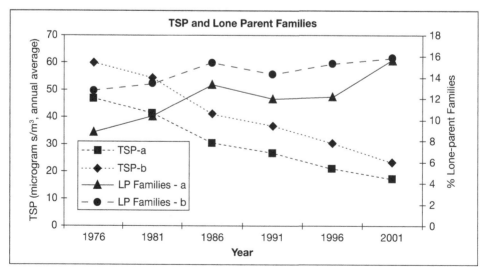

Figure 26.1   Education, lone-parent families and TSP in Vancouver, 1976–2001

families consistently bore the greatest burden of exposure to TSP (albeit with some convergence in 2001).

EJ has of course informed a range of studies of alternative designs, contexts and groups of interest. Kingsley et al. (2014) used a GIS approach to examine the spatial relationship between road infrastructure and 114 644 schools across the US in 2005–6, finding that schools serving predominantly Black students were 18 per cent more likely to be located within 250 metres of a major roadway. Tian et al. (2013) found similar results in their more recent US national-scale study. Often we find that these kinds of relationships are more granular when examined locally. For example, Wu and Batterman (2006) studied these kinds of relationships in Wayne County, Michigan but differentiated traffic (e.g. heavy truck traffic) and road type. They found students of schools near high traffic roads (<150 metres, >5000 trucks/day) were more likely to be Black or Hispanic and to reside in poor neighbourhoods. Similar results have been reported in other regional case studies (e.g. Gunier et al. 2003). Here again we find nuances in relationships but based also on community of interest. In Tampa Bay (Chakraborty 2009) and Miami (Grineski et al. 2013), for instance, race and ethnicity have played a consistent role in conditioning exposure and health outcomes associated with transportation emissions. In Miami specifically, however, predominantly Cuban and Colombian neighbourhoods have had elevated risks whereas neighbourhoods with predominantly Mexican-origin populations have actually had reduced risks.

What is significant about research like this is that it identifies how different groups (note the emphasis on communities, areas or groups rather than individuals) are differentially exposed to air pollution. The Reanalysis Project has the advantage of being a time-continuous picture of relationships. However, these other examples identify how particular segments of our society are exposed to alternative levels of pollution instead of an exposure assignment of a single value for all subjects (individuals, neighbourhoods) in a given region. In our Vancouver example in particular, the persistent systemic nature of alternative exposures speaks directly to the EJ question: that even as overall pollution levels diminished, exposure disparities remained between very different kinds of neighbourhoods. One policy implication of this kind of insight is that it is not sufficient to address only total or average pollution reductions (cf. Lin et al. 2013; Mitchell et al. 2015).

Before turning to the discussion of policy issues that grow out of EJ research, we address what these kinds of advancements have meant on the other side of the dose–response relationship: health outcomes. When we are interested in the health effect of an environmental health risk such as air pollution we want to be sure we are seeing its independent role in our statistical models. This often means we need statistical control for the influence of 'confounding' factors like smoking which can generate similar health outcomes to air pollution. What also interests us in EJ research is the 'effect modification' of lower socioeconomic position (see Figure 26.2). Is being of higher status somehow protective against health impacts? By contrast, are lower status communities more susceptible to health effects? Greater pollution exposure can be thought of as a disproportionately greater threat to health amongst those of lower socioeconomic status. Added to this is the typically compromised health status we know to exist for these individuals and communities (CSDH 2008). Indeed, when combined, we often see an amplified 'triple jeopardy' health impact above and beyond the additive effect of these two sets of health insults. Of course the difference between the concepts of confounding and effect modification may be immeasurably complex. Those of lower status, for instance, are more likely to exhibit risky health behaviours such as smoking and not taking exercise. Indirectly, then, smoking may not amplify the effect of air pollution exposure (whatever that may be, such as asthma) as a confounder but as an indirect status

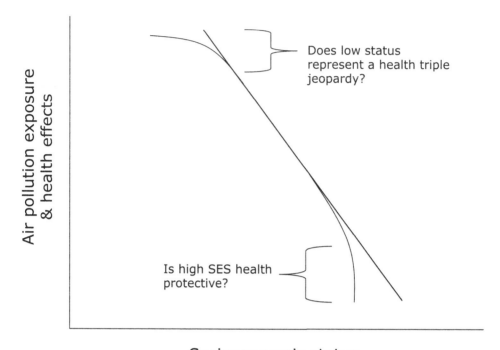

Is high SES health protective?

Does low status represent a health triple jeopardy?

Socioeconomic status

Figure 26.2 Concepts of the SES modification of air pollution health effects

effect modifier. But perhaps this is a limitation of our ability to measure and model these processes. In any case, an example can help us understand how health research has heeded the EJ message.

In their research on Hamilton, Canada, Jerrett et al. (2005) took a similar analytical approach to that used in the Vancouver example presented above. In addition, they were interested in whether chronic exposure to air pollution remained significantly associated with health effects after consideration of a range of socioeconomic variables (e.g. income and wealth variables, education, employment status). They found that across most study subjects in that city in 1991, premature mortality rose by as much as 25 per cent between the least and most polluted neighbourhoods and that, in most analyses, death by cancer and cardio-respiratory disease also rose statistically significantly. When including socioeconomic, demographic and lifestyle variables for census neighbourhoods, they found these factors

> reduced but did not eliminate the pollution effect on mortality. Economic variables representing household income, poverty, and income inequality exerted the most consistent [effect modification] on particle effect .... In models that controlled for multivariate confounding, we also observed elevated point estimates and some significant pollution effects.
>
> *(Jerrett et al. 2005: 2858)*

As with exposure analysis, we also see a diversity of applications in health effects studies targeting EJ-inspired questions. For example, in their study of acute cardiovascular mortality from PM exposure in Phoenix, Wilson et al. (2007) found that socioeconomic status may modify the effect of air pollution with distance from the City's central monitoring station. Also using a time series approach, Montresor-López et al. (2015) focused on short-term ozone exposure and hospital stroke admissions in South Carolina. They found no significant relationship, however the risk of haemorrhagic stroke appeared to be higher (though still not statistically significant) among African Americans. Others have been concerned with the effects of multiple pollutants on various health outcomes and hospitalizations, both in the US and beyond (e.g. Bravo et al. 2015; Son et al. 2012).

While we could list more studies like these what we hope to impart is that the putative health outcomes traditionally associated with air pollution alone are now understood to interact with sociodemographic markers of individuals and communities. One can say the EJ movement and research literature has had its message heard and that we probably have better health models for it. And yet although this kind of evidence may improve our understanding of disease pathways it may, as discussed below, paradoxically muddy our understanding of policy options.

More evidence = policy paralysis?

A useful way to frame the policy (broadly defined) approaches we might take to address the evidence and concepts encountered above is to remind ourselves of the useful framing of EJ issues put forth by Susan Cutter (1995): that EJ can be seen from both 'outcome' and 'process' perspectives. The former is concerned with the distribution of environmental bads; the latter with the processes (e.g. policy, legal, market) that lead to and reproduce unequal and unfair distributions. We apply these alternative concepts to reflect on policy options.

Implicit in our discussion of EJ in air pollution exposure and health is a focus on outcomes. This is no surprise to the epidemiologist or public health clinician. With a focus on health effects in particular, these latter fields have always taken a positivist/quantitative 'outcome' approach to the dose–response relationship. It makes sense that, as EJ came into contact with air pollution epidemiology, outcome studies would flourish. However, whereas traditional or perhaps non-critical (in a social science sense) research in this domain might understandably point to a need for overall air pollution reductions, we have already seen in our Vancouver example that reductions may not achieve the goal that EJ sets: to reduce disparities between the more and less affluent segments of our societies. Mitchell et al. (2015) similarly show that, despite overall air pollution improvements in Britain from 2001 to 2011, pollution levels improved the least or worsened in the most deprived areas (small census zones) depending on the pollutant in question. Indeed, overall reductions are themselves not simply a technical problem. The quality of life and standard of living to which we have grown accustomed necessarily bring with them the regrettable by-products of consumption and pollution. The emissions of private auto transportation are a prime example, rising inexorably against the tide of evidence both on health impacts and our climate.

More complex still is policy on community pollution disparities. Why? Because the relationships we see are infinitely complex. For starters, the nascent EJ-informed air pollution health research is not free of research methods and design issues. A common set of concerns has to do with the ways in which we delimit study areas and examine their associations with air pollution. Ecological fallacy tells us we may define study areas infinitely though of course we typically use a given set of boundaries as areas and communities based on census or other administrative processes. As modifiable areal units, boundary changes can generate different relationship and thereby statistical changes as well. At the same time, the boundaries of our communities of interest are in constant 'dialogue' with the other moving part of our models:

air pollution itself, slipping as it does across regions with no regard for our administrative data collection and aggregation. Strictly speaking, these concerns mean we must always interpret the relationships we find with caution, particularly when using aggregated/areal data: the relationships are always one instance of innumerable possibilities.

Nested within these issues is a further consideration alluded to earlier: that exposure assignment is often made at the place of residence. Time-geography would question: how we characterize exposures as people spend their time indoors as opposed to out of doors; air exchange between indoor and outdoor residential environments; exposures that occur during time spent away from the home such as during the journey to work; and exposures at the workplace, to name a few. Each of these can create unexplained 'noise' in our exposure and effects models and in turn misinform policy. Take for example research on walkable neighbourhoods in Vancouver (e.g. Marshall et al. 2009). While these kinds of areas might encourage physical activity, they might also generate much higher exposure to roadside traffic emissions. In this sense we have a policy conflict in which the desired urban fabric of dense neighbourhoods of mixed land uses, connectedness and commercial success needed for walkability and physical activity also generates roadway traffic and elevated air pollution exposure in the self-same geographies. Notably, since these exposures are away from the home, they exemplify the care needed to interpret much of our air pollution health evidence and the policy solutions it can inform. In recent years we have seen the leading edge of air pollution epidemiology engage other forms of monitoring including time-activity diaries, mobile monitoring and GPS-tracking and monitoring of individual subjects (Glasgow et al. 2016; Smargiassi et al. 2012). These are important steps and it will be interesting to see how this work can be harnessed to answer our EJ questions.

Let us now turn to what Cutter (1995) referred to as 'outcome' EJ issues. Even if the evidence we have can in one way or another be 'picked apart', it will not surprise the reader that civil society should nonetheless act upon what it knows. After all, we do have outcomes generated by a political economy that is unlikely to correct itself; at least not in the near term. Thus we need intervention that addresses how unfair outcomes come to be and are reproduced.

In this context an obvious target is air quality management planning, in which we could aim for the greatest pollution reductions for those with undue exposures and impacts. Operationalized in some way (by barring siting of noxious land uses, imposing limited traffic zones), this orientation targets reductions in the disparity between the most and least polluted areas of interest. Trade-offs and political implications ensue. The most obvious is that the responsible administrative body (e.g. a municipal government) commits to being more 'hot-spot' focused and in a sense less regionally representative. Some communities will come to be relatively less served, including perhaps some of the more advantaged areas that would either 1) see the greatest withdrawal of resources otherwise committed under traditional approaches, or 2) not see an equivalent rise in services if EJ occasions an absolute expansion of programmes. One resource implication flowing from this kind of approach is relatively more or less air quality monitoring, leaving some areas well covered and others less so. If EJ ideals are applied, the respective implication is that relatively less affluent areas see more monitoring; more affluent areas are left to know relatively less about the air they breathe. Beyond monitoring, the implications grow starker still: progressive social and land-use planning could target such processes as road traffic limits and more equitable noxious facility siting.

How receptive are communities and governments to these ideas? This will depend at least on the availability of resources, the severity of air pollution disparities and how willing the wider community is to engage in critical air quality planning and remediation. Existing risk perception and public engagement research suggests this is a long and steep climb. In environmental health in particular, this has often meant we over-respond to the concerns of those of higher socioeconomic status even when they are less burdened by health risks (e.g. Wakefield et al. 2001). The activation of EJ principles within air quality management planning (such as less monitoring in a context of limited resources) will take courage on the part of decision-makers, benevolent-minded citizens and perhaps also potent activism.

This brings us to an alternative to 'top-down' approaches (Marshall 2010). How might communities themselves take action to redress their air quality concerns? Reinforcing other EJ researchers' calls for grassroots approaches, Deacon et al. (2015: 420) note "it is problematic to give the local and wider environmental justice movements over to legal and scientific procedures . . . communities facing perceived environmental injustices should not wait idly by and let academics narrowly define environmental justice issues." The well-known example of the South West Organizing Project (SWOP) appears to support this position. SWOP is a community group founded in the 1980s in New Mexico to fight for environment and health rights in communities (Buzzelli 2008). Faced with obvious air pollution threats, including large-scale point source industrial emissions, the community's "bucket brigade" used simple but effective monitoring equipment to measure, continuously publish (www.swop.net/) and ultimately gain traction over its air quality concerns. Though encouraging, examples like this remind us that communities need to understand why they need to take action in the first instance. We could ask: how might they respond, if at all, to less obvious but equally insidious health threats? In addition, SWOP was aided by outside technical help; expertise necessary even in the simplest forms of environmental monitoring (Ottinger 2009; Bonney et al. 2014). And of course for activist evidence and demands to be heeded we still need a degree of benevolence in the wider civil society. The message, perhaps, is that grassroots movements can effect change but it might take a special coincidence of circumstances.

Where does this leave us? The reader has probably never heard it said that policy is easy. We might say we need a hint of optimism to avoid reflecting on the above discussion as a 'wicked policy problem'. But if EJ succeeds in its aim of greater environmental equity it will have done so on a foundation of sound research. Perhaps it is necessary for a disruptive and radical movement to first expose complexity before a sharper focus can be achieved. Once that happens, we will feel more certain of charting the right kind of progress.

References

Agyeman, J., P. Cole, R. Haluza-Delay, P. O'Riley. (2009). *Speaking for ourselves: Environmental justice in Canada*. Vancouver: UBC Press.

Been, V. (1993). "What's fairness got to do with it? Environmental justice and the siting of locally undesirable land uses." *Cornell Law Review*, vol. 78, pp. 1001–1036.

Bonney, R., J.L. Shirk, T.B. Phillips, A. Wiggins, H.L. Ballard, A.J. Miller-Rushing, J. Parrish. (2014). "Next steps for citizen science." *Science*, vol. 343, 1436–7.

Bravo, M., J. Son, C. Umbelino de Freitas, N. Gouveia, M.L. Bell. (2015). "Air pollution and mortality in São Paulo, Brazil: Effects of multiple pollutants and analysis of susceptible populations." *Journal of Exposure Science and Environmental Epidemiology*, pp. 1–12.

Buzzelli, M. (2007). "Bourdieu does environmental justice? Probing the linkages between population health and air pollution epidemiology." *Health and Place*, vol. 13, pp. 3–13.

Buzzelli, M. (2008). "The political ecology of scale in urban air pollution monitoring." *Transactions of the Institute of British Geographers*, vol. 33, pp. 502–17.

Buzzelli, M., J. Su, N. Le, T. Bache. (2006). "Health hazards and socio-economic status: A neighbourhood cohort approach, Vancouver, 1976–2001." *Canadian Geographer*, vol. 50, 3, pp. 376–391.

Chakraborty, J. (2009). "Automobiles, air toxics, and adverse health risks: Environmental inequities in Tampa Bay, Florida." *Annals of the Association of American Geographers*, vol. 99, 4, pp. 674–697.

CSDH – World Health Organization Commission on the Social Determinants of Health. (2008). *Closing the gap in a generation: Health equity through action on the social determinants of health. Final Report of the Commission on Social Determinants of Health.* Geneva: World Health Organization.

Cutter, S.L. (1995). "Race, class and environmental justice." *Progress in Human Geography*, vol. 19, 1, pp. 111–122.

Deacon, L., J. Baxter, M. Buzzelli. (2015). "Environmental justice: An exploratory snapshot through the lens of Canada's mainstream news media." *Canadian Geographer*, vol. 59, 4, pp. 419–432.

Glasgow, M.L. C.B. Rudra, E.H. Yoo, M. Demirbas, J. Merriman, P. Nayak, et al. (2016). "Using smartphones to collect time–activity data for long-term personal-level air pollution exposure assessment." *Journal of Exposure Science and Environmental Epidemiology*, vol. 26, 4, pp. 356–364.

Grineski, S.E., T.W. Collins, J. Chakraborty. (2013). "Hispanic heterogeneity and environmental injustice: Intra-ethnic patterns of exposure to cancer risks from traffic-related air pollution in Miami." *Population and Environment*, vol. 35, pp. 26–44.

Gunier, R.B., A. Hertz, J. von Behren, P. Reynolds. (2003). "Traffic density in California: Socioeconomic and ethnic differences among potentially exposed children." *Journal of Exposure Analysis and Environmental Epidemiology*, vol. 13, pp. 240–246.

Jerrett, M., M. Buzzelli, R. Burnett, P. DeLuca. (2005). "Particulate air pollution, social confounders, and mortality in small areas of an industrial city." *Social Science and Medicine*, vol. 60, 12, pp. 2845–2863.

Kingsley, S.L., M.N. Eliot, L. Carlson, J. Finn, D.L. MacIntosh, H.H. Suh, G.A. Wellenius. (2014). "Proximity of US schools to major roadways: A nationwide assessment." *Journal of Exposure Science and Environmental Epidemiology*, vol. 24, pp. 253–259.

Krewski, D., R.T. Burnett, D. Goldberg, K. Hoover, J. Siemiatycki, M. Jerrett, M. Abrahamowicz, and W. White. (2000). *Reanalysis of the Harvard Six Cities Study and the American Cancer Society Study of Particulate Air Pollution and Mortality.* Cambridge, MA: Health Effects Institute.

Lin, S., R. Jones, C. Pantea, H. Özkaynak, S. Trivikrama Rao, S. Hwang, V.C. Garcia. (2013). "Impact of NOx emissions reduction policy on hospitalizations for respiratory disease in New York State." *Journal of Exposure Science and Environmental Epidemiology*, vol. 23, pp. 73–80.

Marshall, A. (2010). "Environmental justice and grassroots legal action." *Environmental Justice*, vol. 3, 4, pp. 147–51.

Marshall, J., M. Brauer, L. Frank. (2009) "Healthy neighbourhoods: Walkability and air pollution." *Environmental Health Perspectives*, vol. 117, 11, pp. 1752–1759.

Mennell, M. and K. Bhattacharyya. (2002). "Air quality management" in D. Bates and R. Caton, eds., *A citizen's guide to air pollution.* Vancouver: David Suzuki Foundation, pp. 277–338.

Mitchell, G., P. Norman, K. Mullin. (2015). "Who benefits from environmental policy? An environmental justice analysis of air quality change in Britain, 2001–2011." *Environmental Research Letters*, 10, pp. 1–19.

Montresor-López, J.A., J.D. Yanosky, M.A. Mittleman, A. Sapkota, X. He, J.D. Hibbert, et al. (2015). "Short-term exposure to ambient ozone and stroke hospital admission: A case-crossover analysis." *Journal of Exposure Science and Environmental Epidemiology*, vol. 26, 4, pp. 162–166.

Ott, W.R. (1995). "Human exposure assessment: The birth of a new science." *Journal of Exposure Analysis and Environmental Epidemiology*, vol. 5, 4, pp. 449–72.

Ottinger, G. (2009). "Buckets of resistance: Standards and the effectiveness of citizen science." *Science, Technology and Human Values*, vol. 35, 2, 244–70.

Özkaynak, H., L.K. Baxter, K.L. Dionisio, J. Burke. (2013). "Air pollution exposure prediction approaches used in air pollution epidemiology studies." *Journal of Exposure Science and Environmental Epidemiology*, vol. 23, pp. 566–572.

Pope, C.A., 3rd and Dockery, D.W. (2006). "Health effects of fine particulate air pollution: Lines that connect." *Journal of the Air and Waste Management Association*, vol. 56, pp. 709–42.

Pulido, L. (1996). "A critical review of the methodology of environmental racism research." *Antipode*, vol. 28, 2, 142–59.

Smargiassi, A., A. Brand, M. Fournier, F. Tessier, S. Goudreau, J. Rousseau, M. Benjamin. (2012). "A spatiotemporal land-use regression model of winter fine particulate levels in residential neighbourhoods." *Journal of Exposure Science and Environmental Epidemiology*, vol. 22, 4, pp. 331–8.

Smith, D.M. (1994). *Geography and social justice.* London: Wiley-Blackwell.

Son, J., J.T. Lee, H. Kim, O. Yi, M.L. Bell. (2012). "Susceptibility to air pollution effects on mortality in Seoul, Korea: A case-crossover analysis of individual-level effect modifiers." *Journal of Exposure Science and Environmental Epidemiology*, vol. 22, pp. 227–34.

Tian, N., J. Xue, T.M. Barzyk. (2013). "Evaluating socioeconomic and racial differences in traffic-related metrics in the United States using a GIS approach." *Journal of Exposure Science and Environmental Epidemiology*, vol. 23, pp. 215–222.

Wakefield S., S. Elliott, D. Cole, J. Eyles. (2001). "Environmental risk and (re)action: Air quality, health, and civic involvement in an urban industrial neighbourhood." *Health and Place*, vol. 7, pp. 163–77.

Wilson, W.E., T.F. Mar, J.Q. Koenig. (2007). "Influence of exposure error and effect modification by socioeconomic status on the association of acute cardiovascular mortality with particulate matter in Phoenix." *Journal of Exposure Science and Environmental Epidemiology*, vol. 17, S11–S19.

Wu, Y.C. and S.A. Batterman. (2006). "Proximity of schools in Detroit, Michigan to automobile and truck traffic." *Journal of Exposure Science and Environmental Epidemiology*, vol. 16, 457–470.

Young, I.M. (1990). *Justice and the politics of difference*. New Jersey: Princeton University Press.

27

WATER JUSTICE

Key concepts, debates and research agendas

Leila M. Harris, Scott McKenzie, Lucy Rodina,
Sameer H. Shah and Nicole J. Wilson

Relating freshwater and environmental justice

In this chapter, we focus on a wide range of justice concerns related to freshwater. While we acknowledge social and environmental justice concerns in marine and estuarine systems – including those linked with climate change, environmental flows, contamination, or fish stock depletion – our chapter explicitly attends to injustices related to freshwater, notably for domestic consumption, but also regarding irrigation and another livelihood uses. For drinking water, close to 1.8 billion people regularly rely on contaminated and unsafe sources. As a result, water-borne and diarrhoeal disease remain among the leading causes of ill health and disease globally, causing over half a million deaths per year (WHO 2015a). Women and girls often travel considerable distances to collect water – with an estimated 200 million hours spent *every day* on this core domestic task. This situation poses a suite of physical and safety risks (UNESCO 2015) apart from time losses that take away from rest, leisure, or other productive tasks. Access to water and sanitation is also critical for education, livelihood generation, and other key facets of human development and social well-being (UNESCO 2015; Hall et al. 2013; Crow et al. 2012; Mehta 2014). There are significant challenges related to water quality and needs for livelihood uses. Among them, *de jure* transformations in water rights and *de facto* shifts in water allocations pose significant risks to food, economic, and livelihood security (e.g. Molle and Berkoff 2006). Many concerns related to domestic and irrigation water are considered to be increasingly acute given ongoing hydro-climatological changes associated with shifting climatic baselines, ongoing quality challenges, and competing uses at the agriculture-energy-water nexus (see Chapter 29).

Linked to water's importance, particularly for the world's vulnerable and impoverished, the notion of 'water justice' is one that has garnered considerable analytical and political attention over the past several decades. The concept has been used to interrogate the dispossession occurring with mining and capitalist expansion in Peru and Bolivia (Budds and Hinojosa-Valencia 2012; Perreault 2013), anti-privatization movements from Argentina to Ghana that have worked to resist private sector control and profiteering from water resources (Bennett et al. 2005; Bakker 2010), neoliberal governance shifts that further marginalize First Nations in Canada (Mascarenhas 2007), and key negotiations leading up to the UN adoption of the Human Right to Water and Sanitation in 2010 (Mirosa and Harris 2012).

Key environmental justice concerns relate to the fact that water access and quality are highly unequal, and vary according to a range of social and spatial gradients. Households in some areas may have to spend one quarter or more of their income to acquire water for their daily needs (e.g. work in Ghana by Amenga-Etego and Grusky 2005) – well above the international recommendations of 3–6 per cent of household income to access basic needs (Hutton 2012; UNDP 2010). There are also core justice concerns related to absolute costs, given that on a per unit basis a person with access to high quality or subsidized water may pay 1/10th or even 1/100th of the price for the same amount of water as those living in contexts that do not enjoy such access (Crow 2001). As one example, a resident in well-served parts of Metro Manila pays PHP 10 – PHP 12/m^3 (US$0.22 – US$0.27) for consumption of 10 m^3 as compared to PHP 200/m^3 (US$4.44) for those living in nearby *barangays* (village-level administrations) reliant on deliveries from water carts and tankers (Torio 2016). All told, higher income populations in many parts of the world often enjoy unimpeded access to high quality water at relatively low or negligible cost, while lower income populations often pay a greater proportion of income, as well as more in absolute terms, for water that may be less safe and/or reliable.

Globally, such differences are stark. Consider, for instance, that a resident of the US consumes on average 13 times more water than an average resident of Bangladesh (216 m^3/per capita/year of domestic water in the US, as compared to 16.3 m^3/per capita/year for residents in Bangladesh, data from Hoekstra and Chapagain 2007). Considerations of this type reveal the importance of a justice lens to understand and respond to water challenges, whether unequal domestic access, affordability, or the differentiated effects of water quality, water-related hazards, and water variability for productive needs – all of which we discuss further below. As we also suggest, the flip side is also true – water impinges on a range of justice considerations, from health to economic development, thus meriting consideration.

Water access and quality in relation to social, spatial and hydro-ecological difference

Tracing social and spatial differences – from the neighbourhood to the global scale, and between the global North and South – is one way to investigate water-related inequalities. Water justice considerations for drinking, sanitation, and productive uses must be identified, understood, and assessed in relation to a wide range of social and political factors, including gender, income, indigeneity, and race. Recent works have also increasingly emphasized the ways in which biophysical factors and material conditions, including topographic and environmental conditions, contribute to inequalities (cf. Perreault 2014; Sultana 2011). While we cannot address all of these differences, the following sections provide several illustrations of social, economic, and biophysical differences as key to characterizing and assessing water injustice.

Social, economic and demographic factors

Gender

Gender is frequently highlighted as crucial for unequal water access or changing conditions. Linked to the gendered labour burden of water access for household needs (highlighted above), it has been argued that women may be particularly vulnerable to water pollution, droughts or floods. Specifically, the fact that women are commonly responsible for water fetching, as well as cooking, cleaning, or care of ill or elderly household members suggests that poor water quality or unreliable access might affect women disproportionately, albeit in diverse ways. Sugden et al. (2015) highlight that the out-migration of males in agrarian communities of the Eastern Gangetic Plains (India) produces gendered vulnerability to climate change effects for women belonging to marginal farms and tenant households. In this case, women are particularly vulnerable to drought due to irregular income and lower capacities to invest in off-farm activities (*ibid.*). In work on the arsenic crisis in Bangladesh, Sultana (2011) calls attention to the ways in which contaminated wells result in complex social and health outcomes for women, including emotional distress linked to the inability to provide safe water for their families (thus challenging their ability to meet expectations of being a 'good mother'), and having to rely on tenuous social networks to maintain access to safe water (e.g. by doing favours or begging neighbours who might have access to a safe well).

Highlighting similar themes, Wutich (2009) suggests gender–water linkages are mediated through household roles and livelihoods, referencing the relative stress faced by men and women in urban Cochabamba, Bolivia. In this example, women indicated higher levels of stress on survey responses related to negotiations with vendors over water prices, among other factors. In all of these ways, gender is often considered foundational to water-related equity and justice considerations (see also Alston 2006; Sultana 2014; O'Reilly et al. 2011).

Indigeneity

Indigeneity is another key axis to consider when evaluating water-related justice concerns. This is largely due to the unique history of Indigenous[1] peoples vis-à-vis colonial processes and colonial state practices, which often fail to recognize Indigenous territorial and water rights. These complex histories have often led to compromised water quality and quantity as a result of prioritization of commercial, industrial, and settler uses – often disconnecting Indigenous people, governance practices, and livelihoods from water within (and without) traditional territories (Boelens et al. 2006; Simms 2014). Moreover, changes to water and land access have specific implications for Indigenous peoples due to physical, cultural and spiritual connections to water (Boelens et al. 2006; Wilson 2014). The examples below illustrate unique water justice challenges facing Indigenous peoples including the safety of and access to drinking water as well as governance and control of waters, including source water.

In Canada, drinking water on First Nation reserves (parcels of land set aside under the Indian Act and treaty agreements in Canada for the exclusive use of First Nations (Hanson 2015)) is often of considerably lower quality than for other Canadian residents, resulting in what many have referred to as a 'two-tiered system' (MacIntosh 2008; Simeone 2010). Unsafe drinking water affects both the physical health (i.e. high incidence of illnesses associated with water-borne disease) and the cultural or spiritual well-being of First Nations (First Nations, Métis and Inuit peoples are terms for the Indigenous peoples of Canada), given their complex socio-cultural relationships to water (Basdeo and Bharadwaj 2013; McGregor 2009). One reason for this inequality is the differential governance system – drinking water off-reserve is a provincial responsibility while authority for on-reserve drinking water is the responsibility of the federal government. While significant investments and legislative proposals have been made to remedy the situation, on-reserve drinking water quality remains a persistent problem across the country (Cave et al. 2013). Among the reasons for this persistent problem is the lack of clarity around roles and responsibilities between multiple departments within the federal government and First Nations communities, as well as capacity challenges (Simeone 2010; Simms 2014) – issues that are compounded by the fact that many of the reserves are geographically remote and small in size (Spence and Walters 2012). Broader ontological, epistemological and structural conditions are also critical to these concerns (Simms et al. 2016; Yates et al. 2017).

Other water justice concerns for Indigenous peoples have also been evidenced in a range of contexts across the globe, often linked to territorial rights. In the example of the Central Kalahari Game Reserve (CKGR), Botswana courts allowed Indigenous San people to return to their traditional territories after a forced and highly contested relocation (Morinville and Rodina 2013). While the court ruling enabled the San to return to the reserve, they were not permitted to access traditional water sources – meaning they could not *effectively* return. The situation was eventually rectified through a 2011 legal judgment that upheld that territorial access necessarily also implied water access (*ibid.*, see also McKenzie 2015 for the case of the Yakye Axa people of Paraguay).

Indigenous water justice also invites attention to broader cultural considerations, governance interactions, and source water conditions (rather than a sole focus on domestic drinking water), in addition to longer histories and relationships that condition present day realities. This is the case given that Indigenous livelihood practices – critical to physical and cultural health and identity – are frequently linked with ecosystem conditions (e.g. water quality, quantity and flows, as well as broader ecosystem issues, such as health of fish populations). As one example, Wilson et al. (2015) discuss ways that the Koyukon Athabascan people of Ruby, Alaska express socio-cultural connections to water and the vulnerability associated with ongoing changes. In this case, climate induced hydrologic change reduced the ability of residents of this community to predict the timing of river ice freeze-up, ice thickness, and break-up. This has direct consequences for Indigenous culture, health and livelihoods as residents are less able to access the Yukon River and its tributaries during the winter months. The community also faces diminishing seasonal access to key harvesting areas in addition to increasing travel hazards (see Chapter 40).

Income, race, and other intersectional factors

Highlighting income, poverty, and race as several of the longstanding concerns of environmental justice, scholarship on water has also found that household income is a strong predictor of safe water access (e.g. Dapaah 2014 and Mahama et al. 2014 for case studies of Accra, Ghana, where higher income residents are better able to afford fees for vendors, or storage facilities). Of course, as with any of these categories, they cannot be neatly separated and are best understood as intersectional. Consider the recently highly publicized example of lead contamination in Flint (Michigan, USA). In this case, a budget crisis and ongoing austerity measures led officials to switch the municipal water supply to the Flint River with expectations that this would save the city over $5M. The tragic story of the majority African American city of Flint now serves as paradigmatic of environmental injustice. In brief, this "economic" decision to change the water source to one with more corrosive properties resulted in leaching from old lead pipes – poisoning thousands of residents with lead – known to have long-term health and developmental effects, especially for young children (Lin et al. 2016). As Ranganathan (2016) explains, such examples should not be understood merely as outcomes of racism but as linked to broad political economic shifts and associated disinvestments, including those couched in liberalism (including the associated language of equality). Here, we also see the complex intersections of race, class and place, particularly in the context of austerity and neoliberal governance shifts (cf. Mascarenhas 2007).

Biophysical conditions: flooding, drought, climate change and other material considerations

As several of the above examples help to illustrate, biophysical conditions are another dimension critical to assessing water-related environmental justice concerns, particularly as certain communities might be more vulnerable to ongoing degradation or environmental changes. Many examples show that even as ecological conditions or environmental change are often cast as apolitical, or equity-neutral, all people and places are not affected similarly. As political ecologists, critical hazard researchers, and other scholars have shown, even when cast as 'natural', droughts, floods, or similar events do not have the same effect on all segments of the population – indeed, there is considerable evidence that vulnerability and exposure to risks and hazards are highly inequitable (see Chapter 28). In a key comparative study, Neumayer and Plumper (2007) demonstrated the gender-differentiated effects of floods, droughts and other 'natural' disasters. Statistically evaluating data from 141 countries, they show that women and girls are more likely to suffer morbidity and mortality following a disaster event. This correlation is stronger for bigger events and also in contexts that are considered to be less gender equitable (see Ribot 2009; Nightingale 2015; Watts and Bohle 1993 for other examples of differentiated vulnerabilities to risks and hazards). Returning to the arsenic crisis in Bangladesh – while presumably the geology of arsenic potentially affects all residents equally or at least randomly, it is clear that wealthier residents are better able to dig deeper wells to maintain access to safe water (Sultana 2011).

In the context of climate change, it is well established that many regions anticipate, and are already experiencing, more pronounced extreme events, including drought and flooding. Given that marginalized populations often live on degraded or unsuitable lands, it is clear that low-income settlements may be particularly susceptible to flooding, wastewater pollution, or similar hazards (see Chapter 29). One stark example of this is in Cape Town, South Africa, where there are estimated to be more than 220 informal settlements throughout the city (Mels et al. 2009), 80 per cent of which are located in low-lying, flood-prone areas, such as wetlands or other marginal lands. These areas are particularly vulnerable to the impacts of flooding made worse by a lack of basic services, disaster relief and other key resources (Ziervogel et al. 2014).

'Drought' and water scarcities are also often tied to a range of institutions, discourses, infrastructures, and processes that foreground justice considerations. Mehta (2001) highlights the ways in which discourses of scarcity were naturalized to bring legitimacy to the construction of the Sardar Sarovar Dam in India. While presented as a solution for scarcity, powerful elites benefited from this development while the water needs of the poor went largely ignored. In another example from Israel/Palestine, Alatout (2007) documents that water in the 'land of milk and honey' was initially presented by hydrologists as abundant in ways that supported considerable in-migration of Jews to historic Palestine. Then, after the consolidation of the new state of Israel, the scientific discourse shifted to one of 'scarcity' that supported centralized water management and the formation of centralized state institutions and infrastructures to manage the scarce and important resource. In these ways, we see that hydrological concepts and debates, including notions of 'drought' or 'scarcity,' may be explicitly or implicitly linked to a range of political outcomes, goals, and justice considerations.

Water infrastructures

Much recent work on water justice and equity has also highlighted the importance of infrastructures as key to consolidating differentiated water access, or conditions. In brief, it is suggested that one way to trace networks of power, or inequity, is through mapping the physical infrastructure of water or sanitation, following the pipes, wells, dams, and taps to understand ways that these infrastructures materialize and consolidate inequities (e.g. Barnes 2012; Carse 2012; Kooy and Bakker 2008; Anand 2011; Birkenholtz 2009). For instance, water taps and metering devices have been shown to shape uneven geographies of water access and experiences of marginalization – key themes in work on impoverished and peri-urban settlements in South Africa as documented by Loftus (2006), von Schnitzler (2008) and Rodina (2016). Inequality and marginalization are also strong themes in work on sanitation infrastructure and access (e.g. Morales et al. 2014), large scale damming, development and irrigation infrastructures (e.g. McCully 1996; Harris 2008) and other household infrastructures, whether access to pumps to enable groundwater withdrawals in Rajasthan (Birkenholtz 2009), or water storage in urban Accra (Dapaah 2014).

Conceptual tools and intellectual traditions to understand water justice

Given these considerations and the range of case studies and illustrative examples, it is helpful to briefly survey some of the key literatures and concepts that served to better understand, and respond to, water justice issues. Key traditions and concepts include water security (Cook and Bakker 2012), political ecologies of water (Loftus 2009), and infrapolitics (Anand 2011). In addition, we find inspiration in emergent work from feminist theory on emotions and subjectivities as well as contributions from post-humanism that highlight more-than-human injustices.

Water security has been highlighted in research and policy documents of the past several decades, drawing from interdisciplinary perspectives (Cook and Bakker 2012). Earlier research focused on the connections between secure water provision and national security, autonomy, and stability (Gleick 1993). Work of the past three decades constitutes a paradigmatic shift – as the concept increasingly refers to governance and practice to mitigate and prevent unacceptable water-related risks for food, livelihood, health, ecological, and personal security – all aspects of "human security" (Grey and Sadoff 2007; Garrick and Hall 2014). Linked to water security, "nexus" thinking offers a basis to think through the interconnections between water and agriculture, energy, domestic use and environmental health. Bringing these themes together, a water security framework combined with a 'nexus' approach has the potential to 1) understand how variegated water-related risks confer harm to different segments of the population; and 2) trace the variable effects when considering water-related intervention at the system-scale (e.g. "trade-offs" between water for domestic use or irrigation). Although justice concerns are not always central to water security studies, the concept can be useful to think through the complex and intersecting water requirements across many different uses, users, locales or scales, including the possibility of more adequately acknowledging complex trade-offs in decision making (Bakker 2012).

As many examples cited throughout this chapter make clear, water justice has been a clear theme of studies in political ecology and feminist political ecology – focusing considerable attention on basic needs, livelihood concerns and North–South inequities in water access or quality. While environmental justice studies often focus on 'local' scale concerns, political ecological research seeks to attend to multi-scalar and political economic conditions important for understanding observed environmental changes, or linked inequities – giving explicit consideration to power relationships and structural conditions that underlie those issues. Political ecological understandings have been reflected in diverse water policy circles, and even increasingly in global institutions and policy frameworks. For instance, the 2006 United

Nations *Human Development Report* rejected the dominant paradigm that water stress was a function of physical scarcity. Rather, it argued, "the roots of the crisis in water can be traced to poverty, inequality and unequal power relationships, as well as flawed water management policies that exacerbate scarcity" (p. 5). This statement makes abundantly clear that water access cannot be viewed as an apolitical hydrological process, but rather, as a "hydro-social" process, where water access is reconfigured and reshaped through socio-political processes (Swyngedouw 2004; Linton and Budds 2014). Even as the ideas of water justice appear to be gaining traction in formal debates, many institutions and practices nonetheless maintain economistic, individualistic, or natural science framings of the issues in ways that make justice concerns appear to be secondary at best, or at worst, irrelevant. Indeed, there continues to be strong reliance on hegemonic water governance practices with a focus on privatization, market approaches, or technocratic solutions in ways that sideline or avoid complex justice challenges (Goldman 2007; Harris et al. 2013).

Recent scholarship has also highlighted water in the context of inter-generational and inter-species justice. Among such contributions, feminist theorists have discussed the possibility of a 'watery subjectivity,' highlighting the porosity of bodies, human and non-human, in the ways that water moves through us – literally connecting all life. As Neimanis (2013: 28) argues, attention to water as a metaphorical and material connector between all living beings allows us to understand ourselves, and our bodies, as connected and linked to other peoples, places, animals, and ecologies. She writes:

> Perhaps by imagining ourselves as irreducibly watery, as literally part of a global hydro-commons, we might locate new creative resources for engaging in more just and thoughtful relations with the myriad bodies of water with whom we share this planet.

Contributions of this type are opening new aspects of water justice debates, particularly with recent contributions from historians, literary theorists and other humanities scholars, adding new insights to a field that has historically been dominated by natural sciences and engineering fields (e.g. Chen et al. 2013).

Shifting water governance in light of justice concerns

While environmental justice considerations related to water appear to be increasingly recognized, how to best respond to these challenges is a source of ongoing debate. Water governance broadly refers to processes and practices that shape decision making over water and its uses, including, but not limited to, actors and institutions, as well as formal and informal laws and regulations that govern how water is accessed and used. Governance is critical because of its potential to overcome and respond to key justice considerations, or given the very real possibility that governance structures and processes can also exacerbate justice concerns (see the example of First Nations in Canada above). Foregrounding equity and justice, we might ask: How can we use water, or make decisions related to water, differently to respond to justice considerations, and especially to better meet the needs of those who are particularly vulnerable and underserved? Such a justice orientation is somewhat distinct from, and potentially resistant to, efficiency, cost recovery or other governance priorities that are often emphasized (Goldman 2007).

While much of the above discussion is concerned with distributional justice considerations, it is clear that procedural and compensatory justice, as well as notions of recognition, are all key considerations for water governance. These issues are perhaps addressed most readily through efforts to engender participatory water governance. The full and meaningful engagement of communities in water-related decisions that affect them is often promoted as key to addressing equity concerns, as well as broader sustainability goals linked to notions of effective governance (Goldin 2010). Even with growing agreement on the importance of participatory governance and citizen involvement, there are nonetheless considerable obstacles to meaningful participation, especially in ways that also serve equity aims. Among other considerations, to the degree that water is viewed as an engineering or technical challenge, this can reinforce barriers to broad and inclusive governance, particularly for women or other marginalized populations (Barnes 2013; Goldin 2010). Moreover, participatory governance at times treats certain groups tokenistically, looking only for the physical presence of members, or 'representative' participation, rather than considering whether there is engaged dialogue across communities (which would be key for a meaningful understanding of procedural justice).

Given ongoing debate about how to better respond to these challenges, it is important to maintain a critical perspective in terms of equity claims and how they are enrolled to promote particular governance shifts. The discourse on water privatization, for instance, has been propelled over the past several decades in part based on the idea that efficiencies and cost savings will enable needed investments to be made to extend infrastructure to under-served communities (Bakker 2010). Yet, numerous case study examples have shown that the equity outcomes have frequently not been borne out – for example, for the well-known case of privatization in Cochabamba, Bolivia, equity concerns were paramount in community protests that eventually led to the cancellation of the contract (see Torio 2016, and Harris et al. 2013 for other examples).

Concluding remarks and future research needs

Adequate water quantity, quality, risk considerations, and changing governance practices are all concerns that require careful consideration through an environmental justice lens. Given the manifold concerns related to changing water conditions, we have also suggested that there are ample reasons why water is meaningful for broader discussions of justice. The examples provided in this chapter demonstrate the unevenness and inequities inherent to the history, geography, and ongoing challenges related to water. Given that water is simul-taneously social and biophysical (Bakker 2012; Perrault 2014), understanding and responding to water justice challenges requires targeted thinking at the intersection of social, ecological, technological and political-economic *relations*. We need to continue to develop our theoreti-cal and empirical understanding of the complex systemic and lived experiences of water injustices, including varied perceptions of water-related risks in a changing world. To be better able to respond to these concerns, research and innovation is required to develop governance and policy approaches to meet sustainability and justice goals, particularly in the face of multivalent constraints that often make these goals difficult to achieve.

While not an exhaustive list, we offer several specific areas in need of future water justice research, policy, and practice. Among the governance questions that remain to be addressed are practices that will serve the effective and meaningful implementation of participatory governance, as well as the human right to water. Unfortunately, more often than not, research has served to identify what does not work, with little in the form of governance innovation and evaluation to better achieve these goals. We ask: How can governance pro-cesses be more inclusive, in ways that enable meaningful engagement, or to better address complex trade-offs in ways that are transparent and accountable? For the human right to water, there are many outstanding questions about how governments, NGOs, and civil society can promote its realization given ongoing inequities, complex trade-offs (e.g. between domestic and irrigation water, or financial and capacity obstacles facing many governments), as well as given future uncertainties and challenges (e.g. climate change).

In addition, further research remains critical to better connect questions of injustice across various scales, intersectionally (Indigeneity in relation to material conditions and income, for instance), and in relation to nexus understandings (trade-offs of water–energy–food). In short, understanding and responding to water injustice will also necessarily involve attention to key linkages related to food, energy, health and other sectors. To date, we have not yet been able to solve challenges related to narrow 'water security' definitions, making integrated and complex responses in a context of ongoing uncertainties and variabilities a daunting challenge.

As with other sustainability and justice discussions, much more attention is also needed from humanities, ethics, and arts communities to avoid treatment of these issues as merely 'technical' problems to be solved. Instead, full engagement with a wide range of scholars, policy makers, artists, and community practitioners will provide a stronger foundation from which to respond to the broader 'crisis of imagination' that hinders innovative responses. It is clear to us as water justice scholars that dealing with these challenges requires careful, creative, and innovative thinking in the context of broader questions of what we face as humans, living together with other communities, species, and ecologies in a complex and ever-changing world.

Note

1 While there is no consensus regarding the term Indigenous, capitalization is frequently used to convey respect (Kesler 2015).

References

Alatout, S. (2007), 'State-ing Natural Resources through Law: the codification and articulation of water scarcity and citizenship in Israel', *The Arab World Geographer*, vol. 10, no. 1, pp. 16–37.

Alston, M. (2006), 'The gendered impact of drought', in B.B. Bock & S. Shortall (eds), *Rural gender relations, issues and case studies*, Oxfordshire: CABI Publishing.

Amenga-Etego, R. & Grusky, S. (2005), 'The new face of conditionalities: the World Bank and water privatization in Ghana', in D. McDonald & G. Ruiters (eds), *The age of commodity: water privatization in Southern Africa*, London and Sterling, VA: Earthscan.

Anand, N. (2011), 'Pressure: the PoliTechnics of water supply in Mumbai', *Cultural Anthropology*, vol. 26, no. 4, pp. 542–564.

Bakker, K. (2010), *Privatizing water: governance failure and the world's urban water crisis*, Ithaca and London: Cornell University Press,.

Bakker, K. (2012), 'Water: political, biopolitical, material', *Social Studies of Science*, vol. 42, no. 4, pp. 616–623.

Barnes, J. (2012), 'Pumping possibility: agricultural expansion through desert reclamation in Egypt', *Social Studies of Science*, vol. 42, no. 4, pp. 517–538.

Barnes, J. (2013), 'Who is a water user? The politics of gender in Egypt's water user associations', in L. Harris, J. Goldin, & C. Sneddon (eds), *Contemporary water governance in the Global South: scarcity, marketization, and participation*, London: Routledge.

Basdeo, M. & Bharadwaj, L. (2013), 'Beyond physical: social dimensions of the water crisis on Canada's First Nations and considerations for governance', *Indigenous Policy Journal*, vol. 23, no. 4.

Bennett, V., Davila-Poblete, S. & Nieves Rico, M. (2005), *Opposing currents: the politics of water and gender in Latin America*, Pittsburgh: University of Pittsburgh Press.

Birkenholtz, T. (2009), 'Irrigated landscapes, produced scarcity, and adaptive social institutions in Rajasthan, India', *Annals of the Association of American Geographers*, vol. 99, no. 1, pp. 118–137.

Boelens, R., Chiba, M. & Nakashima, D. (2006), *Water and indigenous peoples*, Paris: UNESCO.

Budds, J. & Hinojosa-Valencia, L. (2012), 'Restructuring and rescaling water governance in mining contexts: the co-production of waterscapes in Peru', *Water Alternatives*, vol. 5, no. 1, pp. 119–137.

Carse, A. (2012), 'Nature as infrastructure: making and managing the Panama Canal watershed', *Social Studies of Science*, vol. 42, no. 4, pp. 539–563.

Cave, K., Plummer, R. & Loë, R. (2013), 'Exploring water governance and management in Oneida Nation of the Thames (Ontario, Canada): an application of the institutional analysis and development framework', *Indigenous Policy Journal*, vol. 23, no. 4.

Chen, C., MacLeod, M. & Neimanis, A. (2013), *Thinking with water*, Montreal and Kingston: McGill-Queen's University Press.

Cook, C. & Bakker, K. (2012), 'Water security: debating an emerging paradigm', *Global Environmental Change*, vol. 22, no. 1, pp. 94–102.

Crow, B. (2001), 'Water: gender and material inequalities in the global South', CGIRS Working Paper Series, Center for Global, International and Regional Studies and the Department of Sociology, University of California, Santa Cruz.

Crow, B., Swallow, B. & Asamba, I. (2012), 'Community organized household water increases not only rural incomes, but also men's work', *World Development*, vol. 40, no. 3, pp. 528–541.

Dapaah, E. (2014), *Water access and governance among indigenous and migrant low income communities in the Greater Accra Metropolitan Area (GAMA), Ghana*, MA Thesis, University of British Columbia.

Garrick, D. & Hall, J.W. (2014), 'Water security and society: risks, metrics, and pathways', *Annual Review of Environment and Resources*, vol. 39, pp. 611–639.

Gleick, P. (1993), 'Water and conflict: fresh water resources and international security', *International Security*, vol. 18, no. 1, pp. 79–112.

Goldin, J.A. (2010), 'Water policy in South Africa: trust and knowledge as obstacles to reform', *Review of Radical Political Economics*, vol. 42, no. 2, pp. 195–212.

Goldman, M. (2007), 'How "'Water for all!'" policy became hegemonic: the power of the World Bank and its transnational policy networks', *Geoforum*, vol. 38, pp. 786–800.

Grey, D. & Sadoff, C.W. (2007), 'Sink or swim? Water security for growth and development', *Water Policy*, vol. 9, no. 6, pp. 545–571.

Hall, R.P., van Koppen, B. & van Houweling, E. (2013), 'The human right to water: the importance of domestic and productive water rights', *Science and Engineering Ethics*, vol. 20, no. 4, pp. 849–868.

Hanson, E. (2015), *Reserves*, University of British Columbia, viewed 24 February 2015 http://indigenousfoundations.arts.ubc.ca/home/government-policy/reserves.html

Harris, L. (2008), 'Water rich, resource poor: intersections of gender, poverty and vulnerability in newly irrigated areas of Southeastern Turkey', *World Development*, vol. 36, no. 12, pp. 2643–2662.

Harris, L.M., Goldin, .JA. & Sneddon, C. (2013), *Contemporary water governance in the Global South: scarcity, marketization, participation*, London: Routledge.

Hoekstra, A.Y. & Chapagain, A.K. (2007), 'Water footprints of nations: water use by people as a function of their consumption pattern', *Water Resources Management*, vol. 21, no. 1, pp. 35–48.

Hutton, G. (2012), *Monitoring "affordability" of water and sanitation services after 2015: review of global indicator options*, paper submitted to the United Nations Office of the High Commission for Human Rights, viewed 8 January 2016, www.wssinfo.org/fileadmin/user_upload/resources/END-WASH-Affordability-Review.pdf

Kesler, L. (2015), *Aboriginal identity & terminology*, University of British Columbia, viewed 24 February 2015, http://indigenousfoundations.arts.ubc.ca/home/identity/aboriginal-identity-terminology.html

Kooy, M. & Bakker, K. (2008), 'Splintered networks: The colonial and contemporary waters of Jakarta', *Geoforum*, vol. 39, no. 6, pp. 1843–1858.

Lin, J., Rutter, J. & Park, H. (2016), 'Events that led to Flint's water crisis' *New York Times*, viewed 15 March, 2016, www.nytimes.com/interactive/2016/01/21/us/flint-lead-water-timeline.html?_r=0

Linton, J. & Budds, J. (2014), 'The hydrosocial cycle: defining and mobilizing a relational-dialectical approach to water', *Geoforum*, vol. 57, no. C, pp. 170–180.

Loftus, A. (2006), 'Reification of the dictatorship of the water meter', *Antipode*, vol. 38, no. 5, pp. 1023–1045.

Loftus, A. (2009), 'Rethinking political ecologies of water', *Third World Quarterly*, vol. 30, no. 5, pp. 953–968.

MacIntosh, C. (2008), 'Testing the waters: jurisdictional and policy aspects of the continuing failure to remedy drinking water quality on First Nations reserves', *Ottawa Law Review*, vol. 39.

Mahama, A.M., Anaman, K.A. & Osei-Akoto, I. (2014), 'Factors influencing householders' access to improved water in low income areas in Accra, Ghana', *Water and Health*, vol. 12, no. 2, pp. 318–331.

Mascarenhas, M. (2007), 'Where the waters divide: first nations, tainted water and environmental justice in Canada', *Local Environment*, vol. 12, no. 2, pp. 565–577.

McCully, P. (1996), *Silenced rivers: the ecology and politics of large dams*, London: Zed Books.

McGregor, D. (2009), 'Honouring our relations: an Anishnaabe perspective', in J. Aqyeman, P. Cole, R. Haluza-DeLay & P. O'Riley (eds), *Speaking for ourselves: environmental justice in Canada*, Vancouver: University of British Columbia Press.

McKenzie, S.O. (2015), 'Yakye Axa v. Paraguay: upholding and framing the human right to water', in Y. Haeck, O. Ruiz-Chiriboga & C.B. Herrera (eds), *The Inter-American court of human rights: theory and practice, present and future*, Cambridge: Intersentia Press.

Mehta, L. (2001), 'The manufacture of popular perceptions of scarcity: dams and water-related narratives in Gujarat, India', *World Development*, vol. 29, no. 12, pp. 2025–2041.

Mehta, L. (2014), 'Water and human development', *World Development*, vol. 59, pp. 59–69.

Mels, A., Castellano, D., Braadbaart, O., Veenstra, S., Dijkstra, I., Meulman, B., et al. (2009), 'Sanitation services for the informal settlements of Cape Town, South Africa', *Desalination*, vol. 248, pp. 330–337.

Mirosa, O. & Harris, L. (2012), 'The human right to water: contemporary challenges and contours of a global debate', *Antipode*, vol. 44, no. 3, pp. 932–949.

Molle, F. & Berkoff, J. (2006), *Cities versus agriculture: Revisiting intersectoral water transfers, potential gains and conflicts*, Colombo, Sri Lanka: Comprehensive Assessment Secretariat, pp. 1–80.

Morales, M., Harris, L. & Öberg, G. (2014), 'Citizenshit: the right to flush and the urban sanitation imaginary', *Environment and Planning A*, vol. 46, no. 12, pp. 2816–2833.

Morinville, C. & Rodina, L. (2013), 'Rethinking the human right to water: water access and dispossession in Botswana's Central Kalahari Game Reserve', *Geoforum*, vol. 49, pp. 150–159.

Neimanis, A. (2013), 'Feminist subjectivity, watered', *Feminist Review*, vol. 103, pp. 23–41.

Neumayer, E. & Plumper, T. (2007), 'The gendered nature of natural disasters: the impact of catastrophic events on the gender gap in life expectancy, 1981–2002', *Annals of the Association of American Geographers*, vol. 97, no. 3, pp. 551–566.

Nightingale, A.J. (2015), 'Adaptive scholarship and situated knowledges? Hybrid methodologies and plural epistemologies in climate change adaptation research', *Area*, vol. 48, no. 1, pp. 41–47.

O'Reilly, K., Halvorson, S.J., Sultana, F. & Laurie, N. (2011), 'Introduction: global perspectives on gender-water geographies', *Gender, Place & Culture*, vol. 16, no. 4, pp. 381–385.

Perreault, T. (2013), 'Dispossession by accumulation? Mining, water and the nature of enclosure on the Bolivian Altiplano', *Antipode*, vol. 45, no. 5, pp. 1050–1069.

Perreault, T. (2014), 'What kind of governance for what kind of equity? Towards a theorization of justice in water governance', *Water International*, vol. 39, no. 2, pp. 233–245.

Ranganathan, M. (2016), 'Thinking with Flint: racial liberalism and the roots of an American water tragedy', *Capitalism Nature Socialism*, vol. 27, no. 3, pp. 17–33.

Ribot, J. (2009), 'Vulnerability does not just fall from the sky: toward multi-scale pro-poor climate policy', in R. Mearns & A. Norton (eds), *Social dimensions of climate change: equity and vulnerability in a warming world*, Washington: The World Bank.

Rodina, L. (2016). 'Human right to water in Khayelitsha, South Africa: lessons from a "lived experiences" perspective', *Geoforum*, vol. 72, pp. 58–66.

Simeone, T. (2010), *Safe drinking water in First Nations communities*, Parliamentary Information and Research Service, viewed 8 January 2016, www.parl.gc.ca/content/LOP/ResearchPublications/prb 0843-e.htm

Simms, B. (2014), *"All of the water that is in our reserves and that is in our territory is ours": colonial and Indigenous water governance in unceded Indigenous territories in British Columbia*, MA Thesis, University of British Columbia.

Simms, R., Harris, L., Bakker, K. & Joe, N. (2016), 'Navigating the tensions in collaborative watershed governance: water governance and indigenous communities in British Columbia, Canada', *Geoforum* vol. 73, pp.6–16.

Spence, N. & Walters, D. (2012), '"Is it safe?" Risk perception and drinking water in a vulnerable population', *The International Indigenous Policy Journal*, vol. 3, no. 3.

Sugden, F., Maskey, N., Clement, F., Ramesh, V. & Philip, A. (2015), 'Agrarian stress and climate change in the Eastern Gangetic Plains: gendered vulnerability in a stratified social formation', *Global Environmental Change*, vol. 29, pp. 258–269.

Sultana, F. (2011), 'Suffering for water, suffering from water: emotional geographies of resource access, control and conflict', *Geoforum*, vol. 42, no. 2, pp. 163–172.

Sultana, F. (2014), 'Gendering climate change: geographical insights', *The Geographical Journal*, vol. 66, no. 3, pp. 372–381.

Swyngedouw, E. (2004), *Social power and the urbanization of water: flows of power*, Oxford: Oxford University Press.

Torio, P. (2016), *Water privatization in Metro Manila: assessing the state of equitable water provision*, PhD Thesis, University of British Columbia.

United Nations Development Programme (UNDP) (2006), *Human Development Report 2006*, viewed 8 January 2016, http://hdr.undp.org/sites/default/files/reports/267/hdr06-complete.pdf

United Nations Development Programme (UNDP) (2010), *The Human Right to Water. UN-Water Decade Programme on Advocacy and Communication*, 8 January 2016, www.un.org/waterforlifedecade/pdf/facts_and_figures_human_right_to_water_eng.pdf

United Nations Educational, Scientific and Cultural Organization (UNESCO) (2015), *Water for women*, viewed 8 January 2016, www.unwater.org/fileadmin/user_upload/worldwaterday2015/docs/Water%20For%20Women.pdf

von Schnitzler, A. (2008), 'Citizenship prepaid: water, calculability, and techno-politics in South Africa', *Journal of Southern African Studies*, vol. 34, no. 4, pp. 899–917.

Watts, M.J. & Bohle, H.G. (1993), 'The space of vulnerability: the causal structure of hunger and famine', *Progress in Human Geography*, vol. 17, pp. 43–67.

Wilson, N.J. (2014), 'Indigenous water governance: insights from the hydrosocial relations of the Koyukon Athabascan village of Ruby, Alaska', *Geoforum*, vol. 57, pp. 1–11.

Wilson, N.J., Walter, M.T. & Waterhouse, J. (2015), 'Indigenous knowledge of hydrologic change in the Yukon River Basin: a case study of Ruby, Alaska', *ARCTIC*, vol. 68, no. 1, pp. 93–106.

World Health Organization (WHO) (2015a), *Drinking-water: key facts*, World Health Organization Media Centre, viewed 8 January 2016, www.who.int/mediacentre/factsheets/fs391/en/

Wutich, A. (2009), 'Intrahousehold disparities in women and men's experiences of water insecurity and emotional distress in urban Bolivia', *Medical Anthropology Quarterly*, vol. 23, no. 4, pp. 436–54.

Yates, J., et al. (2017), 'Multiple ontologies of water: politics, conflict and implications for governance' *Environment and Planning D: Society and Space*.

Ziervogel, G., Waddell, J., Smit, W., & Taylor, A. (2014), 'Flooding in Cape Town's informal settlements: barriers to collaborative urban risk governance', *South African Geographical Journal*, vol. 98, no. 1, pp. 1–20.

28

ENVIRONMENTAL JUSTICE AND FLOOD HAZARDS

A conceptual framework applied to emerging findings and future research needs

Timothy W. Collins and Sara E. Grineski

Introduction

In recent decades, distributive justice issues have become centrally important in risk assessment of environmental risks and hazards. Recognition of social inequalities in the spatial distribution of toxic pollution hazards and risks – first in the United States and then globally – has spawned a vibrant social movement, policy debates, and a large body of research. Under the rubric of environmental justice (EJ) analysis, numerous quantitative spatial studies have focused on examining whether toxic risk burdens are distributed evenly across people and places, or the extent to which racial/ethnic minority, lower class, or other oppressed communities are disproportionately exposed to toxic pollution hazards (see Chapters 25 and 26). Various statistical and spatial analytic techniques (see Chapters 15, 16, 17) have been employed in a variety of locational contexts worldwide. The majority of studies indicate that racial/ethnic minorities, people of low socioeconomic status and other socially marginalized groups experience disproportionate residential exposure to hazards such as air pollution, chemical releases from industrial facilities, and residual risks from inactive hazardous waste sites (Chakraborty et al. 2011; Walker 2012). The impacts of Hurricane Katrina in 2005 and subsequent state response failures catalysed academic inquiry and activism to address social injustices associated with events such as hurricanes and floods. More specifically, concerns regarding the uneven impacts of Katrina on African-American, elderly and low-income residents of New Orleans led to an expansion of empirical EJ research to include the unjust implications of flooding (Bullard and Wright 2009; Colten 2007; Ueland and Warf 2006; Walker and Burningham 2011).

A premise of this chapter is that, while distributive EJ research has produced important knowledge regarding injustices associated with toxic pollution hazards, this body of work is bound by limitations that constrain empirical and normative understandings of EJ implications of flood hazards, despite recent focus in the EJ literature. From an EJ perspective, this chapter first highlights the complex and sometimes counterintuitive findings emerging from relatively recent studies of flood hazards. Next, it highlights some key limitations of conventional distributive EJ research before introducing a conceptual framework with corresponding propositions for enhancing understanding of the complex distributive dimensions of environmental injustices associated with flood hazards.

Environmental justice, social vulnerability and flood hazards: empirical findings

EJ-related studies of floods have yielded ambiguous findings regarding relationships between indicators of social vulnerability/marginality and flood risks. Divergent relationships between indicators of social vulnerability and risks associated with flood hazards have been found for studies focused on disproportionality in pre-flood hazard exposure, versus those focused on disproportionality associated with actual flood events.

One approach to examining EJ implications of flooding is to clarify which people live at greatest risk of flooding or the social characteristics of those residing in locations within or proximate to floodable zones. Such distributive EJ-type studies have not consistently found associations between indicators of social vulnerability/marginality and pre-flood hazard exposure. More than a few quantitative spatial analyses have found, counterintuitively (from a conventional distributive EJ perspective which assumes that socially disadvantaged populations will be exposed to greater risks), that socially advantaged people (i.e. those of low social vulnerability) experience the highest pre-event exposure to flood hazards in particular contexts (Chakraborty et al. 2014; Fielding and Burningham 2005; Grineski et al. 2013; Montgomery and Chakraborty 2013; Ueland and Warf 2006). Of the pre-flood exposure studies focused in the US, we found only one reporting a consistent pattern of areas characterized by high social vulnerability experiencing disproportionately high pre-flood hazard exposure (Burton and Cutter 2008). Notably, this study is based on a measure of spatial flood risks associated with failure-susceptible levees, a variable that integrates (a lack of) flood hazard mitigation (Burton and Cutter 2008), and which points toward issues of process/procedural injustice in the context of flooding (Johnson et al. 2007). Results from other US-based studies have not revealed clear patterns of disproportionate exposure for socially marginal groups (e.g. Mantaay and Maroko 2009), including one focused on spatial associations between neighbourhood socioeconomic composition and flood exposure in Hurricane Katrina (Masozera et al. 2007). In the UK, research reveals that flood risks in inland areas are not inequitably distributed, but that coastal flood risks are borne disproportionately within lower social class areas subject to economic decline via port-based deindustrialization (Fielding 2007; Walker and Burningham 2011; Walker 2012).

On the other hand, far more numerous and methodologically broader ranging studies of flood impacts, response and post-event recovery have typically documented disproportionate impacts for traditional EJ communities. Most such work has not been conducted under the rubric of EJ research, but rather has been conducted using a social vulnerability to hazards/ disasters lens (Rufat et al. 2015). The last three decades have marked the emergence of a social vulnerability perspective on hazards and disasters, which emphasizes the influence of social inequalities on differential risks (Cutter 1996; Wisner et al. 2004). Studies of disaster events associated with flood hazards, as well as a range of hazard types, reveal that disadvantaged social groups are at increased risk of experiencing debilitating damage, uncompensated loss, and long-term suffering. Key characteristics explaining variations in natural disaster impacts are context-dependent, but often include social class, race, ethnicity, gender, age, disability and health status, and immigration and citizenship status – some of the axes of social inequality that EJ research focuses on (Cutter et al. 2003; Wisner et al. 2004).

Vulnerability studies reveal that socially marginalized people have reduced capacities for self-protection in terms of mitigating flood hazards at home sites pre-event; evacuating in response to flooding; returning home or to employment following flood-induced livelihood disruption; and accessing social protection resources to reduce the impacts of flooding such as flood insurance, pre-flood hazard mitigation infrastructure, emergency response information, and post-disaster assistance (Collins 2009, 2010; Elliott and Pais 2006; Maldonado et al. 2016; Mustafa 2005; Pelling 1999). Additionally, such studies indicate that socially vulnerable groups experience the more adverse consequences of flood disasters in terms of morbidity and mortality (Collins et al. 2013; Jimenez et al. 2013; Zahran et al. 2008), which may reflect both their increased exposure to flooding during actual flood events and their reduced access to protective resources.

To summarize, in contrast to the focus of pre-flood studies on the *distributive* dimension of (in)justice in terms of the spatial correspondence between traditional EJ communities and flood hazards, studies of social vulnerability to floods typically emphasize the role of *process-based* inequalities in shaping disproportionate risks for socially marginal groups of people. Thus, from an EJ perspective, the anomalous and divergent findings from distributive and process-based studies of flood hazards are difficult to reconcile. To better integrate these findings within an EJ perspective, the next sections identify limitations of the distributive EJ approach and propose a conceptual framework with specific considerations that analysts should make when examining the EJ implications of flood hazards via future studies.

Two considerations for expanding understanding of the environmental justice implications of flooding

To expand knowledge of the EJ implications of flooding, we must begin from the premise that research on the EJ dimensions of flooding (and the broader quantitative distributive EJ literature) suffers from several limitations. We focus on two limitations here, since they are most relevant to conceptualizing EJ in the context of flood hazards. The distributive EJ literature is founded on 1) an incomplete conception of risk as hazard exposure, which neglects people's capacities to reduce risks; and 2) an underlying assumption that people primarily aim to avoid hazards, which neglects other factors influencing residential decision-making such as locational benefits (see also Chapter 5).

With reference to limitation (1), in order to broaden the EJ conception of risk, it is useful to engage concepts from studies of social vulnerability to hazards/disasters. The social vulnerability to hazards/disasters and EJ research fields are topically related, yet conceptually distinct. A theoretical premise of vulnerability studies is that *risk* is determined partly by human exposure to a *hazard* and partly by people's *social vulnerability*. While there is debate about the meaning and measurement of social vulnerability, the following definition is useful: *"the characteristics of a person or group and their situation that influence their capacity to anticipate, cope with, resist and recover from the impact of a natural hazard"* (Wisner et al. 2004: 11). The recognition that risk is shaped in part by social vulnerability factors that condition people's capacities to mitigate risks is an important advance upon the conception of *risk as hazard exposure* reflected in the distributive EJ literature (Collins 2010). The environmental equity conception of risk supports the expectation that the least powerful groups in society inhabit the most hazardous environments. While findings from most environmental equity studies of *technological* hazards validate that expectation, results from spatial analyses of social patterns of *natural* hazard exposure most often do not (Kates and Haarmann 1992; Maantay and Maroko 2009; Masozera et al. 2007). We argue that the vulnerability studies conception of risk facilitates understanding of complex socio-spatial patterns of exposure to various hazard types – patterns that would be viewed as anomalous from an environmental equity perspective (e.g. the residence of socially elite people in high hazard zones) – because it illuminates capacities that enable people to mitigate risks associated with living in hazardous environments. A strength of the vulnerability perspective is the acknowledgement that people have varying capacities to deploy resources (e.g. from their own reserves or through access to social protections), to enhance security and reduce risks, even under conditions of high hazard exposure.

With regard to limitation (2), distributive EJ research has been limited by its one-dimensional treatment of the environment as hazard. EJ scholars have focused their attention primarily on negative environmental attributes. The EJ expectation that vulnerable people inhabit the most hazardous locations, for example, is rooted in the assumption that environments present only hazards and not benefits to people (Collins 2008, 2009, 2010). Although a bevy of more recent EJ studies have examined distributional injustices associated with positive environmental attributes, e.g. amenities such as tree cover, parks, and green space (Landry and Chakraborty 2009; Wolch et al. 2005), this research has also treated the environment in a one-dimensional manner. Few distributive EJ or social vulnerability analysts, however, have examined the role of locational benefits in the distribution of risks across social groups. A handful of EJ studies suggest that employment opportunities – a type of locational benefit that stems from residence near industrial zones – may influence residential patterns of air pollution exposures (Boone 2002; Oakes et al. 1996). For example, in Baltimore (Maryland, USA), Boone (2002) found that whites historically lived closer to factories than African Americans, which was counterintuitive from an EJ perspective; it turns out that housing proximate to factories was reserved for white workers, while African Americans lived further away, without access to gainful employment. Similarly, a few vulnerability analysts have described how exposure to hazards is influenced by associated locational benefits such as production and livelihood opportunities, accessibility, and amenities (e.g. Pelling 1999; Wisner et al. 2004). However, the influence of locational benefits on exposure to risks has also been a tangential emphasis of the literature on social vulnerability to hazards.

Locational benefits associated with hazards have been a more central focus of scholarship emanating from the disciplines of psychology and economics than they have from the fields most central to EJ scholarship (sociology, geography, public health). The psychology-based risk perception literature reveals that, in the process of making decisions about a place to live, people are influenced by multiple interacting factors, with an important one being the trade-off between the perceived risks and benefits of a location (Slovic 2000). Hazards to which people knowingly expose themselves (assuming they have some agency) are associated with benefits; otherwise, people would generally avoid risky locations. To the extent that the benefits are perceived as exceeding the risks of hazard exposure, people may choose to remain at risk (technically speaking). Technical risk assessments generally find a positive correlation between benefits and hazards; i.e. people tend to accept higher levels of hazard exposure in locations where the associated benefits are high (Slovic et al. 1980). Scenic views and a sense of social exclusivity, for example, entice people to live on highly combustible and geologically unstable slopes in Malibu (California, USA) (Davis 1998). In contrast to technical risk assessments, people's perceptions of benefits and hazards tend to be negatively correlated; residents typically underestimate hazards in environments where associated benefits are perceived to be great, which may lead them, even those with relatively low social vulnerability, to accept higher levels of hazard exposure (Alhakami and Slovic 1994; Siegrist and Cvetkovich 2000; Slovic 1987).

Additionally, economists have explored spatial relationships between hazard exposure, environmental benefits, and land values. Along the coastline, flood risk is often associated with environmental amenities such as ocean views and proximity to beaches. Research shows that proximity to shoreline in the US is highly desirable in residential land markets (Earnhart 2001). Bin and Kruse (2006) found that properties located within the 100-year flood zone (with a one per cent or greater chance of flooding per year) had significantly higher cash values than comparable properties outside of the 100-year flood zone. They hypothesized that this counter-intuitive result may relate to the close relationship between risk and amenity value in coastal settings. In a subsequent study, Bin et al. (2008) statistically isolated the effects of amenity value and flood risk on property values, providing support for that hypothesis.

In sum, analytical consideration of risk mitigation and locational benefits is needed in order to expand understanding of the EJ implications of flood hazards. Next, extending from the literatures on EJ and social vulnerability to hazards/disasters, we present what we see as key elements that should be integrated in a conceptual framework designed to address these two limitations and advance knowledge of socio-spatial influences on exposure to flood hazards, which may be applicable to other types of hazards as well.

Conceptualizing environmental justice implications of flooding

Divisibility refers to the ability to separate (geographically speaking) the risks and benefits associated with a hazard (Kates 1971). Theoretically, divisibility plays a pivotal role in spatial relationships between social vulnerability and hazard exposure. In particular, fundamental differences in *divisibility* may generate divergent social distributions of exposure to flood vs. air pollution hazards. However, divisibility has not been conceptually well integrated in either EJ or vulnerability studies. This relates to the aforementioned lack of consideration of both negative and positive environmental attributes in these two fields. Theoretically, differences exist in divisibility between "natural" (e.g. flood) and "technological" (e.g. sources of toxic air pollution) hazards (see Chapter 26). The benefits associated with many natural hazards must be consumed in place (i.e. they are not geographically separable), whereas the risks/benefits associated with technological hazards are separable by design. Thus, risks/benefits may typically be less divisible for many types of natural hazards and more divisible for many types of technological hazards.

To conceptualize how divisibility plays a pivotal role in distributive aspects of EJ, it is useful to draw a comparison between air pollution – the most studied type of hazard in the EJ literature – and flooding. Figure 28.1 depicts the traditional EJ conceptual model of the relationship between social vulnerability and residential hazard exposure, in which there is assumed to be a positive relationship between social vulnerability and hazard exposure. This model does not account for the divisibility of risks/benefits associated with hazards or people's (differential) capacities to mitigate hazards, and is based largely on EJ scholars' empirical analyses of distributive injustices associated with residential exposure to air pollution and other toxic pollution hazards.

Figure 28.2 depicts our conceptual model of the relationship between social vulnerability and residential exposure to air pollution from major stationary sources of hazardous air pollutants (such as factories), where commodities are produced and distributed for profit and pollution is released into surrounding areas. Nearby residents may be at risk and not benefiting from their proximity to polluting factories, especially when the factories do not provide them with employment. This is an example of how stationary sources of air pollution are characterized by a high degree of divisibility. With respect to environmental hazards in which risks/benefits are highly divisible, proximate locations tend to be less desirable for residential uses and have relatively low land values, which may influence the in-migration of socially

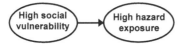

Figure 28.1 Traditional EJ conceptual model of relationship between social vulnerability and residential hazard exposure

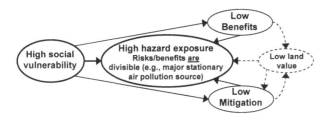

Figure 28.2 Conceptual model of relationship between social vulnerability and residential air pollution exposure

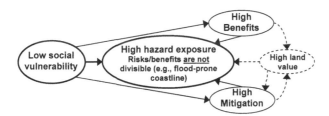

Figure 28.3 Conceptual model of relationship between social vulnerability and residential flood risk

vulnerable residents due to the relatively low cost of housing (Been 1994; Oakes et al. 1996; Hernandez et al. 2015). At the same time, socially vulnerable residents typically lack political power and have reduced capacities to mitigate risks of hazard exposure and, through time, lower value lands near their homes may be continually targeted for development by industries that emit toxic air pollution (Bolin et al. 2005; Cole and Foster 2001; Pulido 2000).

Figure 28.3 depicts our conceptual model of the relationship between social vulnerability and residential flood risk. In contrast to our conceptual model for air pollution, and as our previous points about coastal amenities suggest, risks and benefits are largely indivisible in the case of coastal flood hazards. With respect to hazard types in which risks and benefits are highly indivisible, socioeconomic factors may influence hazard exposure indirectly through the capitalization of amenities in land markets. Socially powerful groups may reside in locations characterized by high levels of hazard exposure as they are able to garner/implement structural and nonstructural forms of mitigation to minimize risks and protect values (Collins 2008, 2009, 2010; Davis 1998).

Recent collaborative work focused on the Miami, Florida metro area illustrates the applicability of our conceptual framework for understanding environmental injustice in the context of flood risk. Metro Miami's population of 6 million is ethnically diverse, with non-Hispanic White residents accounting for only 35 per cent of the total population, and Hispanics and non-Hispanic Blacks, respectively, accounting for 42 per cent and 21 per cent of the population. Miami is also highly residentially segregated along ethnic lines and characterized by some of the highest levels of income inequality in the US. In terms of flood risks, metro Miami is one of the most hurricane-prone urban areas in the world. A study of coastal flood risk ranked Miami as first in asset exposure and fourth in population exposure for cities world-wide (Nicholls et al. 2008). The three counties of metro Miami (Broward, Miami-Dade, and Palm Beach) have been ranked first, second, and third, respectively, for flood-induced property damage in Florida (Brody et al. 2007). Miami is thus a particularly suitable location for analysing the EJ implications of flood hazards.

In alignment with the models presented above, recent studies of metro Miami reveal that distributive relationships between social vulnerability indicators (e.g. lower socioeconomic status, racial/ethnic minority status) and flood risk tend to be opposite from the relationships between social vulnerability indicators and air toxics. That is, greater social vulnerability is typically negatively associated with residential exposure to flooding (Chakraborty et al. 2014; Grineski et al. 2015), while it is generally positively associated with exposure to air pollution risk (Chakraborty et al. 2016; Collins et al. 2015; Grineski et al. 2013, 2015). Evidence suggests that the divergence in social profiles of risk for residential exposure to flooding vis-à-vis air pollution is due in part to the low divisibility of locational benefits within coastal landscapes from the risks of residential flood exposures (as opposed to the relatively higher divisibility of benefits from risks in the case of air pollution). In Miami, floodable landscapes along the coast are amenity rich. Indeed, Miami's urban structure is largely shaped by the value of coastal access. In Miami, indicators of coastal amenity value (e.g. housing cash value, prevalence of seasonal/recreation homes, proximity to public beach access) exhibit robust positive relation-ships with coastal flood risk (Chakraborty et al. 2014; Montgomery and Chakraborty 2015), and a pattern of environmental injustice exists whereby socially vulnerable groups experience constrained access to coastal amenities (e.g. public beaches) (Montgomery et al. 2015). When high risk flood zones in Miami are disaggregated based on inland or coastal zonation, social vulnerability indicators are generally positively associated with inland flood zones and nega-tively associated with coastal flood zones (Chakraborty et al. 2014). This provides further evidence that the counterintuitive social patterning of exposure to coastal flood risk in Miami is partly contingent upon the indivisibility of amenity values from those high risk landscapes. In contrast, inland flood zones – which typically lack the water-based amenities of coastal zones – are characterized by a risk pattern that aligns with traditional EJ expectations.

To fully comprehend the EJ implications of those distributive patterns, we must apply a process-based perspective. While metro Miami is generally at very high risk of flooding, the fact that the most socially powerful residents are concentrated in areas at the greatest risk of coastal flooding suggests that amenity values must outweigh the costs of flood risks. In this context, socially privileged residents (e.g. highly affluent non-Hispanic Whites) are able to externalize risks of coastal living to all US taxpayers through insurance coverage offered by the National Flood Insurance Program (NFIP), which provides highly subsidized premiums to residents in high risk coastal flood zones (US Congressional Budget Office 2007). Socially privileged residents of metro Miami typically choose to live in or near coastal locations because coastal flood risks are mitigated through a variety of public investments (e.g. flood insurance, engineered flood control structures, "beach nourishment" programmes, etc.). Thus, the risks of dwelling within coastal zones are offset by institutionally mediated access to mitigation resources, including flood insurance policies with premiums well below the actuarial costs of flood risks. Indeed, Chakraborty et al.'s (2014) results align with our conceptual model, in that they suggest that flood insurance subsidization through the NFIP in particular may inflate property values and facilitate residential risk-taking, particularly for socially privileged households, because it enables them to externalize risks in their pursuit of coastal amenities.

Recent efforts in the US Congress to systematically increase flood insurance premiums to match actuarial rates has generated resistance from representatives of flood-prone coastal districts as well as home construction and real estate industry groups, who argue that planned rate increases will depress real estate markets in areas where flood insurance is currently heavily subsidized by the NFIP. Although increases in flood insurance premiums would probably affect real estate markets in such locations, this reality underscores the fact that NFIP subsidization has indeed facilitated land speculation, housing development, and residential risk-taking in flood-prone coastal locations.

Some socially vulnerable residents of metro Miami also experience high exposure to flood risks (especially within inland flood zones), which is of particular concern owing to their generally reduced capacities to prepare for, respond to, and recover from flood events (Chakraborty et al. 2014). Miami is one of the most economically unequal metropolitan areas of the US and also perhaps the most flood-prone. Flood exposure similar to that experienced in New Orleans during Hurricane Katrina would generate highly disparate adverse impacts based on prevailing racial/ethnic and economic inequalities. The presence of socially vulnerable groups at risk of inland flooding in greater Miami creates the potential for a disaster of a magnitude unprecedented in US history. Beyond the NFIP, which targets the recovery needs of property owners in particular, protective resources for socially vulnerable groups at risk of flooding in metro Miami are minimally available, and some highly vulnerable groups (e.g. undocumented immigrant renter-occupants) are fortunate if they are able to access any resources to partially meet their recovery needs (Maldonado et al. 2016). Without the implementation of genuine need-based programmes to ameliorate vulnerability among socially marginal residents of flood-prone areas, the risks of a flood disaster of monumental proportions in metro Miami will remain alarmingly high. In sum, environmental injustices in metro Miami appear to be (re)produced as socially privileged groups seek to monopolize access to coastal amenities while the added costs of flood risk mitigation are treated as an externality conveyed to the broader public; in the process, socially vulnerable residents are relegated to areas with air toxics and/or inland flood risks, where they are less able to access protective resources or enjoy coastal amenities.

Conclusion: propositions for deepening understanding

We conclude with three propositions for future research on the EJ implications of flood hazards. First, to improve understanding, a process-based EJ perspective focused on unequal power relations and differential capacities to mitigate risks should be integrated within distributive EJ analysis, which has traditionally been pattern-based and focused on quantifying disproportionate exposure to hazards. To better analyse process–pattern linkages and shed new light on the formation of environmental injustice in the context of flood hazards, distributive EJ analysts should consider the differential capacities of individuals and social groups to mitigate flood risks by deploying or appropriating resources (e.g. from their own reserves or through access to social protections made available through institutions), in order to enhance their security and reduce the threat of loss, even while living amid flood-prone environments. Here, we conceive of mitigation broadly as any structural and nonstructural strategy used to minimize loss and enable recovery from the impacts of hazard exposure (NRC 2006). In the context of flood hazards, risks of damaging exposures can be mitigated through a variety of means. Alterations to the built environment ("structural" mitigation strategies) can reduce physical exposure to flood hazards and "nonstructural" strategies can reduce social vulnerability to flooding. The built environment can be modified through structural mitigation strategies to protect human settlement and reduce exposure to floods in myriad ways, for example through the construction of flood protection structures, through alterations in building design and land cover, and through the designation, restoration, and maintenance of undeveloped public floodways/coastal zones in cities (Alexander 1993). Specific nonstructural strategies to reduce vulnerability to floods include: land-use controls to restrict development in hazardous zones, weather forecasting to predict flood events, emergency response plans to mitigate flood impacts, flood insurance to compensate losses, and post-event aid to enable recovery (Alexander 1993). More broadly, nonstructural strategies to reduce vulnerability include labour laws and access to safe housing, health care, education, and social services. Such entitlements increase people's livelihood security, physical safety, and capacity to cope with flood hazards. Reliance on these mitigation strategies may have dramatic effects upon the actual societal distribution of flood risk as well as the measurement of distributive environmental (in)justice in the context of residential flood risks.

Second, the reality that flood-prone environments typically reflect both positive and negative attributes necessitates that scholars examine the EJ implications of both flood hazards *and* water-based benefits, when making inferences regarding distributive (in)justices. Expanding upon the conceptual model introduced above (Figure 28.3), from an EJ perspective, we must acknowledge the fact that floodable landscapes are often simultaneously environmentally attractive and biophysically dynamic, and, furthermore, the risk instantiates an obstacle to accumulation for elite social groups (i.e. those of the lowest social vulnerability). Maximizing values over the long term (i.e. increasing the exchange value of private property and consuming the use values of coastal environments while averting flood risks) necessitates the enlistment of state and market institutions to redirect appropriated social surplus toward mitigating flood risks. Thus, analysts of the EJ implications of flood risks should direct their attention to the multidimensional character of flood-prone environments (which is necessary since sources of risk and reward are typically not easily separable), with recognition that counterintuitive social patterns of exposure to flood hazards may result from processes founded on highly uneven power relations.

Consider the findings from Collins' (2010) study of a flood disaster at the US–Mexican border. While elite residents on the US side voluntarily exposed themselves to flood hazards in their pursuit of environmental amenities (e.g. scenic arroyo views), they were able to externalize risks by harnessing a disproportionate share of the flood mitigation resources that were redistributed by state and market institutions to support flood preparedness, response, recovery and reconstruction before, during and after the event. It would be erroneous to infer that those social elites suffered environmental injustices even though some experienced acute flood impacts. The case study, in fact, supports precisely the opposite interpretation, i.e. social elites living in scenic, flood-prone environments had continually benefited through environmentally unjust processes (Collins 2010). Such findings highlight limitations to the one-dimensional conception of the *environment as hazard* (or, alternatively, as *amenity*), which is employed (often tacitly) in EJ research.

Third, analysts should integrate distributive and procedural approaches with a focus on pattern–process linkages in order to more fully comprehend environmental injustices in the context of flood hazards, as well as other unevenly distributed environmental risks and resources of EJ concern. The fact that seemingly contradictory social patterns of exposure to hazards can flow from environmentally unjust processes provides support for the point that pattern–process linkages merit closer attention in EJ research. Multiscalar, mixed methods, and comparative research approaches should be integrated in studies of distributive and procedural EJ implications of flooding. In addition to quantitative analyses of aggregated socio-demographic data for census-designated areal units, there is a need to focus on people's decision-making regarding costs and benefits associated with flood-prone residential locations, including their subjectivities and differential material constraints, in order to characterize the environmental injustices they experience (see Chapters 19 and 21). Such micro-scale research demands the inclusion of qualitative methods, which are best suited to uncovering complex, situational and subjective factors that influence experiences of injustice in the context of flood hazards.

Micro-level compositional qualitative research focused on distributional injustices has been underutilized (Hernandez et al. 2015). It holds the potential to more fully explain patterns that quantitative distributive EJ studies have found and, perhaps, to reveal important EJ implications of flooding that have been overlooked. Analysts must also recognize that contemporary patterns of risk exposure across social groups articulate with macro-scale historical-geographical processes (see Chapter 18 on historical approaches). An explicitly multiscalar perspective on environmental injustice, originally advanced in the field of urban political ecology (Swyngedouw and Heynen 2003; Collins 2010), has not been widely adopted by EJ scholars. Given the complex EJ implications of flooding, we believe there is a need for EJ scholars to embrace a multiscalar perspective, since it fosters insights into how distributional injustices are actively (re)produced through the articulation of constraining and enabling forces.

References

Alexander, D. (1993). *Natural Disasters*. New York: Chapman and Hall.

Alhakami, A., & Slovic, P. (1994). A psychological study of the inverse relationship between perceived risk and perceived benefit. *Risk Analysis*, 14, 1085–1096.

Been, V. (1994). Locally undesirable land uses in minority neighborhoods: Disproportionate siting or market dynamics? *The Yale Law Journal*, 103, 1383–1406.

Bin, O., & J.B. Kruse. (2006). Real estate market response to coastal flood hazards. *Natural Hazards Review*, 7, 137–144.

Bin, O., Crawford, T.W., Kruse, J.B., & Landry, C.E. (2008). Viewscapes and flood hazard: Coastal housing market response to amenities and risk. *Land Economics*, 84, 434–448.

Bolin, B., Grineski, S. & Collins, T. (2005). The geography of despair: Environmental racism and the making of South Phoenix, Arizona, USA. *Human Ecology Review*, 12, 155–167.

Boone, C.G. (2002). An assessment and explanation of environmental inequity in Baltimore. *Urban Geography*, 23, 581–595.

Brody, S., Zahran, S., Maghelal, P., Grover, H., & Highfield, W. (2007). The rising costs of floods: Examining the impact of planning and development decisions on property damage in Florida. *Journal of the American Planning Association*, 73, 330–345.

Bullard, R., & Wright, B. (2009). *Race, Place, and Environmental Justice after Hurricane Katrina: Struggles to Reclaim, Rebuild, and Revitalize New Orleans and the Gulf Coast*. Boulder, CO: Westview.

Burton, C., & Cutter, S.L. (2008). Levee failures and social vulnerability in the Sacramento-San Joaquin Delta area, California. *Natural Hazards Review*, 9, 136–149.

Chakraborty, J., Maantay, J., & Brender, J. (2011). Disproportionate proximity to environmental health hazards: Methods, models, and measurement. *American Journal of Public Health*, 101, S27–S36.

Chakraborty, J., Collins, T., Montgomery, M., & Grineski, S. (2014). Social and spatial inequities in exposure to flood risk in Miami, Florida. *Natural Hazards Review*, 15, 04014006.

Chakraborty, J., Collins, T., & Grineski, S. (2016). Cancer risks from exposure to vehicular air pollution: A household level analysis of intra-ethnic heterogeneity in Miami, Florida. *Urban Geography*. doi: 10.1080/02723638.2016.1150112.

Cole, L., & Foster, S.R. (2001). *From the Ground Up: Environmental Racism and the Rise of the Environmental Justice Movement*. New York: New York University Press.

Collins, T. (2008). The political ecology of hazard vulnerability: Marginalization, facilitation and the production of differential risk to urban wildfires in Arizona's White Mountains. *Journal of Political Ecology*, 15, 21–43.

Collins, T. (2009). The production of unequal risk in hazardscapes: An explanatory frame applied to disaster at the U.S.-Mexico border. *Geoforum*, 40, 589–601.

Collins, T. (2010). Marginalization, facilitation, and the production of unequal risk: The 2006 *Paso del Norte* floods. *Antipode*, 42, 258–288.

Collins, T., Jimenez, A., & Grineski, S. (2013). Hispanic health disparities after a flood disaster: Results of a population-based survey of individuals experiencing home site damage in El Paso (Texas, USA). *Journal of Immigrant and Minority Health*, 15, 415–426.

Collins, T., Grineski, S., & Chakraborty, J. (2015). Household-level disparities in cancer risks from vehicular air pollution in Miami. *Environmental Research Letters*, 10, 095008.

Colten, C. (2007). Environmental justice in a landscape of tragedy. *Technology in Society*, 29, 173–179.

Cutter, S.L. (1996). Vulnerability to environmental hazards. *Progress in Human Geography*, 20, 529–539.

Cutter, S.L, Boruff, B.J., & Shirley, W.L. (2003). Social vulnerability to environmental hazards. *Social Science Quarterly*, 84, 242–261.

Davis, M. (1998). *The Ecology of Fear: Los Angeles and the Imagination of Disaster*, New York: Metropolitan.

Earnhart, D. (2001). Combining revealed and stated preference methods to value environmental wetland permit decisions. *Land Economics*, 55, 213–222.

Elliott, J.R., & Pais, J. (2006). Race, class, and Hurricane Katrina: Social differences in human responses to disaster. *Social Science Research*, 35, 295–321.

Fielding, J. (2007). Environmental injustice or just the lie of the land: An investigation of the socio-economic class of those at risk from flooding in England and Wales. *Sociological Research Online*, 12(4), 4.

Fielding, J., & Burningham, K. (2005). Environmental inequality and flood hazard. *Local Environment*, 10, 379–395.

Grineski, S., Collins, T., & Chakraborty, J. (2013). Hispanic heterogeneity and environmental injustice: Intra-ethnic patterns of exposure to cancer risks from vehicular air pollution in Miami. *Population & Environment*, 35, 26–44.

Grineski, S., Collins, T., Chakraborty, J., & Montgomery, M. (2015). Hazardous air pollutants & flooding: A comparative interurban study of environmental injustice. *GeoJournal*, 80, 145–158.

Hernandez, M., Collins, T., & Grineski, S. (2015). Immigration, mobility, and environmental injustice: A comparative study of Hispanic people's residential decision-making and exposure to hazardous air pollutants in Greater Houston, Texas. *Geoforum*, 60, 83–94.

Jimenez, A., Collins, T., & Grineski, S. (2013). Intra-ethnic disparities in respiratory health outcomes among Hispanic residents impacted by a flood. *Journal of Asthma*, 50, 463–471.

Johnson, C., Penning-Roswell, E., & Parker, D. (2007). Natural and imposed injustices: The challenges in implementing 'fair' flood risk management policy in England. *Geographical Journal*, 173, 374–390.

Kates, R. (1971). Natural hazard in human ecological perspective: Hypotheses and models. *Economic Geography*, 47, 438–451.

Kates, R.W., & Haarmann, V. (1992). Where the poor live: Are the assumptions correct? *Environment*, 34, 4–28.

Landry, S.M., & Chakraborty, J. (2009). Street trees and equity: Evaluating the spatial distribution of an urban amenity. *Environment and Planning A*, 41, 2651–2670.

Maantay, J., & Maroko, A. (2009). Mapping urban risk: Flood hazards, race, and environmental justice in New York. *Applied Geography*, 29, 111–124.

Maldonado, A., Collins, T., & Grineski, S. (2016). Hispanic immigrants' vulnerabilities to flood and hurricane hazards in two US metro areas. *Geographical Review*, 106, 109–135.

Masozera, M., Bailey, M., & Kerchner, C. (2007). Distribution of impacts of natural disasters across income groups: A case study of New Orleans. *Ecological Economics*, 63, 299–306.

Montgomery, M., & Chakraborty, J. (2013). Social vulnerability to coastal and inland flood hazards: A comparison of GIS-based spatial interpolation methods. *International Journal of Applied Geospatial Research*, 4, 58–79.

Montgomery, M., & Chakraborty, J. (2015). Assessing the environmental justice consequences of flood risk: A case study in Miami, Florida. *Environmental Research Letters*, 10, 095010.

Montgomery, M., Chakraborty, J., Grineski, S., & Collins, T. (2015). An environmental justice assessment of public beach access in Miami, Florida. *Applied Geography*, 62, 147–156.

Mustafa, D. (2005). The production of an urban hazardscape in Pakistan: Modernity, vulnerability, and the range of choice. *Annals of the Association of American Geographers*, 95, 566–586.

Nicholls, R., Hanson, S., Herweijer, C., Patmore, N., Hallegatte, S., Corfee-Morlot, J., et al. (2008). Ranking port cities with high exposure and vulnerability to climate extremes: Exposure estimates. *OECD Environment Working Papers* 1. OECD Publishing. doi: 10.1787/011766488208.

NRC (National Research Council). (2006). *Facing Hazards and Disasters: Understanding Human Dimensions*, Washington, DC: National Academy Press.

Oakes, J.M., Anderton, D.L., & Anderson, A.B. (1996). A longitudinal analysis of environmental equity in communities with hazardous waste facilities. *Social Science Research*, 25, 125–148.

Pelling, M. (1999). The political ecology of flood hazard in urban Guyana. *Geoforum*, 30, 249–261.

Pulido, L. (2000). Rethinking environmental racism: White privilege and urban development in Southern California. *Annals of the Association of American Geographers*, 90, 12–40.

Rufat, S., Tate, E., Burton, C., & Maroof, A. (2015). Social vulnerability to floods: Review of case studies and implications for measurement. *International Journal of Disaster Risk Reduction*, 14, 470–486.

Siegrist, M., & Cvetkovich, G. (2000). Perception of hazards: The role of social trust and knowledge. *Risk Analysis*, 20, 713–719.

Slovic, P. (1987). Perception of risk. *Science*, 236, 280–285.

Slovic, P. (2000). *The Perception of Risk*, London: Earthscan.

Slovic, P., Fischhoff, B., & Lichtenstein, S. (1980). Facts and fears: Understanding perceived risk. In: Schwing, R., & Albers, Jr., W. (eds.) *Societal Risk Assessment: How Safe Is Safe Enough?* New York: Plenum Press.

Swyngedouw, E., & Heynen, N. (2003). Urban political ecology, justice and the politics of scale. *Antipode*, 35, 898–918.

Ueland, J., & Warf, B. (2006). Racialized topographies: Altitude and race in southern cities. *Geographical Review*, 96, 50–78.

US Congressional Budget Office. (2007). Value of properties in the national flood insurance program. Available: www.cbo.gov/sites/default/files/110th-congress-2007-2008/reports/06-25-flood insurance.pdf [Accessed 27 January 2016].

Walker, G. (2012). *Environmental Justice: Concepts, Evidence, and Politics*, New York: Routledge.

Walker, G., & Burningham, K. (2011). Flood risk, vulnerability and environmental justice: Evidence and evaluation of inequality in a UK context. *Critical Social Policy*, 31, 216–240.

Wisner, B., Blaikie, P., Cannon, T., & Davis, I. (2004). *At Risk: Natural Hazards, People's Vulnerability and Disasters*, 2nd edition, London: Routledge.

Wolch, J., Wilson, J.P., & Fehrenbach, J. (2005). Parks and park funding in Los Angeles: An equity-mapping analysis. *Urban Geography*, 26, 4–35.

Zahran, S., Brody, S.D., Peacock, W.G., Vedlitz, A., & Grover, H. (2008). Social vulnerability and the natural and built environment: A model of flood casualties in Texas. *Disasters*, 32, 537–560.

29

CLIMATE CHANGE AND ENVIRONMENTAL JUSTICE

Philip Coventry and Chukwumerije Okereke

Introduction

The last 25 years have seen climate change become one of the defining international issues of the present time, evolving from scientific warnings of "global warming" to the complex challenge we understand today. Responses are found at every level of global and national governance, and throughout the realms of government, business and civil society. Fundamental to the problem of anthropogenic climate change is determining and reacting to the role of humans in its cause, and dealing with its widespread and unequal effect on human life and wellbeing. As a result, climate change has been labelled a "perfect moral storm" and a "super wicked problem", confounding ethicists and policymakers alike with a seemingly insurmountable challenge to our established understanding of environmental justice and governance (Gardiner 2011; Levin et al. 2012).

Moral philosophers have wrestled with the intricacies of climate change since it first emerged as an international issue, with Shue (1992) noting what he described as the 'unavoidability of justice' in the multilateral decision making process (see also Paterson & Grubb 1992; Bodansky 1993). At the same time, politicians have grappled with the challenge of interpreting the justice implications of climate change and how best to incorporate equity principles in prevailing or emerging governance arrangements. Over the same period, a broad and vocal climate justice movement has emerged from the more established environmental justice community to speak up for the most vulnerable in society and seek greater justice in climate change governance, from the global to the local level. Politicians, meanwhile, have invoked ethical arguments frequently in negotiations and policymaking processes, making use of numerous perspectives from the wide range put forward by ethical scholars (Okereke & Dooley 2010).

In this chapter we will outline the history of climate change as an international issue and how its complex justice elements have been interpreted and utilized by scholars, campaigners and policymakers. In the first section we discuss how scholars have analysed and conceived of the many facets of climate justice. In the second section we explore the intellectual and advocacy links between the movement pursuing global climate justice and the more established environmental justice movement from which it emerged. In the third section we trace the role of climate justice in global governance institutions and the varied positions on justice presented by nation states. Our aim is to provide an overview of these three important and interlinking realms of climate justice, all of which are central to a cohesive understanding of the climate change problem and our collective response to it.

Climate change and justice principles

The scientific theory explaining the effect of atmospheric carbon on global temperatures was first developed in the late nineteenth century, but it was not until the 1980s that the implications of this phenomenon reached the public consciousness and the global political agenda. A strong scientific consensus has since developed regarding the existence of climate change and the role of humans in causing it (Cook et al. 2013; Goldin 2013). An important part of this consensus is the role of the Intergovernmental Panel on Climate Change (IPCC), which was set up in 1988 by the United Nations Environment Programme (UNEP) and the World Meteorological Organization (WMO) to collate and review climate change research and provide the international community with balanced periodic summary assessments. The IPCC's First Assessment Report was published in 1990 and expressed the IPCC's "certainty" that 'emissions resulting from human activities are substantially increasing the atmospheric concentrations of . . . greenhouse gases . . . resulting on average in an additional warming of the Earth's surface' (IPCC 1990a: XI). Demonstrating the consistent message from the scientific community and the strength of consensus, the Fifth Assessment Report, published in 2013–14, stated that 'since the 1950s, many of the observed changes are unprecedented over decades to millennia' (IPCC 2013: 4), and 'It is extremely likely that human influence has been the dominant cause of the observed warming since the mid-20th century' (IPCC 2013: 15; 17).

The emerging scientific consensus played a key role in bringing climate change onto the international agenda and has also contributed to shaping the perspectives on justice taken by those involved in climate change. As well as the creation of the IPCC, 1988 also saw the Toronto Conference on the Changing Atmosphere and NASA's widely reported testimony to the US Congress that climate change is related to fossil fuel emissions (Vanderheiden 2008). These events and the IPCC's First Assessment Report in 1990 helped to facilitate the creation of the United Nations Framework Convention on Climate Change (UNFCCC) during the 1992 Earth Summit in Rio de Janeiro.

In the run up to and as a result of the UNFCCC agreement in 1992, the extent of the problem of anthropogenic climate change and the magnitude of the costs and disruption associated with an effective response became clearer. At the same time, the greater understanding of the existence, causes and potential impact of anthropogenic climate change exposed several complex justice dimensions of the phenomenon. These include huge variation in the contribution and vulnerability to climate change, both between and within nations. Primarily, while the majority of developing countries have contributed little to overall greenhouse gas emissions over the last 150 years, these are the countries expected to suffer the negative impacts of climate change most rapidly. This inequality also has a notable regional aspect, since it soon became clear that countries in tropical regions (mostly developing countries) will experience more severe effects of climate change and are usually less resilient to them (IPCC 1990b, 2014). With these circumstances in mind, philosophers have explored justice principles that can help establish a basis for an effective and equitable global response.

As scientists provided increasingly robust data to show that greenhouse gas emissions from human activity are driving climate change, initial commentators referred to existing legal and ethical principles and suggested that industrialized countries are the polluters who must pay for the environmental damage they have caused (O'Riordan & Cameron 1994). This perspective frames climate change in a similar way to more traditional forms of environmental degradation such as chemical spills, where both the damaged soil or water and the origin of the pollutants are identifiable and attributable. However, it quickly became apparent that this otherwise useful environmental justice principle struggles to deal with the multifaceted climate change problem. A particular challenge was that a significant amount of climate polluting activity took place over the last few hundred years and was carried out by countless different individuals and businesses, and that until recently its implications were not widely or comprehensively understood (Page 2008). Furthermore, at a basic level carbon emissions are not problematic – they only become harmful when multiplied to levels seen after the industrial revolution (Miller 2008).

Another popular ethical principle that came to prominence focuses on the gains from carbon-intensive economic development and is known as the "beneficiary pays" perspective. This principle suggests that a responsibility to assist other countries in dealing with climate change is generated by the disproportionate material benefits that developed countries have obtained through their emissions of fossil fuels, and the resulting harm caused to the environment (Caney 2005). Greater financial and technological resources make developed countries more resilient to the effects of climate change, compounding the likelihood that these effects will be less severe than for developing countries. However, once again this perspective is weakened by the acknowledgement that industrialization took place over two centuries, while the full implications of greenhouse gas emissions have only been fully understood relatively recently. Moreover, industrialization has profited not only those in the West, but also many in developing countries in the form of technologies for water purification, medicine, and more efficient agricultural practices, for example.

Some scholars go further, by exploring the extreme inequalities of wealth, life expectancy and opportunity created by the process of industrialization in the Global North; these inequalities characterize the relationship between developed and developing countries and are tied intimately to colonial history (see Chapters 43 and 46) and asymmetrical geopolitical power relationships (Roberts & Parks 2007). Developed countries have amassed a great "ecological debt", which articulates the reparations owed by the developed world to the developing for generations of exploitative resource extraction and use, international trade, and environmental damage, as well as disproportionate greenhouse gas emissions (the "climate debt") (Roberts & Parks 2009). Exploring these more all-encompassing dimensions of climate justice requires consideration of the hegemonic global economic system, which is underpinned by these inequalities and means governments are constrained in how they conceive of justice and redistribution (Okereke 2010).

Some ethical scholars have sought to disconnect the responsibilities of developed countries from the way their wealth was generated and the benefits they accrued, by appealing to the "ability to pay" principle. This principle appeals to cosmopolitan global justice ideals and suggests that a duty to assist and support developing countries derives directly from the far greater wealth that developed countries possess. Contribution to the problem and the manner in which economic development occurred are irrelevant, since achieving a just outcome is what matters in ethical terms; as a result, this perspective looks at the present and the future, rather than the past (Page 2008). While avoiding complicated responsibility concerns that are politically unpalatable in developed countries, this perspective bypasses guilt and the acceptance of responsibility, which, although difficult for governments to deal with, are important facets of justice in the eyes of most people and cultures.

As well as the challenges of dealing with the costs and burdens of responding to climate change, philosophers have attempted to address the prospect of future greenhouse gas emissions. Anthropogenic climate change is already happening, but, given economic and political realities, emissions will not cease immediately. The implications of how to restrict and allocate future emissions raise issues of entitlement and prompt consideration of whether we have a right to emit, either at individual or national levels (see also Chapter 13). Dividing emissions between each person would give each nation an annual emissions allowance (see Hayward 2007; Caney 2012). Such per capita approaches seem equitable and appeal to developing countries, whose per capita emissions remain relatively low. However, emissions are produced by a huge range of activities and determining which emissions are essential for basic subsistence or for economic development (to which many argue developing countries have a right), is fraught with difficulty (Gardiner 2004). Equity can be framed from a different perspective by starting with current or recent emissions and determining appropriate allocations from there – the "grandfathering" approach. Developing countries argue that this approach constrains their potential economic development and unfairly locks in current economic inequality (Moellendorf 2015).

It is clear from the overview in this section that exploring the ethics of climate change has produced a wide variety of perspectives, and finding a way through the complexity is a challenge for scholars. Nevertheless, being aware of these philosophical perspectives is critical when seeking to understand the development of the climate justice movement and the arguments and demands articulated by its advocates. Similarly, any analysis of the political response to climate change must take account of the ethical basis of positions taken by national governments and negotiating blocs at the global governance level. Dealing with past emissions, the prospect of blame, the financial burdens of climate action and the need for economic development in the Global South are challenges that remain central to the governance of climate change and the future of our planet.

Emergence of the climate justice movement

Climate justice has been the subject of growing scholarly attention, but the academic community has not led its development as an advocacy movement (see Chapters 3 and 4 on social movements). Here the connections between climate justice and environmental justice are important if the grassroots and elite NGO components of the climate justice movement are to be understood. As climate change became increasingly highlighted as an emerging global issue, the various manifestations of injustice that characterize the problem connected with those seen in environmental injustices over the preceding decades. Advocates have responded to greater clarity on the reality of climate change, the uneven geographic profile of its effects, and the disproportionate role played by industrialized countries in causing it. These features of climate change shaped the way campaign organizations took up the cause of climate justice.

Schlosberg and Collins (2014) argue that climate justice developed as a movement directly from the existing environmental justice movement, which itself emerged from the separate environmental protection and social justice movements. Early environmental justice activists in the US drew attention to links between the inferior living conditions and life chances of racial minorities, and various environmental issues such as pollution and toxic waste disposal (see Chapters 25 and 26). Since the scope of the environmental justice movement had already expanded to address local and national issues around the world, as well as transnational environmental problems, the movement could adapt and engage with the complex and interconnected implications of climate change as scientific knowledge became more robust and widely disseminated.

Early formulations of climate justice within the global activist community, such as the Bali Principles of 2002, adopted the historical responsibility perspective and drew directly from earlier articulations of environmental justice within the American activist community. The broad environmental justice movement addressed 'distributive inequity, lack of recognition, disenfranchisement and exclusion, and, more broadly, an undermining of the basic needs, capabilities, and functioning of individuals and communities' (Schlosberg & Collins 2014: 361). All of these facets can be found in the climate justice movement, intertwined with climate change-specific perspectives related to philosophical analysis of the specific injustices of this particularly global problem.

Routledge (2011) takes a perspective focused on the Global South, finding the roots of the climate justice movement in Southern efforts to address the inequalities created and perpetuated by industrialized nations powering their economic development with the engine of capitalism, with its literal and metaphorical reliance on fossil fuels. Links with the philosophical exploration of climate justice are clear here, in terms of historical responsibility and retributive obligations. Routledge focuses on the solidarities formed when place-based struggles become connected through shared injustices and aligned objectives. While climate justice advocates are often motivated by a local issue or injustice, a desire to protect the natural world around them and oppose the power of corporations or governments mean their existing interests, values and reactions align with the objectives of the wider environmental justice movement (Martinez-Alier 2013). In today's globalized world the pursuit of both environmental justice and climate justice are no longer based in a particular region, with individuals, communities and groups around the world advocating climate and environmental justice with vigour and determination (see Chapter 39).

Climate change is an unavoidably global issue, which encourages discussion of global inequalities and established economic and power structures. The concept of "ecologically unequal exchange", for example, illustrates the relationship between the developed and developing world, where the power of industrialized countries has enabled them to create a flow of resources from the Global South to the North (Roberts & Parks 2009). Climate change is bound up with these economic inequalities and the climate justice movement has kept retributive justice at its heart (Schlosberg & Collins 2014), echoing the beneficiary pays perspective and emphasizing the need for industrialized countries to bear the burden of reducing greenhouse gas emissions and funding climate change action in the developing world.

Inequalities are present not only in the causes of climate change, but also in the geographic distribution of its effects and their severity. Although climate change will affect all parts of the world in the end, the greatest adverse impacts are expected in developing countries in and around the tropics (IPCC 2013). Poverty is inextricably linked to climate change, since poverty makes individuals, communities and regions more vulnerable to suffering when precipitation changes or extreme weather events occur, and less resilient to the effects of such phenomena (Parnell 2014) (see also Chapter 28 on flooding). It is important to note that such inequality is not confined to developing countries, though, and can be found within developed countries. For example, studies of the "climate gap" have echoed the original themes of the environmental justice movement by identifying differences in the vulnerability of different racial and socioeconomic groups within states and cities in the US to the effects of climate change (e.g. Shonkoff et al. 2011; Wilder et al. 2016).

In the same vein as environmental justice, the climate justice movement has developed to challenge the existing economic systems, political and institutional structures and power dynamics that create the inequality experienced by countries and communities around the world and exacerbate the effects of climate change (Barrett 2013). Early civil society demands for climate action were simply too stringent for the global establishment to meet, and the 'failure of a more collaborative strategy between major environmental NGOs and the global capitalist managerial class' (see Chapters 38 and 46) contributed to the evolution of the climate justice movement towards targeting the core features of this capitalist establishment (Bond & Dorsey 2010: 287).

The relevance of social and economic inequalities when considering the impact and injustice of climate change highlights the close connection between climate change and economic development. Some have argued that the importance of facilitating development in poor countries should take precedence over the need to curb greenhouse gas emissions, although this perspective can be refuted by pointing out that climate change will harm such countries even as their development continues and considering how emissions can rise in some countries and fall in others (Caney 2011). The need for development remains acute in large portions of the world and as a result this latter consideration has spawned concepts such as "carbon space" and "greenhouse development rights". The division of a scientific cap on total greenhouse gas emissions between the global population is the basis for philosophical perspectives and policy proposals about how to achieve justice when looking both at previous emissions, and establishing how potential future emissions should be distributed amongst nations and between current and future generations (e.g. Hayward 2007; Baer & Fieldmann 2009; Caney 2012). Such per capita approaches tap into egalitarian intuition amongst climate justice advocates and influence many to campaign for developing countries' rights to their fair share of future greenhouse gas emissions – the aforementioned "carbon space" (e.g. Harmeling 2014).

These egalitarian approaches also link to human rights norms and language, which can provide an established basis to underpin climate justice campaigns (see Chapter 13). Rights to develop and rights to emit greenhouse gases are part of the wider climate justice discourse, but the links go further since rights to food, a healthy environment, and ultimately to life, will all be affected by climate change (Moellendorf 2012). Once again, those who currently live without these rights being met are likely to be the most vulnerable to climate change and have the least resources to facilitate lasting adaptation and recovery from specific damaging events. Development planning is evolving to consider the circumstances, rights, and needs of individual communities, all within the context of adaptation to climate change (Ensor 2011). Some governmental and NGO participants in international climate change negotiations have advocated a principle of equitable access to sustainable development to guide how climate change governance can support rather than constrain development (UNFCCC 2012).

The question of development and climate change is a pertinent illustration of the multiple scales at which climate justice is relevant and influential. Justice for communities affected by climate change is at the heart of local, small-scale climate justice advocacy. National and international development strategies reach up to the highest levels of governance and the broadest transnational NGOs. Development is seen as a key way to enhance the resilience of poor communities in the face of impending climate change, and recent developments such as the 2015 Sustainable Development Goals bring the need to combat climate change and ensure sustainable resource use firmly into mainstream development objectives (UN 2015). However, rights to development are unavoidably affected by the existence of climate change and the natural limits of the planet and its ecosystems (Löfquist 2011). How to achieve development in the context of a need to reduce global emissions and the constraints of the natural world has proven difficult to resolve for academics and advocates alike (Okereke & Schroeder 2009; Schlosberg & Collins 2014).

This section has progressed from local injustices and grassroots advocacy to a global perspective and international issues. Climate change goes beyond environmental justice here, by involving all countries in the justice and politics of dealing with a problem that has international causes and effects, and requires co-ordinated international solutions. The local level remains critical, however, in terms of the billions of vulnerable people whose health and livelihoods are threatened by climate change, and the near-infinite variation in their circumstances and the way international policies will affect them and the achievement of justice at the local level. The nature of climate change connects individuals, businesses and governments in the Global North to those in the South, because climate change means greenhouse gas emissions in one region have real and harmful consequences (Moellendorf 2012). This connection is reflected in the international political response to climate change, where justice has played an integral role. It is to this political level that we will turn in the next section.

Climate justice and global governance

The features that make climate change ethics so complicated have had a similar impact in the political sphere, by contributing to varying and strongly held national positions and making institutional and policy design extremely difficult. As early as the first IPCC Assessment Report in 1990 there was acknowledgement that industrialized countries had 'specific responsibilities' and that domestic measures were required because 'a major part of emissions affecting the atmosphere at present originates in industrialized countries where the scope for change is greatest.' The report also suggested that industrialized countries should 'cooperate with developing countries in international action, without standing in the way of the latter's development'; including provision of finance and technology transfer (IPCC 1990b: XXVI). These early expressions of justice and inequality concerns raised by climate change echo the philosophical positions in §1 and civil society advocacy in §2.

The first IPCC report signalled the firm linkage of justice and governance to the scientific evidence and language that provides the impetus for addressing climate change, which in turn ensured ongoing legitimacy for justice concerns within governance institutions (Okereke 2010). As the IPCC noted, the global nature of climate change requires countries to collaborate if they are to successfully deal with a problem that affects each nation differently. The collaboration of nation states in climate change governance has brought ethical arguments right into the centre of the international political negotiating space. While some scholars argue that social justice is necessary for overcoming the political challenges of environmental governance (Paavola 2008), so far seeking justice in policy in light of the complexity of climate justice has proven a central impediment to agreement about an international response to climate change.

As the scientific consensus strengthened and the implications of climate change became clearer, the need for an international political response led to the development of an institutional structure in which to co-ordinate and manage governance activities. The United Nations Framework Convention on Climate Change (UNFCCC) was signed in 1992 at the Earth Summit in Rio de Janeiro, creating a treaty-based foundation for action to reduce greenhouse gas emissions and limit global temperature rise. With near-universal membership and decisions requiring agreement from all parties, the UNFCCC is certainly inclusive in principle, but this design has allowed different national positions on what is or would be fair to feature heavily in discussions, and has generated power dynamics that complicate the realization of climate justice.

From the 1970s onwards various international environmental treaties had been formulated and agreed, and in the late 1980s a global effort to address ozone layer depletion led to the

Montreal Protocol, signed in 1988. A principle of differentiating between different countries' economic circumstances and contributions to the problem had been evolving in these successive environmental treaties; in the Montreal Protocol developing countries were given reduced obligations and substantial additional time to implement the various measures in the treaty (Okereke 2008). Obligations for developed countries (which would largely solve the problem) were dealt with first in negotiations, and then measures for developing countries were addressed. The formulation of this equity principle in the Montreal Protocol was considered successful, and directly influenced its inclusion in the UNFCCC (Hale et al. 2013).

The UNFCCC named this equity principle for the first time; "common but differentiated responsibility" (CBDR) reflected the need for an equitable approach, but the negotiations included all parties together and as a result the measures for industrialized countries became mixed up with the participation of developing countries. Further complexity was added by economic and geopolitical rivalry between developed nations and rapidly developing nations with large economies and rising emissions (Cumberledge 2009). A crude division between developed and developing nations was built into the UNFCCC and Kyoto Protocol, the first treaty based on the UNFCCC to include legally binding obligations, ignoring the significant current and future emissions of large developing nations such as China, India and Brazil. Developing countries were not given any binding commitments, much to the consternation of developed countries, particularly the US Congress and later the Bush Administration, which withdrew from the Kyoto Protocol soon after the treaty was finalized. This can be seen as a misinterpretation of the Montreal Protocol precedent, which created an 'equitable imbalance that threatens the success of the entire mission' (Green 2009: 279).

Rather than constituting a victory for environmental justice in international governance, the clumsy formulation of what had become an important principle of justice in environmental treaties has actually proven a consistent source of disagreement and national rivalry (Brunnée & Streck 2013). Disagreement over how CBDR should be interpreted in the context of climate change focused the governance discourse on the economic burdens of climate change action and how obligations should be allocated, which remains the dominant frame in climate politics (Hoffman 2013). Despite this, 'the UNFCCC remains a uniquely relevant site for the articulation and pursuit of climate effectiveness as well as justice' (Derman 2014: 24).

Nations and negotiating groups consistently invoke contrasting positions on what climate justice looks like, with a range of philosophical perspectives reflected to varying degrees. Some countries, particularly developing and vulnerable nations, share many negotiating perspectives and articulations of climate justice with the wider civil society climate justice movement outlined in the previous section. Certainly demands for finance flows from the developed to the developing world to fund adaptation and mitigation action have been present throughout the negotiations, evolving recently to include "loss and damage" mechanisms to compensate developing countries in the event of damaging severe weather events related to climate change (see Chapters 40, 45 and 46).

Developing countries have consistently argued that developed nations must take the initiative and lead on climate change action. China, for example, has consistently sought to reinforce the divide between developed and developing countries' mitigation responsibilities and called for more ambitious actions on the part of industrialized countries (Stalley 2013). As well as championing the right to development, India has maintained a focus on an equitable per capita approach to emissions that differs from China's focus on historical responsibility (Qi 2011). Although not a homogenous bloc in themselves, small island nations often combine forces and together have been consistently vocal about historical responsibility for climate change. Such nations repeatedly highlight the humanitarian aspect of their vulnerability to climate change and potential loss of entire nations, reinforcing links between climate change and wider global justice issues. Recognizing their vulnerability and the global nature of the climate change problem, island nations seek to stimulate action by larger emitters, developed and developing, rather than focusing on developed countries (Betzold et al. 2012).

Developed countries assert their own perceptions of justice in climate change governance, but again are not homogenous in their views. There is some consistency in the avoidance of the historical responsibility and per capita perspectives; developed countries tend to focus on their greater financial and technical resources rather than the historical background to their ability to pay. On paper, developed countries have accepted historic responsibility in documents such as the UNFCCC (as well as the Montreal Protocol in the ozone context) but are unwilling to accept the full extent of the links between climate change and wider economic imbalances that characterize historic and current relationships between the Global North and South (Boyte 2010). To do so could expose developed countries to financial obligations that would simply not be politically acceptable, and reveal the extent to which climate justice shakes the foundations of the global capitalist hegemonic system.

Countries such as the US appeal explicitly to justice to demand the "fair" participation of large developing economies in climate action (e.g. Bush 2001). Others, such as the European Union (EU), have been more willing to advocate a legally binding climate treaty and back this up with robust domestic measures (Falkner et al. 2010). Despite this greater willingness to take action, there are clear limits to the terms of justice that the EU and other developed countries will accept, mainly in relation to redistributive and compensatory financial mechanisms as well as the familiar, but more varied reticence around emissions reductions.

Justice has been a source of negotiating positions and political disagreement throughout the course of the global institutional governance of climate change. Many perspectives and objectives of climate justice advocates and scholars are present in the negotiating positions of countries within the UNFCCC, albeit often modified for use in the political context. This section is not seeking to provide an exhaustive explanation of the history of this governance, but rather to illustrate how important the global governance of climate change is to an understanding of climate justice. Trends from environmental justice within global governance have influenced the course of climate negotiations, but, since the 1990s, climate justice has moved beyond the realm of environmental justice, with its own specific institutions, language and justice intricacies, and has become a complex aspect of policy processes at the global governance level.

Conclusion

Climate change has become the pre-eminent global environmental issue of the twenty-first century, garnering headlines like no environmental issue before. Its global reach and almost infinitely complex and extensive implications are key to this high profile, and create a situation where a global political response is both necessary and extremely difficult to manage. In this chapter we have outlined the three key realms of climate justice, and sought to illuminate the way in which each is connected to the others but remains separate in function, scope, and in the many ways justice is internally articulated. Climate justice is inherently connected to environmental justice, and each of the three realms either emerged from or were influenced by prior principles or experiences of environmental justice. But as philosophers, policymakers and civil society advocates grapple with climate change and how to distribute emissions cuts equitably, deal justly with historical emissions, and fairly redistribute money to cover the costs of climate action, the challenge of justice is far from resolved. In amongst the discussions and the political manoeuvring are the poorest and most vulnerable in our global society, whose health, livelihoods and very nations are at risk. If these billions of people and the planet we all occupy are to be protected from climate change, the challenge of climate justice must be tackled.

References

Baer, P. & Fieldmann, G. (2009). Greenhouse development rights: Towards an equitable framework for global climate policy. In P.G. Harris (ed.), *The Politics of Climate Change: Environmental Dynamics in International Affairs*. Abingdon, Oxon: Routledge. 192–212.

Barrett, S. (2013). The necessity of a multiscalar analysis of climate justice. *Progress in Human Geography* 37(2): 215–233.

Betzold, C., Castro, P. & Weiler, F. (2012). AOSIS in the UNFCCC negotiations: From unity to fragmentation? *Climate Policy* 12(5): 591–613.

Bodansky, D. (1993). United Nations Framework Convention on Climate Change: A commentary. *Yale Journal of International Law* 18: 451–-58.

Bond, P. & Dorsey, M.K. (2010). Anatomies of environmental knowledge and resistance: Diverse climate justice movements and waning eco-neoliberalism. *Journal of Australian Political Economy* 66: 286–316.

Boyte, R. (2010). Common but differentiated responsibilities: Adjusting the "developing"/"developed" dichotomy in international environmental law. *New Zealand Journal of Environmental Law* 14: 63–101.

Brunnée, J. & Streck, C. (2013). The UNFCCC as a negotiation forum: Towards common but more differentiated responsibilities. *Climate Policy* 13(5): 589–607.

Bush, G.W. (2001). *Text of a Letter from the President to Senators Hagel, Helms, Craig, and Roberts*. The White House (President George W. Bush). [Online]. 13 March 2001. Available at http://georgewbush-whitehouse.archives.gov/news/releases/2001/03/20010314.html [Accessed 9 July 2014].

Caney, S. (2005). Cosmopolitan justice, responsibility, and global climate change. *Leiden Journal of International Law* 18: 747–775.

Caney, S. (2011). Climate change, energy rights, and equality. In D.G. Arnold (ed.), *The Ethics of Global Climate Change*. Cambridge: Cambridge University Press. 77–103.

Caney, S. (2012). Just emissions. *Philosophy & Public Affairs* 40(4): 255–300.

Cook, J., Nuccitelli, D., Green, S.A., Richardson, M., Winkler, B., Painting, R., et al. (2013). Quantifying the consensus on anthropogenic global warming in the scientific literature. *Environmental Research Letters* 8: 1–7.

Cumberledge, S. (2009). Multilateral environmental agreements from Montreal to Kyoto: A theoretical approach to an improved climate change regime. *Denver Journal of International Law and Policy* 37(2): 303–329.

den Elzen, M.G.J., Olivier, J.G.J., Höhne, N. & Janssens-Maenhout, G. (2013). Countries' contributions to climate change: Effect of accounting for all greenhouse gases, recent trends, basic needs and technological progress. *Climatic Change* 121: 397–412.

Derman, B.B. (2014). Climate governance, justice, and transnational civil society. *Climate Policy* 14(1): 23–41.

Ensor, J. (2011). *Uncertain Futures: Adapting Development to a Changing Climate*. Rugby: Practical Action Publishing.

Falkner, R., Stephan, H. & Vogler, J. (2010). International climate policy after Copenhagen: Towards a 'building blocks' approach. *Global Policy* 1(3): 252–262.

Gardiner, S.M. (2004). Ethics and global climate change. *Ethics* 114: 555–600.

Gardiner, S.M. (2011). *A Perfect Moral Storm: The Ethical Tragedy of Climate Change*. Oxford: Oxford University Press.

Goldin, I. (2013). *Divided Nations: Why Global Governance Is Failing and What We Can Do About It*. Oxford: Oxford University Press.

Green, B.A. (2009). Lessons from the Montreal Protocol: Guidance for the next International Climate Change Agreement. *Environmental Law* 39: 253–283.

Hale, T., Held, D. & Young, K. (2013). *Gridlock: Why Global Cooperation Is Failing When We Need It Most*. Cambridge: Polity Press.

Harmeling, S. (2014). *COP20: Building a Fair and Just Climate Deal for the World's Poorest People*. CARE International. [Online]. Available at www.careclimatechange.org/files/Expectations_Paper_FINAL. pdf [Accessed 22 July 2015].

Hayward, T. (2007). Human rights versus emissions rights: Climate justice and the equitable distribution of ecological space. *Ethics and International Affairs* 21(4): 431–50.

Hoffman, M.J. (2013). Climate change. In T.G. Weiss & R. Wilkinson (eds.), *International Organization and Global Governance*. Abingdon, Oxon: Routledge. 605–17.

IPCC (1990a). Policymakers summary. In J.T. Houghton, G.J. Jenkins & J.J. Ephraums (eds.), *Climate Change: The IPCC Scientific Assessment. Working Group I Report for the First Assessment Report of the Intergovernmental Panel on Climate Change*. [Online]. Available at www.ipcc.ch/publications_and_ data/publications_ipcc_first_assessment_1990_wg1.shtml [Accessed 14 April 2016].

IPCC (1990b). Policymakers summary of the response strategies. In *Climate Change: The Response Strategies. Working Group III Report for the First Assessment Report of the Intergovernmental Panel on Climate Change*. [Online]. Available at www.ipcc.ch/publications_and_data/publications_ipcc_first_ assessment_1990_wg3.shtml [Accessed 22 October 2015].

IPCC (2013). Summary for policymakers. In T.F. Stocker, D. Qin, G.-K. Plattner, M. Tignor, S.K. Allen, J. Boschung, et al. (eds.), *Climate Change 2013: The Physical Science Basis. Contribution of Working Group I to the Fifth Assessment Report of the Intergovernmental Panel on Climate Change*. Cambridge, UK & New York, NY: Cambridge University Press.

IPCC (2014). Summary for policymakers. In C.B. Field, V.R. Barros, D.J. Dokken, K.J. Mach, M.D. Mastrandrea, T.E. Bilir, et al. (eds.), *Climate Change 2014: Impacts, Adaptation, and Vulnerability. Part A: Global and Sectoral Aspects. Contribution of Working Group II to the Fifth Assessment Report of the Intergovernmental Panel on Climate Change*. Cambridge, UK & New York, NY: Cambridge University Press.

Levin, K. et al. (2012). Overcoming the tragedy of super wicked problems: Constraining our future selves to ameliorate global climate change. *Policy Sciences* 45: 123–152.

Löfquist, L. (2011). Climate change, justice and the right to development. *Journal of Global Ethics* 7(3): 251–260.

Martinez-Alier, J. (2013). The environmentalism of the poor. *Geoforum* 54: 239–241.

Miller, D. (2008). National responsibility and global justice. *Critical Review of International Social and Political Philosophy* 11(4), 383–399.

Moellendorf, D. (2012). Climate change and global justice. *WIREs Climate Change* 3: 131–143.

Moellendorf, D. (2015). Climate change justice. *Philosophy Compass* 10: 173–186.

Okereke, C. (2008). Equity norms in global environmental governance. *Global Environmental Politics* 8(3): 25–50.

Okereke, C. (2010). Climate justice and the international regime. *WIREs Climate Change* 1(3): 462–474.

Okereke, C. & Dooley, K. (2010). Principles of justice in proposals and policy approaches to avoided deforestation: Towards a post-Kyoto climate agreement. *Global Environmental Change* 20(1): 82–95.

Okereke, C. & Schroeder, H. (2009). How can justice, development and climate change mitigation be reconciled for developing countries in a post-Kyoto settlement? *Climate and Development* 1(1): 10–15.

O'Riordan, T. & Cameron, J. (eds.) (1994). *Interpreting the Precautionary Principle (Vol. 2)*. Abingdon, Oxon: Earthscan.

Paavola, J. (2008). Science and social justice in the governance of adaptation to climate change. *Environmental Politics* 17(4): 644–659.

Page, E.A. (2008). Distributing the burdens of climate change. *Environmental Politics* 17(4): 556–575.

Parnell, S. (2014). *The (Missing) Link: Climate Change Mitigation and Poverty*. Research Paper, Issue 23. Mitigation Action Plans & Scenarios, Cape Town. [Online]. Available at www.mapsprogramme. org/category/publications/papers/ [Accessed 22 July 2015].

Paterson, M. & Grubb, M. (1992). The international politics of climate change. *International Affairs* 68(2): 293–310.

Qi, X. (2011). The rise of BASIC in UN climate change negotiations. *South African Journal of International Affairs* (18)3: 295–318.

Roberts, J.T. & Parks, B.C. (2007). *A Climate of Injustice: Global Inequality, North-South Politics, and Climate Policy*. Cambridge, MA: MIT Press.

Roberts, J.T. & Parks, B.C. (2009). Ecologically unequal exchange, ecological debt, and climate justice: The history and implications of three related ideas for a new social movement. *International Journal of Comparative Sociology* 50(3–4): 385–409.

Routledge, P. (2011). Translocal climate justice solidarities. In J.S. Dryzek, R.B. Norgaard & D. Schlosberg (eds.), *The Oxford Handbook of Climate Change and Society*. Oxford Handbooks Online. Available at: www.oxfordhandbooks.com/view/10.1093/oxfordhb/9780199566600.001.00 01/ oxfordhb-9780199566600-e-26 [Accessed 25 October 2015].

Schlosberg, D. & Collins, L.B. (2014). From environmental to climate justice: Climate change and the discourse of environmental justice. *WIREs Climate Change* 5: 359–374.

Shonkoff, S.B., Morello-Frosch, R., Pastor, M. & Sadd, J. (2011). The climate gap: Environmental health and equity implications of climate change and mitigation policies in California – a review of the literature. *Climatic Change* 109(Suppl. 1): S485–S503.

Shue, H. (1992). The unavoidability of justice. In A. Hurrell & B. Kingsbury (eds.), *The International Politics of the Environment*. Oxford: Oxford University Press. 373–397.

Stalley, P. (2013). Principled strategy: The role of equity norms in China's climate change diplomacy. *Global Environmental Politics* 13(1): 1–8.

UN (2015). Zero draft of the outcome document for the UN Summit to adopt the Post-2015 Development Agenda. United Nations. [Online]. Available at https://sustainabledevelopment.un. org/?page=view&nr=829&type=230&menu=2059 [Accessed 22 October 2015].

UNFCCC (2012). *Report on the Workshop on Equitable Access to Sustainable Development*. United Nations Framework Convention on Climate Change, FCCC/AWGLCA/2012/INF.3/Rev.1. [Online]. Available at http://unfccc.int/resource/docs/2012/awglca15/eng/inf03r01.pdf [accessed 25 October 2015].

Vanderheiden, S. (2008). *Atmospheric Justice: A Political Theory of Climate Change*. New York, NY: Oxford University Press.

Wilder, M., Liverman, D., Bellante, L. & Osborne, T. (2016). Southwest climate gap: Poverty and environmental justice in the US Southwest. *Local Environment: The International Journal of Justice and Sustainability*. [Online]. Available at http://dx.doi.org/10.1080/13549839.2015.1116063 [Accessed 12 April 2016].

30

ENVIRONMENTAL JUSTICE AND LARGE-SCALE MINING

Leire Urkidi and Mariana Walter

Introduction

From the 2000s to the 2010s, the extraction of metals has almost doubled worldwide, passing from 764 000 000 to 1 551 000 000 tons per year (Schaffartzik et al. 2016). In this chapter, we address environmental justice (EJ) in the context of large-scale metal mining activities, an increasingly relevant activity in EJ debates and one of the most polluting activities in the world (EPA 2013).

This chapter approaches the environmental injustice of large-scale metal mining from three complementary angles. First, we outline some of the key biophysical features of large-scale mining activities (e.g. environmental intensity, temporal length of the activity, territorial impacts) and some of the implications for EJ. Second, we examine a central debate: the (spatial and non-spatial) dynamics shaping the distribution of mining burdens and benefits. Third, we address mining struggles, movements and discourses. In this section, we examine the transnational diffusion of the EJ framework among anti-mining movements. We also review some of the central grievances in mining conflicts and EJ discourses. We underline the role played by anti-colonial discourses in mining conflicts given the presence of large transnational corporations and the significance of mining in historical processes of colonization. We also address debates revolving around the current relevance of environmental awareness, resource control, recognition and participation in mining protests, as well as the implication of scaling processes. We conclude by signalling that EJ frameworks express themselves and mutate during mining struggles in complex ways that require approaches able to capture the diverse and dynamic dimensions, scales, contexts and learning processes in motion. Some research gaps and future lines of inquiry are suggested.

Material and territorial features of mining activities

When approaching mining issues, EJ examinations require attention to the biophysical features of the activity and natural resources involved. The biological and geophysical properties of commodities are intertwined with the social relations that are created around them (Perreault 2006). Large-scale modern mining is a highly industrialized activity (e.g. large open pits, chemical and mechanical processes) that largely differs from the imaginary of mining (i.e. underground pick and spade work, alluvial mining) and that can lead to EJ conflicts at different temporal stages of extraction.

One of the particularities of the metal mining production chain is that its initial stages (extraction and processing) are characterized by their large social and environmental costs (Giurco et al. 2010). Moreover, as pressure to extract metals increases and the extraction frontier expands, lower quality deposits (i.e. low grades, toxic minerals present) are exploited, increasing social and environmental pressures (Giurco et al. 2010). The worldwide decline in the quality of ores has direct implications in terms of land intervention of mining activities (e.g. land clearing), as larger and open cast mines have to be built, the amount of water, energy and chemical inputs increases and higher quantities of waste rock are generated (Giurco et al. 2010; Mudd 2007; Prior et al. 2012). Waste rock is especially sensitive when sulfidic material is present, as in many metal ores (Bridge 2004a; Giurco et al. 2010; Mudd 2010). Pollution is caused by the oxidation that naturally occurs in sulfidic minerals, triggering the formation of sulphuric acid and acid drainages and the mobilization of heavy metals. This process has been pointed out as one of the main environmental challenges of the mining industry (Bridge 2004a; Giurco et al. 2010; Government of Australia 2007). Mining-related chemical pollution can also be generated by the release to the environment of reagents added during mineral processing, such as the sulphuric acid that is used for the leaching of copper oxides, or the mercury or cyanide used to process gold.

Precious materials such as gold tend to have the highest generation of overburden, since their high prices make them economically feasible to extract at decreasing qualities/grades, entailing the processing of large amounts of ore and the generation of increasing amounts of waste rock and tailings. The significance of these trends grows as we consider the expansion of the mining frontier to sensitive and critical ecosystems such as tropical and cloud forests, or the very high mountains next to pasturelands and glaciers. These are also often the homes of Indigenous people and peasants (Bebbington 2012b).

It is important to stress that in the case of mining activities, eco-efficiency and technological approaches are limited (Bridge 2004a). For instance, considering an eco-efficiency approach, inputs to the mining process – such as water, energy, or chemical compounds – can be reduced per unit of production, the management of waste can be improved (e.g. better membranes to isolate waste from soil), and mining sites can be rehabilitated (e.g. re-vegetation) (Bridge 2004a). However, the adverse impacts can be reduced per unit of output but not eliminated. This is particularly significant in the context of a global decrease in the quality of deposits that entails the processing of larger amounts of ores to extract the same amount of output. In this vein, Prior and colleagues (2012) suggest that the "peak metal" (the time at which extraction can no longer rise to meet the demand) has more to do with a carefully weighed decision that considers the social and environmental implications of continuing to extract, than a question of existing metal quantities.

A particularity of mining activities is that their related social, cultural and environmental impacts surpass the actual duration and location of extractive activities. EJ conflicts can arise at many spatial and temporal stages of mining activities. From a production chain perspective, conflicts can start before extraction itself. For instance, conflicts may emerge due to problems of access to resources, such as land or water, that are claimed by a company fostering pro-cesses of speculation and dispossession (Bebbington 2012a; Perreault 2013). This occurred in Botswana when the government granted concessions for mineral exploration to diamond companies over an area encompassing the entire ancestral territories of the Gana and Gwi people (San or Bushmen), forcing the relocation of most local inhabitants (Özkaynak et al. 2015). The extraction of diamonds has also been related to wars in Africa since homelessness, injustice, and corruption in the diamond sector provided a cause, or at least a rationale, for some combatants to join insurrections (Le Billon 2008).

Environmental injustices can also emerge during mining operation, in relation to processing, waste management or transportation activities. For instance, the notorious spill of 151 kg of liquid mercury in Choropampa, near the Yanacocha goldmine in the Peruvian Andes in the 2000s, poisoned over 900 villagers and fostered civil resistance. Accidents at some mining waste dams are well known, such as the 2015 disaster in Minas Gerais (Brazil), or the 2000 cyanide spill resulting from the dam break in Bahia Mare, Romania. Sometimes mining operations dump the liquid waste directly into rivers, such as in the Huanuni tin mine in Bolivia. In 2015, one of the largest metal mining companies of the world, Barrick Gold, was accused by a federal prosecutor in Argentina of dumping over 224 000 litres of a cyanide and waste water solution in a local river near its Veladero mine, which has a long history of conflict with local communities (*Buenos Aires Herald* 2015/09/23).

Socio-environmental impacts can also persist for decades or centuries after mine closure. In August 2015, the US Environmental Protection Agency (EPA) reported a polluting spill of three million gallons containing heavy metals when conducting an investigation/treatment at the Gold King Mine, which had been closed since 1923 (US EPA 2015). Acid mine drainage can impact underground and surface water quality during a period that exceeds the productive life of a mine and can even increase with time (Kuipers 2003). The Montana Department of Environmental Quality (1997) determined that acid drainage at the Golden Sunlight mine would continue for thousands of years. Studies conducted by the US government signal that insurance coverage of large-scale metal mines has often been insufficient to handle the full environmental liabilities costs in the long term (US Accountability Office 2011: 8). This is particularly relevant in the case of radioactive metal mines: the large amounts of tailings from abandoned uranium mines entail a potential negative impact, resulting from both the radioactive material and high heavy metal concentrations (Antunes et al. 2007).

The (unjust) distribution of mining impacts

The great environmental, social and territorial impacts of metal mining have placed this activity at the centre of large EJ debates and struggles (Urkidi and Walter 2011; Walter and Urkidi 2016; Özkaynak et al. 2015). The relationship between the extraction of minerals and their geological location shapes a singular debate over the distributive dimensions of the injustices at stake. Is there a disproportionate impact on marginalized social groups or critical ecosystems? If this is the case, why?

In fact, an increasing proportion of mineral exploration and investment expenditures during the 1990s targeted the tropical areas around the globe, reaching ecologically sensitive and/or high value conservation areas (Bridge 2004b). The International Union for Conservation of Nature (IUCN) has issued concerns related to the expansion of the mining, gas and oil frontier in World Heritage Sites, demanding protection for them (IUCN 2011).

Regarding the impact on Indigenous lands, half of the gold mined in the world from 1995 to 2015 (excluding Africa) is estimated to come from these areas (Moody 1996). Recent studies led by scholars and activists point to the high overlap of mining concessions with the land of peasants and Indigenous people in Latin America (Bebbington 2012b). For instance, de Echave (2009, quoted in Bebbington 2012a) estimates that over half of Peruvian peasant communities are affected by mining projects or concessions. This is also the case in Guatemala, where 95 per cent of the existing licences in 2004 had been granted after 2000, and most of them were in Indigenous provinces (Defensoría Q'eqchi 2004).

In Canada, 36 per cent of First Nations communities are located within 50 km of a mine (Hipwell et al. 2002). A report by Mackasey (2000) indicated that there were 160 abandoned mines in the north of Canada, 67 of them having problems of chemical pollution or physical instability and many being located within Aboriginal territories. However, Keeling and Sandlos (2009) argue that these data do not reveal causal or intentional factors but incidental ones, since mining companies located their operations where they found significant mineral deposits and where transportation routes allowed for a profitable exploitation. These authors suggest that it would be difficult to demonstrate the imposition of disproportionate environmental impacts on Aboriginal communities because they form a majority of the population in many areas of Northern Canada.

However, we cannot understand mining developments only in relation to geology, as fiscal and legal frameworks facilitate extraction in some locations that could be considered comparatively unattractive in geological terms. Since 1985, in a context of increasing environmental and labour protection measures in the Global North (Europe, Canada and the US), over 90 states adopted or revised mining laws and codes to increase foreign investments in this sector (Bridge 2004b). For instance, Latin American countries, most of which conducted such reforms, have seen a steep increase of mining investments (Ericsson and Larsson 2013) and related conflicts (OCMAL 2013). Similarly, in the context of a pre-Fukushima global uranium boom, social reaction and stricter environmental regulations in Australia and the US, among others, coincide with a shift of uranium mining activities to poorer countries. Pressure to extract has increased in African countries that have less restrictive legislation, despite having poorer deposits (Conde and Kallis 2012).

Moreover, analysis of the spatial distribution of damages and risks tends to overlook the multiple geographic scales in which particular stories of environmental injustice and industrial development are situated (Keeling and Sandlos 2009). As shown by Bickerstaff and Agyeman (2009), there is a peril of having (spatial) distributive arguments shadowing other legitimate environmental justice concerns. Kurtz (2003) points out that EJ raises difficult questions about the best way to measure environmental inequities across space and to address the nature and spatial extent of both problems and possible solutions.

Furthermore, the EJ distributive debates tend to focus on the allocation of environmental impacts, neglecting the economic distributive injustices at play. There are patterns of (economic, social and environmental) mal-distribution among mining companies, affected communities and socio-political elites. The thesis of the resource curse or the paradox of plenty suggests that countries with an abundance of natural resources, particularly non-renewable ones, tend to perform worse in terms of economic growth and development outcomes than those less dependent on natural resources (Auty 1993). The deterioration of institutions (Sala-i-Martin and Subramanian 2003), socio-economic conditions, and inadequate policies (Arellano Yanguas 2011) have been identified as relevant factors to understand this paradox. It has also been argued that the multiplication of foreign junior companies in exploration stages has affected industrial performance, since those companies tend to engage in corrupt behaviours, especially when operating in weak institutional regions (Dougherty 2015).

Civil society organizations claim that transparency in the extractive is instrumental in overcoming the 'resource curse' through two important functions: empowering citizens and communities to participate in decision-making, and fostering more accountable governments and corporations (McGrath 2014). In the case of transnational companies, it has also been suggested that regulatory frameworks must be developed simultaneously in host countries, in the home countries of the corporations, and at the international level (Canel et al. 2010; see also Chapter 38).

Moreover, while most studies analysing the impact of mining rents and windfalls focus on the national level, little research has examined the local level, where most of the contestation is occurring. A statistical analysis by Arellano Yanguas (2011) in Peru concludes that those regions that have received more mining rents had more socio-environmental conflicts, and that there is no evidence that mining rents benefit the local scale or improve local economic and social well-being indicators. Arellano signals that new conflicts are rising from groups that aim to improve their access to mining rents and benefits, or from unsatisfied expectations generated by mining activities.

Other quantitative and qualitative studies on the local impact of mining royalties point that the availability of royalties does not by itself guarantee growth and improvement of living conditions at the community level (Caselli and Michaels 2009; Perry and Olivera 2009; Van der Ploeg 2008). Mining rent investment tends to benefit urban and departmental (municipal/provincial) levels and men, rather than rural areas and women – particularly if they are poor (Lagos and Blanco 2010; Perry and Olivera 2009; Ward and Strongman 2011). Moreover, mining areas are vulnerable to economic downturns when mining activities are coming to an end or have ended because of the high dependence that is usually created (Hernández 2004; Perry and Olivera 2009).

The multiple scales and dimensions that shape EJ in mining require an approach that rather than focusing on a single scale or single dimension of injustice, considers the multiple and dynamic scales and dimensions at play. In the following section, we argue for approaching mining conflicts from multi-dimensional justice frameworks and multi-scalar and context-dependent perspectives. Moreover, we highlight the relevance of analysing the different socio-economic and cultural inequalities within mining contexts.

Mining conflicts, social movements and environmental justice

The concept of EJ, born in the midst of waste-disposal conflicts, was transferred rapidly to US mining struggles. Mining-related environmental injustices have also been a relevant theme in the transnational diffusion of the EJ framework to Latin America, Canada and South Africa. We must note that we are referring to the diffusion and adoption of the "environmental justice" term. We acknowledge that previous (and ongoing) environmental struggles have expressed similar concerns without explicitly framing them as EJ. Indeed, from our point of view, many of the claims articulated in mining struggles can be understood under an EJ framework that includes non-explicit EJ discourses. There are relevant academic debates about which are the central discourses, grievances and concerns in mining protests and which is the most accurate framework for understanding them. In this section, we approach the relevance of mining in the transnational diffusion of EJ through some examples and review the central discourses deployed in mining conflicts and the related scholar debates.

Mining and the transnational diffusion of environmental justice

Uranium mining was a key concern for many Native Americans who joined the US EJ movement (Rosier 2006). Since the 1970s, Native Americans have been protesting against coal mining and, in the 1980s, uranium mining and the storage of nuclear materials started to dominate their EJ claims (Rosier 2006). After decades of uranium mining in Navajo and Pueblo reservations, the health impacts prompted a protest against 'nuclear colonialism'. With uranium mining as a relevant issue in their agenda, Native American activists took part in the 1991 First People of Color Environmental Leadership Summit, as a multi-racial movement for change (Rosier 2006). There, the self-determination right of Indigenous peoples was asserted in the foundational 'Principles of Environmental Justice'. According to Bullard and Johnson (2000), the 'radioactive colonialism' that still operates on Indian lands is due to the legacy of institutional racism.

In South Africa, mining issues have attracted growing EJ activism in the last 15 years (Madihlaba 2002; see also Chapter 46). The South African EJ movement is highly permeated by anti-racist concerns and mining conflicts are not an exception. Indeed, there are close similarities between the history of the EJ movement in the US in the 1980s and that of South Africa in the 1990s since, in both countries, the history of racial discrimination is central to the environmental discourse (Khan 2002). During the post-apartheid period, several radical organizations acknowledged the relationship between the environment, politics, racial inequality and poverty (Khan 2002) and mining started to be a key theme. In 2012, the Mining and Environmental Justice Community Network of South Africa was created as a network of community-based organizations, whose environmental and human rights are affected, directly or indirectly, by mining-related activities (MEJCON-SA 2012).

In the Australian case, Banerjee (1999) notes that since the earliest times of European invasion, the two activities that led to the greatest dispossession of Aboriginal peoples from their land were mining and pastoral activities (see Chapter 41). The Aboriginal and Northern Australian activists connect the infringement of environmental and cultural rights by uranium mining with the historical European mining pressures (Katona 2006). In Canada, the resistance to the environmental impacts of mining is coupled with the broader themes of cultural survival and sovereignty over traditional lands (Keeling and Sandlos 2009).

Mining and anti-mining movements have also had a central place in the EJ networks emerging since 2000 in Latin America (Urkidi and Walter 2011). In March 2007, a Latin American meeting on Environmental Justice and Mining took place in Oruro, Bolivia (CEPA/OCMAL 2008). In Chile, the inter-regional Network of Action for Environmental and Social Justice was created in 2006, aimed at "resisting and mobilizing against plunder", alluding to past colonial abuses, and at denouncing the impacts and injustices of large-scale mines, plantations, hydropower plants, and other industries mainly driven by transnational corporations and IFIs (Urkidi and Walter 2011). In Argentina, EJ concerns transverse socio-environmental claims. During the past decade, Argentinean socio-environmental struggles and EJ-related claims have expanded with a significant influence of large-scale metal mining conflicts (Urkidi and Walter 2011). In 2015, the first explicit EJ association was born in Argentina (i.e. AJAM) with an emphasis on denouncing extractive industries. The combination of human rights and social justice tradition, environmental concerns, and the experiences of mobilized communities contributed to the development of a (more or less explicit) EJ framework in Latin America (Urkidi and Walter 2011; see also Chapter 43).

Mining activities have been related with EJ claims and grievances worldwide. Moreover, anti-mining networks have played a key role in diffusing EJ discourses (Walter and Urkidi 2016). However, the demands, concerns and discourses of the anti-mining movement differ among regions and with particular cases, opening debates about their specific character.

Mining grievances and environmental justice discourses

Mining activities are related with a large number of EJ-related struggles. In this section, we highlight some key dimensions that appear in the claims and discourses of anti-mining movements around the world and address some relevant scholarly debates. In some cases, we highlight concerns regarding (neo)colonialism, cultural difference and recognition (see Chapter 10). In other cases, environmental justice movements stress grievances related to power imbalances present in mining spaces and the lack of genuine participation mechanisms. In many cases, damages to local livelihoods and the environment are at the origin of local protests. These different concerns usually appear in mining struggles in intertwined and differentiated ways. However, the case-specific meaning of concepts such as livelihood or environment is a contested issue in the literature, as well as the suitability of using some explicative frameworks such as EJ or environmentalism.

As shown in the previous section, grievances related to cultural and racial discrimination and demands for recognition are central in the mining conflicts of many regions. In the case of Indigenous communities, these maintain complex and variable relationships with mining companies, but many groups associate mineral extraction with their historical memory of dispossession and the disruption of traditional lifestyles (Keeling and Sandlos 2009). As Goodall (2006) notes, history is central to questions of how Indigenous peoples understand environmental justice, as the exercise of colonial power has had a spatial expression, through control over space and environment.

According to Banerjee (1999), large-scale colonial mining has devastated many Indigenous communities around the world – Africa, the Asia-Pacific, North and South America – and the trends of deregulation and privatization of the last 20 years have led to a mining industry increasingly resembling the way it looked during colonial times. According to an Australian Aboriginal activist, "the types of benefits the mining companies are talking about is another form of assimilation for Aboriginal people . . .; their relationship with the land and all the values and beliefs that underpin that . . . is being challenged by this development" (Katona 1998, in Banerjee 1999: 25). In Guatemala, for instance, since the protests against the quincentenary of Columbus' arrival, mining contestation is permeated by ethnic claims and concerns about historical justice, due to the historical marginalization of the Maya population (Urkidi 2011).

Demands for recognition and participation also come from non-Indigenous groups in the South that denounce historical and current colonialism and deep power imbalances among Northern companies and Southern societies. These asymmetries are notable in the conflicts and negotiations among corporations, states, communities and civil society organizations (Canel et al. 2010). As in other cases in the world, communities affected by transnational gold mines in Ghana argued that the lack of engagement by government agencies resulted in companies acting in a surrogate governmental capacity (Garvin et al. 2009). In the context of the post-Washington Consensus,[1] companies' responsibility codes and corporate–community agreements are replacing state regulations (Canel et al. 2010), with different results. Garibay et al. (2014) signal that, in Zacatecas (Mexico), the Canadian Goldcorp manoeuvred, with the help of the Mexican State, though coercive strategies and negative reciprocity principles ("get something for nothing with impunity"). In this sense, Keeling and Sandlos (2009) highlight the relevance of studying mining within broader processes of capital accumulation.

Another key grievance in mining protests is related to impacts on local livelihoods. However, the relation between discourses in defence of livelihoods and those in defence of the environment is debated. In a quantitative study analysing the determinants of social conflict in Latin American mining, Haslam and Tanimoune (2016) signalled that livelihood issues are a major concern and defined the problem as the overlap of agriculture and mining and, mainly, the increasing scarcity of agricultural opportunities. Moreover, they showed that poverty was a variable to take into account when analysing the prevalence of protests but that the relation is not linear: poverty intersects with distributional struggles in a complex way (Haslam and Tanimoune, 2016).

Other authors relate livelihood and environmental claims with compensation concerns, pointing to the influence of rent distribution processes in the emergence of mining conflicts. According to Arellano Yanguas (2010), some (apparent) environmental mining conflicts are instead related to the effort of some actors to improve their political leverage and gain access to mining rents. From our Latin American review of 68 mining consultations and our in-depth research in some case studies (Walter and Urkidi 2015), we argue that, in general, these were not among the key collective concerns that led movements and communities to mobilize. However, we do not discard this hypothesis for some of the involved actors. Even within a single case, there are different motives and worries behind those actors that protest against mining projects.

A related debate was developed by Banks et al. when studying mining conflicts in the island of New Guinea. In a review of three mining projects, Banks (2002) argues that Kirsch (1997) and Hyndman (2001) overlooked the influence of the environment in the development of the Ok Tedi conflict in Papua New Guinea, when framing it as "ecological resistance". Kirsch and Hyndman criticized the economically based explanations given by previous observers (Banks 2002). Banks (2002) defends a framework that would look at the way in which control over a range of resources is affected by mining operations, including subsistence resources derived from the natural environment, other material resources such as cash crops and wages, and social, political and cultural resources such as relationships, rights and responsibilities, attachments and identities that maintain the physical, social, and cultural lifeworlds of communities.

For instance, Banks (2002) suggests that, in the Porguera mine in Papua New Guinea, an environmental fix would resolve a series of community concerns over use and abuse of the river system, but it would not address other concerns over regional relationships, decision-making, resource control, and equity (Banks 2002). In the Grasberg mine in Indonesia, evidence did not indicate direct links between the degree of environmental damage and the extent of community protest. However, there seemed to be a link with the loss of autonomy over cultural and subsistence resources (particularly land), as well as human rights (Banks 2002). In the Ok Tedi mine, some communities were not involved in environmental protest as they enjoyed monetary compensation or had moved and had access to monetary resources (Hyndman 2001; Banks 2002). Banks (2002) argues that there are cases where environmental effects are the overwhelming concern because of the massive impact on the community's resource base, but that, in many other, marginalization, oppression, or access to new forms of resources (material or political) were the key issue at play. Banks (2002) notes that there is an agreement among authors that community involvement in decision-making processes is at the root of tensions.

However, Banks (2002) also acknowledges that this debate may become one of semantics, as the environment is understood to be separate from society or not, and argues for dropping the essentially Eurocentric divisions between the environment and the rest of one's daily life. We agree on the need to put this division aside in order to better understand the intertwined environmental and socio-political claims of anti-mining groups. As pointed by Guha and Martinez-Alier (1997), the environmentalisms of the poor originate in social conflicts over control and access to natural resources, since many communities depend on these resources for their material and cultural survival. Indeed, Kirsch (2007), in a later article, highlights that it is risky to reduce Indigenous movements to a binary simplification of choices between the environment and development. He explains that the movement against the Ok Tedi mine, like many other Indigenous movements, sought compensation for the damages to the environment and to limit further pollution of the river, but that they also hoped that the mine would continue to operate, providing them with economic opportunities, albeit not at the cost of the river and rainforest (Kirsch 2007).

We argue that mining conflicts are cut across by different concerns and that the pluralistic approach to EJ developed by Schlosberg (2007), among others, allows capturing the complexity of environmental, political and cultural claims deployed EJ struggles. In the analysis of two notable Latin American mining conflicts (Pascua-Lama in Chile and Esquel in Argentina), we identified the different dimensions of EJ distinguished by Schlosberg (2007) (Urkidi and Walter 2011). The studied movements reclaimed distributional justice (denouncing the unjust allocation of projects and the accumulation of benefits by transnational mining companies), participation in decision-making processes, and the recognition of their culturally and ecologically differentiated needs and livelihoods (Urkidi and Walter 2011). However, we noted that the initial reason behind the defence of local livelihoods was the environmental risk related to metal mining. Similarly, in a later study examining mining consultations in Latin America, we identified that concerns related to the defence of livelihoods, cultural recognition, territorial control, participation and self-determination were central in most of the mining conflicts that led to consultations (Walter and Urkidi 2015).

We have seen that some keywords dominate the debate around mining conflicts in the world. However, it is almost impossible to reduce the mining issue to one specific concern. In this section, we have focused on cases from the South but the heterogeneity can be even further if we consider Northern cases in depth. Indeed, some authors point to the great gap between the Global North and the Global South, when analysing environmental discourses in mining conflicts (Doyle 2002). Even a single conflict cannot be understood under reductionist approaches, but under frameworks that acknowledge the interrelation among environmental, social, cultural, economic and political spheres of life and the complex (and sometimes contradictory) claims of communities and movements. From our point of view, the multidimensional and context-dependent EJ framework offers an inspiring viewpoint from which to examine these struggles.

Finally, studies signal the relevance of examining extractive struggles with a geographical and temporal scale-sensitive approach. As we have analysed in two mining struggles in Latin America, social discourses mutate as conflicts evolve and movements engage in rescaling processes. While some dimensions of justice appeared first (participation and recognition), others (distribution of environmental and socio-economic goods and damages across multiple lines of social discrimination) emerged later, as movements jumped scales, engaging with national and international networks that provided a systemic perspective of the conflicts (Urkidi and Walter 2011). Similarly, other authors have studied the strategic rescaling of movements in extractive conflicts and their successes and risks. On the one hand, Haarstad and Floysand (2007) studied how the ability of the anti-mining movement of Tambogrande (Peru) to draft and re-frame claims in legitimated terms at different scales empowered the community. Kirsch (2007), on the other hand, highlighted the risks of counter-globalization processes for Indigenous movements as their complex ambitions regarding mining projects might be misinterpreted and simplified in those scaling dynamics. These studies point to the need to move from static analysis of social protest to approaches that capture both the dynamic social learning processes at play and the strategic nature of social mobilization.

Conclusion and final notes

This chapter reviewed the environmental injustices of large-scale metal mining activities. We introduced some of the key biophysical features of this environmentally and socially sensitive activity, showing how EJ conflicts emerge at different spatial and temporal moments of extraction. We examined the multiple patterns of maldistribution involved and the socio-environmental injustices underlined by communities and social movements in mining struggles, highlighting the relevance of anti-mining movements in the diffusion of EJ frames. We signalled that the multiple scales and dimensions that shape EJ in mining conflicts require an approach that considers them. We argued for approaching mining conflicts from multi-dimensional justice frameworks and multi-scalar and context-dependent perspectives, signalling the relevance of analysing the different socio-economic and cultural inequalities within mining contexts.

Mining activities offer interesting debates for EJ studies due to their material, territorial and historical particularities. Impacts surpass the duration and location of mining activities and EJ conflicts arise at many spatial and temporal stages. Extractive activities tend to be located in rural, biodiversity-rich and Indigenous areas. Large-scale mining claims tend to be related to (neo)colonization in EJ discourses due to the historical role of mining in processes of colonization and the transnational character of mining companies. These are key EJ features of mining conflicts that could also partially fit other types of activities.

EJ mining studies offers relevant avenues of future research. Social movements worldwide tend to frame and use the "environmental justice" term in differing ways. While in some contexts it is used or avoided given its "legalistic" connotation, in others locally developed terms are preferred to emphasize locally sensitive approaches. It would be relevant to analyse the temporal and spatial evolution of the EJ concept among anti-mining groups. Moreover, further research is needed on how the EJ frame diffuses among networks and movements worldwide. Anti-mining related networks seem to be playing a significant role in this process of diffusion. There is also a research gap regarding the exploration of environmental injustices and power-asymmetries in cases where there are no explicit conflicts (Conde and Kallis 2012). It also seems relevant to examine the medium-term outcomes of successful struggles and negotiation strategies and how EJ views and discourses evolve over time (e.g. uranium mining in Gabon, gold mining in Tambogrande, Peru).

Another line of research emerges from a criticism of the EJ framework and its distributional focus that points to the need to widen EJ approaches. In their study of the 'Save Rosia Montana' anti-mining movement in Romania, Velicu and Kaika (2015) argue that the struggle went beyond the 'traditional' pursuit of justice as a normative idea to be applied through citizen participation or compensation and that it was a process of disruption. They noticed that many local inhabitants were not asking just for redistribution or recognition by institutions; they were questioning the very frameworks at play (development and neoliberalism, for instance). This is also happening in extractive conflicts in Latin America, where local inhabitants refuse to participate in official participation events or event boycott them, and question the 'development paradigm' used to justify mining expansion (Walter and Urkidi 2015). The restrictive nature of mining frameworks is also explained by Blackburn (2009), when he argues that the Indigenous cultural objections to neo/colonialism in Canada are unheard within a political context that champions property as the only social good.

Rosia Montana's activists refused to enter negotiations and consensus-building exercises between a pre-defined set of actors and instead redefined their positions and identities and practised EJ as a transformative act. Activists imagined EJ as an open egalitarian socio-environmental ideal that has to be re-negotiated, re-embodied and performed to change not only the power relations and the set of actors in the process, but also the subjective positions of the actors themselves (Velicu and Kaika 2015). Similarly, many groups involved in extractive conflicts are engaging in broader reflective processes over the idea of 'good living'.

These processes seek to transform identities and subjectivities and promote the reconstitution of the general socio-cultural and environmental project (Gudynas 2011). In this sense and to conclude, a relevant matter that requires more attention is the link between specific EJ struggles and deeper transformative processes. Are environmental justice struggles over mining a way to transit towards more just environments?

Note

1 The Washington Consensus refers to a set of policy reforms fostered by IFIs (mostly based in Washington DC) for the Global South during the 1980s. Regulations aiming at reducing the role of the state and promoting the free market were approved in most Latin American countries.

References

Antunes, S.C., Pereira, R. & Gonçalves, F. (2007). "Acute and Chronic Toxicity of Effluent Water from an Abandoned Uranium Mine." *Archives of Environmental Contamination and Toxicology*, vol. 53(2), pp. 207–213.

Arellano Yanguas, J. (2010). *Local Politics, Conflict and Development in Peruvian Mining Regions*. DPhil thesis. University of Sussex.

Arellano Yaguas, J. (2011). *¿Minería sin fronteras? Conflicto y desarrollo en regiones mineras del Perú*. Lima: Instituto de Estudios Peruanos/Pontificia Universidad Católica del Perú.

Auty, R. (1993). *Sustaining Development in Mineral Economies: The Resource Curse Thesis*. London: Routledge.

Banerjee, B. (1999). "Whose Mine Is it Anyway? National Interest, Indigenous Stakeholders and Colonial Discourses: The Case of the Jabiluka Uranium Mine." *Critical Management Studies Conference* (Postcolonial Stream), Manchester, UK, July 14–16.

Banks, G. (2002). "Mining and the Environment in Melanesia: Contemporary Debates Reviewed." *The Contemporary Pacific*, vol. 14(1), pp. 39–67.

Bebbington, A. (2012a). "Underground Political Ecologies: The Second Annual Lecture of the Cultural and Political Ecology Specialty Group of the Association of American Geographers." *Geoforum*, vol. 43, pp. 1152–1162.

Bebbington, A. (2012b). *Social Conflict, Economic Development and the Extractive Industry: Evidence from South America*. London, New York: Routledge.

Bickerstaff, K. & Agyeman, J. (2009). "Assembling Justice Spaces: The Scalar Networking of Environmental Justice in North-east England." *Antipode*, vol. 41(4), pp. 781–806.

Blackburn, C. (2009). "Differentiating Indigenous Citizenship: Seeking Multiplicity in Rights, Identity, and Sovereignty in Canada." *American Ethnologist*, vol. 36, pp. 66–78.

Bridge, G. (2004a). "Contested Terrain: Mining and the Environment." *Annual Review of Environmental Resources*, vol. 29, pp. 205–259.

Bridge, G. (2004b). "Mapping the Bonanza: Geographies of Mining Investment in an Era of Neoliberal Reform." *Professional Geographer*, vol. 56, pp. 406–421.

Bullard, R.D. & Johnson, G.S. (2000). "Environmentalism and Public Policy: Environmental Justice: Grassroots Activism and Its Impact on Public Policy Decision Making." *Journal of Social Issues*, vol. 56(3), pp. 555–578.

Canel, E., Idemudia, U. & North, L.L. (2010). "Rethinking Extractive Industry: Regulation, Dispossession, and Emerging Claims." *Canadian Journal of Development Studies/Revue canadienne d'études du développement*, vol. 30(1–2), pp. 5–25.

Caselli, F. & Michaels, G. (2009). "Do Oil Windfalls Improve Living Standards? Evidence from Brazil." *Working Paper* 15550. Cambridge: National Bureau of Economic Research.

CEPA/OCMAL. (2008). *Justicia Ambiental y Minería*. Memoria del Encuentro Internacional de Marzo del 2007. Oruro: Centro de Ecología y Pueblos Andinos CEPA/OCMAL.

Conde, M. & Kallis, G. (2012). "The Global Uranium Rush and Its Africa Frontier. Effects, Reactions and Social Movements in Namibia." *Global Environmental Change*, vol. 22, pp. 596–610.

Defensoría Q'eqchi (2004). *Análisis de Concesiones Mineras – Tierras y Culturas Indígenas Amenazadas*. Guatemala: Defensoría Q'eqchi.

Dougherty, M.L. (2015). *By the gun or by the bribe: Firm size, environmental governance and corruption among mining companies in Guatemala*. U4 No 17. Anti-Corruption Resource Centre. CMI, September 2015 No 17.

Doyle, T. (2002). "Environmental Campaigns against Mining in Australia and the Philippines." *Mobilization: An International Journal*, vol. 7(1), pp. 29–42.

EPA. (2013). *2011 Toxic Release Inventory National Overview*. Washington DC: EPA.

Ericsson, M. & Larsson, V. (2013). "E&MJ's Annual Survey of Global Mining Investment Project Survey 2013." *E&MJ Engineering and Mining Journal*. Available at: www.e-mj.com/features/3674-e-mj-s-annual-survey-of-global-metal-mining-investment.html#.WMBDi9ThCt9

Garibay, C., Boni, A., Panico, F. & Urquijo, P. (2014). "Corporación minera, collusión gubernamental y desposesión campesina. El caso de Goldcorp Inc. en Mazapil, Zacatecas." *Desacatos*, vol. 44, pp. 113–142.

Garvin, T., McGee, T.K., Smoyer-Tomic, K.E. & Aubynn, E.A. (2009). "Community–Company Relations in Gold Mining in Ghana." *Journal of Environmental Management*, vol. 90, pp. 571–586.

Giurco, D., Prior, T., Mudd, G.M., Mason, L. & Behrisch, J. (2010). *Peak Minerals in Australia: A Review of Changing Impacts and Benefits*. Sydney: Institute of Sustainable Futures and CSIRO.

Goodall, H. (2006). "Indigenous Peoples, Colonialism, and Memories of Environmental Injustice." In Hood, S., P. Washington, C. Rosier and H. Goodall (eds.), *Echoes from the Poisoned Well: Global Memories of Environmental Injustice*. Oxford: Lexington Books, pp.73–96.

Government of Australia (2007). *Managing Acid and Metalliferous Drainage*. Sydney, Government of Australia.

Gudynas, E. (2011). "Caminos para las transiciones post extractivistas." In Alayza, A. and E. Gudynas (eds.), *Transiciones. Post extractivismo y alternativas al extractivismo en Peru*. Lima: RedGE and CEPES, pp. 187–216.

Guha, R. & Martinez-Alier, J. (1997). *Varieties of Environmentalism: Essays North and South*. London: Earthscan.

Haarstad, H. & Floysand, A. (2007). "Globalization and the Power of Rescaled Narratives: A Case of Opposition to Mining in Tambogrande, Peru". *Political Geography*, vol. 26(3), pp. 289–308.

Haslam, P.A. & Tanimoune, N.A. (2016). "The Determinants of Social Conflict in the Latin American Mining Sector: New Evidence with Quantitative Data." *World Development*, vol. 78, pp. 401–419.

Hernández, G. (2004). "Impacto de las regalías petroleras en el departamento del Meta." *Serie Ensayos sobre Economía Regional*, vol. 14, Villavicencio.

Hipwell, W., Mamen, K.,Weitzner, V. & Whiteman, G. (2002). "Aboriginal People and Mining in Canada: Consultation, Participation and Prospects for Change." *Working Discussion Paper* 4. Ottawa: North-South Institute.

Hyndman, D. (2001). "Academic Responsibilities and Representation of the Ok Tedi Crisis in Postcolonial Papua New Guinea." *The Contemporary Pacific*, vol. 13(1), pp. 33–54.

IUCN (2011). "Mining Threats on the Rise in World Heritage Sites". International Union for Conservation of Nature.

Katona, J. (2006). "The Mirrar Fight for Jabiluka: Uranium Mining and Indigenous Australians to 2004." In Hood, S., P. Washington, C. Rosier and H. Goodall (eds.), *Echoes from the Poisoned Well: Global Memories of Environmental Injustice*. Oxford: Lexington Books, pp. 285–298.

Keeling, A. & Sandlos, J. (2009). "Environmental Justice Goes Underground? Historical Notes from Canada's Northern Mining Frontier." *Environmental Justice*, vol. 2(3), pp. 117–125.

Khan, F. (2002). "The Roots of Environmental Racism and the Rise of Environmental Justice in the 1990s." In McDonald, D.A. (ed.), *Environmental Justice in South Africa*. Ohio University Press, pp. 15–48.

Kirsch, S. (1997). "Is Ok Tedi a Precedent? Implications of the Lawsuit." In Banks, G. and C. Ballard (eds.), *The Ok Tedi Settlement: Issues, Outcomes and Implications*. Canberra: National Centre for Development Studies and Resource Management in Asia Pacific project, pp. 118–140.

Kirsch, S. (2007). "Indigenous Movements and the Risks of Counterglobalization: Tracking the Campaign against Papua New Guinea's Ok Tedi Mine." *American Ethnologist*, vol. 34(2), pp. 303–321.

Kuipers, J. (2003). *Putting a Price on Pollution*. Washington DC: Mineral Policy Center.

Kurtz, H.E. (2003). "Scale Frames and Counter-scale Frames: Constructing the Problem of Environmental Injustice." *Political Geography*, vol. 22, pp. 887–916.

Lagos, G. & Blanco, E. (2010). "Mining and Development in the Region of Antofagasta." *Resource Policy*, vol. 35, pp. 265–275.

Le Billon, P. (2008). "Diamond Wars? Conflict Diamonds and Geographies of Resource Wars." *Annals of the Association of American Geographers*, vol. 98(2), pp. 345–372.

Mackasey, W.O. (2000). *Abandoned Mines in Canada*. WOM Geological Associates Inc. Available at: www.abandoned-mines.org/pdfs/mackasey.pdf

Madihlaba, T. (2002). "The Fox in the Henhouse: The Environmental Impact of Mining on Communities in South Africa." In McDonald, D.A. (ed.), *Environmental Justice in South Africa*. Ohio University Press, pp. 156–167.

Martinez-Alier, J. (2001). "Mining Conflicts, Environmental Justice, and Valuation." *Journal of Hazardous Materials*, vol. 86, pp. 153–170.

McGrath, F. (2014). *Sustainability and Environmental Justice in Resource-rich Countries*. Prague: Bank Watch.

MEJCON-SA. (2012). Mining and Environmental Justice Community Network of South Africa. Centre for Environmental Rights. Available at: http://cer.org.za/programmes/mining/communities/mining-environmental-justice-community-network-south-africa

Montana Department of Environmental Quality. (1997). *Draft Environmental Impact Statement, Golden Sunlight Mine*. Montana.

Moody, R. (1996). "The Lure of Gold – How Golden is the Future?" *Panos Briefing*, vol. 19, Panos Institute.

Mudd, G.M. (2007). "Global Trends in Gold Mining: Towards Quantifying Environmental and Resource Sustainability." *Resource Policy*, vol. 32, pp. 42–56.

Mudd, G.M. (2010). "The Environmental Sustainability of Mining in Australia: Key Mega-trends and Looming Constraints." *Resource Policy*, vol. 35, pp. 98–115.

OCMAL. (2013). Observatorio de Conflictos Mineros de América Latina. Available at: www.conflictosmineros.net

Özkaynak, B., Rodriguez-Labajos, B., Aydın, C.İ., Yanez, I. & Garibay, C. (2015). *Towards Environmental Justice Success in Mining Conflicts: An Empirical Investigation*. EJOLT Report No. 14. ICTA-UAB.

Perreault, T. (2006). "From the Guerra Del Agua to the Guerra Del Gas: Resource Governance, Neoliberalism and Popular Protest in Bolivia." *Antipode*, vol. 38(1), pp. 150–172.

Perreault, T. (2013). "Dispossession by Accumulation? Mining, Water and the Nature of Enclosure on the Bolivian Altiplano." *Antipode*, vol. 45(5), pp. 1050–1069.

Perry, G. & Olivera, M. (2009). "El Impacto del petróleo y a minería en el desarrollo regional y local en Colombia." *Documentos de Trabajo*, No. 2009/06, CAF.

Prior, T., Giurco, D., Mudd, G.M., Mason, L. & Behrisch, J. (2012). "Resource Depletion, Peak Minerals and the Implications for Sustainable Resource Management." *Global Environmental Change*, vol. 22, pp. 577–587.

Rosier, P. (2006). "Fond Memories and Bitter Struggles: Concerted Resistance to Environmental Injustices in Post-War Native America." In Hood, S., P. Washington, C. Rosier and H. Goodall (eds.), *Echoes from the Poisoned Well: Global Memories of Environmental Injustice*. Oxford: Lexington Books, pp. 35–54.

Sala-i-Martin, X. & Subramanian, A. (2003). "Addressing the Natural Resource Curse: an Illustration from Nigeria." *IMF Working Paper* 01/139. Washington, DC: International Monetary Fund.

Schaffartzik, A., Mayer, A., Eisenmenger, N. & Krausmann, F. (2016). "Global Patterns of Metal Extractivism, 1950–2010: Providing the Bones for the Industrial Society's Skeleton." *Ecological Economics*, vol. 122, pp. 101–110.

Schlosberg, D. (2007). *Defining Environmental Justice: Theories, Movements, and Nature*. NYC: Oxford University Press.

Urkidi, L. (2011). "The Defense of Community in the Anti-mining Movement of Guatemala." *Journal of Agrarian Change*, vol. 11(4), pp. 556–580.

Urkidi, L. & Walter, M. (2011). "Concepts of Environmental Justice in Anti-Gold Mining Movements in Latin-America." *Geoforum*, vol. 42, pp. 683–695.

US Accountability Office (2011). *Abandoned Mines. Information on the Number of Hardrock Mines, Costs of Clean up and Value of Financial Assurances*. Highlights of GAO-11-834T.

US EPA (United States Environmental Protection Agency). (2015). Frequent Questions Related to Gold King Mine Response. Available at: www2.epa.gov/goldkingmine/frequent-questions related-gold-king-mine-response.

Van der Ploeg, F. (2008). "Challenges and Opportunities for Resource Rich Economies", *OxCarre Research Paper* No. 2008-05, Oxford.

Velicu, I. & Kaika, M. (2015). "Undoing Environmental Justice: Re-imagining Equality in the Rosia Montana Anti-mining Movement." *Geoforum* (article in press).

Walter, M. & Urkidi, L. (2015). "Community Mining Consultations in Latin America (2002–2012): The Contested Emergence of a Hybrid Institution for Participation." *Geoforum*. http://dx.doi.org/10.1016/j.geoforum.2015.09.007

Walter, M. & Urkidi, L. (2016). "Community Consultations: Local Responses to Large-Scale Mining in Latin America." In De Castro, F., B. Hogenboom, and M. Baud (eds), *Environmental Governance in Latin America*. NYC: Palgrave Macmillan, pp. 287–325.

Ward, B. & Strongman, J. (2011). *Gender-sensitive Approaches for the Extractive Industry in Peru: Improving the Impact on Women in Poverty and their Families*. Washington DC: World Bank Publications.

31

JUSTICE IN ENERGY SYSTEM TRANSITIONS

A synthesis and agenda

Karen Bickerstaff

Introduction

There tends to be an assumption that (transitioning towards) low carbon and renewable energy systems is self-evidently a good thing (Laird 2013). Yet in recent years the social and equity implications of such socio-technical transformation have emerged as a critical, though still relatively underdeveloped, domain of environmental justice research, drawing attention to the structural and material processes driving energy vulnerabilities, the spatial complexity of injustices associated with infrastructure and policy change, and the (re)distribution of political rights and responsibilities for delivering change (Bickerstaff et al. 2013). In this chapter, I highlight key concepts being articulated in *energy justice* research through a focus on low carbon energy system transitions generally and the deployment of nuclear power specifically. In doing so, I explore some critical opportunities for future research, crystallizing around first, the ways in which justice issues are constituted spatially and temporally across the whole lifecycle of the energy system (innovation, through resource extraction, power generation, consumption and managing waste residues) and second, how injustices emerge and evolve through the intersections of technology choices, political contexts and societal expectations.

Human-induced climate change, by its nature, involves the production of injustice (see Chapter 29). The impacts of climate change impose spatially uneven harms on present and future generations and the production of greenhouse gas emissions is also socially and spatially uneven. While these issues of global politics and ethics have been rehearsed over the past two decades, the implications for addressing climate change in a just manner have received far less attention. It is now increasingly apparent that the justice dimensions of mitigating climate change, specifically through decarbonizing energy systems, raise significant public, policy and political challenges that require sustained academic attention – empirically and theoretically (Bickerstaff et al. 2013).

Within this context, the concept of energy justice has proliferated in recent years (e.g. Bickerstaff et al. 2013; Hall et al. 2013; Preston et al. 2013; Walker 2008; Parkhill et al. 2013; Sovacool and Dworkin 2014), with researchers articulating agendas around two central (and often overlapping) themes: the social and spatial distribution of energy poverty and the justice dimensions of particular (low carbon) energy technologies or system components. My attention in this chapter is focused on the latter, the justice consequences of low carbon energy systems. In what follows I reflect on three energy justice concepts that underpin much of the extant research literature: distributional, procedural and recognition justice. In doing so, I draw primarily – though not exclusively – on UK experience and, most extensively, from the case of nuclear power (past and future). Nuclear energy at present forms an important part of the United Kingdom's energy policy and is expected to play a significant role in the delivery of low carbon transitions (DECC 2011; Butler and Simmons 2013). It has been held up in policy discourse as economic, safe and low-carbon – a necessary part of the solution to climate change (e.g. BERR 2008; Butler and Simmons 2013). At the same time nuclear energy is the product of extremely complex technological systems, the nuclear fuel cycle is spatially extensive and sites of extraction, fuel fabrication, power generation and waste disposal have all been the focus of intense controversy. In this respect, my use of nuclear power offers an opportunity to shed light on some of the challenges and ambiguities raised by energy justice research. Whilst I focus on the UK situation, I will draw on wider research and literature to illustrate how issues and processes discussed are applicable to other national contexts.

Distributional justice and energy technologies

Debates about present and future energy technologies have brought questions around the social, spatial and temporal distribution of benefits and burdens associated with the world's energy system to the fore. Here a distinct body of research has addressed the socially regressive impacts of moves towards low carbon energy systems and their costs (e.g. Garman and Aldridge 2015). For example, the UK Feed-in-Tariff (FiT) for microgeneration – a payment made to households or businesses generating their own electricity, proportional to the amount of power generated, and designed to encourage uptake of small-scale renewable energy generation technologies – has been criticized for disproportionately benefiting people in higher-income groups who can afford the upfront capital required to install solar photovoltaic technology (PV). The introduction of the UK FiTs in 2010 is estimated to have cost consumers £6.7 billion (Stockton and Campbell 2011). Stockton and Campbell (2011) question the distributional outcomes of this form of subsidy; whilst the costs are levied on all electricity bills, households' access to the payment and the supplementary benefit of free electricity is limited to those that can pay the upfront capital costs. Investing in these technologies is simply not feasible for financially disadvantaged households. As Walker (2008) suggests, the prospect might be one of middle classes investing available capital, and increasingly realizing returns as capital costs fall, whilst lower income groups remain dependent on increasingly expensive electricity and gas supplies, making them an 'energy underclass'. The point that these authors make is that the potential for microgeneration to improve access to affordable energy for low-income households will not be realized through a model of low carbon transition that focuses on households paying upfront for new technologies; varied delivery and governance mechanisms are needed that recognize the role for local governments, housing associations and energy suppliers (Walker 2008; Stockton and Campbell 2011; Wolsink 2013).

It has also been suggested that the costs of building new nuclear capacity will have a disproportionate impact on low income energy consumers (Garman and Aldridge 2015). In this regard, Mitchell (2011) cogently makes the point that recent UK electricity market reforms, put in place – in part at least – to support private-sector led nuclear new build, will substantially raise the price of electricity. A key element of these reforms, the Contracts for Difference, requires government to commit to pay a premium for low-carbon electricity into the long-term future with energy consumers shouldering the cost via their utility bills. Mitchell argues that with nuclear power, in contrast to other low carbon forms of electricity generation, there is almost no transparency of costs and the government can only negotiate with a couple of nuclear operators, leading to an excessive agreed or 'strike' price.

Distributional inequities have been to the fore in analysing disputes over the siting of energy infrastructures relating to power generation, disposal of waste residues (see Chapter 25) and, to a lesser degree, resource extraction (see Chapters 30 and 31). Justice arguments have tended to centre on siting processes and outcomes in terms of the direct or indirect targeting of vulnerable groups such as Indigenous and economically marginal communities (Blowers and Leroy 1994; Gowda and Easterling 2000; Endres 2009). Blowers (2010) asks the question: "Why dump on us?" in the context of new nuclear power plants in the UK. He notes that all sites listed for nuclear power stations in 2010 were existing nuclear locations, where communities are considered 'peripheral' – that is, characterized by social and economic marginalization: dependency on a dominant employer or sector; political powerlessness; and fatalistic acceptance (of a nuclear presence). He argues that processes of peripheralization are reproduced and reinforced by the repeated focus on these communities as sites for nuclear activities. As Blowers (2010: 157) remarks: "[t]he burden of nuclear risk extending into the far future thus becomes concentrated at a few socially and physically vulnerable locations". Similarly, Gunderson and Rabe (1999: 212), addressing Canadian experience, point to concerns that nuclear facility siting processes tend to rely on single communities where there is 'historic conversance with and economic dependence on nuclear energy'.

Similar observations have been made in relation to other energy system developments such as hydraulic fracturing of rocks to exploit shale gas reserves ('fracking') – described by proponents as providing a lower (rather than low) carbon supply to coal-based production in the short term (Helm 2011). In the US, studies have reported statistically significant correlations between proximity to well sites and environmental and health problems, such as raised levels of methane in drinking water (Clough and Bell 2016). Some research has pointed to breaches of environmental justice – as expressed in the spatial distribution of benefits through ownership of mineral rights (Fry et al. 2015; Clough and Bell 2016) and, for parts of the Marcellus Shale region in the US, evidence that unconventional gas wells are disproportionately located in areas with larger populations living in poverty. In the UK, Cotton et al. (2014: 434) focus on the explicit spatial framing of sites for shale gas exploration, citing the head of one shale gas exploration company who notoriously called for the "'traditional heartlands' of the Midlands and the North to be fracked before the Southeast, in order to save the industry from 'becoming bogged down by a storm of protesting rural areas". As one Lancashire MP subsequently observed: "the North gets the dirty end and the South sucks up all the energy" (cited in Cotton et al. 2014: 434). The authors also comment on the compromised impartiality of local authorities, whereby support for the industry becomes framed as a solution to existing economic deprivation – in essence a form of economic coercion (cf. Blowers and Leroy 1994).

Geopolitical accounts of peripheralization and injustice that tie up different points in global commodity chains can also be observed in existing work – notably the writings of historian Gabrielle Hecht (2012), who narrates the history of uranium mining from an African perspective. She traces how the "invention" of a uranium market in former French colonies such as Gabon and Niger ensured a cheap flow of uranium for Western powers (see also Endres 2009; Taebi and Kloosterman 2008, on the highly uneven distribution of burdens and benefits across the nuclear fuel cycle), whilst large mining companies and post-colonial powers simultaneously played down the (exceptional) health risks of uranium mining, with devastating consequences for the life, work and health histories of miners (see also Chapter 41 on Australia). From the perspective of justice theory, then, Hecht's analysis foregrounds the play of post-colonial relations, structures and institutions in the uneven distribution of nuclear benefits and burdens. With a similar concern for spatialized commodity chains, Mulvaney (2013: 231) highlights how processes and components involved in the production of PV technologies rely on toxic materials and waste flows similar to those in the electronics industries, and how such technological innovations may disproportionately impact workers and communities distant from sites of consumption. The diffuse nature of solar energy also means that to substantially harvest this resource requires tremendous land use footprints, raising conflicts with other uses for land, with possible damage to ecological and cultural resources (Mulvaney 2013).

Procedural justice and energy transitions

Although issues of distributive justice have been central concerns, debates about low carbon energy systems and infrastructure siting are equally permeated by questions of procedural justice – that is, the fairness of, and access to, decision-making procedures (e.g. Cowell et al. 2012; Bickerstaff et al. 2013) (see Chapter 9). So, for instance, much of the existing geographical literature on the siting of energy and/or nuclear facilities has centred on the notion (and critique) of NIMBY (not in my backyard) as a means to understand siting con-flicts and local opposition – typically characterized as locally organized, emotional and self-interested campaigns opposing a locally unwanted land use. These critical accounts make clear that opposition cannot be just dismissed as NIMBY in character but rather is rooted in concerns about the democratic adequacy of planning and consent processes, connected to breaches of social and spatial justice (e.g. Kemp 1990; Blowers 2010). Kemp's (1990) analysis of public concerns around the search for a site to house low and intermediate level radioac-tive waste exemplifies this point very well. Kemp refers to a series of more fundamental technical, socio-economic and environmental issues, beyond the parochialism of the NIMBY label, underpinned by a structural context that reasserted the power of the status quo and failed to acknowledge and take seriously the 'vocabularies of motive' that lay behind public opposition. Hinchliffe and Blowers (2003: 30) similarly position the Decide-Announce-Defend (DAD) approach adopted by Nirex (the organization tasked with finding a site for UK waste, and funded by the nuclear industry) – essentially picking sites in secret and then consulting with communities in the selected areas – as central to the intense opposition leading up to and following identification of West Cumbria as the favoured location for a geological dis-posal facility in the early 1990s (cf. Endres 2009, on the controversy surrounding the Yucca Mountain High Level Waste site in the US). Recent UK and international experience with the development and deployment of carbon capture and storage has similarly suggested that local concerns about the risks of infrastructure may be intensified by their perceived imposi-tion through inaccessible or prejudiced decision-making with poor or limited opportunities for participation (McLaren et al. 2013).

In terms of policy mechanisms for delivering greater procedural justice, much of the social science engagement with energy system transitions embeds an implicit or explicit understand-ing of participation as a key trope of a fair decision-making process and outcome (McLaren et al. 2013). But, too often, these efforts to involve publics have centred on the expression of concerns in relation to decisions being made at downstream stages in innovation and policy processes, and as such with minimal scope to inform decision-making. Reflecting on international experience, McLaren et al. (2013) observe the partial reading of procedural justice in the development and deployment of low carbon energy innovation generally and CCS technologies specifically – with instrumental goals driving the practice of public participation and engagement. The authors note a dominating concern with diagnosing the causes of opposition and, from this, articulating modes of participation that will support socially acceptable siting processes and outcomes.

Emerging work on shale gas policy in the UK (e.g. Cotton et al. 2014) reinforces this point. With regard to locally affected site communities in Lancashire and Suffolk, Cotton et al. (2014) note public concerns that the energy company, Caudrilla's, deployment of extensive local participatory processes represents a form of 'deliberative speak': "a rhetoric of engagement that is not matched by mechanisms to ensure community involvement in decisions" (Cotton et al. 2014: 434). The authors quote a local activist who reflects on the constrained political opportunities available:

> I just think the whole thing is just unfair. And Cuadrilla will come and do big public presentations and they talk about engaging with the community. Excuse me what you've done, you've bunged a couple of hundred quid for somebody to buy some flowers to go outside a village hall somewhere; that is not engaging with the community. So it is that lack of legitimacy that fires me up and inspires me to give up my time.
>
> *(Cotton et al. 2014: 434)*

As such, the policy and research emphasis on promoting (more) participation remains distant from procedural justice issues such as power, voice, access to early decision-making and recognition of difference in fundamental values and beliefs.

The UK government's Managing Radioactive Waste Safely (MRWS) process initiated in 2001, leading up to and beyond a decision regarding how to manage higher activity nuclear wastes (HAW), arguably offered something different, and has been read by many as a radical embracing of a deliberative and participatory style of decision-making by political authorities (e.g. Reiner and Nuttall 2011). As Blowers (2014: 545) comments, through MRWS a new decision-making lexicon came into play "emphasising openness, transparency, engagement, dialogue". An Independent Committee on Radioactive Waste Management (CoRWM) was established in 2003, tasked with conducting a "root and branch exercise to recommend the option, or combination of options, that can provide a long-term solution" (Blowers 2014: 545) with regard to higher activity legacy wastes. As part of this process CoRWM was asked to start from a "blank sheet of paper" in terms of options being considered (for which they were criticized by some, e.g. HLSTC (2004), and embarked on extensive consultation with the public and expert groups (Chilvers and Burgess 2008). CoRWM also placed a great deal of emphasis on intra-and intergenerational ethics (CoRWM 2006). Crucially CoRWM stated that intra-generational equity should be considered in deciding where to locate a waste facility in relation to the needs and interests of those affected. Such thinking led the Committee to recommend a Geological Disposal Facility (GDF) as 'the best available approach', alongside a process by which a site might eventually be selected, emphasizing the idea of volunteerism, whereby communities would be invited to express a willingness to participate in a siting process, with an emphasis on partnership and benefits packages (i.e. incentives to participate and compensation for participation) (Blowers 2014). Nonetheless, in terms of procedural justice, some researchers (notably Chilvers and Burgess 2008) have emphasized the wider policy and institutional contexts (e.g. policy discourse around the role of nuclear power in the UK's future energy mix) in which the analytic-deliberative practices sat, which they suggest served to perpetuate power relations and problem framings, marginalizing particular concerns, knowledges, meanings, and forms of expression.

Recognition

A particular issue that the CoRWM process raises, then, and a central thematic in the energy justice literature, is the (need for) recognition of alternative cultural understandings of risk and the ability of marginalized groups to be heard – to have a voice (Wolsink 2007) – as fundamental principles underpinning procedural justice (see also Chapters 7, 9 and 10). Bell and Rowe (2012), for instance, raise a serious risk that climate mitigation proposals will exacerbate current injustices by further concentrating power and voice in the hands of business interests and more affluent communities. The issue of recognition, and who legitimately should constitute the 'demos', and how, remains a critical ethical challenge – geographically many decisions about climate policies are made by a territorially defined democratic community, yet these policies have effects, spatially and temporally, beyond the boundaries of that community (Bell and Rowe 2012). For Bell and Rowe the most just response is to adopt the principle of proportionality – that power in any decision-making process should be proportional to individual stakes – to increase the power of the least powerful. In this vein, considerable attention, in the context of siting energy system infrastructures, has rested on procedural responses in the form of policies of community volunteerism and compensation (Cowell et al. 2012; Sundqvist 2014 on Swedish and Belgian experience). It has been suggested that community volunteerism expands opportunities for local publics and stakeholders to participate (proportionately) in siting decisions and that compensation measures offer a corrective to unfair distributional outcomes. Others have, however, highlighted the challenges of redistributing power in accordance with relative stakes in this way. Hunhold (2002), in relation to Canadian experience with HAW repository siting, has argued that voluntary approaches are no more successful, due to a typically and excessively local conception of public participation. UK experience with community volunteerism, where efforts to site a GDF focused on the West Cumbria region of England (which hosts the Sellafield complex) between 2008 and 2013, makes a similar point. Although *local government* were identified as the decision-making authority for the host community (DEFRA 2008), there was contestation between different administrative tiers over how this power (and associated right of withdrawal) would be exercised in practice. Whilst the two local 'District' councils voted in 2013 to proceed with the siting process, the regional 'County' authority voted against – essentially bringing the process to an immediate halt (DECC 2014). As the government review of the siting process makes clear, the experience in Cumbria exposed the scalar tensions in establishing decision-making over national infrastructure as essentially a local matter (DECC 2013; Bickerstaff 2012). Research also underlines how benefits packages can be equated with bribery; the likely reality being that only already disadvantaged communities will volunteer to host a HAW repository (Butler and Simmons 2013; Bickerstaff 2012; Gunderson and Rabe 1999) on social (acceptability) rather than (purely) physical suitability criteria – arguably exacerbating the concentration of risk in socially, politically and economically vulnerable places (Blowers 2010).

Post-politics and the management of dissent

Concerns about recognition arguably fit into a more fundamental 'post-political' critique of the democratic adequacy of procedures for low carbon (energy) policy in general (Swyngedouw 2010) and the siting of energy system infrastructure in particular (Johnstone 2014). The 'postpolitical' condition is conceived as that in which politics, identified as 'dissensus' involving competing ideologies of socio-economic trajectories, is effectively foreclosed, establishing a consensual policy framework driven by the dominant ideological convictions of neoliberalism in the guise of a 'value-free', technocratic governance regime (Johnstone 2014). An emerging feature of the post-political condition is the way in which agreement can be negotiated through the technique of appealing to universal themes, such as 'sustainable development', and 'climate change' which encourage, even necessitate, agreement and support in principle (Allmendinger and Haughton 2012; Johnstone 2014). It has been suggested that this drive for technocratic consensus will further marginalize the opportunities for opponents of major planning applications to gain a voice. Post-political arguments emphasize how the management of dissent is characterized by carefully allocating the handling of controversial issues to alternative methods and scales – displacements that are bound up with processes of rescaling decision-making. This critique, which raises fundamental questions about the justice dimensions of major infrastructure siting, has visible purchase in relation to recent UK planning reforms that have reshaped and re-spatialized opportunities for public participation (McLaren et al. 2013; Johnstone 2014; Bickerstaff 2012). Bell and Rowe (2012) note the contradictory political trends in the UK government's responses to the decarbonization agenda: whilst government requires developers to consult local communities over major proposals before planning applications are submitted, the implementation of projects such as wind farms and nuclear power stations is being speeded up, with decisions being taken by ministers rather than local planning authorities.

The UK Planning Act 2008 (HM Government 2008) represents a critical piece of legislation in this respect, part of a radical overhaul of the planning system involving the abandonment of the public inquiry system, and widely seen as the solution to the previous delays that had befallen the siting and construction of new nuclear (Johnstone 2014). These planning changes have effectively passed responsibility for dealing with development consent orders for nationally significant infrastructure from local authorities to a non-departmental public body, to examine applications and make recommendations to government. Through these processes consultation opportunities have been rescaled, between discussion of national policy statements (e.g. in relation to the political need identified for nuclear new build) and local consultations on selected sites – dealing only with the specificities of the particular development in question (Blowers 2010; Johnstone 2014).

Hinkley Point C nuclear power station, the first of eight potential nuclear power sites announced by the UK government to progress through new planning legislation, saw EDF (as the lead implementer) undertake local consultations solely concerned with site specific issues – the environmental *need* for Hinkley C had been established and fixed through national policy. Johnstone (2013) recounts the sheer quantity of events put on by EDF between 2009 and 2012 – demonstrating the high level of visible public 'participation'. However, the scalar separation of national and local issues undoubtedly silenced many substantive concerns about the trajectory of energy policy. In this regard Johnstone (2013) notes how the Major Infrastructure Planning Unit (MIPU, the national body administering planning applications) would not, during the planning hearings, consider 'matters of policy'. Nor was it considered the remit of the MIPU to acknowledge 'external' events that were occurring in the tumultuous world of energy policy – such as the Fukushima disaster in Japan or the UK electricity market reforms. As such, the MIPU proceedings were increasingly seen by NGOs and activist groups as failing to offer an arena for worthwhile participation. The key point here then is that the seemingly uncontested nature of decision-making around nuclear new build was in large part achieved by very explicit processes of rescaling with more or less direct effects in terms of excluding dissenting voices.

We see some similar trends in the context of managing nuclear waste. Reformulated government proposals for implementation of a GDF were set out for consultation in 2013

(DECC 2013) following the withdrawal of West Cumbria from the voluntary siting process. The consultation reiterates the ongoing government commitment to a siting approach based on voluntarism. Yet, at the same time, new proposals sharply reveal efforts to rescale decision-making, to co-opt or neuter a more confrontational mode of politics and to achieve specific planning outcomes. In the consultation document, much is made of the principle of subsidiarity – to guarantee a degree of independence for a lower authority in relation to a higher body or for a local authority in relation to central government (DECC 2013). On this basis, government proposals stated that there should be one representative level of local government that holds the right of withdrawal and has the final decision on proceeding and that that should be the relevant district council in England (DECC 2013), that is, the proximate rather than regional tier of local government. The response to policy failure in West Cumbria therefore appears to be one of re-scaling decision-making responsibilities in order to facilitate a 'democratic' and affirmative outcome. Government has also consigned outstanding problems – the roles and responsibilities of community representatives, how and when a test of public support will be made, and what options there will be for disbursement of community investment – to the deliberations of a new technical expert panel, the Community Representation Working Group (CRWG). Alongside this deferral of complex (and controversial) issues to expert committees, the principle of volunteering is now firmly positioned within an amended Planning Act (2008) which brings a GDF in England within the definition of 'nationally significant infrastructure'. The establishment of a GDF as a nationally significant infrastructure, and the lack of public debate about such a substantive change, has raised concerns that 'in principle' decisions would foreclose debates at an early stage (Blowers 2014: 552). It also sits uncomfortably with the localism principles enshrined in volunteerism – effectively it strips local authorities of the ability to stop a facility being sited within their boundaries (Johnstone et al. 2013).

Conclusions and future agenda

This chapter has sought to review the substantial and diverse body of literature that addresses the justice dimensions of energy system transformation, with a particular focus on nuclear energy and the nuclear fuel cycle. The use of this core case study serves to highlight how existing research tends to address rather narrowly drawn and spatially located components of the energy system, focusing on the benefits and burdens presented by certain fuel types and infrastructure, processes of decision-making and siting, and measures put in place specifically to mitigate and reduce potential injustice. These observations lead to a number of conclusions about the preoccupations of energy justice research and where researchers might wisely invest future efforts.

First, this chapter underlines the importance of research that can develop more holistic and joined up accounts of low carbon energy system transitions across commodity and lifecycle chains. A small but important emerging focus in the literature makes explicit the unevenness of energy system benefits and burdens within and across national boundaries, and how matters of distribution are undercut by entrenched structural relations and exclusions (e.g. Hecht 2012; Endres 2009). As Mulvaney (2013) recognizes in relation to PV technologies, innovations have been treated as a black box, as regards environmental justice, and largely ignored. Mulvaney's work offers a rich insight into how addressing the whole technological supply chain, including institutional innovations, generates a very different set of questions about the unequal distribution of harms from the solar energy transition (also Raman 2013 on the refining of rare earth elements in the use of renewable energy technologies). In this regard we need to find ways of conceptualizing and articulating justice across complex spatial assemblages of humans and non-humans (cf. Miller et al. 2013), to develop multi-sited commodity chain approaches to scrutinize the social and environmental relations that constitute production systems (Mulvaney 2013) – looking upstream as far back as the raw materials deep in the supply chains and forward through the life of commodities and technologies to ultimate disposal or materials recovery.

Second, a linked absence lies with the dearth of work addressing the justice implications of multi-faceted future energy scenarios. So, whilst there has been a huge amount of work on future energy system scenarios (e.g. Ekins et al. 2013) that assess, and seek to optimize the costs (e.g. resources, infrastructures, technologies, taxes) of meeting energy service demands under a range of physical and policy constraints – notably to meet GHG emission reduction targets – we have seen little, if any, meaningful attention to the justice ramifications of holistic scenario trajectories (cf. McLaren et al. 2013). As Miller et al. (2013) observe, technological choices reshape social practices, values, relationships, and institutions, and ways of knowing and living, and in so doing can contribute to creating or reinforcing unequal distributions of power and wealth (also Laird 2013). In this sense future energy scenarios present critical normative questions: about the sorts of socio-technical and political regimes that different energy systems (re)produce; about control, access and affordability and, in turn, the social and spatial patterning of burdens and benefits. This raises the question of how we can conceptualize and empirically scrutinize the processes and outcomes associated with future energy scenarios that combine a diverse range of technologies and policy instruments. Such matters present important challenges not only in terms of the concepts we deploy but also the methods used to quantify, contextualize, map, and deliberate the production and experience of energy justice associated with energy system transformations (see also Chapter 8). There is also an important role for developing stronger comparative method-ologies that recognize, and connect up, different framings of what counts as just or fair. There are useful precedents here in analytic-deliberative research that has sought to use quantitative metrics and outputs as a basis for informed (and critical) public discussion about the societal implications of technical change (Burgess et al. 2007; Chilvers and Burgess 2008; Butler and Simmons 2013; Ottinger 2013) – methods that are demonstrably able to reconcile very different framings of the ethical dimensions of technology, as well as enhance the fairness of decision-making.

Finally, whilst the discussion in this chapter has focused around classic environmental justice concepts of distributional, procedural and recognition justice it has, at the same time, suggested certain limits to the application of these descriptive and interpretative ideas. Arguably these concepts tend to focus research attention on describing the organization of different kinds of benefits and burdens, inclusions and exclusions – focusing on the establish-ment of injustice rather than necessarily foregrounding spatial, political and material causes (Butler and Simmons 2013). Indeed much of the work that has been reviewed in this chapter has not been framed in explicitly environmental justice terms. It is therefore important that future work develops a more eclectic set of theoretical tools to deploy in its critical scrutiny of the dynamics driving, as well as consequences of, energy system transformations. Gabrielle Hecht's (2012) account of uranium exploitation in Africa through the (re)articulation of post-colonial relations offers one notable example of how we might think about and with justice in different ways. In this chapter, I have specifically considered a post-political reading of the shifting democratic landscape around energy policy – the rescaling of decision-making, and the political mobilization of the *urgency* of climate change (mitigation) – which has served to destabilize and neuter opposition or dissensus around certain energy system pathways.

Whilst we might debate how far energy policy in the UK (or elsewhere) can be truly characterized as post-political, such accounts generate some profoundly important questions about how we understand and tackle energy injustice, making clear that choices about fuel sources, and the infrastructures that sustain them, are intensely political, and far from the narrow technical decisions they are often presented to be (Laird 2013). As Mulvaney (2013: 237) reflects, "too often the consequences of climate change render any criticism of low carbon energy moot". This chapter has demonstrated that critical scrutiny of 'low carbon' and 'renewable' energy systems – their social and spatial outcomes as well as the institutional arrangements and political agendas they embed and sustain – is an absolute priority for social science research, if future transitional pathways are to be (as they certainly should be) a powerful force for improving socio-ecological well-being.

References

Allmendinger, P., & Haughton, G. (2012) Post-political spatial planning in England: a crisis of consensus? *Transactions of the Institute of British Geographers*, 37(1), pp. 89–103.

Bell, D., & Rowe, F. (2012) *Are Climate Policies Fairly Made?* York: Joseph Rowntree Foundation.

BERR (Department for Business, Enterprise and Regulatory Reform) (2008) *Meeting the Energy Challenge: A White Paper on Nuclear Power*. London: Department for Business, Enterprise and Regulatory Reform.

Bickerstaff, K. (2012) "Because we've got history here": nuclear waste, cooperative siting, and the relational geography of a complex issue. *Environment and Planning A*, 44(11), pp. 2611–2628.

Bickerstaff, K., Walker, G., & Bulkeley, H. (2013) *Energy Justice in a Changing Climate: Social Equity Implications of the Energy and Low-carbon Relationship*. London: Zed Books.

Blowers, A. (2010) Why dump on us? Power, pragmatism and the periphery in the siting of new nuclear reactors in the UK. *Journal of Integrative Environmental Sciences*, 7(3), pp. 157–173.

Blowers, A. (2014) A geological disposal facility for nuclear waste: if not Sellafield, then where? *Town & Country Planning*, 83(12), p. 546.

Blowers, A. & Leroy, P. (1994) Power, politics and environmental inequality: a theoretical and empirical analysis of the process of "peripheralisation". *Environmental Politics*, 3, pp. 197–228.

Burgess, J., Stirling, A., Clark, J., Davies, G., Eames, M., Staley, K. & Williamson, S. (2007) Deliberative mapping: a novel analytic-deliberative methodology to support contested science-policy decisions. *Public Understanding of Science* 16, pp. 299–322.

Butler, C. & Simmons, P. (2013) Nuclear power and climate change: just energy or conflicting justice? In Bickerstaff, K., Walker, G., & Bulkeley, H. (eds) *Energy Justice in a Changing Climate*. London: Zed, pp. 139–157.

Chilvers, J., & Burgess, J. (2008) Power relations: the politics of risk and procedure in nuclear waste governance. *Environment and Planning A*, 40, pp. 1881–1900.

Clough, E., & Bell, D. (2016) Just fracking: a distributive environmental justice analysis of unconventional gas development in Pennsylvania, USA. *Environmental Research Letters* doi:10.1088/1748-9326/11/2/025001

CoRWM (2006) *Managing Our Radioactive Waste Safely: CoRWM's Recommendations to Government*. London: Committee of Radioactive Waste Management.

Cotton, M., Rattle, I., & Van Alstine, J. (2014) Shale gas policy in the United Kingdom: an argumentative discourse analysis. *Energy Policy* 73, pp. 427–438.

Cowell, R., Bristow, G., & Munday, M. (2012) *Wind Energy and Justice for Disadvantaged Communities*. York: Joseph Rowntree Foundation.

DECC (Department of Energy and Climate Change) (2011) *The Carbon Plan: Delivering Our Low Carbon Future*. London: The Stationery Office.

DECC (2013) *Consultation Review of the Siting Process for a Geological Disposal Facility*. London: The Stationery Office.

DECC (2014) *Implementing Geological Disposal: A Framework for the Long-term Management of Higher Activity Radioactive Waste*. London: The Stationery Office.

DEFRA (Department for Environment, Food and Rural Affairs) (2001) *Managing Radioactive Waste Safely: Proposals for Developing a Policy for Managing Solid Radioactive Waste in the UK*. London: The Stationery Office.

DEFRA (2008) *Managing Radioactive Waste Safely: A Framework for Implementing Geological Disposal.* London: The Stationery Office.

Ekins, P., Keppo, I., Skea, J., Strachan, N., Usher, W., & Anandarajah, G. (2013) *The UK Energy System in 2050: Comparing Low-carbon, Resilient Scenarios.* London: UK Energy Research Centre.

Endres, D. (2009) From wasteland to waste site: the role of discourse in nuclear power's environmental injustices. *Local Environment: The International Journal of Justice and Sustainability* 14(10), pp. 917–937.

Fry, M., Briggle, A., & Kincaid, J. (2015) Fracking and environmental (in)justice in a Texas city. *Ecological Economics* 117, pp. 97–107.

Garman, J., & Aldridge, J. (2015) *When the Levy Breaks: Energy bills, Green Levies, and a Fairer Low-carbon Transition.* London: IPPR.

Gowda, R., & Easterling, D. (2000) Voluntary siting and equity: the MRS facility experience in Native America. *Risk Analysis* 20, pp. 917–930.

Gunderson, W.C., & Rabe, B.G. (1999) Voluntarism and its limits: Canada's search for radioactive waste-siting candidates. *Canadian Public Administration/Administration Publique du Canada* 42(2), pp. 193–214.

Hall, S., Hards, S., & Bulkeley, H. (2013) New approaches to energy: equity, justice and vulnerability: introduction to the special issue. *Local Environment* 18, pp. 413–421.

Hecht, G. (2012) *Being Nuclear: Africans and the Global Uranium Trade.* Cambridge, MA: MIT Press.

Helm, D. (2011) Shale gas and the low carbon transition in Europe. In Hinc, A. (ed.) *The era of Gas: How to use this new potential?* Warsaw: PKN ORLEN, pp. 9–13.

Hinchliffe, S., & Blowers, A. (2003) Environmental responses: radioactive risks and uncertainty. In Blowers, S. and Hinchliffe, S. (eds) *Environmental Responses.* Chichester and Milton Keynes: John Wiley and the Open University, pp. 7–49.

HM Government (2008) *Planning Act 2008.* London: The Stationery Office.

House of Lords Science and Technology Committee (HLSTC) (2004) *5th Report of Session 2003–04 Radioactive Waste Management.* London: The Stationery Office.

Hunhold, C. (2002) Canada's low-level radioactive waste disposal problem: voluntarism reconsidered. *Environmental Politics* 11, 49–72.

Johnstone, P. (2013) *From Inquiry to Consultation: Contested Spaces of Public Engagement with Nuclear Power.* Unpublished PhD thesis, Exeter UK: University of Exeter.

Johnstone, P. (2014) Planning reform, rescaling, and the construction of the post-political: the case of the Planning Act and nuclear power consultation in the UK. *Environment and Planning C: Government and Policy* 32(4), pp. 697–713.

Johnstone P., Gross, M., MacKerron, G., Kern, F., & Stirling, A. (2013) *Response to the DECC Consultation of the Siting Process for a Geological Disposal Facility.* University of Sussex: SPRU.

Kemp, R. (1990) Why not in my backyard? A radical interpretation of public opposition to the deep disposal of radioactive waste in the United Kingdom. *Environment and Planning A*, 22, pp. 1239–1258.

Laird, F.N. (2013) Against transitions? Uncovering conflicts in changing energy systems. *Science as Culture* 22(2), pp. 149–156.

McLaren, D., Krieger, K., & Bickerstaff, K. (2013). Procedural justice in energy system transitions: the case of CCS. In Bickerstaff, K., Walker, G., & Bulkeley, H. (eds) *Energy Justice in a Changing Climate.* London: Zed, pp. 158–181.

Miller, C.A., Iles, A., & Jones, C.F. (2013) The social dimensions of energy transitions. *Science as Culture* 22(2), pp. 135–148.

Mitchell, C. (2011) Nuclear power is the reason for the new energy regulations. *Guardian*, 11 March.

Mulvaney, C. (2013) Opening the black box of solar energy technologies: exploring tensions between innovation and environmental justice. *Science as Culture* 22(3), pp. 214–21.

Ottinger, G. (2013) The winds of change: environmental justice in energy transitions. *Science as Culture* 22(2), pp. 222–229.

Parkhill, K., Demski, C., Butler, C., Spence, A., & Pidgeon, N. (2013) *Transforming the UK Energy System: Public Values, Attitudes and Acceptability – Synthesis Report.* London: UKERC.

Preston, I., White, V., Thumim, J., Bridgeman, T. (2013) *Distribution of Carbon Emissions in the UK: Implications for Domestic Energy Policy.* York: JRF.

Raman, S. (2013) Fossilizing renewable energies. *Science as Culture* 22(2), pp. 172–180.

Reiner, D.M., & Nuttall, W.J. (2011) Geological disposal of carbon dioxide and radioactive waste: a comparative assessment. *Advances in Global Change Research* 44, pp. 295–315.

Sovacool, B.K., & Dworkin, M.H. (2014) *Global Energy Justice: Problems, Principles and Practices.* Cambridge: Cambridge University Press.

Stockton, H., & Campbell, R. (2011) *Time to Reconsider UK Energy and Fuel Poverty Policies?* www.jrf.org.uk/publications/time-reconsider-uk-energy-and-fuel-policies

Sundqvist, G. (2014) "Heating up" or "cooling down"? Analysing and performing broadened participation in technoscientific conflicts. *Environment and Planning A* 46, pp. 2065–2079.

Swyngedouw, E. (2010) Apocalypse forever? Post-political populism and the spectre of climate change. *Theory, Culture & Society*, 27(2–3), pp. 213–232.

Taebi, B., & Kloosterman, J.L. (2008) To recycle or not to recycle? An intergenerational approach to nuclear fuel cycles. *Science and Engineering Ethics* 14(2), pp. 177–200.

Walker, G. (2008) Decentralised systems and fuel poverty: are there any links or risks? *Energy Policy* 36, pp. 4514–4517.

Wolsink, M. (2007) Wind power implementation: the nature of public attitudes: equity and fairness instead of "backyard motives". *Renewable and Sustainable Energy Review* 11, pp. 1188–1207.

Wolsink, M. (2013). Fair distribution of power-generating capacity: justice, microgrids and utilizing the common pool of renewable energy. In Bickerstaff, K., Walker, G., & Bulkeley, H. (eds) *Energy Justice in a Changing Climate*. London: Zed, pp. 116–138.

32

TRANSPORTATION AND ENVIRONMENTAL JUSTICE

History and emerging practice

*Alex Karner, Aaron Golub, Karel Martens
and Glenn Robinson*

Introduction

Transportation systems are fundamental to participation in modern societies. We need effective transportation systems to connect us to opportunities that are distributed throughout space. Access to work, healthy food, medical care, education, recreation, and social interaction are all facilitated by highways, public transit lines, and non-motorized facilities. But this access is not shared equitably across space or demographic groups (Bullard et al. 2004; Ihlanfeldt & Sjoquist 1998; Lucas 2004; Taylor & Ong 1995). Historical and ongoing planning practices and investment decisions have tended to disadvantage those who live near transportation infrastructure and who choose not to or cannot afford to drive (Avila 2014; Golub et al. 2013; Henderson 2006; Mohl 2002). Most often those populations are low-income people and people of colour. In the United States (US), this disconnect runs counter to federal and state law and the stated goals of regional transportation and public transit planners. Similar disparities have been reported in the European and Global South contexts (e.g. Lucas 2012).

This chapter provides an overview of the academic and applied understandings of how justice might best be achieved in the domain of transportation. The emphasis is on the US experience because of its fraught transportation policy and planning history, the subsequent creation of federal laws and regulations focused on public involvement and justice, and the emergence of advocacy and philanthropic organizations aimed at meaningful legal and regulatory enforcement.

We begin by outlining the academic evidence on environmental injustice in transportation related to three dimensions: 1) inequitable access to participation in the planning process; 2) inequitable exposure to localized environmental burdens; and 3) inequitable distribution of the benefits of transportation investments and systems. We also address how justice or fairness should be conceptualized along each dimension. We then turn to the legal and policy framework for achieving (environmental) justice in the US and briefly discuss how transportation planning agencies address environmental justice in their practices. Finally, we turn to the role of civil society. Because of the tenacity of prior injustices, communities, non-profit organizations, and public interest attorneys have begun to create their own frameworks geared towards the achievement of justice. The chapter concludes by synthesizing best practices from the academic literature, agency practice, and community activism to chart a path towards and develop an analytical framework to advance an equitable and just transportation system for the US and beyond.

Environmental justice and transportation

Participation and transportation infrastructure siting

The issue of participation in decision-making has been the subject of extensive study and theorizing and its scope reaches well beyond the environmental justice or transportation literature (e.g. Arnstein 1969; Chess & Purcell 1999; Innes & Booher 2010). Two dimensions of fairness can be distinguished with respect to this area. The first concerns the *level* of participation: the extent to which citizens should be involved in public decision-making. There is widespread agreement that more meaningful forms of public involvement are required, but opinions diverge about what level of participation is morally required and what is practically feasible. The second dimension relates to the question of *who* should able to participate in decision-making. One key issue here is that the delineation of some level of participation does not automatically imply that all citizens have a comparable power to affect outcomes. Some forms of participation may formally be open to all, but still tend to exclude certain groups from meaningful involvement, amongst other reasons because of the skills and resources required for participation (language, technical expertise, public speaking ability, time, etc.) or the need to organize. Thus, merely extending possibilities for citizen participation without affirmative policies or actions to involve those traditionally excluded does not necessarily move the decision-making process towards justice, as particular groups may benefit much more from increased opportunities for involvement than others. Likewise, limiting involvement may do little to create a level playing field, as some actors have easy access to decision-makers irrespective of formal avenues for participation.

Public participation efforts in transportation have evolved from an early period of stark injustice to include some contemporary efforts with promising outcomes aimed at inclusivity and impact. The profound shortcomings of the early efforts were clearly on display during the construction of the interstate highway system, which began in earnest following the passage of the 1956 Federal-Aid Highway Act. Requirements for public participation at that time were virtually non-existent and limited to a single hearing at which representatives of a state department of transportation would announce that a particular highway had been sited and construction was planned. After widespread negative reactions to this policy, the require-ments were increased to two hearings by 1969 (Weiner 2008: 59–60). But the toll on low-income communities and communities of colour had already been taken. As summarized by Mohl (2002), entire African American business districts were decimated (pp. 30–38). In cities across the US, tens of thousands of black, mostly low-income residents were displaced by freeway construction. This type of community destruction was not limited to highways. In the San Francisco Bay Area, construction of the Bay Area Rapid Transit (BART) system wiped out a thriving black business district in West Oakland (Self 2005).

It is difficult to imagine that robust public involvement campaigns would have mitigated the worst of these excesses, but they certainly would have made these issues salient prior to, rather than after, construction. According to Schlosberg (2004), procedural justice – access to the process and an opportunity to have your voice heard – is a key element in achieving a just distribution of outcomes. The oft-cited EJ mantra "We speak for ourselves" embodies a similar sentiment (Cole & Foster 2001). Despite increased emphasis on and development of methods for conducting robust public engagement efforts in the wake of the 1991 Intermodal Surface Transportation Equity Act (ISTEA), problems remain. Aimen and Morris (2012: 1–17) helpfully make a distinction between mere "public involvement" and "meaningful involvement." While the former emphasizes a one-way flow of information from agencies to the public and seeks to manage interactions with the public, meaningful involvement seeks to provide the opportunity for individuals to change the outcome of a particular course of events or a particular project.

There have been promising practices in meaningful involvement, subsequent to the end of major interstate construction in the late 1970s. Although these facilities continue to act as dividing lines in many communities, there are examples where freeway teardowns or reconstructions have resulted in mitigation of early injustices (Mohl 2012). These usually involve robust public engagement efforts. In Oakland, California, the Loma Prieta earthquake destroyed part of Cypress Freeway. Constructed in 1957, the highway divided the vibrant black community of West Oakland. Despite the high costs of construction and right-of-way acquisition, the new alignment avoided the community, reconnecting what had previously been separated (FHWA 2000). Part of the justification for the Atlanta Streetcar project in 2014 was to provide a link for the black community in Downtown Atlanta that had been separated by interstate construction (Ball 2014). Yet despite these successes, projects and plans strongly opposed by advocates and community members, and which tend to benefit wealthier, whiter transportation system users continue to be pursued. In the San Francisco Bay Area, a rail connection to Oakland International Airport opened in 2015 to much fanfare, but the project was earlier found to run afoul of civil rights laws and was opposed by transit equity advocates who argued that the funding could be better spent to meet the needs of transit dependent populations in the region. Clearly, historical patterns of transportation decision-making continue to shape contemporary space, especially for disadvantaged populations in the US (Golub et al. 2013).

Environmental burdens

Transportation systems produce environmental burdens including air, soil, and noise pollution, related health impacts, and traffic safety risks for users as well as non-users. Furthermore, as noted above, transportation infrastructure physically alters urban space, creating barriers between places that were previously connected, or, when placed in dense urban environments, razing entire neighbourhoods (Bullard et al. 2004; Mohl 2012; Schweitzer & Valenzuela 2004).

There seems to be some convergence about the appropriate justice "standard" regarding environmental burdens. The general agreement, both in the academic literature and in US regulations and guidelines, seems to be that burdens from transportation infrastructure as well as other sources of environmental pollution should not be disproportionally carried by disadvantaged populations (e.g. Bullard & Johnson 1997; Mohl 1993; Schweitzer & Valenzuela 2004). More specifically, burdens should be distributed so that each group's share is roughly comparable to its size, with deviations from the ideal of perfect equality between population groups acceptable as long as they remain within reasonable boundaries. While intuitively attractive, the proportionality principle is difficult to apply to transport systems in practice for a number of reasons. First, by their very nature interventions in the transportation system have disparate impacts over space. It may therefore be practically infeasible to avoid disproportional impacts (whether in favour or not of disadvantaged groups). Second, the proportionality standard largely ignores processes of residential mobility. As long as residential location patterns are largely the result of market factors, it may be expected that higher income groups will disproportionately reside in neighbourhoods with low levels of traffic-related pollution and a high quality of life. As long as race, ethnicity and gender are strongly correlated with income levels, it may be expected that these groups will thus carry a disproportionate share of the burdens generated by transportation systems, *even if transportation interventions live up to the proportionality standard*. The proportionality standard is thus by no means beyond scrutiny.

Academic research has shown that people of colour and low-income populations indeed carry a disproportionate share of transport-related burdens. The most blatant injustice has already been discussed above: the displacement of entire neighbourhoods housing mostly poor people of colour to clear the way for massive highway building schemes. These deliberate forms of injustice may largely be a thing of the past, but more subtle forms of injustice are very much present in all major cities in the US. For instance, Rowangould (2013) found that, while almost 20 per cent of the US population lives within 500 m of a road carrying a volume of 25 000 average annual daily vehicle-trips, approximately 24 per cent of the black population and 30 per cent of the Latino population live within that same buffer. Air pollution concentrations are known to be elevated within these short distances from roads, yet regional monitoring stations are not sited to capture near-roadway conditions (Karner et al. 2010). This proximity translates directly into health outcomes including increased risk of cardiovascular and respiratory illness and cancer (Chakraborty 2009; Gauderman et al. 2007). While these patterns may partly be the result of market-driven processes of residential competition and selection, they are a main concern for the environmental justice movement, not least because in neighbourhoods with high concentrations of people of colour, air pollution concentrations routinely exceed regional averages, and other traffic-related impacts including noise, vibration, and safety, can be severe (Morello-Frosch et al. 2001; Karner et al. 2009; Rowangould 2015; see also Martens 2011).

Higher use of non-motorized modes, lower likelihood of automobile ownership, and roadway proximity also combine to create safety risks for low-income and minority populations. Specifically, the risk of injury or death from motor vehicles is higher for people of colour than the population as a whole in the US. Daniels et al. (2002) reported that, overall, blacks account for approximately 40 per cent of all traffic-related injuries in the country, compared to their 13 per cent population share. Walking is particularly dangerous for black Americans. Using data from the Fatality Analysis Reporting System, Hilton (2006) demonstrated that a disproportionate number of non-occupant children killed by motor vehicles are black. Increased risk of death as a pedestrian also extends over the lifetime and affects other people of colour as well. Campos-Outcalt et al. (2003), reporting results from Arizona, found that Latino and black males were 1.33 and 1.75 times more likely to be killed as pedestrians than non-Hispanic whites, respectively.

Benefits of transportation infrastructure: accessibility

Clearly, transportation generates not only (environmental) burdens, but also benefits, and accessibility has been proposed as the fundamental benefit conferred by transportation (Grengs 2015a; Martens 2006; Martens 2012). Accordingly, analyses of social equity in transportation systems often emphasize accessibility metrics both in practice and in the academic literature (e.g. Golub & Martens 2014; Páez et al. 2010). In a transportation context, accessibility refers to the ability to reach desired destinations which are separated in space

(Geurs & van Wee 2004; Handy & Niemeier 1997). Greater accessibility is associated with shorter travel times, lower costs, as well as closer proximity to activity locations.

Some authors have sought to establish criteria by which a particular distribution of accessibility could be judged to be equitable or which could be used to guide planning efforts (e.g. Golub & Martens 2014; Grengs 2015b; Lucas et al. 2015; Martens et al. 2012), but there is little agreement in the literature or in practice about what constitutes a just or fair distribution of accessibility benefits. While academic studies into the patterns of accessibility abound, few are informed by a well-defined justice standard for accessibility. Philosophies of justice have been invoked to develop explicit justice standards in transport (e.g. Beyazit 2011; Mullen et al. 2014; van Wee 2011), drawing on prior work that made initial steps in a similar direction (Bullard & Johnson 1997; Rosenbloom & Altshuler 1977; Schaeffer & Sclar 1980). Golub and Martens (2014) calculate the ratio between automobile and public transit accessibility for a particular area and argue that below a particular threshold an area would experience "access poverty" and transportation disadvantage. Martens et al. (2012) describe how a maximax principle can be used to guide transportation planning, whereby the average accessibility of a population is maximized subject to the constraint that the gap between the least and most accessible groups is held below a maximum acceptable value. More recently, drawing heavily on the contract theory of social justice, Martens (2016) has proposed a sufficiency standard for accessibility, arguing that a transportation system is fair if, and only if, it provides every person with a sufficient level of accessibility. The latter approach resonates clearly in the more qualitative social exclusion literature, which defines transport-related social exclusion as the "process by which people are prevented from participating in the economic, political and social life of the community because of reduced accessibility to opportunities, services and social networks, due in whole or in part to *insufficient* mobility" (Kenyon et al. 2002: 148, emphasis added). This broad definition is, again, intuitively appealing. But while the literature provides some direction on how to define a sufficiency standard, establishing such a standard would be contentious in practice, regardless of its exact form. Thus, while substantial progress has been made exploring possible principles for the distribution of the benefits of transportation, agreement about the most appropriate standard is still far away.

While typically not framed using environmental justice discourse, there is a host of literature studying patterns of accessibility and inaccessibility. Much of this literature focuses on accessibility to employment opportunities. These studies typically apply location-based measures of accessibility and compare job accessibility by car and transit (e.g. Blumenberg & Ong 2001; Kawabata & Shen 2006; Shen 1998). Other work exists that measures disparities in access to healthy foods and the locations of food deserts (e.g. Farber et al. 2014; Widener et al. 2013) as well as access to health care services of various types (e.g. Harrison & Wardle 2005; Martin et al. 2008). These literatures vividly illustrate the vast disparities in accessibility between persons with and without access to a car. Certainly in the US context, public transport services tend to provide very low levels of accessibility, severely inhibiting the ability of car-less households to gain access to employment, health care, education or even healthy food.

Not all studies along these lines are sensitive to typical environmental justice concerns. For instance, much of the work assessing job accessibility tends to ignore the suitability of particular workers for particular jobs. This is especially important in light of the typical structure of US cities. Thus, while "measured accessibility" may be quite high for minority populations living close to the central business district, few of these jobs may actually match their skills or expertise. It is precisely this problem that has been highlighted by the extensive spatial mismatch literature. Given the close correlation between education level and minority status, it can be argued that many accessibility studies are 'colour blind' by failing to take into account the match between workers' abilities and job requirements (see Hu and Giuliano (2017) and Golub and Martens (2014) for counterexamples). While these accessibility analyses can thus tell us a great deal about the interactions between land use and transportation in a particular region, they tell us relatively little about the conditions faced by a particular person or demographic group.

Importantly, travel behaviors and the use of particular pieces of transportation infrastructure are known to differ between demographic groups. Low-income people and people of colour generally own automobiles at lower rates, make shorter trips, and use transit and carpool more readily than higher income, generally whiter populations (Clifton & Lucas 2004; Pucher & Renne 2003). This means that a transportation policy emphasizing highway capacity expansion will tend to disproportionately benefit these non-disadvantaged populations. A historical example is instructive. Part of the outcome of interstate freeway and rapid transit construction in the 1950s and 1960s was to allow relatively wealthy whites to access employment opportunities in central cities while living in suburban locations (Henderson 2006; Pulido 2000). Commonly referred to as "white flight," the construction of transportation infrastructure has been identified as a causal agent in the depopulation of US central cities as regions continued to grow during the second half of the twentieth century (Baum-Snow 2007). This massive investment of public dollars undoubtedly benefited wealthy white populations while exacting a profound cost on people of colour and low income.

Legal and policy frameworks for achieving justice

Recognizing the existence of injustice across multiple issue domains, governmental entities in the US have enacted an array of laws and regulations to prevent further disparate impacts. The threat of legal action, administrative censure, and/or loss of funding are thus powerful tools wielded by environmental justice advocates and activists against transportation agencies and decision-makers. In the absence of supportive public policies, members of the public seeking justice would have far less recourse in the face of potential disparate impacts.

Federal guidance on environmental justice applies to the transportation sector through various regulations promulgated by the US Department of Transportation and its modal agencies including the Federal Highway Administration and the Federal Transit Administration (e.g. Federal Highway Administration 2012; Federal Transit Administration 2012a; Federal Transit Administration 2012b; US Department of Transportation Office of the Secretary 2012; see also Karner & Niemeier 2013; Golub & Martens 2014). These regulations are influenced by and derived from laws such as Title VI of the 1964 Civil Rights Act and executive actions such as the 1994 Executive Order 12898 on environmental justice. In general, agencies that receive federal funding must provide for full and fair participation in transportation planning processes, ensure that the impacts of their actions do not disproportionately affect protected populations, and guarantee that those same populations are not denied the timely receipt of benefits from public investments.

Two types of transportation planning agencies are particularly important for implementing US federal regulations in this area: metropolitan planning organizations (MPOs) and public transit agencies. MPOs are responsible for transportation planning in all urbanized areas that exceed 50 000 in population. Those in larger urban areas have additional authority, with direct control over some funds, and various planning and programming responsibilities. Regions are generally defined by commute patterns and include many different city and county governments. One key mission of an MPO is to address issues that cross the boundary of a single jurisdiction. The planning of a commuter rail system, implementing high-occupancy toll lanes, or providing work trip reduction incentives are all activities that an MPO would undertake. In the wake of interstate highway development, regions are also home to cities and counties with widely differing fiscal resources, a phenomenon that has been identified as regional inequity (Orfield 2002; Pastor et al. 2000). Suburban sprawl and gentrification are two other major issues whose mitigation falls outside of the responsibilities of a single jurisdiction.

Public transit agencies manage the day-to-day operations of public transit systems. In the US, although individual routes may cover their costs, in general, agencies require public subsidies. This is because transit agencies seek to provide both revenue generating service as well as service that is more explicitly oriented towards geographic or social equity (Walker 2012). Transit agencies must assess the equity impacts of their fare and service changes, following guidance laid out by the FTA (Karner & Golub 2015).

Both MPOs and transit agencies play key roles in environmental justice analysis and mitigation. MPOs are important because of the regional nature of travel patterns, mobility, and injustice. There are at least two problems with this emphasis. First, such agencies play mostly a coordination and aggregation role, rather than a leadership role in terms of regional project prioritization (Goldman & Deakin 2000). They are often bound by decisions made at both lower and higher levels of government (e.g. Crabbe et al. 2005). Second, their overarching goal appears to be conflict minimization. Because of this, the analyses undertaken by MPOs have historically not uncovered evidence of injustice at either the project or plan level (Karner 2016; Karner & Niemeier 2013; Sanchez et al. 2003). Transit agencies play a role because transit dependents are a key constituency and they overlap with environmental justice populations. However, these agencies exist in a rather severe fiscal environment; even when new revenues are made available, calls to mitigate congestion often dominate and transit agencies can find themselves left out. Consequently, new funding for transit and new transit projects tend to promote mobility and accessibility for wealthier, whiter transit riders as opposed to transit dependents.

As mentioned above, both agencies assess the equity of their plans and decisions (Karner & Golub 2015; Karner & Niemeier 2013; Martens et al. 2012). Analyses of the patterns of accessibility have become quite common over the past decade. These analyses, however, typically fail to account for the differences in car ownership between various population groups. As a result, they hardly ever result in findings of inequity (Golub & Martens 2014). Moreover, these analyses are typically only weakly linked to decision-making (Karner & Niemeier 2013; Rowangould et al. 2016; Sanchez et al. 2003). For example, an analysis demonstrating that accessibility to destinations by automobile is high in central city areas where the low-income population is also high tells us very little given that low-income people own vehicles at low rates. Other analyses have examined the distribution of transit accessibility across the population and have demonstrated that disadvantaged population groups tend to enjoy relatively high rates of (job) accessibility (Al Mamun & Lownes 2011; Currie 2004; Foth et al. 2013). Yet again, these analyses tell us little about the sufficiency of transit service or its performance relative to the car.

Community-based responses

It seems clear that the agencies tasked with mitigating environmental injustice often conduct analyses that shed little light on the problem. To communities struggling with environmental burdens and often profound disparities in accessibility, the situation can seem intractable. A promising way forward can be found in emerging community-based responses to environmental injustice that are founded on the principle of achieving meaningful public participation in the transportation planning process, consistent with the evolution of public involvement in transportation planning decisions discussed earlier. Although this type of meaningful participation is not a panacea for all transportation injustice, as we will illustrate below, it has generated promising and concrete wins in several planning processes and related to specific projects. Marcantonio and Tepperman-Gelfant (2015) summarize several successful practices in the realm of public participation, noting three factors that appear to be vitally important for success. These include: 1) establishing a shared agenda for what constitutes "success" across diverse stakeholders; 2) rewiring the process so that the agency is not the only entity defining how and when key decisions will be made; and 3) combining "inside" and "outside" tactics, blending participation in formal structures with traditional community organizing and advocacy approaches. These methods have been applied in several cases as of the mid-2010s. Below, we discuss promising examples of the application of these principles in Los Angeles and the San Francisco Bay Area.

In Los Angeles, the expansion of Interstate 710, to accommodate projected growth in truck traffic along a key goods movement route leading north from the ports of Los Angeles and Long Beach, was vehemently opposed by residents proximate to the facility (East Yard Communities for Environmental Justice 2015). They viewed the agency's plan to expand capacity outright as a losing strategy that would only result in induced demand and additional noise and air quality impacts while providing few benefits. Supported by a number of attorneys from the Natural Resources Defense Council and Earthjustice, a coalition of local advocacy organizations developed "Community Alternative 7" proposing no mixed flow capacity expansion, only an increase in capacity along facilities that would be dedicated to heavy vehicle use. Agency analysis demonstrated that the community alternative would meet the stated project needs at lower cost and with fewer environmental and social impacts.

In the San Francisco Bay Area, the Six Wins Coalition united around multiple related, but previously disparate ideas including affordable housing, local transit service, public health, and displacement (Marcantonio & Karner 2014). Again supported by foundation funding and public interest attorneys, the coalition proposed a community-defined alternative – entitled the "Equity, Environment, and Jobs" (EEJ) scenario – after the MPO declined to include it in an earlier round of modelling. The alternative directly spoke to needs that had been identified by community members in advance and was aggressively pursued and advocated for in formal venues such as board and advisory committee meetings. Key public meetings where votes on the EEJ scenario were held were heavily populated by supporters. Surprisingly, once the alternative was simulated using standard transportation modelling approaches, it was found to outperform the agency's preferred alternative. By running more frequent local transit service and locating affordable housing near job centres, the environmental impacts of the EEJ scenario were much lower than the preferred plan while key environmental justice goals were met simultaneously.

Both of these cases illustrate that community-led processes can surface alternatives that otherwise would not be considered and which can lead to improved plan and project performance as well as superior distributive outcomes. Yet in both cases, substantial resources were required to develop the alternative plans, including those required to hire personnel to engage with agencies and to provide technical assistance and advice on plan design. One final emerging best practice for moving towards environmentally just outcomes is thus the provision of funding for groups seeking to engage in planning processes. Such practices are not without precedent. In their 2014 regional transportation plan, the Fresno County Council of Governments provided small grants ($1000 – $3000) to community-based organizations to support their engagement with the process.

Conclusion and future directions

The early history of transportation planning was rife with unjust practices, creating a system in which low-income travellers, people of colour, and transit dependent people were substantially disadvantaged relative to whiter, wealthier, riders. Mid-twentieth century legal and regulatory requirements to address environmental injustice in transportation, while leading to incidental success, have not led to fundamental changes in policy and planning actions. Consequently, historical disparities in burdens and benefits persist.

The demographics of the US are changing, and new challenges have presented themselves to transportation planners in terms of evolving passenger travel and goods movement patterns, aging infrastructure, funding constraints, the threat of climate change, and a growing appreciation for the potential environmental, public health and social consequences of transportation systems and services and transportation decisions. Many of these changes offer the promise of advancing environmental justice goals. But in order to deliver results, environmental justice analyses, of both burdens and benefits, have to become much more rigorous. Since both the burdens and benefits of transportation investments vary across a variety of social and geographic dimensions, it is of particular importance that they are not only examined in the aggregate, but also in terms of their incidence upon particular populations or communities.

Furthermore, it is vitally important that environmental justice analyses become a key component of the transportation planning process and are carried out *before* policies and plans are created. There are few signs that this change will happen from the top down, so continuing involvement and pressure from communities will be required to make the case for rigorous, well-timed analyses that can inform planning and decision-making. Indeed, the most promising opportunities for achieving just transportation systems appear to be originating from communities themselves in partnership with skilled attorneys and researchers and with support from philanthropic foundations. Empowering these types of efforts runs counter to existing practice but may hold the greatest hope for success in delivering state, regional, and local transportation plans capable of redressing past injustices.

References

Aimen, D. & Morris, A., 2012. *Practical Approaches for Involving Traditionally Underserved Populations in Transportation Decisionmaking*, Washington, DC: National Cooperative Highway Research Program.

Al Mamun, S. & Lownes, N., 2011. Measuring service gaps. *Transportation Research Record: Journal of the Transportation Research Board*, 2217, pp. 153–161.

Arnstein, S.R., 1969. A ladder of citizen participation. *Journal of the American Institute of Planners*, 35(4), pp. 216–224.

Avila, E., 2014. *The Folklore of the Freeway*, Minneapolis, MN: University of Minnesota Press.

Ball, J., 2014. Modern Streetcars Return to Atlanta. In H.F. Etienne & B. Faga, eds. *Planning Atlanta*. APA Press.

Baum-Snow, N., 2007. Did highways cause suburbanization? *The Quarterly Journal of Economics*, 122(2), pp. 775–805.

Beyazit, E., 2011. Evaluating social justice in transport: Lessons to be learned from the capability approach. *Transport Reviews*, 31(1), pp. 117–134.

Blumenberg, E. & Ong, P., 2001. Cars, buses, and jobs: Welfare participants and employment access in Los Angeles. *Transportation Research Record: Journal of the Transportation Research Board*, 1756, pp. 22–31.

Bullard, R.D. & Johnson, G.S. eds., 1997. *Just Transportation: Dismantling Race & Class Barriers to Mobility*, Gabriola Island, BC: New Society Publishers.

Bullard, R.D., Johnson, G.S. & Torres, A.O. eds., 2004. *Highway Robbery: Transportation Racism & New Routes to Equity*, Cambridge, MA: South End Press.

Campos-Outcalt, D., Bay, C., Dellapena, A. & Cota, M.K., 2003. Motor vehicle crash fatalities by race/ethnicity in Arizona, 1990–96. *Injury Prevention*, 9(3), pp. 251–256.

Chakraborty, J., 2009. Automobiles, air toxics, and adverse health risks: Environmental inequities in Tampa Bay, Florida. *Annals of the Association of American Geographers*, 99(4), pp. 674–697.

Chess, C. & Purcell, K., 1999. Public participation and the environment: Do we know what works? *Environmental Science & Technology*, 33(16), pp. 2685–2692.

Clifton, K.J. & Lucas, K., 2004. Examining the empirical evidence of transport inequality in the US and UK. In K. Lucas, ed. *Running on Empty: Transport, Social Exclusion and Environmental Justice*. Bristol, UK: Policy Press.

Cole, L.W. & Foster, S.R., 2001. *From the Ground Up: Environmental Racism and the Rise of the Environmental Justice Movement*, New York: NYU Press.

Crabbe, A.E., Hiatt, R., Poliwka, S.D. & Wachs, M., 2005. Local transportation sales taxes: California's experiment in transportation finance. *Public Budgeting & Finance*, 25(3), pp. 91–121.

Currie, G., 2004. Gap analysis of public transport needs: Measuring spatial distribution of public transport needs and identifying gaps in the quality of public transport provision. *Transportation Research Record: Journal of the Transportation Research Board*, 1895, pp. 137–146.

Daniels, F., Moore, W., Conti, C., Perez, L.C.N., Gaines, B.M., Hood, R.G., et al., 2002. The role of the African-American physician in reducing traffic-related injury and death among African Americans: Consensus report of the National Medical Association. *Journal of the National Medical Association*, 94(2), pp. 108–118.

East Yard Communities for Environmental Justice, 2015. I-710 Corridor Project. Available at: http://eycej.org/campaigns/i-710/.

Farber, S., Morang, M.Z. & Widener, M.J., 2014. Temporal variability in transit-based accessibility to supermarkets. *Applied Geography*, 53, pp. 149–159.

Federal Highway Administration, 2012. Nondiscrimination: Title VI and Environmental Justice. Available at: www.fhwa.dot.gov/environment/environmental_justice/facts/index.cfm [Accessed July 8, 2013].

Federal Transit Administration, 2012a. *Circular FTA C 4702.1B: Title VI Requirements and Guidelines for Federal Transit Administration Recipients*, Washington, DC: US Department of Transportation.

Federal Transit Administration, 2012b. *Circular FTA C 4703.1: Environmental Justice Policy Guidance for Federal Transit Administration Recipients*, Washington, DC: US Department of Transportation.

FHWA, 2000. Cypress Freeway Replacement Project. Available at: www.fhwa.dot.gov/environment/environmental_justice/case_studies/case5.cfm.

Foth, N., Manaugh, K. & El-Geneidy, A.M., 2013. Towards equitable transit: examining transit accessibility and social need in Toronto, Canada, 1996–2006. *Journal of Transport Geography*, 29, pp. 1–10.

Gauderman, W.J. et al., 2007. Effect of exposure to traffic on lung development from 10 to 18 years of age: A cohort study. *The Lancet*, 369(9561), pp. 571–577.

Geurs, K.T. & van Wee, B., 2004. Accessibility evaluation of land-use and transport strategies: Review and research directions. *Journal of Transport Geography*, 12(2), pp. 127–140.

Goldman, T. & Deakin, E., 2000. Regionalism through partnerships? Metropolitan planning since ISTEA. *Berkeley Planning Journal*, 14(1), pp. 46–75.

Golub, A., Marcantonio, R.A. & Sanchez, T.W., 2013. Race, space, and struggles for mobility: Transportation impacts on African Americans in Oakland and the East Bay. *Urban Geography*, 34(5), pp. 699–728.

Golub, A. & Martens, K., 2014. Using principles of justice to assess the modal equity of regional transportation plans. *Journal of Transport Geography*, 41, pp. 10–20.

Grengs, J., 2015a. *Advancing Social Equity Analysis in Transportation with the Concept of Accessibility*, Ann Arbor, MI: University of Michigan Population Studies Center.

Grengs, J., 2015b. Nonwork accessibility as a social equity indicator. *International Journal of Sustainable Transportation*, 9(1), pp. 1–14.

Handy, S.L. & Niemeier, D.A., 1997. Measuring accessibility: An exploration of issues and alternatives. *Environment and Planning A*, 29(7), pp. 1175–1194.

Harrison, W.N. & Wardle, S.A., 2005. Factors affecting the uptake of cardiac rehabilitation services in a rural locality. *Public Health*, 119(11), pp. 1016–1022.

Henderson, J., 2006. Secessionist automobility: Racism, anti-urbanism, and the politics of automobility in Atlanta, Georgia. *Urban Geography*, 30(2), pp. 293–307.

Hilton, J., 2006. *Race and Ethnicity in Fatal Motor Vehicle Traffic Crashes 1999–2004*, Washington, DC: National Highway Traffic Safety Administration.

Hu, L. & Giuliano, G., 2017. Poverty concentration, job access, and employment outcomes. *Journal of Urban Affairs*, 39(1), pp. 1–16.

Ihlanfeldt, K.R. & Sjoquist, D.L., 1998. The spatial mismatch hypothesis: A review of recent studies and their implications for welfare reform. *Housing Policy Debate*, 9(4), pp. 849–892.

Innes, J.E. & Booher, D.E., 2010. *Planning with Complexity: An Introduction to Collaborative Rationality for Public Policy*, New York: Routledge.

Karner, A., 2016. Planning for transportation equity in small regions: Towards meaningful performance assessment. *Transport Policy*, 52, pp. 46–54.

Karner, A. & Golub, A., 2015. Comparing two common approaches to public transit service equity evaluation. *Transportation Research Record*, 2531, pp. 170–179.

Karner, A. & Niemeier, D., 2013. Civil rights guidance and equity analysis methods for regional transportation plans: A critical review of literature and practice. *Journal of Transport Geography*, 33, pp. 126–134.

Karner, A., Eisinger, D., Bai, S. & Niemeier, D., 2009. Mitigating diesel truck impacts in environmental justice communities: Transportation planning and air quality in Barrio Logan, San Diego, California. *Transportation Research Record: Journal of the Transportation Research Board*, 2125, pp. 1–8.

Karner, A.A., Eisinger, D.S. & Niemeier, D.A., 2010. Near-roadway air quality: Synthesizing the findings from real-world data. *Environmental Science & Technology*, 44(14), pp. 5334–5344.

Kawabata, M. & Shen, Q., 2006. Job accessibility as an indicator of auto-oriented urban structure: A comparison of Boston and Los Angeles with Tokyo. *Environment and Planning B: Planning and Design*, 33(1), pp. 115–130.

Kenyon, S., Lyons, G. & Rafferty, J., 2002. Transport and social exclusion: Investigating the possibility of promoting inclusion through virtual mobility. *Journal of Transport Geography*, 10(3), pp. 207–219.

Lucas, K. ed., 2004. *Running on Empty: Transport, Social Exclusion, and Environmental Justice*, Bristol, UK: The Policy Press.

Lucas, K., 2012. Transport and social exclusion: Where are we now? *Transport Policy*, 20, pp. 105–113.

Lucas, K., van Wee, B. & Maat, K., 2015. A method to evaluate equitable accessibility: Combining ethical theories and accessibility-based approaches. *Transportation*, 43(3), pp. 473–490.

Marcantonio, R. & Karner, A., 2014. Disadvantaged communities teach regional planners a lesson in equitable and sustainable development. *Poverty & Race*, 23(1), pp. 5–12.

Marcantonio, R. & Tepperman-Gelfant, S., 2015. Seizing the Power of Public Participation. Available at: http://povertylaw.org/article/Seizing-the-Power.

Martens, K., 2006. Basing transport planning on principles of social justice. *Berkeley Planning Journal*, 19(1), pp. 1–17.

Martens, K., 2011. Substance precedes methodology: On cost–benefit analysis and equity. *Transportation*, 38(6), pp. 959–974.

Martens, K., 2012. Justice in transport as justice in accessibility: Applying Walzer's "Spheres of Justice" to the transport sector. *Transportation*, 39(6), pp. 1035–1053.

Martens, K., 2016. *Transport Justice: Designing Fair Transportation Systems*, New York and London: Routledge.

Martens, K., Golub, A. & Robinson, G., 2012. A justice-theoretic approach to the distribution of transportation benefits: Implications for transportation planning practice in the United States. *Transportation Research Part A: Policy and Practice*, 46(4), pp. 684–695.

Martin, D., Jordan, H. & Roderick, P., 2008. Taking the bus: Incorporating public transport timetable data into health care accessibility modelling. *Environment and Planning A*, 40(10), pp. 2510–2525.

Mohl, R., 2002. *The Interstates and the Cities: Highways, Housing, and the Freeway Revolt*, Poverty and Race Research Action Council. Available at: www.prrac.org/pdf/mohl.pdf.

Mohl, R.A., 1993. Race and space in the modern city: Interstate-95 and the black communities in Miami. In A.R. Hirsch & R.A. Mohl, eds. *Urban Policy in Twentieth-Century America*. New Brunswick, NJ: Rutgers University Press, pp. 100–158.

Mohl, R.A., 2012. The expressway teardown movement in American cities: Rethinking postwar highway policy in the post-interstate era. *Journal of Planning History*, 11(1), pp. 89–103.

Morello-Frosch, R., Pastor, M. & Sadd, J., 2001. Environmental justice and Southern California's "Riskscape": The distribution of air toxics exposures and health risks among diverse communities. *Urban Affairs Review*, 36(4), pp. 551–578.

Mullen, C., Tight, M., Whiteing, A. & Jopson, A., 2014. Knowing their place on the roads: What would equality mean for walking and cycling? *Transportation Research Part A: Policy and Practice*, 61, pp. 238–248.

Orfield, M., 2002. *American Metropolitics: The New Suburban Reality*, Washington, DC: Brookings Institution Press.

Páez, A., Mercado, R.G., Farber, S., Morency, C. & Roorda, M., 2010. Relative accessibility deprivation Indicators for urban settings: Definitions and application to food deserts in Montreal. *Urban Studies*, 47(7), pp. 1415–1438.

Pastor, M. et al., 2000. *Regions That Work: How Cities and Suburbs Can Grow Together*, Minneapolis, MN: University of Minnesota Press.

Pucher, J. & Renne, J.L., 2003. Socioeconomics of urban travel: Evidence from the 2001 NHTS. *Transportation Quarterly*, 57(3), pp. 49–77.

Pulido, L., 2000. Rethinking environmental racism: White privilege and urban development in Southern California. *Annals of the Association of American Geographers*, 90(1), pp. 12–40.

Rosenbloom, S. & Altshuler, A., 1977. Equity issues in urban transportation. *Policy Studies Journal*, 6(1), pp. 29–40.

Rowangould, G.M., 2013. A census of the US near-roadway population: Public health and environmental justice considerations. *Transportation Research Part D: Transport and Environment*, 25, pp. 59–67.

Rowangould, G.M., 2015. A new approach for evaluating regional exposure to particulate matter emissions from motor vehicles. *Transportation Research Part D: Transport and Environment*, 34, pp. 307–317.

Rowangould, D., Karner, A. & London, J., 2016. Identifying environmental justice communities for transportation analysis. *Transportation Research Part A: Policy and Practice*, 88, pp. 151–162.

Sanchez, T.W., Stolz, R. & Ma, J.S., 2003. *Moving to Equity: Addressing Inequitable Effects of Transportation Policies on Minorities*, Cambridge, MA: The Civil Rights Project at Harvard University.

Schaeffer, K.H. & Sclar, E., 1980. *Access for All: Transportation and Urban Growth*, New York: Columbia University Press.

Schlosberg, D., 2004. Reconceiving environmental justice: Global movements and political theories. *Environmental Politics*, 13(3), pp. 517–540.

Schweitzer, L. & Valenzuela, A., 2004. Environmental injustice and transportation: The claims and the evidence. *Journal of Planning Literature*, 18(4), pp. 383–398.

Self, R.O., 2005. *American Babylon: Race and the Struggle for Postwar Oakland*, Princeton University Press.

Shen, Q., 1998. Location characteristics of inner-city neighbourhoods and employment accessibility of low-wage workers. *Environment and Planning B: Planning and Design*, 25(3), pp. 345–365.

Taylor, B.D. & Ong, P.M., 1995. Spatial mismatch or automobile mismatch? An examination of race, residence and commuting in US metropolitan areas. *Urban Studies*, 32(9), pp. 1453–1473.

US Department of Transportation Office of the Secretary, 2012. Updated Environmental Justice Order 5610.2(a). *Federal Register*, 77(91), pp. 27534–27537.

van Wee, B., 2011. *Transport and Ethics: Ethics and the Evaluation of Transport Policies and Projects*, Northampton, MA: Edward Elgar.

Walker, J., 2012. *Human Transit: How Clearer Thinking about Public Transit Can Enrich Our Communities and Our Lives*, Washington, DC: Island Press.

Weiner, E., 2008. *Urban Transportation Planning in the United States: History, Policy, and Practice*, New York: Springer.

Widener, M.J., Farber, S., Neutens, T. & Horner, M.W., 2013. Using urban commuting data to calculate a spatiotemporal accessibility measure for food environment studies. *Health & Place*, 21, pp. 1–9.

33

FOOD JUSTICE

An environmental justice approach to food and agriculture

Alison Hope Alkon

West Oakland, California has approximately 40 000 residents, but lacks a full-scale grocery store. Historically, the neighbourhood was one of the few places where African Americans could buy homes, and though recent gentrification has brought many new residents, it remains predominantly black and low-income. Life expectancy here in the "flatlands" is approximately 10 years less than in the whiter and more affluent hills (Haley et al. 2012). Violence is certainly one cause, but more pervasive, if less dramatic, are the elevated rates of diet related conditions such as heart disease and diabetes. In addition, there are few opportunities for community-led economic development. The few businesses that aren't frightened away by the neighbour-hood's tough reputation are largely owned by non-residents. It's hard to find a job, let alone a career.

This bleakness is the result of structural inequalities that have first created and then hamstrung black urban neighbourhoods across the US. Well-documented processes of racial inequalities in urban economic development include redlining, differential policing, and systemic underinvestment (Massey and Denton 1993; Squires and Kubrin 2006). The de-industrialization of the United States since the late 1970s (Wilson 2011), a result of the spatial reconfigurations of capital due to neoliberal globalization (Harvey 2005), exacerbated already existing racial inequalities by way of massive capital flight, job loss, and economic devastation in the largely black working-class areas of major cities (Wacquant 1995; McClintock 2011). More recently, the 2008 economic crisis has driven black unemployment in these comm-unities to nearly 50 per cent, levels last seen during the Great Depression, and showing little signs of recovery (Walker 2010). In the Bay Area, gentrification and rising rents create additional barriers to both residential stability and community-led economic development.

Since 2008, Mandela Foods Cooperative has stood as an exception to these trends. A small grocery store, Mandela features produce from regional black and Latino farmers, as well as bulk and packaged foods. It is cooperatively owned by four African American neighbourhood residents (though one recently could not afford a rental increase and was forced to move to a neighbouring city). Mandela Foods is also supported by Mandela Marketplace, a non-profit organization which provides the co-op with access to credit, technical assistance and training: resources that entrepreneurs from wealthier communities can often mobilize through banks and social networks.

Mandela seeks to address health disparities by providing neighbourhood residents with access to the "healthy, culturally appropriate food" that the neighbourhood otherwise lacks. It also works to create economic opportunities for community members to participate in the process of supplying that food. In the words of Oakland's Mandela Marketplace's Executive Director Dana Harvey, "It's not about plopping a grocery store down in a community. It's about engaging and resourcing a community to solve their problems, and own those solutions" (personal communication).

Mandela Foods Cooperative is a prime example of an organization working towards "food justice." While the movement is still relatively young, and a shared definition of food justice is still emergent, one attempt to delineate the term comes from geographer and activist Rasheed Hislop (2014: 19), who describes it as "the struggle against racism, exploitation, and oppression taking place within the food system that addresses inequality's root causes both within and beyond the food chain." The food justice movement consists mainly of local, community-based projects devoted to increasing access to fresh produce and creating economic opportunities in low-income communities of colour, though the movement is beginning to expand in some exciting ways.

As with environmental justice, a burgeoning academic literature on food justice seeks to understand the circumstances that create food injustices, to amplify and substantiate activist claims, and to analyse movement strategies and goals. These scholars view food justice as the confluence of three previous areas of research: environmental justice, alternative agriculture, and food studies. Food justice scholarship argues that food is an important lens through which we can comprehend social, racial and environmental injustices, and that an emphasis on inequalities is a fruitful way to understand the production, distribution and consumption of food (Alkon and Agyeman 2011; Gottlieb and Joshi 2010). This framework is most relevant to scholarship and activism in North America, and to a limited extent, Europe. In the Global South, parallel desires have been raised through the concept of food sovereignty, which is discussed below (Via Campesina 2002).

In practice, projects and policies designed to achieve food justice have cohered around two goals: increasing access to healthy food among marginalized communities and the establishment of community control over food and agricultural systems. Food justice, then, becomes a way to emphasize and amplify the roles of low-income communities and communities of colour in efforts to reform industrial food systems and create local alternatives.

In this chapter, I offer an overview of food justice focused on its roots in environmental justice, alternative agriculture and food studies. Food justice scholarship and activism combine alternative agriculture's critique of industrial food with environmental justice's emphasis on the distribution of benefits and harms and food studies' attention to the links between food and identity. I will conclude by speculating on future directions for theory and practice.

Food justice and environmental justice: examining impacts and benefits

The environmental justice movement has cohered around the well-substantiated claim that low-income people and people of colour bear a disproportionate share of the burden of environmental degradation. Low-income people and people of colour are more likely to live in neighbourhoods dominated by toxic industries and diesel emissions, or in rural areas burdened by the pesticides and dust that result from agribusiness (United Church of Christ 1987, 2007). Many environmental justice studies are largely quantitative and epidemiological in nature, aiming to prove disproportionate exposure (Pastor et al. 2006; Mohai and Saha 2003), and to link such exposure to public health outcomes (Israel et al. 2005; Petersen et al. 2003). The academic literature has also followed communities as they have mobilized against incinerators (Cole and Foster 2000), petrochemical facilities (Allen 2004) and toxic waste (LaDuke 2004).

The food justice movement parallels environmental justice in many ways. Both rely on an institutional concept of racism consistent with those of anti-racist activists and the academic literature (Sbicca 2012). Popular approaches to racism in the US tend to assume that an individual must consciously make a biased decision based on race. *Institutional racism*, on the other hand, occurs when institutions such as government agencies, the military or the prison system adopt policies that exclude or target people of colour either overtly or in their effects (see also Chapter 2). From the perspective of environmental justice scholars and activists, the disproportionate burden of toxics borne by low-income people and people of colour need not be linked to intentional discrimination (though the US Environmental Protection Agency has tried to define it in this limited way) if the outcome is a disproportionate burden borne by communities of colour. Similarly, food justice activists point to a variety of institutional policies, such as the US Department of Agriculture's discrimination against African American farmers or the supermarket industry's practice of charging lower prices in suburban vs. urban locations, through which communities of colour have been systematically disadvantaged. An institutional approach claims that racial and economic inequalities are built into the zoning ordinances, mortgage requirements and other policies that determine how industries, human communities, and goods and services come to exist in particular places (Massey and Denton 1993; Lipsitz 1998; Pellow 2002; McClintock 2011). Such institutional understandings of racism do not see it as separable from class, but rather as producing and being produced by economic inequalities. Activists from both movements tend to emphasize issues of race, although they do so with the knowledge that communities of colour are disproportionately poor. Moreover, they have linked food injustices to issues of worker exploitation, immigration and the prison-industrial complex (Alkon and Guthman 2017).

Foundationally, environmental justice scholarship establishes, and the movement seeks to rectify, the disproportionate locating of environmentally hazardous land uses in communities of colour and low-income communities. Food justice activists are primarily concerned not with too many environmental bads, but with access to the goods of environmental sustainability, particularly fresh produce and organic food. Often working closely with activists, food justice scholars have highlighted many of the barriers that make it more difficult for low-income people and people of colour to access local and organic food as both producers and consumers. For example, their work has illustrated the processes through which farmers of colour have been disenfranchised, which range from discrimination by the USDA to forced relocation to immigration laws barring land ownership by particular ethnic groups (Gilbert et al. 2002; Norgaard et al. 2011, Minkoff-Zern et al. 2011). In response, scholars such as Monica White (2017) and Priscilla McCutcheon (2013) work to capture the history of black agriculture in the United States, demonstrating its essential vitality and its importance for civil rights and other social movements. These projects, and others like them, depict agriculture as a proud tradition in communities of colour, connected to both everyday life and essential political work. In addition, food justice scholars call attention to the hardships faced by farmworkers on both organic and conventional farms (Brown and Getz 2011), and highlight efforts these workers make to improve their own food access and working conditions (Minkoff-Zern 2012; Mares and Peña 2011).

Food justice scholars also examine inequalities through the lens of consumption, and argue that low-income communities and communities of colour face a variety of obstacles to the consumption of local and organic food, and that supporters of sustainable agriculture have not done enough to bridge these barriers. For example, locally grown and organic foods tend to be more expensive than conventional alternatives, especially with regard to canned and packaged items. It is, of course, quite difficult for low-income people to afford these foods, particularly in the context of escalating housing and healthcare costs (Alkon et al. 2013; Lea and Worsley 2005).

Another important barrier is the relative lack of available fresh produce – let alone locally grown and organic options – in low-income communities and communities of colour (Wrigley 2002; Morland et al. 2002). Scholars refer to areas lacking fresh food as "food deserts," though activists are critical that the desert imagery naturalizes this political and economic process, and that too much emphasis on the presence or absence of supermarkets results in the offering of incentives to chain supermarkets rather than addressing root causes such as racism and poverty (Holt-Giminez 2013). Nonetheless, there is a fair amount of agreement that the lack of available fresh produce is an obstacle to its consumption. Scholars have traced the political and economic processes through which food retailers and other purveyors of necessary resources have abandoned low-income urban communities (McClintock 2011) and described community-based responses (White 2010; Bradley and Galt 2013).

In addition to opposing the permitting of locally unwanted land uses, environmental justice activists have also organized for *procedural justice*, which encompasses the rights of all affected communities to be included in environmental decision making (Fletcher 2004; Shrader-Frechette 2002; also Chapter 9). This concern is embedded in the concept of food sovereignty, which is a cornerstone of food justice activism. Originally coined by peasant rights movements in the Global South, food sovereignty is a community's "right to define their own food and agriculture systems" (Via Campesina 2002). Like procedural justice, food sovereignty moves beyond the distribution of benefits and burdens to call for a greater distribution of power in the management of food and environmental systems.

As food justice activists translate peasant demands for sovereignty into an urban US context, they have expanded US food movements' goals beyond support for local/organic farming to more broadly include the lived realities of those communities most harmed by industrial food systems. Food justice activists have highlighted the disproportionate positive coverage received by whites in the organic food and farming scene, and even quantified disparities in funding and political connections between white-led and people of colour-led organizations (Reynolds 2014). While it has become common for activists to declare the necessity of marginalized communities being, in the words of one Oakland activist "at the table, instead of just on the table" (cited in Alkon 2012), successfully advocating for these communities entails a variety of insider and outsider roles, as support from large foundations and the USDA have become more commonplace (Broad 2016).

In sum, like the EJ literature, food justice scholarship locates inequalities in the wider political, economic and cultural systems that produce both environmental degradation and racial and economic inequality. It also highlights the ways in which communities of colour resist these conditions by creating local and organic food systems and practices. Its focus on food access mirrors the environmental justice movement's emphasis on dispropor-tionate harms while food sovereignty presents a food system-specific notion of procedural justice. Food justice activism and scholarship has roots in environmental justice, but com-bines this with insights yielded by critiques of alternative agriculture and the wider field of food studies.

A critique of alternative agriculture

In its desire to create sustainable alternatives to environmentally destructive processes and practices, food justice borrows from and has much in common with the movement to create alternative agriculture, an umbrella term that encompasses local, slow, organic and other attempts to create more sustainable food. The two diverge, however, because alternative agriculture pays scant attention to how the harms of industrial agriculture are distributed, and who has access to and sovereignty over sustainable food systems.

In the United States, social movements to reform and transform the food movement cohere mainly around desires for improved personal and ecological health. Though there are earlier antecedents, particularly among Indigenous peoples and those farmers and gardeners who could not afford chemical inputs, Don Worster (2005) traces the ecological critique of US agriculture to the 1930s, when a movement for "permanent agriculture" arose in response to depressed agricultural prices and the ecological effects of the Dust Bowl. This movement resurfaced and gained ground with the growth of the counterculture in the 1960s (Belasco 1993). During that time, opposition to the state and mass production, as well as a budding environmental ethic, led to a desire for foods that were less processed, and for farming methods that countered the chemical dependency of industrial agriculture. Young, countercultural types went "back to the land" in search of this more organic lifestyle, forming communes or developing individual homesteads. Others stayed in the cities and purchased food from these new farmers through co-ops, health food stores, and food conspiracies (buying clubs in which members pooled money for bulk purchases from nearby suppliers). These new farmers and their supporters were animated by a belief that organic farming was a dance between the farmer and the forces of nature, and that cooperation between the two would offer not only sustenance, but a model for an alternative society (Belasco 1993). Early adherents hoped that as the movement grew, demand for local and organic food produced in alternative food systems would cause it to replace the dominant, industrial model.

Today the movement for alternative agriculture is one force driving the explosive growth in markets for local and organic foods. Retail sales of organic products in the United States were only $3.6 billion in 1997, but reached $21.1 billion in 2008, and organic acreage more than doubled between 1997 and 2005 (Dimitri and Oberholtzer 2009). Natural food stores were the primary distributor in 1997, but by 2008, nearly half of this food was purchased in chain supermarkets (Dimitri and Oberholtzer 2009). Big-box stores like Wal-Mart and Safeway, which the counterculture once labelled as inherently contrary to organic philosophies, are now major retailers of organic products. Local, decentralized distribution grew as well. For example, the number of farmers' markets in the United States quadrupled from less than 2000 in 1994 to more than 8000 in 2013 (USDA 2013).

Much of the early social science scholarship on sustainable food systems and food movements seeks to document and celebrate successes and make the case for movement goals (see, for example, Lyson 2004; Hassanein 2003; Magdoff et al. 2000). Scholars argued that industrial agriculture is environmentally, socially, and economically destructive. Ecologically, industrial agriculture depends on mechanization and monoculture, creating increased reliance on fossil fuel-based inputs, decreasing soil fertility, and, through erosion, polluting rivers and streams (Altieri 2000). Economically, reliance on inputs creates economic advantages for those with ready supplies of capital, encouraging the consolidation of small farms into agribusiness corporations (Clapp and Fuchs 2009). Such consolidation is aided by the US Farm Bills (Rausser 1992), which favour large-scale growers and for which agribusiness lobbies heavily (Liebman 1983). Thus, despite a cultural rhetoric depicting farmers as self-reliant and independent (Bradley 1995; Bell 2004), industrial agriculture's dependence on government subsidies provides evidence that the industry is not economically viable (Kent and Meyers 2001). This consolidation has had devastating economic and social effects on rural communities (Bell 2004; Goldschmidt 1978[1947]). Farmers are incentivized to undermine their own security – both social and economic – as well as the productive capabilities of the land (Bell 2004).

But perhaps the harshest social effects are felt by farmworkers. Despite numerous attempts to organize, these immigrant populations are often confined to seasonal work for which they are paid minimal wages (McWilliams 2000[1939]; Daniel 1981; Pulido 1996). Farmworkers are particularly vulnerable to pesticide poisoning and, ironically, often lack steady access to healthy food (Nash 2007; Harrison 2011; Brown and Getz 2011). Industrial agriculture, on the other hand, benefits greatly from what Taylor and Martin (1997: 855) call "the immigrant subsidy in US agriculture," as they pay scant wages and benefits to their workers. Farmworkers' struggles and working conditions have received some attention from environmental justice scholars (Pulido 1996; Harrison 2011), and this is a foundational area of overlap between food and environmental justice.

While earlier generations of scholars helped to support these claims, a more recent vein of critical work has begun to think through the problems and contradictions embodied by alternative agriculture efforts within the United States. Some scholarship is directly critical of sustainable agriculture activism for the ways that whiteness infiltrates its narratives and practices. Although US agriculture developed on the backs of people of colour, and people of colour have since been present in agriculture in a variety of ways, the increased recognition of farmers provided by the alternative agriculture movement tends to amplify the presence of whites while eliding the contributions of communities of colour.

However, scholarly descriptions of the whiteness of alternative foods do not only refer to the presence or absence of bodies. According to Ruth Frankenberg's foundational work, whiteness "carries with it a set of ways of being in the world, a set of cultural practices often not named as 'white' by white folks, but looked upon instead as 'American' or 'normal'" (Frankenberg 1993: 4, see also Saldanha 2006; Slocum 2006; Kobayashi and Peake 2000; also Chapter 2). This whiteness can inhibit the participation of people of colour in alternative food systems, and can constrain the ability of those food systems to meaningfully address inequality. Thus, such whiteness may prevent alternative food movements, despite their growing popularity, from contributing to food justice.

Guthman, for example, argues that phrases common to the sustainable agriculture movement, such as "getting your hands dirty in the soil" and "looking the farmer in the eye," point to "an agrarian past that is far more easily romanticized by whites than others" (2008b: 394). Given the disenfranchisement of so many African American, Native American, Latino/a, and Asian American farmers (Romm 2001), Guthman argues that it is likely these phrases do not resonate with communities of colour in the ways intended by their primarily white orators. This cultural barrier can suggest to low-income communities and communities of colour that sustainable agriculture is not for them, especially when combined with the lack of available organic and local produce in their neighbourhoods.

A second scholarly critique of alternative food activism revolves around the concept of neoliberalism. Neoliberalism is a political economic philosophy that asserts that human well-being can best be achieved if the so-called "free" market is allowed to function with little to no intervention from the state (Harvey 2005: 2). Prominent social scientists have argued that current modes of food activism may explicitly oppose aspects of neoliberalism in their discourses, but their practices tend to embrace it, primarily by "relying on markets rather than the state to pursue change" (Harrison 2008: 163–64; see also Guthman 2008a; Allen

2008; Brown and Getz 2008). The dominance of such strategies has prompted Michael Pollan (2006) to refer to food activism as a "market-as-movement," in which supporters "vote with our forks" for the kind of food system we want to see. This ideal moves away from long-standing social movement strategies pursuing state-mandated protections for labour, the environment, and the poor, and posits individual entrepreneurialism and consumer choice as the primary pathways to social change. Activists encourage one another to build and support alternative food businesses, and believe that change will come through shifting market demand. These food justice and neoliberalism critiques are interrelated, as strategies pursued through the market, such as starting a business or buying particular kinds of goods, are by definition less accessible to low-income people (Alkon 2012).

There is much to recommend the alternative agriculture movement, particularly its critique of corporate agribusiness and emphasis on ecological production. But much as the environmental justice movement has been critical of environmentalists' lack of attention to the distribution of environmental hazards, food justice scholars and activists have attended to questions concerning the disparate impacts of industrial agriculture and access to the benefits provided by sustainable alternatives.

Food studies: race and identity

Food justice scholarship draws upon the broader field of food studies to incorporate a theoretically sophisticated understanding of race that will be useful to many researchers interested in environmental justice. Studies of food justice offer an excellent opportunity to enrich our understanding of racial identity formation because food is deeply intertwined with both personal and cultural notions of who we are. In a now classic piece, Winson (1993) refers to food as an "intimate commodity" that is literally taken inside the body and imbued with heightened significance. It is not only a physiological necessity; food practices – what scholars often call *foodways* – are manifestations and symbols of cultural histories and proclivities. As individuals participate in culturally defined proper ways of eating, they perform their own identities and memberships in particular groups (Douglass 1996). Food informs individuals' identities, including their racial identities, in a way that other environmental justice issues – such as energy, water, garbage, etc. – do not.

Many food justice projects attempt to tap into this deep relationship between food and cultural identity in order to develop explicitly race-conscious responses to issues of food access and food sovereignty. Norgaard et al. (2011), for example, offer what they term a "racialized environmental history" of the Karuk tribe's access to their central ancestral food, salmon. This access has been disrupted by the construction of dams on the Klamath River, with disastrous health consequences for tribal members. The authors emphasize the importance of salmon to Karuk physical, spiritual and emotional health. Similarly, Teresa Mares and Devon Peña (2011) and Laura-Anne Minkoff-Zern (2014a, b) all worked with Latino/a immigrant community gardeners, whose deep knowledges of cultivation and cooking help to maintain not only their histories and heritages, but also their resilience against the poverty and exploitation they face as farmworkers.

The relationship between food and identity cannot, however, be reduced to simplistic assertions that certain kinds of people eat certain kinds of food. Food studies teaches that processes of individual and collective identity formation are hardly static, but develop fluidly in accordance with particular cultural moments and movements. Food is culturally essential, yet the relationship between food and culture is not deterministic and cannot be over-simplified. For example, Hayes-Conroy's work on school gardens and Slow Food Convivia in Berkeley, California and Nova Scotia, Canada describes the ways in which particular groups have projected different notions of good food onto particular cultural moments, and how these can vary across social difference (2014). Similarly, the work of black vegan activists and chefs such as Bryant Terry and Breeze Harper seeks to create plant-based cuisines that speak to their black identities and experiences (Terry 2009; Harper 2010).

Food justice advocates such as those at Mandela Food Cooperative also argue that links between food and cultural identity can serve as sources of wealth for their communities, providing culturally relevant underpinnings for economic development (for other examples see White 2010; Sbicca 2012). This is the same argument that food scholar Psyche Williams Forson (2006) made historically, as she examined the practice of selling chicken by black women, though now the kind of food being sold has shifted. Creating good jobs in the food system has been an essential component of later food justice activism, which has prompted increased alliances with farmworker organizations and labour unions working to improve pay and conditions across the food system (Alkon and Guthman 2017; Myers and Sbicca 2015; Minkoff-Zern 2014b).

Broadly, scholars have established a relationship between food, social status and cultural identity that is extremely fluid and certainly not proscriptive or straightforward. This complexity, however, is precisely the reason why studies of food have so much to offer to the environmental justice literature, which has thus far rarely engaged with social science understandings of race that conceptualize it not as biologically determined, but as learned, performed and practised through social interaction (cf. Pulido 1996, 2000; Park and Pellow 2004, 2011; Sze 2006; also Chapter 2). Moreover, the field of food studies has devoted significant attention to the relationships between food and cultural identity, but has rarely examined the structural context in which these relationships occur. In the contemporary US, this context consists of an environmentally and socially destructive centralized agribusiness system in which race and class inform inequalities of material resources and decision-making power. Neither has the food studies literature addressed the power of deeply held and meaningful foodways to inspire culturally resonant social movements against this agribusiness system. It is only in this newly emergent body of work on food justice that the racialized political economy of food production and distribution meets the cultural politics of food consumption.

Future directions

In just a few short years, the burgeoning literature on food justice has brought together key concepts from environmental justice, alternative agriculture and food studies to attempt to understand how inequalities affect the production, distribution and consumption of food. It has incorporated the environmental justice literature's emphasis on the distribution of environmental toxins and benefits, and used this as a lens to critique the alternative foods movement, which has previously explored this all-important aspect of the food system. Simultaneously, it borrows from food studies scholars' sophisticated understandings of the relationship between food, race and identity to produce a nuanced, constructed notion of race. Food justice activism began primarily as community based alternatives that were rooted in and attempted to serve communities of colour. Future directions include a broadening of strategies and a move beyond efficacy and evaluation.

First, projects have varied in the degree to which they are truly rooted in the communities they seek to serve. Some are started by local residents, while others are initiated by well-intentioned outsiders. Claims of cooptation often fly fast and furious between organizations, especially when forced to compete for scant funding. Researchers have often concluded, along with activists, that community control is essential for success. But more sophisticated work is beginning to examine the effects of insider versus outsider leadership (Reynolds and Cohen 2016) or to analyse the ways that insider-ness and outsider-ness are constructed (Broad 2016). These are but a few strategies through which researchers can move beyond questions of evaluation and efficacy to better understand the complexities of social life.

In addition, scholars have begun to move beyond a narrow focus on individual, community-based projects toward analyses of collective campaigns that foster the aims of food justice. Scholars have examined efforts to address pesticide drift (Harrison 2011), ban pesticides and GMOs (Alkon and Guthman 2017; Eaton 2017) and improve the wages and working conditions of those who labour throughout the food system (Minkoff-Zern 2014a, b; Lo and Koenig 2017). This scholarship helps to strengthen links between supporters of alternative food and food justice projects and those who have historically preferred more confrontational strategies (Sbicca 2017).

Beyond labour, food justice activists have begun exploring connections to a variety of progressive social issues and researchers have found these intersections to be fertile ground for scholarship. Labour, as described above, is one key area, and food justice activists have been working with labour unions and other workers' groups ranging from the Coalition of Immokalee Workers (a farmworker organization) to the OUR Wal-Mart struggle to organize workers at the nation's largest food distributor. In addition, two prominent food justice scholars are each researching the ways in which farms served as an essential asset for the Civil Rights Movement (White 2017; McCutcheon 2013). Another movement with recent ties to food justice is prison reform. Several food justice projects in Oakland have begun to train former inmates for food-related jobs as a way to provide opportunities not often afforded this population (Sbicca 2017). Additionally, a number of scholars are examining the links between food justice and migration. A soon-to-be released edited volume called *Food Across Borders* (DuPuis and Mitchell) contains several chapters to this effect, including Mares' work on how migration policy affects workers on Vermont Dairy farms and Minkoff-Zern's analysis of how immigrant farmworkers transition to farm ownership.

Lastly, North American food justice activists are increasingly connecting with allies rooted in peasant movements for food sovereignty throughout the Global South. Food First, a non-profit policy research organization has developed food sovereignty tours that bring US activists into contact with activists across the globe, and is also a founding member of the US Food Sovereignty Alliance. In addition, North American communities increasingly discuss their struggles in terms of sovereignty and independence, even as they engage the state and corporations around issues of land use. Each of these areas is an important future direction for food justice scholarship. Clearly food justice scholarship and activism has roots in the environmental justice and alternative agriculture movements, but its wings touch a variety of progressive causes.

References

Alkon, A.H. 2012. *Black, White and Green: Farmers' Markets, Race and the Green Economy.* Athens, GA: UGA Press.

Alkon, A.H. and J. Agyeman. 2011. *Cultivating Food Justice: Race, Class and Sustainability.* Cambridge, MA: MIT Press.

Alkon, A. and J. Guthman. 2017. *The New Food Activism.* Berkeley, CA: UC Press.

Alkon, A., D. Block, K. Moore, K. Gillis, N. DiNuccio, and N. Chavez. 2013. "Foodways of the Urban Poor." *Geoforum* 48: 126–35.

Allen, P. 2004. *Together at the Table: Sustainability and Sustenance in the American Agrifood System.* University Park, PA: Penn State University Press.

Allen, P. 2008. "Mining for Justice in the Food System: Perceptions, Practices, and Possibilities." *Agriculture and Human Values* 25: 157–61.

Altieri, M. 2000. "Ecological Impacts of Industrial Agriculture and the Possibilities for Truly Sustainable Farming." In *Hungry for Profit*, ed. F. Magdoff, J. Bellamy Foster and F. Buttel, 77–92. New York: Monthly Review Press.

Belasco, W. 1993. *Appetite for Change: How the Counterculture Took on the Food Industry.* Ithaca, NY: Cornell University Press.

Bell, M. 2004. *Farming for Us All: Practical Agriculture and the Cultivation of Sustainability.* State College, PA: Penn State University Press.

Bradley, K.J. 1995. "Agrarian Ideology: California Grangers and the Post-World War II Farm Policy Debate." *Agriculture History* 69(2): 240–56.

Bradley, K. and R.E. Galt. 2013. "Practicing Food Justice at Dig Deep Farms & Produce, East Bay Area, California: Self-determination as a Guiding Value and Intersections with Foodie Logics." In "Subversive and Interstitial Food Spaces," special issue, *Local Environment: The International Journal of Justice and Sustainability* 19(2): 172–86.

Broad, G. 2016. *More than Just Food.* Berkeley, CA: UC Press.

Brown, S. and C. Getz. 2008. "Privatizing Farm Worker Justice: Regulating Labour Through Voluntary Certification and Labeling." *Geoforum* 39: 1184–96.

Brown, S. and C. Getz. 2011. "Farmworker Food Insecurity and the Production of Hunger in California." In *Cultivating Food Justice: Race, Class, and Sustainability*, ed. A. Alkon and J. Agyeman. Cambridge, MA: MIT Press.

Clapp, J. and D. Fuchs. 2009. *Corporate Power in Global Agrifood Governance.* Cambridge, MA: MIT Press.

Cole, L. and S.R. Foster. 2000. *From the Ground Up.* NY: NYU Press.

Daniel, C.E. 1981. *Bitter Harvest: A History of California Farmworkers.* Ithaca, NY: Cornell University Press.

Dimitri, C. and L. Oberholtzer. 2009. "Marketing US Organic Foods: Recent Trends from Farms to Consumers." *Economic Information Bulletin* No. (EIB-58). www.ers.usda.gov/publications/eib-economic-information-bulletin/eib58.aspx#.UtA_4UQYt2c (accessed 1/10/2014).

Douglass, M. 1996. *Purity and Danger: An Analysis of the Concepts of Purity and Taboo.* NY: Taylor.

DuPuis, M., D. Mitchell and M. Garcia. Forthcoming. *Food Across Borders: Production, Consumption and Boundary Crossing in North America.* NJ: Rutgers.

Eaton, E. 2017. "How Canadian Farmers Fought and Won the Battle Against GMO Wheat." In *The New Food Activism*, ed. A. Alkon and J. Guthman. Berkeley, CA: UC Press, pp. 55–79.

Fletcher, T.H. 2004. *From Love Canal to Environmental Justice: The Politics of Hazardous Waste on the Canada–US Border.* Ontario, Canada: Broadview Press.

Forson, P.W. 2006. *Building Houses out of Chicken Legs.* Chapel Hill: UNC Press.

Frankenberg, R. 1993. *White Women, Race Matters: The Social Construction of Whiteness.* Minneapolis: University of Minnesota.

Garcia, M., M. DuPuis and D. Mitchell. 2016. *Food Across Borders.* University of Arizona Press.

Gilbert, J., G. Sharp and S. Felin. 2002. "The Loss and Persistence of Black-owned Farms and Farmland: A Review of the Research Literature and Its Implications." *Southern Rural Sociology* 18: 1–30.

Goldschmidt, W. 1978[1947]. *As You Sow.* New York: Harcourt, Brace, 1947. Reprinted Montclair, NJ: Allanheld, Osmun.

Gottlieb, R. and A. Joshi. 2010. *Food Justice.* Cambridge, MA: MIT Press.

Guthman, J. 2008a. "Neoliberalism and the Making of Food Politics in California." *Geoforum* 39: 1171–83.

Guthman, J. 2008b. "Bringing Good Food to Others: Investigating the Subjects of Alternative Food Practice." *Cultural Geographies* 15(4): 431–47.

Haley, A.D., E. Zimmerman, S. Wolf and B. Evans. 2012. "Neighbourhood-level determinants of life expectancy in Oakland." Technical report, Center on Human Needs. Virginia Commonwealth University. Richmond, VA. Available from www.societyhealth.vcu.edu/media/society-health/pdf/PMReport_Alameda.pdf (accessed 12/2/2015).

Harper, A.B. 2010. *Sistah Vegan: Food, Identity, Health, and Society: Black Female Vegans Speak.* Herndon, VA: Lantern Books.

Harrison, J.L. 2008. "Lessons Learned from Pesticide Drift: A Call to Bring Production Agriculture, Farm Labour, and Social Justice Back into Agrifood Research and Activism." *Agriculture and Human Values* 25(2): 163–67.

Harrison, J.L. 2011. *Pesticide Drift and the Pursuit of Environmental Justice*. Cambridge, MA: MIT Press.

Harvey, D. 2005. *A Brief History of Neoliberalism*. New York: Oxford University Press.

Hassanein, N. 2003. "Practicing Food Democracy: A Pragmatic Politics of Transformation." *Journal of Rural Studies* 19(1): 77–86.

Hayes-Conroy, J. 2014. *Savoring Alternative Food*. NY: Routledge.

Hislop, R. 2014. "Reaping Equity Across the USA: FJ Organizations Observed at the National Scale." MA thesis. University of California-Davis.

Holt-Giminez, E. 2013. "Land Grabs Versus Land Sovereignty: Food First Backgrounder." www. foodfirst.org/en/Land+grabs+vs+land+sovereignty (accessed 1/10/2014).

Israel, B.A., E.A. Parker, Z. Rowe, A. Salvatore, M. Minkler and J. Lopez. 2005. "Community-based Participatory Research: Lessons Learned from the Centers for Children's Environmental Health and Disease Prevention Research." *Environmental Health Perspectives, 113(10)*, 1463–1471.

Kent, J. and N. Meyers. 2001. *Perverse Subsidies: How Tax Dollars Can Undercut the Environment and the Economy*. Manitoba, Canada: IISD Press.

Kobayashi, A. and L. Peake. 2000. "Racism out of Place: Thoughts on Whiteness and an Anti-racist Geography in the New Millennium." *Annals of the Association of American Geographers*, 90: 392–403.

LaDuke, W. 2004. *Indigenous People, Power and Politics: A Renewable Future for the Seventh Generation*. Minneapolis, MN: Honor the Earth.

Lea, E. and T. Worsley. 2005. "Australians' Organic Food Beliefs, Demographics and Values." *British Food Journal* 107(10/11): 855–69.

Liebman, E. 1983. *California Farmland: A History of Large Agricultural Landholdings*. Totowa, NJ: Rowman & Allanheld.

Lipsitz, G. 1998. *The Possessive Investment in Whiteness*. Pennsylvania: Temple University Press.

Lo, J. and B. Koenig. 2017. "Food Workers and Consumers Organizing Together for Food Justice." In *The New Food Activism*, ed. A. Alkon and J. Guthman. Berkeley CA: UC Press.

Lyson, T.A. 2004. *Civic Agriculture: Reconnecting Farm, Food and Community*. Boston: Tufts University Press.

Magdoff, F., J.B. Foster and F. Buttel, eds. 2000. *Hungry for Profit: The Agribusiness Threat to Farmers, Food, and the Environment*. New York: Monthly Review Press.

Mares, T.M. and D. Peña. 2011. "Environmental and Food Justice: Toward Local, Slow, and Deep Food Systems." In *Cultivating Food Justice: Race, Class, and Sustainability*, ed. A. Alkon and J. Agyeman, 197–220. Cambridge, MA: MIT Press.

Massey, D.S. and N.A. Denton. 1993. *American Apartheid: Segregation and the Making of the Underclass*. Cambridge, MA: Harvard University Press.

McClintock, N. 2011. "From Industrial Garden to Food Desert: Unearthing the Root Structure of Urban Agriculture in Oakland, California." In *Cultivating Food Justice: Race, Class, and Sustainability*, ed. A. Alkon and J. Agyeman, 89–120. Cambridge, MA: MIT Press.

McCutcheon, P. 2013. "Returning Home to Our Rightful Place: The Nation of Islam and Muhammad Farms." *Geoforum* 49: 61–70.

McWilliams, C. 2000. *Factories in the Field*. Berkeley, CA: UC Press.

Myers, J. and J. Sbicca. 2015. "Bridging Good Food and Good Jobs: From Secession to Confrontation within Alternative Food Movement Politics." *Geoforum* 61: 17–26.

Minkoff-Zern, L.-A. 2012. "Knowing 'Good Food': Immigrant Knowledge and the Racial Politics of Farmworker Food Insecurity." In "Race, Space, and Nature," special issue, *Antipode: A Radical Journal of Geography*.

Minkoff-Zern, L.-A. 2014a. "Subsidizing Farmworker Hunger: Food Assistance Programs, Farmworker Gardens, and the Social Reproduction of California Farm Labour." *Geoforum* 57: 91–98.

Minkoff-Zern, L.-A. 2014b. "Hunger Amidst Plenty: Farmworker Food Insecurity and Coping Strategies in California." *Local Environment: The International Journal of Justice and Sustainability*. Special issue on Interstitial and Subversive Food Spaces. Vol. 19(2).

Minkoff-Zern, L.-A., N. Peluso, J. Sowerwine and C. Getz. 2011. "Race and Regulation: Asian Immigrants in California Agriculture." In *Cultivating Food Justice: Race, Class, and Sustainability*, ed. A. Alkon and J. Agyeman, 65–86. Cambridge, MA: MIT Press.

Mohai, P. and R.K. Saha. 2003 "Reassessing Race and Class Disparities in Environmental Justice Research Using Distance-Based Methods" Paper presented at the annual meeting of the American Sociological Association, Atlanta Hilton Hotel, Atlanta, GA. Online PDF 2009-05-26 from www.allacademic.com/meta/p107607_index.html

Morland K., S. Wing, A.D. Roux and C. Poole. 2002. "Neighbourhood Characteristics Associated with the Location of Food Stores and Food Service Places." *American Journal of Preventive Medicine* 22: 23–29.

Nash, L. 2007. *Inescapable Ecologies: A History of Environment, Disease, and Knowledge.* Berkeley, CA: University of California Press.

Norgaard, K., R. Reed and C. Van Horn. 2011. "A Continuing Legacy: Institutional Racism, Hunger, and Nutritional Justice on the Klamath." In *Cultivating Food Justice: Race, Class, and Sustainability*, ed. A. Alkon and J. Agyeman, 23–46. Cambridge, MA: MIT Press.

Park, L.S.-H. and D. Pellow. 2004. "Racial Formation, Environmental Racism, and the Emergence of Silicon Valley." *Ethnicities* 4(3): 403–424.

Park, L.S.-H. and D. Pellow. 2011. *Slums of Aspen.* NY: NYU Press.

Pastor, M., R. Morello-Frosch and J. Sadd. 2006. "Breathless: Air Quality, Schools, and Environmental Justice in California." *Policy Studies Journal* 34(3): 337–362.

Pellow, D.N. 2002. *Garbage Wars: The Struggle for Environmental Justice in Chicago.* MA: MIT Press.

Petersen, D., M. Minkler, V. Vasquez Breckwich, and A. Baden 2006. "Community-based Participatory Research as a Tool for Policy Change: A Case Study of the Southern California Environmental Justice Collaborative." *Review of Policy Research* 23(2): 339.

Pollan, M. 2006. "Voting with Your Fork." *New York Times.* http://pollan.blogs.nytimes.com/2006/05/07/voting-with-your-fork/ (accessed 8/6/2013).

Pulido, L. 1996. *Environmentalism and Economic Justice: Two Chicano Cases from the Southwest.* Tucson: University of Arizona Press.

Pulido, L. 2000. "Rethinking Environmental Racism: White Privilege and Urban Development in Southern California." *Annals of the Association of American Geographers* 90(1): 12–40.

Rausser, G.C. 1992. "Predatory Versus Productive Government: The Case of US Agricultural Policies." *Journal of Economic Perspectives* 6(3): 133–57.

Reynolds, K. 2014. "*Disparity Despite Diversity: Social Injustice in New York City's Urban Agriculture System*". *Antipode 47(1): 240–259.*

Reynolds, K. and N. Cohen. 2016. *Beyond the Kale.* Athens, GA: UGA Press.

Romm, J. 2001. "The Coincidental Order of Environmental Justice." In *Justice and Natural Resources*, ed. K. Mutz. Washington, DC: Island Press.

Saldanha, A. 2006 "Re-ontologizing Race." *Environment and Planning D Society and Space* 24(1): 9–24.

Saxton, D. 2015. "Strawberry Fields as Extreme Evironments: The Ecobiopolitics of Farmworker Health." *Medical Anthropology.* 34(2): 166–83.

Sbicca, J. 2012. "Growing Food Justice by Planting an Anti-oppression Foundation: Opportunities and Obstacles for a Budding Social Movement." *Agriculture and Human Values* 29(4): 455–66.

Sbicca, J. 2017. "Resetting the Good Food Table: Labour and Food Justice Alliances in Los Angeles." In *The New Food Activism*, ed. A. Alkon and J. Guthman. Berkeley, CA: UC Press.

Shrader-Frechette, K. 2002. "Environmental Justice: Creating Equality, Reclaiming Democracy." New York: Oxford University Press.

Slocum, R. 2006. "Anti-racist Practice and the Work of Community Food Organizations." *Antipode* 38(2): 327–49.

Squires, G.D. and C.E. Kubrin. 2006. *Privileged Places: Race, Residence, and the Structure of Opportunity.* Boulder, CO: Lynne Rienner.

Sze, J. 2006. *Noxious New York: The Racial Politics of Urban Health and Environmental Justice.* Boston: MIT Press.

Taylor, E.J. and P.L. Martin. 1997. "The Immigrant Subsidy in US Agriculture: Farm Employment, Poverty, and Welfare." *Population and Development Review* 23(4): 855–74.

Terry, B. 2009. *Vegan Soul Kitchen: Fresh, Healthy, and Creative African-American Cuisine.* New York: Da Capo Press.

United Church of Christ. 2007. *Toxic Waste and Race at Twenty.* New York: United Church of Christ.

United Church of Christ. 1987. *Toxic Wastes and Race in the United States: A National Report on the Racial and Socio-economic Characteristics with Hazardous Waste Sites.* New York: United Church of Christ.

United States Department of Agriculture (USDA). 2013. "Farmers Markets and Local Food Marketing." www.ams.usda.gov/AMSv1.0/ams.fetchTemplateData.do?template=TemplateS&leftNav=Whole saleandFarmersMarkets&page=WFMFarmersMarketGrowth&description=Farmers Market Growth. (accessed 3/17/2014).

Via Campesina. 2002. "Food Sovereignty." Flyer distributed at the World Food Summit +5, Rome. Available at viacampesina.org. (accessed 6/5/2009).

Wacquant, L.J.D. 1995. "The Ghetto, the State, and the New Capitalist Economy." In *Metropolis: Center and Symbol of Our Times*, ed. P. Kasinitz, pp. 418–49. New York: New York University Press.

Walker, D. 2010. "The Unreported Economic Depression in Black America." Alternet.org. November 29. www.alternet.org/speakeasy/2010/11/29/the-unreported-economic-depression-in-black-america.

White, M.M. 2010. "Shouldering Responsibility for the Delivery of Human Rights: A Case Study of the D-Town Farmers of Detroit." *Race/Ethnicity: Multidisciplinary Global Perspectives* 3(2): 189–211.

White, M.M. 2017. "'A pig and a garden': Fannie Lou Hamer and the Freedom Farms Cooperative.' *Food and Foodways* 25(1): 20–39.

Wilson, W.J. 2011. *When Work Disappears: The World of the New Urban Poor*. New York: Vintage Books.

Winson, A. 1993. *The Intimate Commodity*. Toronto: Garamond Press.

Worster, D. 2005. *Dust Bowl*. Oxford: Oxford University Press.

Wrigley, N. 2002. "'Food deserts' in British Cities: Policy Context and Research Priorities." *Urban Studies* 39(11): 2029.

34

ENVIRONMENTAL CRIME AND JUSTICE

A green criminological examination

Michael J. Lynch and Kimberly L. Barrett

Criminologists paid little attention to environmental crime and justice until the 2000s following Lynch's (1990) argument concerning developing a criminological specialization focused on environmental crime, law and justice called 'green criminology' (GC). Several early GC studies addressed environmental justice (EJ) by examining environmental racism and proximity to environmental hazards (see below). In the GC literature, however, concerns for other forms of environmentally related justice issues such as harms against non-human animals/species, green victimization of the public, and studies of the punishment of green/environmental offenders also have implications for EJ research. Taking a broad view, Holifield (2001) noted that EJ was difficult to precisely define due to historical, political and cultural variations. We would add to that observation academic variability, as GC calls attention to how harms against ecosystems generate a variety of environmental injustices that affect not only humans, but also nonhuman species.

To illustrate these points, we begin with a brief review of the definition of green crime, its connection to EJ, and review several primary areas of GC research. In each area, we draw attention to how that research addresses traditional and expanded notions of EJ.

Defining green crime

Generally, green crimes are legal or illegal acts (Walters 2010a), defined either as violations of laws, as social harms, or as scientifically identified harms (Lynch and Stretesky 2014) that involve injurious behaviours to humans, nonhuman animals/species and ecosystems (Brisman and South 2013a). As Brisman and South note, damage to ecosystems and wildlife destruction may result from legally permissible behaviours (e.g. pollution, timber harvesting and mining). Because environmental law permits a wide range of ecologically destructive behaviours, green criminologists argue that it is the kind and extent of harm produced by environmentally destructive behaviours rather than their legal standing that makes those behaviours green crimes. From a political economic perspective, Stretesky et al. (2013a: 2) argue that green crimes are "acts that cause or have the potential to cause significant harm to ecological systems for the purpose of increasing or supporting production." In the latter view, green crimes are also identified as behaviours which "scientific evidence suggests may cause significant ecological destruction." In the latter case, scientific evidence of harm rather than legal standards define the existence of green crime. For example, while mountaintop removal mining (MRM) is legally permissible once the appropriate permits are acquired, MRM can cause extensive ecological destruction to forest ecosystems and wildlife which can be conceptualized as a green crime based on scientific evaluations of MRM ecological harms.

Green criminologists often define green justice/injustice using philosophical positions on 'animal rights,''ecological rights' and 'ecosystem rights' (Beirne 2007; Benton 2007). White (2007) notes that these GC perspectives on justice can be divided into two categories: 1) those that relate to EJ rights that affect humans including social injustices such as differential exposure to environmental toxins and hazards (as suggested by Lynch and Stretesky 1998); and 2) those related to ecological justice, or rights issues for nonhuman animals and ecosystems. Brisman (2008) suggests that the concept of green justice/injustice is related to the green concept of crime/harm, and that where there is green crime, green injustice exists, an idea that extends to the oppression of nonhuman animals.

Here, we extend this argument noting that green justice and EJ intersect when ecological injustice against nature or nonhuman animals also creates conditions of environmental injustice for humans. In making that connection, we draw on Holifield's (2001) argument that the definition of EJ can vary historically, politically and contextually. For example, oil extraction on Indigenous lands in developing countries is not only an example of ecological harm/injustice, but also illustrates the ways in which ecological injustice promotes EJ concerns for Indigenous peoples (see Chapters 40 and 41). In this case, Indigenous peoples are made proximate to toxic wastes from oil extraction through oil spills and groundwater contamination, and have reduced access to the healthy ecosystems they employ for survival. Ecological injustice may also affect humans by creating environmental injustice through ecological damage that involves harms to nonhuman animals. We can use the term 'ecological-induced-environmental injustice' (EEJ) to identify this condition. The sections that follow provide examples of how green criminologists address EJ/EEJ.

Environmental justice as green injustice

The study of environmental justice – or inequality in exposure to environmental hazards and unequal protection under environmental laws related to a community's class, race and/ or ethnic composition – has been an important sociological and activist subject since the early 1980s (Bullard 1983). That subject was also a concern raised during GC's development (Lynch and Stretesky 1998, 1999; Stretesky and Hogan 1998; Stretesky and Lynch 1998, 1999, 2002). GC is concerned with unequal exposure to environmental hazards for two reasons: 1) as an extension of traditional criminological studies of race and class biases in criminal justice processes; and 2) in relation to GC's expanded concept of victimization – that is, as an example of how green crimes and unequal exposure to toxic hazards produce injustice and public health crimes (White 2003).

Empirical studies related to EJ in GC explore several issues. Stretesky and Lynch (2002) performed one of the first EJ studies to examine the effect of public school racial and class composition on proximity to hazardous waste sites, finding a race effect after controlling for class composition. Using geographic data on hazardous waste sites and community characteristics over time (1970, 1980 and 1990), Stretesky and Hogan (1998) found that after controlling for class effects, Black and Hispanic communities were closer to hazardous waste sites than other communities in Florida. In two studies of the effect of community race, class and ethnicity on penalties given to petroleum companies that violated environmental laws, Lynch et al. (2004a, 2004b) found evidence that refineries in Black, Hispanic and low income communities received lesser punishments, indicating differential legal treatment of environ-mental law violators related to the characteristics of the communities they victimized. These studies not only contribute to the EJ knowledge base, but also suggest that the intersection of racial, ethnic and class variability in EJ outcomes can be explained as part of a broader political economic approach to green crime and injustice, as noted below.

Political economy and green crime

Green criminology is currently more applied than theoretical. By 'applied' we mean that green criminologists tend to use case studies and examples of green crime and justice, and have been less concerned with developing a widely shared theoretical perspective. Nevertheless, there are a few theoretical orientations which green criminologists employ.

Lynch (1990) initially proposed GC as an extension of radical criminology which specifically includes a focus on political economic explanations of crime and justice. Various alternatives to that view have been proposed. Beirne (1999, 2007, 2009, 2014) makes a strong case for green criminological theories that address harms to nonhuman animals, and others have developed analyses derived from ecosystem rights approaches (Barnett 1999; Benton 1998; White 2003). Green cultural criminology draws on concepts from cultural criminology in an effort to examine "contested terrains" or cultural spaces that emerge around the definition and study of green crimes, including the definition of green crime and transgressions of those definitions, resistance to cultural definitions of green crime, and media constructions of green crime, and in doing so can also address EJ concerns (Brisman and South 2013b). Finally, as noted, green criminologists incorporate EJ theories that explore how race, ethnicity and class as social structures impact where green crime occurs and affect responses to those crimes (see sections on environmental justice and punishment of green offenders).

We prefer political economic explanations of green crime/justice and the use of treadmill of production (ToP) theory and ecological Marxism (Lynch et al. 2013; Stretesky et al. 2013a). That view draws ecological Marxists' observations that capitalism is and must be an ecologically destructive force that produces ecological disorganization (Foster 1992, 1997, 1999, 2000; Foster and Burkett 2008; Foster et al. 2010; see also Chapter 6). In that view, the inherent growth imperative of capitalism requires a constantly expanding stream of raw material inputs from ecological withdrawals, and generates an expanding stream of ecological additions or pollutants. Ecological withdrawals and additions promote ecological destruction and disorganization, and in this sense the growth imperative of capitalism and the stability/growth needs of nature are said to be in contradiction with one another (Foster 1992, 2000), with each increase in capitalism causing a contraction of nature. Ecological withdrawals/additions interfere with nature's cycles and reproductive abilities, and cause nature to become increasingly disorganized over time. Criminologically speaking, then, ecological withdrawals/additions can be viewed as crimes against nature from the perspective of nature, which includes the needs of all species that rely upon nature and the ability of nature to reproduce itself and the conditions for life on earth (Lynch et al. 2013; Stretesky et al. 2013a). Those crimes against nature can also be interpreted as forms of injustice, and it is difficult to disentangle green crimes from green injustice.

This brief summary does not address the intricate details and complexities of a political economic explanation of green crime and justice which also includes reference to the problem of metabolic rift, urban–rural ecological contradictions within capitalism, and class conflict (Foster 1999; Foster et al. 2010). Also relevant are scientific theories related to the earth's planetary boundaries (Rockström et al. 2009) and green criminological discussions of how planetary boundary analysis and green crime and justice intersect and relate to political economy (Long et al. 2014; Lynch et al. 2013; Lynch and Stretesky 2014). In short, taking a political economic approach, green criminologists have addressed how the capitalist ToP generates green crimes associated with ecological withdrawals/additions caused by production. This is a complex process that plays out in different ways within nations and at the global level. Nevertheless, this view also suggests that the distribution of green crimes impacts the distribution of EJ. As noted, some green criminological studies address traditional EJ justice issues related to class/race/ethnic differentials in exposure to pollution and their effect on the punishment of environmental offenders empirically. Some green criminological studies expand the concept of EJ by examining how green harms produce ecological injustice for non-human species (defined earlier as EEJ). Less developed are green criminological studies that address EJ issues for Indigenous/Native Peoples (Lynch and Stretesky 2012).

Ecological withdrawal as green crimes

As noted, ecological withdrawals involve the extraction of raw materials from nature for manufacturing commodities or supplying energy. These withdrawals have direct ecological consequences, and also promote ecological-environmental injustice.

Treadmill of production theory (Schnaiberg 1980) argues that ecological withdrawals required for capitalism's expansion facilitates continuous growth of production and consumption as well as the disorganization of nature. Numerous empirical studies examine the effect of the ToP on ecological destruction (e.g. Clark et al. 2012 on coal consumption; Jorgenson 2003 and York et al. 2003 on ecological footprints). Green criminologists have also examined how the ToP explains ecological destruction and creates crimes/injustice against ecosystems and their inhabitants (Lynch et al. 2013; Stretesky et al. 2013a).

Drawing attention to the connection between harms caused by ecological withdrawals and EJ, van Solinge (2010) noted that extensive legal and illegal deforestation in the Amazon creates detrimental effects for Native Peoples who are forced to move from their traditional homes and increasingly come into contact with modern civilization, which quickly erodes traditional ways of life and threatens the existence of Native Peoples. Moreover, deforestation destroys the local ecosystem, plants, wildlife and other natural resources (e.g. waterways) which Native Peoples in the Amazon rely upon for survival. In this sense, both legal and illegal deforestation serve as examples of how green crimes facilitate not only ecological destruction, but environmental injustice for Native Peoples. These green crimes also generate EJ protests by Native Peoples over deforestation and oil and mineral exploration (van Solinge 2013), and involve conflicts over land use, important EJ concerns for Native Peoples (van Solinge 2014; Zaitch et al. 2014).

Ecological additions, green crime and green injustice

ToP theory also draws attention to ecological additions or pollutants that result from production. As with ecological withdrawals, ToP theory posits that continuous expansion of the capitalist ToP increases ecological additions and expands ecological disorganization. Green criminologists have drawn attention to this approach and it relates to EJ (Stretesky et al. 2013a, 2013b; Lynch et al. 2013; Lynch and Stretesky 2014; Long et al. 2012; Greife and Stretesky 2013).

Lynch et al. (2013; Lynch and Stretesky 2014) used the ToP approach to argue that green criminologists should define ecological additions as green crimes that promote ecological disorganization. From the perspective of ecological Marxism, ecological additions can be conceptualized as crimes against nature – as behaviours that harm the organization, operation and functions of the natural ecological world. Crimes involving ecological additions are widespread and include everyday outcomes such as the emission of industrial pollutants into the air, water and land. Accumulation of pollutants over time can create large industrial waste hazards and toxic waste sites, and can pose EJ concerns. For example, in the US there are 1163 acknowledged Superfund Sites – the most hazardous of the known toxic waste sites, and prior research indicates that these sites are unequally distributed, having adverse effects on low-income and minority communities (Stretesky and Hogan 1998). Such areas exist world-wide, as ecological additions were spread by the growth of global capitalism, and due to these routine emissions, pollutants are now globally ubiquitous. As Lynch and Stretesky (2014) note, in the US routine ecological additions emitted by industries averaged 24 billion pounds of *reported* toxic emissions annually under the US EPA's Toxic Release Inventory program over the past decade – which probably significantly underestimates the true extent of these emissions.

As other green criminologists note, ecological additions also involve green crimes related to radioactive waste (Walters 2007), and oil and other forms of chemical pollution (Ruggiero and South 2013b). In the view of green criminologists, these ecological additions constitute "dirty collar crimes" (Ruggiero and South 2010a), crimes of the economy (Ruggiero and South 2013a) and the routine "toxic crimes of capitalism" (Lynch and Stretesky 2001; Ruggiero and South 2013b) that now undermine the stability of the global ecosystem. These green crimes cause much more victimization than the crimes criminologists ordinarily study (e.g. homicide, rape, burglary, etc., see section on green victimization) and also pose specific EJ problems, including those for Native Peoples.

From the above, it should be evident that green crimes also have EJ dimensions. From studies of EJ we know that pollution is unevenly distributed (e.g. Chakraborty 2012, 2004). This suggests that ecological additions are unevenly distributed, and when linked to ToP theory also suggests that the uneven distribution of ecological additions and the forms of EJ that emerge stem from the structural and geographic organization of capitalism. This is a complex issue with different dimensions across nations and cannot be easily summarized.

The scope of green victimization

Green criminologists also draw attention to the scope of green victimization, which appears in various forms and affects not only humans, but nonhuman life forms (Beirne 1999), including ecosystems (South 2007; van Solinge 2010). Green criminologists suggest that all these forms of environmental injustice are worthy of discussion (White 2007).

It is difficult to translate victimization of nonhuman entities (nonhuman animals, plants, ecosystems, etc.) into quantitative counts given the broad scope of victimization and the large number of individual entities victimized (Jarrell et al. 2013; Lynch 2013). It is, however, possible to estimate the extent of human green victimization because data relevant to this task are more readily available (e.g. estimates of the number of people exposed to air and water pollution, or who live near toxic waste sites). These estimates have been compared to estimates of street crime (e.g. robbery, rape, larceny, motor vehicle theft, homicide, etc.) victimization to illustrate that green victimization is more widespread than the forms of victimization criminologists typically study. Criminologists have spent significant effort counting human victims of street crime (victimology), and helped develop surveys for estimating street crime victimization in the 1970s (in the US, National Crime Victimization Survey (NCVS), initiated in 1973).

Prior studies have compared green crime and street crime victimization estimates (Jarrell et al. 2013; Lynch 2013). Lynch and Barrett (2015) compared violent street crime victimization to one form of green victimization – exposure to small particles from coal-fired power plants (CFPP). Drawing on earlier research (Stretesky and Lynch 1998), Lynch and Barrett argued that CFPP pollution is a form of corporate environmental violence that causes widespread green victimization for humans across the US. Using evidence from the 2010 Clean Air Task Force study of exposure to CFPP pollution, Lynch and Barrett compared CFPP exposure estimates to street crime victimization. The results of their comparison noted that green victimization from CFPP small particle matter pollution causes nearly 53 per cent more deaths in the US than homicide – a point illustrating the importance of examining the deleterious green victimization effects of pollution.

The above estimate relates only to green victimization produced by a very specific source: small particle matter emitted by CFPPs. Using broader estimates of green victimization, Jarrell et al. (2013) and Lynch (2013) compared exposure to chemical pollution using air, land and water pollution exposure estimates as a measure of green victimization to estimates of victimization from criminal assaults as an equivalent form of harm. Both studies found that the likelihood of green victimization was *millions of times* more likely than street crime victimization. While much more likely, one could argue that the potentially small levels of environmental exposure associated with green victimization from air/land/water pollution is trivial compared to some forms of street crime victimization (e.g. a violent assault). However, the routine nature of exposure to environmental pollutants can mean they accumulate in the body, leading to the promotion of diseases which in the long run may lead to debilitating diseases and even death.

Criminological studies of green victimization have relevance to the study of EJ. These studies indicate the widespread nature of green victimization from ecological additions. Because these green victimizations are also persistent, and persistently exposed populations often include those of low income, and racial and ethnic minorities – a point empirical studies on the distribution of pollution have long made (e.g. Sexton et al. 1993) – there is a clear link between green crimes that result from exposure to ecological additions and EJ.

Crimes and injustice affecting nonhuman animals

Green criminologists have devoted significant attention to crimes affecting nonhuman species, especially wildlife poaching. Poaching laws, created in the fourteenth century to protect the property rights of the landed classes, extended to ownership of wildlife on privately owned land (Eliason 2012). These laws were designed to protect property rights rather than wildlife, and it was not until the industrial era that poaching laws emerged to protect wildlife from extinction (Eliason 2012). With respect to the concept of ecological-environmental justice, one must consider that poaching laws can have adverse impacts on poor/Indigenous peoples' access to ecological resources, and while these laws may protect wildlife species (Ngoc and Wyatt 2013; Wyatt 2014a) and limit animal abuse and trafficking (Sollund 2013a), they also impose conditions that adversely affect the survival of poor/Indigenous peoples. The latter issue has not been sufficiently examined by green criminologists, who instead tend to focus on the effectiveness of poaching regulations and the effects of poaching on nonhuman species survival, as noted below.

Several green criminological studies examine the effects of poaching on neotropical parrot species (Clarke and Rolf 2013; Pires and Clarke 2011, 2012; Pires and Petrossian 2015). Studies have also examined how access opportunities (e.g. fish prevalence and value of fish species) and fishing/fish poaching laws and their enforcement might control this problem (Petrossian 2015; Petrossian and Clark 2014; Petrossian et al. 2015). As Hauck (2007) argues, illegal fishing and mismanagement of fisheries also impacts humans by threatening human and environmental security due to declining availability of ecological resources – which is an EJ issue consistent with our concept of ecological-environmental justice.

Other studies draw attention to how deficient social control/regulation of poaching and illegal wildlife trade affects species survival (Nurse 2013; Wellsmith 2011; on elephant poaching ban see Lemieux and Clarke 2009). Green criminologists note that while poaching is an important concern, it is not the sole cause of species endangerment, and attention must also be directed to how anthropogenic forces such as the continued expansion of the global capitalist ToP also contribute to species decline and biodiversity loss and generate injustice for nonhuman animal species (Lynch et al. 2015; see also Sollund 2013b). Missing in these studies, however, is a direct discussion of how poaching regulations designed to limit species decline create ecological-environmental injustice for Native Peoples, and how the law might be reconceptualized to address this concern.

The punishment of green offenders

Criminologists often study the punishment/sentencing of street criminals, but rarely pay direct attention to the punishment of green offenders (e.g. corporations that violate environmental laws). Only a handful of green criminological studies address the punishment of green offenders. Some of those studies address EJ questions related to equity in punishment when offences victimize communities with different racial, ethnic and class characteristics. Others address EJ questions indirectly by examining whether penalties for environmental offences deter environmental offenders and slow ToP ecological additions.

Addressing the latter issue, Stretesky et al. (2013b) examined whether the large penalties applied to the top 25 penalized companies between 2003 and 2010 (N = 158) affected future compliance with environmental laws. Drawing on ToP and deterrence theory – the latter approach asserting that punishment deters future illegal behaviour – the researchers analysed whether large penalties imposed by the US EPA changed toxic emission behaviour in the future. Theoretically, the question is whether EPA-imposed penalties slow the ToP and ecological additions. Overall, the results from various models controlling for other factors that affect toxic release by corporations suggest that penalties did not slow the ToP for the companies that received the largest penalties. One of the key factors driving toxic releases was the volume of toxic releases in the prior year, indicating that the ToP continued to drive production and pollution upward despite the imposition of environmental penalties. Thus, the researchers concluded that large state-imposed penalties did not slow the ToP, and that there is a reason to rethink the role penalties are believed to play in constraining ecological destruction. These results also suggest that because penalties do not slow the ToP, they would also be ineffective responses to environmental injustice. That is, since penalties for environmental violations do not alter the behaviour of polluting corporations, there is little reason to expect that targeting corporations that pollute low income/minority communities with enhanced penalties or more frequent penalties would change their behaviour significantly and lessen the extent of environmental injustice.

In another test of ToP theory related to the social control of corporations, Long et al. (2012) examined whether political campaign contributions made by core coal ToP corporations affected enforcement actions taken against those companies. ToP theory suggests that treadmill actors use their economic and political power to affect laws and regulations, and Long et al.

suggest that one way in which this can happen is through political campaign contributions. The researchers found that political donations significantly increased for companies just prior to the conclusion of an enforcement event (odds ratio = 6.36), and that corporations appear to employ political campaign contributions to affect how much punishment they receive for environmental crimes. The EJ implications of these findings require further investigation. For example, future studies can examine whether corporations are more or less likely to contribute to political campaigns to affect their punishment depending on the race/class/ethnic composition of the communities their offence affects.

In another study that employed ToP theory and has more direct EJ implications, Stretesky and Lynch (2011) examined penalties meted out to 100 coal mining facilities between 2002 and 2008. As Stretesky and Lynch noted, prior research suggests that the level of poverty and inequality around a facility affects the penalty received by a corporation, and that the higher the rate of poverty and the greater the per cent minorities surrounding an area with an environmental violation, the lower the financial penalty received (Atlas 2001; Lavelle and Coyle 1992; but see Ringquist 1998). Stretesky and Lynch found that 35 per cent of the mines in the study had at least one environmental violation, and that four variables affected whether mines were more likely to have an environmental violation: the number of EPA permits required, the number of facility inspections, the type of coal mining operation, and the size of the coal mine. When a measure of area poverty was added to the model, it was significantly related to increased odds of an environmental violation occurring. The social control variable in this case – inspections – was found to be negatively related to poverty, so that as area poverty diminished, facilities were more likely to be inspected, meaning that those living in poorer communities receive less environmental justice.

Conclusion

Green criminology, conceived 25 years ago, examines a range of issues that occur at the intersection between the study of environmental and criminological issues. As noted, it employs rich philosophical traditions that define the rights of nonhuman species and ecosystems, and political economic analysis to draw attention to green crime and injustice. In GC, green crimes also produce green injustices that affect various species. As a result, green criminologists take a broad view of environmental injustice and also draw attention to the ways in which ecological/green crimes generate interconnected forms of environmental justice for humans, nonhuman species and ecosystems.

Power differentials and injustice represent a key theme across areas that green criminologists explore. For example, the non-random spatial distribution of ecological withdrawals and ecological additions illustrates major inequalities with respect to who suffers and who benefits from these production processes. Consistent with environmental justice work in other disciplines, green criminologists have observed that communities with high percentages of minority residents as well as communities with high rates of poverty endure more severe environmental injustices (e.g. reduced penalties for environmental violations, increased frequencies of environmental crimes) compared to higher-income, predominantly white communities. Green criminologists have also integrated green criminology with ToP and ecological Marxist perspectives, emphasizing the role of political economy in environmental degradation, including examining ecological disorganization as crime and a form of injustice.

For those outside of criminology, the key lesson learned from GC's examination of EJ include attending to how ecological destruction generates injustice not simply for humans, but for ecosystems and nonhuman species, and how these forms of injustice intersect and affect one another. While criminologists have learned a great deal from geographers, sociologists and public health researchers, we hope this review stimulates researchers in those disciplines to consider other dimensions of the problems posed by environmental justice.

References

Atlas, M. (2001). "Rush to judgment: an empirical analysis of environmental equity in U.S. Environmental Protection Agency enforcement actions." *Law & Society Review*, vol. 35, pp. 633–682.

Barnett, H. (1999). "The land ethic and environmental crime." *Criminal Justice Policy Review*, vol. 10, no. 2, pp. 161–191.

Beirne, P. (1999). "For a nonspeciesist criminology: animal abuse as an object of study." *Criminology*, vol. 37, no. 1, pp. 117–148.

Beirne, P. (2007). "Animal rights, animal abuse and green criminology." In Beirne, P. and N. South (eds), *Issues in green criminology: confronting harms against environments, humanity and other animals*, Oxford, UK: Willan, pp. 55–86.

Beirne, P. (2009). *Confronting animal abuse: law, criminology and human–animal relationships*. Totowa, NJ: Rowman and Littlefield.

Beirne, P. (2014). "Theriocide: naming animal killing." *International Journal for Crime, Justice and Social Democracy*, vol. 3, pp. 50–67.

Benton, T. (1998). "Rights and justice on a shared planet: more rights or new relations?" *Theoretical Criminology*, vol. 2, no. 2, pp. 149–175.

Benton, T. (2007). "Ecology, community and justice: the meaning of green." In Beirne, P. and N. South (eds), *Issues in green criminology: confronting harms against environments, humanity and other animals*. Oxford, UK: Willan, pp. 3–30.

Brantingham, P. J. and Brantingham, P.L. (eds). (1981). *Environmental criminology*. Beverly Hills, CA: Sage Publications.

Brisman, A. (2008). "Crime-environment relationships and environmental justice." *Seattle Journal of Social Justice*, vol. 6, pp. 727–907.

Brisman, A. (2014). "Of theory and meaning in green criminology." *International Journal for Crime, Justice & Social Democracy*, vol. 3, no. 2, pp. 22–35.

Brisman, A. and South, N. (2013a). "Introduction: horizons, issues, and relationships in green criminology." In South, N. and A. Brisman (eds), *The Routledge international handbook of green criminology*, Oxon, UK: Routledge, pp. 1–24.

Brisman, A. and South, N. (2013b). "A green-cultural criminology: an exploratory outline." *Crime, Media, Culture*, vol. 9, no. 2, pp. 115–135.

Bullard, R.D. (1983). "Solid waste sites and the Black Houston community." *Sociological Inquiry*, vol. 53, pp. 273–288.

Chakraborty, J. (2004). "The geographic distribution of potential risks posed by industrial toxic emissions in the US." *Journal of Environmental Science and Health, Part A*, vol. 39, no. 3, pp. 559–575.

Chakraborty, J. (2012). "Cancer risk exposure to hazardous air pollutants: spatial and social inequalities in Tampa Bay, Florida." *International Journal of Environmental Health Research*, vol. 22, no. 2, pp. 165–183.

Clark, B., Jorgenson, A.K. and Auerbach, D. (2012). "Up in smoke: the human ecology and political economy of coal consumption." *Organization & Environment*, vol. 25, no. 4, pp. 452–469.

Clarke, R.V. and Rolf, A. (2013). "Poaching, habitat loss and the decline of neotropical parrots: a comparative spatial analysis." *Journal of Experimental Criminology*, vol. 9, no. 3, pp. 333–353.

Eliason, S. (2012). "From the King's deer to a capitalist commodity: a social historical analysis of the poaching law." *International Journal of Comparative and Applied Criminal Justice*, vol. 36, no. 2, pp. 133–148.

Foster, J.B. (1992). "The absolute general law of environmental degradation under capitalism." *Capitalism, Nature, Socialism*, vol. 3, no. 3, pp. 77–82.

Foster, J.B. (1997). "The crisis of the Earth: Marx's theory of ecological sustainability as a nature-imposed necessity for human production." *Organization & Environment*, vol. 10, no. 3, pp. 278–295.

Foster, J.B. (1999). "Marx's theory of metabolic rift: classical foundations for environmental sociology." *American Journal of Sociology*, vol. 105, no. 2, pp. 366–405.

Foster, J.B. (2000). *Marx's ecology: materialism and nature*. NY: New York University Press.

Foster, J.B. and Burkett, P. (2008). "Classical Marxism and the second law of thermodynamics: Marx/ Engels, the heat death of the universe hypothesis, and the origins of ecological economics." *Organization & Environment*, vol. 21, no. 1, pp. 3–37.

Foster, J.B., Clark, B. and York, R. (2010). *The ecological rift: capitalism's war on the Earth*. NY: Monthly Review Press.

Greife, M.B. and Stretesky, P.B. (2013). "Treadmill of production and state variations in civil and criminal liability for oil discharges in navigable waters." In South, N. and A. Brisman (eds), *Routledge international handbook of green criminology*, London: Routledge, pp. 150–166.

Hauck, M. (2007). "Non-compliance in small-scale fisheries: a threat to security." In Beirne, P. and N. South (eds), *Issues in green criminology:confronting harms against environments, humanity and other animals*, Oxon, UK: Willan, pp. 270–289.

Holifield, R. (2001). "Defining environmental justice and environmental racism." *Urban Geography*, vol. 22, no. 1, pp. 78–90.

Jarrell, M.L., Lynch, M.J. and Stretesky, P.B. (2013). "Green criminology and green victimization." In Arrigo, B. and H. Bersot (eds), *The Routledge handbook of international crime and justice studies*, London: Routledge, pp. 423–44.

Jorgenson, A.K. (2003). "Consumption and environmental degradation: a cross-national analysis of the ecological footprint." *Social Problems*, vol. 50, no. 3, pp. 374–394.

Lavelle, M. and Coyle, M. (1992). "Unequal protection: the racial divide in environmental law." *National Law Journal*, vol. 21, pp. S1–S11.

Lemieux, A.M. and Clarke, R.V. (2009). "The international ban on ivory sales and its effects on elephant poaching in Africa." *British Journal of Criminology*, vol. 49, no. 4, pp. 451–471.

Long, M.A., Stretesky, P.B. and Lynch, M.J. (2014). 'The treadmill of production, planetary boundaries and green criminology." In Sapiens, T., R. White and M. Kluin (eds), *Environmental crime and its victims*, Devon, UK: Ashgate, pp. 263–276.

Long, M.A., Stretesky, P.B., Lynch, M.J. and Fenwick, E. (2012). "Crime in the coal industry: implications for green criminology and treadmill of production." *Organization & Environment*, vol. 25, no. 3, pp. 328–346.

Lynch, M.J. (1990). "The greening of criminology: a perspective for the 1990s." *The Critical Criminologist*, vol. 2, no. 3, pp. 3–4,11–12.

Lynch, M.J. (2013). "Reflecting on green criminology and its boundaries: comparing environmental and criminal victimization and considering crime from an eco-city perspective." In South, N. and A. Brisman (eds), *The Routledge international handbook of green criminology*, London: Routledge, pp. 43–57.

Lynch, M.J. and Barrett, K.L. (2015). "Death matters: victimization by particle matter from coal fired power plants in the US, a green criminological view." *Critical Criminology*, vol. 23, no. 3, pp. 219–234.

Lynch, M.J. and Stretesky, P.B. (1998). "Uniting class and race with criticism through the study of environmental justice." *The Critical Criminologist*, vol. 1, pp. 4–7.

Lynch, M.J. and Stretesky, P.B. (1999). "Clarifying the analysis of environmental justice: further thoughts on the critical analysis of environmental justice issues." *The Critical Criminologist*, vol. 9, no. 3, pp. 5–8.

Lynch, M.J. and Stretesky, P.B. (2001). "Toxic crimes: examining corporate victimization of the general public employing medical and epidemiological evidence." *Critical Criminology*, vol. 10, no. 3, pp. 153–172.

Lynch, M.J. and Stretesky, P.B. (2012). "Native Americans, social and environmental justice: implications for criminology." *Social Justice*, vol. 38, no. 3, pp. 34–54.

Lynch, M.J. and Stretesky, P.B. (2014). *Exploring green criminology: toward a green criminological revolution*. Surrey, UK: Ashgate.

Lynch, M.J., Stretesky, P.B. and Burns, R.G. (2004a). "Determinants of environmental law violation fines against oil refineries: race, ethnicity, income and aggregation effects." *Society and Natural Resources*, vol. 17, no. 4, pp. 333–347.

Lynch, M.J., Stretesky, P.B. and Burns, R.G. (2004b). "Slippery business: race, class and legal determinants of penalties against petroleum refineries." *Journal of Black Studies*, vol. 34, no. 3, pp. 421–440.

Lynch, M.J., Long, M.A., Barrett, K.L. and Stretesky, P.B. (2013). "Is it a crime to produce ecological disorganization? Why green criminology and political economy matter in the analysis of global ecological harms." *British Journal of Criminology*, vol. 55, no. 6, pp. 997–1016.

Lynch, M.J., Long, M.A. and Stretesky, P.B. (2015). "Anthropogenic development drives species to be endangered: capitalism and the decline of species." In Sollund, R.A. (ed.), *Green harms and crimes: critical criminology in a changing world*, NY: Palgrave-Macmillan, pp. 117–146.

Ngoc, A.C. and Wyatt, T. (2013). "A green criminological exploration of illegal wildlife trade in Vietnam." *Asian Journal of Criminology*, vol. 8, no. 2, pp. 129–142.

Nurse, A. (2013). "Privatising the green police: the role of NGOs in wildlife law enforcement." *Crime, Law and Social Change*, vol. 59, no. 3, pp. 305–318.

Petrossian, G.A. (2015). "Preventing illegal, unreported and unregulated (IUU) fishing: a situational approach." *Biological Conservation*, vol. 189, pp. 39–48.

Petrossian, G.A. and Clarke, R.V. (2014). "Explaining and controlling illegal commercial fishing: an application of the CRAVED theft model." *British Journal of Criminology*, vol. 54, no. 1, pp. 73–90.

Petrossian, G.A., Weis, J.S. and Pires, S.F. (2015). "Factors affecting crab and lobster species subject to IUU fishing." *Ocean & Coastal Management*, vol. 106, pp. 29–34.

Pires, S.F. and Clarke, R.V. (2011). "Sequential foraging, itinerant fences and parrot poaching in Bolivia." *British Journal of Criminology*, vol. 51, no. 2, pp. 314–335.

Pires, S.F. and Clarke, R.V. (2012). "Are parrots CRAVED? An analysis of parrot poaching in Mexico." *Journal of Research in Crime and Delinquency*, vol. 49, no. 1, pp. 122–146.

Pires, S.F. and Petrossian, G.A. (2015). "Understanding parrot trafficking between illicit markets in Bolivia: an application of the CRAVED model." *International Journal of Comparative and Applied Criminal Justice*, vol. 40, no. 1, pp. 63–77.

Ringquist, E. (1998). "A question of justice: equity in environmental litigation, 1974–1991." *Journal of Politics*, vol. 60, pp. 1148–1165.

Rockström, J., Steffen, W., Noone, K., Persson, Å., Chapin, F.S., Lambin, E.F., Lenton, T.M. et al. (2009). "A safe operating space for humanity." *Nature*, vol. 461, no. 7263, pp. 472–475.

Ruggiero, V. and South, N. (2010a). "Green criminology and dirty collar crime." *Critical Criminology*, vol. 18, no. 4, pp. 251–262.

Ruggiero, V. and South, N. (2010b). "Critical criminology and crimes against the environment." *Critical Criminology*, vol. 18, no. 4, pp. 245–250.

Ruggiero, V. and South, N. (2013a). "Green criminology and crimes of the economy: theory, research and praxis." *Critical Criminology*, vol. 21, no. 3, pp. 359–373.

Ruggiero, V. and South, N. (2013b). "Toxic state–corporate crimes, neo-liberalism and green criminology: the hazards and legacies of the oil, chemical and mineral industries." *International Journal for Crime, Justice and Social Democracy*, vol. 2, no. 2, pp. 12–26.

Schnaiberg, A. (1980). *The environment: from surplus to scarcity*. NY: Oxford University Press.

Sexton, K., Gong, H., Bailar, J.C., Ford, J.G., Gold, D.R., Lambert, W.E. and Utell, M.J. (1993). "Air pollution health risks: do class and race matter?" *Toxicology and Industrial Health*, vol. 9, no. 5, pp. 844–878.

Sollund, R. (2013a). "Animal trafficking and trade: abuse and species injustice." In Westerhuis, D., R. Walters and T. Wyatt (eds), *Emerging issues in green criminology: exploring power, justice and harm*, New York: Palgrave-MacMillan, pp. 72–92.

Sollund, R. (2013b). "Oil production, climate change and species decline: the case of Norway." In White, R. (ed.), *Climate change from a criminological perspective*, NY: Springer, pp. 135–147.

South, N. (2007). "The 'corporate colonisation of nature': bio-prospecting and bio-piracy and the development of green criminology." In Beirne, P. and N. South (eds), *Issues in green criminology: confronting harms against environments, humanity and other animals*, Oxford, UK: Willan, pp. 230–247.

Stretesky, P.B. and Hogan, M.J. (1998). "Environmental justice: an analysis of superfund sites in Florida." *Social Problems*, vol. 45, no. 2, pp. 268–287.

Stretesky, P.B. and Lynch, M.J. (1998). "Corporate environmental violence and racism." *Crime, Law and Social Change*, vol. 30, no. 2, pp. 163–184.

Stretesky, P.B. and Lynch, M.J. (1999). "Environmental justice and the prediction of distance to accidental chemical releases in Hillsborough County, Florida." *Social Science Quarterly*, vol. 80, no. 4, pp. 830–846.

Stretesky, P.B. and Lynch, M.J. (2002). "Environmental hazards and school segregation in Hillsborough County, Florida, 1987–1999." *The Sociological Quarterly*, vol. 43, no. 4, pp. 553–573.

Stretesky, P.B. and Lynch, M.J. (2011). "Coal strip mining, mountain top removal and the distribution of environmental violations across the United States, 2002–2008." *Landscape Research*, vol. 36, no. 2, pp. 209–230.

Stretesky, P.B., Long, M.A. and Lynch, M.J. (2013a). *The treadmill of crime: political economy and green criminology*. Abingdon, UK: Routledge.

Stretesky, P.B., Long, M.A. and Lynch, M.J. (2013b). "Does environmental enforcement slow the treadmill of production? The relationship between large monetary penalties, ecological disorganization and toxic releases within offending corporations." *Journal of Crime and Justice*, vol. 36, no. 2, pp. 235–249.

van Solinge, T.B. (2010). "Deforestation crimes and conflicts in the Amazon." *Critical Criminology*, vol. 18, no. 4, pp. 263–277.

van Solinge, T.B. (2013). "Equatorial deforestation as a harmful practice and a criminological issue." In White, R. (ed), *Global environmental harm*, Devon, UK: Willan, pp. 20-35.

van Solinge, T.B. (2014). "Researching illegal logging and deforestation." *International Journal for Crime, Justice and Social Democracy*, vol. 3, no. 2, pp. 36–49.

Walters, R. (2007). "Crime, regulation and radioactive waste in the United Kingdom." In Beirne, P. and N. South (eds), *Issues in green criminology: confronting harms against environments, humanity and other animals*, Oxford, UK: Willan, pp. 186–205.

Walters, R. (2010a). "Eco crime." In Muncie, J., D. Talbot, and R. Walters (eds), *Crime: local and global*, Cullompton, UK: Willan, pp. 174–208.

Walters, R. (2010b). "Toxic atmospheres air pollution, trade and the politics of regulation." *Critical Criminology*, vol. 18, no. 4, pp. 307–323.

Wellsmith, M. (2011). "Wildlife crime: the problems of enforcement." *European Journal on Criminal Policy and Research*, vol. 17, no. 2, pp. 125–148.

White, R. (2003). "Environmental issues and the criminological imagination." *Theoretical Criminology*, vol. 7, no. 4, pp. 483–506.

White, R. (2007). "Green criminology and the pursuit of social and ecological justice." In Beirne, P. and N. South (eds), *Issues in green criminology: confronting harms against environments, humanity and other animals*, Oxford, UK: Willan, pp. 32–54.

White, R. (2008). *Crimes against nature: environmental criminology and ecological justice*. Cullompton, UK: Willan.

White, R. (2010). "A green criminology perspective." In McLaughlin, E. and T. Newburn (eds), *The SAGE handbook of criminological theory*, Beverly Hills, CA: Sage, pp. 410–426.

Wyatt, T. (2014a). "Non-human animal abuse and wildlife trade: harm in the fur and falcon trades." *Society & Animals*, vol. 22, no. 2, pp. 194–210.

Wyatt, T. (2014b). "The Russian Far East's illegal timber trade: an organized crime?" *Crime, Law and Social Change*, vol. 61, no. 1, pp. 15–35.

York, R., Rosa, E.A. and Dietz, T. (2003). "Footprints on the earth: the environmental consequences of modernity." *American Sociological Review*, vol. 68, no. 2, pp. 279–300.

Zaitch, D., van Solinge, T.B. and Müller, G. (2014). "Harms, crimes and natural resource exploitation: a green criminological and human rights perspective on land-use change." In Bavinck, M., L. Pellegrini and E. Mostert (eds), *Conflicts over natural resources in the global south*, Boca Raton, FL: CRC Press, pp. 91–108.

35

URBAN PARKS, GARDENS AND GREENSPACE

Jason Byrne

Introduction

Why do some people enjoy better environmental quality than others? In an era of global rapid urbanization, planetary-scale environmental changes, and increasing social polarization, finding answers to this question is a compelling environmental justice concern. Much environmental justice research has focused on how marginalized and vulnerable communities (e.g. people of colour and low-income earners) have been disproportionately exposed to environmental harms, including landfills, polluting factories and contaminated sites (see Chapters 25 and 26). Since the late 1970s, a growing body of research has also considered how environmental benefits (e.g. safe, clean and healthy air, water and food) might be unequally distributed among urban populations. Urban greenspaces are an example.

The term 'greenspace' refers to a wide variety of natural and human-modified areas including remnant landscapes (e.g. forests, national parks), as well as parklands (e.g. urban parks, botanic gardens, cemeteries and commons) and green infrastructure (e.g. street trees, green alleys, green roofs and green walls etc.) (Wolch et al. 2014). For the purpose of this chapter the term 'public open space' in included in this definition, recognizing there are distinctions (i.e. some open spaces such as squares or plazas may not be very green; some parks are private).

Research on urban greenspace has generally demonstrated that marginalized and vulnerable communities (e.g. the urban poor, oppressed ethno-racial groups) have less access to parks, forests and community gardens. While much of this research has been concentrated in North America, recent research in Europe, Asia, Latin America (see Chapter 44), Australia, the Middle East and South Africa has confirmed similar disparities (acknowledging regional differences) (Boone et al. 2009; Astell-Burt et al. 2014). Even where the spatial distribution of greenspace appears equitable, researchers have found disparities in greenspace quality (facilities, condition) and/or levels of service provision (e.g. events, security). These patterns have been observed in cities within the developing and the developed world. Additionally, some scholars contend that certain greenspaces may be implicitly coded as the preserve of dominant ethno-racial groups, functioning to exclude others (Byrne 2012; Finney 2014). And patterns of funding for greenspace acquisition and management may reflect and entrench ethno-racial and income-based disparities (Wolch et al. 2005).

This chapter concisely overviews the history, methods and theories of environmental injustice research on urban greenspace, touching upon some issues related to activism. The chapter examines current debates and controversies, and provides insights into some of the challenges facing both researchers and policy-makers. Attention is given to intersecting issues of environmental inequality such as climate injustice, food insecurity and ecosystem services, functions and benefits. The chapter concludes by pointing to emerging issues and future research directions, highlighting the strengths and weaknesses of different conceptual understandings and methodological approaches.

Types of urban greenspace and why they matter

A wide variety of different types of greenspaces can be found in cities. Many greenspaces such as urban parks, playgrounds, and sporting fields originated alongside the institution of urban planning and are usually regarded as hallmarks of liveable places. Together with urban forests, community gardens, city farms, playing fields, golf courses, promenades, wildlife reserves, protected areas (e.g. water catchments), botanical gardens, regional parks, zoos, squares, street verges and stream- and riverbanks (to list a few), these spaces comprise a verdant tapestry that suffuses most cities (Byrne and Sipe 2010).

However, this diversity of urban greenspace belies a now indisputable fact – not all urban residents have the same levels of greenspace access, nor do they enjoy the same standard of facilities or the same level of service provision (e.g. sporting programmes, festivals, cultural activities etc.). Moreover, under pressures related to global urbanization, many cities are struggling to preserve existing greenspaces; sometimes they are regarded as secondary concerns (Lin et al. 2015; Rupprecht et al. 2015). Many greenspaces may simply fail the contemporary needs of diverse urbanites. These issues are environmental justice concerns.

Much of the recent environmental justice greenspace literature has focused on the myriad 'ecosystem services' afforded to urban residents (Lin et al. 2015). While a detailed review of ecosystem services, functions and benefits is beyond the scope of this chapter, it is useful to briefly consider some of the benefits that urban greenspace is said to provide. For instance, scholars have found that urban greenspaces can foster increased levels of physical activity (e.g. walking, cycling), thus providing health-regulation functions (Mavoa et al. 2015). Greenspaces such as parks and urban forests can improve socialization, aid mental restoration, reduce stress and combat anxiety and depression (Wolch et al. 2014; Lin et al. 2015; Thompson et al. 2012). Research shows that residents need not *use* greenspace to derive some of these benefits. Just a view of urban greenery may improve mental health (Kaplan 2001). Limited greenspace access may thus harm some urban populations (Kuo 2001). There are also environmental repercussions if generations of children grow up in cities without biodiverse landscapes (Rupprecht et al. 2015). And greenspaces may have an important role to play in limiting some climate change impacts (e.g. heat, flooding, wind) (Jim et al. 2015) and food insecurity (Wolch et al. 2014). The design, allocation and management of urban greenspaces raise important justice considerations, across scales, in diverse settings and among different social groups.

Conceptions of justice in greenspace research

Leisure researchers have long observed ethno-racial and income-based differences in how people use parks and other recreation spaces (Johnson et al. 1998), proffering three explanations. First, people's income can determine their residential location and hence their physical access to greenspace, their ability to travel, and their capacity to afford some recreational activities (e.g. golf, polo or tennis) (More 2002). Second, cultural differences may shape recreation preferences; different ethno-racial groups may prefer different types of greenspace based on their shared values (Floyd and Johnson 2002). And third, park administrators and urban service providers (e.g. planners) can influence the provision of facilities (e.g. basketball hoops) and maintenance levels (Wicks and Crompton 1986), and hence which recreational opportunities are available to diverse communities.

Over the past three decades, a growing cadre of scholars has observed that such user-focused explanations are too simplistic. They ignore historical and spatial processes affecting 'who gets what', and hence how and why greenspaces are used in different ways. Urban greenspaces are often inequitably spatially and socially distributed. Affluent and socially dominant communities seemingly enjoy better levels of access, better quality facilities, and better-maintained facilities than those in so-called disadvantaged areas (McIntyre et al. 1991). These patterns are found in both developed and developing countries (Lara-Valencia and García-Pérez 2015; Willems and Donaldson 2012; Omer and Or 2005; Yin and Xu 2009; Yasumoto et al. 2014). The key questions are not so much about what patterns are observable but ask which factors and processes determine disparities in greenspace availability and accessibility, and why are these outcomes inequitable and/or unjust? Studies on parks, urban forests and community gardens have provided some answers.

Parks

Most greenspace scholarship has examined environmental inequalities stemming from park accessibility. Early research was focused on service provision, which has remained a strong theme within the literature. Since the late 1970s, following political economy analyses illuminating social injustices in built environments, scholars have investigated whether different urban populations receive the same level of park access. Access has been defined in multiple ways, however, such as the number of parks in a given locale, the size or area of these parks, the physical distance from houses (e.g. 400 metres – straight line vs. street network), or the amount of park space per capita (e.g. 1 ha per 1000 residents). Conceptions of equity have been premised on the notion that irrespective of class, race, ethnicity or other axes of difference (e.g. disability, gender), everyone should have the same level of access to parks (Lucy 1981; Nicholls 2001).

As Gordon Walker (2009) has observed for other environmental justice studies, this 'first generation' of greenspace research focused on the unequal distribution of environmental benefits or 'outcome equity'. The conception of justice here – *distributive justice* – is spatial; all urban residents should have the same amount of park space. But distributive justice does not fully account for disparities in service provision. Much of the literature has tended to treat parks as homogenous entities (i.e. same size, design, facilities, vegetation type and cover etc.). Analyses of park distribution often fail to account for the heterogeneity of park spaces and for historical, institutional, political or social reasons why some communities have better park access than others (Byrne and Wolch 2009).

For example, Carolyn Finney demonstrated how a history of racial discrimination in the United States (US) resulted in the unequal distribution of facilities such as parks within communities of colour (Finney 2014). Legislation once banned people of colour from using White park spaces; non-Whites had separate facilities – typically fewer in number and of inferior quality (Byrne and Wolch 2009). Moreover, vicious attacks by White supremacists made people of colour fearful of venturing into naturalistic or wildland greenspaces, such as national parks, leaving a legacy of unequal use (also see Butler and Richardson 2015).

Brownlow's (2006) study of Fairmont Park in Philadelphia highlighted similar processes at work. He found that a reduction in park funding and management, associated with local government fiscal constraints, disproportionately impacted non-Whites. Neighbourhoods of colour surrounding Fairmont Park experienced increasingly violent crimes as vegetation management deteriorated. A second conception of justice – recognition justice – is at work here. Premised on the understanding that "space . . . is constructed by and through social practices" (Walker 2009: 615), recognition justice perspectives show how processes of denigration, disrespect and violence systematically condition people and places. To undo the harmful impacts of these processes (termed 'process equity') requires understanding how historical practices have configured contemporary social and biophysical environments.

Burgess et al. (1988) were among the first scholars to identify how social practices can shape access to greenspace benefits (and costs). Although their study was not explicitly framed as environmental justice research, they considered many issues that have since shaped the research agenda. For example, they observed that multiple axes of difference (i.e. gender, age, disability) can shape people's experiences of diverse greenspaces: women, older people and people of colour often fear urban wildlands (see Figure 35.1). People's perceptions of greenspaces may thus affect their choices to use them – or not (Rossi et al. 2016).

Some studies have shown how park management can be exclusionary. Park signs, language, and online sources of park information can discriminate against non-native speakers, people with reduced access to computers or those who are illiterate (or blind). A study of an urban national park in Los Angeles revealed that Latinos were disenfranchised from park information

Figure 35.1   Jogger in Westerpark, Amsterdam. Researchers have found that some people of colour feel unwelcome or afraid in certain green spaces. (Source: author)

because much of the promotional material was only available in English, giving the impression that it was an 'Anglos-only' park (Byrne et al. 2009; Byrne 2012). Even where the *amount* of parkland appears to be equitable, discriminatory practices embedded in park management can limit the type and quality of facilities available to marginalized and vulnerable communities. Congested parks, for example, effectively have lower levels of service provision (Boone et al. 2009; Sister et al. 2010).

Fewer studies have examined participatory democracy in park design, provision and management. A recent review (Fors et al. 2015) noted that residents are seldom able to meaningfully contribute to the process or outcomes of park design and management. Here a third dimension of justice is at work – participatory and procedural justice (Walker 2009) (see Chapter 9). Among the few studies of participation in park design, Smiley et al. (2016) found that people of colour in the US expressed a preference for better maintained parks with good quality facilities, and a desire for more effective participation, over simply better park access (also see Abbasi et al. 2016 for the UK). Besides ethno-racial composition and socio-economic disadvantage, scholars have also found that disabled people are seldom involved in park design; all-abilities parks are rare (Evcil 2010; Seeland and Nicolè 2006; Woolley 2013). And a growing number of studies have begun to consider how children, teenagers and older people may be systematically excluded from some parks because designers rarely consider their needs (Gardsjord et al. 2014; Gearin and Kahle 2006; Gilliland et al. 2006; Veitch et al. 2007; Kaczynski et al. 2009). But parks are not the only form of greenspace where environmental inequalities occur.

Urban forests

Environmental justice researchers have also examined inequalities in the distribution of urban trees, urban forests, and roadside and spontaneous vegetation. Many studies have found a close correlation between trees and income: poor and disadvantaged places generally have fewer trees and/or less canopy coverage. Some studies have also found ethno-racial disparities in tree cover (Landry and Chakraborty 2009), where ethno-racially marginalized or vulnerable neighbourhoods tend to have less canopy cover and fewer trees – even on publicly managed land (e.g. street verges, parks). And while the research has been dominated by North American studies, examples from Australia, Brazil, South Africa and China reveal similar inequities. Some scholars have noted an interaction effect between tree cover, income, ethno/racial background, education and home-ownership (Landry and Chakraborty 2009; Pham et al. 2012). Tree canopy cover is often positively associated with higher income neighbourhoods where residents have better education, higher levels of home-ownership and are predominantly White. Some scholars note that tree canopies are inherited from previous generations – as most trees take a long time to grow. Others have pointed to the influence of climate and water availability (Schwarz et al. 2015). Yet research across cities of varying age, in older and younger neighbourhoods, traversing different densities and in different climatic zones often reveals the same disparities. A crucial question then is what can explain these findings?

Like parks, urban trees have been found to provide a wide array of benefits including modulating ambient temperatures, intercepting pollution, reducing storm-water runoff, fostering physical activity (i.e. walkability) and improving mental health. Benefits such as temperature modulation are becoming increasingly important as climate change impacts cities. Scholars have offered several explanations for disparities in access to tree benefits. Wealthier residents can afford to live in greener neighbourhoods due to their purchasing power (Heynen et al. 2006; Perkins et al. 2004). Neighbourhoods with better canopy coverage and more trees usually have higher property values. Wealthier residents also tend to be better educated; higher education levels appear to be associated with better knowledge of tree benefits, so affluent residents may seek out greener neighbourhoods or plant more trees (Schwarz et al. 2015). Researchers have noted that affluent residents can afford higher management costs of trees (e.g. watering, pruning) and lower-income residents may be circumspect about negative tree impacts (e.g. windthrow, pavement uplift, asthma). Wealthier residents tend to be better politically incorporated, thus able to leverage elected representatives and government departments to plant and maintain trees in their neighbourhoods (Landry and Chakraborty 2009). And affinity for trees may be linked to cultural or class predilections. But observed disparities cannot be fully explained by cultural affinities, class proclivities, perceived returns on investment and the like. Nor can differences be fully accounted for due to neighbourhood age or population density. Many observed disparities occur on public lands, and similar patterns of inequality occur for both remnant and ornamental vegetation. Community gardens are instructive.

Community gardens

International studies, but particularly those in the US, have found associations between the location of community gardens and spatial and social disadvantage (Milbourne 2012). In many countries, these gardens are found in inner city areas, characterized by impoverished and often immigrant communities (Saldivar-Tanaka and Krasny 2004; Jermé and Wakefield 2013). Community gardens are typically established by disadvantaged groups, acting informally, who cooperate to collectively use public or private land to grow food (Guitart et al. 2012). Community gardens are cooperatively shared, unlike allotment gardens, which although public, only have plots for individual cultivation. Often community gardens have evolved over time; gardeners typically self-organize to grow food, removing trash from landscapes blighted by economic and political disinvestment.

In contrast to parks and trees, community garden research has tended to address issues other than accessibility and distributional equity. Many community garden studies have occurred at the neighbourhood scale, focusing on a smaller number of gardens, as opposed to municipal or metropolitan-scale park and tree studies. Moreover, community gardens tend to be spaces of production, not consumption. International research has found that for some neighbourhoods, these gardens fulfil subsistence and livelihood functions, enabling gardeners to sell their produce for profit, such as gardens in South Africa, Brazil, the Philippines, Zambia and Mexico (Guitart et al. 2012). And research has demonstrated that poverty alleviation is a primary function of community garden development, because gardeners can often grow fresh produce cheaper than it can be purchased.

Much of the community garden literature has highlighted issues surrounding insecure tenure. Although some gardens have been in operation for decades, because they are located on private land or neglected municipal land, when a municipal government or landowner decides to pursue development, gardeners are at risk of dispossession. There are many examples (e.g. from New York and Los Angeles) where gardens have been shut down, and gardeners forcibly evicted as land is redeveloped for other purposes (Irazabal and Punja 2009; Staeheli et al. 2002). Some studies have found a cultural politics of nature at work, where at-risk gardens have been rescued by non-profit organizations, only to be forced to comply with their visions of greenspace (i.e. manicured flowerbeds, formal facilities, reduced cultivation) (Eizenberg 2012). Much of the research has been framed around the 'rights to the city'.

Unlike parks and street trees, community gardens have been found to be important in developing social capital among marginalized and disadvantaged populations because they bring people together for knowledge exchange and skills development (Saldivar-Tanaka and Krasny 2004). These gardens also enable migrant communities to retain cultural connections through the food plants they grow. For instance, a study of what was grown among three gardens with culturally homogenous members in Toronto, Canada (Sri Lankan, African and Chinese), found that plants grown were related to gardeners' country of origin (Baker 2004). However, an important issue now salient in the literature is how community gardens established on brownfield sites may inadvertently expose gardeners to carcinogenic substances from soil contamination (e.g. hydrocarbons or heavy metals). There are examples from cities such as Oakland in California, Berlin in Germany, and Nairobi in Kenya where food safety concerns have been raised (McClintock 2012; Gallaher et al. 2013; Säumel et al. 2012). Paradoxically, gardens developed to nurture and support vulnerable populations may be slowly poisoning them.

Current debates, controversies and emerging research

Research has advanced beyond basic considerations of service distribution (i.e. accessibility, proximity and location) to consider factors explaining *how* the character of greenspaces (e.g. design, landscaping, facilities, funding) may affect both who uses these spaces and the potential ecosystem service benefits they derive from them. And the scope of greenspaces considered has expanded beyond parks, forests and community gardens to more recently consider beaches, informal green spaces (e.g. railway corridors and street verges), and so-called 'blue spaces' (Rupprecht et al. 2015; Montgomery et al. 2015). There has also been a growing awareness of inter-linkages between multiple environmental justice issues and multiple greenspace types. For example, eco-gentrification (see Chapter 36), where actions to improve neighbourhood greenspace (e.g. increased tree cover) can have pernicious consequences, such as increasing property values (Wolch et al. 2014). The eco-gentrification of greenspaces to improve the international competitiveness of cities and attract knowledge workers can dispossess vulnerable people from their neighbourhoods, turning greenspace systems into revenue generating mechanisms (Wolch et al. 2014). Efforts to demonstrate the financial worth of greenspaces – as an argument for their preservation – may be co-opted by neoliberal ideology, and contribute to the commodification of greenspaces (Heynen et al. 2006).

Concerns about healthy cities and urban populations are also related to concerns about the financial viability of cities – and their greenspaces. Healthy cities are said to be able to attract footloose capital and skilled populations of professionals, while reducing the financial burden of sedentary and unwell populations. Greenspaces are readily co-opted into a neoliberal logic. Local ecologies are reimagined as 'natural capital', providing a range of functions, services and benefits. Consequently, greenspaces are increasingly being required to demonstrate that they can 'pay their way'. Otherwise, they may be deemed surplus to a city's needs and sold to bolster municipal coffers. There is a real risk here that marginalized and vulnerable populations, like surplus parklands, are recast as 'liabilities'.

Another issue with the ecosystem services, functions and benefits approach is that it often neglects the costs/disbenefits associated with some types of greenspace. These costs may be disproportionately concentrated in some places, and may disproportionately harm marginalized and vulnerable communities (although research is presently underdeveloped) (Roy et al. 2012). For example, re-greening cities using the wrong species could result in higher levels of pollen, leading to a higher incidence of asthma. It might also heighten the risk of wildfire, or wind-throw during storms. Urban greening could lead to higher populations of species regarded as 'pests', leading to damage to infrastructure, housing and potentially to disease. We lack studies on these important topics. Do biodiversity benefits of urban forests accrue to some urban populations and not others (Shanahan et al. 2014)? And humans are not the only inhabitants of cities. Ecological justice suggests that it is important we make space in cities (and in our research) for non-human others.

A final problem is that researchers have tended to treat environmental justice communities as passive victims, robbing them of agency in engaging with these complex issues and working through solutions. Some of the community garden literature has pointed to a White, liberal tendency to disempower environmental justice communities by forcing them to comply with particular ideals and ideologies of greenspace. By unquestioningly accepting the logic of ecosystem services, researchers risk naturalizing this neoliberal approach. Greenspaces may be valued only for their services and functions. This ignores insights from the place-attachment literature showing that urban residents develop strong affinities for some greenspaces, irrespective of their economic value (see Chapter 47). There are few if any greenspace studies examining issues related to Indigenous peoples, such as the protection of sacred sites or native title conflicts around parks or forests. And the increasingly sophisticated technologies and analyses characterizing newer greenspace research may potentially alienate communities, activists and policy-makers, who are unable to access, understand and/or use complex statistics and indices. Moreover, it keeps knowledge in the hands of experts, who use reductionist ontologies to negotiate the politics of greenspace preservation and management, ignoring and marginalizing local knowledge and social and environmental innovations.

Conclusion

This chapter has overviewed the many environmental justice considerations entailed in urban greenspace research. It has concisely assessed the research, discussed key conceptual and methodological concerns, and explored the multiple dimensions of greenspace provision and use. However, there are still some areas that have been poorly theorized and studied, and the research agenda remains incomplete. For example, we now know that access to greenspace is critical for human physical and mental health. But as Low and Gleeson (1998) observed some two decades ago, environmental justice research is often deeply anthropocentric. We tend to forget that we share our cities with a stunning array of other species – and that we ourselves are colonies of organisms. A human-centred perspective can blind us to other conceptions of justice such as ecological justice (Baxter 2005), and can prevent us from understanding complex socio-ecological interrelationships between humans and the biogeochemical world on which we depend (Byrne et al. 2014).

As Julian Agyeman and others (2002) have pointed to in their conception of *just sustainabilities* (see Chapter 14), environmental justice researchers and theorists must better understand how socio-ecological systems generate different types of injustice and how in turn remedies must demonstrate a sophisticated approach to problem-solving, lest they substitute one form of injustice for another. For example, efforts to retrofit green infrastructure to deprived neighbourhoods can trigger eco-gentrification, displacing marginalized and vulnerable communities (Wolch et al. 2014). Efforts to stop sprawl and accommodate burgeoning urban populations by increasing residential densities may destroy pockets of urban greenspace and concentrate vulnerable populations in hotter places, forcing up energy expenditure and increasing heat-related mortality and morbidity (Byrne and Portanger 2014).

Building community gardens to improve food security may inadvertently expose disadvantaged populations to soil contamination or may function as a type of middle-class welfare unless managed properly.

Methodological issues are important too. Over time, environmental justice research on urban greenspace has become increasingly sophisticated, partly in response to criticisms levelled at earlier research (Landry and Chakraborty 2009). Improvements in technology and better access to data have overcome some limitations. Researchers have increasingly combined remote sensing (satellite data, high resolution aerial photography and LiDAR) with population census data, and using Geographic Information Systems have interrogated spatial relationships and developed indices of disadvantage. Yet such broad-scale analysis can mask local inequalities.

The appropriate unit of analysis is important. Most researchers now recognize the modifiable areal unit problem, where the area chosen for analysis (e.g. census block, neighbourhood) can affect the findings of the research (see Chapter 16). Researchers have also adopted advanced spatial statistics to overcome problems with spatial autocorrelation (see Chapter 17) and analyses have begun to consider a much wider range of explanatory variables. Nonetheless, there still remains some debate about differences in problem definition and differences in the operationalization of 'equity'. For example, different results are found if a researcher investigates perceived as opposed to physical access to greenspace, or uses straight line (Euclidean distance) versus network distance. And different patterns of inequality emerge if researchers assess distance, area, facilities, funding, crowding, temporal concerns, and/or vegetation cover. Studies must employ a wider range of methods to manage such variability.

While quantitative studies have burgeoned in recent years, there are limitations to what they can reveal; they are less adept at explaining why inequities occur. More recently there have been signs that qualitative approaches are being adopted (see Chapters 19, 24). For example, residents have been interviewed about their attitudes and values towards trees (Kirkpatrick et al. 2012), and tree managers about their policies and practices (Kirkpatrick et al. 2013). Archival research, focus groups, time diaries and text analyses are beginning to illuminate some of the processes that drive observed disparities. One of the key weaknesses of much of this research, though, has been the absence of a temporal or historical dimension (Maroko et al. 2009 is a notable exception). And much of the research to date has been at a broader scale.

To date, little greenspace research has questioned codification of standards for greenspace provision. Few scholars have challenged the metrics for assessing 'who gets what and how'. While a good deal of effort has been expended on investigating whether urban populations are properly serviced by prescribed greenspace standards (such as x hectares per capita, or y acres per 1000 residents), few scholars and practitioners have paused to question where these standards originated, how they were derived, and whether they are appropriate (Byrne and Sipe 2010). Nor have we done a particularly good job of assessing the quality and character of greenspaces. Not all parks are the same, nor are all community gardens environmental equals. Street trees and urban forests are composed of many species – with specific benefits and costs – yet scholars have tended to overlook their heterogeneity. And rarely have scholars questioned different conceptions of fairness and justice that underpin appraisals of greenspace provision and use. Which types of nature are being promulgated, for whom, why and how? What purposes do they serve? Whose interests are being met? Who gets to decide what is appropriate and how can the broader community participate in greenspace decision-making, if at all? Race, ethnicity and income are only some of the axes of difference that can (re)produce environmental inequalities and researchers need to become more attentive to others – including gender, disability, sexual preference, and even species. Finally, few studies have considered integrated metropolitan greenspace systems – parks, trees, gardens, liminal spaces and the like. There is still much to learn about greenspace and environmental equity.

References

Abbasi, A., Alalouch, C. and Bramley, G. 2016. Open space quality in deprived urban areas: user perspective and use pattern. *Procedia–Social and Behavioral Sciences*, 216, 194–205.

Agyeman, J., Bullard, R.D. and Evans, B. 2002. Exploring the nexus: bringing together sustainability, environmental justice and equity. *Space and Polity*, 6(1), 77–90.

Astell-Burt, T., Feng, X., Mavoa, S., Badland, H.M. and Giles-Corti, B. 2014. Do low-income neighbourhoods have the least green space? A cross-sectional study of Australia's most populous cities. *BMC Public Health*, 14(1), online.

Baker, L.E. 2004. Tending cultural landscapes and food citizenship in Toronto's community gardens. *Geographical Review*, 94(3), 305–325.

Baxter, B. 2005. *A Theory of Ecological Justice*. Abingdon, Oxon: Routledge.

Boone, C.G., Buckley, G.L., Grove, J.M. and Sister, C. 2009. Parks and people: an environmental justice inquiry in Baltimore, Maryland. *Annals of the Association of American Geographers*, 99(4), 767–787.

Brownlow, A. 2006. An archaeology of fear and environmental change in Philadelphia. *Geoforum*, 37(2), 227–245.

Burgess, J., Harrison, C.M. and Limb, M. 1988. People, parks and the urban green: a study of popular meanings and values for open spaces in the city. *Urban Studies*, 25, 455–473.

Butler, G. and Richardson, S. 2015. Barriers to visiting South Africa's national parks in the post-apartheid era: black South African perspectives from Soweto. *Journal of Sustainable Tourism*, 23(1), 146–166.

Byrne, J. 2012. When green is White: the cultural politics of race, nature and social exclusion in a Los Angeles urban national park. *Geoforum*, 43(3), 595–611.

Byrne, J. and Portanger, C. 2014. Climate change, energy policy and justice: a systematic review. *Analyse & Kritik*, 36(2), 315–343.

Byrne, J. and Sipe, N. 2010. *Green and Open Space Planning for Urban Consolidation – A Review of the Literature and Best Practice*. Brisbane: Griffith University, Urban Research Program.

Byrne, J., Sipe, N. and Dodson, J. 2014. *Australian Environmental Planning: Challenges and Future Prospects*. New York: Routledge.

Byrne, J. and Wolch, J. 2009. Nature, race, and parks: past research and future directions for geographic research. *Progress in Human Geography*, 33(6), 743–765.

Byrne, J., Wolch, J. and Zhang, J. 2009. Planning for environmental justice in an urban national park. *Journal of Environmental Planning and Management*, 52(3), 365–392.

Eizenberg, E. 2012. The changing meaning of community space: two models of NGO management of community gardens in New York City. *International Journal of Urban and Regional Research*, 36(1), 106–120.

Evcil, N. 2010. Designers' attitudes towards disabled people and the compliance of public open places: the case of Istanbul. *European Planning Studies*, 18(11), 1863–1880.

Finney, C. 2014. *Black Faces, White Spaces: Reimagining the Relationship of African Americans to the Great Outdoors*. Chapel Hill, NC: University of North Carolina Press.

Floyd, M.F. and Johnson, C.Y. 2002. Coming to terms with environmental justice in outdoor recreation: a conceptual discussion with research implications. *Leisure Sciences*, 24, 59–77.

Fors, H., Molin, J.F., Murphy, M.A. and van den Bosch, C.K. 2015. User participation in urban green spaces – for the people or the parks? *Urban Forestry & Urban Greening*, 14(3), 722–734.

Gallaher, C.M., Kerr, J.M., Njenga, M., Karanja, N.K. and WinklerPrins, A.M. 2013. Urban agriculture, social capital, and food security in the Kibera slums of Nairobi, Kenya. *Agriculture and Human Values*, 30(3), 389–404.

Gardsjord, H., Tveit, M. and Nordh, H. 2014. Promoting youth's physical activity through park design: linking theory and practice in a public health perspective. *Landscape Research*, 39(1), 70–81.

Gearin, E. and Kahle, C. 2006. Teen and adult perceptions of urban green space in Los Angeles. *Children Youth and Environments*, 16(1), 25–48.

Gilliland, J., Holmes, M., Irwin, J.D. and Tucker, P. 2006. Environmental equity is child's play: mapping public provision of recreation opportunities in urban neighbourhoods. *Vulnerable Children and Youth Studies*, 1(3), 256–268.

Guitart, D., Pickering, C. and Byrne, J. 2012. Past results and future directions in urban community gardens research. *Urban Forestry & Urban Greening*, 11(4), 364–373.

Heynen, N., Perkins, H.A. and Roy, P. 2006. The political ecology of uneven urban green space: the impact of political economy on race and ethnicity in producing environmental inequality in Milwaukee. *Urban Affairs Review*, 42(1), 3–25.

Irazabal, C. and Punja, A. 2009. Cultivating just planning and legal institutions: a critical assessment of the South Central Farm struggle in Los Angeles. *Journal of Urban Affairs*, 31(1), 1–23.

Jermé, E.S. and Wakefield, S. 2013. Growing a just garden: environmental justice and the development of a community garden policy for Hamilton, Ontario. *Planning Theory & Practice*, 14(3), 295–314.

Jim, C., Lo, A.Y. and Byrne, J.A. 2015. Charting the green and climate-adaptive city. *Landscape and Urban Planning*, 138, 51–53.

Johnson, C.Y., Bowker, J.M., English, D.B.K. and Worthen, D. 1998. Wildland recreation in the rural South: an examination of marginality and ethnicity theory. *Journal of Leisure Research*, 30, 101–120.

Kaczynski, A.T., Potwarka, L.R., Smale, B.J. and Havitz, M.E. 2009. Association of parkland proximity with neighborhood and park-based physical activity: variations by gender and age. *Leisure Sciences*, 31(2), 174–191.

Kaplan, R. 2001. The nature of the view from home psychological benefits. *Environment and Behavior*, 33(4), 507–542.

Kirkpatrick, J., Davison, A. and Daniels, G. 2012. Resident attitudes towards trees influence the planting and removal of different types of trees in eastern Australian cities. *Landscape and Urban Planning*, 107(2), 147–158.

Kirkpatrick, J.B., Davison, A. and Harwood, A. 2013. How tree professionals perceive trees and conflicts about trees in Australia's urban forest. *Landscape and Urban Planning*, 119, 124–130.

Kuo, F.E. 2001. Coping with poverty: impacts of environment and attention in the inner city. *Environment and Behavior*, 33(1), 5–34.

Landry, S.M. and Chakraborty, J. 2009. Street trees and equity: evaluating the spatial distribution of an urban amenity. *Environment and Planning A*, 41(11), 2651–2670.

Lara-Valencia, F. and García-Pérez, H. 2015. Space for equity: socioeconomic variations in the provision of public parks in Hermosillo, Mexico. *Local Environment*, 20(3), 350–368.

Lin, B., Meyers, J. and Barnett, G. 2015. Understanding the potential loss and inequities of green space distribution with urban densification. *Urban Forestry & Urban Greening*, 14(4), 952–958.

Low, N. and Gleeson, B. 1998. *Justice, Society, and Nature: An Exploration of Political Ecology*. London: Routledge.

Lucy, W. 1981. Equity and planning for local services. *Journal of the American Planning Association*, 47(4), 447–457.

Maroko, A.R., Maantay, J.A., Sohler, N.L., Grady, K.L. and Arno, P. 2009. The complexities of measuring access to parks and physical activity sites in New York City: a quantitative and qualitative approach. *International Journal of Health Geographics*, 8(34), 1–23 (online).

Mavoa, S., Koohsari, M.J., Badland, H., Davern, M., Feng, X., Astell-Burt, T. and Giles-Corti, B. 2015. Area-level disparities of public open space: a geographic information systems analysis in Metropolitan Melbourne. *Urban Policy and Research*, 33(3), 306–323.

McClintock, N. 2012. Assessing soil lead contamination at multiple scales in Oakland, California: implications for urban agriculture and environmental justice. *Applied Geography*, 35(1), 460–473.

McIntyre, N., Cuskelly, G. and Auld, C. 1991. The benefits of urban parks: a market segmentation approach. *Australian Parks and Recreation*, 27(1), 11–18.

Milbourne, P. 2012. Everyday (in) justices and ordinary environmentalisms: community gardening in disadvantaged urban neighbourhoods. *Local Environment*, 17(9), 943–957.

Montgomery, M.C., Chakraborty, J., Grineski, S.E. and Collins, T.W. 2015. An environmental justice assessment of public beach access in Miami, Florida. *Applied Geography*, 62, 147–156.

More, T.A. 2002. The marginal user as the justification for public recreation: a rejoinder to Crompton, Driver and Dustin. *Journal of Leisure Research*, 34(1), 103–118.

Nicholls, S. 2001. Measuring the accessibility and equity of public parks: a case study using GIS. *Managing Leisure*, 6(4), 201–219.

Omer, I. and Or, U. 2005. Distributive environmental justice in the city: differential access in two mixed Israeli cities. *Tijdschrift voor Economische en Sociale Geografie*, 96(4), 433–443.

Perkins, H.A., Heynen, N. and Wilson, J. 2004. Inequitable access to urban reforestation: the impact of urban political economy on housing tenure and urban forests. *Cities*, 21(4), 291–299.

Pham, T.-T.-H., Apparicio, P., Séguin, A.-M., Landry, S. and Gagnon, M. 2012. Spatial distribution of vegetation in Montreal: an uneven distribution or environmental inequity? *Landscape and Urban Planning*, 107, 214–224.

Rossi, S.D., Pickering, C.M. and Byrne, J.A. 2016. Not in our park! Local community perceptions of recreational activities in peri-urban national parks. *Australasian Journal of Environmental Management*, 23(3), 245–264.

Roy, S., Byrne, J. and Pickering, C. 2012. A systematic quantitative review of urban tree benefits, costs, and assessment methods across cities in different climatic zones. *Urban Forestry & Urban Greening*, 11(4), 351–363.

Rupprecht, C.D., Byrne, J.A., Ueda, H. and Lo, A.Y. 2015. 'It's real, not fake like a park': Residents' perception and use of informal urban green-space in Brisbane, Australia and Sapporo, Japan. *Landscape and Urban Planning*, 143, 205–218.

Saldivar-Tanaka, L. and Krasny, M.E. 2004. Culturing community development, neighborhood open space, and civic agriculture: the case of Latino community gardens in New York City. *Agriculture and Human Values*, 21(4), 399–412.

Säumel, I., Kotsyuk, I., Hölscher, M., Lenkereit, C., Weber, F. and Kowarik, I. 2012. How healthy is urban horticulture in high traffic areas? Trace metal concentrations in vegetable crops from plantings within inner city neighbourhoods in Berlin, Germany. *Environmental Pollution*, 165, 124–132.

Schwarz, K., Fragkias, M., Boone, C.G., Zhou, W., McHale, M., Grove, J.M., et al. 2015. Trees grow on money: urban tree canopy cover and environmental justice. *PloS One*, 10(4), e0122051.

Seeland, K. and Nicolè, S. 2006. Public green space and disabled users. *Urban Forestry & Urban Greening*, 5(1), 29–34.

Shanahan, D., Lin, B., Gaston, K., Bush, R. and Fuller, R. 2014. Socio-economic inequalities in access to nature on public and private lands: a case study from Brisbane, Australia. *Landscape and Urban Planning*, 130, 14–23.

Sister, C., Wolch, J. and Wilson, J.P. 2010. Got green? Addressing environmental justice in park provision. *GeoJournal*, 75(3), 229–248.

Smiley, K.T., Sharma, T., Steinberg, A., Hodges-Copple, S., Jacobson, E. and Matveeva, L. 2016. More inclusive parks planning: park quality and preferences for park access and amenities. *Environmental Justice*, 9(1), 1–7.

Staeheli, L.A., Mitchell, D. and Gibson, K. 2002. Conflicting rights to the city in New York's community gardens. *GeoJournal*, 58(2–3), 197–205.

Thompson, C.W., Roe, J., Aspinall, P., Mitchell, R., Clow, A. and Miller, D. 2012. More green space is linked to less stress in deprived communities: evidence from salivary cortisol patterns. *Landscape and Urban Planning*, 105(3), 221–229.

Veitch, J., Salmon, J. and Ball, K. 2007. Children's perceptions of the use of public open spaces for active free-play. *Children's Geographies*, 5(4), 409–422.

Walker, G. 2009. Beyond distribution and proximity: exploring the multiple spatialities of environmental justice. *Antipode*, 41(4), 614–636.

Wicks, B.E. and Crompton, J.L. 1986. Citizen and administrator perspectives of equity in the delivery of park services (USA). *Leisure Sciences*, 8(4), 341–365.

Willems, L. and Donaldson, R. 2012. Community neighbourhood park (CNP) use in Cape Town's townships. *Urban Forum*, 23(2), 221–231.

Wolch, J., Wilson, J.P. and Fehrenbach, J. 2005. Parks and park funding in Los Angeles: an equity-mapping analysis. *Urban Geography*, 26(1), 4–35.

Wolch, J.R., Byrne, J. and Newell, J.P. 2014. Urban green space, public health, and environmental justice: the challenge of making cities 'just green enough'. *Landscape and Urban Planning*, 125, 234–244.

Woolley, H. 2013. Now being social: the barrier of designing outdoor play spaces for disabled children. *Children & Society*, 27(6), 448–458.

Yasumoto, S., Jones, A. and Shimizu, C. 2014. Longitudinal trends in equity of park accessibility in Yokohama, Japan: an investigation into the role of causal mechanisms. *Environment and Planning A*, 46(3), 682–699.

Yin, H. and Xu, J. 2009. Spatial accessibility and equity of parks in Shanghai. *Urban Studies*, 6, 71–76.

36

URBAN PLANNING, COMMUNITY (RE)DEVELOPMENT AND ENVIRONMENTAL GENTRIFICATION

Emerging challenges for green and equitable neighbourhoods

Isabelle Anguelovski, Anna Livia Brand,
Eric Chu and Kian Goh

Introduction

Starting in the late 1970s, environmental justice (EJ) protests not only targeted toxic sites and other Locally Unwanted Land Uses (LULUs), they were also – even if implicitly – directed at providing historically marginalized neighborhoods with new or restored environmental goods and amenities. As a result, EJ mobilizations have come to support poor and minority residents in their fights for new environmental goods, including recreational spaces, parks, urban gardens, affordable fresh food, healthy housing, and improved waste management systems (Gottlieb 2005; Bullard 2007; Anguelovski 2014).

Despite these advancements, urban EJ scholarship is now at a critical juncture. The improved environmental quality of many urban distressed areas is making them newly attractive to private investors who are increasingly buying previously disinvested or abandoned buildings and transforming them into high end condominium complexes. Gradually, economically vulnerable residents (often people of color) are displaced as wealthier (often whiter) residents move in to enjoy new urban lifestyles and associated amenities that former residents had fought for (Anguelovski 2015). In many instances, municipal policymakers and politicians officially back neighborhood revitalization efforts to fulfil a broader environmental agenda that espouses sustainable urban forms such as compactness, density, mixed uses, walkability, diversity, and greening into neighborhood planning (Jabareen 2006). Their agenda also increasingly embodies new urban planning approaches such as "Complete Streets" (Brand 2015).

This process of land revaluation exemplifies what critical urban scholars have named "green gentrification" or "ecological gentrification" (Dooling 2009; Checker 2011). While municipalities present an apparently apolitical, technical agenda such as "greening" or "sustainability" (defined as the biophysical environment benefiting from ecological improvements), as has often happened in the United States, they end up sponsoring projects that result in inequitable outcomes. Increasingly, cities worldwide are engaging in green planning or smart growth agendas, with trends emerging in Canada (Bunce 2009; Jones and Ley 2016; Quastel et al. 2012; Rosol 2013), Europe (Inroy 2000), and Asia (Schuetze and Chelleri 2015). Ecological gentrification can also occur in the aftermath of low-income residents and people of color organizing to improve the environmental quality of their neighborhood (Anguelovski 2014). As a result, EJ activists now face critical challenges as they adjust their strategies in response to emerging development conflicts and affordability concerns (Faber and Kimelberg 2014; Anguelovski 2015).

In this chapter, we examine how so-called urban greening and sustainability approaches can produce or exacerbate environmental inequities and how select urban communities are beginning to organize against new injustices. Our chapter focuses on the US and examines diverse examples of urban environmental inequity produced in the context of neighborhood greening and community (re)development. In our analysis of community-based environmental revitalization in Boston, sustainable community planning in New Orleans, and disaster resilience planning in New York City, we find that historically marginalized neighborhoods are doubly exposed in the same space to environmental impacts and environmental privilege, that is the disproportionate access to environmental goods and amenities from which upper income classes and whites benefit while more marginalized groups are excluded (Park and Pellow 2011).

A new paradox for environmental justice activists

The evolution of the urban environmental justice agenda

Early EJ scholars who examined inequities in exposure to contamination and health risks (Holifield et al. 2009) revealed that, compared to white and rich communities, minority and low-income populations suffered greater environmental harm from Locally Unwanted Land Uses (LULUs), including waste sites, disposal facilities, transfer storage, incinerators, refineries, and other contaminating industries (Bryant and Mohai 1992; Bullard 1990; Downey and Hawkins 2008; Lerner 2005; Sze 2007; Mohai et al. 2009). Thirty years later, despite significant local victories and changes to federal policies with the passing of the 1994 Executive Order, statistics on inequitable risks, exposure, and impacts are still revealing inequitable trends (Bullard et al. 2007; Wilson et al. 2012; Chakraborty 2012; Brown 2013; Mohai and Saha 2015). A critical manifestation of environmental racism has been the recent water contamination conflict in Flint, Michigan, a city where 40 per cent of the residents live in poverty and 57 per cent of the population is black.

During the 2000s, EJ scholars started documenting inequalities in the allocation of environmental services such as street cleaning, public transportation, and waste collection. They demonstrated that access to trees, parks, public recreation areas, and park maintenance differed by ethnic and income groups (Dahmann et al. 2010; Pham et al. 2012), with low-income and minority groups having less access to green space than privileged residents (Landry and Chakraborty 2009; Heynen et al. 2006; Hastings 2007). The causes of such inequities go back to a long history of racial discrimination in housing rentals or sales, redlining by insurance companies or lending institutions, and restrictive covenants (Self 2003; Sugrue 2005; Hillier 2003), which have all contributed, for instance, to the fact that black residents tend to live in areas with smaller and poorly equipped or maintained parks (Boone et al. 2009).

Recent research has also focused on the development of advocacy coalitions for green, affordable, and healthy housing (Loh and Eng 2010). Demands for healthy housing have been supported by coalition building for federal funding for green jobs, training for environmental sustainability projects, or revenue redistribution from utility companies for weatherizing buildings (Fitzgerald 2010; Song 2014). Some scholars have examined community rebuilding and place attachment in environmental justice projects (Agyeman 2013; Anguelovski 2014), emphasizing the importance of place-remaking and the creation of safe havens in distressed urban communities.

Yet today, the process of transforming formerly degraded and abandoned inner city neighborhoods into greener and more environmentally safe ones is accompanied by a process of rebranding, in which neighborhoods seen as abandoned ghettoes become cultural artefacts and sites of revitalization in which developers can invest (Hyra 2008; Boyd 2008; Brand 2015). Narratives of former blight and current cultural vitality combine to enable the "creative destruction" of these neighborhoods (Anguelovski 2015). Here, environmental restoration and improvements are linked to new forms of inequality and gentrification.

Studies in land and real estate economics are indeed pointing at the existence of strong correlations between land clean-up; investment in parks or open space, waterfront redevelopment, and ecological restoration; and changes in demographic trends and neighborhood property values (Anguelovski 2015). For instance, removing sites from the Superfund list in the US has seemed to result in a 26 per cent increase in mean household income and a 31 per cent increase in the proportion of residents with a college degree (Gamper-Rabindran and Timmins 2011). If contaminated land decreases its value, it also over time can create a rent gap for investors (Bryson 2013). Much of this process takes place as part of sustainability planning to make cities greener and more liveable (Checker 2011; Anguelovski and Carmin 2011; Bai et al. 2010; Wheeler and Beatley 2009). This cycle creates what is now known as environmental gentrification, that is "the implementation of an environmental planning agenda related to public green spaces that leads to the displacement or exclusion of the most economically vulnerable human population while espousing an environmental ethic" (Dooling 2009: 630).

Within cities, public investments for urban sustainability materialize in new environmental amenities such as parks, waterfronts, or playgrounds (Dooling 2009; Hagerman 2007), which makes neighborhoods more attractive and can create new engines of economic growth and revitalization (De Sousa 2003; Low et al. 2005; Dooling 2009; Quastel 2009). Yet, redevelopment projects presented under the principles of sustainability or smart growth actually neglect to consider their impacts on working class residents and minorities (Rosan 2012; Pearsall 2008). Recent urban resilience projects have been further shown to affect or displace poor communities, while in other cases protecting and prioritizing elite groups at the expense of the urban poor (Anguelovski et al. 2016). Yet, a truly comprehensive vision of sustainability should take increases in vulnerability much more seriously (Mueller and Dooling 2011) and aim for building a just city (Connolly and Steil, 2009).

Environmental gentrification highlights the contradictions inherent in restoring urban nature through municipal greening agendas, and often utilizes a moral discourse of win-win benefits for all (Checker 2011). This contradiction questions traditional planning approaches that view green spaces as tools for social reform, public health improvement, and economic development (Dooling 2009). Municipal initiatives for sustainable and compact cities also mirror a form of capital accumulation that embodies the "depoliticization of the social" (Madra and Adaman 2014). Today, local sustainability seems to be a constituent of capital accumulation and enables the survival of capitalism (Keil 2007; Gibbs and Krueger 2007). As Swyngedouw (2007) argues, sustainability illustrates a post-political, post-democratic, and a-conflictual turn in which neoliberal governance regimes advance local greening projects while ending possibilities for a real politics of the environment and a real debate about the purposes and impacts of such projects (Anguelovski 2015). Sustainability has come to be an engineering endeavor that eludes core urban questions at the intersection of racial inequalities, social hierarchies, and environmental privilege.

New targets for environmental justice activism: municipal sustainability and greening agendas

In response, EJ activists are increasingly protesting against projects perceived as contributing to environmental gentrification and displacement (Pearsall and Anguelovski 2016). Some of their targets are smart growth policies in the context of neighborhood revitalization and upgrading (Tretter 2013) and other greening programs (Battaglia et al. 2014). Resistance has also been emplaced in "Complete Streets" planning processes which municipalities use in order to make neighborhoods more walkable and liveable for pedestrians and bikers (Agyeman and Zavetoski 2015). In other words, while traditional EJ research suggested that residents are fixed in their neighborhood and cannot easily escape toxic industries or waste sites, more recent studies of EJ activism reveal the emergence of numerous resistance movements against gentrification and displacement. Such an emphasis shows how much EJ groups are still fighting a collective imaginary that equates green with *white* and *middle (or upper) class* (Anguelovski 2015). This dynamic is reinforced by the ways in which the urban political economic regimes are structured and the fact that discourses of sustainability and environmentalism are rarely questioning the inequalities that underpin them.

With processes of green gentrification in mind – processes that can inadvertently be triggered by resident-led environmental improvement initiatives – activists are now forced to connect the pursuit of environmental justice to other claims linked to more incremental and territorial approaches for environmental revitalization, including greater access to affordable housing, protection of place identity, creation of new types of economic development, and a refusal to accept green amenities at any price (Anguelovski 2015). They often defend, for instance, the idea of a "just green enough" strategy for their neighborhood (Curran and Hamilton 2012). They embrace its industrial fabric and activities, rather than considering them as traditional LULUs, and ask for greening strategies to be accompanied by housing trust funds and a commitment of public officials against real estate speculation (Wolch et al. 2014). In other cases, residents decide to raise their voice at community meetings to advocate, among others, for the maximum height of new buildings and for the designation of their area as a historic district (even though historic districts can also become unaffordable) (Pearsall 2012). Last, they have also built alliances with middle-class residents (Hamilton and Curran 2013; Curran and Hamilton 2012).

In the next section, we illustrate these historical evolutions and emerging tensions between the right to a cleaner and safer urban environment and the fear of displacement and exclusion through three examples in Boston, New Orleans, and New York.

Three tales of urban environmental justice tensions

From environmental blight to attractive environmental haven in Dudley, Boston

Dudley is a central neighborhood situated in the district of Roxbury in Boston, with a majority of long-time low-income African-American, Cape Verdean, and Latino residents. In the early 1980s, more than 1300 neighborhood lots were vacant in an area of 1.5 square miles, and the neighborhood had become the center of contractor-led dumping offences (Medoff and Sklar 1994). Dudley was filled with contaminating industries, including auto-repair shops, scrap metal dealers, and truck storage facilities (Layzer 2006; Shutkin 2000). In addition, supermarkets, community centres, parks, and recreational facilities were scarce (Anguelovski 2014).

Early restoration efforts in the 1980s focused on the denunciation and clean-up of contaminated land. For instance, in 1985, the Dudley Street Neighborhood Initiative (DSNI) organized a comprehensive clean-up campaign, "Don't Dump on Us," and in 1987 successfully advocated for the closure of two illegal trash transfers by the Boston Public Health Department (Anguelovski 2014). DSNI then obtained the power of eminent domain over a triangle of 64 acres of abandoned land and created a community land trust to transform the blighted area into a dynamic neighborhood controlled by the community (Medoff and Sklar 1994).

Residents and local organizations fought against environmental toxics, while also dedicating attention to open space, parks, and playground rehabilitation and development. For instance, they obtained funding from the City of Boston for the construction of green spaces such as the Dudley Town Common (built in 1993) and the Dennis Street Park (inaugurated in 2008), which sits on the land trust. Together with organizations such as the Boston Schoolyard Initiative, they transformed neglected schoolyards into productive environments for safe and creative learning and playing. Community organizations such as Project Right successfully advocated for and participated in the creation of multi-purpose facilities offering athletic, academic, prevention, and intervention services (Anguelovski 2014).

Since the beginning of Dudley's revitalization, residents, community organizations such as the DSNI, and EJ non-profits such as The Food Project and Alternatives for Community and the Environment (ACE) spent much of their effort on the redevelopment of vacant and/or contaminated lots for large-scale urban agriculture. In addition to dozens of community gardens, Dudley hosts three urban farms and a greenhouse run by The Food Project, which every year produces 250 000 pounds of food for donation, youth-driven food enterprise, and farmers' markets. It also has a non-profit bakery, the Haley House Bakery, which offers healthy food options, organizes healthy eating and cooking classes for at-risk kids, and hires formerly incarcerated residents to run its kitchen.

With more than 650 parcels of vacant land having been cleaned up and redeveloped since 1985, Dudley has become a more liveable, green, and attractive neighborhood. Since the renovation of the historic Ferdinand Dudley Square and overall environmental improvements, Dudley Square has received the attention of new businesses, investors, and real estate agencies. Gentrification pressures are growing. Average condo sale prices have increased about 12 per cent from $179 000 in 2011 to $287 000 in 2014[1]. In June 2015, a developer proposed a plan for a 25-storey mixed residential and office tower without any affordable units. With DSNI's former director John Barros as the City of Boston's new Chief of Economic Development, the municipality is trying to encourage Dudley's revitalization through "development without displacement" and through residents' investment and control of community resources – but it is aware that such a path is full of pitfalls.

With affordability concerns in mind, community development corporations (CDCs) have partnered with DNSI to offer green affordable housing to residents. For instance, in 2009, Dorchester Bay Economic Development Corporation (EDC) completed the construction of Dudley Village, with 50 green affordable rental units, in which residents also helped build a park and playground. Projects such as this contribute to both physical and mental health improvements, reconcile the community with its land and landscape, create safe havens for residents where they are protected from stigmas and trauma, and allow families affected by environmental trauma to have access to secure, healthy, and affordable homes (Anguelovski 2014). In addition to being a powerful tool for community-planned and community-centered investment, DSNI has also put much effort into building affordable homes on its land trust, Dudley Neighbors Incorporated, which has proved to be a secure form of tenure and land management to preserve the neighborhood from (green) gentrification.

Since 2011, new business models have also taken root in Dudley, bringing together job creation and training, community wealth creation, and environmental justice. Partnerships between non-profits, small businesses, and residents are facilitating integrated food networks by bringing together growers (e.g. City Growers or The Food Project), processors and trans-formers (e.g. numerous small businesses participating in the food incubator Commonwealth Kitchen), retail (e.g. Dorchester Food Coop, Davies Market, etc.), restaurants and catering non-profits (e.g. Haley House, Dudley Dough, Davies Market, and Fresh Food Generation), waste management (e.g. CERO), and land (e.g. Dudley Land Trust). Through a partnership called the Dudley Real Food Hub (DRFH), community members work together to address complex development issues, increase affordable fresh produce, provide healthier restaurants and other foodscapes, offer fresh school food, reuse vacant lots for growing food, develop community-owned food businesses and jobs, and help create new ventures around food waste collection, management, and recycling.

In sum, Dudley illustrates EJ organizing in a community affected by environmental blight, rebuilt by safe havens, and now devising environmental solutions to ensure stability and eco-nomic development opportunities for families living in the neighborhood and threatened by gentrification pressures. It illustrates the complexity encountered by planners and comm-unity organizations as they seek to create community wealth mechanisms and ventures in a way that supports the long term ability of most residents – not just a few business entrepre-neurs – to remain in their neighborhood and benefit from the environmental amenities that they helped create and restore.

Sustaining the Lower Ninth Ward in New Orleans

The Lower Ninth Ward, located east of the Industrial Canal in New Orleans, is made up of two neighborhoods, Holy Cross and the Lower Ninth Ward, that are nearly surrounded by water. Due largely to its impoverished and majority black population, the neighborhood has historically suffered from environmental racism, degradation, and abandonment.

Pre-Hurricane Katrina, in 2000, a snapshot of the Lower Ninth Ward shows a pre-dominantly black community of nearly 20 000 residents, with whites living to a large extent on better-drained, higher ground near the river. High rates of poverty, transportation vulner-abilities, as well as high rates of home ownership characterized the neighborhood. As a community that has always had to make do with less, residents employed "sustainable" survival strategies, using the land as an economic and material resource (Haymes 1995; Pile

& Keith 1993). Historically, the combination of larger lots and lower income families gave rise to small local farming efforts. The neighborhood also has a robust history of EJ and political activism – local movements for school desegregation and to prevent widening the Industrial Canal shaped residents' ongoing fight for city investments (even in the wake of Hurricane Katrina).

Following Hurricane Katrina in 2005, responding to calls by planners, developers, and city officials to shrink the footprint of the city (Nelson et al. 2007), and to comments by local politicians that neighborhoods like the Lower Ninth Ward had to prove their viability in order to be rebuilt, residents used a combination of self-initiated planning processes, participation in city-led planning processes, and protests to secure their place in the city's future (Brand 2007). Many interpreted early city-led planning efforts, such as the Urban Land Institute's recommendations to prohibit return to the northern portion of Lower Ninth Ward, as part of a larger effort to permanently displace the city's majority black population (Nelson et al. 2007; Brand 2007). Resident-driven recovery plans such as the *Sustainable Restoration Plan*,[2] and *A People's Plan for Overcoming the Hurricane Katrina Blues* (Reardon et al. 2009) articulated sustainable development strategies to increase economic and spatial equality and promoted redistributive agendas that could address the deeply racialized poverty and inequality that pervaded the area before Katrina. While Lower Ninth Ward residents also participated in city-led planning efforts, many felt that these efforts avoided the critical question of how the city would get rebuilt and for whom.

The tensions between government-led planning and resident-driven green development highlight the challenges of sustainable development and EJ in an impoverished community. On the one hand, much of the government-led efforts have reinforced inequality and made an equitable return to the neighborhood more difficult. The city's direct prohibitions on return in the Lower Ninth Ward shaped the neighborhood's initial slow pace of recovery.[3] While rebuilding efforts were hampered by succession paperwork, Louisiana's post-Katrina homeownership recovery program, the Road Home program, reinforced the historically patterned land devaluation in communities of color by equating rebuilding assistance with pre-Katrina values.[4]

On the other hand, resident-driven development, including planning and housing activism, local gardening and food servicing, environmental reclamation efforts in Bayou Bienvenue, and strategic alliances with national partners such as the Make It Right Foundation and Global Green, "have" shaped emerging (though tenuous) elements of sustainable development. Local residents define sustainable redevelopment as physical (green buildings, solar panels, reduced carbon impact, etc.), economic (community land banking, green jobs, and urban agriculture), cultural (community, cultural, and religious spaces), and environmental (activism against environmental injustice and for environmental reclamation and natural lines of defense). Across organizations in the Lower Ninth Ward (including the Lower Ninth Ward Center for Sustainable Engagement and Development, Lower Ninth Ward Neighborhood Empowerment Network, the Holy Cross Neighborhood Association, and Common Ground Relief), this approach has shaped more holistic efforts to address historical inequities. Housing redevelopment projects such as Make It Right and Global Green have focused on affordable, green housing and green infrastructure that have primarily served local residents.

Despite this groundswell, the city remains committed to a redevelopment agenda focused on increasing the economic value of land and attracting creative class residents. For instance in 2015, despite overwhelming criticism by local residents about the scale and the land use being unreflective of the neighborhood context, the City Council approved a condominium redevelopment of the historic Holy Cross School site along the river. House values are on the rise, fueling residents' concerns over land grabbing and the possible spillover gentrification from the Bywater and Marigny neighborhoods on the western side of the Industrial Canal. A citywide planning effort to improve stormwater drainage, the *Urban Water Plan*,[5] which was supported by city funding but did not include citizen participation, has all but ignored any equity lens in its promotion of market-based environmental planning and green infrastructure (Fisch 2014). Its vision for the Lower Ninth Ward includes a new blue-way along Bayou Bienvenue where a large number of homes have already been rebuilt. While heightened awareness of local stormwater issues and new design approaches to ameliorate localized environmental vulnerability is laudable, missing is a deeper discussion of the sorts of tradeoffs made in choosing environment over equity (Campbell 2016; Krumholz 1982; Agyeman 2013; Brand 2015).

While post-Katrina resident activism in the Lower Ninth Ward has managed to ensure that projects such as Make It Right bring direct benefits to local residents, without this type of policy structure and equity lens this project could have easily promoted gentrification. In contrast, rising housing prices, which prey upon historically low land values, increase gentrification pressures that already concern local residents. The Lower Ninth Ward's developmental and environmental vulnerability is heightened by policy and planning processes long steeped in environmental racism that view lower-income communities of color as logical sites of either degradation and neglect or gentrification and capital accumulation (Brand 2015). This bifurcated view offers little room for planners hoping to make more equitable and sustainably just distributional and development decisions for communities that have long suffered from disinvestment and environmental inequities.

Environmental and social resilience intertwine in Brooklyn, New York

Red Hook, in Brooklyn, New York, sits between waterfront port infrastructure on the west and the elevated Brooklyn–Queens Expressway on the east. The neighborhood is home to disparate worlds: art galleries and cafes along Van Brunt Street; the Red Hook Ball Fields, famous for Latin American food trucks; the Gowanus Canal, a federal Superfund site; and the Red Hook Houses, the largest New York City Housing Authority (NYCHA) public housing project in Brooklyn comprising 30 buildings over 39 acres. Historically an industrial port of immigrant communities and dockworkers, the neighborhood underwent decline in the 1970s and 1980s. Red Hook has been at the forefront of EJ struggles since the early 1990s, when activists organized against a proposed waste facility in the neighborhood (Sze 2007). More recently, residents lost the fight against the building of a cement factory.

In October 2012, Red Hook was one of the city's worst hit neighborhoods during Superstorm Sandy. Residents in the Red Hook Houses lost power for weeks. But in the days following the storm, Red Hook emerged as a hub of grassroots post-disaster recovery. Staff from the Red Hook Initiative (RHI), a local community organization, Occupy Sandy and other volunteers, and government officials worked together to organize recovery actions. RHI played a particularly central role, serving as command center and soup kitchen for post-disaster recovery. Posts on social media sites further coordinated volunteering and fundraising efforts around the Red Hook efforts (RHI 2013; Schmeltz et al. 2013).

RHI provides health and education workshops and job training to neighborhood youth, and hires staff primarily from within the public housing community. The executive director of RHI notes that for those who know the neighborhood well, confronting the impacts from Sandy required the same skills and experiences as tackling problems like poverty and youth homelessness – knowing community members, assessing a situation, and developing an action plan. She refers to this as "building on the social capital" of the neighborhood. Yet, she also highlights the confusion stemming from a *lack* of local knowledge in recovery efforts. For example, she recounts one of the organization's board members getting into a truck with an official from the Federal Emergency Management Agency (FEMA) because he did not know where to go.[6]

Red Hook has been the focus of several post-Sandy resiliency initiatives, including the New York Rising Community Reconstruction program and the Rebuild By Design international competition. However, local community leaders expressed ambivalence about these initiatives. An architect and resident of Red Hook and a member of the Red Hook NY Rising planning committee stresses the positive aspects of state-led community planning efforts, including the numerous educational training sessions that illustrated key issues and potential solutions. At the same time, she notes the problems brought on by the unequal stakes of community members, in particular the lack of agency and influence among NYCHA residents and their lack of "empowerment over their environment."[7] RHI's executive director is more direct about NY Rising's shortcomings, citing the incongruity of including public housing residents in discussions about long-term infrastructure such as sea walls and floodgates when, two years after Sandy, they still breathe in fumes from temporary boilers.[8]

This lack of representation of lower-income residents, who account for approximately 50 per cent of the 12,400 residents in Red Hook, exposes the challenges of understanding systemic barriers in post-disaster recovery planning, and the missed opportunity to build on grassroots social support infrastructures. Even though Red Hook is slated to receive FEMA funding for NYCHA housing repairs, the funds are slow to come. NYCHA's CEO also talks about adding security doors and cameras, "upgrades" that are more aligned with the surveillance of residents than their empowerment or resilience. So while Red Hook continues to receive media attention – and with the recovery work of community organizations well recognized (see Red Hook-NYRCR 2014) – these planning processes still do not appear to engage effectively across disparate stakeholders.

Despite such challenges, experiences from Red Hook also highlight the possibilities when community social relationships are better represented in post-disaster response. RHI launched the Red Hook Wi-Fi project with the Open Technology Institute (OTI) in late 2011, providing free Internet access and local network applications such as real-time bus tracking and New York Police Department (NYPD) relations surveys (RHI 2012). Sandy left much of Red Hook without electricity, but Red Hook Wi-Fi remained operational. FEMA officials and volunteers set up additional routers to extend the network to support recovery efforts. RHI and OTI now train neighborhood youth as "digital stewards" to install and manage networking equipment, effectively tying a social program that promotes education and job creation to future environmental resilience.

Red Hook has for many years avoided the worst impacts of gentrification due to its relative physical isolation. However, as the neighborhood continues to receive federal recovery funds and awaits an integrated flood protection infrastructure system (see NYC SIRR 2013), it increasingly confronts new development pressures that are positioned to take advantage of these resiliency improvements. For example, a controversial for-profit private school and new townhouses are under construction, and ambitious plans have been proposed for a privately developed "innovation district" along the waterfront. In this process, the role of strong community networks among public housing residents, so critical in the immediate aftermath of Sandy, risks being neglected, or at best treated as a footnote to these larger initiatives.

In sum, Post-Sandy Red Hook does not lack resiliency initiatives. However, state and municipal planning processes have suffered from the challenges of addressing systemic socio-spatial inequality. They instead appear poised to perpetuate conditions of marginalization and enable further unequal urban development and resilience-fueled gentrification.

Concluding remarks

Early EJ movements denounced policymakers, regulators, and firms for treating working-class and minority residents as if they do not deserve to live in healthy neighborhoods, and for allowing "brown" LULUs to be sited in their neighborhoods. Exclusion and displacement in the process of creating greener, more sustainable, and more resilient neighborhoods might well be the twenty-first century embodiment of this process, where the urban poor and people of color end up in less healthy and liveable neighborhoods. Marginalized groups are thus facing double exposure to environmental impacts and environmental gentrification, often in the same space.

In the mid- and long-term, the impacts of combined processes of green redevelopment, gentrification, and displacement might be more severe in terms of community vulnerability compared to toxic waste sites. Our case examples illustrate how municipal green plans and projects – and even community-based neighborhood greening – can promote exclusion and unequal development. For low-income and minority residents, practices and discourses of urban greening increasingly signal to them that they might become excluded from the environmental, health, and social benefits of new parks or ecological corridors, which are then transformed into new forms of LULUs: GreenLULUs (Anguelovski 2015).

Urban planners must face the fact that greening projects might trigger or accelerate environmental gentrification, encroachment, and un-affordability for the city's most vulnerable residents because greening has become the new "urban frontier" for speculators and investors taking advantage of a "green gap." In the mid- and long-term, will green and resilient neighborhoods only be for middle (or upper) class and white residents? If being green equates to having access to new environmental goods in a sustained way, then cities will be faced with this reality.

The more recent manifestations of EJ activism call for continuing research on greening cities and greening practices. To date, there has been no quantitative study of the racial and social impact and the magnitude of displacement in cities implementing so-called "green" agendas. Additionally, urban economists should examine the role of real estate dynamics and municipal financialization (Weber 2010) in contributing to green gentrification and speculation. Questions remain about whether sustainability projects in marginalized neighborhoods can positively redistribute access to environmental amenities throughout the city. More critically, EJ activists and scholars must challenge the apolitical and technocratic discourse of sustainability to reassert its social and political dimensions and residents' right to the city. In the end, just cities must create more equitable environmental outcomes for low-income groups and communities of color by fostering bottom-to-bottom social networks, inclusive decision-making mechanisms, and engagement in equity-based planning.

Notes

1 See http://bostinno.streetwise.co/2015/02/12/roxbury-real-estate-gentrifying-boston-innovation-districts/
2 See http://davidrmacaulay.typepad.com/SustainableRestorationPlan.pdf
3 See www.datacenterresearch.org/reports_analysis/neighborhood-recovery-rates-growth-continues-through-2015-in-new-orleans-neighborhoods/

4 In 2010, a federal judge ruled that the Road Home program was discriminatory against black homeowners.
5 See http://livingwithwater.com/blog/urban_water_plan/about/
6 Interview by the author, Brooklyn, NY, September 13, 2014.
7 Interview by the author, Brooklyn, NY, August 22, 2014.
8 Interview by the author, Brooklyn, NY, September 13, 2014.

References

Agyeman, J. (2013) *Introducing Just Sustainabilities*, London: Zed Books.

Agyeman, J. & Zavetoski, S. (2015) *Incomplete Streets: Processes, practices, and possibilities*, New York: Routledge.

Anguelovski, I. (2014) *Neighbourhood as Refuge: Environmental justice, community reconstruction, and place-remaking in the city*, Cambridge: MIT Press.

Anguelovksi, I. (2015) 'From toxic sites to parks as (green) LULUs? New challenges of inequity, privilege, gentrification, and exclusion for urban environmental justice', *Journal of Planning Literature*, vol. 31, no. 1, February, pp. 23–36.

Anguelovski, I. & Carmin, J. (2011) 'Something borrowed, everything new: innovation and institutionalization in urban climate governance', *Current Opinion in Environmental Sustainability*, vol. 3, pp. 169–175.

Anguelovski, I., Shi, L., Chu, E., Gallagher, D., Goh, K., Lamb, Z., et al. (2016) 'Equity impacts of urban land use planning for climate adaptation: critical perspectives from the Global North and South', *Journal of Planning Education and Research*, vol. 36, no. 3, September, pp. 333–348.

Bai, X., Roberts, B. & Chen, J. (2010) 'Urban sustainability experiments in Asia: patterns and pathways', *Environmental Science & Policy*, vol. 13, pp. 312–325.

Battaglia, M., Buckley, G., Galvin, M. & Grove, J.M. (2014) 'It's not easy going green: obstacles to tree-planting programs in East Baltimore', *Cites and the Environment*, vol. 7.

Boone, C., Buckley, G., Grove, M. & Sister, C. (2009) 'Parks and people: an environmental justice inquiry in Baltimore, Maryland', *Annals of the Association of American Geographers*, vol. 99, pp. 767–787.

Boyd, M.R. (2008) *Jim Crow Nostalgia: Reconstructing race in Bronzeville*, Minneapolis: University of Minnesota Press.

Brand, A.L. (2007). 'Rebuilding the right to return. toward a framework of social and spatial justice in New Orleans', *Critical Planning*, no. 14.

Brand, A.L. (2015) The most complete street in the world: a dream deferred and coopted. In Agyeman, J. & Zavetoski, S. (eds.) *Incomplete Streets: Processes, practices, and possibilities*. New York: Routledge.

Brown, P. (2013) *Toxic Exposures: Contested illnesses and the environmental health movement*, New York: Columbia University Press.

Bryant, B.I. & Mohai, P. (1992) *Race and the Incidence of Environmental Hazards : A time for discourse*, Boulder: Westview Press.

Bryson, J. (2013) 'The nature of gentrification', *Geography Compass*, vol. 7, pp. 578–587.

Bullard, R. (1990) *Dumping in Dixie: Race, class, and environmental quality*, Boulder: Westview Press.

Bullard, R.D. (2007) *Growing Smarter: Achieving liveable communities, environmental justice, and regional equity*, Cambridge: MIT Press.

Bullard, R.D., Mohai, P., Saha, R. & Wright, B. (2007) *Toxic Wastes and Race at Twenty: 1987–2007*. United Church of Christ Justice and Witness Ministries.

Bunce, S. (2009) 'Developing sustainability: sustainability policy and gentrification on Toronto's waterfront', *Local Environment*, vol. 14, pp. 651–667.

Campbell, S. (2016) Green cities, growing cities, just cities? Urban planning and the contradictions of sustainable development. In Fainstein, S.S. & DeFilippis, J. (eds.) *Readings in Planning Theory*. Malden, MA: Wiley-Blackwell. pp 214–240.

Chakraborty, J. (2012) 'Cancer risk from exposure to hazardous air pollutants: spatial and social inequities in Tampa Bay, Florida', *International Journal of Environmental Health Research*, vol. 22, pp. 165–183.

Checker, M. (2011) 'Wiped out by the "Greenwave": environmental gentrification and the paradoxical politics of urban sustainability', *City & Society*, vol. 23, pp. 210–229.

Connolly, J. & Steil, J. (2009) Can the Just City be built from below: brownfields, planning, and power in the South Bronx. In Marcuse, P. (ed.), *Searching for the just city: debates in urban theory and practice*. London; New York: Routledge.

Curran, W. & Hamilton, T. (2012) 'Just green enough: contesting environmental gentrification in Greenpoint, Brooklyn', *Local Environment*, vol. 17, pp. 1027–1042.

Dahmann, N., Wolch, J., Joassart-Marcelli, P., Reynolds, K. & Jerrett, M. (2010) 'The active city? Disparities in provision of urban public recreation resources', *Health & Place*, vol. 16, pp. 431–445.

De Sousa, C.A. (2003) 'Turning brownfields into green space in the City of Toronto', *Landscape and Urban Planning*, vol. 62, pp. 181–198.

Dooling, S. (2009) 'Ecological gentrification: a research agenda exploring justice in the city', *International Journal of Urban and Regional Research*, vol. 33, pp. 621–639.

Downey, L. & Hawkins, B. (2008) 'Race, income, and environmental inequality in the United States', *Sociological Perspectives*, vol. 51, pp. 759–781.

Faber, D. & Kimelberg, S. (2014) Sustainable urban development and environmental gentrification: the paradox confronting the U.S. environmental justice movement. In: Hall, H.R., Robinson, C. & Kohli, A. (eds.) *Uprooting Urban America: Multidisciplinary perspectives on race, class & gentrification*. New York: Peter Lang Publishers.

Fisch, J. (2014) Green infrastructure and the sustainability concept: a case study of the Greater New Orleans Urban Water Plan. Thesis. University of New Orleans, Department of Planning and Urban Studies

Fitzgerald, J. (2010) *Emerald Cities: Urban sustainability and economic development*, New York: Oxford University Press.

Gamper-Rabindran, S. & Timmins, C. (2011) 'Hazardous waste cleanup, neighbourhood gentrification, and environmental justice: evidence from restricted access census block data', *The American Economic Review*, vol. 101, pp. 620–624.

Gibbs, D.C. & Krueger, R. (2007) Containing the contradictions of rapid development? New economic spaces and sustainable urban development. In: Krueger, R. & Gibbs, D.C. (eds.) *The Sustainable Development Paradox: Urban politial economy in the United States and Europe*. London: Guilford Press.

Gottlieb, R. (2005) *Forcing the Spring: The transformation of the American environmental movement*, Washington, DC: Island Press.

Hagerman, C. (2007) 'Shaping neighbourhoods and nature: urban political ecologies of urban waterfront transformations in Portland, Oregon', *Cities*, vol. 24, pp. 285–297.

Hamilton, T. & Curran, W. (2013) 'From "Five Angry Women", to "Kick-ass Community": gentrification and environmental activism in Brooklyn and beyond', *Urban Studies*, vol. 50, pp. 1557–1574.

Hastings, A. (2007) 'Territorial justice and neighbourhood environmental services: a comparison of provision to deprived and better-off neighbourhoods in the UK', *Environment and Planning C*, pp. 896–917.

Haymes, S.N. (1995) *Race, Culture, and the City: A pedagogy for Black urban struggle*. Albany, NY: State University of New York.

Heynen, N., Perkins, H. & Roy, P. (2006) 'The political ecology of uneven urban green space', *Urban Affairs Review*, vol. 42, pp. 3–25.

Hilier, A. (2003) 'Redlining and the Home Owners' Loan Corporation', *Journal of Urban History*, vol. 29, pp. 394–420.

Holifield, R., Porter, M. & Walker, G. (2009) 'Spaces of environmental justice: frameworks for critical engagement', *Antipode*, vol. 41, pp. 591–612.

Hyra, D.S. (2008) *The New Urban Renewal: The economic transformation of Harlem and Bronzeville*, Chicago: University of Chicago Press.

Inroy, N.M. (2000) 'Urban regeneration and public space: the story of an urban park', *Space and Polity*, vol. 4, pp. 23–40.

Jabareen, Y.R. (2006) 'Sustainable urban forms: their typologies, models, and concepts', *Journal of Planning Education and Research*, vol. 26, pp. 38–52.

Jones, C.E. & Ley, D. (2016) 'Transit-oriented development and gentrification along Metro Vancouver's low-income SkyTrain corridor', *The Canadian Geographer/Le Géographe canadien*, vol. 60, pp. 9–22.

Keil, R. (2007) Sustaining modernity, modernizing nature: the environmental crisis and the survival of capitalism. In: Krueger, R. & Gibbs, D. (eds.) *The Sustainable Development Paradox: Urban political economy in the United States and Europe*. London: Guilford Press.

Krumholz, N. (1982) 'A retrospective view of equity planning in Cleveland, 1969–1979', *Journal of the American Planning Association*, Spring, pp. 163–174.

Landry, S. & Chakraborty, J. (2009) 'Street trees and equity: evaluating the spatial distribution of an urban amenity', *Environment and Planning A*, vol. 41, pp. 2651–2670.

Layzer, J. (2006) *The Environmental Case: Translating values into policy*, Washington, DC: CQ Press.

Lerner, S. (2005) *Sacrifice Zones: The front lines of toxic chemical exposure in the United States*, Cambridge, MA: MIT Press.

Loh, P. & Eng, P. (2010) *Environmental Justice and the Green Economy: A vision statement and case studies for just and sustainable solutions*. Boston: Alternatives for Community and the Environment.

Low, S., Taplin, D. & Scheld, S. (2005) *Rethinking Urban Parks: Public space and cultural diversity*, Austin: University of Texas Press.

Madra, Y.M. & Adaman, F. (2014) 'Neoliberal reason and its forms: de-politicisation through economisation', *Antipode*, vol. 46, pp. 691–716.

Medoff, P. & Sklar, H. (1994) *Streets of Hope: The fall and rise of an urban neighbourhood*, Boston: South End Press.

Mohai, P., Pellow, D. & Roberts, J.T. (2009) 'Environmental justice', *Annual Review of Environment and Resources*, vol. 34, pp. 405–430.

Mohai, P. & Saha, R. (2015) 'Which came first, people or pollution? Assessing the disparate siting and post-siting demographic change hypotheses of environmental injustice', *Environmental Research Letters*, vol. 10.

Mueller, E. & Dooling, S. (2011) 'Sustainability and vulnerability: integrating equity into plans for central city redevelopment', *Journal of Urbanism: International Research on Placemaking and Urban Sustainability*, vol. 4, pp. 201–222.

Nelson, M., Ehrenfeucht, R. & Laska, S. (2007) 'Planning, plans and people: professional expertise, local knowledge and governmental action in post-Hurricane Katrina New Orleans. *Cityscape*, vol. 9(3), pp. 23–52.

NYC Special Initiative for Rebuilding and Resiliency (SIRR). (2013) A stronger, more resilient New York. City of New York Office of the Mayor. Retrieved from http://s-media.nyc.gov/agencies/sirr/SIRR_singles_Lo_res.pdf

Park, L.S.H. & Pellow, D. (2011) *The Slums of Aspen: Immigrants vs. the environment in America's Eden*, New York: New York University Press.

Pearsall, H. (2008) 'From brown to green? Assessing social vulnerability to environmental gentrification in New York City', *Environment and Planning C*, vol. 28, pp. 872–886.

Pearsall, H. (2012) 'Moving out or moving in? Resilience to environmental gentrification in New York City', *Local Environment*, vol. 17, pp. 1013–1026.

Pearsall, H. & Anguelovski, I. (2016) Contesting and resisting environmental gentrification: responses to new paradoxes and challenges for urban environmental justice. *Sociological Research Online*, vol. 21(3), 6.

Pham, T.-T.H., Apparicia, P., Séguin, A.M., Landry, S. & Gagnon, M. (2012) 'Spatial distribution of vegetation in Montreal: an uneven distribution or environmental inequity?', *Landscape and Urban Planning*, vol. 107, pp. 214–224.

Pile, S. & Keith, M. (1993) *Place and the Politics of Identity*. New York: Routledge.

Quastel, N. (2009) 'Political ecologies of gentrification', *Urban Geography*, vol. 30, pp. 694–725.

Quastel, N., Moos, M. & Lynch, N. (2012) 'Sustainability-as-density and the return of the social: the case of Vancouver, British Columbia', *Urban Geography*, vol. 33, pp. 1055–1084.

Reardon, K., Green, R., Bates, K.K. & Kiely, R.C. (2009). 'Commentary: overcoming the challenges of post-disaster planning in New Orleans: lessons from the ACORN Housing and University Collaborative', *Journal of Planning Education and Research*, vol. 28, pp. 391–400.

Red Hook Initiative (RHI). (2012). Recovering from Sandy: RHI WiFi Project. Brooklyn, NY: Red Hook Initiative.

Red Hook Initiative (RHI). (2013). Red Hook Initiative: a community response to Hurricane Sandy. Brooklyn, NY: Red Hook Initiative.

Red Hook – NY Rising Community Reconstruction Program (NYRCR). (2014) Red Hook NY Rising community reconstruction plan. NY Rising Community Reconstruction Program. Retrieved from https://stormrecovery.ny.gov/sites/default/files/crp/community/documents/redhook_nyrcr_plan_20mb_0.pdf

Rosan, C.D. (2012) 'Can PlaNYC make New York City "greener and greater" for everyone?: sustainability planning and the promise of environmental justice', *Local Environment*, vol. 17, pp. 959–976.

Rosol, M. (2013) Vancouver's "EcoDensity" planning initiative: a struggle over hegemony?', *Urban Studies*, vol. 50, pp. 2238–2255.

Schmeltz, M.T., González, S.K., Fuentes, L., Kwan, A., Ortega-Williams, A. & Cowan, L.P. (2013) 'Lessons from Hurricane Sandy: a community response in Brooklyn, New York', *Journal of Urban Health*, vol. 90(5), pp. 799–809.

Schuetze, T. & Chelleri, L. (2015) 'Urban sustainability versus green-washing: fallacy and reality of urban regeneration in downtown Seoul', *Sustainability*, vol. 8, pp. 33.

Self, R.O. (2003) *American Babylon: Race and the struggle for postwar Oakland*, Princeton: Princeton University Press.

Shutkin, W. (2000) *The Land That Could Be: Environmentalism and democracy in the twenty-first century*, Cambridge: MIT Press.

Smith, N. (1996) *The New Urban Frontier: Gentrification and the revanchist city*, New York: Routledge.

Song, L.K. (2014) 'Race, transformative planning, and the just city', *Planning Theory*, Online.

Sugrue, T. (2005) *The Origins of the Urban Crisis: Race and inequality in postwar Detroit*, Princeton: Princeton University Press.

Swyngedouw, E. (2007) Impossible Sustainability and the Postpolitical Condition. In: Krueger, R. & Gibbs, D.C. (eds.) *The Sustainable Development Paradox: Urban political economy in the United States and Europe*. London: Guilford Press.

Sze, J. (2007) *Noxious New York: The racial politics of urban health and environmental justice, Environmental justice in America: A new paradigm*, Cambridge: MIT Press.

Tretter, E.M. (2013) 'Contesting sustainability: "SMART growth" and the redevelopment of Austin's Eastside.'*International Journal of Urban and Regional Research*, 37(1), 297–310.

Weber, R. (2010) 'Selling city futures: the financialization of urban redevelopment policy.' *Economic Geography*, vol. 86, pp. 251–274.

Wheeler, S.M. & Beatley, T. (2009) *The Sustainable Development Reader*, London: Routledge.

Wilson, S.M., Fraser-Rahim, H., Williams, E., Zhang, H., Rice, L., Svendsen, E. & Abara, W. (2012) 'Assessment of the distribution of toxic release inventory facilities in metropolitan Charleston: an environmental justice case study', *American Journal of Public Health*, vol. 102, pp. 1974–1980.

Wolch, J.R., Byrne, J. & Newell, J.P. (2014) 'Urban green space, public health, and environmental justice: the challenge of making cities "just green enough"', *Landscape and Urban Planning*, vol. 125, pp. 234–244.

37

JUST CONSERVATION

The evolving relationship between society and protected areas

Maureen G. Reed and Colleen George

Introduction

The past few decades have demonstrated rapid growth in land and marine areas designated as protected areas globally. The United Nations' (UN) 2014 list of protected areas reported 32.9 million km^2 protected, compared to 2.4 million km^2 in 1962 (Deguignet et al. 2014). Historically, nature conservation – conceptually and practically – has focused on identifying incentives and means for protecting, restoring, and enhancing the natural environment without much consideration to the social implications of such efforts (Sandler and Pezzullo 2007). The conservation movement has been driven by Western elite values of nature that emphasize the importance of protecting pristine landscapes, waterscapes, and habitat for endangered species (Redford 2011; Lele 2011). Today, however, protected areas are not solely about the protection of nature. In practice, protected areas are now expected to make much broader contributions to human society including maintaining critical ecosystem "goods and services" such as water, food, carbon storage; mitigating climate change; alleviating poverty; and providing opportunities for economic development (Watson et al. 2014).

Originally, environmental justice scholars focused less on the establishment of protected areas and more on the inequitable distribution of effects of *industrial development* across geographic areas and social groups. More recently, however, environmental justice scholars have raised multiple critiques including how striving for the biocentric ideal of nature preservation has reproduced colonial or neocolonial practices, neglected Indigenous rights and knowledge systems, marginalized local peoples, and perpetuated sharp divides of privilege and inequality (e.g. West et al. 2006). Wolfgang Sachs has called this "ecology without equity" (Sachs 2013: 22). These scholars ask key questions: whose nature do we seek to protect?; who reaps the benefits and who bears the costs of our choices?; what systems of governance will contribute to protecting biological diversity and cultural integrity? and finally, who gets to decide? (e.g. Sachs and Santarius 2007; Sandler and Pezzullo 2007). These questions draw attention to the need for conservation with a justice agenda. Hence, issues posed by environmental justice scholars involving the distribution of benefits and costs and the procedures established for allocation and management of lands and resources are as relevant to conservation as they are to urban and industrial environmental problems.

This chapter reflects on the emergence and application of "just conservation" in relation to protected areas. We focus on the changing emphasis from setting aside protected areas solely to conserve biological diversity to including multiple objectives and ensuring that protected areas also provide opportunities for people to be involved in their establishment and management. We begin by explaining selected international initiatives that have begun to include equity considerations within a conservation agenda. Next, we provide a framework for assessing procedural equity. We then examine the evolution of the world network of biosphere reserves (BRs), created under the Man and Biosphere (MAB) programme of the United Nations Educational, Cultural, and Scientific Organization (UNESCO) and consider its application in Canada. As a country with 18 BRs established between 1978 and 2016, Canada's BRs demonstrate evolution in practice including the challenges of meeting procedural justice considerations. We close by considering the implications of these challenges for the broader agenda of "just conservation".

Reconciling conservation and environmental justice: evidence through international priorities

Inherent tensions exist between biodiversity protection and environmental justice. Where environmental justice focuses on procedural and distributive justice for humans, ecological justice, advocated for by conservation scientists, focuses on bringing justice to the natural world (Shoreman-Ouimet and Kopnina 2015). Dobson (2007) suggests these priorities may be incompatible, as the considerable cultural differences between parties advocating for each side create tensions that cannot be reconciled. Schlosberg (2007) disagrees, suggesting that these issues are at their foundation issues of justice.

Contemporary efforts for biodiversity conservation are becoming much more conscious of environmental justice issues, by seeking to mitigate the costs of conservation to local people and to reconcile goals of conservation and development (Martin et al. 2013). At the international level, academics, government officials, practitioners, and activists have worked collaboratively for almost half a century to include equity considerations within conservation initiatives. International calls have worked to reconcile environmental and economic imperatives and draw attention to the interdependence of biological diversity, human livelihoods, poverty and well-being. The 1972 UN Stockholm Conference on the Human Environment, the 1980 World Conservation Strategy, the 1992 Earth Summit (where the UN Convention on Biological Diversity [CBD] was established), and the 2000 UN Millennium Development Goals explicitly connected environmental degradation with poverty and human rights and contributed to increased action to establish protected areas that simultaneously address concerns for social equity. The 2000 Millennium Development Goals suggested that protected areas should contribute to poverty reduction and sustainable development (Sachs 2005). The (2005) Millennium Ecosystem Assessment argued that continuous environmental degradation adversely affects rights of individuals and communities to basic needs such as water, food, health and life itself.

Armed with these observations and aspirations, in 2010, the IUCN, with seven other international NGOs, formed the Conservation Initiative on Human Rights to ensure conservation actions upheld human rights, enhanced capability for implementation, and adopted appropriate accountability measures. This "rights-based approach" acknowledges that the establishment of protected areas and the formulation and implementation of management plans must balance conservation goals with other legitimate rights and interests. It also requires stakeholders to be involved throughout the process to ensure their interests are identified and addressed based on the best information available, including both scientific and traditional knowledge (Grieber et al. 2009). Advocates of the approach also acknowledged the importance of diversity by concluding that "applying the rights-based approach will not necessarily give the same result in each situation, but it should ensure that decisions and actions are not only legal but also legitimate" (Grieber et al. 2009: 109).

The CBD has also sought to explicitly address social equity considerations. The CBD's Program of Work on Protected Areas, first agreed to in 2004, contains goals that include promoting equity and benefit sharing and enhancing and securing the involvement of Indigenous and local communities. In 2010, the tenth Conference of the Parties (COP) to the CBD adopted the Strategic Plan for Biodiversity 2011–2010, known as the Aichi Targets. Of the 20 targets, Target 11 calls for an increase in terrestrial, coastal and marine areas, "especially areas of particular importance for biodiversity and ecosystem services [be] conserved through *effectively and equitably managed*, ecologically representative and well connected systems of protected areas" (Leadley et al. 2014: 259). Woodley et al. (2012: 29) argued that "effective and equitable management means that protected areas management includes the need and rights of stakeholders as a fundamental part of management".

Importantly, the IUCN has also established several categories of protected areas that allow for a range of governance regimes. Category I are strictly protected areas that limit human use, while Categories V and VI provide for human settlements and recognize a range of governance systems and actors including NGOs, Indigenous peoples, and co-management arrangements. Furthermore, IUCN now recognizes new forms of protected areas such as Indigenous Peoples' and Community Conserved Territories and Areas (ICCAs) that are created to support both biological and cultural diversity. These new types of arrangements are buttressed by increasing emphasis on management effectiveness of protected areas. For example, the IUCN and UNEP generated a guidebook for assessing management effectiveness that includes evaluating opportunities for benefit sharing among affected local communities (Leverington et al. 2010). These kinds of initiatives have supported the introduction of the new social compact theme at the 2014 World Parks Congress emphasizing "effective and just conservation, and social and ecological connectivity" (IUCN 2014: 38).

Informing conservation through environmental justice scholarship

Clearly, individuals and institutions directly involved in biodiversity conservation and protected area planning are now working on ways to develop strategies that jointly meet conservation and human development goals. While environmental justice scholars have long focused on the distribution of costs and benefits of human use of the environment (Schlosberg 2007), their more recent assessments have come to focus on how *procedures* – governance processes – have contributed to inequalities (e.g. Martin et al. 2013; also Chapter 9, this volume). In other words, to fairly address issues associated with the distribution of outcomes, scholars and activists argue that decision-making *processes* must be designed to take into account the opinions and perspectives of those affected. Some have identified process as being foundational because it is through unfair processes that distributions of protected areas and management schemas are formulated. Drawing from those who inform justice theory (e.g. Fraser 2001; Nussbaum 2001, 2011; Sen 1999; Young 1990), Schlosberg (2007) described three procedural imperatives that must be addressed to advance environmental justice: *recognizing* those typically disenfranchised, seeking *participation* from broad social groups, and building the *capabilities* of communities (see also Chapters 9, 10, and 11).

An important consideration under "recognition" is not simply the physical recognition of peoples such as those who are Indigenous or local to a region. Rather, "recognition is about seeking equality between different ways of knowing the world . . . or . . . being reflexive regarding whose culture is privileged and respected" (Martin et al. 2013: 124). This is a more fundamental perspective in that it acknowledges that there may be significant cultural differences among those involved in a conservation agenda, including differences in cosmologies, ways of knowing, and what constitutes just distribution or procedures. The domination of certain ways of knowing and doing may entrench hierarchies of status that obscure some local cultural and historical understandings and "perpetuate failures of recognition through the extension of popular, dominant conceptions of environmental sustainability" (Martin et al. 2013: 128). Hence, a focus on recognition may tell us a lot about who gets invited to participate, who makes the invitation, and whether or how local capabilities are degraded, built or enhanced. In combination, the distributive and procedural insights offered through environmental justice theory urge a reconsideration of who is involved in protected area governance, as well as how protected areas are designated, managed, and how their mandates and priorities are defined. In the next section we document the evolution of BRs, as they have long been promoted as institutions designed to simultaneously address conservation and development objectives.

BRs: moving toward just conservation in practice?

In 1976, the MAB programme began to designate the first in the world network of BRs. In that year, 52 BRs were created in eight countries; by April 2016, UNESCO reported 669 BRs in 120 countries (www.unesco.org/new/en/natural-sciences/environment/ecological-sciences/biosphere-reserves/). Like protected areas more generally, BRs have grown rapidly in number, and have refined their purpose and implementation strategies. A review of their 40-year experience, however, also reveals that the three imperatives of procedural justice are exceedingly difficult to achieve in practice.

BRs were conceived from the outset as something other than strictly protected areas: they were to be multi-purpose sites designed to foster understanding of human impacts on ecological and cultural systems. Each BR was to contain a core protected area, a buffer zone with uses (typically recreational and research) that were compatible with maintaining the biological values of the core area, and a transition zone that allowed for more intensive human activities. In Canada, the buffer zone is frequently called a zone of cooperation. The classical configuration resembled a fried egg, allowing for scientists to better understand how human activities influenced 'natural' ecosystems (Figure 37.1). This zonation system would then allow for explicit experimentation and learning about how humans affected biodiversity. The variety of sites would offer opportunities for comparative research and thereby demonstrate their value for research.

During the first years of implementation, BRs primarily served as field sites for scientists. The original strategy was to select representative ecosystems based on a global classification of biogeographical provinces created by biologist Miklos Udvardy (1975). While this ideal was never realized, BRs were initially designated where scientists and civil servants could demonstrate to MAB that sites had a strong potential for conservation, some form of legal protection, and a history of, or facilities for, research (Batisse 1982). Natural scientists determined what problems were significant and how they were to be addressed. Scientists were then expected to transmit the knowledge they gained to managers and policy-makers. Although early documentation speaks of *human*–environment interactions, early researchers were steeped in epistemologies of natural science, and thus not well positioned to consider the cultural, social, and economic contexts that contributed to those changes (Reed 2009). Local and/or Indigenous peoples had little or no say in site selection nor were they significantly involved in the operations of the BRs.

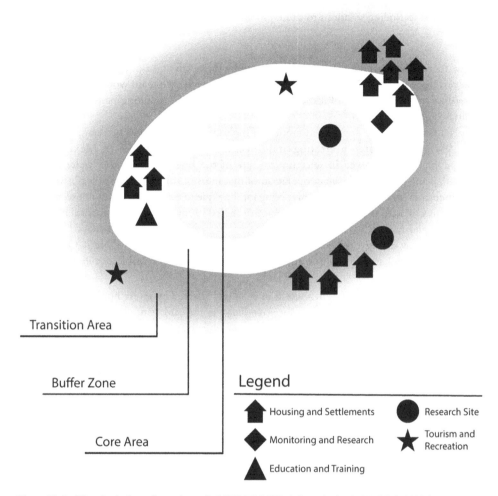

Figure 37.1 The classical configuration of a UNESCO BR (adapted after UNESCO 2000 by Jamie Nunn)

In the late 1970s, the term "integrated development" emerged in documents describing the mission of BRs (UNESCO–ICSU 1981: 36). BRs were to function through partnerships established at local, national and international levels. Academic partnerships were to emerge across disciplinary divides by identifying real-world problems and attracting the required expertise across the spectrum of academic disciplines. National committees were composed mainly of scientists from universities or national research institutions, with representatives of relevant public and private bodies. These committees would, ideally, be able to establish research priorities and funding mechanisms. While this overall governance structure suggested the exclusion of local people, UNESCO-sponsored reports continued to suggest that the involvement of local people was critical to the success of BRs. For example, the same report stated:

> The people living in the area should be listened to so that the significance of their claims, the nature and magnitude of the problems affecting them, the changes that they

desire, and so on, can be clearly grasped. The work carried out in the BR should then be geared to finding solutions to those problems and improving the overall situation.

(UNESCO–ICSU 1981: 21)

Clearly, there was an appetite for research inspired by the needs and aspirations of local people. However, there was not clarity about what members of a research team would establish direct partnerships with members of local communities and how the resulting partnerships would be initiated and maintained.

Only a few BRs were considered models of inclusive participation. For example, Halffter (1981: 95) praised efforts in creating and governing the Mapimi BR in Mexico where "local people were involved in selecting the site of the reserve, under the leadership of a local cattleman, and through the cooperation of members of the Ejido de la Flor". Although a few other cases were also noted for involving local people (e.g. Pays d'Enhaut in Switzerland, Waterton in Canada), in general, BRs in practice shared in the criticisms of classical protected areas (Reed and Massie 2013).

In 1995, a new strategy, the Seville Strategy and Statutory Framework (UNESCO 1996: 5), officially introduced a new function to the BR programme – sustainable development.

> Rather than forming islands in a world increasingly affected by severe human impacts, they [BRs] can become theatres for *reconciling* people and nature, they can bring knowledge of the past to the needs of the future, they can demonstrate how to over-come the problems of the sectoral nature of our institutions. In short, BRs are much more than just protected areas.
>
> *(our emphasis)*

From that point onward, founding documents indicate that maintaining cultural diversity and local livelihoods and protecting ecological goods and services to meet human needs were significant elements of the rationale for the creation of BRs (e.g. UNESCO 1996, 2000, 2008). Additionally, meeting these goals required local people to become more directly involved in the nomination of BR sites and the implementation of its mandate (UNESCO 2000, 2008; Bouamrane 2006). For example, the international strategic plan, the Lima Action Plan 2016-2025, seeks to ensure that BR activities are open and participatory by ensuring that all stages from planning and implementation to review of BRs account for local and Indigenous practices (UNESCO 2016).

Practice was intended to match the concept. The 1995 Statutory Framework made BRs subject to periodic review of their effectiveness. Attention to zonation and governance capability became the primary criteria for judging effectiveness of BRs in practice (Reed and Egunyu 2012). Zonation concerns reflected the original mandate to understand (and address) the impact of various levels of human use of landscape. Interest in governance reinforced the importance of social context and the involvement of local people in decision-making (Price et al. 2010). By 2007, a background paper declared:

> In the past 15 years, a shift from a research-driven to a management-driven programme has taken place in MAB as a result of the need to focus on identifying management solutions at the local level. This was accomplished to a certain extent at the expense of scientific research and monitoring.
>
> *(UNESCO 2007: 9)*

By 2015, sustainable development had become the *raison d'être* for BRs. The first strategic action area of the 2015 MAB Strategy (UNESCO 2015: 12) declares that "the World Network of Biosphere Reserves [is] comprised of effectively functioning models for sustainable development". In the same document, biodiversity conservation is intended to support the sustainable development function and earlier ideas about "training" have been expanded to include education, learning and capacity building. The new strategy also explicitly speaks to "sustainability science" as a key mechanism to generate, communicate and share knowledge from a range of individuals and groups and to adopt multiple ways of learning and sharing in order to expand lessons learned beyond a single group or sector.

In some ways, the creators of BRs were ahead of their time as they sought in the 1970s to better understand the connection between biological diversity and human well-being and to reconcile the potential contradictions. By emphasizing the learning dimension of BRs, they also sought to include local people, particularly in developing countries, in research and training activities. But BRs also reflected the conceptual and practical limitations of their creators. Despite the high-level rhetoric, in most cases, *scientists* trained in western epistemologies and research traditions remained primarily responsible for designation of BRs, research within BRs, and transmission of acquired knowledge to local people. In the early years, at least, the recognition of local people as knowledge holders and resource managers did not emerge and their participation was limited. Capacity building was conceived in terms of training opportunities for local peoples rather than as opportunities for co-production of knowledge or co-management of sites. Hence the contribution of BRs to the procedural aspects of just conservation was more aspirational than operational.

The BR experience in Canada

Canada's experience related to BRs has mirrored international priorities and challenges. Canada's early BRs reflected a strong conservation and research agenda. Mont Saint Hilaire (established 1978) and Long Point BR (established 1986) were justified by their long history of scientific research (Figure 37.2). Waterton BR (established 1979) had greater local involvement where the concept of a BR was appealing to the Superintendent of the national park that would come to form its core protected area. He believed that a designation might help the park to do a better job of working with its rural neighbours. The four BRs established in the late 1980s were developed in large part through planning and direction from the national MAB Committee, suggesting a top-down managerial strategy for the creation of BRs during this time period.

Over time, the creation and management of BRs became the purview of local communities. During the period from 2000 to 2016, twelve BRs were designated in Canada, each with a strong local catalyst. One that was created in 2016 emerged from a self-governing Indigenous community – the first BR of its kind in the world. Additionally, the original six BRs also began to strengthen their connections to their local communities through revisions to their governance structures and initiation of new projects.

BRs in Canada have adopted a multi-stakeholder governance structure designed to recognize the variety of perspectives, interests and values required to inform their broadened mandate, as well as offer an opportunity for collaboration, knowledge-sharing, and capacity building. Participants include municipal, provincial and federal governments; representatives of natural resource industries (forestry, fisheries, mining, agriculture, ecotourism) and environmental organizations; academic and/or government researchers and "members-at-large". Many BRs have seats available for Indigenous representatives. Each biosphere in

Figure 37.2 The location and designation dates of Canadian BRs 2016 (map credit: Canadian Biosphere Reserves Association)

Canada has a slightly different local arrangement, with different mechanisms for identifying representatives. In many cases, board members are prescribed by the terms of reference of each BR. Municipal and Indigenous leaders from the region are examples. Some resource sector representatives may also be appointed because of the position they hold. In some cases, members are nominated from the general public. Typically, these people have been actively involved in establishing the BR. Academic or scientific members of the board (who have often conducted research in the region) have become involved in this way. Frequently, BR boards have a combination of appointed and elected members (George 2015).

In Canada and abroad, research questions posed by social scientists and the application of social science research methods emerged in the post-1995 period. The terminology describing BRs shifted from learning laboratories to learning sites (see Schultz and Lundholm 2010). Community-based research involving formal or informal partnerships between researchers and local residents was also initiated, with the idea that scientists, managers, and local communities might coordinate efforts to better understand how to achieve conservation of biodiversity and sustainability (e.g. Schultz and Lundholm 2010; Stoll-Kleemann et al. 2010; Mendis-Millard and Reed 2007). This shift is clearly evident in Canada where closer links between social scientists and practitioners became established through the 2000s allowing local people to become more central participants by helping to shape research questions and methods (see Mendis-Millard and Reed 2007). In some cases, natural scientists have also worked more closely with local residents. In Clayoquot Sound, a scientific symposium was held in 2011 to exchange lessons learned between scientists and practitioners regarding the preceding 15 years of ecosystem management. In Mont St. Hilaire, studies on the protection of forest corridors and ecological services involved local residents in community restoration projects (Godmaire et al. 2013). Hence, research conducted in the 2000s introduced strategies that included practitioners, residents, and scholars in research practice and exchange of knowledge.

Representation in governance is one of the most critical challenges facing BRs. Because BRs in Canada hold no legislative or regulatory authority, they are not a formal threat to interest groups and can, therefore, serve as politically neutral open forums (Pollock 2009). As mentioned earlier, BRs in Canada continue to utilize stakeholder models to help promote access to a number of interest groups. The mid-1990s saw a significant increase in the level of volunteerism, with communities taking ownership of their local BR. The primary opportunity for members to participate in BRs is by serving as a member of the board. Board members are able to participate in information sharing, knowledge generation and decision-making in a multi-perspective and interdisciplinary environment; however, there are drawbacks associated with this model. A number of boards have very low turnover, discouraging new members. Some boards in Canada have struggled to achieve active participation from board members and those board members who are very involved experience burnout and frustration. Beyond the board, opportunities for broader community participation in these organizations are typically limited to workshops or events put on by the BR, signing up for email updates, or 'liking' Facebook pages. Though common among small volunteer-based, non-profit organizations, this phenomenon does not promote active participation.

An enduring challenge, but growing opportunity, lies in the recognition and inclusion of Indigenous peoples and knowledge (see also Chapters 40 and 41). Two-thirds of BRs in Canada are located in regions with significant resident Indigenous populations and traditional territories. Additionally, the Canadian Constitution and treaties recognize the rights of Indigenous peoples in environmental and resource management. As United Nations organizations, BRs are also required to uphold the UN Declaration on the Rights of Indigenous

Peoples. Nevertheless, a national survey of BRs in 2011 determined that Indigenous organizations participated in the events of eight BRs, and partnered with seven. Yet, only three of the then-15 BRs indicated that they maintained communication with Indigenous organizations about their activities and only two BRs reported having Indigenous representatives on their management boards. One of these, the Clayoquot Sound Biosphere Trust Society's Board, has equal representation from Nuu-Chah-nulth (Indigenous) Nations and other local communities. Most BRs in Canada have not demonstrated this level of focus on local or Indigenous membership on their boards although as noted below, one restructured its board to become a round table and one BR is now wholly led by Indigenous peoples. Nevertheless, many boards continue to be dominated by researchers and government officials representing elite environmental interests. Furthermore, one might reflect on whether a stakeholder board structure is one that is conducive to recognition and participation of Indigenous ways of knowing and decision-making.

BR organizations have recognized this gap and started to take steps to address it. For example, the Canadian-MAB committee (composed of six individuals) introduced an Indigenous representative in 2012. Through his guidance, new BR nominations have been required to demonstrate how Indigenous peoples have been included in determining the rationale and structure of the proposed reserve. In 2013, a regional conference of BRs (EuroMAB) devoted several sessions to inclusion of Indigenous peoples and established an on-going working group. In 2014, several Canadian BRs met at Clayoquot Sound to learn from the Clayoquot experience strategies, challenges, and rewards of improved Indigenous inclusion. In that same year, one BR that had struggled to maintain positive working relationships dissolved its board structure and created a round table that has since gained greater legitimacy among local Indigenous people. Canadian Indigenous people raised the issue of Indigenous rights and knowledge systems during the 2016 World Congress of BRs. Consequently, the resulting Action Plan (2016–2025) now contains stronger language acknowledging the rights of Indigenous peoples, although placing Indigenous knowledge on equal footing with scientific knowledge was not achieved. The Annual General Meeting of Canadian biosphere reserves that followed the World Congress included the first national-level round table discussion of the membership in how to strengthen relationships with Indigenous peoples across all activities of BRs. In 2017, BRs from across Canada will work with Indigenous leaders, academic researchers and government representatives in a 3-day meeting devoted to developing an action plan devoted to co-building sustainability and reconciliation.

In summary, Canada and MAB's history of providing conditions for the recognition and participation of Indigenous people within the governance system of BRs has been uneven. The nominal participation of Indigenous peoples on governing boards has been low as has been the recognition of Indigenous ways of knowing and doing within most BR organizations and across the national and international networks. Steps to address this challenge in Canada are in their infancy. However, renewed commitments and tangible changes are beginning to take root. Canadian BR practitioners are beginning to rethink their governance strategies to make their actions towards sustainable development more in line with procedural aspects advocated for in environmental justice theory. Both academics and practitioners agree that moving towards 'just' conservation in BRs will require longstanding efforts to build and strengthen relationships among all involved.

Conclusion

Conservation has historically focused on biocentric imperatives of nature protection; however, the act of conservation is undeniably a social and political process, and decisions made about how protected areas are designated and managed have associated social and economic impacts, with burdens traditionally falling on local and Indigenous groups. With the broadening mandate of protected areas (provide ecosystem services, contribute to poverty alleviation), as well as the increasing attention being paid to the distributional and procedural injustices associated with protected area designation and management, there is a need to develop models for conservation that reconcile environmental protection and social justice concerns.

The need for these efforts has been enforced by a number of international calls to promote social equity through conservation (IUCN, CBD). Although some governance models have been developed to address these goals, few have been deemed successful in addressing both the social and environmental imperatives (Martin et al. 2013). Conservationists in the natural sciences acknowledge their limited perspectives (e.g. Redford 2011) and identify the need to engage alternative voices in discussions to identify strategies able to holistically and sustainably address conservation issues. Indeed, greater interdisciplinary training and collaboration is needed to develop resilient conservation strategies that promote social justice while conserving biodiversity (Redford 2011). Those with understandings of environmental justice can help to inform both distributive and procedural discussions, connecting to issues of recognition, participation, and capabilities.

UNESCO BRs offer an integrated model that supports interdisciplinary innovations for conservation and protected area management. As reflected in their 40-year history, their mandates have evolved to consider local perspectives and incorporate sustainability objectives and capacity building. However, BRs also demonstrate a rather bumpy heritage with respect to procedural justice including recognition, participation, and capacity building. The increasing appetite for Indigenous engagement in Canadian BRs offers opportunities for learning, capacity building and modelling both within individual BRs and across the international network.

It is important that environmental justice scholars, practitioners, and activists become more involved in and vocal about promoting inclusive and representative discussions that drive the mandates and designations of conservation initiatives, including BRs. International organizations including the IUCN, UNEP, and UNESCO provide broad directives that encourage issues of social justice to be considered in conservation planning. However, the questions that have long been posed by environmental justice scholars and activists remain: whose nature do we seek to protect?; who reaps the benefits and who bears the costs of our choices?; what systems of governance will contribute to protecting biological diversity and cultural integrity? and finally, who gets to decide?

Acknowledgements

Funding for this research was provided by the Social Sciences and Humanities Research Council. We could not contemplate this research without the tremendous support from our research partners. We are also grateful to the editors of this handbook for constructive comments on earlier drafts and to Zac Carreiro for his final editorial polish.

g for

References

Batisse, M. (1982). "The Biosphere Reserve: A Tool for Environmental Conservation and Management." *Environmental Conservation* vol. 9, pp. 101–111.

Bouamrane, M. (ed.) (2006). *Biodiversity and Stakeholders: Concertation Itineraries*. Technical Note 1. Paris, UNESCO.

Deguignet, M., Juffe-Bignoli, D., Harrison, J., MacSharry, B., Burgess, N. and Kingston, N. (2014). *2014 United Nations List of Protected Areas*. Cambridge, UK: UNEP-WCMC.

Dobson, A. (2007). "Social justice and environmental sustainability: ne'er the twain shall meet?" In Agyemand, J., Bullard, R., and Evans, B. (eds), *Just Sustainabilities: Development in an Unequal World*, Cambridge, MA: MIT Press, pp. 83–95.

Fraser, N. (2001). "Recognition without Ethics?" *Theory, Culture and Society* vol. 18, no. 1–2, pp. 21–42.

George, C. (2015). *Exploring Just Sustainability in a Canadian Context: An Investigation of Sustainability Organizations in the Canadian Maritimes*. PhD Thesis, University of Saskatchewan, Saskatoon, Canada.

George, C. & Reed, M.G. (2015). "Operationalizing just sustainability: Towards a model for place-based governance." *Local Environment*. Available: DOI: 10.1080/13549839.2015.1101059

Godmaire, H., Reed, M.G., Potvin, D. and Canadian BRs. (2013). *Learning from Each Other: Proven Good Practices in Canadian BRs*. [Online]. Ottawa: Canadian Commission for UNESCO. pp. 66. Available: http://unesco.ca/en/home-accueil/biosphere [19 September 2016]

Grieber, T., Janki, M., Orellana, M., Savaresi, A. and Shelton, D. (2009). "A Rights-based Approach in Protected Areas: Conservation with Justice, a Rights-based Approach." *IUCN Environmental Law and Policy Paper No. 71*: Gland, Switzerland. pp. 87–109.

Halffter, G. (1981). The Mapimi Biosphere Reserve: Local Participation in Conservation and Development. *Ambio*, vol. 10, no. 2/3, pp. 93–96.

International Union for Conservation of Nature (IUCN). (2014). *World Parks Congress 2014 Summary Report*. IISD Reporting Services. Available: www.iisd.ca/iucn/wpc/2014/ [19 September 2016]

Leadley, P.W., Krug, C.B., Alkemade, R., Pereira, H.M., Sumaila, U.R., Walpole, M., et al. (2014). "Progress towards the Aichi Biodiversity Targets: An Assessment of Biodiversity Trends, Policy Scenarios and Key Actions." *Secretariat of the Convention on Biological Diversity*, Montreal, Canada. Technical Series 78.

Lele, S. (2011). "Rereading the Interdisciplinary Mindscape: A Response to Redford." *Oryx*, vol. 45, no. 3, pp. 331–332.

Leverington, F., Costa, K.L., Courrau, J., Pavese, H., Nolte, C., Marr, M., et al. (2010). *Management Effectiveness Evaluation in Protected Areas: A Global Survey*: 2nd edition. Gland, Switzerland, and Cambridge: IUCN.

Martin, A., McGuire, S. and Sullivan, S. (2013). "Global Environmental Justice and Biodiversity Conservation." *The Geographical Journal*, vol. 179, no. 2, pp. 122–131.

Mendis-Millard, S. & Reed, M.G. (2007). "Understanding Community Capacity Using Adaptive and Reflexive Research Practices: Lessons from Two Canadian BRs." *Society and Natural Resources*, vol. 20, pp. 543–559.

Nussbaum, M.C. (2001). *Women and Human Development: The Capabilities Approach*, vol. 3, New York: Cambridge University Press.

Nussbaum, M.C. (2011). *Creating Capabilities*. Boston: Harvard University Press.

Pollock, R. (2009). *The Role of UNESCO BRs in Governance for Sustainability: Cases from Canada*. Unpublished PhD Thesis. Peterborough, ON: Trent University.

Price, M.F., Park, J.J. and Bouamrane, M. (2010). "Reporting Progress on Internationally Designated Sites: The Periodic Review of BRs." *Environmental Science & Policy*, vol. 23, pp. 549–557.

Redford, K. (2011). "Misreading the Conservation Landscape." *Oryx*, vol. 45, no. 3, pp. 324–330.

Reed, M.G. (2009). "A Civic Sort-of Science: Addressing Environmental Managerialism in Canadian BRs." *Environments*, vol. 36, no. 3, pp. 17–35.

Reed, M.G. & Egunyu, F. (2012). "Management Effectiveness in UNESCO BRs: Learning from Canadian Periodic Reviews." *Environmental Science & Policy*, vol. 25, pp. 107–117.

Reed, M.G. & Massie, M. (2013). "What's Left? Canadian BRs as 'Sustainability in Practice.'" *Journal of Canadian Studies/Revue D'études Canadiennes*, vol. 47, no. 3, pp. 200–225.

Sachs, J.D. (2005). "Investing in Development: A Practical Plan to Achieve the Millennium Development Goals (Millennium Project)," Report to the UN Secretary-General. London: Earthscan.

Sachs, W. (2013). "Liberating the World from Development." *New Internationalist*. March, pp. 22–27.

Sachs, W. & Santarius, T. (2007). *Fair Future: Resource Conflicts, Security and Global Justice*. Canada: Fernwood Publishing.

Sandler, R.D. & Pezzullo, P.C. (2007). *Environmental Justice and Environmentalism: The Social Justice Challenge to the Environmental Movement*. Cambridge, MA: MIT Press.

Schlosberg, D. (2007). *Defining Environmental Justice: Theories, Movements and Nature*. Oxford: Oxford University Press.

Schultz, L. & Lundholm, C. (2010). "Learning for Resilience? Exploring Learning Opportunities in BRs." *Environmental Education Research*, vol. 16, pp. 645–663.

Sen, A. (1999). *Development as Freedom*. Oxford: Oxford University Press.

Shoreman-Ouimet, E. & Kopnina, H. (2015). "Reconciling Ecological and Social Justice to Promote Biodiversity Conservation." *Biological Conservation*, vol. 184, pp. 320–326.

Stoll-Kleeman, S., de la Vega-Leinert, A.C. and Schultz, L. (2010). "The Role of Community Participation in the Effectiveness of BR Management: Evidence and Reflections from Two Parallel Global Surveys." *Environmental Conservation*, vol. 37, pp. 227–238.

Udvardy, M. (1975). "A Classification of the Biogeographical Provinces of the World." Prepared as a contribution to UNESCO's Man and the Biosphere Programme Project No. 8, IUCN Occasional Paper No. 18. Morges, Switzerland: International Union for Conservation of Nature and Natural Resources.

UNESCO, MAB, UNESCO–ICSU. (1981). Conference and Exhibit, Ecology in Practice: "Establishing a scientific basis for land management," Paris, 22–29 September 1981: Theme 3: providing a basis for ecosystem conservation, Communication 3/1: The BR concept: its implementation and its potential as a tool for integrated development. By Michel Maldague, Laval University, Quebec Canada SC–81/WS/56. (copy located in Francis fonds, Wilfrid Laurier University Archives).

UNESCO. (1996). *BRs: The Seville Strategy and the Statutory Framework of the World Network*. Paris: UNESCO. Available: http://unesdoc.unesco.org/images/0010/001038/103849Eb.pdf [19 September 2016]

UNESCO. (2000). *Solving the Puzzle: The Ecosystem Approach and BRs*. Paris: UNESCO.

UNESCO. (2007). 3rd World Congress of BRs: Biosphere Futures, "UNESCO BRs for Sustainable Development." Background Paper to the Palacio Municipal de Congresos, Madrid. SC–08/CONF.401/5. Paris. October 25, 2007.

UNESCO. (2008). *Madrid Action Plan for BRs 2008–2013*. Paris: UNESCO.

UNESCO. (2015). *MAB Strategy 2015–2015*. Paris: UNESCO.

UNESCO. (2016). *Lima Action Plan for UNESCO's Man and the Biosphere (MAB) Programme and its World Network of Biosphere Reserves (2016-2025)*. Paris: UNESCO.

Watson, J.E.M., Dudley, N., Segan, D.B. & Hockings, M. (2014). "The Performance and Potential of Protected Areas." *Nature*, vol. 515, pp. 67–73.

West, P., Igoe, J. & Brockington, D. (2006). "Parks and Peoples: The Social Impact of Protected Areas." *Annual Review of Anthropology*, vol. 35, pp. 251–277.

Woodley, S., Bertzky, B., Crawhall, N., Dudley, N., Londoño, J.M., MacKinnon, K., et al. (2012). "Meeting Aichi Target 11: What Does Success Look Like for Protected Area Systems?" *Parks*, vol. 18, pp. 23–36.

Young, I. (1990). *Justice and the Politics of Difference*. Princeton NJ: Princeton University Press.

PART IV

GLOBAL AND REGIONAL DIMENSIONS OF ENVIRONMENTAL JUSTICE RESEARCH

38

FREE-MARKET ECONOMICS, MULTINATIONAL CORPORATIONS AND ENVIRONMENTAL JUSTICE IN A GLOBALIZED WORLD

Ruchi Anand

Introduction

Approximately thirty years after the environmental justice movement took shape in the United States, its theoretical and practical underpinnings have become a rhetorically inherent part of law and policy at the national and international levels. This chapter forwards the argument that the economic backdrop of globalization and free market capitalism accompanied by an explosion in the number of multinational corporations (MNCs) obstructs the enforcement of environmental justice in powerful ways.

Human welfare, justice, and equity concerns are trumped by forces of the market which provide economic rationalizations for all such charges, evading political responsibility (Porritt 2012). Such a tyranny of the market where all ills are justified in the name of economics "is extending its reach across the planet like a cancer, colonizing ever more of the planet's living spaces, destroying livelihoods, displacing people, rendering democratic institutions impotent, and feeding on life in an insatiable quest for money" (Korten 1999: 14). This cancer of the unleashed free market does not stop at state borders but spills over in ways that make questions of international environmental justice pertinent.

At the national level, the lure of the global capitalist system that promises economic growth and development allows us to understand why the problem of low-income and minority communities being disproportionately impacted by environmental burden is slighted. At the international level, it allows us to understand why less-developed countries suffer disproportionate environmental burdens as compared to developed ones (Anand 2004). In less-developed countries, it is the poorest of the poor who bear the biggest burdens. The economic logic and justification of the free market hurts the possibility and intensity with which we respond to environmental justice claims. At the international level, the "unholy trinity" (Peet 2003), namely the WTO, the IMF and the World Bank, promote the 'free' and uninterrupted flow of the market, discouraging barriers even at the expense of justice. While the *laissez faire* market remains incapable of rendering justice, neither the United

Nations nor the international laws that emanate from it are adequate to regulate the ensuing environmental justice issues that arise from the functioning of MNCs.

Since MNCs have the fluidity to move production and the associated 'environmental bads' between countries without much legal oversight, environmental justice issues are side-tracked as 'collateral damage.' The marginalization, both procedural (in their ability to participate in the decision-making processes) and distributive (in their dealing with disproportionate environmental burdens in the form of toxic wastes, health risks, pollution, and unsafe jobs) that low-income and minority populations face is worsened manifold when the perpetrators are MNCs. This is partly due to the legal difficulties in determining accountability, responsibility, compensation for damages and solutions for redress when environmental justice claims arise. The complex interlinkages between governments, international organizations and MNCs in a neoliberal global context make solutions cumbersome to envision and enforce. Such deprivations enhance structural inequality which has been the primary concern of environmental justice and are often the result of market, legal or public policy failures at the national and international levels. One of the first casualties of such failures is democracy, due to the distancing of voices from the most marginalized populations of the world. The associated casualties are justice and the human rights of poorer marginalized people to be safe from disproportionate environmental harms. Free market capitalism is not a win-win game but has winners and losers. This clearly negates the hypothesis of the world being flat (Friedman 2005) and reiterates the hypothesis of the world as spiky (Florida 2005).

This chapter explores the bleak potential for the realization of environmental justice goals in the wake of an explosion of MNCs that have begun to shape global governance, and is organized in the following manner. The first section discusses the growth in size and influence of MNCs, which are the embodiments of the globalized capitalist economy. The next section explores some legal questions about MNCs in global governance and how their activities can be legislated by national and international laws. The final section concludes on a pessimistic note by reiterating how environmental justice is impacted adversely by the neoliberal agenda and the growth of MNCs.

Multinational corporations and environmental justice and the logic of neoliberalism

Multinational corporations are at the epicentre of this globalized world symbolized by free markets, privatization, deregulation and foreign direct investment (FDI). They have a decisive role in how environmental justice plays out nationally and internationally or as Schlosberg (2013) calls it, horizontally and vertically. An MNC is "an enterprise that engages in foreign direct investment (FDI) and owns or controls value-added activities in more than one country" (Dunning 1992: 3). Also referred to as multinational enterprises (MNEs), transnational corporations (TNCs) and transnational enterprises (TNEs), they differ in their degrees of multi-nationality or transnationality owing to criteria such as the number of foreign affiliates or subsidiaries abroad, the number and kind of countries they operate in, the proportion of overseas assets, revenues and profits, and the international nature of their employees, owners and staff, to name a few (Spero and Hart 1997). They also differ in terms of the industry they specialize in (e.g. extracting raw material, high technology products, petroleum, banking, insurance, retail, telecommunications), their revenues, capital, level of technology, organizational structures and levels of corporate social responsibility (Spero and Hart 1997; Bantekas 2004). Regardless of the kind of MNC, questions of global justice abound with their activities which cross geographical, social, political, cultural and educational boundaries (Mujih 2012).

It is estimated that approximately 100 000 MNCs exist today. They account for a fourth of the global gross domestic product (GDP). MNCs produce revenues that surpass the overall GDP of many states (Mikler 2013). Multinational corporations have become the principal players in world trade. According to Mujih (2012: 83), more MNCs than countries fill the positions of the largest economies in the world – "of the largest economies in the world in 2001, 51 were MNCs, only 49 were countries." The most powerful MNCs surpass the wealth of developing and some developed countries alike. For example in 2001, "General Motors was bigger than Ireland, New Zealand and Hungary combined, Royal Dutch Shell was bigger than Norway; Mitsubishi had sales greater than the gross domestic product of Indonesia, while Ford was bigger than South Africa" (Mujih 2012: 84). These facts and figures are resonated by the UN World Investment Report, World Bank and UN Development Program sources. Many of the most powerful MNCs originate in industrialized countries although non-industrialized countries also feature on the list (Graham and Woods 2006). The geometric explosion of MNCs with a simultaneous weakening of the state makes questions of social responsibility more pertinent than ever before.

MNCs have developed numerically and financially, continuously searching to relocate their undertakings through FDI to places with the most competitive conditions, i.e. lower costs, low wages and weak environmental and regulatory legislation. These places, typically in less-developed countries (LDCs), also happen to be in countries that depend heavily on the presence of MNCs for jobs, technical expertise, research and development, financial investment, improvement in balance of payments, an increase in tax revenue, entrepreneurship, improved quality of goods and services, competition and the related advantages of an increase in dynamism (Amao 2013; Spero and Hart 1997; Graham and Wood 2006). This has been referred to as the 'benign model' by Moran (1999).

At the same time, MNCs have drawn a lot of negative attention for their role in environmental damage, human rights violations and a lack of social responsibility, raising the concerns of environmental justice (Spero and Hart 1997). MNCs have been accused of neo-colonialism, pressuring governments of weaker countries to open their markets to competition, technological fraud, a lack of accountability, and undermining social, economic and human rights. This has been referred to as the 'malign model' (Moran 1999). The 'malign' model questions the ability of MNCs to increase capital in the host countries and criticizes MNCs for benefiting from lax political and legal infrastructure through the host countries' lack of economic strength and bargaining leverage. This model argues that MNCs generate by-products of environmental degradation (through the over-exploitation of natural resources, greenhouse gas emissions, toxic and hazardous waste production and water pollution), worker exploitation and human rights violations that they are not held accountable or liable for (Moran 1999; Bhagwati 2012).

The paradox that is inherent in this situation is that the competitiveness of a country comes from a simultaneous erosion of regulations in a globalized world. Although MNCs bring in many economic advantages, they are associated with a host of social and environmental problems, particularly the disadvantaged in LDCs. MNCs tend to relocate to countries having weak regulatory regimes with 'enforcement gaps' (Zerk 2010: 66). Taking any stringent action against MNCs could mean that host states jeopardize millions of dollars, leading to a 'lack of incentive' by host states to act against MNC operations (Morimoto 2005: 145). Yet, not doing anything and continuing with lax regulatory regimes implies violating the right to justice for the most downtrodden and affected people. The impunity of MNCs at the national level can be explained by less-developed countries prioritizing their dependence on foreign direct investment over ensuring environmental justice to their populations.

As a result of this conflict of interest, MNC transactions confront several issues of environmental justice. It is attractive and lucrative for MNCs to seek short-term benefits by using these facets of weak legislation to their advantage. In some cases, governments of developing countries deliberately maintain the lack of regulation in order to be competitive with other developing countries for FDI. In such cases, there is a dangerous "race to the bottom" by potential hosts of MNCs in developing countries (Graham and Woods 2006) at the same time as there is a practice of 'divide and rule' by MNCs to create discord in communities by offering monetary incentives and compensation for anticipated problems (Greyl et al. 2013). In such situations where economic factors are the ultimate decision-makers, the natural environment can quickly become an irreparable casualty. Allegations of unfairness or injustice are countered by the question, "what's fairness got to do with it?" (Been 1993).

For example, a critique of the disproportionate siting of locally unwanted land uses (LULUs) in poor neighbourhoods is countered by four objections from free-market advocates (Been 1993). The first, the '*causation objection*', disassociates the siting process from the disproportionate environmental burdens afflicted on the poor and persons of colour. This objection claims that community demographics had nothing to do with siting decisions. It is the housing and job markets that led low income and minority populations to move in to neighbourhoods with LULUs. The second, the '*mobility objection*', implies that property value would decline as a result of LULU siting, the rich would move out and the poorer people would move in, recreating the pattern of disproportionate siting. The third, the '*aggregation objection*', encourages a calculation of not merely the burdens but certain benefits that may accrue from siting proposals before one can calculate fair-siting. The fourth, namely the '*free market objection*,' contends that allocations of environmental 'goods' and 'bads' should be made through the free markets. Put simply, a free market is a market-based economy that is based on supply and demand and where governmental intervention or control is nonexistent or minimal. Justice is one of the prime victims of such a market where monetary gain justifies every claim of injustice.

The underlying belief is that the free market will deliver the socioeconomic welfare programmes that the state and international institutions were accustomed to undertaking. This anti-state and pro-market position or 'neo-liberalism' can be argued to be the "dominant paradigm of development" (Önis 1995: 97). The neoliberal agenda is what is dominant in most advanced capitalist countries to a lesser or greater extent. Many developing countries in Asia, Africa and Latin America follow suit, be it by choice or not. Organizations such as the IMF, the World Bank and the WTO regulate global trade and finance and propagate the neoliberal agenda in their workings (Haque 1999: 203–204). Harvey (2005: 33) claims that "neoliberalization has meant, in short, the financialization of everything" including environmental justice.

We are witnessing the commodification of environmental resources in a system that is fuelled by the logic of the free market with the goal being to maximize profits regardless of the destruction of the planet. In *The Great Transformation*, Polanyi (2001: 3) affirms the impossibility of a 'self-adjusting market system' and provides a powerful and compelling critique of free market liberalism, referring to it as a 'stark utopia.' The neoliberal strategy has "become hegemonic as a discourse" (Cervantes 2013: 28), permeating education, media, state and international institutions, environmental justice and sustainable development.

What does 'sustainable development' mean in the face of neoliberalism? Clearly, the main pillars of the market-driven perspective are incompatible with the principal goals of sustainable development. Haque (1999: 199) states:

> These neoliberal policies are likely to expand industrialization (which causes environmental pollution), globalize consumerism (which encourages consumption of

environmentally hazardous products), multiply the emission of CO_2 and CFCs (worsening the greenhouse effect and ozone layer depletion), overexploit natural resources (depleting nonrenewable resources), and increase the number of urban poor and rural landless (forcing them to build more slums and clear more forests), and therefore, threaten the realization of sustainable development objectives.

As a result of neoliberal policies of economic growth, we are witnessing increased pollution, erosion, global warming and desertification at unstoppable rates, leading us to a path of development that is unsustainable and often unjust (Cervantes 2013). The question that begs to be answered is 'how are the benefits and burdens of such a policy distributed?' The answer, for those who occupy the powerful seats of decision-making, lies in the market.

In the case of the transboundary export of hazardous wastes from OECD to non-OECD countries, where developing countries were receiving a disproportionate amount of hazardous waste, questions of international justice and equity began to surface. Lawrence Summers (1992: 82), the former Chief Economist of the World Bank, in his infamous statement (see Chapter 5, Exhibit 1), said that "the economic logic of dumping a load of toxic waste in the lowest-wage country is impeccable." His statement in the form of a memorandum made on December 12, 1991, gave three economic reasons for the export of hazardous wastes:

> First, the costs of pollution depend on earnings foregone through death or injury; these costs are lowest in the poorest countries ... Second, costs rise disproportionately as pollution increases, so shifting pollution from dirty places to clean ones reduces costs. Third, people value a clean environment more as their income rises, if other things are equal, costs fall if production moves from rich to poor ones.
>
> *(Summers 1992: 82)*

Clearly, economic and market based explanations, as catalogued in Summers' statement above, are incapable of addressing the social and humanitarian variables required for rendering environmental justice to the poor and underprivileged of the world. Divorced from ethical and social justice concerns, the current system of neoclassical economic thinking stands for competition, accumulation and profit-making (Gonzalez 2001; Gonzalez 2015). This free-market globalized system raises a host of ethical questions (Singer 2002) but has already expanded nationally, internationally and transnationally. Be it the case of the export of toxic e-waste to poor countries by Tesco or Barclays, chocolate production or the bottled water business by Nestlé, the violation of basic human rights by Walmart, oil spills by oil giants such as Exxon-Mobil and Chevron, environmental pollution by Dow Chemicals or the monopolization of the world's seed supply by Monsanto, the logic remains the same, 'profits over people' (Chomsky 1999).

This 'profits over people' globalized system is motored by the IMF, World Bank and WTO, which together constitute a "nascent global state" (Chimni 2004: 1). These institutions undermine the democracy and independence of sovereign states, particularly developing countries (by replacing their national laws by "uniform global standards in order to remove the barriers to capital accumulation at the global level" (Chimni 2004: 7). This institutional backdrop distances states (LDCs more than rich ones) from decision-making, privileging the transnational capitalist class (TCC), the corporate elite or the *Davos* class (Buxton 2014). This class of people who benefit from such a system consists of the owners of transnational capital, corporate executives, bankers, brokers, financial experts and media managers (Chimni 2004). A globalized world with a growing number of MNCs has changed the playing field of environmental justice by increasing social, economic, political and environmental inequities worldwide (Agyeman and Carmin 2011; Hurrell and Woods 1995; Falk 1999; Hertz 2001; George 2014). It begs the question of whether (or not) MNCs are subjects of international law.

Regulating activities of multinational corporations

The status of MNCs in international law is an ongoing and inconclusive debate (Alvarez 2011; Pentikainen 2012). Are MNCs subjects of international law? Do they possess international legal personality? Do MNCs have the same rights and duties as other subjects of international law? Traditionally, states are the main subjects of international law and MNCs were regulated by national laws – either in the states where they operate (host state) or/and the state that they come from (home state). Today, we are confronting the question of whether or not MNCs also have legal personality in the same manner as states do (Muchlinski 2007). Most international scholars would argue that MNCs do not possess international legal personality under international humanitarian or criminal law (Bismuth 2010). Other scholars argue that MNCs are subjects of international law (Hendrawan 2014). Still others talk of degrees of international legal personality (Mayer and Jebe 2010; Osofsky et al. 2012).

In these different responses to the question, MNCs continue to be powerful business entities worldwide contributing to issues of justice and fairness and violations of human rights. The issue that is of prime concern to environmental justice scholars has been to explore what avenues are possible if and when there are violations of environmental justice transnationally by MNCs. As we have discussed previously in this chapter, MNCs have faced charges of violating human rights for monetary purposes such as damaging the environment cross-nationally at the same time as endangering the basic human rights of people inside and across national borders. How, if at all, can MNCs be held to laws, standards and codes of conduct, preventively or post-operation, in the event that they violate the rights of people they impact? There is a serious 'protection gap' and a 'governance gap' in the operations of MNCs that have significant rights without corresponding obligations in cases of violations of human rights laws (Ruggie 2010). The mechanisms that exist for regulating the worldwide conduct of MNCs and holding them accountable if human rights abuses should occur are 1) self-regulation by MNCs; 2) regulation by and from inside the host state; 3) regulation from the home state where the company originates; and 4) MNCs' codes of conduct at the international level (Marquez 2015).

Self-regulation implies that MNCs uphold themselves to human rights standards. Concealed is the assumption that MNCs have adequate incentives to respect human rights when juxtaposed against profit-making concerns. Advocates of this approach argue that MNCs care about their reputation, inspiring them to embrace social consciousness as a strategy for inviting more business among socially responsible clients. It is in the interest of MNCs to embrace corporate social responsibility which is not necessarily at odds with their profit-making goals (McLeah 2006). Critics of this approach highlight the voluntary nature of self-imposed codes of conduct, which gives MNCs the choice of whether or not to follow them. Codes are often written in vague language and are never enforceable. Critics argue that, in the majority of cases, it is more lucrative for MNCs to produce goods unethically and those MNCs would never trade off profit for social causes. MNCs that adopt ethical business codes agreeing to respect all internationally recognized human rights are less than half of the total number (BHRRC 2016). Self-regulation can be viewed as a complementary means of MNC regulation, not an alternative to legal enforcement.

Regulation from the host state is another mechanism to make MNCs abide by the national and international laws governing human rights. In developed countries with established legal and political systems, MNCs tend to abide by the national laws. In the event of violations, MNCs are held accountable. However, when MNCs operate in LDCs, regulating the behaviour of MNCs to conform to human rights codes does not work as well. This is due to several reasons: weak or non-existent legislation; a lack of will to enforce laws that do exist; corruption amongst state officials who may receive bribes from MNCs to allow them to violate laws; and the use of weak human rights standards as a comparative advantage to attract MNCs (Stiglitz 2007; Ruggie 2010).

Regulation from the home state is another possible mechanism for monitoring the activities of MNCs for their conformity with human rights codes of conduct. Parent companies of the majority of MNCs are located in the US, European Union or Japan, where standards of living are higher, and human rights and legal systems are upheld to a greater extent than in LDCs (UNCTAD 2009). The accountability and subsequent liability of MNCs to victims, in the event that there are negative consequences, is still unclear and limited. Problems of MNC accountability are further aggravated since the parent companies and their subsidiaries have separate legal personalities. Additionally, "neither the 'home' state—where the parent company is established—nor the 'host' country of the subsidiary's location exercise complete control over the functioning of the whole entity of the MNC" (Namballa 2014: 182). There is also reluctance by home states to regulate their MNCs under their country's domestic law in order not to disadvantage their own MNCs' competitiveness with other countries, and this can harm this strategy's success (Joseph 2000).

Having assessed the problems in regulating MNCs at the level of the state, a discussion of *regulatory mechanisms for MNCs at the international level* is needed. Although several initiatives exist at the international level which attempt to regulate the activities of MNCs, they remain largely unregulated and defy corporate responsibility because the measures remain voluntary and so lack any potential for enforcement. These codes of conduct include the Guidelines for Multinational Enterprises (1976); the Tripartite Declaration of Principles Concerning Multinational Enterprises and Social Policy (1977); the UN Draft Code on Transnational Corporations (1982); the Global Compact (1999); the UN Norms on the Responsibility of Transnational Corporations and other Business Enterprises with Regard to Human Rights (2003); Norms on the Responsibilities of Transnational Corporations and Other Business Enterprises with Regard to Human Rights (2005); the IFC Performance Standards on Environmental and Social Sustainability (2006); the Guiding Principles on Business and Human Rights: Implementing the United Nations 'Protect, Respect and Remedy' Framework adopted by the UN Human Rights Council (2011); and the UN Human Rights Council to adopt a binding international instrument to govern MNCs (2014) (Abrahams 2004; Marquez 2015).

Despite all these international legal instruments, the impunity of MNCs with respect to human rights and environmental justice violations continues. These guidelines and principles could not ensure an enforcement mechanism because MNCs are not legally bound by these commitments. MNCs may exercise voluntary choices to respect human rights codes of conduct but they cannot be bound to them because they are not enforceable. On the one hand, these regulations have been lauded as steps in the right direction while on the other hand, they have been criticized for giving an 'illusion of regulation' when in fact there are none that are binding (Chesterman 2010).

Conclusion

The ambiguous legal status of MNCs allows them to function in a way that is unaccountable to international law, evading responsibility in cases of violations of human rights or questions of environmental justice. Economics and profits drive their choices of location, commodity and how they choose to operate. In a famous article Milton Friedman (1970: 126) writes, quoting his own 1962 book *Capitalism and Freedom*, "There is one and only one social responsibility of business – to use its resources and engage in activities designed to increase its profits." Karen Bell (2015: 1) asks, "Can the capitalist economic system deliver environmental justice? Or is it the capitalist system, itself, that has been at the root of the environmental and social crises we now face?" The answer is probably a bit of both.

Environmental Justice Organizations, Liabilities and Trade (EJOLT) has published an inventory and an atlas of environmental injustices and conflicts, demonstrating how local political ecologies are progressively becoming multinational and interconnected. In all of these environmental justice conflicts, local communities charged MNCs with environmental 'bads' and human rights violations. With 1672 cases, the long list involves transnational charges related to nuclear energy, gas, climate justice, biomass and land conflicts, mining, industrial waste and oil (https://ejatlas.org/company). Today, the culprits among multinationals that appear to have the most cases against them are (EJOLT): Royal Dutch Shell (42 cases), Nigerian National Petroleum Corporation (41 cases), Chevron Corporation (32 cases), Shell Petroleum Development Company (27 cases), Nigeria Agip Oil Company (22 cases), Monsanto Corporation (20 cases), Anglo Gold Ashanti (18 cases), ExxonMobil Corporation (15 cases), BHP Billiton (14 cases), ENEL Group (14 cases), Sacyr (12 cases), Barrick Gold Corporation (12 cases), Sinohydro Corporation Limited (11 cases), Endesa (11 cases), Total (11 cases), Dow Chemical Company (10 cases), Agip Group (9 cases). These cases involve low-income and minority populations of less-developed, resource-rich countries including specific Indigenous groups that have faced incursions on their land by MNCs to procure resources such as oil (Rourke and Connolly 2003).

Neo-liberal capitalism has become the unchallenged custodian of allocating environmental goods and 'bads', privileging the logic of the so-called 'free market' over any other contender, environmental justice included. The logic of the free market is flawed. It is 'not-so-free' after all and has created, maintained and aggravated the gulf between the rich and the poor in the world. The profit motives associated with every conceivable relationship have led to the commodification of sustainable development and environmental justice. The future of environmental justice, national and international, looks rather bleak in the face of the spreading culture of capitalism. The question is whether free-market capitalism can have a soul when its motives are profit or whether any such effort would merely be what Milton Friedman called "hypocritical window dressing"?

As a future direction for research, it could be interesting to run a correlation between the wealth generated by MNCs and the numbers of environmental justice cases held against them. One could hypothesize that the greater the profits generated by MNCs, the greater the case for environmental justice. If there are examples of MNCs that generate wealth through free-market capitalism but are not charged for violating EJ, lessons need to be drawn and shared from a legal, institutional and EJ perspective.

References

Abrahams, D. (2004). "Regulating Corporations: A Resource Guide." *United Nations Research Institute for Social Development (UNRISD)*. Geneva. www.unrisd.org

Agyeman, J. and Carmin, J. (2011). "Introduction: Environmental injustice beyond borders." In Carmin, J., and Agyeman, J. (eds.), *Environmental inequalities beyond borders: local perspectives on global injustices*, Cambridge, MA: MIT Press, pp. 1–16.

Alvarez, J.E. (2011). "Are corporations 'subjects' of international law?" *Santa Clara Journal of International Law*, vol. 1, pp. 1–36.

Amao, O. (ed.) (2013). *Corporate social responsibility, human rights and the law: Multinational corporations in developing countries*. Routledge Research in Corporate Law, Reprint Edition. Abingdon, UK: Routledge.

Anand, R. (2004). *International environmental justice: A north–south dimension*. Aldershot, UK: Ashgate Publishing.

Anderson, S. and Cavanagh, J. (2000). "Top 200: The rise of global corporate power." *Corporate Watch*, 2000, www.globalpolicy.org/component/content/article/221/47211.html

Bantekas, I. (2004). "Corporate social responsibility in international law." *Boston University International Law Journal*, vol. 22, pp. 309–347.

Been, V. (1993). "What's fairness got to do with it? Environmental justice and the siting of locally undesirable land uses." *Cornell Law Review, vol.* 78, pp. 1001–1085.

Bell, K. (2015). "Can the capitalist economic system deliver environmental justice?" In *Environmental Research Letters*, vol. 10, no. 12. Available at http://iopscience.iop.org/article/10.1088/1748-9326/10/12/125017/pdf (Last accessed 25 September 2016).

Bhagwati, J.N. (2012). "Multinational corporations and development: Friends or foes?" In Tihanyi, L., Devinney, T.M., and Pedersen, T. (eds.), *Institutional theory in international business and management (Advances in International Management*, vol. 25). Bingley, UK: Emerald Group Publishing Limited, pp. 5–14.

Bismuth, R. (2010). "Mapping a responsibility of corporations for violations of international humanitarian law: Sailing between international and domestic legal orders." *Denver Journal of International Law & Policy*, vol. 38, no. 2, pp. 203–226.

Business and Human Rights Resource Center (BHRRC) (2016). "Company policy on human rights." Retrieved September 12, 2016 from www.business-humanrights.org/

Buxton, N. (2014). "Introduction." In Buxton, N. (ed.), *State of Power 2014*. Available at www.tni.org/stateofpower2014.

Cervantes, J. (2013). "Ideology, neoliberalism and sustainable development, human geographies." *Journal of Studies and Research in Human Geography*, vol. 7, no. 2, pp. 25–34.

Chesterman, S. (2010). "Lawyers, guns, and money: The governance of business activities in conflict zones." *Chicago Journal of International Law*, vol. 11, pp. 321–342.

Chimni, B.S. (2004). "International institutions today: An imperial global state in the making." *European Journal of International Law*, vol. 15, no. 1, pp. 1–37.

Chomsky, N. (1999). *Profit over people: Neoliberalism and global order*. New York: Seven Stories Press.

Dunning, J.H. (1992). *Multinational enterprises and the global economy*. Reading, MA: Addison-Wesley.

Falk, R. (1999). *Predatory globalization: A critique*. Cambridge, UK: Polity Press.

Florida, R. (2005). "The world is spiky." *The Atlantic*, vol. 296(3), pp. 48–51.

Friedman, M. (1970). "The social responsibility of business is to enhance its profits." *New York Times Magazine*, vol. 32(13), pp. 122–126.

Friedman, T.L. (2005). *The world is flat: A brief history of the twenty-first century*. New York: Farrar, Straus & Giroux.

George, S. (2014). "State of corporations: The rise of illegitimate power and the threat to democracy." In Buxton, N. (ed.), *State of Power 2014*. Available at www.tni.org/stateofpower2014.

Gonzalez, C.G. (2001). "Beyond eco-imperialism: An environmental justice critique of free trade." *Denver University Law Review*, vol. 78, pp. 979–1016.

Gonzalez, C. (2015). "Environmental justice, human rights and the global south." *Santa Clara Journal of International Law*, vol. 13, pp. 151–195.

Graham, D. and Woods, N. (2006). "Making corporate self-regulation effective in developing countries." *World Development*, vol. 34, no. 5, pp. 868–883.

Greyl, L., Ojo, G.U, Williams, C., Certoma, C., Greco, L., Ogbara, N., and Ohwojeheri, A. (2013). "Digging deep corporate liability: Environmental justice strategies in the world of oil." *Report*,

No. 9. Available at: www.cid.org.nz/assets/Key-issues/Enviroclimate-change/2013.-Environmental-justice-strategies-in-the-world-of-oil.pdf (Last accessed January 25, 2016).

Haque, S. (1999). "The fate of sustainable development under neo-liberal regimes in developing countries." *International Political Science Review*, vol. 20, no. 2, pp. 197–218.

Harvey, D. (2005). *A brief history of neoliberalism*. Oxford: Oxford University Press.

Hendrawan, D. (2014). "Multinational corporations as the subject of international law." February 17, 2014. Available at SRN: http://ssrn.com/abstract=2397007 or http://dx.doi.org/10.2139/ssrn.2397007

Hertz, N. (2001). *The silent takeover: Global capitalism and the death of democracy*. London: Heinemann.

Hurrell, A. and Woods, N. (1995). "Globalisation and inequality." *Millennium: Journal of International Studies*, vol. 24, no. 3, pp. 447–470.

Joseph, S. (2000). "An overview of the human rights accountability of multinational enterprises." In Kamminga, M., and Zia-Zarifif, S. (eds.), *Liability of multinational corporations under international law*, The Hague: Kluwer Law International.

Korten, D.C. (1999). *The post-corporate world: Life after capitalism*. West Hartford CT and San Francisco CA: Kumarian Press and Berrett-Koehler Publishers.

McLeah, F. (2006). "Corporate codes of conduct and the human rights accountability of transnational corporations: A small piece of a larger puzzle." In DeSchutter, O. (ed.), *Transnational corporations and human rights*, Portland OR: Hart Publishing, pp. 219–240.

Marquez, D.I. (2015). "Legal avenues for holding multinational corporations liable for environmental damages in a globalized world." *ARACE – Direitos Humanos em Revista*, Ano 2, Numero 3, Setembro 2015, pp. 58–74. Available at https://arace.emnuvens.com.br/arace/article/download/53/38 (Accessed September 11, 2016).

Mayer, D. and Jebe, R. (2010)."The legal and ethical environment for multinationals." In O'Toole, J. and Mayer, D. (eds.), *Good business: Exercising effective and ethical leadership*, New York: Routledge.

Mikler, J. (2013). "Global companies as actors in global policy and governance." In Mikler, J. (ed.), *The handbook of global companies*, Oxford: Wiley-Blackwell, pp. 1–16.

Moran, T.H. (1999). *Foreign direct investment and development: The new policy agenda for developing countries and economies in transition*. Washington, DC: Institute for International Economics.

Morimoto, T. (2005). "Growing industrialization and our damaged planet: The extraterritorial application of developed countries' domestic environmental laws to transnational corporations abroad." *Utrecht Law Review*, vol. 1, no. 2, pp. 134–159.

Muchlinski, P.T. (2007). *Multinational enterprises & the law*. 2nd edition, International Law Library. Oxford: Oxford University Press.

Mujih, E. (2012). *Regulating multinationals in developing countries: A conceptual and legal framework for corporate social responsibility*. Farnham, UK and Burlington, VT: Gower.

Namballa, V.C. (2014). "Global environmental liability: Multinational corporations under scrutiny." *Exchanges: The Warwick Research Journal*, vol. 1, no. 2, pp. 181–205. Available at: http://exchanges.warwick.ac.uk/index.php/exchanges/article/view/26 (Last accessed: January 25, 2016).

Önis, Z. (1995). "The limits of neoliberalism: Toward a reformulation of development theory." *Journal of Economic Issues*, vol. 29, no. 1, pp. 97–119.

O'Rourke, D. and Connolly, S. (2003). "Just oil? The distribution of environmental and social impacts of oil production and consumption." *Annual Review of Environment and Resources*, vol. 28, no. 1, pp. 587–617.

Osofsky, H.M., Baxter-Kauf, K., Hammer, B. Mailander, A., Mares, B., Pikovsky, A., et al. (2012). "Environmental justice and the BP deepwater horizon spill." *NYU Environmental Law Journal*, vol. 20, pp. 99–198.

Peet, R. (2003). *Unholy trinity: The IMF, World Bank and WTO*. Malaysia: Zed Books.

Pentikainen, M. (2012). "Changing international 'subjectivity' and rights and obligations under international law: Status of corporations." *Utrecht Law Review*, vol. 8, no. 1, pp. 145–154.

Polanyi, K. (2001). *The great transformation: The political and economic origins of our time*. Boston, MA: Beacon Press.

Porritt, J. (2012). *Capitalism: As if the world matters*. Limited edition, London: Routledge.

Ruggie, J. (2010). "Protect, respect and remedy: The UN framework for business and human rights." In Baderin, M. and Ssenvonjo, M. (eds.), *International Human Rights Law*, England: Ashgate Publishing Limited, pp. 519–538.

Schlosberg, D. (2013). "Theorizing environmental justice: The expanding sphere of a discourse." *Environmental Politics*, vol. 22, no. 1, pp. 37–55.

Singer, P. (2002). *One world: The ethics of globalization.* Terry Lectures Series. New Haven, CT: Yale University Press.

Spero, J.E. and Hart, J.A. (1997). *The politics of international economic relations.* New York: St Martins Press.

Stiglitz, J.E. (2007). "Multinational corporations: Balancing rights and responsibilities." *Proceedings of the American Society of International Law,* vol. 101, pp. 3–60.

Summers, L. (1992). "Let them eat pollution." *The Economist* (London, England), Saturday, February 8, 1992, issue 7745, p. 82.

UNCTAD (2009). "World investment report 2009: Transnational corporations, agricultural production and development." New York–Geneva: United Nations.

Zerk, J.A. (2010). "Extraterritorial jurisdiction: Lessons for the business and human rights sphere from six regulatory areas." *Corporate Social Responsibility Initiative Working Paper* No. 59. Cambridge, MA: John F. Kennedy School of Government, Harvard University. Available at www.hks.harvard. edu/m-rcbg/CSRI/publications/workingpaper_59_zerk.pdf (Last accessed 25 September 2016).

39

GLOBALIZING ENVIRONMENTAL JUSTICE

Radical and transformative movements past and present

Leah Temper

Berta's struggle was not only for the environment, it was for system change, in opposition to capitalism, racism and patriarchy.

This quote comes from a statement delivered by the family of Berta Caceres, a Honduran Indigenous, female, environmental activist who was assassinated in her home on March 3, 2016, while this essay was being penned. Berta, a leader of the Lenca Indigenous people, and General Coordinator of the Civic Council of the Indigenous and Popular Organizations of Honduras (COPINH), was best known for her struggle against the Agua Zarca dam, for which she won the prestigious Goldman Environmental Prize in 2015. The case became a global symbol because after COPINH staged a blockade for over a year, they managed to get the world's largest dam company, Chinese State-owned Sino-Hydro, to abandon the project.

Berta's death, unfortunately, is not an extraordinary occurrence. In 2015, environmental activists were killed at the rate of three per week (Global Witness 2016). Nor was it a surprise – she was a known target. What was unexpected perhaps, was the scale and breadth of the globalized reaction to her death. Her tragic assassination triggered an unprecedented global response demanding justice for Berta and an end to impunity and corruption of extractive projects. The call came from Indigenous and human rights groups, LGBTQ activists, those contesting extractivism including mining and hydropower, climate justice activists, and feminists. I argue that in this out-pouring of solidarity, we can begin to trace the contours of an incipient and increasingly coherent global movement for environmental justice.

One dimension of the EJ movement's globalization has been its diffusion throughout the world. Environmental justice struggles are taking place across the Global North and Global South. They include movements of agrarian resistance against land-grabbing led by La Via Campesina, climate justice movements fighting dirty coal plants and pipelines that are rallying to "Leave the Oil in the Soil", as well as those contesting new processes of commodification of nature through mechanisms such as REDD, carbon credits and bio-diversity offsets; movements defending rights of access, such as pastoralists claiming access to watering holes and waste-pickers to revalorized waste resources being commodified by "green" incinerators. Over the past three years, the Environmental Justice Atlas (EJatlas), a project I co-direct with Joan Martinez-Alier, has traced the outline of this incipient global movement for environmental justice through its localized manifestations (Figure 39.1). The EJatlas (www.ejatlas.org) currently tracks almost 1900 ecological conflicts around the world (as of October 2016). Each point represents one conflict defined as a local mobilization against an environmentally destructive activity or policy (Figure 39.2). Each case or data-sheet is documented by an organization involved in the resistance, or an activist scholar, and outlines the history, actors, impacts, claims and outcomes of the conflict as well as photos, references, and links (for a description of the project see Temper et al. 2015).

While these cases are collected under the banner of environmental justice, it should be noted that many of the peasant and urban movements documented may not explicitly adopt an environmental justice discourse. There is no global unified environmental justice movement *per se* (Martinez-Alier et al. 2016), but rather a plurality of movements and struggles with their own agency and ideological hybridity (Baviskar 2005) that are increasingly networked and brought into strategic alliances. Environmentalism is thus one representation of several contending subjective meanings attached to struggles, along with gender, class, caste, ethnic and nationalist meanings. This networking and alliance-building between struggles constitutes a second dimension of the EJ movement's globalization.

Recent scholarship has thus grappled with the question of how to define "global environmental justice" (Schlosberg 2009, 2013; Martinez-Alier et al. 2016; Sikor & Newell 2014; Agyeman 2014; Walker 2009). This work has made productive contributions regarding the dimensions of justice and inequality across both locations and the global trans-national institutions and interconnections that join them, continually highlighting the plurality of justice norms across diverse cultural, social and environmental contexts. Following Guha and Martinez-Alier's seminal work (1997) that introduced the "environmentalism of the poor" it has been generally accepted that there is an important distinction in terms of origins and forms of articulation between the ways in which environmental action characteristically expresses itself in the North and the South. While Northern and particularly US environmentalism is deemed to be aesthetic and moral, rooted in the politics of consumption, Southern environmentalisms are understood to be about survival and subsistence, the ecosystem people against the omnivores (Gadgil & Guha 1995).

Here I would like to attempt to move past this distinction, agreeing with Newell (2005) that the generic categories of North and South are increasingly inadequate for the study of global environmental politics. As local political ecologies are becoming increasingly trans-national and interlinked across geographies, new points of convergence are manifold as campaigns around biofuels and food sovereignty, land-grabbing, climate and food justice simultaneously address sectors such as agriculture, energy generation, water management and financial markets, demanding action at a global scale of governance.

This chapter contributes to clarifying the nature and shape of the global environmental movement, drawing on literature on the history of environmental protest as well as on the empirical evidence from the Global Environmental Justice Atlas. Through these narratives, as well as literatures from political ecology, environmental philosophy, eco-feminism, ecological economics, history, anthropology and sociology, this chapter aims to distill some of the core characteristics, unresolved tensions and relevant future lines of enquiry of an emerging radical and transformative global EJ movement. These include a focus on ecological justice that takes into account relations with nonhuman nature and navigates the tension between conservation and livelihoods, considered in the first section; a global materialist

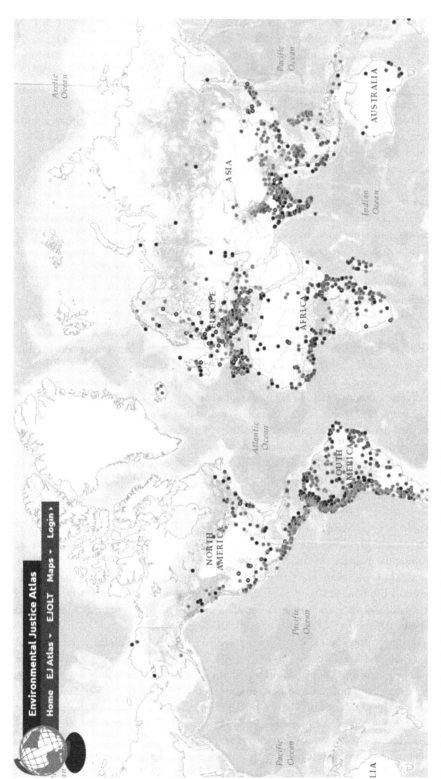

Figure 39.1 The online platform of the Environmental Justice Atlas (www.ejatlas.org)

Figure 39.2 A case in the Environmental Justice Atlas: Yanacocha mine, Peru

perspective that questions the structural basis of the economy, taken up in the second; and the increasing globalization and interconnectedness between struggles and their simultaneously oppositional and constructive politics, illustrated in the third. The article concludes with avenues for future research.

By drawing on historical as well as current examples, I hope to demonstrate that EJ, while only recently defined as such, has a long and rich history around the world that far predates Western environmentalism as we know it. This brings attention to the need for diverse and situated understandings of environmentalism, and for going beyond northern theoretical understandings of environmental justice to one informed by non-Eurocentric epistemologies and ontologies.

Within this diversity, this chapter outlines defining characteristics and tensions that have endured over time while also highlighting how a globalizing world has served to transform environmentalism through an emerging planetary consciousness.

Exhibit A: justice for human and nonhuman nature – the eco-dharma of the Bishnois

Environmental justice has generally been considered an anthropocentric endeavour, primarily concerned with the displacement of environmental risks onto parties not implicated in their production. This framing may imply that the activities themselves do not need to be questioned, only the distribution of risks. In contrast, ecological justice has been used to refer to the justice of the relationship between humans and the rest of the natural world (Low & Gleeson 1998), and calls for addressing the ecological quality of our practices (Stevis 2000). While environmental justice has traditionally positioned itself counter to an environmentalism of conservation and wilderness protection (Guha & Martinez-Alier 1997), the global brand of EJ transcends and dissolves some of these distinctions – pointing to the inseparability between justice for nature and justice for humans.

To illustrate this blurring of the distinction between environmental and ecological justice, we can turn to what is perhaps the oldest recorded protest event in environmental history, the story of the Bishnoi collective martyrdom against deforestation in India in 1730 CE. The campaign was led by Amrita Devi, the first known tree-hugger and eco-feminist, and a follower of the Bishnoi, an offshoot sect of Hinduism, based on the teachings of Guru Jambhesvara (Jain 2010).

When the Maharaja Abhayasingh of Jodhpur demanded wood from his subjects to build a fortress and the Bishnoi community defied his orders he sent soldiers to their village to cut down the Khejari trees which grew in the area. When Amrita Devi came to know about it, she faced the soldiers and proclaimed that anyone wishing to cut a tree would have to first cut through her, saying: *Sar santey rookh rahe to bhi sasto jaan* (If a tree is saved even at the cost of one's head, it's worth it). It is said that weaponless, she hugged the nearest Khejari, and the axe-wielding soldiers cut through her neck and the tree, severing both. Her daughters followed in her footsteps and also lost their heads. In response, the community decided that for every tree cut, one Bishnoi volunteer would sacrifice his or her life. When the Maharaja was informed of the events, he ordered the felling to be stopped. Yet by this time, 363 Bishnois had been martyred (Dwivedi 2005).

The story is probably preserved for posterity because the Maharaja, in recognition of their courage, apologized and issued a royal decree, engraved on a copper plate, prohibiting the cutting of trees and hunting of animals within the boundaries of Bishnois villages. The case is exceptional in that despite the bloodshed, it represents a rare record of historical success in what we may term proto-environmental justice activism. The territory remains a protected area until today, while the Bishnois' struggle has gone on to inspire other movements in India, particularly the Chipko Andolan "forest huggers" movement in the Himalayas led by women (Shiva & Bandyopadhyay 1986) and can thus be seen as an important influence for global EJ (see also Chapter 48).

The Bishnoi case highlights the often spiritual and religious aspect of environmental defence, which they term an 'eco-dharma'. It was inspired by the tenets of Guru Jambhesvara (1451–1536) who, after witnessing drought that led to decimation of the local wildlife for meat, had a vision where he saw people quarrelling with nature and destroying the environment that sustained them. He realized that humans have to sustain the environment around them in order for nature to sustain the humans. His 29 tenets of the religion include everyday practices of ecological activism aimed to enable the flourishing of life and a synthesis between ecological, social and community health in harsh desert conditions. Informed by this philosophy we can say that the Bishnoi defence of the forest went beyond a conservationist ethic, instead emanating from a religious reverence for nonhuman life, and can also be understood as the defence of the underlying basis of their social ecology.

The Bishnoi philosophy can be seen as an early example of EJ activism that appears to transcend anthropocentric and eco-centric approaches to justice. Increasingly this perspective is gaining prominence demonstrated by a push to progressively broaden the claimants of justice to include nonhuman nature. The Principles of Ecological Justice from the First People of Color Leadership Conference, UN Declaration on the Rights of Indigenous Peoples, the Indigenous Environmental Network, the Bolivian "rights of Mother Nature" initiative, as well as numerous other declarations on environmental and climate change by Indigenous peoples support this view (Figueroa 2011; Maldonado et al 2013; Kopnina 2014).

Such initiatives are informed by Indigenous environmental epistemologies and often rely on a narrative of reciprocity and relational ontologies based on natural law, which defines the relationship and responsibility between people and the environment. As Indigenous EJ scholar Winona LaDuke (1994; ix) explains, "All parts of the environment—plants, animals, fish, or rocks—are viewed as gifts from the Creator. These gifts should not be taken without a reciprocal offering, usually tobacco or *saymah*, as it is called in the Ojibwa language."

On the surface, there may appear to be a contradiction between an ecological perspective based on the rights of mother nature, and issues of access represented by the livelihood approach of the environmentalism of the poor. Schlosberg (2009), for example, eschews an a priori hierarchy among species, considering various models of political deliberation among a plurality of parties, including nonhumans. However, scholars making such claims tend to underestimate the potential for serious conflict among the various dimensions of and claimants to ecological justice and the political implications of "subjectifying" nature. When it is humans against nature, who comes first?

This question was brought to the members of the 17th World Congress of the Anthropological and Ethnological Sciences in 2013, when the motion 'Justice for people should come before justice for the environment' was debated (Kopnina 2014). Those in favour, including Amita Baviskar (2005), pointed to the social costs of nature preservation and how conservation is frequently a tool of neo-colonialism insensitive to local cultures, arguing that anthropologists have a duty to uphold human (economic and social) rights and Indigenous entitlements against Western environmentalists.

Those against the motion countered that humans, other species, and the material world are bound together in communal processes of production and reproduction that are interdependent and that only a recognition of this interconnectedness and a consideration of the rights of nonhuman species simultaneous with the social justice movement holds the promise for long-term sustainability and justice (Kopnina 2014). At the end, the attendees overwhelmingly supported this position.

This tension between justice for humans and justice for nonhuman nature, between conservation and access, is not easily resolved and remains an ongoing challenge for the practice, policy and study of global environmental justice. For example, EJ activists have been vocal supporters of new legislation that aims to give rights to nature such as the implementation of Ecuador's constitution on the rights of Mother Nature (Fish 2013) and a more recent initiative to give the Te Urewera National Park in New Zealand juridical status similar to corporations under the stewardship of the Maori (Magallanes 2015). Such initiatives have been lauded for integrating Indigenous cosmo-visions and ontologies that grant subjectivity to nature. However, the translation of such rights on paper into real rights is fraught with difficulties and the politics of socionatural hybridity remain very much under construction.

Granting rights to nature, while discursively powerful, still calls for someone to defend those rights. Accepting that nonhuman beings have interests that need to be taken into account in public decisions necessitates accepting that these interests will be voiced by human agents. As O'Neill (2006: 262) says, "The nonhuman natural world does not speak to us. Neither does nature listen." This leads him to the question: Who speaks for nature? With what legitimacy? Nature is not a single and homogenous entity but a complex ecosystem with competing interests among organisms at the local scale, as well as between scales. Further, nature cannot elect a representative nor represent itself.

One potentially productive avenue when deciding who can represent nature is to draw on eco-feminist epistemology which draws a parallel between domination over women and other marginalized groups and domination of nature, contending that nature and women are both socially constructed as something "other" to have power over (Mies & Shiva 1993). Feminist scholars suggest that the way forward is to work towards the dismantling of dualisms such as man/woman and culture/nature and instead bring attention back to the distributions of power and privilege at play (Mies & Shiva 1993; Plumwood 2004). A feminist lens directs our attention away from questions of whether nature's rights are being harmed towards the underlying power dynamics, asking "what specific social/institutional configurations make it such that . . . many do not have the resources to attempt to 'live in integrity' with nature?" (Davion 2002: 57). Recast this way, even the soldiers who severed Amrita's head were acting in the way they thought they were supposed to behave within their cultural and structural constraints. This perspective calls attention to the reality that since nature cannot represent itself, many will speak for the environment, and such voices need to be heard within concrete contexts of power and privilege.

The story of the Bishnoi illustrates a rare historical victory for sub-altern voices, albeit at a price. It also serves as an early example of how engagement with local identities and meanings, norms and existing taboos can inform environmental management based on widespread participation and deliberative democracy and informed by an ethics of care. This does not foreclose the political act of relating to nature, but requires an expression of the kind of relationships we want to build and the power relationships embedded in them.

Exhibit B: social metabolism and the environmental philosophy of Tanaka Shozu

Another key concept relevant to understanding global EJ is an understanding of how local issues and struggles are affected by larger-scale processes and vice versa. Relevant frameworks for analysing these dynamics include social metabolism and the related concepts of unequal ecological exchange and ecological debt. The concept of social metabolism refers to the physical throughput of the economic system, in terms of the energy and materials associated with economic activities, as either direct or indirect inputs or wastes (Fischer-Kowalski & Haberl 2007). Geographically uneven and socially unequal metabolic processes are key to understanding environmental inequality, which in turn both reflects and reinforces overt forms of hierarchy and exploitation. EJ informed by ecological economics brings attention to the metabolic patterns driving environmental change and gives further insights into how uneven flows of matter and energy and transformations in the extraction and provision of natural resources characterize different socio-ecological transitions (Fischer-Kowalski 1997; Fischer-Kowalski & Haberl 2007) and lead to novel forms of social contestations.

The earliest recorded case in the Environmental Justice Atlas is that of the Ashio copper mine in Japan (1903). It provides a fruitful illustration of early EJ activism driven by new socio-metabolic transformations. While environmentalism in the West emerged primarily out of a concern for preserving nature and wildlife, early movements in Japan were very much focused on the local social and health impacts associated with new forms of industrial pollution, similar to the US EJ movement (Imura & Schreurs 2005).

The leader of the movement against the mine, Tanaka Shozo (1841–1913), widely acknowledged as Japan's "peasant environmentalist" (Martinez-Alier 2009), can also be heralded as one of the country's first public EJ activists and thinkers. He played a dual role in mobilizing and participating in direct action as well as representing the villagers in the National Diet in Tokyo. Yet the prescience and significance of his ecological thought, laid out in detail by Stolz (2006, 2007), remains under-appreciated and is worth examining as it holds continuing relevance for global EJ today.

Tanaka developed a sophisticated ecological theory of society informed by his experience of activism, based on the twin processes of nature: "poison" (*doku*) and "flow" (*nagare*) (Stolz 2006). When the government concluded that the solution to the mine pollution was flood control and the construction of a dam, it instituted a massive re-engineering of the watershed, signalling "the beginning of the Japanese state's systematic intervention in nature". Tanaka, the anti-mining activist, became a water justice activist and went on to develop his fundamental River Law (*konponteki kasenho*), foreshadowing recent studies such as that of the World Commission on Dams (2000) by over a century, by pointing to the intertwined environmental and social harms that come from ignoring the dictates of an active nature in the name of absolute human agency (Stolz 2006):

> For Tanaka, because motion was inherent in nature, the state's policies of control through constriction and manipulation of the rivers' currents would not have the desired effect of wholly controlling the river. On the contrary, they would result in a harmful "backflow" or "reversal of flow" (gyakuryu) as the river confronted the concrete banks, sluices, and reservoirs and reversed itself, resulting in flooding upstream. (This is precisely what happened.) ... As Tanaka's thought developed, doku came to describe not only the presence of toxins in the watershed's fields, but also the horizon of beneficial human intervention in nature.
>
> *(Stolz 2007: 4)*

Tanaka showed how the accumulation of harm from bad environmental policy moved from the environmental to the social realm, making the link between environmental issues and social justice at the heart of EJ. He demonstrated how the increase in *doku* led inevitably to social and political repression, including the repression of the anti-mine petitioners, the taming of the river through damming and an escalating cycle of displacement of externalities culminating in the destruction of Yanaka village. Tanaka stated, "the mine poison problem has mutated; it has become the theft and destruction of homes" (Arahata 1999: 8, quoted in Stolz 2006).

Tanaka's ideas about flow and poison foreground the global perspective on EJ today, in the acknowledgement of how impacts at one scale radiate outwards. His description of *doku* can be seen as a precursor to Georgescu-Roegen's work on entropy and the economic process (1987), where he demonstrated how societies generate order by continually importing low-entropy matter-energy from the environment and exporting high-entropy matter-energy back to the environment. Georgescu-Roegen's work has been extended by scholars such as Hornborg (1992) to explain how environmental inequality is a result of both uneven access to the basic sources of low entropy energy – solar and terrestrial or stored energy (such as coal and oil) – as well as the uneven flows of high entropy wastes (such as carbon dioxide and other pollutants) that are discharged from the economic metabolic system and displaced onto the most vulnerable populations. This socio-metabolic perspective brings to the fore how conflicts over oil extraction, coal and fossil fuels (of which 352 have been documented to date in the EJatlas), and the sea-level rise, droughts, floods and tornadoes associated with climate chaos, are opposite sides of the same coin. Global environmental and climate justice activists increasingly bring attention to this interconnectedness between access to resources and risk, and from sources to sinks.

Climate chaos represents the most cogent example of the unequal transfer of entropy at a global scale. Thus, a Black Lives Matter protest in Britain in September 2016 blocked London City Airport, arguing that "the climate crisis is a racist crisis" because seven out of the ten countries most impacted will be in Africa. Activists deploy concepts such as the ecological debt from North to South, a concept born in Latin America in 1991 (Martinez-Alier et al. 2014), and "ecologically unequal exchange" to draw attention to how the high material standards of developed countries are dependent on net transfers of materials and energy from the periphery to the industrial centre; while the less developed countries and regions exporting the resources experience a net increase in entropy (disorder), leading to environmental disruption and degradation, as natural resources and traditional social structures are dismembered (Hornborg 1992).

Across the world, those protesting against environmental injustice increasingly go beyond asking for compensation or access to resources, and instead question the basic configuration of the global socio-metabolic system based on a false growth imperative and on prices that do not reflect environmental costs. In this way, the discourse of global EJ goes beyond a mere quantitative question of distribution or of reducing the size of the economy and its material and energy throughput. It includes calls for a complete transformation of the economic system. A central element of this transformation is recognition of the reproductive and care labour undertaken by women, as well as nature, which are both considered free gifts to capital (Salleh 2010). Thus it can be said that while the US Environmental Protection Agency considers that EJ will be achieved when "everyone enjoys the same degree of protection from environmental and health hazards", global EJ assumes a more radical position, claiming that only the restructuring of dominant economic models, social relations and institutional arrangements can address social, political, economic and environmental inequities.

Exhibit C: globalizing resistance – the Unist'ot'en camp

A global EJ perspective highlights how localized processes of resistance are intertwined with processes at larger scales brought about by expanding capitalist markets, multiple forms of commoditization, and global flows of resources and information. Those engaged in place-based struggles can contest the "spaces of flows" of global capitalism through a variety of strategies. This may include "scale jumping", or forging alliances with actors at different spatial scales and with differential access to networks of institutional, financial, and political support, to externalize their political claims (Smith 1996).

At the same time, local movements motivated in part by material demands, such as those of the environmentalism of the poor, are increasingly linking their struggles to impacts and processes occurring at much broader scales and of global concern, while they propose new alternatives. The degrowth movement, the global movement for food justice, and Indigenous activists and their allies are not mobilizing due to concern for their own livelihood and security alone. Instead, they share the concern that environmental pressures wrought by capitalism, its crises, and its "spatial and temporal fixes" are threatening the material basis of the entire planet. Global environmental justice can thus be said to be motivated by a new planetary consciousness and understanding of the interlinked nature of geographies of environmental injustice. This networking between struggles and movements constitutes the global EJ movement.

The case of the Unist'ot'en camp of the Wet'suwet'en is instructive. The Wet'suwet'en First Nation territory spans over 22 000 km in northwestern British Columbia, Canada, and lies directly in the path of several proposed gas and oil pipelines. Since 2010, the Unist'ot'en clan, members of the Wet'suwet'en Nation, have been reoccupying and re-establishing themselves on their ancestral lands in opposition to these projects. They have set up a camp on the GPS coordinates of the pipeline route and refuse to allow any pipelines to cross their territory, which they term as "occupied and un-ceded" as the tribe has never signed a treaty with the Government of Canada.

The First Nation claims they have not been properly consulted, and objects that the pipeline contributes to expanding shale gas extraction through hydraulic fracturing ('fracking'), which uses and destroys enormous volumes of fresh water. They also claim that they are operating in solidarity with neighbouring communities who want to stop all pipelines, reverse climate change, shut down tar sands and oppose what they claim are false solutions to climate change: carbon marketing; carbon, boreal and biological offsets; and reducing emissions from deforestation and forest degradation (REDD).

The position of the Unist'ot'en camp is counter to a NIMBY approach, which would confine their resistance to concerns for distributive impacts on their territory alone. They are uncompromising; neither the Pacific Trails Pipeline nor any other pipeline will be allowed through their territories. This complete opposition to all pipelines – existing, proposed or approved to expand – has generated wide support for the camp, which has become a symbol against extractivism. This networked resistance is a defining characteristic of global environmental justice today, as groups working on global climate change create alliances with local place-based movements to demand structural transformation of the economy. As Combes describes of the fracktivist movement (in Temper et al. 2013), while part of the struggle keeps a local defensive attitude, another line is focused more on proactive work to broaden the mobilization beyond the locality, integrating discourses on energy democracy, sovereignty and climate justice. Similarly, we see global anti-incineration movements forming coalitions with local waste-picker and recycling movements. They point to how they reduce greenhouse gas emissions through the reuse of materials and are fighting for recognition of their significant environmental contributions in recycling and the role this plays in climate change mitigation and a healthy economy.

While the Unist'ot'en checkpoint disrupts flows of capital and blocks the movement of resources out of the territory, it simultaneously creates a space where the people may practise and assert their sovereignty and enact what they view as their sacred responsibility to all life within their territory. For example, the Unist'ot'en have erected an Indigenous healing centre within the pathway of multiple pipelines meant to cross their territory (Temper et al. 2015). As Mel Basil, a member of the collective, says: "I don't have a right to these fish – I have a responsibility to this river and I will not let that responsibility be diminished."

The recent announcement of the Canadian Government revoking the licence of the Enbridge oil pipeline largely in response to Indigenous protest, including that of the Unist'ot'en camp, attests to how this power of responsibility can indeed be exercised from below. It demonstrates how environmental justice activism contributes to re-ordering and reshaping social and material relations, and the spatial organization and distribution of risk, pollution, entropy, and environmental inequities for some time to come, and at a global scale.

Conclusion

> We are Nature Defending Itself
>
> (activist slogan from 2015 Paris Climate Summit)

This chapter has drawn from both historical and emblematic cases from the Environmental Justice Atlas to describe key features of an emerging global environmental justice discourse and a globalizing movement. The tools may have changed across time – today the response to Berta's death is globally transmitted and shared through digital communications. However, violence against nature and women is an omnipresent motif, from Amrita to Berta, mostly unchanged in shape and form. In Ashio, the dam was a means to stem and control the polluted floodwaters from the mine; meanwhile the Agua Zarca dam that COPINH is resisting is one among 48 dams planned or underway on their lands, primarily for hydro-power for mining projects. An ironic reversal. A reflection on these key historical and current examples demonstrates both the permanence and the novelty in how popular environmental activism expresses itself, the values that drive it, the response from the state and power-holders, and the impacts of activism across time and space.

Numerous scholars have contended that climate change and a convergence of crises over food, fuel, energy and finance are contributing to an "environmentalization of social issues" (Acselrad 2010). While EJ movements are sometimes painted as defensive and reactionary, I argue that then as now, global EJ displays characteristics of being both confrontational and constructive. Counter to Swyngedouw's (2010) concern that environmental questions are becoming part of a new "post-political" consensus, this chapter suggests that global environmental justice offers the possibility to position environmental questions within a larger critique of unequal social relations and the capitalist globalization process they are embedded in. According to authors such as Acosta (2009) and Svampa (2013), the escalation of socio-environmental conflicts has often been accompanied by the emergence of new forms of political mobilization and civic participation focusing on defending the commons, biodiversity, and the environment. They lead to the introduction of new forms of governance across multiple scales as well as new forms of participatory democratic decision-making. Concrete examples include the spread and diffusion of popular consultations as regards mining projects in Latin America (Walter & Urkidi 2015).

A globalized EJ calls for the need for intercultural communication and acceptance of other worldviews and a plurality of ways of understanding nature. In this way it reminds us that conflicts over the environment are epistemic struggles wherein other forms of the political, other economies, other knowledges are produced and theorized and hegemonic worldviews are questioned and reformulated (Escobar 2016). And that new knowledge practices are created through processes of struggle (Temper & Del Bene 2016), through the reversal of enclosures of land and resources, and through the active defence of existing and new commoning practices and social relations as communities understand and contest the mechanisms of their oppression (Casas-Cortés et al. 2008; Brownhill et al. 2012). Engagement in processes of struggle offers a space to exercise a citizenship and participation that is often denied, and for the expression of plural values that may conflict in ways that simply cannot be neatly or even satisfactorily resolved. The practice of environmental justice through struggle opens up this space for contesting claims, and for competing definitions of nature and justice to be expressed and debated.

The EJatlas and the current 1850 cases (and growing) provide empirical material for a research agenda that contributes to understanding how inequalities are shaped through socio-metabolic transformations in the economy and how they are contested, and to what outcomes. Further documentation and analysis via a comparative political ecology holds significant promise for extending both the praxis and theory of environmental justice and geographical scholarship.

Further, the EJatlas offers fruitful opportunities for comparative research delving into the shape of global EJ and the differences and similarities between environmentalisms across locations, as well as on how local political contexts shape strategies and action-forms. For example, we may ask: Are environmentalist concerns in some countries more likely to be expressed in a contentious manner? What is the role of disruptive protest and how does escalation of protest activities lead to differing outcomes? How are different languages of valuation such as identity politics deployed? How is the technical language of Western environmentalism (increasingly used for strategic reasons) combined with arguments about identity and culture, and to what effect? How do new forms of action appear and how are they diffused across time and space? Which of the sustained actions achieve their goals and why?

Finally, while the EJatlas clearly illustrates what environmental injustice is and contributes to understanding the mechanisms of enclosure, exclusion and externalization through which it operates, a question demanding further study is whether the positive ideals of EJ exist in practice. If so, what does it look like and when and how does it obtain? What is the relationship between EJ and transformations to sustainability? How and when have communities been able to put their alternative practices and visions into place?

An examination of global EJ through the stories of struggle can help point towards practicable paths to equitable and sustainable society–nature relationships. At the same time, an ecologically informed global resistance under the banner of EJ can serve as a convergence issue where diverse movements challenging diverse forms of oppression, including racism, sexism, speciesism and physical and epistemic violence can coalesce and potentially where new solidarities and alliances can be born.

References

Acosta, A. (2009). *La Maldición de la Abundancia*. Quito: Abya Yala.

Acselrad, H. (2010). "Ambientalização das lutas sociais-o caso do movimento por justiça ambiental". *Estudos avançados*, vol. 24, no. 68, 103–119.

Agyeman, J. (2014). "Global environmental justice or le droit au monde?" *Geoforum*, vol. 54, pp. 236–38.

Agyeman, J., Bullard, R.D., & Evans, B. (2003). *Just Sustainabilities: Development in an unequal world*. London: Earthscan.

Baviskar, A. (2005). "Red in Tooth and Claw? Looking for Class in Struggles over Nature" in Ray, R. and Katzenstein, M.F. (eds.), *Social movements in India: Poverty, power, and politics*. Oxford: Rowman and Littlefield, pp. 161–178.

Brownhill, L., Turner, T.E., & Kaara, W. (2012). "Degrowth? how about some "de-alienation?" *Capitalism Nature Socialism*, vol. 23, no. 1, pp. 93–104.

Casas-Cortés, M.I., Osterweil, M., & Powell, D.E. (2008). "Blurring boundaries: recognizing knowledge-practices in the study of social movements." *Anthropological Quarterly*, vol. 81, pp. 17–58.

Chatterton, P., Featherstone, D., & Routledge, P. (2013). "Articulating climate justice in Copenhagen: antagonism, the commons, and solidarity." *Antipode*, vol. 45, no. 3, pp. 602–620.

Davion, V. (2002). "Ecofeminism, Integrity and the Earth Charter: A Critical Analysis" in Miller, P. and Westra, L. (eds.), *Just ecological integrity: The ethics of maintaining planetary life*. Oxford: Rowman & Littlefield.

Dwivedi, O.P. (2005). "Satyagraha for Conservation: Awakening the Spirit of Hinduism" in Gottlieb, R.S. (ed.), *This sacred earth: Religion, nature, environment*. London: Routledge, pp. 145–157.

Escobar, A. (2016). "Thinking-feeling with the earth: territorial struggles and the ontological dimension of the epistemologies of the south." *AIBR. Revista de Antropología Iberoamericana*, vol. 11, pp. 11–32.

Figueroa, R.M. (2011). "Indigenous peoples and cultural losses" in Dryzek, J.S., Norgaard, R.B. and Schlosberg, D. (eds.), *The Oxford handbook of climate change and society*. Oxford: Oxford University Press, pp. 232–250.

Fischer-Kowalski, M. (1997). "Society's metabolism: origins and development of the material flow paradigm." *From Paradigm to Practice of Sustainability*, vol. 21, p. 15.

Fischer-Kowalski, M. & Haberl, H. (eds.) (2007). *Socioecological transitions and global change: Trajectories of social metabolism and land use*. Cheltenham, UK and Northampton, MA: Edward Elgar Publishing.

Fish, L. (2013). "Homogenizing community, homogenizing nature: an analysis of conflicting rights in the rights of nature debate." Stanford University. Available at: http://web.stanford,edu/group/journal/cgi-bin/wordpress/wp-content/uploads/2013/06/Fish->pdf.

Gadgil, M. & Guha, R. (1995). *Ecology and equity: The use and abuse of nature in contemporary India*. New Delhi: Penguin Books.

Georgescu-Roegen, N. (1971). *The entropy law and the economics process*. Cambridge MA: Harvard University Press.

Global Witness (2016). "On Dangerous Ground." www.globalwitness.org/en/reports/dangerous-ground/

Guha, R. & Martinez-Alier, J. (1997). *Varieties of environmentalism: Essays north and south.* London: Earthscan.

Hornborg, A. (1992). "Machine fetishism, value, and the image of unlimited good: towards a thermodynamics of imperialism." *Man*, vol. 27, pp. 1–18.

Hornborg, A. (1998). "Towards an ecological theory of unequal exchange: articulating world system theory and ecological economics." *Ecological Economics*, vol. 25, pp. 127–136.

Hornborg, A. (2005). "Footprints in the cotton fields: the industrial revolution as time–space appropriation and environmental load displacement." *Ecological Economics*, vol. 59, pp. 74–81.

Imura, H. & Schreurs, M.A. (eds.). (2005). *Environmental policy in Japan.* Cheltenham, UK and Northampton, MA: Edward Elgar.

Jain, P. (2010). "Bishnoi: an eco-theological 'new religious movement' in the Indian desert." *Journal of Vaishnava Studies*, vol. 19, no. 1, pp. 1–20.

Jain, P. (2011). *Dharma and ecology of Hindu communities: Sustenance and sustainability.* Burlington: Ashgate.

Karan, P. & Suganuma, U. (eds.). (2008). *Local environmental movements: A comparative study of the United States and Japan.* Lexington, Kentucky: University Press of Kentucky.

Kopnina, H. (2014). "Environmental justice and biospheric egalitarianism: reflecting on a normative-philosophical view of human–nature relationship." *Earth Perspectives*, vol. 1, no. 1, pp. 1–11.

LaDuke, W. (1994). "Foreword" in Gedicks, A. *The new resource wars: Native and environmental struggles against multinational corporations.* Boston: South End Press.

Low, N. & Gleeson, B. (1998). "Situating justice in the environment: The case of BHP at the Ok Tedi Copper Mine." *Antipode*, vol. 30, no. 3, pp. 201–226.

Magallanes, C.J. (2015). "Nature as an Ancestor: Two Examples of Legal Personality for Nature in New Zealand" *VertigO - la revue électronique en sciences de l'environnement* [online], consulted septembre 2016. URL : http://vertigo.revues.org/16199 ; DOI : 10.4000/vertigo.16199

Maldonado, J.K., Shearer, C., Bronen, R., Peterson, K., & Lazrus, H. (2013). "The impact of climate change on tribal communities in the US: displacement, relocation, and human rights." *Climatic Change*, vol. 120, no. 3, pp. 601–614.

Martinez-Alier, J. (2003). *The environmentalism of the poor: A study of ecological conflicts and valuation.* Cheltenham, UK: Edward Elgar.

Martinez-Alier, J. (2009). "Social metabolism, ecological distribution conflicts, and languages of valuation." *Capitalism Nature Socialism*, vol. 20, no. 1, pp. 58–87.

Martinez-Alier, J., Anguelovski, I., Bond, P., Del Bene, D., Demaria, F., Gerber, J.F., et al. (2014). "Between activism and science: grassroots concepts for sustainability coined by EJOs." *Journal of Political Ecology*, vol. 21, pp. 19–60.

Martinez-Alier, J., Temper, L., Del Bene, D., & Scheidel, A. (2016). "Is there a global environmental justice movement?" *The Journal of Peasant Studies*, vol. 43, pp. 731–755.

McNeill, J.R. (2001). *Something new under the sun: An environmental history of the twentieth-century world.* New York: WW Norton & Company.

Mies, M. & Shiva, V. (1993). *Ecofeminism.* London. Zed Books.

Muradian, R., Walter, M., & Martinez-Alier, J. (2012). "Hegemonic transitions and global shifts in social metabolism: implications for resource-rich countries. Introduction to the special section." *Global Environmental Change*, vol. 22, no. 3, pp. 559–567.

Newell, P. (2005). "Race, class and the global politics of environmental inequality." *Global Environmental Politics*, vol. 5(3), pp. 70–94.

O'Neill, J. (2006). "Who speaks for nature?" in Haila, J. and Dyke, C. (eds.), *How nature speaks: The dynamics of the human ecological condition.* Durham NC: Duke University Press, pp. 261–78.

Plumwood, V. (2004). "Gender, eco-feminism and the environment." In White, R. (ed.), *Controversies in environmental sociology.* Cambridge: Cambridge University Press.

Salleh, A. (2010). "From metabolic rift to 'metabolic value': reflections on environmental sociology and the alternative globalization movement." *Organization & Environment*, vol. 23, no. 2, pp. 205–219.

Schlosberg, D. (2009). *Defining environmental justice: Theories, movements, and nature.* Oxford: Oxford University Press.

Schlosberg, D. (2013). "Theorising environmental justice: the expanding sphere of a discourse." *Environmental Politics*, vol. 22, pp. 37–55.

Shiva, V. & Bandyopadhyay, J. (1986). "The evolution, structure, and impact of the Chipko move-ment." *Mountain Research and Development*, vol. 6, pp. 133–142.

Sikor, T. & Newell, P. (2014). "Globalizing environmental justice?" *Geoforum*, vol. 54, pp. 151–57.

Smith, N. (1996). "Spaces of vulnerability: the spaces of flows and the politics of scale." *Critique of Anthropology*, vol. 16, pp. 63–77.

Stevis, D. (2000). "Whose ecological justice?" *Strategies: Journal of Theory, Culture & Politics.* vol. 13, no. 1, pp. 63–76.

Stolz, R. (2006). "Nature over nation: Tanaka Shōzō's fundamental river law." *Japan Forum*, vol. 18, no. 3, pp. 417–437.

Stolz, R. (2007). "Remake politics, not nature: Tanaka Shozo's philosophies of 'poison' and 'flow' and Japan's environment." *The Asia-Pacific Journal/Japan Focus*, vol. 5, no. 1, pp. 1–9.

Svampa, M. (2013). "'Consenso de los Commodities' y lenguajes de valoración en América Latina". *Nueva sociedad*, vol. 244, pp. 30–46.

Swyngedouw, E. (2010). "Apocalypse forever? Post-political populism and the spectre of climate change." *Theory, Culture & Society*, vol. 27, no. 2–3, pp. 213–232.

Temper, L. & Bliss, S. (2015). "Decolonising and decarbonising: how the Unist'ot'en are arresting pipelines and asserting autonomy." in Temper, L. and Gilbertson, T. (eds.), *Refocusing resistance to climate justice: COPing in, COPing out and beyond Paris*. Barcelona: EJOLT report no. 23.

Temper, L. & Del Bene, D. (2016). "Transforming knowledge creation for environmental and epistemic justice." *Current Opinion in Environmental Sustainability*, vol. 20, pp. 41–49.

Temper, L., Yánez, I., Sharife, K., Ojo, G., Martinez-Alier, J., CANA, et al. (2013). "Towards a post-oil civilization: Yasunization and other initiatives to leave fossil fuels in the soil." *EJOLT Report*, no. 6. Accessed at www.ejolt.org/wordpress/wpcontent/uploads/2013/05/130520_EJOLT6_Low2.pdf

Temper, L., Del Bene, D., & Martinez-Alier, J. (2015). "Mapping the frontiers and front lines of global environmental justice: the EJAtlas." *Journal of Political Ecology*, vol. 22, pp. 256–278.

Walker, G. (2009). "Globalizing environmental justice: the geography and politics of frame context-ualization and evolution." *Global Social Policy*, vol. 9, no. 3, pp. 355–82.

Walter, M. and Urkidi, L. (2015). "Community mining consultations in Latin America (2002–2012): the contested emergence of a hybrid institution for participation. *Geoforum*. www.sciencedirect.com/science/article/pii/S0016718515002602

World Commission on Dams. (2000). *Dams and development: A new framework for decision-making: The report of the World Commission on Dams*. Cape Town: World Commission on Dams.

40

ENVIRONMENTAL JUSTICE FOR A CHANGING ARCTIC AND ITS ORIGINAL PEOPLES

Alana Shaw

The Arctic is currently warming twice as fast as anywhere else in the world and the approximately 400,000 Indigenous peoples who dwell in this region, such as the Saami, Nenets, Evenks, Chuckchi, and Inuit are quickly realizing that their ancestral territories have essentially become 'ground zero' for global climate change (Arctic Council 2015; Reiss 2012). Arctic peoples are disproportionately affected by climate change not only in terms of the early impacts that they are already experiencing but also due to the significant threats that these rising temperatures present to the maintenance of their traditional lifeways (Trainor et al. 2007). As Native scholars like Robyn (2002) have contended, the lifeways of many Indigenous communities are themselves informed by understandings of the natural world that are predicated on radically different ideals from those held by dominant Western society. Both Robyn and environmental justice (EJ) theorist Schlosberg (2007) have therefore suggested that an inability to give meaningful consideration to such cultural difference is often a key source of injustice within environmental management and development scenarios that involve Indigenous peoples.

This chapter builds on the work of scholars such as Schlosberg and Robyn in order to argue that an environmentally just future for a changing Arctic and its original peoples is contingent upon the disruption of such cycles of procedural marginalization. In order to accomplish this, those decision-making structures, which are now being established to respond to the issues associated with regional climatic change, must possess the capacity to recognize and engage with both the traditional values and the contemporary perspectives of Arctic peoples. A geographic interest in the idea of nature as a social construction is featured here in support of this argument. As such, it works to emphasize how the dismissal of other ways of knowing and relating to nature is an act of ideological domination that can have profound material and social consequences for those societies who do not recognize the 'Great Divide' between culture and nature (Braun & Castree 2001; Latour 1993). In addition, as part of the process of recognizing cultural difference, it is equally important to attend not only to the traditional values and worldviews of Indigenous peoples, but also to their current realities (Whyte 2011). Such an awareness helps to highlight how the limited economic and political opportunities that are often available to Indigenous communities are the by-product of a colonial politics of nature that has long sought to dominate both nature and Native peoples (Blaser 2004; Ishiyama 2003).

The realities of a warming Arctic have so far manifested themselves through a number of shifting environmental conditions, which have begun to challenge the ability of Arctic peoples to continue to safely pursue their traditional subsistence way of life. Examples of these changes include increased rates of coastal erosion and melting permafrost, a reduction in the quantity and quality of foraging opportunities for the reindeer herds maintained by the Saami and many Russian peoples, and a decline in year-round sea ice coverage that has affected the timing and success of seasonal hunting activities, all of which are well documented in the literature on Arctic climate change (ACIA 2004; Cochran et al. 2013; Ford et al. 2006; ICC c. 2004; Tyler et al. 2007; see also Chapter 29).

Arctic peoples face additional pressures in terms of what Banerjee (2013) has called the 'Arctic paradox.' In this instance, the changing environmental circumstances with which they must now contend, which are largely the product of global extraction projects, have also ushered in the possibility of a new era of commercial activity and resource exploitation in the Far North. One example is the offshore drilling plans recently pursued by Shell Oil in the Chukchi Sea. The continuing loss of summer sea ice could also facilitate the opening of the Northwest Passage, a long-desired trade route around the top of Canada and Alaska, to general ocean traffic as early as 2020. When this occurs, it is predicted that Arctic shipping lanes could host as much as 25 per cent of the earth's total sea commerce (Reiss 2012). Future scenarios also include the possibility of increased levels of marine tourism throughout the region and the eventual opening up of commercial fisheries in the Arctic Ocean once a scientific baseline on available fish stocks has been established (Brigham 2014). Many of these development scenarios, which are now made feasible by rising temperatures, present an additional threat to Indigenous lands and lifeways and as such they may well represent a 'new round of dispossession' just beginning to unfold in the region (Cameron 2012).

The following section of this chapter begins to more fully elucidate how the hegemonic supremacy accorded to Western ways of understanding nature has often led to the dispossession of Indigenous lands and has worked against the abilities of Native communities to pursue their own self-determined lives. Two concise examples from the Alaskan Arctic are then employed in order to more clearly illustrate how a contemporary politics of nature, which privileges Western conceptualizations of the natural world, has begun to intersect with the realities of a changing climate to produce distinct consequences for the Arctic's original peoples. These scenarios illustrate how crucial it is to attend to the traditional perspectives of Arctic peoples if they are to be equitably represented within the decision-making processes that are likely to shape their futures.

The next section then emphasizes the need to recognize how complex national histories of colonial domination and cultural assimilation have meant that contemporary playing fields of power are rarely equal. This is especially so when it comes to the knowledge negotiations that are inherent to any cross-cultural forms of communication and decision-making (Ishiyama 2003; Nadasdy 2005). From the development pressures within Russia that overshadow climate change as the most pressing concern currently identified by the Nenet, to the economic and political positioning of the Saami of Norway that constrains their abilities to successfully adapt to changing environmental conditions, it is important to attend to the heterogeneous realities of Arctic Indigenous existence today.

Finally, this chapter concludes with a consideration of the unsuccessful nature of the legal challenges recently raised by Arctic peoples regarding the unjust impacts of climate change. Such difficulties further suggest that environmental justice in this region will require a willingness on the part of Western society to think beyond the nature–culture divide in order to explore those alternative types of relationships that could help us all to reimagine how we might "responsibly inhabit our complex socioecological worlds" (Braun 2002: 10).

Environmental justice, the nature–culture divide and Arctic peoples

As Di Chiro (1998) has noted, a defining feature of environmental justice activism and scholarship was the early realization that the nature–culture dichotomy has been linked to a long history of environmental racism and colonial oppression. This awareness facilitated a significant break from the mainstream environmental movement, which generally sought to preserve pristine wilderness and was therefore seen as less responsive to the everyday struggles of many marginalized peoples. EJ theorists thus became interested in examining the linkages between injustice and how certain hegemonic framings of nature have served to marginalize other knowledge systems and ways of interacting with our environments.

Latour's understanding of the divide constructed by Western societies between culture and nature helps to highlight these linkages and their particular importance for Indigenous peoples. Latour (1993: 39) has shown how this construct allowed Western societies to give up the seemingly "ridiculous constraints of their past which required them to take into account the delicate web of relations between things and people." Such a divide also led to the belief that their ability to see an external nature as it 'truly is' meant that Western knowledge was far superior to all other 'partial' knowledge systems (Demeritt 2001; Latour 1993). Western society's increasing ability to exert control over the natural world therefore rationalized a similar domination of Indigenous peoples (Blaser 2004) as the forced assimilation of Native societies was posited as a 'well-meaning' effort to civilize their cultures. This was despite the fact that such interventions attempted to fundamentally reorder the relationships that they had long held with their lands in ways that ultimately compromised the economic and political powers of their people (Li 1999).

As critical geographers like Braun and Castree (2001) have argued, once the constructed nature of this divide is recognized, it becomes possible to analyse the social and material effects of this privileging of one particular way of knowing nature. The very idea of nature as a discrete resource to be managed through a process of 'rational allocation' is, in fact, entirely at odds with the perspectives held by most Indigenous peoples (Cajete 2000; Harvey 1996). As an example, the Inuit of the Canadian Arctic largely see nature as a set of relations and the beluga whales they have traditionally hunted are understood to share the same social space as humans. As hunters, their actions have always been guided by a respect for the sentience of these whales that persists today regardless of any managerial ranking of the animal's value as either a scarce or abundant 'resource' (Tyrrell 2007).

Western science and society's preoccupation with the domination of nature has also encouraged the production of new technologies that have only served to intensify many people's sense of alienation from an inanimate nature (Cajete 2000). Conversely, while Native peoples have often been willing to adopt new tools and technology, it is usually done as a means of maintaining their connections to the natural world (Ingold 2000). In the context of a changing climate, this has meant that new machinery such as front end loaders have saved Iñupiat whalers on the North Slope of Alaska from having to drag the body of the bowheads they hunt over the sand and gravel that is exposed during the fall whaling season. This allows them to continue to treat the animal in the "right way" so that such technology is seen as a means of maintaining appropriate relations with the whales (Sakakibara 2010: 1008). These same whalers have, however, been critiqued for these adaptions in ways that leave little room for the dynamic cultural evolution that is needed to successfully respond to the realities of a changing climate (Heimbuch 2011).

Overall, EJ scholars Schlosberg (2007) and Robyn (2002) have argued that contemporary resource management and development scenarios often have environmentally unjust consequences because they are predicated on a peculiarly Western way of apprehending the world that is fundamentally in conflict with traditional Indigenous perspectives. These scenarios represent a form of cultural imperialism that has worked to push those ideals that do not conform to the conceptually dominant nature–culture divide into further obscurity (Perry & Robyn 2005). As Schlosberg (2007) explains, a refusal to engage with the complexities of Native knowledge systems, which includes their unique conceptualizations of nature, represents a form of cultural misrecognition (see also Chapter 10). Such misrecognition invariably leads to the procedural marginalization of Indigenous peoples within the very decision-making structures that are meant to act as the forum for their concerns. Such superficial levels of inclusion, in turn, often foment intercultural miscommunications and mistrust. Such scenarios should therefore be rightly understood as 'epistemological acts of violence' that deny "the legitimacy of other ways of knowing and managing nature" (Goldman & Turner 2011: 17).

Arctic peoples are especially vulnerable to the politics of nature surrounding regional climate change because its associated environmental issues and developmental pressures both represent a significant risk to their subsistence lifestyles. Like most Native societies found throughout the world, the Arctic's original peoples have sustained a deep connection with their ancestral lands (and with each other) through their subsistence practices. These relations have, in turn, shaped their collective identities as a people (Freeman et al. 1998; Kassam 2009; Trainor et al. 2007). Industrial incursions on these lands clearly represent a threat to the maintenance of these relationships. Yet even without the presence of any new development, the unpredictability of current weather patterns has already begun to unsettle these connections. Many subsistence hunters, like the Inuit of Arctic Bay, Canada, have come to question the ongoing accuracy of their traditional knowledge, which has long served them as a guide on how to operate safely in their environments (Ford et al. 2006). Gwich'in elders in Alaska have also discussed the difficulties associated with the transmission of their knowledge to younger generations because of the increasing changes they have noted in the land and animal migration patterns (Alexander et al. 2013). This has left some with the feeling that they are becoming "strangers in their own land" (ACIA 2004: 94).

The possibility that environmental conditions could continue to deteriorate to the point that it becomes necessary to relocate their communities is also of great concern given the significant nature of the cultural disruptions that are likely to result. Yet, many countries have largely abandoned the idea of mitigating climate change and have instead chosen an 'adaption only' focus (Loring 2013). The fact that most adaptation strategies tend to rely on relocation as the primary option (Tsosie 2007) works against Indigenous interests because it ignores the fact that Arctic peoples such as the Iñupiat understand themselves as living *in* rather than just *off* their land. "There can be no separation, no amputation" (Gallagher 2001: 28) without significant risk to their own cultural being. The cultural costs of relocation are, however, rarely considered alongside its economic costs so that most adaptation strategies continue to promote removal as the most reasonable plan (Adger et. al 2011; Tsosie 2007).

An adaption-only strategy is also convenient in that it does little to consider how Western society's instrumental view of nature as an external set of resources that exists for humanity's benefit (Harvey 1996) has prompted the ongoing exploitation of our global environment. It has also consistently rationalized the degradation of Indigenous lands as an unfortunate by-product of economic growth (Grinde & Johansen 1995). A singular focus on adaptation clearly serves to perpetuate these processes as mere 'business as usual' (Bullard 1992) in ways that tend to imply that the ever-increasing industrialization of a warming Arctic is the only viable path towards further 'progress' in this region (Loring 2013).

The vulnerabilities of Arctic peoples to climate change are also affected by their powers (or lack thereof) to participate in its associated decision-making processes and the colonial histories that have generally led to their political and economic marginalization within the nation-states in which they now dwell (Abate & Kronk 2013; Shearer 2013). The emergent issues raised by global climate change tend to both reflect and increase pre-existing social inequalities (Mohai et al. 2009) and many Native communities, in particular, have very limited resources and governing capacities after sustained periods of colonial dispossession (Cronin & Ostergren 2007). Within the Arctic, most adaptation strategies have been framed around the issue of addressing 'local capacity', while the colonial acts of systemic inequity that have worked and still continue to work to diminish these capacities have generally been overlooked. It is therefore critically important to attend to those processes that are now attempting to address these limited capacities, if the patterns of ideological and structural domination that created such vulnerabilities in the first place are to be unsettled (Cameron 2012; Marino & Ribot 2012).

A warming Arctic thus presents a fundamental challenge in terms of crafting management and policy decisions that contend with the many environmental and social issues it raises in a manner that reflects the multitude of ways that people know and relate to Arctic natures. A look towards the Alaskan Arctic helps to further highlight the difficulties associated with this task as attempts are now being made to respond to the changing environmental conditions that are threatening the safety of Native communities on the coast. The Arctic offshore drilling programme in the Chukchi Sea is also one of the first examples of the type of industrial development made much more realizable by rising temperatures in the region. Illuminating how an Arctic politics of nature has informed these scenarios works to reveal how justice in this region will necessarily entail the "extension or reconstruction of expertise beyond modern scientific knowledge, to include traditional, cultural, and alternative forms of knowledge and representations of nature" (Schlosberg 2007: 198).

Cultural misrecognition and procedural injustice

Kivalina and Shishmaref are two Iñupiat villages in the Alaskan Arctic facing the very real possibility that their communities will need to be relocated due to the high rates of coastal erosion and extreme flooding that they are now experiencing as a result of climate change. In the specific context of Shishmaref, Marino (2012) has shown how federal decision-making processes have contributed to both the current vulnerability of residents to these environmental phenomena and their lack of representation within the procedures intended to address these issues. She highlights how it was only at the federal government's insistence that the seasonally nomadic Iñupiat permanently settled on this barrier island, which their ancestors always insisted would eventually disappear. The high costs associated with moving the infrastructure that has since been established means, however, that residents of this economically isolated village would never be able to undertake such a project on their own.

Yet, as Marino further explains, residents have not felt as if they have been well represented in the processes surrounding their village's potential relocation because these procedures have been dominated by bureaucratic agencies. In relation to the governmental meetings held around these issues, residents have become increasingly frustrated by the fact that they have been asked the same questions or repeatedly told the same things, but have been left with little sense that anything has changed as a result. The federal government's response to this issue, which was partly produced by its failure to heed the prior warnings of the Iñupiat, seems to be merely perpetuating the marginalization of local peoples within those administrative processes that will profoundly affect their futures.

Kivalina's residents have expressed a similar level of frustration with their experiences regarding the possible relocation of their village and they have come to challenge "the legitimacy of the decision-making process and social structures that allow such decisions to be made without involvement of those most intimately concerned" (Shearer 2013: 212). These two communities are not alone in their concern regarding appropriate representation as this issue also appears in the context of the Arctic offshore drilling programme in the Chukchi Sea, which was pioneered by Shell Oil up until their 2015 announcement that they would no longer pursue these efforts. In this case, Iñupiat communities throughout Alaska's North Slope expressed a sense of disillusionment with those forums that were meant to elicit their perspectives and concerns regarding this development. The former President of the Native Village of Point Hope, Caroline Cannon, stated that she felt as if they were not full participants in the decision-making process where community members had "to repeat ourselves over and over again, both to the same audiences that don't seem to hear or respect what we are saying, and to the constantly shifting audiences that come to our community to talk about oil and gas industrial development" (Thompson et al. 2013: 326).

The privileging of Western understandings of nature as an external object open to human manipulation also worked to exacerbate the concerns of Iñupiat residents within these processes. Shell Oil consistently claimed that its use of the best technology available would allow it to handle any potential issues that came up in its offshore operations (Reiss 2012). However, most Indigenous societies believe that it is an act of great hubris to try and control nature (Nadasdy 2011). The Iñupiat's intimate knowledge of the mercurial Arctic environments in which they have long dwelled has, in fact, taught them that it is only by adhering to the laws of nature rather than trying to transcend them through technical mastery that one can hope to safely operate in these environments (Kassam 2009). It is therefore worth noting that such concerns were not unfounded since Shell did encounter several rather serious technical setbacks in its initial forays into the Arctic offshore drilling theatre, although it rationalized its ultimate withdrawal largely based on the economic infeasibility of extraction given current market prices for oil (Martinson 2015).

The lack of consideration given to the Iñupiat's concerns regarding Shell's perceived overconfidence in its technology is strikingly similar to the ways in which Shishmaref elders and their belief that the island would eventually disappear were not taken seriously. Both the 'positional superiority' that Western knowledge has long been afforded over other knowledge systems and the emphasis it places on scientific expertise and technological capabilities as the sole arbiters of environmental issues and debates (Willems-Braun 1997; Said 1994) have worked in concert to efface the perspectives of those people who have lived in these regions for millennia. These acts clearly reflect a pattern of ideological domination and dismissal, which explains the profound sense of frustration and lack of representation felt by the Iñupiat within these decision-making processes. Such procedural injustice is further compounded by complex colonial histories through which Native peoples have become politically and economically disenfranchised participants within these larger bureaucratic structures. How these histories continue to inform the abilities of Arctic peoples to actively participate in the climate change discourse and decision-making processes that will shape the future of this region is therefore another key area of concern.

Recognition and the contemporary realities of Arctic peoples

As Forbes and Stammler (2009) have pointed out in the context of Nenet reindeer herders of the Yamal Peninsula, it is important that the lived realities of those communities impacted by climatic change are not lost in the overarching discourse on this issue. This is partly because the idea of 'climate change' is itself an 'exogenous construct' that does not always find resonance with the very people most affected by its consequences. For herders like the Nenet who maintain a direct and daily relationship with their lands, the environment is understood to always be undergoing constant and even sometimes dramatic change. Concrete discussions regarding their abilities to adapt to shifting weather conditions, which impact their immediate realities, are thus inherently more meaningful to them than any engagement with the idea of a changing climate itself.

In addition, while climate change is producing new environmental issues and development pressures in the Arctic, they are often only the latest manifestations of historic processes of resource extraction that have long sacrificed Indigenous lands in the name of 'progress' and economic growth (Grinde & Johansen 1995; Robyn 2002). This means that climate change is not always the most pressing issue for those communities who are already embedded in capitalist economies in which they are operating at a distinct disadvantage. In the case of the Nenet, they usually identify the continuing encroachment of oil and gas activities on their traditional herding grounds as their paramount concern (Forbes & Stammler 2009).

In terms of recognitional justice, it is critical to attend to the 'situational particularities' of each group, which includes not only their unique customs and worldviews but also the colonial histories and processes that have shaped their current economic and political realities (Whyte 2011). In truth, not all Arctic peoples possess an equal ability to influence the decision-making processes that affect them. Inuit in Canada can be seen as slightly better off than the Iñupiat of Alaska due to the many co-management scenarios that were included in their land settlement agreements, which provide them with some measure of political power over their lands (Trainor et al. 2007; Zellen 2008). In turn, Indigenous groups in the Arctic's western countries are considered to have much more political clout than those in post-Soviet Russia, where a sustained period of socioeconomic instability meant that little progress was made at the national level in regards to Indigenous rights (Forbes & Stammler 2009).

However, even the contemporary capacities of supposedly 'better-off' peoples such as the Saami of Norway are constrained by regional histories, which have involved their structural integration into the economies and governments of 'well-meaning' colonial powers. Here we find that the Saami's abilities to adapt their herding techniques in response to climate change are hindered by a combination of governmental policies and development pressures, which are themselves the product of Western ideals that do not reflect their own values and traditions. Such issues include everything from the loss of foraging habitat through the encroaching development of oil and gas pipelines and railways, to governmental regulation of the price of reindeer meat, which has stagnated their economy and reduced their overall autonomy, and reindeer husbandry laws that conflict with the Saami's customary practices (Tyler et al. 2007).

Overall, environmental justice for Arctic peoples is contingent upon supporting their abilities to exert control over their environments in order to lead their own self-determined lives (Adger et al. 2011; ICC c. 2004; Tsosie 2007). Without this control, their unique values and lifeways invariably become enmeshed in Western systems of bureaucracy, which operate according to fundamentally different ways of understanding the natural world and humanity's rightful place in it. If a warming Arctic is to increasingly become a place where these worldviews collide, it is imperative that appropriate decision-making structures be created that are able to engage with cultural difference as a matter of justice. At the same time, it is important to recognize that any such knowledge negotiations are likely to occur in the context of larger political environments in which Native peoples are still marginalized (Fernandez-Gimenez et al. 2006). Without careful consideration given to the lingering legacies of colonial epistemologies and structures of power (Wainwright 2008), the recognition of cultural difference can itself quickly become but a "cheap gift of political and economic inclusion" (Coulthard 2014: 173) where the terms of engagement are always laid out in the interest of the dominant parties who seek to maintain the status quo.

Arctic futures

Arctic peoples have not been willing to accept the limiting role of passive victims within the overarching context of global climate change. However, their initial attempts to contest the injustices that they are experiencing have not been promising in terms of their ability to assert their sovereign rights within this realm. In 2005, the Inuit Circumpolar Council (ICC), which advocates for the cultural interests of all Inuit, filed a petition with the Inter-American Commission on Human Rights claiming that the United States was violating the rights of Arctic peoples by refusing to limit its greenhouse gas emissions. This was despite the fact that the US had shown through prior signatory actions that it was well aware of the nature of the harm that these emissions were causing. The petition sought to force the US to adopt emission limits and to consider the impacts of global warming on the Inuit in order to begin to develop appropriate adaptation strategies. The petition was, however, rejected because the ICC's charges were seen as lacking sufficient evidence to constitute a clear violation of their rights (Shearer 2013; Tsosie 2007).

In 2008, two groups of plaintiffs representing Kivalina (as a city and as a Tribe) filed suit in a US District Court against 24 oil, electricity, and coal companies, which they considered to be the most significant producers of the greenhouse gas emissions that had caused the climatic conditions that now imperilled their village. They sought almost $400 million in damages, which was the approximate price tag associated with their community's potential relocation. This suit was also eventually dismissed on the grounds that it was seen as a political rather than a legal issue. In addition, it was argued that because there were so many greenhouse gas emitters located across the globe, it was not possible to trace the issues they were now experiencing to any of the specific actions of the named defendants (Shearer 2013; Native Village of Kivalina v. ExxonMobil et al. 2009).

The underlying message in both cases appears to be that because climate change is everyone's problem, it ends up being no one's. This way of thinking, of course, belies the fact that it is Western society's preoccupation with increasing mastery over an external nature that has largely engendered these issues. Meanwhile, the lifeways of most Indigenous peoples mean that they are not likely to have significantly contributed to such change. As Mohai, Pellow, and Roberts (2009: 420) have argued, in terms of climate change it is important to remain clear about "who suffers most its consequences, who caused the problem, who is expected to act, and who has the resources to do so."

While the ability to creatively adapt to shifting environmental and social circumstances is a hallmark of Arctic societies, the rapidity and magnitude of regional climate change has placed a significant strain on even these well-established capacities (ACIA 2004; Nelson 1969). In terms of thinking about the future it becomes important to consider whether "our society wants to continue with an Arctic that, as a social artifact, reflects a culture of social injustice and fear of controversy" (Loring 2013: 8) or one that supports the basic human rights of its original peoples to continue to pursue their own lifeways.

Arctic peoples must be able to participate in the larger decision-making processes that affect their futures in ways that acknowledge "their authority to protect and promote their ways of life" (ICC c. 2004). EJ theorists are well suited to contribute to this task by helping to find a place for traditional Indigenous knowledge and understandings of nature within the bureaucratic decision-making structures now shaping the future of the Arctic. Critiquing those ongoing processes that merely cloak themselves with the false patina of cultural inclusivity will remain necessary. Yet, it is also time to begin to envision what a truly intercultural space of deliberation would look like where no one worldview is allowed to dominate the other.

Ultimately, we must ask ourselves: How can we begin to move beyond the nature–culture divide in order to ensure that alternative visions for the Arctic, which are rooted in ideas of interconnectedness rather than alienation, and mutual respect and reciprocity rather than domination, be given their due consideration alongside the current push to open up the region to a new round of large-scale development and potential dispossession? Here it is important to recognize that there are other ways of being in this world that fundamentally reject the idea that such 'business as usual' scenarios necessarily entail such high environmental and social costs. The challenge that remains is to find a way for a multitude of perspectives to be meaningfully represented at the decision-making table so that we as a global society can collectively craft a future for a changing Arctic that both reflects the traditional values and engages with the contemporary realities of its original peoples.

References

Abate, R.S. & Kronk, E.A. (2013). 'Commonality among unique indigenous communities: An introduction to climate change and its impacts on indigenous peoples.' *Tulane Environmental Law Journal*, vol. 26, pp. 179–195.

ACIA (2004). *Impacts of a warming climate: Arctic climate impact assessment*. Cambridge University Press, Cambridge, UK.

Adger, W.N., Barnett, J., Chapin F.S. III, & Ellemor, H. (2011). 'This must be the place: Underrepresentation of identity and meaning in climate change decision-making.' *Global Environmental Politics*, vol. 11, no. 2, pp. 1–25.

Alexander, D., Savage, M., & Gilbert, M. (2013). 'We'll fight to protect the Gwich'in homeland and our way of life', in S. Banerjee (ed.), *Arctic voices: Resistance at the tipping point*, Seven Stories Press, New York.

Arctic Council (2015). *Indigenous peoples of the Arctic countries*, viewed 15 October 2015, www.arctic-council.org/images/PDF_attachments/Maps/indig_peoples.pdf

Banerjee, S. (2013). 'From Kolkata to Kaktovik en route to Arctic voices: Something like an introduction', in S. Banerjee (ed.), *Arctic voices: Resistance at the tipping point*, Seven Stories Press, New York.

Blaser, M. (2004). 'Life projects: Indigenous peoples' agency and development', in M. Blaser, H.A. Feit, & G. McRae (eds), *In the way of development: Indigenous peoples, life projects, and globalization*, Zed Books, London.

Braun, B. (2002). *The intemperate rainforest: Nature, culture, and power on Canada's west coast*, University of Minnesota Press, Minneapolis.

Braun, B. & Castree, N. (2001). 'Preface', in N. Castree & B. Braun (eds), *Social nature: Theory, practice, and politics*, Blackwell Publishers Ltd., Malden.

Brigham, L.W. (2014). 'The fast-changing maritime Arctic', in B. Zellen (ed), *The fast-changing Arctic: Rethinking Arctic security for a warmer world*, University of Calgary Press, Calgary.

Bullard, R. (1992). 'Environmental blackmail in minority communities', in B. Bryant & P. Mohai (eds), *Race and the incidence of environmental hazards*, Westview Press, Boulder.

Cajete, G. (2000). *Native science: Natural laws of interdependence*, Clear Light Publishers, Santa Fe.

Cameron, E.S. (2012). 'Securing indigenous politics: A critique of the vulnerability and adaption approach to the human dimensions of climate change in the Canadian Arctic', *Global Environmental Change*, vol. 22, pp. 103–114.

Cochran, P., Huntington, O.H., Pungowiyi, C., Tom, S., Chapin, F.S. III, Huntington, H.P., et al. (2013). 'Indigenous frameworks for observing and responding to climate change in Alaska', *Climatic Change*, vol. 120, pp. 557–567.

Coulthard, G.S. (2014). *Red skin, white masks: Rejecting the colonial politics of recognition*, University of Minnesota Press, Minneapolis.

Cronin, A. & Ostergren, D. (2007). 'Tribal watershed management: Culture, science, capacity and collaboration', *The American Indian Quarterly*, vol. 31, no. 1, pp. 87–109.

Demeritt, D. (2001). 'Being constructive about nature', in N. Castree & B. Braun (eds), *Social nature: Theory, practice, and politics*, Blackwell Publishers Ltd, Malden.

Di Chiro, G. (1998). 'Environmental justice from the grassroots: Reflections on history, gender, and expertise', in D. Faber (ed.), *The struggle for ecological democracy: Environmental justice movements in the United States*, The Guilford Press, New York.

Fernandez-Gimenez, M.E., Huntington, H.P., & Frost, K.J. (2006). 'Integration or co-optation? Traditional knowledge and science in the Alaska Beluga Whale Committee', *Environmental Conservation*, vol. 33, no. 4, pp. 306–315.

Forbes, B.C. & Stammler, F. (2009). 'Arctic climate change discourse: The contrasting politics of research agendas in the West and Russia', *Polar Research*, vol. 28, pp. 28–42.

Ford, J.D., Smit, B., & Wandel, J. (2006). 'Vulnerability to climate change in the Arctic: A case study from Arctic Bay, Canada', *Global Environmental Change*, vol. 16, pp. 145–160.

Freeman, M.M.R., Bogoslovskaya, L., Caulfield, R.A., Egede, I., Krupnik, I.I., & Stevenson, M.G. (1998). *Inuit, Whaling, and Sustainability*, AltaMira Press, Walnut Creek.

Gallagher, H.G. (2001). *Etok: A story of Eskimo Power*, Vandamere Press, Clearwater.

Goldman, M.J. & Turner, M.D. (2011). 'Introduction', in M.J. Goldman, P. Nadasdy, & M.D. Turner (eds), *Knowing nature: Conversations at the intersection of political ecology and science studies*, University of Chicago Press, Chicago.

Grinde, D.A. & Johansen, B.E. (1995). *Ecocide of Native America: Environmental destruction of Indian lands and peoples*, Clear Light Books, Santa Fe.

Harvey, D. (1996). *Justice, nature, and the geography of difference*, Blackwell Publishers Inc., Malden.

Heimbuch, H. (2011). 'Lightning strikes twice for Wainwright whaling captain', *Alaska Dispatch*, 5 November, viewed 1 December 2013, www.adn.com/article/lightning-strikes-twice-wainwright-whaling-captain

ICC (c. 2004). *Responding to global climate change: The perspective of the Inuit Circumpolar Conference on the Arctic Climate Impact Assessment*, viewed 27 October 2015, www.inuitcircumpolar.com/responding-to-global-climate-change-the-perspective-of-the-inuit-circumpolar-conference-on-the-arctic-climate-impact-assessment.html

Ingold, T. (2000). *The perception of the environment: Essays on livelihood, dwelling and skill*, Routledge, London.

Ishiyama, N. (2003). 'Environmental justice and American Indian tribal sovereignty: Case study of a land-use conflict in Skull Valley, Utah', *Antipode*, vol. 35, no. 1, pp. 119–139.

Kassam, K.S. (2009). *Biocultural diversity and indigenous ways of knowing: Human ecology in the Arctic*, University of Calgary Press, Calgary.

Latour, B. (1993). *We have never been modern*, C. Porter (trans), Harvard University Press, Cambridge.

Li, T.M. (1999). 'Compromising power: Development, culture, and rule in Indonesia', *Cultural Anthropology*, vol. 14, no. 3, pp. 295–322.

Loring, P.A. (2013). 'Are we acquiescing to climate change? Social and environmental justice considerations for a changing Arctic', in F.J. Mueter, D.M.S. Dickson, H.P. Huntington, J.R. Irvine, E.A. Logerwell, S.A. Maclean, et al. (eds), *Responses of Arctic marine ecosystems to climate change*, University of Alaska, Fairbanks.

Marino, E. (2012). 'The long history of environmental migration: Assessing vulnerability construction and obstacles to successful relocation in Shishmaref, Alaska', *Global Environmental Change*, vol. 22, pp. 374–381.

Marino, E. & Ribot, J. (2012). 'Special issue introduction: Adding insult to injury: Climate change and the inequities of climate intervention', *Global Environmental Change*, vol. 22, pp. 323–328.

Martinson, E. (2015). 'Shell calls off its multibillion-dollar mission in Alaska's Arctic', *Alaska Dispatch News*, 28 September, viewed 1 October 2015, www.adn.com/article/20150928/shell-abandons-offshore-oil-exploration-alaskas-arctic

Mohai, P., Pellow, D., & Roberts, T. (2009). 'Environmental justice', *The Annual Review of Environment and Resources*, vol. 34, pp. 405–430.

Nadasdy, P. (2005). 'The anti-politics of TEK: The institutionalization of co-management discourse and practice', *Anthropologica*, vol. 47, no. 2, pp. 215–232.

Nadasdy, P. (2011). 'Application of environmental knowledge: The politics of constructing society/ nature', in M.J. Goldman, P. Nadasdy, & M.D. Turner (eds), *Knowing nature: Conversations at the intersection of political ecology and science studies*, University of Chicago Press, Chicago.

Native Village of Kivalina v. ExxonMobil Corp et al. (ND Cal, 2009). Retrieved from www.leagle. com/decision/In FDCO 2020100405364.xml/NATIVE VILLAGE OF KIVALINA v. EXXONMOBIL CORP.

Nelson, R.K. (1969). *Hunters of the northern ice*, The University of Chicago Press, Chicago.

Perry, B. & Robyn, L. (2005). 'Putting anti-Indian violence in context: The case of the Great Lakes Chippewa of Wisconsin, *American Indian Quarterly*, vol. 29, no. 3/4, pp. 590–625.

Reiss, B. (2012). *The Eskimo and the oil man: The battle at the top of the world for America's future*, Business Plus, New York.

Robyn, L. (2002). 'Indigenous knowledge and technology: Creating environmental justice in the twenty-first century,' *The American Indian Quarterly*, vol. 26, no. 2, pp. 198–220.

Said, E.W. (1994). *Orientalism*, 25th anniversary edn, Vintage Books, New York.

Sakakibara, C. (2010). '*Kiavallakkikput agviq* (into the whaling cycle): Cetaceousness and climate change among the Iñupiat of Arctic Alaska', *Annals of the Association of American Geographers*, vol. 100, no. 4, pp. 1003–1012.

Schlosberg, D. (2007). *Defining environmental justice: Theories, movements, and nature*, Oxford University Press, Oxford.

Shearer, C. (2013). 'From Kivalina: A climate change story,' in S. Banerjee (ed.), *Arctic voices: Resistance at the tipping point*, Seven Stories Press, New York.

Thompson, R., Ahtuangaruak, R., Cannon, C., & Kingik, E. (2013), 'We will fight to protect the Arctic Ocean and our way of life', in S. Banerjee (ed.), *Arctic voices: Resistance at the tipping point*, Seven Stories Press, New York.

Trainor, S.F., Chapin, F.S. III, Huntington, H.P., Natcher, D.C., & Kofinas, G. (2007). 'Arctic climate impacts: Environmental injustice in Canada and the United States', *Local Environment*, vol. 12, no. 6, pp. 627–643.

Tsosie, R. (2007). 'Indigenous people and environmental justice: The impact of climate change', *University of Colorado Law Review*, vol. 78, pp. 1625–1677.

Tyler, N.J.C., Turi, J.M., Sundset, M.A., Strøm Bull, K., Sara, M.N., Reinert, E., et al. (2007). 'Saami reindeer pastoralism under climate change: Applying a generalized framework for vulnerability studies to a sub-arctic social-ecological system', *Global Environmental Change*, vol. 17, pp. 191–206.

Tyrrell, M. (2007). 'Sentient beings and wildlife resources: Inuit, beluga whales and management regimes in the Canadian Arctic', *Human Ecology*, vol. 35, no. 5, pp. 575–586.

Wainwright, J. (2008). *Decolonizing development: Colonial power and the Maya*, Blackwell Publishing, Malden.

Whyte, K.P. (2011). 'The recognition dimensions of environmental justice in Indian Country', *Environmental Justice*, vol. 4, no. 4, pp. 199–205.

Willems-Braun, B. (1997). 'Buried epistemologies: The politics of nature in (post) colonial British Columbia', *Annals of the Association of American Geographers*, vol. 87, no. 1, pp. 3–31.

Zellen, B.S. (2008). *Breaking the ice: From land claims to tribal sovereignty in the Arctic*, Lexington Books, Lanham.

41

ENVIRONMENTAL INJUSTICE IN RESOURCE-RICH ABORIGINAL AUSTRALIA

Donna Green, Marianne Sullivan and Karrina Nolan

This chapter explores the disproportionate impact mining has on Aboriginal Australians. Aboriginal Australians represent the world's longest continuous living culture, and comprise about 2.5 per cent of the Australian population, although in Northern Australia, this percentage rises to about one third. Large mine sites occur across Australia, with many mine sites located on, or near, land owned by Aboriginal Australians. Until this last decade, mining was one of the significant contributors to the Australian economy. Mining, and the processing of natural resources, carries social, environmental and health costs that are not fully accounted for by the mining companies. Aboriginal Australians are therefore disproportionately impacted by these industries due to mining occurring on or near their lands, by the imbalance of power relationships, social and legal disadvantage, and poverty.

The legacy of colonization

The unequal relationship that currently exists between Indigenous and non-Indigenous Australians (Indigenous is a term which includes Aboriginal and Torres Strait Islanders, the two distinct groups of first nation people in Australia) is a direct result of the British invasion in 1788, and the subsequent persecution of Indigenous people since that time (Arthur 2005). Over 200 years later, no treaty has been signed between these parties, making Australia one of the few countries not to have an official negotiated settlement between its first people and its colonizers (Havemann 1999).

Further exacerbating this fraught situation is a legal framework that is set up to exclude and discriminate against Indigenous Australians. An example of this ongoing tension can be seen in recent decades by the enactment of the much-maligned Northern Territory National Emergency Response legislation, which required the suspension of the Racial Discrimination Act (Altman 2007). This legacy of legal and social discrimination and exclusion has resulted in tremendous gaps in social, economic and health outcomes between Indigenous and non-Indigenous Australians. For example, Indigenous households are nearly two and a half times as likely to be in the lowest income bracket, and four times less likely to be in the top income bracket as non-Indigenous households. Additionally, nearly half of all Indigenous children are living in jobless families; this is three times higher than for all Australian children. Indigenous Australians are hospitalized for cardiovascular diseases at 1.7 times the rate of other Australians, and life expectancy at birth is estimated to be 67 years for Indigenous males and 73 years for Indigenous females, representing gaps of 11.5 and 9.7 years, respectively, compared with all Australians (AIHW 2011). In recent years, state and federal government policies have been implemented in an attempt to address some of these gaps, however, progress remains slow. These were part of a suite of ongoing state and federal government policies implemented in an attempt to 'close the gap' (Altman 2009), that is, to reduce the disadvantage between Indigenous and non-Indigenous Australians seen in a range of social, economic and health indicators (AIHW 2011; Sutton 2001).

Land ownership and native title

Within the lifetime of the current generation, progress has been made for Indigenous people regarding land ownership. The Aboriginal Land Rights Act (ALRA) passed in 1976 in the Northern Territory, and the more recent Federal Native Title Act, enacted in 1993, recognize that Indigenous people have rights and interests to their land stemming from their traditional laws and customs (Macintyre et al. 2000). This Act clarifies how Indigenous land ownership, called 'Native Title' can be sought, protected and recognized by the courts. In just over 20 years since this Act was passed, Indigenous people have claimed over three million square kilometres of northern Australia. Today, the Indigenous Estate currently covers 48 per cent of the country's north (NNTT 2016). The overwhelming majority of this land is held by Aboriginal Australians, and for the remainder of this chapter, we will focus on the environmental injustice that they are experiencing on their land.

Northern Australia is sparsely populated with many remote 'outstation' Aboriginal communities living on, or very near, vast ore bodies, many of which have been intensely explored and mined by colonizers since the early 1900s (Arthur 2005; Brereton et al. 2009). Early exploration was carried out with little negotiation or consent from the land's traditional owners (traditional owners are the legally recognized Indigenous owners of the area). Since enactment of Native Title, there is a more established process of negotiation, usually carried out through local Lands Councils, statutory representative bodies representing relevant local Indigenous people (O'Faircheallaigh & Corbett 2005). The recognized traditional owners hold land through communal titles called Land Trusts. If a company wants to explore or mine, they must work through the relevant Land Council to negotiate with the appropriate traditional owner group(s). This process has been shown to cause unanticipated difficulties, with significant knowledge and power imbalances occurring between mining companies and traditional owner groups – unsurprisingly leading to poor outcomes for the latter (O'Faircheallaigh 2008). Given that there are many mineral-rich areas on or near Aboriginal lands, a disproportionate burden of social and environmental damage associated with these resource-intensive industries has occurred for generations.

Government service substitution

In 1996, the Federal Senate Select Committee was asked:

> What other community in Australia is asked to mine their country to purchase what is their basic human right? There is no other community in Australia except the Aboriginal community who are being asked to mine their country to improve their health, housing and education.
>
> *(Christophersen 1996)*

Amidst the declining mining boom, Indigenous people living in areas near existing and proposed mines remain for the most part extremely disadvantaged (Langton 2008). Until the early 1980s, State and Territory governments made a condition of a lease that mining companies provide, or significantly contribute to, local infrastructure such as houses, schools, shops or community or sport centres in addition to the freight and port facilities that the mining operation would require (O'Faircheallaigh 2004).

However, as a result of these conditions, governments have often reduced their role as service providers. In addition, the promise of local employment by the mining companies is often largely unfulfilled, and communities are increasingly seeing the environmental impacts of resource extraction on their land. It would seem that the modern form of 'negotiation' and agreement-making which was meant to generate extensive community benefits packages and subsequent economic development, has failed to deliver for the majority of Aboriginal people (O'Faircheallaigh 2004). For example, in the Northern Territory, the mining industry contributed nearly a quarter of the gross Territory product in 2007–08 (Carroll 2003; NTG 2010). In 2003, over three quarters of this production was contributed by four mines (Gove, Groote Eylandt, McArthur River and Ranger – the last two of which are discussed in the case studies below).

Australian environmental justice studies

Environmental justice research is not well established in Australia, with only a handful of studies carried out in recent decades (EDO 2012). One of the first qualitative studies was produced by the Australian Conservation Foundation (Low & Gleeson 1999). This report highlighted various apparent case studies of environmental justice for low socio-economic groups in Victoria, as well as for Aboriginal people in the Northern Territory. Since that time, a handful of other qualitative environmental justice studies have been published (Byrne et al. 2009; Gross 2007; Heyworth et al. 2009; Lloyd-Smith & Bell 2003; Millner 2011). Unsurprisingly, Australia has no enforceable environmental justice regulations. The closest guidance comes from the legally unenforceable Intergovernmental Agreement on the Environment (DoE 1992), which states that all Australians should "enjoy the benefit of equivalent protection from air, water and soil pollution . . . wherever they live."

The only national level quantitative environmental justice assessment of air pollution linked the spatial distribution emissions associated with industrial air pollution sources derived from the National Pollution Inventory, Australia's pollutant release and transfer register (NPI 2016), to Indigenous status and social disadvantage characteristics of communities – derived from Australian Bureau of Statistics indicators. The study revealed a clear national pattern of environmental injustice based on the locations of industrial pollution sources, as well as volume, and toxicity of air pollution released at these locations (Chakraborty & Green 2014). Communities with the highest number of polluting sites, emission volume, and toxicity-weighted air emissions indicated significantly greater proportions of Indigenous population, and higher levels of socio-economic disadvantage. The quantities and toxicities of industrial air pollution were significantly higher in communities with the lowest levels of educational attainment and occupational status. A study of modelled NO_2 pollution, used as a proxy for urban air pollution from traffic, found similar disproportionate air pollution results for lower socio-economic status communities (Knibbs & Barnett 2015).

Case studies

The majority of Indigenous people in Australia live in major cities and rural areas. For this reason, we have selected case studies that explore the specifics of the Indigenous experience of environmental injustice in a large inland city, Broken Hill, which has a disproportionately high proportion of Indigenous people, as well as two more remote case studies where industrial pollution is occurring on, or immediately adjacent to Indigenous estates.

Each Australian state and territory has its own laws governing monitoring of legal emission levels of certain ambient air pollutants. Because of this lack of uniformity, significant social inequity in exposure to industrial point source air pollution arises (Taylor et al. 2014a). Mining and mineral processing operations have created a number of point source air pollution hotspots across the country (Taylor et al. 2014b). Communities located near these sites suffer disproportionately high exposure to air pollution and consequent adverse health outcomes (Munksgaard et al. 2010).

One significant industrial air pollutant is lead, and the associated toxic compounds released during its mining and smelting. Lead inhalation or ingestion adversely affects the neurological, cardiovascular, immune, endocrine, and reproductive systems (NTP 2012). Since the earliest operations of Australian mines and smelters over a century ago, occupational lead poisoning has been a significant public health concern (Blainey 1968). Non-occupational exposures among communities living near these sites are currently an even greater public health problem. Recent estimates highlight that up to 100 000 Australian children have suffered permanent developmental damage as a result of preventable exposure to lead (Taylor et al. 2013). However, verifiable evidence of the spatial extent and effects of this problem is difficult to obtain, because of a lack of independently verified emissions data, and highly complicated and inconsistent pollution monitoring regulations (Weng et al. 2012). For this reason, two of our case studies focus on leaded locations with high percentages of Indigenous residents.

Broken Hill, New South Wales

Rising out of the red sandy desert in western New South Wales, one of the world's richest lead, zinc and silver deposits was first discovered in 1884. Since that time, there has been over a century of continuous mining in the region, with over 200 million tons of ore mined from both surface and underground mines, producing over 20 million tons of lead, nearly as much zinc and over 30 000 tons of silver (Mudd 2007). The main city of this region, Broken Hill, which grew up around the mines, has a current population of 19 000. The eponymous Broken Hill Proprietary (BHP), now BHP-Billiton, is one of the world's largest mining multinationals.

Today, only 10 per cent of Broken Hill residents work in the mining sector (ABS 2014), although mining is still dominant and inescapable, forever memorialized by the massive piles of capped tailings that tower over the town. The dry and dusty conditions and frequent winds that stir up lead contaminated dust which collects in gardens, playgrounds and homes, are a constant feature of life in the town. Parents are reminded to have their children tested for lead at each immunization visit, and the Health Department recommends that parents constantly mop, sweep and vacuum their homes to try to reduce their children's exposure to lead (Child and Family Health Centre 2014).

Health concerns related to lead mining have always been present in the town. Almost as soon as mining and smelting began, many workers became ill – a local doctor interviewed as part of an official investigation estimated that as many as three-quarters of all mineworkers were 'leaded', with some dying from lead poisoning (Thompson 1893). This remarkably thorough investigation into occupational lead poisoning conducted in the early 1890s documented environmental and community health risks, noting that lead contamination in Broken Hill might be harming cows, cats, dogs and birds. The investigators identified multiple pathways of exposure for the community, including inhalation of lead in air and as re-suspended dust, ingestion of lead from rainwater, and ingestion from lead particles deposited on hands. An examination of children at city schools found that those attending school closest to the mines and smelters were 'pallid' and that the school's collected rainwater contained large amounts of lead (Thompson 1893: 17).

Though attempts to address pollution problems from current operations and legacy heavy metal pollution have been made, recent studies on environmental contamination found elevated concentrations of lead in air, and lead (and other heavy metals such as arsenic and cadmium) on surfaces in children's playgrounds, dust and soil (Taylor et al. 2014b; Dong et al. 2015). Despite the long-standing concerns about health effects, childhood lead exposure in Broken Hill continued throughout the twentieth century and into the twenty-first. Simultaneously, scientific advances in understanding lead poisoning have increased concerns about low levels of lead exposure (NTP 2012). The Australian government considers 5 ug/dL the blood lead level (BLL) at which public health investigation should begin. In 2014, about half of all children tested in Broken Hill had BLLs 5 ug/dL or higher (Lesjak & Jones 2015).

Broken Hill is home to a sizable Aboriginal population; approximately 16 per cent of the city's 0–4 year old residents are Aboriginal. Significant disparities are evident when data on BLLs are broken down by Aboriginal status: the prevalence of elevated blood lead levels is more than two-fold higher in Aboriginal children. In 2014, of the children tested, 76 per cent of Aboriginal children had BLLs 5 ug/dL or higher, while among non-Aboriginal children 37 per cent had BLLs 5 ug/dL or higher. Aboriginal children were also more likely to have BLLs of 10 ug/dL or higher – 43 per cent, compared to 20 per cent of non-Aboriginal children. At higher blood lead levels these disparities persist and are even more significant with Aboriginal children over five times more likely than non-Aboriginal children to have BLLs between 15 and 29 ug/dL (Lesjak & Jones 2015).

Higher average BLL among Aboriginal children have been measured consistently since blood lead screening began in Broken Hill in 1991. Numbers of Aboriginal children screened have varied over the years, reaching a peak of 120 in 2004, then decreasing to a low of 37 in 2009. After concerted efforts to screen more children, and working more closely with an Aboriginal community health care provider, screening rates have improved in recent years. However, for all years since 1991, Aboriginals have had higher mean BLL than non-Aboriginals. The 2014 geometric mean blood lead level among Aboriginal children (7.5 ug/dL) is higher than both the NHMRC reference level, and the mean for non-Aboriginal children (5.2 ug/dL) (Lesjak & Jones 2015).

Unfortunately, the persistence of higher BLLs among Aboriginal children in Broken Hill, while a clear environmental justice issue, is not surprising to experts who study lead as an environmental health problem. Markowitz and Rosner (2002: 137) have called lead a "paradigmatic" poison, "that linked industrial and environmental disease in the first two-thirds of the twentieth century." In other words, by understanding lead, we can understand the myriad other toxicants which originate at mines and factories and often work their way into the environment through industrial pollution, causing harm to human health. Lead is also paradigmatic for environmental justice issues. Though lead can affect people of all social and economic groups, it too frequently disproportionately affects low-income children and children of color (Landrigan et al. 2010).

Low-income and socially disadvantaged people have fewer resources to address environmental health threats that are not properly managed or controlled by government. A qualitative study that included Indigenous people in Broken Hill pointed to challenges they faced in addressing lead health risks. Some said they were not able to get help with addressing sources of lead exposure in their homes that required financial resources – they pointed specifically to "fixing cracks in ceilings, planting grass over dirt yards" (Thomas et al. 2013). Lead exposure has often been framed as an issue of parental responsibility, and the notion that parents in industrial communities can, and should, be able to protect their children from lead exposure by following recommendations of public health experts (mostly having to do with house cleaning and hygiene), has been perpetuated both by government and industry. Unfortunately, there is little evidence to support this belief, but this notion has very much been part of the discourse at Broken Hill; when children are found to have elevated blood lead levels, parents may feel stigmatized and be less likely to participate in public health interventions (Condon-Paoloni 2005; McGee 1998).

McArthur River Mine, Northern Territory

Home to over 1000 people, the town of Borroloola in the Gulf of Carpentaria, Northern Territory, is situated on the traditional country of Yanyula people. The mine site itself is situated on the McArthur River, for which four Aboriginal clan groups, the Mara, Yanyula, Garawa and Gurdanji, have responsibility. These traditional owners have protected this land for thousands of years, and they continue to hunt, fish, hold ceremony and care for sacred sites located there, including sites of great cultural significance, such as the Rainbow Serpent Dreaming site. The story of this site is about the serpent that used to travel along the river creating storms and cyclones.

The McArthur River runs through this land and it is considered the lifeblood of the area by the traditional owners (Young 2015). However, despite consistent opposition from local people, one of the world's largest lead and zinc mines, the McArthur River Mine (MRM), operates within the bed of this river, about 50 km downstream of Borroloola. Prior to the mining activity, Aboriginal elders said they used to drink from the river, but now they know it is too polluted to do that (Green & Kerins 2015).

Environmental and cultural problems

McArthur River Mine, which was given approval to develop and begin commercial operations in 1995, is currently owned by Glencore, an Anglo-Swiss multinational mining company. In its 21 years of operation as an underground, and then open cut mine, it has routinely violated regulations covering the discharge of contaminated water into the river, and regulations around the operation of large dams and the management of toxic waste rock (Erias Group 2015; Mellor 2015). These violations are documented in the annual independent monitoring reports, which began in 2008, and were imposed as a condition of the contested open pit approval (Young 2015).

In 2002, in order to access more material, specifically to allow mining of the ore body below the original river, the company proposed converting the underground mine into an open pit by diverting a 5.5 km stretch of the McArthur River. There was strong opposition to this from the community as well as the Northern Territory Government's Environment Minister. The Minister, referring to an assessment by the Northern Territory Environment Protection Agency (EPA), felt there was significant uncertainty over the environmental risks

(Young 2015). The EPA had concerns about the diversion of the river, noting that it could pose significant risks in terms of contaminated seepage from the mining and milling operations entering the ground water as well as the disruption of wildlife and visible damage to country (Young 2015). However, despite these concerns, five years later, the diversion went ahead.

Uncle Jack Green, a senior Garawa elder and cultural advisor, was one of the community leaders opposing the river's diversion due to the fact that it cut right through the Rainbow Serpent Dreaming cultural site. Uncle Jack Green explains that the river holds the song of the Rainbow Serpent and that the diversion had a devastating impact culturally: "That song's about two places, there were two rainbows. He lies across here and they cut him in half – and this is very important to Aboriginal people. It hurts a lot of Aboriginal people on McArthur" (Green & Kerins 2015).

Since the diversion there have been a series of escalating environmental problems that have been inadequately dealt with by MRM. In October 2014, the mine's independent monitor released a damning report into the mine's management, for example highlighting the seepage from the tailings storage facility which led to heavy metal contamination of waterways including Surprise Creek, a tributary of the McArthur River (Erias Group 2015).

Another major concern is related to the spontaneous combustion of the waste rock dump in 2013 (Bardon 2016). The fire was caused by reactive chemistry in the mine's waste rock dump when over 200 million tonnes of strongly pyritic material was exposed to water and oxygen. The company had significantly underestimated the percentage of reactive material present in the waste rock dump (Bardon 2014), and had failed to classify and store waste rock according to regulatory guidelines to prevent acidic reactions from occurring. The toxic sulphur smoke plume was visible from over 30 km away, and the fire burned for over a year (Mellor 2015). MRM attempted to contain the fire by covering the smouldering area with a 0.6m thick clay cap. But this cap was improperly compacted – exposing it to erosion, air and water, and so the acid forming rock below risks leaching into the groundwater and surrounding floodplains (Green & Kerins 2015). The independent monitoring report identified that as a result of this activity, waters in the region would most likely become polluted, impacting the surrounding ecosystem and, indirectly, the Aboriginal people fishing in the river (Green & Kerins 2015; Erias Group 2015).

The 2014 independent monitor's report noted that the mine site was being poorly managed, and both it and EPA questioned whether there was any adequate solution to managing the waste, and how it would cope with the additional half billion tonnes of pyritic waste rock that continued mining would bring to the surface. Despite these concerns, an expansion of the mine to double its production and continue until 2036 was approved. Interim approval to build a second waste rock dump and continue mining was granted in March 2016 despite Glencore failing to demonstrate that it had any effective solution to managing the huge volumes of reactive waste rock already on site.

Community, culture, country

From the outset most of the Aboriginal community have been concerned about the impacts of the mine damaging their country, which is so critical to their identity and way of life. Community members are also very clear that the mine has problems that must be fixed. Aunty Nancy McDinny, a Garawa/Yanyula traditional owner, says of the mining company, "We don't trust it anymore, it's poison. They should tell us the truth". Following revelations about the scale of the reactive waste rock and water contaminations risks, a meeting was called where a letter was presented to Northern Territory Government signed by over 150 Traditional Owners which called for production at the mine to stop until a solution to the mine's environmental problems could be found, for local people to be included on the mine's monitoring programme and for mining-damaged land to be rehabilitated, with local people taking on some of those jobs.

At this meeting, some community members said that they were worried that if they expressed their opposition to the mine, the company would withdraw services immediately. Unattributed notices were distributed around Borroloola following the meeting warning residents that services would be withdrawn if opposition to the mine intensified. These concerns echoed previous reports about the reduction in government services due to corporate substitution of vital social and educational services, for example as noted by Mellor (Mellor 2015: 9), who documents concerns over the clinic and other health and education programmes that the mine contributions are vital to maintain. While the mine is required to provide a community benefit trust fund of $1.5M per year to the community, the distribution of this money is controlled by Glencore and the Northern Territory Government. The result of this arrangement is that community services, such as the art centre, sea rangers and even basic healthcare and educational services are politically influenced by Glencore and lack independence to criticize or speak out against damage caused by the mine.

As a response, in 2016, a delegation of traditional owners went to Glencore's annual general meeting to discuss their concerns. They are also conducting their own independent research with unaligned researchers to find out the real extent of the pollution on their land, air and water.

Ranger uranium mine, Northern Territory

Great sandstone escarpments rise out of densely forested plains in Kakadu National Park. The Park is inscribed on the World Heritage list for its environmental and cultural values, the latter related to the Park's traditional owners, the Mirarr, who believe that they have a cultural obligation to protect their land for future generations. The Mirarr still practise their traditional culture on their land: they continue to hunt, fish and collect bush food within the Park's boundaries (Scambary 2013). Their primary concern regarding protecting their country today is over the damage caused to it by uranium mining. Kakadu is the only place in the world where government and industry have developed mining at a site – the Ranger uranium mine, run by Energy Resources of Australia (ERA) – completely surrounded by a World Heritage-listed park.

Toxic leaks and spills at Ranger

In 2002, a technical officer who worked at the Ranger environmental laboratory between 1993 and 1998 sent a statement to the Northern Territory and Federal governments in which he provided evidence that the Ranger mine was causing, and failing to accurately report, environmental damage (Kyle 2002). What is now clear, is that early reported leaks of acidic and low grade radioactive material were not isolated incidents, but part of a pattern. Since 1988, over 120 leaks and spills have occurred at Ranger, many of which were not made public until well after the incident had occurred – if at all (Senate 2003). In his statement, Kyle concluded: ". . . having demonstrated its incompetence, insouciance, and unwillingness to employ best practice in the management of mining a dangerous substance in a sensitive area, Ranger had breached its licence conditions and behaved as an unsuitable operator and an irresponsible corporate citizen" (Senate 2003: 71). The federal Senate Committee report agreed with that view, noting that the Ranger mine had "a pattern of underperformance and non-compliance" when it came to protecting the environment, a view shared by many of the Aboriginal corporations representing traditional owners in the mine site area (Senate 2003: 9).

Despite these warnings, toxic leaks continue to occur. The most recent was in 2013, when over 1.3 million litres of radioactive acidic slurry spilled out of a holding tank (NTG 2016). Despite a report into the incident which found that the company's management of process safety and its corporate governance did not meet expected standards, the chief executive of the Department of Mines and Energy decided that "it is not in the public interest to lay a charge against the mining company" (NTG 2016). This appeared to be the final insult to the Mirarr, who advised ERA that they could not consider any possible further extension to their Authority to mine in the Ranger Project Area, but wanted it rehabilitated so it could be incorporated back into Kakadu National Park (GAC 2015).

Few researchers or bureaucrats have investigated how the ongoing leaks and spills at Ranger have affected human health. Despite local Aboriginals particularly being at risk, the Northern Territory government does not regularly monitor the health impacts from these activities. In 2006, the first exploratory study of cancer rates for Aboriginals living in the mining region found almost a doubling in the overall cancer incidence rate (Tatz et al. 2006). However, these health costs, or the social, cultural and environmental costs that uranium mining creates, are not properly factored into government policy or industry licence provisions.

Traditional owners: land rich but dirt poor

Not all Indigenous communities oppose the idea of uranium mining on their land. But what they do object to is being excluded from key decisions about how and where the mining is allowed to happen, especially when it will have serious impacts on their way of life, whether by damaging their health, their land, or just as importantly, by affecting their ability to carry on cultural practices that have evolved over thousands of years. In areas where uranium mining has occurred, mining royalties have been paid as compensation to the traditional owners. Yet those royalties have been only a fraction of the profits extracted from the ground. They have come at a cost of damage to sacred sites, loss of land, access to hunting, fishing and collection of bush food, social disruption resulting from an influx of outsiders, and heightened social tensions created by unequal distribution of benefits from mining. And as the local Aboriginal communities have learnt through bitter experience, the price of mining royalties often includes cutbacks to their already meagre provision of housing, education and medical services. In the end, many already disadvantaged communities end up having to fund basic services that other Australians take for granted (O'Faircheallaigh 2006).

Uranium is found in many parts of Australia. The most economically recoverable deposits are found in the Northern Territory and South Australia, together comprising over one third of the world's known low-cost uranium supplies (Geosciences Australia 2013). Uranium mining could have the potential to enhance the ability of traditional owners to gain greater control over their lives, or it could perpetuate their existing disadvantage. But it is not only the cultural 'costs' that need to be considered: uranium mining appears 'cheap' because the environmental impacts, from radioactive leaks from tailings dams to the massive amounts of water used in mining operations, aren't fully accounted for by the companies that take profits from selling the uranium.

Discussion

The preceding case studies illustrate the intersection of poverty, power imbalance, environmental pollution and cultural, historical and legal distinctiveness in Aboriginal communities confronting the environmental and human health impacts of mining on or near their lands. In contrast to the US, the limited Australian environmental justice research means that studies such as these tend to be isolated, with little coherent, in-depth assessment, either nationally or at a state level, of the extent and nature of disproportionate environmental exposures. Such research is essential to fulfil the aims of the Intergovernmental Agreement on the Environment to ensure 'equivalent protection' for all Australians, and to inform policy that would prohibit disproportionate exposure to environmental hazards based on race or socio-economic status.

The environmental justice issues concerning mining that are currently unfolding on Aboriginal land are complex and not easily resolved. Documenting environmental and public health harms from mining is currently hampered by lack of funding and the remoteness of some of the locations leads to an 'out of sight, out of mind' mentality. Mining may also provide some economic benefits to remote Aboriginal communities, which may have limited opportunities for making money from tourism or art. In others, particularly in more remote areas, mining is often seen as one of the few productive options that may provide jobs and the cash needed to buy food and pay for electricity and other basic needs. Even though mining jobs are frequently well paid, mines tend not to employ a substantial number of people due to their high level of automation (O'Faircheallaigh 1986). Despite the claims by mining companies of increased jobs for Aboriginal people, few Aboriginals are employed long term: most are hired on low skilled contract work.

What many Aboriginal communities have found after mining consents have been given is a gradual erosion of culture through increasing numbers of non-Aboriginal and non-local people living on their land, destruction of their sacred sites and a withdrawal of government funded services (Scambary 2013). Royalty payments are meant to be 'compensatory' and are not meant to affect the provision of public services. However, many Aboriginal communities have seen how these services have been scaled back when they have received mining royalties.

Over the last decade, research has highlighted the lack of long-term economic benefits accruing to local Aboriginal communities from these, and other, mining activities (Scambary 2013). Despite these findings, the disadvantage remains, with mining being presented as the only economic option or 'choice' for many remote communities experiencing limited economic development and government services.

Mining on Aboriginal lands is a complex issue that must be decided by Aboriginal people themselves. However, they need a legal framework that will protect their interests and protect the environment on which they depend. Existing frameworks that are already used by Indigenous people, such as the United Nations declaration on the Rights of Indigenous Peoples, could be better implemented, thus ensuring and re-affirming Indigenous people's rights to protect and manage lands and to be able to exercise their social, political and economic rights. In 2009, Australia endorsed this declaration, although there has been limited acknowledgement of this in national or state policy to date. Because of history, discrimination and the inherent power differences between Lands Councils and multinational mining companies, the following procedures must be strengthened. These comprise federal legal protections for Aboriginal communities negotiating with mining companies covering royalties, funding to Aboriginal groups to conduct their own environmental and health monitoring, and/or to hire independent outside experts so that communities can participate equally in highly technical decision-making, and ensuring that government cannot reduce services and thereby increase dependence on mining companies.

To advance the aim of the Intergovernmental Agreement of 'equivalent protection' for all Australians, there is a pressing need for research on environmental justice issues affecting low-income Australians, as well as research on environmental justice issues specific to Indigenous Australians.

Funding for collaborative research and practice networks dedicated to community-based participatory research approaches to environmental justice issues could help to advance the goal of the Intergovernmental Agreement by linking research with policy and practice. Such research must take into account the unique historical, social, and economic context of Indigenous people in Australia, with specific research foci defined by Indigenous communities themselves. Relevant topics for environmental justice research include the environmental, health and social impacts of resource extraction, approaches to sustainable economic development, the health and social impacts of climate change, and strategies to mitigate climate change impacts on Indigenous communities.

References

Altman, J. 2007, *The Howard government's Northern Territory intervention: Are neo-paternalism and Indigenous development compatible*, vol. 16, Centre for Aboriginal Economic Policy Research, Canberra.

Altman, J. 2009, *Beyond closing the gap: Valuing diversity in Indigenous Australia*, vol. 54, Centre for Aboriginal Economic Policy Research, Canberra.

Arthur, B. 2005, *Macquarie atlas of Indigenous Australia*, PanMcMillan, Australia.

Australian Bureau of Statistics (ABS) 2014, *Region Profile for Broken Hill*, viewed 9 March 2016, http://stat.abs.gov.au/itt/r.jsp?RegionSummary&region=11250&dataset=ABS_NRP9_LGA&geoconcept=REGION&datasetASGS=ABS_NRP9_ASGS&datasetLGA=ABS_NRP9_LGA&regionLGA=REGION&regionASGS=REGION

Australian Institute of Health and Welfare (AIHW) 2011, *The health and welfare of Australia's Aboriginal and Torres Strait Islander people, an overview 2011*, AIHW, Canberra.

Bardon, J. 2014, 'McArthur River mine's burning waste rock pile sparks health, environmental concerns among Gulf of Carpentaria Aboriginal groups', *ABC*, 27 July, viewed 4 March 2016, www.abc.net.au/news/2014-07-27/mcarthur-river-mine-gulf-of-carpentaria-anger-smoke-plume/5625484

Bardon, J. 2016, 'The race to avert disaster at the NT's McArthur River Mine', *ABC Radio Background briefing*, 12 February, viewed 3 March 2016, www.abc.net.au/radionational/programs/background-briefing/the-race-to-avert-disaster-at-the-nts-mcarthur-river-mine/7159504

Blainey, G. 1968, *The rise of Broken Hill*, Macmillan, Australia.

Brereton, D., Klimenko, V., Cote, C. & Evans, R. 2009, *The minerals industry and land and water development in northern Australia*, ch. 8, Northern Australia Land and Water Science Review (full report) October 2009, Queensland.

Byrne, J., Wolch, J. & Zhang, J. 2009, 'Planning for environmental justice in an urban national park', *Journal of Environmental Planning and Management*, vol. 52, no. 3, pp. 365–392.

Carroll, J. 2003, *Future direction of the mining industry in the Northern Territory*, Department of Business, Darwin.

Chakraborty, J. & Green, D. 2014, 'Australia's first national level quantitative environmental justice assessment of industrial air pollution', *Environmental Research Letters*, vol. 9, no. 4, 044010.

Child and Family Health Center 2014, *Lead in Broken Hill*, viewed 9 March 2016, www.leadnsw.com.au/#!lead-in-broken-hill/c20gz

Christophersen, J. 1996, *Committee Hansard*, 3 September, pp. 507, viewed 5 March 2016, www.aph.gov.au/Parliamentary_Business/Committees/Senate/Former_Committees/uranium/report/d07

Condon-Paoloni, D. 2005, *Weighing the risks: Individualisation, trust and risk in three lead contaminated communities in Australia*, PhD, University of Wollongong, viewed 9 March 2016, http://ro.uow.edu.au/theses/772/

Department of the Environment (DoE) 1992, *Intergovernmental Agreement on the Environment*, Commonwealth of Australia, Canberra.

Dong, C., Taylor, M., Kristensen, L. & Zahran, S. 2015, 'Environmental contamination in an Australian mining community and potential influences on early childhood health and behavioural outcomes', *Environmental Pollution*, vol. 207, pp. 345–356.

Environmental Defenders Office (EDO) 2012, *Environmental Justice Project: Final report*, Victoria: Environmental Defenders Office, viewed 29 Feb 2016, https://envirojustice.org.au/downloads/files/law reform/edo vic environmental justice report.pdf

Erias Group 2015, *Independent monitor environmental performance annual report McArthur River Mine 2014*, Report No. 01164A_3_v2

Geosciences Australia 2013, *Uranium*, viewed 5 March 2016, www.ga.gov.au/scientific-topics/minerals/mineral-resources/uranium#heading-4

Green, J. & Kerins, S. 2015, *Developing the North: A case study from the Gulf Country*, occasional seminar, Australian Institute of Aboriginal and Torres Strait Islander Studies, Canberra.

Gross, C. 2007, 'Community perspectives of wind energy in Australia: The application of a justice and community fairness framework to increase social acceptance', *Energy Policy*, vol. 35, no. 5, pp. 2727–2736.

Gundjeihmi Aboriginal Corporation (GAC) 2015, *GAC Media Statement*, 15 October, viewed 5 March 2016, www.mirarr.net/media_releases/gac-media-statement

Havemann, P. 1999, *Indigenous peoples' rights in Australia, Canada, & New Zealand*, Oxford University Press, USA.

Heyworth, J., Reynolds, C. & Jones, A. 2009, *A tale of two towns: Observations on risk perception of environmental lead exposure in Port Pirie and Esperance, Australia*, viewed 4 March 2016, http://ro.uow.edu.au/cgi/viewcontent.cgi?article=1366&context=medpapers

Jack Green 2016, video recording, Seed, Borroloola.

Knibbs, L. & Barnett, A. 2015, 'Assessing environmental inequalities in ambient air pollution across urban Australia', *Spatial and Spatio-temporal Epidemiology*, vol. 13, pp. 1–6.

Kyle, G. 2002, *Statement on the three issues pertaining to environmental monitoring at the ERA Ranger uranium mine, Jabiru, NT*, viewed 4 March 2016, www.environment.gov.au/system/files/resources/6fa33e3b-9f47-4d01-91b4-d891554ded67/files/ssr171-statement.pdf

Landrigan, P., Rauh, V. & Galvez, M. 2010, 'Environmental justice and the health of children', *The Mount Sinai Journal of Medicine*, vol. 77, no. 2, pp. 178–187.

Langton, M. 2008, 'Poverty in the midst of plenty: Aboriginal People, the 'resource curse' and Australia's mining boom', *Journal of Energy and Natural Resources Law*, vol. 26, no. 1, pp. 31–65.

Lesjak, M. & Jones, T. 2015, *Lead health report: Children less than 5 years old in Broken Hill*, viewed 9 March 2016, http://media.wix.com/ugd/ba9a86_457498a9c1f94c5ca8472fd64f0e0371.pdf

Lloyd-Smith, M. & Bell, L. 2003, 'Toxic disputes and the rise of environmental justice in Australia', *International Journal of Occupational and Environmental Health*, vol. 9, no. 1, pp. 14–23.

Low, N. & Gleeson, B. 1999, *One earth: Social & environmental justice*, ACF, Victoria.

Macintyre, S., Atkinson, A., Lake, M., Pons, X. & Macintyre, S. 2000, *A concise history of Australia*, Cambridge University Press, Australia.

Markowitz, G. & Rosner, D. 2002, *Deceit and denial: The deadly politics of industrial pollution*, University of California Press, Berkeley.

McGee, T. 1998, 'The social context of responses to lead contamination in an Australian community: Implications for health promotion', *Health Promotion International*, vol. 13, no. 4, pp. 297–306.

Mellor, L. 2015, 'Poison or poverty? Glencore's blackmail of Borroloola', *Mining Monitor*, vol. 5, pp 7–10.

Millner, F. 2011, 'Access to environmental justice', *Deakin Law Review*, vol. 16, no. 1, pp. 189–208.

Mudd, G. 2007, *The sustainability of mining in Australia: Key production trends and their environmental implications for the future*, viewed 9 March 2016, www.protestbarrick.net/downloads/1_SustMining-Aust-aReport-Master.pdf

Munksgaard, N., Taylor, M. & Mackay, A. 2010, 'Recognising and responding to the obvious: The source of lead pollution at Mount Isa and the likely health impacts', *Medical Journal of Australia*, vol. 193, no. 3, pp. 131–132.

Nancy McDinney 2016, video recording, personal interview, Borroloola.

National Native Title Tribunal (NNTT) 2016, *Indigenous estates and determinations*, viewed 29 Feb 2016, www.nntt.gov.au/Maps/Indigenous Estates and Determinations A1L.pdf

National Pollutant Inventory (NPI) 2016, *National pollutant inventory*, viewed 29 Feb 2016, http://npi.gov.au

National Toxicology Program (NTP) 2012, *NTP Monograph: Health effects of low-level lead*, viewed 9 March 2016 https://ntp.niehs.nih.gov/ntp/ohat/lead/final/monographhealtheffectslowlevellead_newissn_508.pdf

Northern Territory Government (NTG) 2010, *Northern Territory Government 2009–10 Budget*, NTG, Darwin.

Northern Territory Government (NTG) 2016, *Completed investigation into failure of leach tank 1 Ranger Uranium Mine*, NTG, Darwin. http://mediareleases.nt.gov.au/mediaRelease/18275

O'Faircheallaigh, C. 1986, 'The economic impact on Aboriginal communities of the Ranger Project: 1979–1985', *Australian Aboriginal Studies*, vol. 2.

O'Faircheallaigh, C. 2004, 'Evaluating agreements between Indigenous peoples and resource developers in Honour Among Nations?' *Treaties and Agreements with Indigenous People*, Melbourne University Press, Australia.

O'Faircheallaigh, C. 2006, 'Aborigines, mining companies and the state in contemporary Australia: A new political economy or "business as usual"?' *Australian Journal of Political Science*, vol. 41, no. 1, pp. 1–22.

O'Faircheallaigh, C. 2008, 'Negotiating cultural heritage? Aboriginal–mining company agreements in Australia', *Development and Change*, vol. 39, no.1, pp. 25–51.

O'Faircheallaigh, C. & Corbett, T. 2005, 'Indigenous participation in environmental management of mining projects: The role of negotiated agreements', *Environmental Politics*, vol. 14, no. 5, pp. 629–647.

Scambary, B. 2013, *My country, mine country: Indigenous people, mining and development contestation in remote Australia*, CAEPR Monograph 33, viewed 5 March 2016, www.oapen.org/search?identifier=459939

Senate 2003, *Inquiry into Environmental Regulation of Uranium Mining*, Senate Environment, Communications, Information Technology and the Arts References Committee, Parliament of Australia.

Sutton, P. 2001, 'The politics of suffering: Indigenous policy in Australia since the 1970s', *Anthropological Forum*, vol. 11, no. 2, pp. 125–173.

Tatz, C., Cass, A., Condon, J. & Tippett, G. 2006, *Aborigines and uranium: Monitoring the health hazards*, AIATSIS Research Discussion Paper, No. 20.

Taylor, M., Lanphear, B. & Winder, C. 2013, 'Eliminating childhood lead toxicity in Australia: A call to lower the intervention level', *Medical Journal of Australia*, vol. 199, no. 5.

Taylor, M., Davies, P., Kristensen, L. & Csavina, J. 2014a, 'Licenced to pollute but not to poison: The ineffectiveness of regulatory authorities at protecting public health from atmospheric arsenic, lead and other contaminants resulting from mining and smelting operations', *Aeolian Research*, vol. 14, pp. 35–52.

Taylor, M., Mould, S., Kristensen, L. & Rouillon, M. 2014b, 'Environmental arsenic, cadmium and lead dust emissions from metal mine operations: Implications for environmental management, monitoring and human health', *Environmental Research*, vol. 135, pp. 296–303.

Thomas, S., Boreland, F. & Lyle, D. 2013, 'Improving participation by Aboriginal children in blood lead screening services in Broken Hill, NSW', *New South Wales Public Health Bulletin*, vol. 23, no. 12, pp. 234–238.

Thompson, A. 1893, *Report of Board appointed to inquire into the prevalence and prevention of lead poisoning at the Broken Hill Silver–Lead Mines*, viewed 9 March 2016, https://archive.org/details/b21365015

Weng, Z., Mudd, G., Martin, T. & Boyle, C. 2012, 'Pollutant loads from coal mining in Australia: Discerning trends from the National Pollutant Inventory', *Environmental Science & Policy*, vol. 19, pp. 78–89.

Young, A. 2015, 'McArthur River Mine: Monumental regulatory mess', *Land Rights News*, Northern Edition, January, pp. 11–15.

42

ENVIRONMENTAL JUSTICE ACROSS BORDERS

Lessons from the US–Mexico borderlands

Sara E. Grineski and Timothy W. Collins

Environmental injustice in transnational context: a world-systems perspective

Over the last four decades, social scientists have recognized the transnational shifting of environmental burdens, in terms of the hazards associated with production and waste disposal, from powerful to less powerful states (Pellow 2007; Kellenberg 2012; Clapp 2001). This global transference of risk has been framed as an environmental injustice because it is directly related to a global system of stratification whereby powerful core states (e.g. United States) and transnational corporations are able to impose their economic will on less powerful (semi)peripheral states (e.g. Mexico) (Pellow 2007; Frey 2003). (Semi)peripheral states are saddled with environmental burdens because they are marginalized within the global political economic order and represent a path of less resistance (Pellow 2007). For example, core countries are able to externalize residents' consumption-based environmental costs, which is leading to increased deforestation in (semi)peripheral countries (Jorgenson 2010). Similarly, waste imports worldwide have been shown to increase in countries whose environmental regulations are lower than those of their trading partners (Kellenberg 2012).

World systems theory, as conceived originally by Immanuel Wallerstein, offers a historical perspective on development and social change over a long time scale that can further understandings of the transference of environmental burdens across borders. A world systems perspective suggests that the capitalist system simultaneously produces and responds to uneven development (Jones 1998). By design, the capitalist system creates ecological and social harms in specific places through a self-reinforcing mechanism of increasing rates of production and consumption (i.e. the 'treadmill of production' (Gould et al. 2008)).

The polarization created by this unevenness is necessary for the maintenance of the system as a whole (Wallerstein 1974). Core nations are able to develop at the expense of the periphery, extracting surplus (Simpson 1990) and externalizing environmental hazards such as toxic waste (Pellow 2007). Peripheral regions are unable to prevent this because of their relative weakness, which relates to a lack of resources and an inability to unite internal interests. Peripheral states do not have the same level of internal autonomy over their economic affairs as core states; they can be seen as operating in the interests of the core (Simpson 1990). The semiperiphery lies between the core and its periphery, in both geographic and social terms. It is core with respect to periphery but periphery with respect to core. It exploits and is exploited and is a necessary structural element of the world economy because it can deflect political pressure that periphery groups might otherwise direct against core states (Wallerstein 1974).

In what follows, we introduce the US–Mexico border region. The region is a microcosm for processes of uneven development worldwide but they are brought into stark relief here, where the externalization of environmental risks from a core to a semiperipheral state is easily visible. Numerous EJ studies have been done on this border region and this body of work enables us to use this border region as an exemplar of EJ issues created by international borders. The chapter summarizes the environmental injustices occurring in the region at two scales: between the US and Mexico and then within border communities on each side. Then, we offer a series of hypotheses, informed by the US–Mexico border case, for how distributional environmental injustices might be expected to play out in core, semiperiphery and periphery states worldwide before concluding with a discussion of normative concerns.

The US–Mexico borderlands

The US–Mexico border is a place where the core and semiperiphery meet, and this political economic context has inscribed patterns of environmental injustice across regional socio-environmental landscapes. The international border stretches nearly 2 000 miles along the Rio Grande/Bravo River from San Diego, California/Tijuana, Baja California Norte and the Pacific Ocean in the west to the Gulf of Mexico in Brownsville, Texas/Matamoros, Tamaulipas in the east (Figure 42.1). Socially, the border region, defined as 100 km on each side of the international line, is home to more than 14 million residents, with about 7.3 million living in the US and 6.8 million in Mexico. Projected population growth rates in the region exceed national averages. If current trends continue, the border population could increase by an additional 4.6 million people by the year 2020 (Environmental Protection Agency 2012).

The physical geography of the border region exacerbates its environmental problems. The region is generally semiarid, with more rainfall on the coasts, and the dry conditions heighten concerns about water and air quality (Liverman et al. 1999). The interior border region has been identified as a heat vulnerability "hot spot" (Reid et al. 2009). The rapid urbanization and industrialization of the region alongside the arid climate and predictions of severe precipitation decreases and temperature increases under global climate models have contributed to serious concerns about social vulnerability to climate change (Wilder et al. 2010).

Industrial production and its associated environmental consequences have a long history in the border region. The Bracero Program, started by the US government in 1942, legalized the migration of Mexican workers into the US to replace those serving in World War II. When the program ended in 1964, several hundred thousand Mexican workers were returned to Mexican border cities. In an attempt to alleviate overcrowding and unemployment, the Mexican government created the Border Industrialization Program to promote industrial development and employment (Liverman and Vilas 2006). As a result, the *maquiladora* industry grew tremendously. In 1970, Mexico had 72 factories, and by 1979, it had 620. Currently, there are approximately 3000 *maquiladoras* (i.e. export-oriented final assembly plants) in Mexico's northern border region. These mostly US-owned transnational corporations import needed equipment and raw materials tax free.

The passage of the North American Free Trade Agreement (NAFTA) in 1994 enabled continued growth in the *maquiladora* sector along the Mexican side of the border because it

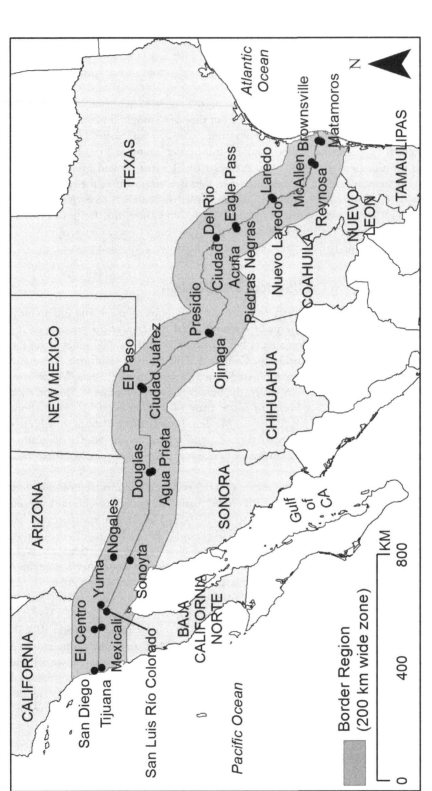

Figure 42.1 US–Mexico border region

reduced tariff barriers to trade (Frey 2003). Some have argued that growth in the *maquiladora* sector post-NAFTA has added to environmental degradation and health risks in the region (Williams and Homedes 2001). Evidence to support this includes violations of environmental laws by foreign-owned companies made possible through poor enforcement, a lack of adequate environmental legislation, and a weak institutional framework in Mexico (Roberts and Thanos 2003). Others have shown that foreign-owned *maquiladoras* are actually cleaner and more responsive to environmental regulations than are locally owned industries (Liverman and Vilas 2006; Contreras et al. 2006). While there is a debate as to the extent that *maquiladoras* damage the environment, there is consensus that the overall growth of industrial activity along the border has caused environmental degradation and amplified health risks (Liverman and Vilas 2006; Schatan and Castilleja 2005).

Environmental injustices between the United States and Mexico

It has been argued that nowhere is the movement of hazardous production processes from core to (semi)peripheral states more readily visible than along the US–Mexico border (Grineski and Juárez-Carillo 2012). Here, hazardous production processes are transferred to Mexico through transnational corporations operating *maquiladoras*; 67 of the top 100 *maquiladoras* in Mexico are owned by US companies (Grineski and Juárez-Carillo 2012). A study of the cross-border cities of Ciudad Juárez (Chihuahua) and El Paso (Texas) showed that the level of industrial hazard, measured by the density of facilities, was 24 times higher in the average Ciudad Juárez neighbourhood in the year 2000, as compared to the average neighbourhood in El Paso, Texas (Grineski and Collins 2010), representing a clear injustice between the two countries. The difference in the level of hazard between the two cities is evidence of the importance of uneven power relations between core and (semi)peripheral nations in determining risk disparities. This risk disparity reflects a world systems-scale environmental injustice, since unequal political-economic relations between the US and Mexico have produced divergent trajectories of hazardous industrialization. Specifically, de-industrialization in El Paso combined with a *maquiladora* boom in Juárez has resulted in a highly disparate cross-border industrial risk profile. The NAFTA appears to have accentuated these divergent trajectories in El Paso and Juárez (Grineski and Collins 2010). The lower level of industrial risk experienced by El Pasoans relative to *Juárenses* is largely a product of the unequal positioning of the two cities in the world economic order, as corporations based in the US (and other core states) have raced along lowest production cost paths by moving factories to cities in (semi)peripheral states (Grineski and Collins 2010).

The growth in industrial employment opportunities on the Mexican side has drawn increasing numbers of migrants to the border, where the lack of affordable high-quality housing for them is a serious issue. Development of neighbourhoods without access to piped water or sewage infrastructure has been linked to the mass migration of rural residents to northern Mexican cities due to the labour opportunities in *maquiladoras* and Mexico's shift to corporate agricultural practices. Poor migrants have settled in fringe areas due to the lack of affordable housing elsewhere. The lack of public infrastructure in these areas relates to several factors, including the fact that *maquiladora* profits have been largely appropriated by global and local elites rather than re-invested into infrastructure development, which could have addressed poor living conditions for many *maquiladora* workers (McDonald and Grineski 2012; Collins 2010). The extent of the problem is reflected in the disparity in access to piped water between El Paso and Juárez. In Juárez, 24 per cent (69 866 households) of housing units were without piped water indoors as per the 2000 census, and in El Paso the percentage was 1.5 (3068 households) (McDonald and Grineski 2012). Parallel disparate exposures to flood hazards – which may also be viewed in terms of environmental injustice – exist between El Paso and Juárez, with marginalized residents of informal peri-urban Juárez settlements experiencing the greatest flood risks (Collins 2010).

A US-owned thermoelectric plant that was sited near Mexicali (capital of Baja California) in 2003 to generate power primarily for southern California provides another example of inter-border environmental injustice that was studied using qualitative methods. This plant is located in Mexico because of the streamlined permitting process, limited political space for popular resistance, lower wages, lower land costs, and a more favourable political and investment community (Carruthers 2008). Another advantage for the US is that Mexico must bear the burden of the plant's substantial air pollution externalities. Despite the fact that electric power plants are the largest source of toxic air pollution in North America and the Mexicali Valley–Salton Sea cross-border air shed was already recognized as seriously polluted, Mexican authorities approved the installation of this plant in 2001 (Carruthers 2008). After intense public pressure to meet California air standards, the US-based corporation installed nitrogen oxide scrubbers at the plant to reduce air emissions. However, when activists demonstrated because the scrubbers were only partially installed, the US Department of Energy pressured the plant to comply with US law, and the plant was closed. Nonetheless, the plant re-opened in 2004 after learning that the scrubbers were not needed to comply with Mexican law (Carruthers 2008).

As seen in the Mexicali case, the combination of lower production costs and lax environmental policies and enforcement attracts industry to (semi)peripheral contexts. As a Tijuana environmental activist reported:

> Government oversight is poor. There aren't enough inspectors. There is no obligatory inspection scheme, only a voluntary one, and inspections are arranged in advance, with no surprise visits. We have seen gradual deterioration in the urban communities where the factories are located.
>
> *(Godoy 2011, n.p.)*

Her perspective has been echoed by a federal study of environmental regulations in Mexican *maquiladoras* which showed that only 38 per cent of the plants were in full compliance with environmental standards and that the trend is toward increasingly fewer inspections (Schatan and Castilleja 2005).

The lack of compliance with standards should not come as a surprise when one considers the institutional and technical capabilities for inspection and enforcement along the Mexican border. There are many challenges that make environmental enforcement difficult, which are likely to be characteristic of semiperipheral contexts more generally. These include strong local and core interests preventing enforcement; vague and weak regulations; underpaid enforcement agents who are easily bribed; a lack of clarity regarding responsibility for environmental enforcement; municipalities' lack of monitoring equipment and technical knowledge; and excessive work demands on staff (Montalvo Corral 2004). Only 54 per cent of the 200 plants that Schatan and Castilleja (2005) surveyed in Tijuana, Juárez and Mexicali had an active environmental policy. This is in spite of the fact that 89 per cent were foreign-owned with parent companies having significant environmental protection measures in place at their domestic facilities (Schatan and Castilleja 2005). On the one hand, these factors negatively influence environmental quality and quality of life for local people; on the other, they make Mexico attractive to transnational corporations seeking to reduce production costs by relocating to a non-core context. Additionally, the factories owned by core companies have the power to quell community resistance to their operations at the local level due to the steady supply of labour available along the border, close relationships with municipal authorities, and the vast economic resources of their transnational parent companies, which far outweigh local resources for social mobilization (Morales et al. 2012).

While less related to the transference of industrial risks, the US and Mexican sides of the border are unevenly socially vulnerable to the impacts of climate change, with the Mexican side experiencing greater risks and reduced capacities to respond. Climate change predictors for the western US–Mexico border region include temperature increases of 3–5°C by 2100, with a possible 58 per cent decrease in precipitation. When comparing climate change-related hazard exposure and social marginality between El Paso and Juárez, indicators were generally higher in Juárez as compared to El Paso. Specifically, peak neighbourhood heat exposures and flood risks were significantly higher in Juárez, as were indices of social marginality, which suggest that Juárez would be at an increased risk during a climate change-related event due to its population characteristics as compared to El Paso (Grineski et al. 2012).

Environmental injustice within Mexican and United States border communities

On the Mexican side

In addition to these transnational injustices present between the US and Mexican sides of the border, there are finer scale injustices occurring within border communities, including those on the Mexican side. Many of the studies have focused on the inequities in exposure to risks from *maquiladoras*. This is because *maquiladoras* create air, soil and water pollution through their activities. They generate toxic chemical wastes, which are sometimes spilled or improperly disposed of (Sanchez 1990). For example, toxic effluents in drainage ditches flowing from *maquiladoras* into neighbouring communities have been documented to contain contaminants in excess of US Environmental Protection Agency standards (Williams and Homedes 2001). The *maquiladoras* also have the potential to be sites for acute industrial accidents, such as the 2013 explosion at a Juárez *maquila* that killed eight and injured many others. A study of children living in six Mexican cities determined that children with the highest levels of flame retardants in their blood serum resided in an urban and industrial area, as compared to the children living in a rural area, near a landfill, or in an urban but not industrial area (Pérez-Maldonado 2009). A recent air monitoring campaign in Tijuana (Bei et al. 2013; Minguillon et al. 2014; Shores et al. 2013) revealed sharp, extremely high, and regularly occurring peaks of black carbon occurring around midnight near a municipal park, a pollutant with known respiratory and cardiovascular effects. These peaks were comparable to occupational exposure in workplaces dominated by diesel exhaust exposure and the authors believe they were caused by clandestine industrial activity (Shores et al. 2013). This evidence suggests that industrial land uses (associated primarily with the *maquiladora* sector) in northern Mexico pose health risks for proximate residents.

In Nogales (Sonora), researchers found significantly higher proportions of educated and affluent residents living within 500 m and 1000 m buffers of hazardous waste-generating facilities (using correlation analysis), with no significant differences in the proportions of recent immigrants living within as compared to outside the buffers (Lara-Valencia et al. 2009). In Tijuana (Baja California), researchers found that formal residential development (measured as the proportions of occupied housing units with no dirt floors, electric lights, piped water, sewage infrastructure, refrigerator, and washing machine) was significantly associated with industrial park density, with formal development accounting for the significant effect of higher socioeconomic status on greater industrial density. Higher proportions of female-headed households were also significantly associated with industrial park density (Grineski et al. 2016). *Maquiladora* employees may also face residential risks from living near their places of employment. A survey of 767 *maquiladora* workers in Tijuana revealed that 20 per cent of workers lived in the same neighbourhood as or in a neighbourhood adjacent to the factory where they worked (Kopinak and Barajas 2002).

Ciudad Juárez (Chihuahua), the fifth largest city in Mexico, has been examined more from an EJ perspective than other Mexican border cities. In Juárez, both socially advantaged and disadvantaged residents are at increased risk from environmental hazards, depending on the hazard of focus. Similar to the patterns in Nogales (Lara-Valencia et al. 2009), neighbourhoods with higher mean levels of education (higher socioeconomic status), lower proportions of migrants, and lower proportions of young children had significantly higher *maquiladora* densities (Grineski and Collins 2010). This is because in Juárez, more affluent residents live in the central city to access social, infrastructural and market-based services (e.g. secondary schools, hospitals, paved roads, sewer systems, and retail) (Grineski and Collins 2010). In contrast to typical US cities, Mexican cities like Juárez tend to have well-developed central cities served by basic infrastructure (e.g. piped water, sewage treatment, paved roads, and electricity), whereas the fringes (where suburbs would be in an American city) contain informal settlements of low-rise, self-constructed homes with reduced access to this infrastructure, that impose health and environmental risks on residents (Graham et al. 2004).

Transnational corporate owners of *maquiladoras* choose to set up operations in the developed areas, on high rent land, because it has either been offered as an incentive by state officials (Zeisel et al. 2006) or it is otherwise relatively inexpensive from a global market perspective (it is a bargain compared to land in the US Rust Belt). Such industrial siting decisions are made possible by the inequalities that exist between core-based transnational corporations and (semi)peripheral regions. The clustering of formal development in the central city means that this area is also home to high concentrations of affluent residents and transnational industrial facilities. In contrast, domestic industries (e.g. brick kilns) cannot afford to compete in a transnational land market and thus are sited in less privileged areas, where more socially marginalized residents live (Grineski et al. 2010).

Given that *maquiladoras* are not the only environmental hazard facing Juárez residents, researchers have also examined risks associated with climate change (Grineski et al. 2015; Collins et al. 2013), brick kilns and the rail line (Grineski et al. 2010). Related to climate change and its intersection with socio-economic forces, many Juárez neighbourhoods face situations of triple exposure, in which residents have suffered due to the conjoined effects of the global recession, drug war violence, and extreme heat since 2008. The economic downturn in 2008 hit the city of Juárez hard, resulting in a loss of more than 80 000 jobs (Kolenc 2010). A steep rise in narco-violence began after the onset of the global economic recession. Between 2008 and 2012, over 10 000 people were killed (Alvarado 2012) and between 2007 and 2009, 230 000 residents out-migrated from the city (Velázquez Vargas 2012). These demographic changes were reflected in neighbourhood census data; over 75 per cent of Juárez neighbourhoods experienced decreasing population density between 2000 and 2010 and the average neighbourhood saw a 40 per cent increase in the proportion of older adults (Grineski et al. 2015). Neighbourhoods with greater drops in population density and increases in the proportion of older residents between 2000 and 2010 were at significantly higher risk in extreme heat, as were neighbourhoods with lower social class. In Juárez, triple exposures were associated with a climate gap as the most endangered lower socio-economic status and increasingly older-aged populations remained in neighbourhoods from which high proportions of residents had fled (Grineski et al. 2015). Meanwhile, over the same period of time, El Paso ranked as one of the safest cities in terms of violent crime in the US.

Risk associated with brick kilns represents a clear case of environmental injustice in Juárez. There are an estimated 369 brick kiln burns per month in Juárez and approximately 391 kg of air contaminants are released during every burn (Romo et al. 2004). Brick kilns are owned by local residents, and the majority of the bricks are sold domestically (Romo et al. 2004). Those of lower social class and a disproportionate number of children reside in neighbourhoods with higher brick kiln density (Grineski et al. 2010). The disproportionate presence of children near the kilns, which is related to the family-owned nature of the business, is a serious health concern, given the air pollution generated by the low-technology wood- and sawdust-fired kilns.

Environmental injustices are also found in relation to the rail line. Residence near the rail line is especially risky in Juárez because one of the world's most dangerous chemical plants (Solvay) ships hydrofluoric acid along it (Morales et al. 2012) and areas near the rail line are disproportionately inhabited by poor residents (Grineski et al. 2010). In addition to posing an acute risk, the rail line is noisy and a personal safety hazard for residents and pedestrians.

While these studies document objective risks faced by border residents, it is also important to consider how Mexican border residents subjectively experience their socio-environments. In-depth analysis in one Matamoros neighbourhood located on a former landfill, next to an operating landfill, revealed that neighbourhood residents did not see themselves as victims (Johnson and Niemeyer 2008). Residents, many of whom were migrants from poorer communities in southern Mexico, understood Matamoros to offer them the possibility of relatively high paying jobs and better educational opportunities for their children. Some interviewed residents earned three to four times more than they had in rural southern Mexico and saw residence in the relative affluence of northern Mexico as a symbol of their rising social status. Residents were proud of transforming empty dumpland into a home and of raising themselves out of absolute poverty. While troubled by ubiquitous trash in their yards, they also used trash to create better shelters (Johnson and Niemeyer 2008). As Johnson and Niemeyer (2008) stated, "clearly, the frame through which [neighbourhood] residents view their lives is more expansive than the monolithic picture most Northern environmentalists have of people living in border [neighbourhoods]" (p. 378).

Based on the EJ studies conducted along the Mexican side of the US–Mexico border, we can conclude that despite the lack of association between lower socioeconomic status and greater levels of hazards, which is commonly found in core contexts, Juárez and other Mexican cities along the US border face numerous environmental injustices that are related to the region's position in the global economic order.

On the US side

Far fewer quantitative studies of unequal risks have been conducted in US border cities. The lack of quantitative studies belies a tradition of community organizing for environmental justice in US border communities. However, several quantitative studies have been conducted in El Paso (Texas), the sister city of Ciudad Juárez. When considering the environmental amenity of neighbourhood parks, there were no significant differences in the number of parks in El Paso neighbourhoods based on median income or the percentage of those foreign-born. However, low income and high percentage foreign-born neighbourhoods contained parks with more safety and quality concerns than did high income and low percentage foreign-born neighbourhoods (Kamel et al. 2014). When relating residential risk from factories to census data, El Paso neighbourhoods with lower socioeconomic status and higher proportions of residents who had recently migrated to the city had significantly higher densities of factory-related hazards (Grineski and Collins 2010).

Increased health risks, in the form of residential exposure to hazardous air pollutants (HAPs) that are known to cause cancer, are also found for El Paso neighbourhoods with higher than average percentages of Spanish speakers with limited English-language proficiency, foreign-born residents and non-US citizens (Collins et al. 2011). The cancer risks from HAPs for lower class Hispanics (based on poverty status and education) were much greater than the risks for upper class Hispanics, whereas for whites, class was not a significant determiner of risk. Findings from this study generally suggest that mutually reinforcing disadvantages multiply cancer risk disparities from HAPs for Hispanics in El Paso County, while disadvantages associated with class, age and gender status have little influence on cancer risks from HAPs for whites (Collins et al. 2011). El Paso schoolchildren also experience reductions in grade point average (GPA) associated with higher levels of HAPs surrounding their home site (Clark-Reyna et al. 2015) and school sites (Grineski et al. 2016). Effects on GPA were particularly strong for non-road mobile sources of pollution, including the airport, a US Army base and numerous rail yards related to the movement of rail freight between the US and Mexico. The importance of non-road mobile HAPs was also found in another El Paso study whereby non-road mobile HAPs were the strongest predictor of children's hospitalization rates from respiratory infections as compared to other risk sources (Grineski et al. 2013).

However, peri-urban socially marginalized residents along the US side of the border are also subject to an environmental injustice that is more commonly found in periphery countries: the lack of access to piped water and sewage systems. This is a critical injustice, given the generally high levels of infrastructure accessible to residents living in the US. US residents lacking access to water reside in the low-income *colonias* (Bath et al. 1998). *Colonias* in the US are unincorporated settlements characterized by high poverty rates and substandard living conditions. In El Paso County in 2010, 86 472 residents lived in 321 *colonias*; 56 of these *colonias*, housing 5 529 residents, lacked access to piped drinking water. These residents rely on water delivery trucks and store their water in tanks; however, this water typically has low levels of chlorine and is sometimes contaminated (McDonald 2012). This approach is substantially more expensive than relying on a public utility. Surveyed *colonia* households living without piped water paid 308 per cent more per gallon for water than did nearby water utility customers (McDonald 2012). The lack of access to water represents a clear case of environmental injustice, especially considering the racist history of water policy-making in El Paso which has led *colonia* residents to be excluded from this basic right (Bath et al. 1998). Taken together, these findings for El Paso demonstrate a traditional pattern of environmental injustice along the US side of the border, whereby disadvantaged residents face increased risks.

Future research and applicability to other transnational contexts

Knowledge of the forces leading to environmental injustices along the US–Mexico border may be extended to enhance understanding of environmental injustices in other transnational contexts via future studies. The border context sensitizes us to the transfer of production-based hazards especially from core to (semi)periphery locales and how this is enabled by a lack of regulation; differences between domestic and foreign firms in terms of their ability to secure desirable land for operations outside of the core; the vulnerabilities of low-paid labourers; the variability in urban development trajectories between world-system contexts; and how position in the world system shapes production-related risks for local residents. With increasing global social and economic unevenness and interconnectedness, most local contexts are inscribed by transnational processes, to greater and lesser degrees, even though it may not be as obvious as at the border of two unequal countries.

Table 42.1 offers hypotheses for patterns of distributional environmental injustices within core, semiperiphery and periphery contexts, extending from the US–Mexico border case as well as published literature on environmental injustice related to production processes in other localities. Table 42.1 also offers hypotheses about distributional injustices related to primary and secondary production processes connected to both domestic and foreign ownership. EJ analysts should test these hypotheses in a variety of international contexts, ranging from core to periphery across different types of ownership and production processes, in future studies. This will contribute to the development of a better understanding of processes leading to environmental injustice with the ultimate goal of ameliorating those injustices.

Table 42.1 Hypothesized patterns of distributive environmental injustice in core, semiperiphery and periphery countries by type of industry

	Environmental regulations	*Economic characteristics*	*Hypotheses related to distributional EJ: primary production*		*Hypotheses related to distributional EJ: secondary production*	
			Domestic	*Foreign*	*Domestic*	*Foreign*
Core	Higher levels of regulation and some enforcement	Diverse economy, including primary and secondary production, service activities, and research and development	Poor employees [1, 2] (H_1) and poor rural residents [3, 4] (H_2) face greatest risks		Poor face greater industrial risks [5-9] (H_9)	
Semi-periphery	Moderate level of regulation and limited enforcement [10, 11]	Mix of primary and secondary production by domestic and foreign firms	Poor subsistence farmers face greatest risks (H_3)	Poor employees face greatest risks [12] (H_4)	Poor residents face greatest risks [13] (H_{10}); employees also face risks	More affluent residents face greatest risks [13-16] (H_{11}); employees face risks

(continued)

Table 42.1 Hypothesized patterns of distributive environmental injustice in core, semiperiphery and periphery countries by type of industry *(continued)*

	Environmental regulations	Economic characteristics	Hypotheses related to distributional EJ: primary production		Hypotheses related to distributional EJ: secondary production	
			Domestic	Foreign	Domestic	Foreign
Periphery	Lower levels of regulation and limited enforcement [10]	Highly extractive; primary production; some foreign investment	Poor subsistence farmers face greatest risks (H$_5$)	Poor employees (H$_6$), Indigenous landholders (H$_7$), and the rural poor (H$_8$) face greatest risks [12, 17]	Less applicable	

Notes:

1. Arcury, T.A., et al., Pesticides present in migrant farmworker housing in North Carolina. *American Journal of Industrial Medicine*, 2014. **57**(3): pp. 312–322.
2. Runkle, J., et al., Occupational risks and pregnancy and infant health outcomes in Florida farmworkers. *International Journal of Environmental Research and Public Health*, 2014. **11**(8): pp. 7820–7840.
3. Mirabelli, M.C., et al., Race, poverty, and potential exposure of middle-school students to air emissions from confined swine feeding operations. *Environmental Health Perspectives*, 2006. **114**(4): pp. 591–596.
4. Wilson, S.M., et al., Environmental injustice and the Mississippi hog industry. *Environmental Health Perspectives*, 2002. **110**: pp. 195–201.
5. Chakraborty, J. and D. Green, Australia's first national level quantitative environmental justice assessment of industrial air pollution. *Environmental Research Letters*, 2014. **9**(4).
6. Dale, L.M., et al., Socioeconomic status and environmental noise exposure in Montreal, Canada. *BMC Public Health*, 2015. **15**: p. 205.
7. Laurian, L. and R. Funderburg, Environmental justice in France? A spatio-temporal analysis of incinerator location. *Journal of Environmental Planning and Management*, 2014. **57**(3): pp. 424–446.
8. Buzzelli, M. and M. Jerrett, Racial gradients in ambient air pollution exposure in Hamilton, Canada. *Environment and Planning A*, 2004. **36**(10): pp. 1855–1876.
9. Downey, L., Environmental inequality in metropolitan America in 2000. *Sociological Spectrum*, 2006. **26**(1): pp. 21–41.
10. Schroeder, R., et al., Third World environmental justice. *Society & Natural Resources*, 2008. **21**(7): pp. 547–555.
11. Schatan, C. and L. Castilleja, *The Maquiladora Electronics Industry and the Environment along Mexico's Northern Border* Commission for Environmental Cooperation, Editor. 2005: Montréal, Quebec, Canada.
12. Cifuentes, E. and H. Frumkin, Environmental injustice: case studies from the South. *Environmental Research Letters*, 2007. **2**(4).
13. Grineski, S.E., et al., No Safe Place: Environmental hazards and injustice along Mexico's northern border. *Social Forces*, 2010. **88**(5): pp. 2241– 2266.
14. Sabapathy, A., S. Saksena, and P. Flachsbart, Environmental justice in the context of commuters' exposure to CO and PM10 in Bangalore, India. *Journal of Exposure Science and Environmental Epidemiology*, 2015. **25**(2): pp. 200–207.
15. Schoolman, E.D. and C.B. Ma, Migration, class and environmental inequality: exposure to pollution in China's Jiangsu Province. *Ecological Economics*, 2012. **75**: pp. 140–151.
16. Lara-Valencia, F., et al., Equity dimensions of hazardous waste generation in rapidly industrialising cities along the United States–Mexico border *Journal of Environmental Planning and Management*, 2009. **52**(2): pp. 195–216.
17. Maiangwa, B. and D.E. Agbiboa, Oil multinational corporations, environmental irresponsibility and turbulent peace in the Niger Delta. *Africa Spectrum*, 2013. **48**(2): pp. 71–83.

Normative concerns

Particular challenges must be tackled in order to promote environmental justice in spaces defined by transnational processes. A critical challenge is the transnationally operating 'treadmill of production' (Gould et al. 2008). This process calls into question the efficacy of the environmental justice movement in the US. In the capitalist world system, basic contradictions undermine the conditions for production through environmental degradation and damage to the health of workers, to the point at which this spills over into the social arena, leading to mass resistance from environmental and/or labour movements (O'Connor 1988). When oppositional movements emerge, the capitalist 'treadmill' searches for new, often (semi)peripheral spaces to exploit, rather than restructuring production along "sustainable" lines.

Indeed, the success achieved by social movements in core states has led many core-based corporate polluters to move their hazardous operations to less controlled and more profitable places (Pellow 2007). The political success of the early EJ movement in the US has inadvertently helped rescale patterns of environmental injustice in exposure to toxics from the domestic to the transnational level (Low and Gleeson 1998). Thus, borders can be strategically exploited to facilitate accumulation for transnational corporations and to circumvent social resistance. In the process, spaces defined by international boundaries and/or transnational processes (e.g. export processing zones) have emerged as particularly vulnerable to socioenvironmental exploitation.

The pursuit of a globally responsive EJ agenda within this context, where international borders are being strategically manipulated to facilitate corporate profits and are (re)producing environmental injustice as a result, is complex. As a first step, EJ activists must recognize that, in their efforts to reduce local risks, they may inadvertently transfer risks to less powerful regions. Thus, we echo the call for the abandonment of a parochial EJ ethic (i.e. not in my back yard or NIMBY) and the adoption of a cosmopolitan one (i.e. not on planet earth or NOPE) as a necessary step toward more globally successful EJ activism (Pellow 2007; Low and Gleeson 1998). Once a cosmopolitan EJ ethic is broadly embraced, the focus must then move toward organizing effective opposition. A particular challenge comes from the discourse emanating from transnational corporations, states and media actors, which drives a wedge between the interests of workers and environmentalists by framing issues in binary terms of 'jobs/development vs. environment' trade-offs. EJ activists must take care not to unintentionally legitimize and perpetuate this discursive frame. To achieve greater success, those concerned with transnational EJ must strive to bring together activists engaged in labour and environmental movement organizing. This unity should be forged based on the shared recognition that most projects presented as transnational economic development opportunities are in fact highly exploitative of both ecological environments and working people. It is difficult to imagine such issues being effectively addressed without EJ activist leadership, since recognition of the connections between labour and environmental exploitation is foundational to the EJ movement.

References

Alvarado, E. (2012). "Juárez death toll surpasses 10,000". *Texas Monthly Daily Post*, May 10.

Arcury, T.A., Lu, C., Chen, H. & Quandt, S.A. (2014). "Pesticides present in migrant farmworker housing in North Carolina." *American Journal of Industrial Medicine*, Vol. 57, pp. 312–322.

Bath, C.R., Tanski, J.M. & Villarreal, R.E. (1998). "The failure to provide basic services to the colonias of El Paso County: A case of environmental racism?" In: Camacho, D.E. (ed.) *Environmental Injustices, Political Struggles: Race, Class, and the Environment*. Durham, NC: Duke University Press. pp. 125–137.

Bei, N.F., Li, G.H., Zavala, M., Barrera, H., Torres, R., Grutter, M., et al. (2013). "Meteorological overview and plume transport patterns during Cal-Mex 2010." *Atmospheric Environment*, Vol. 70, pp. 477–489.

Buzzelli, M. & Jerrett, M. (2004). "Racial gradients in ambient air pollution exposure in Hamilton, Canada." *Environment and Planning A*, Vol. 36, pp. 1855–1876.

Carruthers, D. V. (2008). "Where local meets global: Environmental justice on the US–Mexico border." In: Carruthers, D.V. (ed.) *Environmental Justice in Latin America: Problems, Promise and Practice*. Cambridge: MIT Press. pp. 136–160.

Chakraborty, J. & Green, D. (2014). "Australia's first national level quantitative environmental justice assessment of industrial air pollution." *Environmental Research Letters*, Vol. 9, p. 044010.

Cifuentes, E. & Frumkin, H. (2007). "Environmental injustice: Case studies from the South." *Environmental Research Letters*, Vol. 2, p. 045034.

Clapp, J. (2001). *Toxic Exports: The Transfer of Hazardous Wastes from Rich to Poor Countries*. Ithaca, NY: Cornell University Press.

Clark-Reyna, S.E., Grineski, S.E. & Collins, T.W. (2015). "Residential exposure to air toxics is linked to lower grade point averages among school children in El Paso, Texas, USA." *Population and Environment*, Vol. 37, pp. 319–340.

Collins, T. (2010). "Marginalization, facilitation, and the production of unequal risk: The 2006 Paso del Norte floods." *Antipode*, Vol. 42, pp. 258–288.

Collins, T., Grineski, S., Chakraborty, J. & McDonald, Y. (2011). "Understanding environmental health inequalities through comparative intracategorical analysis: Racial/ethnic disparities in cancer risks from air toxics in El Paso County, Texas." *Health and Place*, Vol. 17, pp. 335–344.

Collins, T.W., Grineski, S.E., Ford, P., Aldouri, R., Aguilar, M.D.L.R., Velazquez-Angulo, G., et al. (2013). "Mapping vulnerability to climate change-related hazards: Children at risk in a US–Mexico border metropolis." *Population and Environment*, Vol. 34, pp. 313–337.

Contreras, Ó.F., Carrillo, J., Garcia, H. & Olea M., J. (2006). "Desempeño laboral de las maquiladoras: Una evaluación de la seguridad en el trabajo." *Frontera Norte*, Vol. 18, pp. 55–86.

Dale, L.M., Goudreau, S., Perron, S., Ragettli, M.S., Hatzopoulou, M. & Smargiassi, A. (2015). "Socioeconomic status and environmental noise exposure in Montreal, Canada." *BMC Public Health*, Vol. 15, p. 205.

Downey, L. (2006). "Environmental inequality in metropolitan America in 2000." *Sociological Spectrum*, Vol. 26, pp. 21–41.

Environmental Protection Agency. (2012). *Border 2020: US–Mexico Environmental Program* [Online]. Washington DC. Available: www.epa.gov/border2020 [Accessed 17 October 2016].

Frey, R.S. (2003). "The transfer of core-based hazardous production processes to the export processing zones of the periphery: The Maquiladora centers of northern Mexico." *Journal of World Systems Research*, Vol. IX, pp. 317–354.

Godoy, E. (2011). *MEXICO: Maquiladora Factories Manufacture Toxic Pollutants* [Online]. Rome. Available: www.ipsnews.net/2011/08/mexico-maquiladora-factories-manufacture-toxic-pollutants/ [Accessed 3 August 2015].

Gould, K., Pellow, D. & Schnaiberg, A. (2008). *The Treadmill of Production: Injustice and Unsustainability in the Global Economy*. Boulder, CO: Paradigm.

Graham, J., Gurian, P., Corella-Barud, V. & Avitia-Diaz, R. (2004). "Peri-urbanization and in-home environmental health risks: The side effects of planned and unplanned growth." *International Journal of Hygiene and Environmental Health*, Vol. 207, pp. 447–454.

Grineski, S.E. & Collins, T.W. (2010). "Environmental injustice in transnational context: Urbanization and industrial hazards in El Paso/Ciudad Juárez." *Environment and Planning A*, Vol. 42, pp. 1308–1327.

Grineski, S.E. & Juárez-Carillo, P.M. (2012). "Environmental injustice in the U.S.–Mexico border region." In: Lusk, M., Staudt, K. & Moya, E. (eds.) *Social Justice in the U.S.–Mexico Border Region*. Tucson, AZ: University of Arizona. pp. 179–198.

Grineski, S.E., Collins, T.W., Romo, L. & Aldouri, R. (2010). "No safe place: Environmental hazards and injustice along Mexico's northern border." *Social Forces*, Vol. 88, pp. 2241– 2266.

Grineski, S.E., Collins, T.W., Ford, P., Fitzgerald., R.M., Aldouri, R., Velázquez-Angulo, G., et al. (2012). "Climate change and environmental injustice in a bi-national context." *Applied Geography*, Vol. 33, pp. 25–35.

Grineski, S.E., Collins, T., Chakraborty, J. & McDonald, Y. (2013). "Environmental health injustice: Exposure to air toxics and children's respiratory hospital admissions." *The Professional Geographer*, Vol. 65, pp. 31–46.

Grineski, S.E., Collins, T.W., McDonald, Y. J., Aldouri, R., Aboargob, F., Eldeb, A., et al. (2015). "Double exposure and the climate gap: Changing demographics and extreme heat in Ciudad Juárez, Mexico." *Local Environment*, Vol. 20, pp. 180–201.

Grineski, S.E., Collins, T.W. & Aguilar, L.R. (2016). "Environmental injustice along the US–Mexico border: Residential proximity to industrial parks in Tijuana, Mexico." *Environmental Research Letters*, Vol. 10, p. 095012.

Johnson, M.A. & Niemeyer, E.D. (2008). "Ambivalent landscapes: Environmental justice in the US–Mexico borderlands." *Human Ecology*, Vol. 36, pp. 371–382.

Jones, A. (1998). "Re-theorising the core: A 'globalized' business elite in Santiago, Chile." *Political Geography*, Vol. 17, pp. 296–318.

Jorgenson, A.K. (2010). "World-economic integration, supply depots, and environmental degradation: A study of ecologically unequal exchange, foreign investment dependence, and deforestation in less developed countries." *Critical Sociology*, Vol. 36, pp. 453–477.

Kamel, A.A., Ford, P.B. & Kaczynski, A.T. (2014). "Disparities in park availability, features, and character-istics by social determinants of health within a US–Mexico border urban area." *Preventive Medicine*, Vol. 69, pp. S111–S113.

Kellenberg, D. (2012). "Trading wastes." *Journal of Environmental Economics and Management*, Vol. 64, pp. 68–87.

Kolenc, V. (2010). "Maquilas dodge the violence: Juárez plants hurt more by recession than drug violence." *El Paso Times*, March 7.

Kopinak, K. & Barajas, M.D.R. (2002). "Too close for comfort? The proximity of industrial hazardous wastes to local populations in Tijuana, Baja California." *Journal of Environment and Development*, Vol. 11, pp. 215–246.

Lara-Valencia, F., Harlow, S.D., Lemos, M.C. & Denman, C.A. (2009). "Equity dimensions of hazard-ous waste generation in rapidly industrialising cities along the United States–Mexico border." *Journal of Environmental Planning and Management*, Vol. 52, pp. 195– 216.

Laurian, L. & Funderburg, R. (2014). "Environmental justice in France? A spatio-temporal analysis of incinerator location." *Journal of Environmental Planning and Management*, Vol. 57, pp. 424–446.

Liverman, D.M. & Vilas, S. (2006). "Neoliberalism and the environment in Latin America." *Annual Review of Environment and Resources*, Vol. 31, pp. 327–363.

Liverman, D.M., Varady, R.G., Chavez, O. & Sanchez, R. (1999). "Environmental issues along the US–Mexico border: Drivers of change and responses of citizens and institutions." *Annual Review of Energy and Environment*, Vol. 24, pp. 607–643.

Low, N. & Gleeson, B. (1998). *Justice, Society and Nature: An Exploration of Political Ecology*, London: Routledge.

Maiangwa, B. & Agbiboa, D.E. (2013). "Oil multinational corporations, environmental irresponsibility and turbulent peace in the Niger Delta." *Africa Spectrum*, Vol. 48, pp. 71–83.

McDonald, Y.J. (2012). *Lacking a Connection to a Community Water System: Water Quality and Human Health Impacts in El Paso Colonias*. M.A., University of Texas at El Paso.

McDonald, Y.J. & Grineski, S.E. (2012). "Disparities in access to residential plumbing: A binational comparison of environmental injustice in El Paso and Ciudad Juarez." *Population and Environment*, Vol. 34, pp. 194–216.

Minguillon, M.C., Campos, A.A., Cardenas, B., Blanco, S., Molina, L.T. & Querol, X. (2014). "Mass concentration, composition and sources of fine and coarse particulate matter in Tijuana, Mexico, during Cal-Mex campaign." *Atmospheric Environment*, Vol. 88, pp. 320–329.

Mirabelli, M.C., Wing, S., Marshall, S.W. & Wilcosky, T.C. (2006). "Race, poverty, and potential exposure of middle-school students to air emissions from confined swine feeding operations." *Environmental Health Perspectives*, Vol. 114, pp. 591–596.

Montalvo Corral, C. (2004). "Challenges for cleaner production in international manufacturing subcontracting: The case of the Maquiladora industry in northern Mexico." *Frontera Norte*, Vol. 16, pp. 69–99.

Morales, O., Grineski, S.E. & Collins, T.W. (2012). "Structural violence and environmental injustice: The case of a US–Mexico border chemical plant." *Local Environment*, Vol. 17, pp. 1–21.

O'Connor, J. (1988). "Capitalism, nature, socialism: A theoretical introduction." *Capitalism, Nature, Socialism*, Vol. 1, pp. 11–38.

Pellow, D.N. (2007). *Resisting Global Toxics: Transnational Movements for Environmental Justice*. Cambridge: MIT Press.

Pérez-Maldonado, I.N., Ramírez-Jiménez, M.R., Martinez-Arevalo, L.P., Lopez-Guzman, O.D., Athanasiadou, M., Bergman, A., et al. (2009). "Exposure assessment of polybrominated diphenyl ethers (PBDEs) in Mexican children." *Chemosphere*, Vol. 75, pp. 1215–1220.

Reid, C.E., O'Neill, M.S., Gronlund, C.J., Brines, S.J., Brown, D.G., Diez-Roux, A.V. & Schwartz, J. (2009). "Mapping community determinants of heat vulnerability." *Environmental Health Perspectives*, Vol. 117, pp. 1730–1736.

Roberts, J.T. & Thanos, N.D. (2003). *Trouble in Paradise: Globalization and Environmental Crises in Latin America*. London: Routledge.

Romo, L., Cervera, L.E. & Córdova, G. (2004). "Estudio urbano ambiental de las ladrilleras en el municipio de Juárez." *Revista Estudios Fronterizos*, Vol. 5, pp. 9–34.

Runkle, J., Flocks, J., Economos, J., Tovar-Aguilar, J.A. & McCauley, L. (2014). "Occupational risks and pregnancy and infant health outcomes in Florida farmworkers." *International Journal of Environmental Research and Public Health*, Vol. 11, pp. 7820–7840.

Sabapathy, A., Saksena, S. & Flachsbart, P. (2015). "Environmental justice in the context of commuters' exposure to CO and PM10 in Bangalore, India." *Journal of Exposure Science and Environmental Epidemiology*, Vol. 25, pp. 200–207.

Sanchez, R.A. (1990). "Health and environmental risks of the Maquiladora in Mexicali." *Natural Resources Journal*, Vol. 130, pp. 163–186.

Schatan, C. & Castilleja, L. (2005). *The Maquiladora electronics industry and the environment along Mexico's northern border*. Montréal, Quebec, Canada : Commission for Environmental Cooperation. Available: www3.cec.org/islandora/en/item/2232-maquiladora-electronics-industry-and-environment-along-mexicos-northern-border-en.pdf

Schoolman, E.D. & Ma, C.B. (2012). "Migration, class and environmental inequality: Exposure to pollution in China's Jiangsu Province." *Ecological Economics*, Vol. 75, pp. 140–151.

Schroeder, R., Martin, K.S., Wilson, B. & Sen, D. (2008). "Third World environmental justice." *Society & Natural Resources*, Vol. 21, pp. 547–555.

Shores, C.A., Klapmeyer, M.E., Quadros, M.E. & Marr, L.C. (2013). "Sources and transport of black carbon at the California–Mexico border." *Atmospheric Environment*, Vol. 70, pp. 490–499.

Simpson, G.R. (1990). "Wallerstein's World Systems Theory and the Cooke Islands: A critical examination." *Pacific Studies*, Vol. 14, pp. 73–94.

Velázquez Vargas, M.D.S. (2012). "Desplazamientos forzados: Migración e inseguridad en Ciudad Juárez, Chihuahua." *Estudios Regionales en Economía, Población y Desarrollo*, Vol. 7, pp. 4–21.

Wallerstein, I.M. (1974). *Modern World System, Volume 1*. New York: Academic Press.

Wilder, M., Scott, C.A., Pablos, N.P., Varady, R.G., Garfin, G.M. & McEvoy, J. (2010). "Adapting across boundaries: Climate change, social learning, and resilience in the U.S.–Mexico border region." *Annals of the Association of American Geographers*, Vol. 100, pp. 917–928.

Williams, D.M. & Homedes, N. (2001). "The impact of the Maquiladoras on health and health policy along the US–Mexico border." *Journal of Public Health Policy*, Vol. 22, pp. 320–337.

Wilson, S.M., Howell, F., Wing, S. & Sobsey, M. (2002). "Environmental injustice and the Mississippi hog industry." *Environmental Health Perspectives*, Vol. 110, pp. 195–201.

Zeisel, K., Paredes, N. & Brunelle, D. (2006). *The North American Free Trade Agreement (NAFTA): Effects on Human Rights* [Online]. Paris: International Federation for Human Rights. Available: www.fidh.org/IMG/pdf/Mexique448-ang2006.pdf [Accessed 10 July 2008].

43

THE DAWN OF ENVIRONMENTAL JUSTICE?

The record of left and socialist governance in Central and South America

Karen Bell

Environmental justice embraces the concept and movement for a healthy and safe environment; equal access to environmental resources alongside avoidance of environmental harms; and meaningful participation in environmental decision-making. Hence, it encompasses the substantive (outcomes and quality), distributive (equality and equity), recognition (respect and inclusion) and procedural (decision-making and access to information) aspects of justice (see Bell 2014). According to this understanding, despite ongoing and strenuous efforts, environmental justice seems as elusive as it did when the term was first coined in the 1980s. As outlined in previous work, I have come to the conclusion that this is because environmental justice is consistently undermined by capitalism with its drives toward inequality and irrationality (Bell 2014).

Looking towards Central and South America, we have something of a natural experiment, which we could use to test this theory. Central and South American citizens, more than those of any other continent, have increasingly elected left wing and socialist governments over the last twenty years. This so-called 'pink tide' allows us to see the extent to which marginalizing and minimizing capitalist processes within a nation state does, indeed, further environmental justice. To this end, this chapter considers whether, under these governments, environmental justice has been enhanced or whether it remains more of a rallying cry than a reality. This surge of left-wing governance is a distinctive feature of the region and it seems that the approach to environmental justice issues may provide a model for achieving environmental justice elsewhere. However, I discuss the tensions that exist between the ideals professed and their practical implementation and the extent to which the successes and failures might be attributable to capitalism's presence or absence. Hence, the chapter does not intend to cover the entire continent (see Chapters 42 and 44 for other perspectives) but, rather, to focus on the discourses and practices of the five key socialist governments in the region – Cuba, Venezuela, Bolivia, Ecuador and Nicaragua – in the context of their membership of the left wing 'ALBA' alliance (Alianza Bolivariana para los Pueblos de Nuestra América or 'The Bolivarian Alliance for the Peoples of Our America'). These five countries together make up the main players in the ALBA alliance, an organization formed to advance social, political, and economic cooperation between left-leaning Central and South American states. Apart from Cuba, which has pursued largely socialist policies since 1959, all were elected as socialist governments relatively recently following lengthy neoliberal periods of governance. Venezuela was the first to take a radical change of direction when the United Socialist Party of Venezuela (PSUV), led by Hugo Chávez, swept to power in 1997. Following this, in 2005, the MAS (Movimiento al Socialismo – Movement toward Socialism) secured electoral victory with Evo Morales as its leader in Bolivia. Shortly after, Ecuador and Nicaragua both embraced socialism in 2006/2007 with the election of governments led by Rafael Correa and José Daniel Ortega, respectively. Since these elections, the countries have embarked on major programmes to regain state control of the economy, effectively changing their economies from predominantly free market to mixed, with strong state management. Socialist NGOs and citizen movements have driven these processes and now work alongside the state to promote and help implement the left wing policies that they campaigned for.

In 2004 ALBA was formed through an initial agreement between Venezuela and Cuba. Since then, the organization has grown to include eleven full member states which, in order of joining, are: Venezuela, Cuba, Bolivia, Nicaragua, Dominica, Ecuador, Antigua and Barbuda, Saint Vincent and the Grenadines, Saint Lucia, Grenada, and the Federation of Saint Kitts and Nevis. (Honduras was formerly a member but the 2009 coup put in place a right-wing government that decided to withdraw from the alliance.) ALBA also organizes sub-regionally, having outposts in numerous other countries of Central and South America, as well as Africa and North America. For example, some of the FMLN controlled local authorities in El Salvador are affiliated to ALBA. The organization has also involved social movements such as Via Campesina (Harris and Azzi 2006). However, the main driving force emanates from the Central and South American countries discussed here.

ALBA is an acronym which, in the dominant language of these countries (Spanish), means 'dawn'. The name is seen, by some, to be significant, implying the dawning of a new reality (e.g. Cole 2008). ALBA's vision is explicitly anti-imperialist, anti-neoliberal, and sometimes anti-capitalist; and dedicated to building an alternative form of relationship based on the principles of solidarity, co-operation and complementarity (Muhr 2010). It maintains a vision of social welfare and mutual economic aid, promoting the rights of Indigenous peoples, protection of the environment, social participation, solidarity, exchange of resources, and respect for diverse cultures and identities. This is facilitated through organizing in ways which foster independence from neoliberal states and institutions (Muhr 2010). For example, ALBA has its own currency, the Sucre, and its own Bank of ALBA. With this currency, the countries involved no longer have to use a foreign currency for international exchanges, enabling some independence from the World Bank and the International Monetary Fund; and circumventing conditionality which interferes in the countries' domestic affairs. ALBA also enables trade to take place between countries without the use of money as an intermediary at all, so that, for example, some of the earliest 'trade' was in Cuban doctors who were sent to Venezuela in exchange for oil imports (Muhr 2010).

Although ALBA does not have a specific policy on the achievement of environmental justice, its declarations and projects are focused on a number of related areas, including: food, energy, literacy, health, sanitation and water; as well as eradicating poverty and reducing inequality. Hence, at international conferences ALBA states have consistently pushed for the recognition and rights of Indigenous and other marginalized communities; binding and stringent environmental regulations; reform of the global financial system and its architecture; transformation of patterns of consumption and production; the need for new development indicators that transcend the limitations of GDP; and the need for a new international economic order based on the principles of equality, sovereignty, and cooperation.

Whilst promoting a common ideology, ALBA does not aim to restrict heterogeneous national approaches, even where they seem to contradict the common principles. Hence, in the ALBA countries, policies and programmes reflect national contexts though all emphasize lifting people out of poverty, improving health, meeting basic needs, ensuring access to land and water, defending traditional practices, appropriate infrastructure developments, resource sovereignty and international climate justice. The next section looks at the particular aspirations of these ALBA nations and their achievements in more detail, focusing on four aspects of environmental justice: building healthy and safe environments; constructing more equitable and equal societies; recognition of formerly marginalized groups; and enabling participatory decision-making structures.

Healthy and safe environments

In terms of creating healthy and safe environments, the ALBA countries tend to approach this holistically. One of the key themes that has emerged from Bolivia and Ecuador which has now been integrated into the wider ALBA discourse, is the philosophy of Living Well ('Vivir Bien' or 'Buen Vivir' in Spanish, 'Suma Qamaña' in Aymara, 'Sumaj Kawsay' in Quechua, and 'Ñande Reko' in Guaraní). Rooted in the worldview of Andean Indigenous groups, the concept of Living Well describes a communal and ecologically balanced approach to addressing multiple environmental crises, whilst meeting human needs and achieving equality. It aspires to living in harmony with other human beings and nature in relationships of service and reciprocity; and subsumes all economic and social objectives to the protection of ecosystems, which are considered to be the foundation for the accomplishment of all social goals (Radcliffe 2012). Importantly, Living Well implies that we are part of a whole, so that we cannot live well if other humans do not, or at the expense of our environment. Living Well inherently critiques the accepted need for economic growth, and moves towards the goal of meeting needs and satisfying rights. Under this ethos, the economy is based not on the profit motive, but on respect and care for humans and the rest of nature in a spirit of solidarity.

According to the new Bolivian constitution, all development projects should now be evaluated through a lens of Living Well. The approach was further strengthened with the passing of the 'Framework Law of Mother Earth and Integral Development for Living Well' in 2012. This national legislation establishes 11 new rights for nature, including: the right to life and to exist; the right to continue vital cycles and processes free from human alteration; the right to pure water and clean air; the right to balance; the right not to be polluted; the right to not have cellular structures modified or genetically altered; and the right not to be affected by mega-infrastructure and development projects that affect the balance of ecosystems. Similarly, under the Correa government in Ecuador, a new constitution was developed based on Living Well as 'the roadmap for the construction of a utopia' (Acosta 2010). The constitution establishes and guarantees related rights, including free universal health care and education to graduate level; food, energy, water, housing, a healthy environment; and economic sovereignty and participation in decision-making. The Ecuadorian constitution also grants rights to nature. Ecuador's national development plan (National Plan for Living Well) has three pillars: To transform the economy from reliance on the primary sector, especially oil, towards a more service and knowledge based economy; to reduce poverty and inequality through developing efficient public services, education, health care and social security; and to establish a more participatory form of democracy by enhancing citizen involvement at all levels of governance (SENPLADES 2009, 2013).

Although these countries are some of the world's smallest contributors to climate change, they are already being severely impacted, experiencing rising temperatures, melting glaciers and more frequent extreme weather events, including floods, droughts, frosts and mudslides. For example, glaciers, major providers of fresh water, are retreating at an alarming rate. The ALBA bloc have taken a principled position in the United Nations climate change negotiations, pushing for a binding, ambitious and justice-based agreement, including advocating climate reparations from the global North to the South. One of their best known interventions occurred during the Fifteenth Conference of the Parties to the United Nations Framework Convention on Climate Change, held in Copenhagen in 2009. The ALBA group denounced the lack of tangible commitments by developed countries and prevented the signing of an agreement which they perceived as leading to climatic catastrophe. Bolivia then went on to host the 35 000-strong conference named the 'World People's Summit on Climate Change and the Rights of Mother Earth'. The summit resulted in the 'People's Agreement', which reiterated the need for a 1°C limit on temperature rises and reaffirmed the demand for climate reparations. It specifically stated that, in order to redress developed countries' historical responsibility for climate change, they should give 6 per cent of their respective GDPs to climate adaptation funds, to be managed by national governments at the United Nations, not the World Bank.

The ALBA countries and organizations have also been at the forefront of rejecting market approaches to resolving environmental problems (Bull and Aguilar-Støen 2015). They reject these approaches as both apolitical and unjust since, without a rational and fair decision-making process, they allow the wealthiest to continue to pollute and harm the environment. In particular, the ALBA nations have voiced their concerns about the commodification of nature, indefinite economic growth and privatization of public goods, proclaiming:

1. That nature is our home and is the system of which we form a part, and that therefore it has infinite value, but does not have a price and is not for sale.
2. Our commitment to preventing capitalism from continuing to expand in the spheres that are essential to life and nature, being that this is one of the greatest challenges confronting humanity.
3. Our absolute rejection of the privatization, monetization and mercantilization of nature, for it leads to a greater imbalance in the environment and goes against our ethical principles.
4. Our condemnation of unsustainable models of economic growth that are created at the expense of our resources and the sovereignty of our peoples.
5. Only a humanity that is conscious of its present and future responsibilities, and states with the political will to carry out their role, can change the course of history and restore equilibrium in nature and life as a whole.

(ALBA 2010)

Hence, market mechanisms such as Payments for Ecosystems Services (PES) projects are rejected on global and national environmental justice grounds. As several authors point out, PES can result in increased competition for control over the ecosystems that provide valuable services, possibly leading to displacement and dispossession of communities and 'green grabbing' (e.g. Fairhead et al. 2012; Sikor and Newell 2014).

The tendency for the ALBA nations to solve environmental problems and injustices outside of the market has often led to radical and effective solutions. For example, in 2006, as part of an international comparison of ecological footprints, Cuba was found to be the only country in the world meeting high standards of human development within the ecological limits of one planet (WWF 2006). This achievement appears to be because it has been able to take flexible and rational decisions, based on current needs, such as largely rejecting industrialized agriculture in favour of organic food production (Levins 2005) and prioritizing renewable energy sources alongside energy conservation (Guevara-Stone 2009). Levins (2005: 14) argues that these changes were possible because of socialism, since "Nobody was pushing pesticides or mechanization to make a profit . . . Socialism made ecological choices more likely." Cuba's 'Revolución Energética' (energy revolution), which began in 2006, is a particularly good example of an environmental policy which also addresses environmental justice. This initiative endeavoured to save energy and use more sustainable sources. The programme included replacing household appliances with more efficient and safer equipment, supplied free or at low cost to the entire population. Hence, as these examples indicate, the ALBA bloc countries work towards the establishment of substantive environmental justice through adopting novel policies which would be difficult to implement under more capitalist structures.

Though many formal policies, legislative proposals and training programmes are still in the process of being set up, we can already see some indication of environmental and social achievements resulting from the Living Well paradigm. For example, in Bolivia, since the MAS government came to power and particularly with the initiation of the Living Well policy, a number of environmental harms, including energy use and Co_2 and greenhouse gas emissions, have been stabilized or reduced (see Tables 43.1, 43.2 and 43.3).

Greenhouse gas emissions have particularly decreased since 2010 (see Table 43.3). Moreover, where Bolivia formerly had one of the highest deforestation rates in the world, this has dropped dramatically since 2010: by 64 per cent (Andersen 2014). Fuentes (2015) points out that 2010 is the year when the government officially opposed carbon offset schemes,

Table 43.1 Energy use (kg of oil equivalent per capita)

	2004	2005	2006	2007	2008	2009	2010	2011	2012	2013	2014
Bolivia	662	697	834	704	737	759	756	779	831		

Source: World Bank 2015 energy use (kg of oil equivalent per capita) http://data.worldbank.org/indicator/EG.USE.PCAP.KG.OE/countries?page=1

Table 43.2 CO_2 emissions (metric tons per capita)

	2004	2005	2006	2007	2008	2009	2010	2011	2012	2013	2014
Bolivia	1.5	1.3	1.6	1.3	1.4	1.5	1.5	1.6			

Source: World Bank 2015 CO_2 emissions per capita http://data.worldbank.org/indicator/EN.ATM.CO2E.PC?page=1

Table 43.3 GHG data: Total GHG emissions including land-use change and forestry ($MtCO_2e$ million metric tons of carbon dioxide equivalent)

	2004	2005	2006	2007	2008	2009	2010	2011	2012
Bolivia	127.56	121.77	128.03	132.80	129.82	128.78	154.85	140.28	136.47

Source: CAIT Climate Data Explorer. 2015. Washington, DC: World Resources Institute. Available online at: http://cait.wri.org.

set up a state body to protect forest areas, and put large areas of forest under the management of local Indigenous people – all programmes that fit with the principles of Living Well.

Equality and equity

ALBA states address the distributive aspect of environmental justice through aiming to ensure that all social groups can access necessary environmental resources. This requires economically empowered and equal societies. They have used macroeconomic and social policies based on universality and equitable access to achieve this. For example, in Cuba, these measures include limiting wage inequality in the public sector; keeping prices for goods and services low; assuring equal and affordable access to essential food and consumer goods; and extending social security, welfare, sports and cultural activities to the entire population for free, or at very low cost (Bell 2014). In addition, Cuba does not demonstrate the distributional environmental injustice built upon residential segregation that is evident in capitalist countries, because there is no social-spatial segregation by race and class. This classic pattern of spatial segregation had also existed in Cuba before the revolution, but has been greatly alleviated by the general programmes designed to reduce poverty, as outlined above, as well as a number of specific housing laws and policies; which made housing free or affordable in any area of the city or country (Taylor 2009). Most importantly, the lack of a legal land or housing market, whilst allowing mobility through housing 'swaps', has largely prevented segregation according to income – though this policy has now been removed as a result of the desire of the majority of the population in a nationwide consultation in 2011.

Following in the footsteps of Cuba, the other countries discussed here have seen rapid and dramatic drops in poverty, deprivation and inequality since their left governments came to power (see, for example, World Bank data in Table 43.4). For example, in Bolivia, poverty levels have fallen by about a quarter; illiteracy, which stood at approximately 14 per cent in 2006, has been eradicated (UNESCO 2009); and lands that were not being used productively have been partially redistributed to the Indigenous peasantry. Similarly, in Venezuela, multiple social programmes (known in Venezuela as 'missions') have significantly reduced poverty and also improved nutrition, eradicated illiteracy, expanded the number of pension recipients, increased university enrolment, provided new housing, reduced unemployment, and provided free medical care to low-income groups for the first time.

Hence, by focusing on reducing poverty and inequality in general, the ALBA countries have enhanced the distributional aspects of environmental justice, as their less well-off citizens are more able to access environmental resources and services, such as food, energy, transport and green space.

Table 43.4 Poverty levels* during the most recent period of socialist governance

	Start	*Current*
Venezuela	50.0% (1998)	32.2% (2013)
Bolivia	60.6% (2005)	39.1% (2013)
Ecuador	36.7% (2007)	22.5% (2014)
Nicaragua	42.5% (2009)	29.6% (2014)

Source: World Bank 2015 (no data available for Cuba from this source).
Nearest dates available.

*Defined here as percentage of the population living on less than $1.9 a day
PPP (purchasing power parity)

Recognition and participation in environmental decision-making

The ALBA countries discussed here have also initiated programmes that support the procedural and recognition elements of environmental justice. For example, in Venezuela, grassroots organization and citizen participation in decision-making have dramatically increased since the socialist government came to power. Public participation with regards to environmental problems and the management of natural resources is encouraged through the creation of community structures such as the Technical Committees for Water; Energy Committees; Conservation Committees, Community Councils, Health Committees and Urban Land Committees. The Venezuelan government's programmes include the Shared Environmental Management programme; the Tree Mission (Misión Arbol); and the Energy and Transport Revolution Mission (Misión Revolución Energética y Transporte). This level of popular organizing is unknown in Venezuela's history and rare globally. These organizations have proliferated in part because the government has funded them generously, enabling local people to address and solve their local problems with regard to transport, housing and the environment, strongly favouring environmental justice.

Similarly, in Bolivia, the MAS government places a strong focus on participatory democracy, which runs alongside a representative system. Hence, Bolivia's National Development Plan contains substantial mechanisms to improve participation in governance structures (Gobierno de Bolivia 2006), as does the 2009 Constitution. The new constitution also highlights the 'pluri-national' nature of the Bolivian state, thus recognizing its 36 Indigenous nations. It specifies a number of Indigenous rights, including that of self-government and self-determination, collective landownership, community involvement in the economy, the

Figure 43.1   Meeting on living well and women's rights – La Paz 2014 (photo by author)

need for prior consultation on matters affecting them (including where non-renewable natural resources are concerned), and the right to benefit from such activities.

In Cuba, a vast network of institutions and organizations play a part in influencing environmental decisions, including mass organizations and environmental NGOs (Bell 2011, 2014). Though the country is generally portrayed in the West as having few, if any, of the ingredients for inclusive decision-making of any kind, in fact Cuba is organized as a system of direct democracy which includes and requires participatory decision-making, election of representative delegates, a vibrant civil society, accountable governance and a willingness to respond to demands for change. Proposals for serious reforms of general government policy are circulated widely and discussed extensively in local branches of mass organizations, schools, workplaces and universities, before being put to referendum or opinion channelled into National Assembly debates (Raby 2006; Kapcia 2008; Bell 2011). Nicaragua has also followed the above models of direct democracy, setting up a system similar to that of the 'Communal Councils' of Venezuela, and the 'Popular Power' assemblies of Cuba where local people make local decisions, often on environmental matters. In Nicaragua, these instruments are called Citizens' Councils and they are similarly organized in a tiered structure through the municipalities, departments, and autonomous regions. A recent achievement of the Councils in Venezuela was a ban on transgenic (GMO) seed and protection of local seeds from privatization. This was a product of direct participatory democracy because the law was developed through a deliberative partnership between members of the country's National Assembly and a grassroots coalition of eco-socialist, peasant, and agroecological oriented organizations and institutions (Camacaro et al. 2016).

Tensions and limitations

The above outline describes how ALBA and some of its member states are developing policies and discourses that encourage and, in some cases, actualize environmental justice. However, as with many paths towards an ideal, the process is far from smooth. Tensions and constraints limit and undermine the achievement of these countries' aspirations.

In particular, the ALBA countries struggle to overcome their legacies of colonialism which has, in all cases, permanently altered them economically, environmentally and socially by orientating them toward primary export production, stifling Indigenous industry and tying them to a dependency on external trade. This has had implications for the choices they have made, pressurizing them to compromise on their principles and, at times, even to act contrary to them. For example, as a result of colonialism, Cuba did not develop a diverse economy and so, when the socialist Council for Mutual Economic Assistance (COMECON) collapsed and the US blockade on Cuba was subsequently tightened in 1989, it needed to begin to trade with the capitalist world. Following extensive consultation with the population, this led to the use of market-based policies so as to 'use capitalism to save socialism' (Taylor 2009: 4). These policies impacted on environmental justice through the development of activities that undermined ecology, equality and sovereignty: tourism, joint ventures with foreign capital, intensifying exploitation of natural resources for export (especially nickel), and increasing the availability and promotion of consumer goods in order to capture dollars sent as remittances (Bell 2011, 2014).

The under-development resulting from their colonial histories means that these countries are often desperate to find ways of increasing income so as to address the needs of their populations. This, too, undermines environmental justice, by limiting policy options. Hence, with continued dependence on foreign aid and markets and the lack of domestic funding for social and environmental policies, the ALBA countries have at times conceded to Western policies that they might, otherwise, have fully rejected. Nature commodification, green capitalism and other market rhetoric has been resisted in the ALBA bloc discourse yet some of the ALBA countries are now integrating these approaches into their environmental policies, albeit to a very limited extent and within a context of deep controversy. For example, the Nicaraguan government has set up some Payments for Ecosystems Services (PES) initiatives (Van Hecken et al. 2015), even though these are widely seen as a neoliberal political instrument: a technical fix that enables capitalist expansion whilst resulting in detrimental social and environmental outcomes (e.g. McAfee 2012). In 2013, the country received US$ 3.6 million from a World Bank REDD-Readiness fund (World Bank 2013). Nicaragua's dependence on foreign aid provided by large donor organizations such as the World Bank, the United Nations Environment Programme (UNEP), and the Inter-American Development Bank (IADB) make it difficult for the country to avoid adopting the environmental policies promoted by those organizations. Similarly, as a result of colonialism which undermined diversification of the economy, the Venezuelan economy became overly dependent on oil and, consequently, is challenged by so-called *Dutch disease*. This means that, as revenues increase in a dominant sector (in this case, oil), the nation's currency becomes strong relative to currencies of other nations, resulting in the nation's other exports becoming too expensive for other countries to buy, and imports becoming cheaper. Hence, apart from oil, almost all goods are cheaper to import than to produce in Venezuela. This undermines efforts to promote domestic production and has been part of the reason for the recent scarcity of everyday consumer goods. It also ties Venezuela to further production of oil, even though it is ecologically harmful.

The need for short-term income has meant that these countries have not been able to move away from the extractivist base to their economies. Though they have used their resource wealth to fund the social programmes that have in some respects enabled greater environmental justice for their populations, this has environmental impacts, locally, nationally and globally. For example, Bolivia has used funds from the nationalization of natural resources, in particular hydrocarbons, and increased the taxes from those companies that remained in private hands from an average 18 per cent of profits to as much as 82 per cent (Postero 2010). These funds have been used to improve sanitation and clean water services across the country as well as to fund poverty alleviation schemes. However, the mining projects upon which this revenue is based will inevitably have severe environmental impacts, including possible distributional justice impacts (see Chapter 30). Similarly, Ecuador, in order to achieve its social development goals, is planning to increase extractive mining activities in the short term, the idea being to 'use the extraction of raw materials in order to stop the extraction of raw materials' (SENPLADES 2013: 48). Correa argues that expanding the extraction of natural resources is only a short term strategy to obtain the much needed capital for social programmes until the economy becomes more diverse and knowledge based (Walsh 2010). He previously has endeavoured to establish a basis for not pursuing extractivism with the Yasuní initiative, wherein Ecuador offered to leave vast oil reserves underground in the Yasuní national park in return for payments of $ 3.6 billion from the international community. This scheme was, itself, a fundamentally market-based environmental programme and it did not attract the funding hoped for (Pellegrini et al. 2014). Hence, the country now continues with extractivism to fund its social programmes until it is in a position to do otherwise. In the case of Cuba, when it lost its COMECON trading partner and turned to policies generally associated with capitalism, these measures, especially tourism and the intensive exploitation of natural resources, had a negative influence on Cuba's environment and society (Taylor 2009).

Although, superficially, some of the contradictions seem to be about whether the provision of services and economic development should override ecological considerations, these tensions could also be seen as arising from the difficulty in deciding how to balance the needs of diverse social groups. For many, decolonization can be accomplished only through indus-trialization accompanied by redistribution. The politics surrounding the construction of a road through the Isiboro Ségure Indigenous Territory and National Park (TIPNIS) in the central lowlands of Bolivia is illuminating in this regard. The international media have focused on the protests and the supposed hypocrisy of the government, in particular of Evo Morales, in wishing to build a road through a sensitive ecosystem and at the expense of the rights of Indigenous people, considering the road to be mainly a means of facilitating hydro-carbon exploration and extraction. However, far from the situation being that of Indigenous people in opposition to the government, as the media tended to portray it, the TIPNIS situation was characterized by conflicts between the different social movements. These tend to take very different positions with regard to a number of environmental and social issues, in part due to the history of their formation. Some, often supported by international NGOs, anthropologists, and religious groups, have organized specifically around identity, ethnicity and culture, for example, the Indigenous Federation of Eastern Bolivia (CIDOB). Others focused more on class or a critique of capitalism, for example the predominantly Aymara and Quechua highland campesinos, such as the National Federation of Bolivian Campesino Women-Bartolina Sisa (FNMCB-BS). Consequently, the government set up a consultation process regarding whether the road should be constructed through TIPNIS and if the area should be designated as untouchable. The vast majority of those who chose to participate – 54 of 69 communities – agreed to the road. The benefits of the road would be felt across the park, the region and the country but the costs would impact on specific groups locally.

It should also be appreciated that all of these countries, even Cuba, have had insufficient time to correct the legacy of almost 500 years of colonial and neoliberal rule. Their depend-ence on exploiting their natural resources to maintain the consumerist lifestyles of many in the West is part of a wider pattern in which, according to Dependency Theory (Frank 1967) and World Systems Theory (Wallerstein 2004), the global economy has long been structured around the mass extraction of resources in the periphery nations of South America, Africa and Asia for consumption in Europe and the United States. However, the ALBA bloc countries emphasize the difference between extraction that contributes to welfare and that which provides profits to corporations.

It is also important to recognize that, where these countries have tried to assert their independence and take a different approach from that of the West, they have been subjected to varying degrees of coercive intervention from the dominant capitalist states. For example, in the case of Cuba, the US blockade has affected the country in terms of the availability and cost of environmentally useful technology, as well as a general shortage of resources. For example, problems with water supply have been linked to a shortage of replacement parts for the distribution network, originally built using US components (American Association for World Health 1997). In addition, the US blockade seems to have conditioned and limited Cuba's decision-making processes. Since the revolution in 1959, Cuba has been the recipient of relentless US efforts to overthrow the government. As well as an ongoing blockade, this has included proactive sponsorship of opposition groups; stated attempts to use academic work to destabilize the Cuban system; a crusade of disinformation; and a campaign of aggres-sion, including sabotage, terrorism, an invasion, attempted assassinations of the leadership, biological attacks and hotel bombings (Saney 2004; Kapcia 2008). Some argue this has led to a siege mentality, whereby it is difficult to take risks, admit mistakes and rectify them (e.g. Kapcia 2008). Similar challenges have faced the other four ALBA nations discussed here. Some environmentalists on the left have criticized the governments of Venezuela, Bolivia, Ecuador and Nicaragua for not taking more transformative redistributive measures, instead of relying on extraction to fund their social programmes. However, it needs to be understood that these governments are in a difficult position with regard to their economic elites, who could orchestrate (and have orchestrated) socioeconomic repercussions (even a possible coup) in the event of a deeper economic redistribution. Hence, they have only redistributed to a limited extent. As I write this, we have just seen the socialist government in Venezuela lose ground in the latest national election. Maduro supporters argue that this happened as a result of the chaos which ensued from a systematic plan amounting to a 'soft coup' which included violence as well as economic attacks (speculation, hoarding, cross-border contraband). The opposition groups behind these activities had been supported by the US State Department, in part through USAID.

We can see, therefore, that the goals of ALBA are not easy to fulfil. Though ALBA was specifically set up to help these countries overcome these undermining tendencies, it is as yet not strong enough to enable them to fully resist Western domination.

The dawn?

Does the Central and South American left herald the dawn of environmental justice, with solidaristic alliances such as ALBA sparking a new paradigm for democracy, ecology and equity? This brief overview of the environmental justice-related achievements in these countries would seem to suggest that this is possible. Yet these governments, as with countries throughout the Global South, are implicated in, and constrained by, myriad regional and global economic forces that make environmental justice a difficult proposition. Even so, they are perhaps the most promising countries globally in terms of the policies that are in place for achieving environmental justice. There is much to be researched and debated in relation to these issues.

The ALBA countries looked at here have attained some environmental justice successes as a result of policies based on meeting basic needs, prioritizing equality of outcomes, innovative environmental programmes, participatory democracy, re-localizing production and consumption, and sharing more. This indicates the importance of a commitment to political and social equality, as well as a healthy environment, in order to achieve environmental justice. The causes of the injustices that continue to occur in these countries seemed to be a result of a complex interplay between a history of colonialism; a shortage of income; and the political hostility of US and dominant powers. These factors are all elements of capitalism. This seems to reinforce the idea that capitalism is a fundamental factor that should not be ignored in any analysis of environmental justice. Some might argue that the realization of what could be called the eco-socialist agenda of the ALBA bloc will require a more complete socio-economic transformation in the long run. The ALBA countries have the potential to lead a new reality which could include such a complete transformation and, as such, may herald a new dawn for environmental justice. Even though they are often thwarted and compromised, the ALBA nations do at least propose to achieve a society that is both ecologically sustainable and socially just. We should remember that dawn always begins with a tiny, imperceptible chink of light which does little to counter the darkness but gradually, gently, and inevitably, it comes to light up the entire sky.

References

AAWH (American Association for World Health) (1997). *The Denial of Food and Medicine: The Impact of the US Embargo on Health and Nutrition in Cuba.* http://archives.usaengage.org/archives/studies/cuba.html

Acosta, A. (2010). "El Buen (con) Vivir, una utopía por (re)construir: Alcances de la Constitución de Montecristi", OBETS. *Revista de Ciencias Sociales*, vol. 6 (1), pp. 35–67.

ALBA (2010). *Bolivarian Alliance for the Americas: Treaty of Commerce of the People (ALBA-TCP)*, La Paz, Plurinational State of Bolivia, Nov 3–5, 2010.

Andersen, L.E. (2014). *Mission Accomplished or Too Good to Be True?* INESAD (Institute of Advanced Development Studies), La Paz, Bolivia.

Bell, K. (2011). "Environmental Justice: Lessons from Cuba". School for Policy Studies. Bristol, University of Bristol. PhD.

Bell, K. (2014). *Achieving Environmental Justice: A Cross National Analysis.* Bristol: Policy Press.

Bull, B., Aguilar-Støen, M. (Eds.), (2015). "Environmental Politics in Latin America: elite dynamics, the left tide and sustainable development". *Routledge Studies in Sustainable Development.* Routledge: New York.

Camacaro, W., Mills, R.B. and Schiavoni, C.M. (2016). "Venezuela Passes Law Banning GMOs, by Popular Demand." Venezuelaanalysis.com http://venezuelanalysis.com/analysis/11798

Cole, K. (2008). "ALBA: A Process of Concientización." *The International Journal of Cuban Studies*, vol. 2 (December), pp. 31–40.

Fairhead, J., Leach, M. and Scoones, I. (2012). "Green Grabbing: A New Appropriation of Nature?" *Journal of Peasant Studies*, vol. 39 (2), pp. 37–261.

Frank, A. (1967). *Capitalism and Underdevelopment in Latin America.* New York: Monthly Review Press.

Fuentes, F. (2015). "Why the Media Distorts Bolivia's Environmental Record." Telesur. www.telesurtv.net/english/opinion/Why-the-Media-Distorts-Bolivias-Environmental-Record-20150722-0016.html

Gobierno de Bolivia (2006). *Plan Nacional de Desarrollo.* La Paz, Government of Bolivia.

Guevara-Stone, L. (2009) "La Revolucion Energetica: Cuba's Energy Revolution." *Renewable Energy World Magazine*, vol. 12.

Harris, D. and Azzi, D. (2006). ALBA "Venezuela's Answer to "Free Trade": The Bolivarian Alternative for the Americas." Occasional Paper. *Focus on the Global South.* www.focusweb.org/node/1087.

Kapcia, A. (2008). *Cuba in Revolution: A History since the Fifties.* London: Reaktion Books.

Levins, R. (2005). "How Cuba Is Going Ecological." *Capitalism Nature Socialism*, vol. 16 (3), pp. 7–25.

McAfee, K. (2012). "The Contradictory Logic of Global Ecosystem Services Markets." *Development and Change*, vol. 43, pp.105–131.

Muhr, T. (2010). "Nicaragua: Constructing the Bolivarian Alliance for the Peoples of Our America (ALBA)." In *Globalization and Transformations of Social Inequality*, ed. U. Schuerkens, 115–134. Routledge.

Pellegrini, L., Arsel, M., Falconí, F. and Muradian, R. (2014). "The Demise of a New Conservation and Development Policy? Exploring the Tensions of the Yasuní ITT Initiative". *The Extractive Industries and Society*, vol. 1, pp. 284–291.

Postero, N. (2010). "The Struggle to Create a Radical Democracy in Bolivia." *Latin American Research Review*, vol. 45, pp. 59–78.

Raby, D.L. (2006). *Democracy and Revolution: Latin America and Socialism Today.* London: Pluto Press.

Radcliffe, S.A. (2012). "Development for a Postneoliberal Era? Sumak Kawsay, Living Well and the Limits to Decolonisation in Ecuador." *Geoforum*, vol. 43 (2), pp. 240–249.

Saney, I. (2004). *Cuba: A Revolution in Motion.* London: Zed Books.

SENPLADES (Secretaría Nacional de Planificación y Desarrollo) (2009). *Plan Nacional para el Buen Vivir 2009–2013. Construyendo un Estado Plurinacional e Intercultural.* Quito: SENPLADES.

SENPLADES (Secretaría Nacional de Planificación y Desarrollo) (2013). *Plan Nacional para el Buen Vivir 2013–2017. Buen Vivir. Todo el mundo mejor.* Quito: SENPLADES.

Sikor, T. and Newell, P. (2014). "Globalizing Environmental Justice?" *Geoforum*, vol. 54, pp. 151–157.

Taylor, J.H.L. (2009). *Inside El Barrio: A Bottom-up View of Neighborhood Life in Castro's Cuba.* Sterling, VA: Kumarian Press.

UNESCO (2009). "Results Achieved by the Republic of Bolivia in Eradicating Illiteracy as a Potentially Valuable Experience in UNESCO's Efforts During the United Nations Literacy Decade (2003–2012)." http://unesdoc.unesco.org/images/0018/001818/181816e.pdf

Van Hecken, G., Bastiaensen, J. and Huybrechs, F. (2015). "What's in a Name? Epistemic Perspectives and Payments for Ecosystem Services Policies in Nicaragua." *Geoforum*, vol. 63, pp. 55–66.

Wallerstein, I. (2004). *World-Systems Analysis: An Introduction*. Durham, NC: Duke University Press.

Walsh, C. (2010). "Development as Buen Vivir: Institutional Arrangements and (de)Colonial Entanglements", *Development*, vol. 53 (1), pp. 15–21.

World Bank (2013). *Nicaragua: Forest Carbon Partnership Facility (FCPF) Readiness Preparation Grant Project*. Washington DC: World Bank Group.

World Bank (2015) "Data, Poverty" http://data.worldbank.org/topic/poverty WWF (2006). *Living Planet* Report. Gland, Switzerland: World Wildlife Fund.

44

URBAN ENVIRONMENTAL (IN)JUSTICE IN LATIN AMERICA

The case of Chile

Alexis Vásquez, Michael Lukas, Marcela Salgado and José Mayorga

Introduction

In 2014, Latin America and the Caribbean (LAC) was the second most urbanized region in the world with approximately 80 per cent of their population living in urban areas (United Nations [UN] 2014). In the 1950s and 1960s an urban explosion took place through rural–urban migration and rapid and uncontrolled urbanization. In the context of import substitution policies and associated industrialization processes, millions of farmers migrated to the cities in search of new employment opportunities. Evidently, this arrival of people in a vertiginous volume and velocity exceeded the capacity of the urban economy to absorb this workforce and of the local and national governments to meet the growing demand for housing and basic services (Romero 2007). The new urban residents had no other option but to occupy peripheral areas informally, by self-constructing slums that were isolated and often exposed to natural (floods and landslides) and anthropogenic (landfills and polluting industries) hazards. This problem still persists today, as 160 million people, that is almost 30 per cent of Latin America's urban population, live in informal settlements (Jiménez 2015).

Despite the persistent problem of access to formal housing, urbanization in Latin America has had positive effects that have increased the income of residents and improved access to health and education. In the last 20 years, the Human Development Index in Latin America has increased by 10 per cent and the poverty rate reduced by half between 1990 (12 per cent) and 2012 (6 per cent) (Jiménez 2015). Nevertheless, hiding behind these figures is one of the most serious social problems that defines the intensity of environmental injustice: the enormous inequality in the distribution of wealth that is characteristic of the region. Latin America continues to be the most unequal region in the world (Jiménez 2015).

Explosive growth, informality and inequality have created serious environmental problems that more specifically result from the lack of provision of basic services, such as garbage collection, potable water or sewage systems. Social movements have emerged that have put these environmental problems on the public agenda in general, and the urban environmental justice agenda in particular, the latter being manifest in an increasing number of socio-environmental conflicts in the region.

According to the Environmental Justice Atlas (2016), the most important conflicts in the region have been those related to solid waste management, access to water, and the impacts of polluting industry. Conflicts over waste management in cities have focused on the transport and final disposal of waste into open dumps or landfills in urban areas. Examples include the social movements opposed to the "Doña Juana" landfill in Bogota-Colombia, and the open dump operated by the company Tersa del Golfo in the Province of Mexico. Social movements have debated the management of and equitable access to water in relation to projects such as the "Alto Maipo" hydroelectric plant in Santiago de Chile, the operation of the Sogamoso hydroelectric plant in Colombia, and the privatization of water in the city of Guayaquil in Ecuador. Finally, activism has developed around the pollution impacts that industries generate, such as the struggle of communities near industries that use asbestos in Bogota-Colombia, or the struggle against the production of cement in the communities of Hidalgo-México.

While social and political activism on urban environmental justice has been consolidating in recent years (see also Chapter 43), research on urban environmental justice in Latin America is still under-developed. Publications on urban environmental justice in Latin America started in 2003, and to date there are only 21 publications in total (according to searches in Scopus, Web of Science, Redalyc, Scielo and Google Scholar). In these, there are different thematic emphases among the different countries of the region and different methodological approaches are applied. In general, the scientific research in the region is highly concentrated in Mexico and Chile, although other countries are involved as well.

Research has been conducted in Mexico using concepts from urban environmental justice as a framework. The main focus of the research led by Grineski and Collins (2008, 2009, 2012, 2015) has been about social environmental differences present on the border between Mexico and the United States (see Chapter 42), identifying that in cities such as Juárez and Tijuana, there is greater exposure of the population to hazards, both natural (floods and landslide) and anthropogenic (chemical emissions, air pollution from industry), than in cities such as El Paso, located in Texas.

Research in Chilean cities has focused on how distinct socio-economic groups are unequally affected by environmental risks (Vásquez & Salgado 2009 and Romero et al. 2010), air pollution (Romero et al. 2013), temperature and heat islands (Romero et al. 2010) and environmental impacts from urban expansion (Henriquez et al. 2009). There has also been, to a lesser extent, concern about inequitable access to urban green spaces (Vásquez & Salgado 2009; Reyes & Figueroa 2010; Vásquez et al. 2016).

In Argentina there are studies about exposure to pesticides in urban areas (Berger 2009), unequal access to public services (García 2003), distribution of environmental quality indicators between different cities (Celemin 2012) and environmental suffering (Auyero & Swistun 2009). In Bolivia, there are two studies about unequal access to water (see Chapter 27) in an urban environment (Bustamante 2012; Crespo 2009). In Montevideo, Uruguay, a study has focused on the effects of lead contamination in the city (Renfrew 2007), and there has been a study in Cuba on community actions to improve the urban environment (Anguelovski 2013) and in Colombia on air pollution (Romero et al. 2013).

In summary, urban environmental justice research in Latin America is recent, scarce and concentrated on the classic distributional aspects of environmental problems, such as natural hazards and pollution, and environmental goods, such as green spaces and public services, among different social groups. Another characteristic of these studies is the dominant interest in working with socio-economic classifications to assess inequalities over other social classifications based on ethnic, age or gender aspects.

In order to go into more detail and depth, over the rest of the chapter we examine the case of Chile, first presenting some background on its urban development model before then investigating more specifically the Chilean environmental justice agenda. In concluding the chapter we synthesize the current state of the urban environmental justice agenda in Chile by looking at gaps in research and practice, and propose directions for future research.

Major development issues of Chilean cities and the processes behind them

Chile is one of the most urbanized countries in the continent and in the last few decades has been undergoing a process of 'metropolization' (Hidalgo et al. 2009). While the capital, Santiago, is the undisputed economic and political centre of the country, in recent decades other metropolitan regions have formed, such as Valparaiso-Viña del Mar on the coast of the central zone and Concepción-Talcahuano in the south. At the same time, there are several intermediate cities that in the last two decades have experienced growth rates higher than those of the large cities (Henríquez 2014). In many cases, this growth dynamic is associated with productive and extractive activities characteristic of the Chilean development model. In that context, there are important urban development processes concentrated around activities such as aquaculture, forestry, mining and agribusiness. Both these activities and associated urbanization processes are steadily increasing the pressure on natural resources such as water, air and land. Together with the negative impacts of these interventions, a more organized and strengthened activist community has generated more frequent environmental conflicts during the last decade.

What has left its mark on the way cities are developing in Chile is the neoliberal urban policy that was adopted under the military dictatorship in 1979 and has since seen only minor revisions. Today Chile's accelerated urban growth processes are poorly regulated and planned, produce strong social-environmental segregation patterns and respond primarily to the profitability criteria of financial capital.

Apart from still being one of the most neoliberal countries in the world, due to its geographical conditions Chile is also one of the countries with the greatest levels of exposure to natural hazards, such as volcanic eruptions, earthquakes and tsunamis (see further discussion below). As a consequence, in recent years there have been major socio-natural disasters that have revealed the weaknesses of a territorial planning process subsumed to the market. Another problem related to Chile's neoliberal urbanism is unequal access to environmental goods and services, which adds an environmental dimension to social segregation. A good example of this is the disappearance, reduction and deterioration of vegetation, including croplands, natural areas and wetlands (Smith & Romero 2007) that is associated with the uncontrolled growth of urban areas and resulting spatial fragmentation.

Development of academic research, public policy and civil society mobilization

In response to the conflicts and tensions emerging around Chile's urban development model and strongly neoliberal politics, academic research, public policy and civil society have become increasingly engaged in questions of urban environmental justice. There are three main issues in Chilean cities that have characterized the urban environmental justice agenda: 1) socio-natural disasters; 2) contamination and pollution; and 3) green spaces.

Socio-natural disasters

Chile has seen a series of major events in recent decades: a volcano eruption in Patagonia in 2006; one of the five strongest earthquakes (Mw=8.8) ever recorded worldwide in central-southern Chile in 2010; a disastrous fire in the port city of Valparaiso in 2014; and storms, landslides and earthquakes in different locations in 2015. All of these events have left important lessons with regard to urban environmental justice, the two most important being: 1) an unequal exposure to threats by different social groups as well as differences in the levels of vulnerability (Vargas 2002; Bankoff 1999); and 2) a lack of recognition and participation of communities as valid interlocutors when defining prevention plans and reconstruction measures when disaster occurs. In Chile it is evident that political action and governance processes, including non-decision making, in many instances tend to increase levels of vulnerability and the unequal distribution of risks, thus promoting scenarios of environmental injustice (Sandoval et al. 2015).

In post-disaster scenarios in Chile – in particular during reconstruction processes – associative community spaces have been shaped and built upon people's grievances and demands for the exercise of citizenship rights in the affected territory (see also Chapter 28 on flooding). Ugarte and Salgado (2014) investigated these issues in Patagonia, where in May 2008 a volcano eruption severely damaged the small city of Chaitén. Residents were forced to evacuate the city, regardless of local knowledge or family and community organization, which displaced families to bigger cities with very different prevailing lifestyles and social structures compared to the small and rural Patagonian city. Subsequently, without any participative processes, the city was declared uninhabitable by the government, and people were banned from entering because it was considered too risky. As an alternative, a new Chaitén was proposed to be established in a location close to the former one. However, this process was strongly questioned by the inhabitants, who claim not to have been properly consulted, positioning themselves in open opposition to the relocation. In this scenario, the spatially and politically displaced inhabitants began to organize and generate resistance (Ugarte & Salgado 2014; Sandoval et al. 2015). In this case it is possible, therefore, to observe the emergence of collective demands focused on the reconstruction of the community and its territory, moving from individual concerns to what Schlosberg (2013) has identified as the articulation of collective and community environmental justice issues.

In a similar vein, post-disaster community organization for recognition and participation, after the magnitude 8.8 earthquake and tsunami of 2010, included claims about irregularities in granting subsidies, delays in the construction of poor quality housing and unjustified expropriation of land and forced displacement. This unleashed a high level of discontent in the population (Fuentes & Shüler 2014; Imilan & Fuster 2014; Ugarte et al. 2015), resulting in demonstrations against the solutions provided by the government and a demand for greater participation in reconstruction processes. Various social organizations converged in the National Movement for Fair Reconstruction (NMFR), which has taken on the task of reporting irregularities and channelling demands that have been highly critical of the public–private and essentially market-based solutions the government has provided. Based on the ideas proposed by Harvey (2013), the actions of the NMFR (and organizations involved in it) can be interpreted as a claim to the 'right to the city', that involves the freedom to make and remake ourselves and our cities, beyond individual interests. Therefore, the actions and motivations of NMFR tend towards overcoming the idea of an organization directly linked to one particular territory, developing instead the idea and practice of networked territories (Ugarte et al. 2015). In this context, demands are identified that are not only individual, but that operate under a territorial logic of community in cases where it is the very functioning of the city that has been affected (Schlosberg 2013).

In summary, the demands for fair reconstruction showed that the recognition of affected populations as stakeholders is required in order to bring justice to post-disaster settings. However, as Schlosberg (2013) points out, it is crucial to realize how the concept of justice is used, understood, articulated and demanded. In this regard, what is denounced in the cases we have reviewed is deficient participation and the lack of knowledge about the needs, projects, and desires of citizens in the decision making processes. This demands greater levels of recognition and participation of citizens in the management of their territories and in the mitigation and restoration processes following from the impacts of extreme natural events.

Contamination and pollution

In Chile, as in other countries, problems of environmental contamination have been at the forefront of the environmental justice agenda since its beginning, in terms of both political mobilization and academic research. In fact, a considerable part of the numerous socio-environmental conflicts that Chile has witnessed in the last two decades has been attached to problems of urban pollution, which in turn is the result of the manner in which the neoliberal development model intertwines with urbanization processes and patterns. In the following, three issues of environmental contamination are briefly explained: 1) air pollution; 2) environmental contamination disasters; and 3) the emerging concept of 'sacrifice zones'.

In an important number of Chilean cities, air pollution (see Chapter 26) has been a longstanding problem (Meléndez 1991). Although the main sources responsible for atmospheric contamination are transportation, industry and the use of firewood (Romero et al. 2010), the relative importance of pollutants in different cities across the country varies and depends on both the city's economic base and geographic conditions. Also, within the cities affected by contamination problems an unequal socio-spatial distribution can be observed which transforms into a distributive environmental justice problem and is also discussed in those terms. In one of the few existing studies, Romero et al. (2010) determine for the city of Santiago a geography of atmospheric contamination (of PM_{10}) that coincides with the historic pattern of socio-economic segregation: in the richer and much greener eastern part of the city, contamination with particulate matter (PM_{10}) is lower than in the poorer western part of the city. The authors (Romero et al. 2010: 59) thus conclude that there is the "existence of levels of environmental injustice that would require mitigation and compensation measurements specifically targeted at balancing the spatial distribution of climates and contaminants". However, they also highlight that new real estate dynamics – gated communities for the middle and upper classes being installed in formerly poor neighbourhoods – lead to increasing socio-environmental heterogeneity, which makes the development of justice-based mitigation and compensation mechanisms more complex. On the political level a first milestone with regard to air pollution was set when the government echoed both the scientific evidence and the mobilization of NGOs and approved the first Plan for Prevention and Atmospheric Decontamination of the Metropolitan Region (PPAD), which was enacted in 1997/1998 and actualized in 2004 and 2010. While in its first version the PPAD focused on diminishing PM_{10} concentration, the actualization that has been followed since 2015 (called "Santiago Respira") focuses on the reduction of finer particulates ($PM_{2.5}$). Furthermore, in its "Decontamination Plans, Strategy 2014–2018", the government proposes a range of actions with a view to decontamination in mid-sized cities, especially in the south of Chile (Fundación Terram 2014). Thus, while the issue of contamination is being addressed on the policy level, it is not being done with explicit consideration of environmental justice claims.

Environmental contamination disasters are another face of urban contamination, which is closely related to the extractive development model. Here a simultaneously tragic and important event has been the disaster in the Nature Sanctuary of the Cruces River – near to the city of Valdivia – that according to Sepúlveda (2011) "marked a before and after in the environmental history of Chile". The disaster began in 2004 with the discharge of effluent from the pulp mill of the Arauco forestry company into the Cruces River, which led to the death of hundreds of black-necked swans and the degradation of an important wetland. For several years, a local citizens' environmental movement confronted the polluting company, which belonged to one of the largest economic groups in the country, but it denied responsibility. Because of the 'agency' of the black-necked swans (in the sense of actor-network theory, see Sepúlveda & Villarroel 2012), the grotesque denial of responsibility from the company, and the ongoing work of the NGOs, the conflict received significant media attention throughout the country. In fact, the development of the conflict gave a major impetus to the environmental institutions of the country, which were modernized substantially in 2010. As a result, in 2013 one of the recently installed environmental courts (see Chapter 12) convicted the company, giving important recognition to the infringed ecosystem and the affected communities, principally those related to the collapsed tourist industry of the region (Sepúlveda 2011). Although there has been undeniable progress in terms of the modernization of the environmental institutional landscape, in the environmental justice movement there is agreement that modernization did not take the opportunity to remedy the lack of citizen participation, which is one of the most important issues.

Considering the unequal socio-spatial distribution of the contamination caused by the development model of the country, the term 'sacrifice zones' has been applied. In Chile these are understood as territories and human settlements (usually small and medium-sized cities) that are environmentally devastated by the cumulative effects of industrial developments. An emblematic case of such a sacrifice zone is the Bay of Quintero, where a large number of polluting industries are concentrated and where in 2014 38 000 litres of oil were discharged into the sea. From an explicit environmental justice perspective, the influential NGO Fundación Terram (2014: 13) determines that the "areas that are highly contaminated as a result of the current development model, the lack of public policies and the negligence of authorities, constitute sacrifice zones of fundamental rights of the communities that live there". These zones in general show high levels of poverty, a marked weakness in basic services such as potable water, deficient public health systems and poor citizen participation in territorial decision making. Faced with this reality, in 2014 the first conclave of sacrifice zones took place where mayors from the Puchuncaví, Quintero, Tocopilla, Huasco and Coronel municipalities met and decided to push towards the creation of the "Union of Sacrifice Zone Municipalities". Several meetings followed and a list of demands was presented, one important point being the strengthening of regulations to diminish PM_{10}-related air pollution, which severely affects the sacrifice zones.

Green spaces

Concern for urban green spaces (see Chapter 35) has emerged only recently in the environmental justice agenda in Chile. Once basic needs such as health, education and housing have been satisfied, at least in general and quantitative terms, concerns about the quality of urban life have slowly moved to aspects related to the neighbourhood, such as infrastructure, public spaces and green areas. Interest in the unequal distribution of urban green spaces in Chile essentially began to appear in the second half of the 2000s, when studies began to measure and understand the degree of unequal access to these spaces. Using purely statistical approaches, these studies have concentrated on the distributive dimension of environmental justice from the early stages of research.

These investigations have shown that there is a markedly unequal distribution of green spaces in Chilean cities, especially in Santiago, where most research has been done. For instance, Escobedo et al. (2006) show that municipalities of low socio-economic status have less urban tree cover than municipalities of high socio-economic status. In addition, higher-income municipalities have a greater diversity of species, and fewer but larger trees (De la Maza et al. 2002; Escobedo et al. 2006). Reyes and Figueroa (2010) add that municipalities of lower socio-economic status have fewer green spaces and less access to large public green areas. Vásquez (2009) and Vásquez and Salgado (2009) found that this type of unequal distribution of green areas can also be observed at an intra-communal level, especially in peripheral municipalities where the fragmented pattern described by Borsdorf and Hidalgo (2010) for Latin American cities is most strongly manifested. As a consequence, a major deficit of green spaces has arisen, especially in the residential complexes developed by the state's social policies, which have focused on the massive construction of a large number of houses of low quality, at a high density, with almost no courtyards or public green areas.

Few studies have addressed such issues in other Chilean cities. However, the Ministry of Environment (2011) indicates that, in general, there is an increasing availability of green spaces in cities following a north–south gradient. In a similar vein, Vásquez and Salgado (2009) found that the close relationships between socio-economic status and availability of green areas identified in the municipality of Peñalolén in Santiago does not exist in San Pedro de la Paz in Concepción, in the south of the country. This may be because of the reduced need for irrigation to maintain urban green areas due to the rainier climate that exists in southern Chile. The difference in purchasing power and relative or municipal budget available to irrigate green spaces thus becomes more important in central and northern Chile, where arid and semi-arid conditions predominate.

This unequal distribution of urban green spaces occurs both in private and in public spaces. For private space the explanation is that small houses have very small courtyards and that these are normally used to expand housing rather than being kept as gardens; in the case of public green spaces, the causes are given by the low availability of municipal resources for the creation of green spaces and especially for maintaining them.

Until now, there have been limited efforts from public policy to moderate these inequalities. One of the initiatives created for this is the Urban Parks programme of the Ministry for Housing and Urban Development (MHUD), which acts to create urban parks in municipalities that have a lack of green areas. These parks are financed by resources from the MHUD and thus not from the municipal coffers, hence large parks can be established in municipalities that would not have sufficient resources for their creation and maintenance. At the same time, the Quiero Mi Barrio programme from MINVU (Ministerio de Vivienda y Urbanismo, Ministry for Housing and Urbanism) aims to make improvements in the urban environment of degraded neighbourhoods of low socio-economic status, commonly including the construction or restoration of green areas within these interventions.

Meanwhile, in the sphere of civil society, NGOs and community organizations concerned with urban green spaces have begun to defend and reclaim them. There are at least two foundations (Mi Parque and Cultiva) that aim to create and rehabilitate public green areas in poor neighbourhoods, using participatory design techniques (see Chapter 23) and community and municipal commitment. These foundations operate with donations from members and companies, where the latter make their contributions through their corporate social responsibility programmes or through compensation for environmental impacts. Although there is a lack of mid- and long-term impact assessments of these initiatives addressing environmental justice, positive outcomes in terms of recuperating green spaces can already be observed.

In addition, neighbourhood organizations have emerged in response to an unmet demand for urban spaces that can contribute to improving the quality of life. These movements have emerged in poor neighbourhoods, many of which were developed by the social housing programme of the state. These movements have been mostly directed at protecting green spaces threatened by road or real estate projects, or at the acquisition of vacant lots to transform them into plazas or community gardens (Opazo and Jaque 2014).

Across these cases, the broad and diverse concern for urban green spaces may be associated with a resignification of what Chilean society understands by the environment, which now encompasses environmental conditions as they are experienced every day where one lives or works.

Synthesis and final reflections

As this chapter has shown, the urban environmental justice agenda in Latin America (see also Chapter 43) and in Chile may be still incipient but seems to be in a gradual process of consolidation. The cities in this region have specific characteristics compared to other regions of the world, particularly considering the effects of the nexus of explosive urbanization, informality, segregation and inequality. While on a general level this applies to Chile as well, here informality is less an issue while socio-spatial segregation and inequality go deeper than in other countries of the region. This translates into environmental justice being a particularly pressing problem in the highly segregated and unequal Chilean urban areas.

In this context it makes sense that Chile stands out in Latin America in terms of the scientific production of work on urban environmental justice, which includes studies not only on the distributive dimension of environmental threats, but also on environmental amenities. Chilean civil society has become increasingly concerned about both the distributive dimension in terms of social groups' vulnerability to socio-natural disasters and pollution problems and the inequality of access to urban environmental assets such as green and blue spaces. This could suggest interesting future directions for the urban environmental justice agenda in other Latin American countries, extending beyond the distribution of environmental burdens once they overcome the persistent problem of lack of basic infrastructure, sanitation and services. Looking the other way around, Chile with its free-market based development ideology (see Chapter 6) can benefit from lessons delivered by more progressive political regimes in the region. In Ecuador and Bolivia, for instance, environmental justice claims are, at least implicitly, part of the conceptions of the '*Sumak kawsay*' or '*buen vivir*', the philosophy of 'good living'. This ancestral Andean cosmovision is based on a harmonious relationship between man, community and nature and in important ways influenced the new constitution of Ecuador in 2008. In this, on the one hand, the social and cultural rights of Indigenous communities were strengthened, while on the other, the rights of nature itself were codified in constitutional terms for the first time.

Chilean society and especially its political elites still have a long way to go in politically recognizing its development model as a structural cause of urban environmental injustice. Rather, in Chile the environmental justice agenda currently is driven by a civil society that sometimes explicitly and sometimes implicitly includes the environmental justice issues of distribution, recognition and participation as part of the comprehensive process of strengthening democracy in post-dictatorship Chile.

In terms of academia, the brief review of literature on the Latin American and Chilean research agendas has shown that there is little work that explicitly alludes to the concept of environmental justice, but there are many more studies that do that implicitly, particularly in the areas of socio-natural disasters, contamination and pollution, and green areas. For academia to contribute further to the urban environmental justice agenda, not only is more research needed that addresses environmental issues from the explicit perspective of justice, but also approaches whose theoretical and methodological focus capture the full complexity of environmental justice questions. In this sense, it is important to strengthen but also to go beyond the distributive dimension and address processes and problems of participation and recognition for communities affected by injustice (see Chapters 9 and 10). This would incorporate variables for understanding both the structural and subjective dynamics that are found behind the production of the unequal distribution of risks, services and environmental amenities. Studies are needed that approach the intersubjective construction of 'justice' and the 'environment' in conflict situations and in civil society actions, as has been seen in post-disaster reconstruction planning or citizen mobilization around environmental contamination events. In addition, only a better understanding of the effect of the scarcity and weakness of institutionalized spaces for participation on the trajectories of citizen actions can illuminate how to articulate them effectively into formal structures.

On a more consensual level there is a need for more and better instances of dialogue between the political, academic and civil society spheres in order to provide feedback and align the respective agendas on environmental justice in the country. Research could be more participative and qualitative (see Chapters 23 and 24), fostering the demands emanating from affected communities and, at the same time, being a source of accessible and local information for these communities. A bottom-line demand and goal is to recognize citizens as valid actors and interlocutors in the political sphere when designing mitigation and management strategies in the context of, for instance, socio-natural disasters, pollution events and access to green areas.

However, if the current extractive development model and the processes of unplanned urban growth which are maintaining or deepening current levels of socio-spatial inequality are not tackled, environmental injustice problems are highly likely to remain in place. Given this scenario, the challenge is to move forward in shaping institutional mechanisms that safeguard environmental justice as a civil right that should have a constitutional character. This is a particularly relevant discussion in Chile given the current national political debate about the possible content of a new constitution designed under a democratic regime.

References

Anguelovski, I. 2013, 'New directions in urban environmental justice: Rebuilding community, addressing trauma, and remaking place', *Journal of Planning Education and Research*, no. 33, pp. 160–175.

Auyero, J. & Swistun, D. 2009, *Flammable: environmental suffering in an Argentine shantytown*. Oxford University Press, Oxford.

Bankoff, G. 1999, 'A history of poverty: The politics of natural disasters in the Philippines, 1985–95', *The Pacific Review*, vol. 12, no. 3, pp. 381–420.

Berger, M. 2009, 'Poblaciones expuestas a agrotóxicos: autoorganización ciudadana en la defensa de la vida y la salud, Ciudad de Córdoba, Argentina', *Physis*, no. 20, pp. 119–143.

Borsdorf, A. & Hidalgo, R. 2010, 'From Polarization to Fragmentation: Recent Changes in Latin American Urbanization'. In P. Lindert & O. Verkoren (eds.), *Decentralized Development in Latin America*, Springer. Netherlands.

Bustamante, R. & Medieu, A. 2012, 'Struggling at the borders of the city: Environmental justice and water access in the southern zone of Cochabamba, Bolivia', *Environmental Justice*, no. 5, pp. 89–92.

Celemin, J. 2012, 'Calidad ambiental y nivel socioeconómico: su articulación en la región metropolitana de Buenos Aires', *Scripta Nova*, no. 441.

CEPAL 2010, *Terremoto en Chile Una primera mirada al 10 de marzo de 2010*, Santiago.

Crespo, C. 2009, 'Privatización del agua y racismo ambiental en ciudades segregadas. La empresa Aguas del Illimani en las ciudades de La Paz y El Alto (1997–2005)', *Anuario de Estudios Americanos*, no. 66, pp. 105–122.

De la Maza, C., Hernández, J., Bown, H., Rodríguez, M. & Escobedo, F. 2002, 'Vegetation diversity in the Santiago de Chile urban ecosystem?', *Arboricultural Journal*, vol. 26, pp. 347–357.

Environmental Justice Atlas, 2016. *EJAtlas | Mapping Environmental Justice*. [online] Available at: https://ejatlas.org/.

Escobedo, F., Nowak, D., Wagner, J., De la Maza, C., Rodríguez, M., Crane, D. &, Hernández, J. 2006, 'The socioeconomics and management of Santiago de Chile's public urban forests', *Urban Forestry & Urban Greening*, vol. 4, no. 3, pp. 105–114.

Fuentes, L. & Shüler, U. 2014, *La política social de Mercalli: El terremoto y la oportunidad de los empresarios*, CEIBO, Santiago.

Fundación Terram 2014, *Balance Ambiental 2014. Sacrificando Chile Por la Inversión. Presidenta: ¿Zonas de sacrificio o justicia ambiental?*, Santiago, viewed 20 November 2015, www.terram.cl/wpcontent/uploads/2015/01/BALANCE_AMBIENTAL_TERRAM_2014.pdf

García, M. 2003, *El desigual acceso a servicios públicos urbanos: Brechas sociales y riesgo ambiental en el caso de Tandil, Argentina*, Universidad Nacional del Centro de la Provincia de Buenos Aires, Buenos Aires.

Grineski, S. & Collins, T. 2008, 'Exploring patterns of environmental injustice in the Global South: Maquiladoras in Ciudad Juarez, Mexico', *Population and Environment*, no. 29, pp. 247–270.

Grineski, S., Collins, T. & Romo, M. 2009, 'Vulnerability to environmental hazards in the Ciudad Juarez (Mexico)–El Paso (USA) metropolis: A model for spatial risk assessment in transnational context', *Applied Geography*, no. 29, pp. 448–461.

Grineski, S., Collins, T., Ford, P., Fitzgerald, R., Aldouri, R., Velázquez-Angulo, et al. 2012, 'Climate change and environmental injustice in a bi-national context', *Applied Geography*, no. 33, pp. 25–35.

Grineski, S., Collins, T. & Romo, M. 2015, 'Environmental injustice along the US–Mexico border: Residential proximity to industrial parks in Tijuana, Mexico'. *Environ*, no. 1, pp. 1–10.

Harvey, D. 2013, *Ciudades rebeldes: Del derecho a la ciudad a la revolución urbana*, Ediciones Akal, Buenos Aires.

Henríquez, C. 2014, *Modelando el crecimiento de ciudades medias, hacia un desarrollo urbano sustentable*, Ediciones UC, Santiago.

Henríquez, C., Arenas, F., Romero, H. & Azócar, G. 2009, 'Justicia socio-ambiental y sostenibilidad en el crecimiento de las ciudades medias de Chillán y Los Ángeles (Chile)'. In *UNESCO, Las Ciudades Medias o Intermedias en un mundo globalizado*. University of Lleida, Lleida Càtedra.

Hidalgo, R., De Mattos, C. & Arenas, F. 2009 (eds.), *Chile: Del país urbano al país metropolitano*, Pontificia Universidad Católica de Chile, Santiago.

Imilan, W. & Fuster, X. 2014, *Terremoto y tsunami post 27F: El caso de Constitución, Arauco y Llico*, Informe del Observatorio de la Reconstrucción, Instituto de la Vivienda, Facultad de Arquitectura y Urbanismo, Universidad de Chile, Santiago.

Jiménez, J. (ed.) 2015, *Desigualdad, concentración del ingreso y tributación sobre las altas rentas en América Latina*, Libros de la CEPAL, N° 134 (LC/G.2638-P), Santiago de Chile, Comisión Económica para América Latina y el Caribe (CEPAL).

Liberturn de Duren, N. (ed.) 2014, *Urbanización rápida y desarrollo: Cumbre de América Latina y China*, Banco Interamericano de Desarrollo, Washington, DC.

McDonald, Y. & Grineski, S. 2011, 'Disparities in access to residential plumbing: A binational comparison of environmental injustice in El Paso and Ciudad Juarez', *Population and Environment*, no. 34, pp. 194–216.

Mehta, L., Allouche, J. & Walnycki, A. 2013, 'Global environmental justice and the right to water: The case of peri-urban Cochabamba and Delhi', *Geoforum*, no. 54, pp. 158–166.

Meléndez, T. 1991, *Salvemos Santiago . . . y salvemos Chile. Los problemas ambientales de Chile y sus alternativas de solución*, Ercilla, Santiago.

Ministry of Environment 2011, *Informe del estado del medio ambiente 2011*, Santiago de Chile.

Molina, M., Romero, H., Sarricolea, P. 2007, 'Características socio ambientales de la expansión urbana de las Áreas metropolitanas de Santiago y Valparaíso'. In R. Hidalgo, C. de Mattos & F. Arenas (eds.), *Chile: Del país urbano al metropolitano*, Geolibros PUC, Santiago.

Moore, S. 2009, 'The excess of modernity: Garbage politics in Oaxaca, México', *The Professional Geographer*, no. 61, pp. 426–437.

Norman, L., Villareal, M., Lara, F., Yuan, Y., Nie, W., Wilson, S., et al. 2012, 'Mapping socio-environmentally vulnerable populations' access and exposure to ecosystem services at the U.S.-Mexico borderlands', *Applied Geography*, no. 34, pp. 413–424.

Opazo, T. & Jaque, J. 2014, *La pelea por el área verde. La Tercera.* October 18. Retrieved from www.latercera.com/noticia/tendencias/2014/10/659-600763-9-la-pelea-por-el-area-verde.shtml

Renfrew, D. 2007. 'Justicia ambiental y contaminación por plomo en Uruguay'. In R. Sonnia (ed.), *Antropología social y cultural en Uruguay*, Unesco, Montevideo.

Reyes, P. & Figueroa, I. 2010, 'Distribución, superficie y accesibilidad de las áreas verdes en Santiago de Chile'. *EURE*, vol. 36, no. 109, pp. 89–110.

Romero, H., Fuentes, C. & Smith, P. 2010. 'Ecología política de los riesgos naturales y de la contaminación ambiental en Santiago de Chile: Necesidad de justicia ambiental', *Scripta Nova*, no. 331.

Romero, H., Irarrázaval, F., Opazo, D., Salgado, M. & Smith, P. 2010, 'Climas urbanos y contaminación atmosférica en Santiago de Chile', *Revista EURE*, vol. 36, no. 109, pp. 35–62.

Romero, L. 2007, 'Are we missing the point? Particularities of urbanization, sustainability and carbon emissions in Latin American cities'. *Environment and Urbanization*, vol. 19, no. 1, pp. 159–175.

Romero, P., Qin, H., & Borbor, M. 2013, 'Exploration of health risks related to air pollution and temperature in three Latin American cities', *Social Science & Medicine*, vol. 83, pp. 1–9.

Sandoval, V., Boano, C., González, C. & Albornoz, C. 2015, 'Explorando potenciales vínculos entre resiliencia y justicia ambiental: El caso de Chaitén, Chile', *Magallania (Chile)*, vol. 43, no. 3, pp. 37–49.

Schlosberg, D. 2013, 'Theorising environmental justice: The expanding sphere of a discourse', *Environmental Politics*, vol. 22, no. 1, pp. 37–55.

Sepúlveda, C. 2011, 'Celco y el desastre de Valdivia: la hora de la verdad', *El Mostrador*, 11 April, viewed 11 September 2015, www.elmostrador.cl/noticias/opinion/2011/04/11/celco-y-el-desastre-de-valdivia-la-hora-de-la-verdad.

Sepúlveda, C. & Villarroel, P. 2012, 'Swans, conflicts, and resonance local movements and the reform of Chilean environmental institutions', *Latin American Perspectives*, vol. 39, no. 185, pp. 181–200.

Smith, P. & Romero, H. 2007, 'Efectos del proceso de urbanización sobre la calidad ambiental de los humedales del área metropolitana de Concepción'. *Anales de la Sociedad Chilena de Ciencias Geográficas*, Santiago, pp. 245–250.

Soltero, E., Mama, S. & Pacheco, A. 2015, 'Physical activity resource and user characteristics in Puerto Vallarta, Mexico', *RETOS. Nuevas Tendencias en Educación Física, Deporte y Recreación*, no. 28, pp. 203–206.

Ugarte, A. & Salgado, M. 2014, 'Sujetos en emergencia: Acciones colectivas de resistencia y enfrentamiento del riesgo ante desastres; el caso de Chaitén, Chile'. *Revista INVI*, vol. 29, no. 80, pp. 143–168.

Ugarte, A., Salgado, M. & Fuster, X. 2015, *Emergencia de sujeto político y experiencias de acción colectiva en desastres socionaturales: Análisis de casos en Santiago, Constitución y Chaitén, Chile.* In C. Arteaga & R. Tapia (eds.), *Desastres socionaturales y vulnerabilidad social en Chile y Latinoamérica*, Editorial Universitaria, Santiago.

United Nations 2014, *World Urbanization Prospects 2014: Highlights.* United Nations Publications. www.compassion.com/multimedia/world-urbanization-prospects.pdf

Vargas, J. 2002, 'Políticas públicas para la reducción de la vulnerabilidad frente a los desastres naturales y socio-naturales'. *CEPAL, Serie medio ambiente y desarrollo no. 50*, Santiago.

Vásquez, A. 2009, 'Vegetación urbana y desigualdades socio-económicas en la comuna de Peñalolén, Santiago de Chile: Una perspectiva de justicia ambiental', Master thesis, Universidad de Chile, Santiago.

Vásquez, A., Devoto, C., Giannotti, E. & Velásquez, P. 2016. 'Green infrastructure systems facing fragmented cities in Latin America – Case of Santiago, Chile. *Procedia Engineering*, vol. 161, pp. 1410–1416.

Vásquez, A. & Romero, H. 2007, 'Desigualdades socioeconómicas en la comuna de Peñalolén, una perspectiva de Justicia Ambiental', *Anales Sociedad Chilena de Ciencias Geográficas*, Santiago, pp. 273–277.

Vásquez, A. & Salgado, M. 2009. 'Desigualdades socioeconómicas y distribución inequitativa de los riesgos ambientales en las comunas de Peñalolén y San Pedro de la Paz. Una perspectiva de justicia ambiental', *Revista de Geografía Norte Grande*, no. 43, pp. 95–110.

45

ENVIRONMENTAL JUSTICE IN NIGERIA

Divergent tales, paradoxes and future prospects

Rhuks T. Ako and Damilola S. Olawuyi

Introduction

The oil rich Niger Delta exemplifies circumstances wherein the proper conceptualization, understanding and pursuit of environmental justice determine the outcomes of natural resource exploitation. For about half a century, multinationals have, in joint ventures with the Federal Government of Nigeria, exploited oil from the Niger Delta, deriving huge revenues for both parties. Oil, which accounts for close to 90 per cent of exports and roughly 75 per cent of the country's consolidated budgetary revenues, has reportedly yielded over $1 trillion for the country (World Bank 2015). In contrast, host communities have over the years consistently decried the negative impacts of the oil industry which they have to bear without any real sustainable positive benefits.

These include well documented incidences of environmental pollution, particularly oil spills and gas flaring, that contribute significantly to the negative impacts the oil industry has had on the Niger Delta environment (Watts 2012). While the frequent oil spills contaminate the soil, vegetation, aquatic resources and groundwater, flaring of gas involves the emission of huge volumes of poisonous flares and gas that pollute the air and return to land via acid rain. The gas flares also artificially light up the environment all day and night while making loud noises, contributing to noise pollution. Other negative environmental consequences of the oil industry include "canalization, dredging, large-scale effluent release, mangrove clearance, massive pollution of surface and groundwater" (Watts 2012: 2) as well as seismic operations and activities related thereto (Amnesty International 2009).

Since the inhabitants of the region rely almost exclusively on environmental resources, mainly as farmers and/or fishermen, their sources of livelihood are impacted by the above environmental consequences of oil exploration and production. In addition, there are health impacts on the human population, livestock, the fauna and flora as well as other species that flourish in the Delta's rich and diverse environment. Simply put, oil-related activities adversely affect the socio-economic wellbeing and health of the inhabitants of the region, as well as by extension their human rights (Ako 2015; Obi 2009, 2014; Human Rights Watch 1999).

The inequitable distribution of environmental benefits and disadvantages has become a source of conflict and the resultant violence has threatened the state, the oil industry and the host communities. In response to the violence, the state has adopted several initiatives to quell the insurgency in the Niger Delta region, including militarization, increased spending on community development, and most recently the amnesty programme, and the oil companies have improved their community outreach programmes and corporate social responsibility strategies (Ako and Omiunu 2013; Ako 2012; Idemudia 2011). Despite these, the region has continued to experience, and remains susceptible to, violent conflicts ostensibly due to the evident environmental injustices. Thus it is imperative to address the environmental justice deficit through a holistic reappraisal of the relevant legal and institutional framework.

This chapter argues, based on the Niger Delta experience, that Nigeria requires a more strategic, focused and holistic framework on environmental justice to replace the current haphazard approach. It is posited that a focal agency be mandated to promote the institutionalization of environmental justice in governance. This focal institution will take the lead on coordinating relevant institutions to develop, evaluate, monitor and assess the performance of government departments and agencies with respect to the tenets of environmental justice. Also, it will provide leadership by assisting government departments to develop the skills and resources they need to promote participation; accountability; equality and non-discrimination; access to information; and access to justice in environmental issues.

The chapter is in five sections, including this introduction. The second section highlights the concept of environmental justice within the context of a resource-rich developing country such as Nigeria. It also discusses the key actors in Nigeria's oil industry to provide a background to understand their roles in the environmental justice debate. The third section highlights, with examples, the role that the legal and institutional framework regulating the oil industry plays in promoting environmental (in)justice. The fourth section discusses the fundamental drawbacks of the current state of inchoate environmental justice methodology, particularly highlighting the problems related to the haphazard approach adopted vis-à-vis institutions and regulations. The fifth section concludes by suggesting that a focal institution should be created to take the lead on the proper coordination and implementation of environmental justice issues and initiatives in Nigeria.

Environmental justice in Nigeria: conceptualization and key actors

Environmental justice is conceptualized differently in developed countries and developing countries. Even among countries in the same category, environmental justice is conceptualized according to the specific circumstances the country in question faces. Hence, while in the US issues of racial equality are fundamental, this is less so in the UK, where socio-economic parity is more emphasized (Ako 2013). With regards to the difference between developed and developing countries, Beinart and McGregor (2003: 2) note that while some environmentalists use the environmental justice platform to emphasize the responsibility to future generations for the wellbeing of the planet, 'Africanists' by contrast consider issues such as access to resources as the critical issue for communities.

With oil being the main natural resource in Nigeria, the discourse on environmental justice is often centred on the oil-rich Niger Delta, where oil exploration and exploitation activities have instigated and/or exacerbated environment-related injustices for about half a century, with the underlying issue being the (in)equitable distribution of 'goods' and 'bads' generated from the oil industry. Obiora's (1991: 477) description of environmental justice: "not simply as an attack against environmental discrimination, but as a movement to rein in and subject corporate and bureaucratic decision-making, as well as relevant market processes, to democratic scrutiny and accountability" adequately captures the Nigerian situation.

Briefly, while revenues in taxes and royalties accrue to the state, the oil companies make profit from exploiting oil. However, the oil producing communities aver, and there is ample evidence that lends credence to their claims, that they bear the varied negative environmental consequences without commensurate benefits accruing to them (Ukeje 2001; Human Rights Watch 1999; Pegg 1999; Osaghae 1995). The above state of affairs is contrary to one of the underlying elements of the environmental justice paradigm in developing countries, which focuses on the need for laws and institutions to address human rights and social impacts of resource utilization on host communities, especially poor and vulnerable groups (Olawuyi 2015b; Oluduro 2014; Ako 2013).

The oil industry's primary stakeholders – the host communities, oil companies and government – have in some way contributed to or even exacerbated the conditions that culminate in environmental injustices. The Federal Government has failed in its primary responsibility of ensuring a legal framework exists to adequately hold the oil industry accountable, mainly due to its joint (and conflicting) role as an operator and regulator of the industry. In addition, the Federal Government has the largest stake in the operation of the oil industry, as it holds majority shares in various joint-venture agreements with oil multinationals and also regulates the industry via the development of legal regimes and institutions to enforce them.

With the Federal Government being the majority shareholder in the joint ventures, it will bear a significant portion of the losses where recorded. Thus it is in the government's best interests that the ventures record profits. This is a fundamental reason why the Federal Government prioritizes the protection of this economic interest over its mandate to ensure the proper regulation of the oil industry. The repercussions of this approach have included environmental pollution and degradation, abuse of human rights and violent conflicts, all of which contribute to the environmental injustices that plague the region. The perceived failure of the oil companies to adopt best practices in their business operations in Nigeria – ranging from implementing environmentally friendly operations to effective delivery of corporate social responsibility objectives – has put them in the category of propagators of environmental injustices in the eyes of their host communities (Ako et al. 2009).

The host communities are themselves not exempt from being implicated in the exacerbation of environmental injustices in the region, as some of their responses to the alleged injustices inflicted on them by the state and oil multinationals have worsened the situation. Briefly, the responses from host communities can be broadly categorized into two: non-violent and violent. The former, initially anchored by community elders and representatives, and later on by non-governmental movements – the latter typically local or community based organizations acting in collaboration with international organizations – rely on negotiations, intellectual publications, protests and litigation, amongst others. The most notable example would be the activities of the Movement for the Survival of the Ogoni People (MOSOP), which reframed resource governance struggles in rights-based language and approaches (Ako 2015; Olawuyi 2015b). In contrast, militant groups such as the Niger Delta People's Volunteer Force (NDPVF) led by Mujahid Dokubo-Asari and the Niger Delta Vigilante (NDV) led by Ateke Tom adopted violent contestations as their *modus operandi* ostensibly due to the failure of the peaceful approach to yield desired results (Ako 2013a; Ikelegbe 2011).

The difference in the approach adopted by the host communities is important to the environmental justice discourse, as the objectives and impacts of the violent approach, with the benefit of hindsight, clearly were not altruistic. Briefly, the MOSOP model that quickly became the *modus operandi* for the entire region, concentrated efforts on mass mobilization and education of the people to appreciate the human rights implications of the oil industry on their environment and lives (Ako 2015). Thus the focus was on both state and corporate accountability for environmental wrongs, the need for a protective and responsive legal and institutional framework to protect the environment and human rights of host communities and effective local participation in the management of the oil industry. In other words, the model was built on popular participation of the affected general public following their intellectual empowerment to appreciate their prevalent circumstances and the need for change.

Militancy, in contrast, became a tool of agitators who argued that the peaceful means of engagement was not yielding palpable benefits for the Niger Delta region (Ikelegbe 2011). Without going into the details of the origin, rise and particular actors of militancy in the Niger Delta region, it has become obvious that these actors rode on the environmental justice bandwagon rhetoric not as a means to emancipate the inhabitants of the Niger Delta, but as a means to achieve personal aggrandizements and individual gratification. Many of the top echelon of the militant groups, leveraging the Federal Government's amnesty programme, became government contractors and apologists. Exploiting the new positions they occupied, they engaged in sharp business practices and made enormous amounts of money illicitly in exchange for 'peace'.

Their foot soldiers reaped the benefits of 'their labour' by way of receiving payment of monthly allowances and study scholarships (vocational and educational) to institutions at home or abroad. Meanwhile, the lives of the supposed beneficiaries of the struggle – the ordinary inhabitants of the region – have not changed significantly, and not positively anyway. The demands made by the militants including the amendment of the relevant regulatory framework, implementation of the Niger Delta Master Plan and environmental clean-up, amongst others, were abandoned as they gave up their arms in exchange for personal emoluments and gratification. If anything, the ordinary inhabitants of the region suffered further debilitation of their socio-economic existence and human rights due to militancy-induced violent conflicts that pervaded the region.

Unfortunately, the inhabitants of the region appear to be getting set for another round of violence as militants are once again threatening and targeting oil installations. This time, the attacks are linked to a former militant leader who is on the run from the judicial system. He has had a bench warrant issued against him and a court order for his arrest, for failing to attend a court summons in respect to forty counts of an alleged 34 billion Naira (approximately US$ 100 million) fraud levelled against him and nine others by the Economic and Financial Crimes Commission (Oladimeji 2016). On reflection, the threat of attacks linked to the embattled former militant leader reinforces the argument put forward in this chapter that militancy, riding on the wave of fighting for the 'environmental justice' of the inhabitants of the Niger Delta region, in fact exclusively (or primarily) benefited its principal actors.

Environmental justice in Nigeria's oil industry: laws and institutions

There is a plethora of laws regulating the oil industry in Nigeria. Without the need to rehash the comprehensive regulatory framework, this section will make a couple of assertions. First is that the legal framework has been identified as a fundamental cause of the violent conflicts that have besieged the oil-rich region for over half a century (Ako 2009, 2011, 2013). Second is that the operators of the oil industry – the state and oil multinationals – have circumvented relevant laws and further aggravated the feelings of environmental injustice held by the region's inhabitants (Emeseh 2006; Ako and Oluduro 2013). Nonetheless, the state has created institutions with the responsibility of promoting elements of environmental justice as understood in the context of developing countries (Olawuyi 2015a).

The last decade has witnessed the enactment of environmental laws and the establishment of institutions that have a bearing on promoting environmental justice in Nigeria, specifically the oil industry. These include the following: National Oil Spill Detection and Response Agency (NOSDRA) (Establishment) Act (Laws of the Federation of Nigeria (LFN) 2006, c N157); Associated Gas Reinjection Act (LFN 2004, c A25); Nigeria Delta Development Commission Act (LFN 2004, c N87); Environmental Impact Assessment Act (LFN 2004, c E12); and National Environmental Standards Regulatory and Enforcement Agency (NESREA) (Establishment) Act (LFN 2004, c N164), amongst others.

However, there has been laxity (sometimes purposely) with regards to the harmonization and implementation of these laws to promote their practical efficacy. For instance, the Associated Gas Re-Injection Act of 1979 has been amended several times to continue to permit oil companies to flare gas following their failures to develop gas utilization projects as envisaged by the Act. Thus in 2016, 32 years after the law was initially passed, several amendments to it legalize gas flaring subject to the payment of affordable fines of $3.50 for every 1000 standard cubic feet of associated gas flared (Ako and Oluduro 2013).

At the institutional level, the government has over the years created a ministry to oversee the environment as well as developmental agencies with a specific focus on the Niger Delta region. The environment ministry now exists at both federal and state levels with broad mandates including: formulating and implementing environmental policies; preparing the necessary action plans for environmental protection; promoting conservation and the sustainable use of natural resources; promoting cooperation in environmental science and conservation technology; and cooperating with agencies, departments, statutory bodies, and research agencies on matters relating to the protection of the environment and the conservation of natural resources. In addition, there are federal agencies – NOSDRA and the Department for Petroleum Resources (DPR) – that contribute to environmental management with a specific focus on the oil industry.

There are, however, serious limitations to these agencies' capacity to regulate the environment as highlighted in the following section. Thus the environmental impacts of the oil industry continue to adversely affect the socio-economic wellbeing and health of Niger Delta communities with inter-generational consequences. With primary economic activities (farming and fishing being the traditional sources of livelihood) negatively affected by the polluted environment, community inhabitants can barely afford medical and educational services. Thus the younger generations are confined to recurrent poverty and poor health despite the rich environmental endowments of the Niger Delta region.

In recognition of the multiple negative impacts the oil industry has on the host communities and the need for a more equitable distribution of the positive benefits, the Federal Government created the Niger Delta Development Commission (NDDC). The primary responsibility of the institution is to speed up the infrastructural development of the Niger Delta by investing funds derived from oil revenues directly in the region. NDDC funding is in addition to the constitutional 13 per cent derivation that is paid to all oil-producing states (S. 162(2) Constitution of the Federal Republic of Nigeria (CFRN) 1999). However, the state governments are not under any legal obligations to invest any portion of these funds in the host communities. That notwithstanding, state governments have created development commissions styled after the NDDC to contribute to the development of the host communities.

Unfortunately, these state commissions, such as the NDDC, are embroiled in the malaise of corruption, embezzlement, mismanagement of funds, nepotism, and alleged underfunding, all contributing to their underperformance (Johnson 2012; Onyeose 2011; Efeizomor 2010). The creation of a full-fledged ministry at the federal level – the Ministry of Niger Delta Affairs – to cater specifically to the region has also not yielded palpable results. In fact, a past minister declared that the region remains bedevilled by massive environmental degradation and huge infrastructure deficit due to financial constraints faced by the ministry (Soriwei 2012).

The summary of the situation of the Niger Delta suggests that the notion of environmental justice is idealistic (Ako 2013). The issues of environmental pollution, oil-related human rights violations, and uneven distribution of oil revenues, amongst others, are not simply consequences of an inadequate regulatory framework *simplicita*. The alliance between the Federal Government and the oil multinationals and the absence of active public participation of host communities in the management of the oil industry are salient factors that further explain the pervasive state of environmental injustice in the Niger Delta region (Ako 2014). The challenge is not in the realization that environmental injustices exist but in fashioning a solution to the problem that has lasted decades with repercussions that have threatened and continue to threaten Nigeria's political unity and socio-economic development.

The protracted injustices in the region persist primarily due to the government's failure to adopt a holistic framework within which it can assess the problems and develop a strategy that recognizes and addresses them. Arguably, numerous ad hoc government inquiry panels and committees on the Niger Delta have been hindered by their limited terms of reference and/or capacity to conceptualize the issues within an environmental justice framework. Thus recommendations from these panels and committees have been made without fully appreciating the wide range of issues that have fed into the underlying cause of violent conflicts that have besieged the region. The creation of multiple (and parallel) development agencies, despite their history of failures, is one indicator of the limited conception of the cause of conflicts in the Niger Delta. Hence, it is imperative to properly understand and (re)conceptualize the problems in the Niger Delta.

At the root of this reconceptualization should be a more expansive understanding of environmental justice in Nigeria that focuses on identifying, addressing, and mitigating human rights impacts of resource development projects that result in poor and vulnerable populations suffering disproportionate adverse impacts of resource development (UNDP 2014). Practically, this approach means that existing resource regimes would be reformed to include elements of active public participation; accountability; equality and non-discrimination; access to information; and access to justice.

Environmental justice in practice: paradoxes and challenges

As noted in the previous section, the conceptualization and understanding of the notion of environmental justice is imperative to fully appreciate the factors that underlie oil-related conflicts in the Niger Delta, and a prerequisite to their long-term resolution. The spasmodic approach to the repercussions of environmental injustice in the Niger Delta, rather than a holistic approach, is responsible for the multiplicity of (uncoordinated) initiatives and institutions that have not yielded overall positive results. This section highlights the drawbacks of the piecemeal methodology that the government has adopted, to further the argument that it is essential that a holistic approach to environmental justice is taken for peace and sustainable development in the oil-rich Niger Delta region.

The first drawback of the fragmented approach is the lack of sustained government commitment to fully and continually engage with the root causes of conflicts in the Niger Delta region – presumed to be primarily environmental and developmental matters. Thus successive governments have embarked on initiatives in accordance with the visions and priorities of the government in power. An example is the succession of development institutions created by successive governments: Niger Delta Development Board (1961–1972); Niger Delta River Basin Development Authority (1972–1983); Oil Minerals Producing Areas Development Commission (1992–2000); Niger Delta Development Commission (2000 to date); and more recently, the Ministry of Niger Delta Affairs (2008 to date). Although all the above bodies were primarily mandated to hasten the physical development of the Niger Delta region, different administrations sought to score political points by creating their own agencies.

Similarly, successive governments instituted inquiry panels and commissions, all to determine the root causes of conflicts in the Niger Delta and to recommend lasting solutions. The recommendations of this plethora of reports have been haphazardly implemented at best, once again raising the question of government consistency. Notably, in 2008, the Federal Government constituted the Technical Committee on the Niger Delta (TCND). The Committee had three key mandates:

- to collate, review, and distil the various reports, suggestions and recommendations on the Niger Delta from the Willinks Commission Report (1958) to the present and give a summary of the recommendations necessary for government action;
- to appraise the summary recommendations and present detailed short, medium and long term suggestions for the challenges in the Niger Delta; and
- to make and present to government any other recommendations that will help the Federal Government achieve sustainable development, peace, human and environmental security in the Niger Delta region (Report of the Technical Committee on the Niger Delta (2008) available online: www.waado.org/nigerdelta/niger_delta_technical_com/ NigerDeltaTechnicalReport.pdf).

However, as with previous reports that this committee collated, key recommendations that relate to environmental justice remain unimplemented. These include, for example, that the government ensure that regulations are made to compel oil companies to have insurance bonds by 2010; strengthen independent regulation of oil pollution and work towards an effective Environmental Impact Assessment mechanism (EIA); make the enforcement of critical environmental laws a national priority; expose fraudulent environmental cleanups of oil spills and prosecute operators; and to end gas flaring by December 31, 2008 as previously ordered by the Federal Government.

Secondly, and linked to the previous point, is the lack of independence of institutions whose remit falls within promoting elements of environmental justice. For example, the Department of Petroleum Resources that has the statutory responsibility of ensuring compliance to petroleum laws, regulations and guidelines in the oil industry is under the supervision of the Nigerian National Petroleum Corporation (NNPC). The NNPC manages the government's interests in the oil industry including joint ventures with oil multinationals. Thus, a paradoxical situation exists wherein a supervisee is mandated to oversee its supervisor. Similarly, it may be argued that the government ministry and agencies that oversee the environment that are under the supervision of the presidency lack the independence to effectively perform their functions, especially where they conflict with the economic interests of the state. This is particularly so with regards to the oil industry, where environmental justice concerns are still considered inimical to optimal revenue.

In essence, the environment ministry and agencies lack the requisite level of independence to serve effectively as 'watchdogs' to monitor and report on the sustainability of government operations and policies vis-à-vis the oil industry. The Canadian model, for example, provides an alternative to the Nigerian system. In Canada, the Commissioner of Environment and

Sustainable Development is independent and has legislative powers to conduct performance audits and to launch independent assessment to determine whether Federal Government departments are meeting their sustainable development objectives, and responding to environmental petitions from the public. Arguably due to the fact that the Commissioner is not a cabinet level member, incumbents are able to pursue their mandates with all independence and to report to the public adequately without undue control and/or fear of reprimand.

This level of institutional independence is imperative if environmental justice is to take a foothold in Nigeria, where the government has been perennially accused of colluding with international oil companies to stifle virtually all the elements of environmental justice (Human Rights Watch 1999). In other words, the architecture of institutions that can deliver on effectively integrating environmental justice into governance, especially in the oil industry, needs to be reassessed and revitalized.

Thirdly, the lack of institutional coordination is a key barrier to the growth of environmental justice in Nigeria. There are over fifty ministries, agencies, departments and other institutions across federal, state and local governments in Nigeria with the mandate to monitor, enforce, and report on environmental protection and the sustainability of government operations and policies. In addition, there are legislative committees at federal and state levels that also play an important role in reviewing and evaluating government policies and programmes, especially as they relate to the tenets of environmental justice. Without a clear institutional framework that highlights a coordinating office, the efforts of the various offices with mandates that implicate environmental justice are often duplicated or left undone.

The consequences of the lack of a coordinated approach to environmental justice, such as lack of conceptual clarity, financial waste and non-delivery of benefits, amongst others, underscore the need for a focal body or office for environmental justice. Such a body will provide holistic evaluation, review and monitoring of a wide range of environmental justice issues ranging from environmental concerns to the enthronement of active public participation, access to justice in environmental matters, participatory development, protection of cultural heritage and the investigation of land grabs.

Conclusion: future prospects for environmental justice architecture in Nigeria

The conceptualization of environmental justice is well underway in Nigeria, with a growing literature, institutions and interest groups that explore its meaning and essence (Ako 2009, 2011, 2013, 2014; Obiora 1991; Olawuyi 2015b). However, the practical awareness, understanding and implementation of the concept are still developing, despite the glaring impacts of environmental injustices occasioned by the oil industry. One of the key reasons for the stunted development of the environmental justice concept in Nigeria, as highlighted in this chapter, is that regulatory and institutional approaches have been largely uncoordinated. Consequently, there is a proliferation of laws and duplicative institutions on environmental justice, with little or no progress in terms of practical results.

It is imperative that environmental justice considerations become integral to governance in Nigeria, especially if its oil industry is to survive. Cuts to oil production occasioned by restive host communities, as is currently being experienced, amidst dwindling oil prices, do not augur well for the national economy. Also, conflict mode has drawbacks for the host communities and the oil industry with severe negative implications for the sustainable development of the region and industry. Thus Nigeria has to better reflect and embed environmental justice in governance by transcending the prevailing haphazard model to a more strategic and autonomous framework that is independently implemented and consistently pursued irrespective of the government in power.

It has been suggested that a body be mandated to focus on the institutionalization of environmental justice in policy-making as well as performance monitoring and evaluation to promote the integration of the elements of the concept into mainstream governance. As a model of governance, the appointment of a focal future generations institute or officer will give a higher profile to environmental justice issues in Nigeria. By taking such a bold step of appointing a distinct environmental justice officer or institution, the Nigerian government will be sending a strong message to business enterprises and stakeholders that resource utilization, investments, planning and development must not jeopardize human rights. Such commitment on the part of government would also encourage government agencies and even private sector corporations to develop internal environmental justice programmes.

The adoption of the United Nations 2030 Agenda for Sustainable Development places further impetus on the Nigerian government to promote environmental justice and the imperative of inclusive governance. Goal 16 that specifically encourages national authorities to ensure responsive, inclusive, participatory and representative decision-making at all levels provides a veritable basis to remove all barriers to the just and equitable management of natural resources in Nigeria. With the 'why' question having been examined and detailed in the literature, focus has to shift to the 'how' question. In essence, 'how' can Goal 16 that promotes the essence of environmental justice be realized in Nigeria's extractive industry (broadly speaking), taking into cognisance – especially from the experience of the oil industry highlighted above – the deep-rooted economic, political, territorial and socio-cultural considerations and implications. Interrogating this critical question will contribute to the unearthing of the fundamental issue of how the elements of environmental justice can be infused into national action plans, policies, initiatives and activities in a holistic manner. This will provide a basis for stakeholders' understanding, information sharing and decision-making, a prerequisite for the sustainable development of environmental resources.

References

Ako, R. (2009). "Nigeria's Land Use Act: An anti-thesis to environmental justice." *Journal of African Law*, vol. 53, pp. 289–304.

Ako, R. (2011). "Resource exploitation and environmental justice in developing countries: The Nigerian experience." In Botchway, F. (ed.), *Natural resource investment and Africa's development*, Cheltenham: Edward Elgar, pp. 72–104.

Ako, R. (2012). "Re-defining corporate social responsibility (CSR) in Nigeria's post-amnesty oil industry." *African Journal of Economic and Management Studies (Special Issue on CSR)* vol. 3, pp. 9–22.

Ako, R. (2013). *Environmental justice in developing countries: Perspectives from Africa and the Asia-Pacific.* Oxon: Routledge.

Ako, R. (2013a). "Militia emancipation or public participation? The quest for sustainable development in the Niger Delta Region." In Etekpe, A. & I. Ibaba (eds.), *Trapped in violence: Niger Delta and the challenges to conflict resolution and peace building*, Port-Harcourt: University of Port-Harcourt Press, pp. 83–112.

Ako, R. (2014). "Environmental justice in Nigeria's oil industry: Recognizing and embracing contemporary legal developments." In Percival, R., J. Lin & W. Piermattei (eds.), *Global environmental law at a crossroads*. Cheltenham: Edward Elgar, pp. 160–176.

Ako, R. (2015). "A lega(l)cy unfulfilled: Reflections of the Wiwa-led MOSOP and the localization of human rights." *The Extractive Industries and Society*, vol. 2, pp. 628–631.

Ako, R. & O. Oluduro (2013). "Bureaucratic rhetoric of climate change in Nigeria: International aspiration versus local realities." In Maes, F., A. Cliquet, W. du Plessis & H. McLeod-Kilmurra (eds.), *Linkages between climate change and biological diversity*, Cheltenham: Edward Elgar, pp. 3–31.

Ako, R. & O. Omiunu (2013). "Amnesty in the Niger Delta: Vertical movement towards self-determination or lateral policy shift?" *The Journal of Sustainable Development Law and Policy*, vol. 1, pp. 86–99.

Ako, R., L. Obokoh & P. Okonmah (2009). "Forging peaceful relationships between oil-companies and host communities in Nigeria's Delta Region: A stakeholder's perspective to corporate social responsibility." *Journal of Enterprising Communities*, vol. 3, pp. 205–216.

Amnesty International (2009). *Nigeria: Petroleum, pollution and poverty in the Niger Delta*. London: Amnesty International Publications.

Beinart, W. & J. McGregor (eds.) (2003). *Social history and African environments*. Oxford: James Currey.

Efeizomor, V. (2010). 'Delta uncovers N8bn fraud in DESOPADEC', This day Live, 3 August 2010, www.thisdaylive.com/articles/delta-uncovers-n8bn-fraud-in-desopadec/81868/ accessed 28 November 2012.

Emeseh, E. (2006). "The limitations of law in promoting synergy between environment and development policies in developing countries: A case study of the petroleum industry in Nigeria." *Journal of Energy and Natural Resources Law*, vol. 24, pp. 574–606.

Emeseh, E. (2011). "The Niger Delta crisis and the question of access to justice." In Obi, C. & S. Rustad (eds.), *Oil and insurgency in the Niger Delta: Managing the complex politics of petro-violence*, London: Zed Books, pp. 55–70.

Human Rights Watch (1999). *The price of oil: Corporate responsibility and human rights violations in Nigeria's oil producing communities*. Washington DC: Human Rights Watch/Africa.

Idemudia, U. (2011). "Corporate social responsibility and the Niger Delta conflict: Issues and prospects." In Obi, C. & S. Rustad (eds.), *Oil and insurgency in the Niger Delta: Managing the complex politics of petro-violence,* London: Zed Books. pp. 167– 183.

Ikelegbe, A. (2011). "Popular and criminal violence as instruments of struggle in the Niger Delta." In Obi, C. & S. Rustad (eds.), *Oil and insurgency in the Niger Delta: Managing the complex politics of petro-violence,* London: Zed Books, pp. 125–135.

Johnson, D. (2012). "EFCC arrests OSOPADEC boss." *Vanguard* (Nigeria) newspaper 21 March 2012, www.vanguardngr.com/2012/03/efcc-arrests-osopadec-boss/ accessed 13 June 2012.

Obi, C. (2009). "Nigeria's Niger Delta: Understanding the complex drivers of violent oil-related conflict." *Africa Development*, vol. 34, pp. 103–128.

Obi, C. (2014). "Oil and conflict in Nigeria's Niger Delta region: Between the barrel and the trigger." *The Extractive Industries and Society*, vol. 1, pp. 147–153.

Obiora, L. (1991). "Symbolic episodes in the quest for environmental justice." *Human Rights Quarterly*, vol. 21, pp. 464–512.

Oladimeji, R. (2016). 'N34bn fraud: Tompolo asks court to vacate arrest warrant', *The Punch* (Nigeria) newspaper 1 February 2016, www.punchng.com/n34bn-fraud-tompolo-asks-court-to-vacate-arrest-warrant/ accessed 18 February 2016.

Olawuyi, D. (2015a). *Principles of Nigerian environmental law*. Ado-Ekiti: Afe Babalola University Press.

Olawuyi, D. (2015b). "The emergence of right-based approaches to resource governance in Africa: False start or new dawn?" *Journal of Sustainable Development Law and Policy*, vol. 16, pp. 15–28.

Oluduro, O. (2014). "Oil exploitation and human rights violations in Nigeria's oil producing communities." *Afrika Focus*, vol. 25, pp. 160–166.

Onyeose, C. (2011). "GEJ sacks NDDC board: President Jonathan sacks the board of the NDDC over large-scale fraud", *Daily Times* (Nigeria) 14 September 2011, http://dailytimes.com.ng/article/gej-sacks-nddc-board accessed 21 March 2012.

Osaghae, E. (1995). "The Ogoni uprising: Oil politics, minority agitation and the future of the Nigerian state." *African Affairs*, vol. 94, pp. 325–344.

Pegg, S. (1999). "The cost of doing business: Transnational corporations and violence in Nigeria." *Security Dialogue*, vol. 30, pp. 473–484.

Soriwei, F. (2012). 'Niger Delta ministry: Postponing real development', *Punch* (Nigeria) newspaper, 3 November 2012, www.punchng.com/politics/niger-delta-ministry-postponing-real-development/ accessed 10 December 2012.

Ukeje, C. (2001). "Oil communities and political violence: The case of ethnic Ijaws in Nigeria's Delta region." *Terrorism and Political Violence*, vol. 13, pp. 15–36.

United Nations Development Program (UNDP) (2014). *Environmental justice: Comparative experiences in legal empowerment* (UNDP 2014), www.undp.org/content/dam/undp/library/Democratic%20Governance/Access%20to%20Justice%20and%20Rule%20of%20Law/Environmental-Justice-Comparative-Experiences.pdf accessed 1 March 2016.

Uwafiokun, I. (2011). "Corporate social responsibility and the Niger Delta conflict: Issues and prospects." In Obi, C. & S. Rustad (eds.), *Oil and insurgency in the Niger Delta: Managing the complex politics of petro-violence*, London: Zed Books, pp. 167–183.

Watts, M. (2012). "Sweet and sour: The curse of oil in the Niger Delta." In Butler, T., D. Lerch and G. Wuerthner (eds.), *The energy reader: Overdevelopment and the delusion of endless growth*, Healdsburg, CA: Watershed Media, Chapter 28.

World Bank (2015) Nigeria overview, www.worldbank.org/en/country/nigeria/overview accessed 12 February 2016.

46

SUB-IMPERIAL ECOSYSTEM MANAGEMENT IN AFRICA

Continental implications of South African environmental injustices

Patrick Bond

Introduction

The South African government was democratized in 1994, but since then has not responded effectively to either inherited or new environmental injustices (Cock 2016; Satgar 2016; Bond 2016a). The resulting footprint of ecological destruction reaches thousands of kilometers north into the African continent. Both Pretoria-based politics and Johannesburg-based economics are responsible for such extreme damage to the continent's environmental sustainability, as to warrant the label 'sub-imperialist.' As the theory of sub-imperialism (Marini 1965; Harvey 2003; Bond and Garcia 2015) would suggest, this occurs in at least three ways:

- first, South Africa's long settler-colonial traditions of facilitating ultra-exploitative ecosystems management as promoted by imperial powers, namely the avaricious use of free environmental space for the purpose of *externalizing pollution* (i.e. without paying environmental and social liabilities especially where these have racial, gender and class bias);
- second, South Africa's homegrown-neoliberal systematization of that power through *intensified ecological modernization*, in the form of supposedly-corrective 'Green Economy' governance strategies that commodify nature so as to save it; and
- third, the role of South African corporations in amplifying capital-nature exploitation in other African settings, aided by Pretoria's diplomatic, financial and military support.

The era of neoliberalism since 1994 affected both global and local elite strategies for environmental governance: 'market solutions for market problems.' As this strategy became generalized in the early twenty-first century thanks to the United Nations, the World Bank and associated institutions, the metabolism of capital-nature relations intensified. This was partly a function of the rhythms of capitalist crisis and global uneven development – most obviously China's role in rapidly shifting both capital accumulation and pollution from West to East – and specifically the 2002–11 commodity super-cycle.

But the minerals and petroleum upturn and subsequent 2011–15 crash left a devastating impact upon African economies and environments. The incentive for firms to produce higher volumes of commodities was one driving force before 2011, but also as the price fell, many firms' shareholders demanded even higher volumes of output to make up for lower prices, which in those instances intensified the already extreme metabolism of extraction. With the heightened metabolism, researchers began noticing upturns in social protests across Africa (ACLED 2016; African Development Bank 2016), as described in the conclusion.

The combination of neoliberal economic policies (imposed since the early 1980s) and resource-extractive dependency (especially since 2002) left Africa much more exposed than it should have been, by the time of the commodity price downturn. Many minerals and petroleum fell more than 50 per cent in price after 2011, with the most dramatic declines during 2015. Once-powerful multinational corporations dependent upon commodities lost more than 85 per cent of their London Stock Exchange share value from their 2009–11 peak to 2015–16 trough, including Glencore (87 per cent), Anglo American (94 per cent) and Lonmin (99 per cent). By mid-2016 the World Bank had downgraded the continent's overall annual GDP growth from the 2000–10 period's 5.4 per cent average to the 2010–15 period's 3.3 per cent average to just 1.6 per cent for 2016, the lowest in two decades.

Such over-exposure not only reflected the Bretton Woods Institutions' policy power, with its export-oriented dogmatism. It was also a function of Pretoria's sub-imperial location as the legitimator and often amplifier of imperial power on the continent. For example, the 2001 New Partnership for Africa's Development (NEPAD) which Pretoria – joined by leaders from Nigeria, Algeria and Senegal – pushed into the African Union, was described by the US State Department as "philosophically spot on" because it relegitimized orthodoxy (Gopinath 2003; Bond 2005). NEPAD soon housed the Program for Infrastructure Development in Africa, a trillion-dollar strategy mainly aimed at providing roads, railroads, pipelines and bridges that, like the colonial era, largely emanate from mines, oil/gas rigs and plantations, and are mainly directed towards ports. Electricity generation is overwhelmingly biased towards multinational corporate mining and smelting needs.

Simultaneously, environmental injustices associated with helter-skelter extraction of minerals (see Chapter 30) and petroleum still worsen under imperial control, with the increasing dimension of sub-imperial legitimation. The 'Africa Rising' myth followed logically (e.g. Perry 2012), as the continent's much higher twenty-first century Gross Domestic Project (GDP) initially appeared to justify the renewed export orientation and overall commitment to liberalization (no matter how misleading the GDP data).

Part of the critique of sub-imperialism is Pretoria's role in multilateral processes that are objectively unjust in relation to Africa, and that are becoming more so over time. In addition to a variety of world and regional economic summits, since 1994 South Africa hosted:

- the 1998–2001 World Commission on Dams, which two leading South African water experts subsequently fatally undermined (Briscoe 2010; Muller 2014);
- the 2002 World Summit on Sustainable Development, which generalized the practice known as 'neoliberal nature' (Bond 2002);
- the 2011 United Nations Framework Convention on Climate Change (UNFCCC) climate conference, which was celebrated by the US State Department (Stern 2011) for ending core justice principles (Bond 2012); and
- the 2016 Convention on International Trade in Endangered Species of Wild Fauna and Flora (CITES), where South Africa joined a few owners of large stockpiles of rhino horn and elephant ivory to attempt to overturn trade bans, against a huge majority of countries which wanted them strengthened (Lunstrum and Bond 2016).

Most importantly, in Paris at the UNFCCC's decisive 2015 summit, a South African chaired the G77 delegation, arguing there that "climate apartheid" required a strong global response, and yet the South African government celebrated the outcome notwithstanding such fatal deficiencies that the world's leading climate scientist, James Hansen, labelled the Paris outcome "bullshit" (Bond 2016b). At the same moment, destructive sub-imperial power over Africa was reflected in South Africa's role in other controversial multilateral bodies, such as the World Trade Organization's December 2015 Nairobi summit (which decisively attacked poor countries' food sovereignty) and the 2010–15 restructuring of International Monetary Fund (IMF) voting power, leaving Africa profoundly disempowered (e.g. Nigeria losing 41 per cent of its share). Pretoria represented Africa in the Brazil-Russia-India-China-South Africa (BRICS) network, which aimed to have the latter play a 'gateway' function in Africa for capital's benefit (Bond and Garcia 2015).

In a context of global governance that is so adverse to Africa's interests, it is not surprising that South Africa's micro-economic, socio-political and environmental interventions are also contributing to the continent's problems, as documented in the next section.

South African sub-imperialism, from corporate to military to financial

The framing for environmental injustices that best reflects the South African economy's and state's power in Africa is the theory of sub-imperialism (Bond and Garcia 2015). Prior to 1994, there were three modes of apartheid's 'constellation of states' in the region that exemplified this power:

- the pull of inexpensive male migrant labour from regional sites to the South African mines, fields and factories, dating to the late nineteenth century;
- the extension of South African mining conglomerates up-continent, starting with Rhodes' British South African Company in 1890; and
- the security capability of the Pretoria regime, especially in fighting border wars (Angola, Namibia and Mozambique) that left millions dead in the course of attacks against African National Congress (ANC) guerrilla operations in neighbouring countries, especially during the era of armed, decolonization struggles in Southern Africa from the 1960s to 1980s.

But after 1994, instead of offering cessation and reparations to the region, the new ANC rulers in Pretoria mainly amplified existing sub-imperial power relations. Legalization of (apartheid-era) resident migrant workers' status in South Africa went hand-in-hand with wider-scale crises in the sub-region that brought in millions more economic refugees from Malawi, Mozambique, Zambia and Zimbabwe, and political refugees from the eastern Democratic Republic of the Congo, Burundi, Rwanda, Somalia, Sudan and Zimbabwe. Together, one impact was a lowering of the cost of unskilled and semi-skilled labour power which, together with intensified internecine competition amongst poor people in township retail and housing markets, generated xenophobic tendencies amongst South Africans.

South African over-accumulation processes date to the 1970s, but the early-1990s capitalist crisis helped facilitate a political power transfer that included the continental legitimation of Johannesburg and Cape Town corporations (Bond 2014). They began to move much more decisively up-continent following the apartheid-era investment drought, at a time when most whites were shunned in the rest of Africa. After 1994, social, economic and environmental change followed new waves of South African retailers, financiers, cellphone operators, infrastructure construction firms, tourism companies and mining houses. The latter sector was often characterized by extreme exploitation of nature and society:

- At the eastern DRC's Mongbwalu mine, Johannesburg-based AngloGoldAshanti closely collaborated – in a context of approximately six million civilian deaths and extreme environmental damage – with notorious warlords (the Nationalist and Integrationist Forces), and when criticized by Human Rights Watch (2005) in *The Curse of Gold*, the firm's CEO Bobby Godsell (2005) remarked, "Our central purpose is to find and mine gold profitably ... [but] mistakes will be made."
- The Johannesburg mining house Mvelaphanda and associates (e.g. Mark Willcox and Walter Hennig) of a leading ANC politician, Tokyo Sexwale, were implicated during the US Securities and Exchange Commission's prosecution of a major ($39 billion) New York financier, Och-Ziff, which in 2016 pled guilty to mining- or petroleum-related bribery of state officials from Libya, the DRC, Chad, Niger, Guinea and Zimbabwe (AmaBhungane 2016).
- The diamond mining house De Beers was charged by two academics, Khadija Sharife and Sarah Bracking (2016), with systematic misinvoicing of more than $2.8 billion in diamonds over a seven year period, including batches from Botswana and Namibia.
- A year after South African President Jacob Zuma took power, his nephew Khulubuse and lawyer Michael Hulley gained access – thanks to assistance from Zuma's ally DRC President Joseph Kabila and the Israeli 'blood diamonds' tycoon Dan Gertler – to a $10 billion oil concession at Lake Albert that originally had been controlled by Ireland's Tullow (Congoleaks 2011).

These were just a few of the highest-profile adverse incidents involving South African mining corporations in Africa. In many other cases associated with resource extraction from Africa, Illicit Financial Flows ('IFFs') were specifically national in character. An example is the documented tax avoidance strategies of Lonmin, Angloplats and Impala Platinum, discovered in 2014 by labour-aligned researchers during the five-month mineworkers' strike, but similar allegations were made against Implats' Zimbabwe subsidiary. More generally, such IFF profits are mainly drawn from minerals and oil ripped from the African soil. In 2015 Global Financial Integrity measured the IFFs alone from 2004 to 2013 as costing $21 billion per year in South Africa and $18 billion in Nigeria, while Sub-Saharan Africa as a whole lost 6.1 per cent of GDP annually to IFFs, more than 50 per cent higher than the rate for poor countries in general (Kar and Spanjers 2015: 8–9, 23).

These environmentally unjust outflows are not merely the result of accounting gimmicks; they are enforced by the sheer might of regional military power. Pretoria's security capacity was tested in the case noted above involving Khulubuse Zuma's eastern DRC investment, worth $10 billion (although in 2016 he claimed to be near penniless when faced with repayment of overdue salaries to his bankrupted Aurora mining house workers). In 2013 at Jacob Zuma's request, the South African National Defence Force (SANDF) deployed more than 1300 'peace-keeping' troops near his nephew Khulubuse's oil stake. A variety of scandals beset the force there, including wild drunken (and sexual) rampages by the SA troops, who also ignored a 2016 massacre by warlords just a kilometre from the SANDF base (Allison 2016).

Pretoria's military commitment to defending mining and petroleum interests in the DRC echoed the earlier deployment (by Presidents Thabo Mbeki and Zuma) of hundreds of SANDF troops in Bangui, Central African Republic (CAR). That decision followed a 2006 deal for diamond market monopoly control signed by Mbeki and the CAR dictator François

Bozizé. By early 2013, socio-political and ethnic tensions in CAR had become extreme, as the French government and military switched support from Bozizé to opposition rebels. But as Bozizé was ousted by a coup during a March 2013 Bangui firefight – a few days before the BRICS had their Durban heads-of-state summit – there were 15 SANDF fatalities. Some SANDF survivors were furious with their role as mercenaries; they told senior Johannesburg reporters (Hosken and Mahlangu 2013): "Our men were deployed to various parts of the city, protecting belongings of South Africans. They were the first to be attacked . . . outside the different buildings – the ones which belong to businesses in Jo'burg."

Lesotho's attempted army coup in mid-1998 was another example, as the SANDF killed two dozen Basotho soldiers at the wall of the Katse Dam that supplies Johannesburg and Pretoria with water, fearing the rebels would blow up the dam (Bond 2002). Added to South Africa's periodic military incursions up-continent, these political, economic and environmental influences contribute to a sense that leading officials in Pretoria amplify rather than reduce the continent's degradation.

This is especially true in light of Pretoria's cooperation with the US military's Africa Command (AFRICOM), which is active in dozens of countries. AFRICOM bears testimony to Washington's overlapping desire to maintain control over core African conflict sites amidst rising Islamic fundamentalism from the Sahel to Kenya, which are, coincidentally, theatres of war in the vicinity of large petroleum reserves (Turse 2014). US Air Force strategist Shawn Cochran (2010: 111) relayed the words of a 'US military advisor to the African Union' whom he interviewed in 2009, "We don't want to see our guys going in and getting whacked . . . . We want Africans to go in." Jacob Zuma (2014) was content with this arrangement, for at a 2014 African leaders' summit in Washington, he announced, "As President Obama said, the boots must be African."

Financing is another area where sub-imperial relationships would be important. The BRICS New Development Bank entered into a collaborative arrangement with the World Bank in 2016, and its much-anticipated launch of a regional centre in South Africa soon afterwards had a precedent: financing by the Development Bank of Southern Africa (DBSA). In 2013–15 the DBSA was given an additional $2 billion in working capital in large part to pave the way into the sub-continent. Many major DBSA corporate beneficiaries in Johannesburg and Cape Town have played a predatory role in African sites of extreme environmental injustice, and by 2016, 14 per cent of DBSA assets were in Africa (outside South Africa), with plans for an additional $400 million in semi-privatized infrastructure investment.

However, profound contradictions soon emerged. The main DBSA regional lender was Mo Shaik, the former head of state intelligence and brother of a man convicted of bribing Jacob Zuma in the late-1990s arms deal, and of another state official most responsible for that notorious $5 billion deal. By 2015, Shaik was disarmingly open about the limitations of his brief. In a talk to the main strategic leadership seminar held at South Africa's foreign ministry (at which this author took notes), he conceded that, in the rest of the continent, he sometimes "felt like an economic hit man . . . [selling] projects they don't need and they can never pay for, but my job is to sell them these projects." He named the largest proposed infrastructure work on earth, the Congo River's Inga Hydropower Project ("who would invest $80 billion in the DRC?") and other random development finance disasters from the Sahel to Southern Africa (Bond 2016c).

Given the unreliability of African conditions for multinational corporate investment and development finance, the DBSA and BRICS New Development Bank will stumble to fulfil the sub-imperial function. They will require much closer cooperation with corporations from Brazil, Russia, China and India. The BRIC bloc had welcomed South Africa in 2010, and in 2013 in Durban, the heads of state planned more of the 'gateway' role that the New Development Bank might play in Africa (Bond 2016c). In this and other settings, South Africa often attempts to 'speak for' Africa in multilateral strategies; e.g., it is the only African state in the G20. But this means in several crucial environmental management functions – climate change, mining, commercial agriculture, wildlife, timber and water (ranging from oceans to household sanitation) – South Africa exhibits power relations based on state policy and regulatory processes biased towards pro-corporate standpoints.

South Africa's pollution and natural capital depletion

South Africa's own economy – which since the early 2000s has been increasingly directed by transnational corporations and international commodities and financial markets – generates exceptionally high levels of greenhouse gas emissions, minerals-related toxics, land and bulk water degradation, solid waste, and maritime pollution. The results include myriad injustices in the way climate change, air pollution, declining land capacity, and household water constraints affect low-income black people. A new 'Blue Economy' oceans management strategy features oil and gas exploration, beachfront tourism and commercial mariculture (in contrast to subsistence fishing), and will exacerbate environmental injustice locally but potentially also in neighbouring sites.

The spillover impacts of all these policies and practices on regional African and even global ecologies is increasingly obvious, especially in terms of climate change, trans-boundary rivers, commodified and genetically-modified food provision, and oceans. Indeed, since the end of apartheid in 1994, various aspects of South African ecosystems management have degenerated: water and soil quality, greenhouse gas contributions to global warming, fisheries and other aspects of the maritime environment, industrial toxics, genetic modification, Acid Mine Drainage, fracking and offshore oil and gas drilling. The South African Government Communications and Information Service (2007) conceded that in the state's most thorough self-assessment, a dozen years after liberation, there was "a general decline in the state of the environment."

There were many ways in which this localized ecological destruction was experienced regionally, in Africa. As a prime example of an overly large carbon footprint, the threat of water scarcity and water table pollution worsened dramatically after 1994 in the country's main megalopolis, Gauteng. With 12 million residents stretching from the Vaal steel and petro-chemical complex through Johannesburg to northern Pretoria's sprawling townships, Africa's largest industrial and commercial complex required new mega-dams known as the Lesotho Highlands Water Project. The first two, Katse and Mohale, were built during the late 1990s, and featured the world's highest-profile construction-corporation corruption cases, the (uncompensated) displacement of Indigenous people, loss of scarce fertile farmland due to dam impounding, and other destructive environmental consequences downriver.

Moreover, the extremely high costs of Lesotho water transfer across the mountains (five times the existing marginal price) deterred consumption by the poorest people in Gauteng townships, since they paid a disproportionate share (see also Chapter 27). One result was an upsurge of social protest and Africa's main early-2000s "water war," between Soweto residents and their municipal supplier (outsourced to a Paris water company, Suez, whose construction subsidiary was one of the firms prosecuted for corruption in Lesotho). This followed higher prices and a commercialized system of water delivery via pre-payment meters and water-limiting sanitation. The wealthiest urban (mainly white) families continued to enjoy swimming pools and English gardens in Johannesburg and Pretoria, which meant that in some of the most hedonistic suburbs water consumption was 30 times greater each day than in low-income townships (some of whose residents continue doing gardening and domestic work for whites) (Bond 2002).

In spite of market-based management strategies in water catchment areas, there remains a growing need for state command-and-control systems (e.g. water rationing) to shift systems associated especially with commercial agricultural irrigation, mining degradation of the water table, the cooling of coal-fired power plants, commercial timber's impact on groundwater, as well as a few million high-income urban South Africans' hedonistic lifestyles in even Lesotho-supplied Johannesburg. Neither traditional Riparian rights (water drawn from owned land) nor market-centric water pricing strategies have proven effective in incentivizing change. Nor has a tokenistic "Free Basic Water" strategy adopted nationally in 2001 supplied sufficient water to poor people, as witnessed by ubiquitous illegal water connections across the low-income areas of South Africa.

There are other examples of wide-ranging environmental injustices within South Africa, including numerous unresolved conflicts over access to energy (highly class segregated), natural land reserves (where in mineral-rich sites, displacement of Indigenous people continues), deleterious impacts of economic activity on biodiversity, insufficient protection of endangered species, and state policies favouring genetic modification for commercial agriculture. Marine regulatory systems have become overstressed and hotly contested by European and East Asian fishing trawlers, as well as by local medium-scale commercial fishing firms fending off new waves of small-scale black rivals. Expansion of gum and pine timber plantations, largely for pulp exports to East Asia, remained extremely damaging, not only because of grassland and organic forest destruction – leading to soil adulteration and far worse flood damage downriver, such as Mozambique suffered in 2000–1 – but also due to the spread of alien invasive plants into water catchments across the country. A constructive state programme, "Working for Water", slowed but did not reverse their growth.

Thanks to accommodating state policies, South African commercial agriculture remained extremely reliant upon fertilizers and pesticides, with Genetically Modified (GM) crops increasing across the food chain while virtually no attention was given to expanding organic farming. The huge US agri-corporation Monsanto used Makhatini Flats in KwaZulu-Natal as a high-profile 'success story' to draw small (black) farmers into GMO cotton after 1999, but by 2005, the strategy had become an embarrassment. Monsanto and the Gates Foundation advanced pro-GM advocacy to the rest of the continent from South Africa, which proved an easy site to establish the strategy, given the preponderance of medium- to large-scale farming (by white landholders and corporations). Tellingly, although the Monsanto Bt maize variety MON801 failed to halt pest resistance in South Africa it was marketed elsewhere in Africa through the Insect Resistant Maize for Africa strategy (African Centre for Bio-Safety 2015).

The South African government's failure to prevent toxic dumping and incineration (see Chapter 25) led to a nascent but portentous group of mass tort (class action) lawsuits. The victims included asbestos and silicosis sufferers who worked in or lived close to the country's mines, and by 2016 they were on the verge of winning tens of millions of dollars in damages. Other legal avenues and social activism were pursued by residents who suffered persistent pollution in extremely toxic pockets such as South Durban and, just south of Johannesburg, the industrial sites of Sasolburg, Secunda (the world's single biggest emission site for CO_2) and Steel Valley. In these efforts, the environmental justice movement almost invariably fought both corporations and Pretoria. From 1994, progressive-sounding environmental statements were regularly made by officials, yet in reality they regularly downplayed ecological crimes (a Green Scorpions anti-pollution team did finally emerge but with subdued powers that barely pricked). Exemplifying Pretoria's approach was the 2015 decision to avoid enforcement of an Air Quality Act on grounds that the largest polluting corporations required further exemptions.

The explanation for Pretoria's lack of concern about environmental injustices at home and abroad is obvious: a national economic structure in which wealth and political influence remained vested in the so-called Minerals-Energy Complex. One result, obvious to even the World Bank by 2000, was the way in which South Africa's reliance upon non-renewable resource extraction gave the country a *net negative per capita income*, once adjustment to standard GDP is made. The typical calculation of 'growth' adds GDP for the extraction and sale of minerals, but does not take into account pollution or depletion of mineral wealth (i.e. the transaction only receives a credit, not a debit in spite of the mineral's depletion).

The Gaborone Declaration (2012) – signed by officials from Botswana, Gabon, Ghana, Kenya, Liberia, Mozambique, Namibia, Rwanda, South Africa and Tanzania – recognized "the limitations that GDP has as a measure of well-being and sustainable growth." The signatories committed to "integrating the value of natural capital into national accounting and corporate planning." But as such corrections begun to be made, it became apparent that South Africa (and most of Africa) suffered a net *dis*accumulation of wealth. The World Bank's (2011) *Changing Wealth of Nations* calculated a 25 per cent drop in South Africa's natural capital and by 2008, according to the Bank (2011), the average South African was losing $245 per person per year. Although methodologies are subject to debate, the overall message is fairly straightforward, namely that even relatively industrialized South Africa had become overly dependent upon natural resources. The more platinum, gold, coal and other metals are dug from the soil, the poorer South Africa becomes. Many other African countries had far worse ratios of net to gross wealth.

Climate injustice in South Africa

The extraction, transport and smelting of minerals and metals remain South Africa's major contributor to climate change (see Chapter 29), mainly because of the extremely high amounts of electricity required. Most of the three dozen corporations comprising South Africa's Energy Intensive Users Group (EIUG) are the largest mining and smelting firms, which together consume 44 per cent of the electricity supplied by Eskom, a parastatal which relied 93 per cent upon coal-fired generation during the early twenty-first century. Although South Africa has the world's third greatest potential for harnessing solar capacity, and although a rapid roll-out of private-supplied renewable energy – wind and solar – raised the proportion of the country's grid to more than 10 per cent from 2010 to 2016, Eskom's chief executive Brian Molefe announced in 2016 that he would no longer purchase further amounts. In search of more 'base load' (for the EIUG's benefit) he would instead begin a nuclear energy procurement process that would in the short term cost the company at least $10 billion (along with anticipated corruption, not to mention longer-term cost estimates of $100 billion).

Yet aside from long-delayed maintenance, the country's short-term grid expansion was unnecessary in the period after 2015 due to the severity of the commodity price crash. South Africans had feared a lengthy period of potential electricity black-outs ('load-shedding'), a phenomenon that began in 2006 because Eskom had delayed new construction following a 1980s–90s oversupply of as much as one third (in turn premised on a high gold price follow-ing a 1979–81 blip). At the time, excessive demand from EIUG mining and smelting firms was apparent, continuing after the 2009 economic crisis until the commodity super-cycle peaked in 2011.

This is not an unusual configuration in 'resource-cursed' Africa, where vast amounts of electricity are delivered via high-tension cables to multinational corporate mining houses for the sake of extraction and capital-intensive smelting (McDonald 2008). Meanwhile, below those wires, most African women slave over fires to cook and heat households: their main energy source is a fragile woodlot, their transmission system is their (and often a child's) back, and their energy consumption is often done while coughing thanks to dense particulates in the air. The transition from HIV-positive status to full-blown AIDS is just one opportunistic respiratory infection away, again with gendered implications for care-giving. In addition to these hazards, in 2006 Christian Aid (2006) estimated that 182 million Africans were at risk of premature death due to climate change this century.

Amongst the conventional neoliberal strategies to address climate change is carbon trading, especially the Clean Development Mechanism (Bond 2012). Moreover, also in the interests of 'internalizing externalities' (such as pollution), South Africa is gradually implementing a carbon tax, but one far lower than what is required to switch high-carbon economic activities towards post-carbon trajectories. In 2014, the Treasury's proposed tax was cut from the equivalent of a $5.50/tonne to $14.90/tonne range which "would be both feasible and appropriate to achieve the desired behavioural changes and emissions reduction targets," to much lower levels: "When the tax-free threshold and additional relief are taken into account, the effective tax rate will range between $0.89 and $3.55 per ton of CO_2e (and zero for Agriculture and Waste)" (South African Treasury 2014). And even more beneficial to corporations, "one of the ways to recycle the expected carbon tax revenue is by reducing other taxes. One such tax that could be reduced is the existing electricity levy on electricity produced from non-renewable sources (e.g. coal) and nuclear energy."

By 2015, in spite of leadership of the G77 at the UNFCCC Paris summit, Pretoria's delegates defended a (voluntary) Intended Nationally Determined Contribution that was deemed by the NGO Climate Action Tracker (2015) to be 'inadequate' (the lowest level of ambition). At the time, in spite of its 2009 announcement of post-2020 greenhouse gas emission cuts of 34 per cent below "growth without constraints" (an imprecise concept), the South African state was:

- building two of the world's four largest coal-fired powerplants for an estimated $20 billion at Kusile and Medupi, with a third planned, each anticipated to add 35 million tonnes of CO_2 to the air;
- planning the export of 18 billion tonnes of coal from the northeastern South African provinces via the Richards Bay terminal, following $20 billion in new rail line expansion (the single highest priority infrastructure project in the country's National Development Plan);
- preparing for an $18 billion port-petrochemical complex expansion in South Durban (which in 2016 was delayed until 2030 though it remained the second highest priority infrastructure project) and a new $6 billion heavy-oil refinery in another port city, Port Elizabeth;
- offering shale-gas fracking exploration rights to South African, Norwegian and US firms in the fragile Drakensburg mountain range and arid Karoo region;
- hastily supporting coal mines in ecologically-sensitive Mpumalanga province (the most fertile land and a tourist zone), even ignoring water licensing requirements, as well as a controversial coal mine on the border of the ancient Mapungubwe national heritage site;
- investing in Carbon Capture and Storage technology, which aims to compress carbon dioxide from the petro-chemical and energy complex into potentially unstable

underground storage sites, even though its boosters were rapidly retreating from Norwegian and US pilot projects, and in spite of the fact that it violates the Precautionary Principle, increases energy to produce power by 25 per cent, is an unproven technology, is at least a decade away from implementation, and prolongs the extraction of coal;

- building or refurbishing ten World Cup stadiums and revising local economic strategies towards sports tourism in spite of the universally acknowledged 'white elephant' character of these investments;
- doubling the capacity of the Durban–Johannesburg oil pipeline, including its redirection from the traditional route (through white suburbs) so that black peri-urban residential areas suffer pipeline breakage risks, at a cost which was ultimately at nearly triple the original estimates;
- subsidising the national airliner with more than $100 million annually as it proved incapable of turning a profit; and
- approving offshore oil and gas exploration drilling prospects to ExxonMobil, Norway's Statsoil and other controversial firms in the dangerous Agulhas Current (near Durban) as part of re-envisaging South Africa's 3000 km-coastline share of the Indian and Atlantic Oceans through the 'Operation Phakisa' Blue Economy strategy.

These and other features of capital accumulation and state infrastructure mega-project construction have far-reaching implications for African climate and water management. Blowback can be expected, for once the desertification of neighbouring states, food crop failure and sustained droughts – or sometimes extreme floods as were witnessed in 2000–01 in Mozambique – hit the continent with full force, often combined with localized warfare, the result will be a profusion of 'climate refugees': migrants who cross the long South African border in a.desperate search for livelihoods. One facet of this blowback is the likely amplification of local working-class xenophobia, which already due to a post-apartheid influx of political and economic refugees, occurred in three major upsurges that left dozens dead and thousands displaced in 2008, 2010 and 2015. These are the indirect aspects of South Africa's prolific environmental injustices, amongst many more direct aspects of sub-imperialism.

Conclusion

The effects of South African sub-imperial economic, financial, infrastructural and military activity – especially in the interests of the extractive industries – include severe pollution and environmental degradation. What can be done by way of mitigation? There are usually two routes that open up once the damage is recognized: managerial 'ecological modernization' strategies strongly premised on market rationality; and environmental justice campaigning. The countervailing strategies for environmental protection favoured by global, by South African and by most African national leaders are typically based upon market principles (with some important exceptions such as banning trade in elephant ivory and rhino horns, in which African countries defeated CITES proposals by Swaziland, South Africa, Namibia and Zimbabwe in 2016). But to date, like carbon trading, they do not appear to be effective (Bond 2012).

The one area of market-related valuation of ecosystems in which implementation might lead to economic transformation is natural capital accounting, but attempts to adjust GDP calculations in this manner – e.g. the Gaborone Declaration of 2012 – were stymied, with no progress to report four years later. As noted above, South Africa recorded a net negative wealth once non-renewable resource depletion is considered, and for the rest of Africa, the situation is even worse. The World Bank's (2014) *Little Green Data Book* recorded 88 per cent of Sub-Saharan African countries suffering net negative wealth accumulation in 2010, because a net 12 per cent of GDP was lost once this adjustment is made. This suggests an indisputable *economic* case for leaving minerals and petroleum underground, until Africa's resource curse is ended.

What can be done to halt the uncompensated depletion of wealth, to address climate change properly (e.g. with systematic demands for 'climate debt' reparations to be paid to African climate victims) and to prevent South Africa and its BRICS allies from adopting explicitly sub-imperial accumulation strategies? These tendencies might be dampened by global capitalist crisis, but the ebb and flow of accumulation often has the effect of intensifying extraction: as commodity prices fall, increased volume is demanded by shareholders who insist on steady net revenue.

On the other hand, there is always the bottom-up factor: class and eco-social struggle. During this stage of increasingly intense extractivism, there have never been more recorded African protests. The African Development Bank (AfDB) commissions measurements based upon journalistic data, and these suggest that major public protests rose from an index level of 100 in 2000 to nearly 450 in 2011. Instead of falling back after the Arab Spring – especially acute in Tunisia, Egypt and Morocco – the index of protests rose higher still, to 520 in 2012, as Algeria, Angola, Burkina Faso, Chad, Gabon, Morocco, Nigeria, South Africa and Uganda maintained the momentum of 2011 (African Development Bank 2013). In 2013, the index rose still higher, to 550 (African Development Bank 2014). In 2014 it fell back just slightly, but as in the earlier years, the main causes of protest were socio-economic injustices (African Development Bank 2015). For 2015, the Sussex University Armed Conflict Location and Event Data Project (2016) found even higher levels of dissent. There are all manner of reasons, but according to Agence France Press and Reuters reports, the vast majority since 2011 were over inadequate wages and working conditions, low quality of public service delivery, social divides, state repression and lack of political reform (African Development Bank 2015: xvi). But a fair share of the turmoil in Africa prior to the 2011 upsurge took place in the vicinity of mines and mineral wealth (ACLED 2016). As the super-cycle is now definitively over and as corporate investment more frantically loots the continent (as argued below), the contradictions may well lead to more socio-political explosions.

The scale of the problems that have emerged, and worse problems that lie ahead, does not only offering sobering considerations about Africa's adverse power relations and the limits of existing managerialist strategies. When viewed from below, they also allow for optimism that once society begins to recognize the threats, a more visionary approach can be established by some of the continent's leading environmental justice organizations. That in turn is likely to entail direct confrontation with South Africa and a more active search for solidarity from progressive South African environmentalists. As Naomi Klein (2014) has posited, *This Changes Everything*. The climate crisis can force a broader recognition of society's need for radical restructuring in virtually all capitalist sub-systems of social reproduction, in areas including energy, transport, agriculture, urbanization, production, consumption, disposal and financing. To these it will be important to add: the system of imperial power that maintains Africa as a victim of environmental injustice, one in which South Africa continues to play a sub-imperial facilitating role.

References

African Centre for Bio-Safety (2015). *Africa Bullied to Grow Defective Bt Maize: The Failure of Monsanto's MON810 Maize in South Africa*. Johannesburg.

African Development Bank (2013). *African Economic Outlook*. Tunis.

African Development Bank (2014). *African Economic Outlook*. Tunis.

African Development Bank (2015). *African Economic Outlook*. Tunis.

African Development Bank (2016). *African Economic Outlook*. Tunis.

Allison, S. (2016). "South African peacekeepers accused of failing to prevent DRC massacre." *Daily Maverick*, 21 January.

AmaBhungane (2016). "US probe links Cape tycoon to mining rights bribery in Africa", *Mail & Guardian*, 18 August.

Armed Conflict Location and Event Data Project (ACLED) (2016). *Conflict Trends*. www.acleddata.com

Bond, P. (2002). *Unsustainable South Africa*. London: Merlin Press.

Bond, P. (Ed.) (2005). *Fanon's Warning*. Trenton: Africa World Press.

Bond, P. (2012). *Politics of Climate Justice*. Pietermaritzburg: University of KwaZulu-Natal Press.

Bond, P. (2014). *Elite Transition*. London: Pluto Press.

Bond, P. (2016a). "South Africa's next revolt?: Eco-socialist opportunities," in L. Panitch and G. Albo (Eds), *Socialist Register 2017*, New York: Monthly Review Press.

Bond, P. (2016b). "Who wins from 'climate apartheid'?: African climate justice narratives about the Paris COP21." *New Politics*, Winter, pp. 122–129.

Bond, P. (2016c). "BRICS banking and the debate over sub-imperialism." *Third World Quarterly*, 37, 4, pp. 611–629.

Bond, P. and A. Garcia (2015). *BRICS: An Anti-capitalist Critique*. London: Pluto Press.

Briscoe, J. (2010). "Overreach and response." *Water Alternatives*, 3, 2, pp. 399–415.

Christian Aid (2006). *The Climate of Poverty: Facts, Fears and Hope*. London: Christian Aid.

Climate Action Tracker (2015). "South Africa." London, 2 October. climateactiontracker.org/countries/southafrica.html

Cochran, S. (2010). "Security assistance, surrogate armies, and the pursuit of US interests in Sub-Saharan Africa." *Strategic Studies Quarterly*, 4, 1, pp. 111–152.

Cock, J. (2016). "An eco-socialist order in South Africa," *Review of African Political Economy*, Radical Agendas #5, 18 January.

Congoleaks (2011). "Democratic Republic of the Congo Report." congoleaks.blogspot.com/2011/11/initial-drc-report.html

Gaborone Declaration (2012). "Gaborone declaration for sustainability in Africa." Gaborone, 12 May. www.gaboronedeclaration.com

Godsell, B. (2005). "Doing business in a conflict zone." *Business Day*. 3 June.

Gopinath, D. (2003). "Doubt of Africa," *Institutional Investor Magazine*, May.

Harvey, D. (2003). *The New Imperialism*. Oxford: Oxford University Press.

Hosken, G. and I. Mahlangu (2013). "We were killing kids." *Sunday Times*, 31 March. www.timeslive.co.za/local/2013/03/31/we-were-killing-kids-1.

Human Rights Watch (2005). *The Curse of Gold*. London.

Kar, D. and J. Spanjers (2015). "Illicit financial flows from developing countries: 2004–2013." Washington, DC: Global Financial Integrity, December.

Klein, N. (2014). *This Changes Everything*. Toronto: Knopf.

Lunstrum, E. and P. Bond (2016). "Militarising game parks and marketing wildlife are unsustainable strategies." *Pambazuka*, 22 September.

Marini, R.M. (1965). "Brazilian interdependence and imperialist integration." *Monthly Review* 17, 7, pp. 14–24.

McDonald, D. (Ed.) (2008). *Electric Capitalism: Recolonising Africa on the Power Grid*. Pretoria: HSRC Press.

Muller, M. (2014). "A more useful agenda for water management." *New Water Policy and Practice*, 1, 1.

Perry, A. (2012). "Africa rising." *Time*, 3 December.

Satgar, V. (2016). "Where to for South Africa's Left?" *Review of African Political Economy*, Radical Agendas #6, 27 January.

Sharife, K. and S. Bracking (2016). "Diamond pricing and valuation in South Africa's extractive political economy." *Review of African Political Economy,* published online, http://dx.doi.org/10.1080/03056244.2016.1177504

South African Government Communication and Information Service (2007). "The state of our environment should remain under a watchful eye." Pretoria, 29 June. Accessed 15 February 2015, www.info.gov.za/speeches/2007/07062911151001.htm

South African Treasury (2014). "Carbon offsets paper." Pretoria. www.treasury.gov.za/public%20 comments/CarbonOffsets

Stern, T. (2011). "Durban wrap-up." Email to Hillary Clinton, 13 December. https://wikileaks.org/ clinton-emails/emailid/24887C05784614

Turse, N. (2014). "Africom becomes a war-fighting combatant command," *TomDispatch*, 13 April. www.tomdispatch.com/blog/175830/tomgram%3A_nick_turse,_africom_becomes_a_%22 war-fighting_combatant_command%22

World Bank (2011). *The Changing Wealth of Nations*. Washington: The World Bank Group.

World Bank (2014). *Little Green Data Book 2014*. Washington: The World Bank Group.

Zuma, J. (2014). "President Zuma welcomes US-Africa Leaders' Summit outcomes," The Presidency, Pretoria, 7 August, www.thepresidency.gov.za/content/president-zuma-welcomes-us-africa-leaders%E2%80%99-summit-outcomes

47

ENVIRONMENTAL JUSTICE AND ATTACHMENT TO PLACE

Australian cases

David Schlosberg, Lauren Rickards and Jason Byrne

Introduction

Australia illustrates a particular arena of environmental justice (EJ) theorizing that warrants closer scrutiny and development – the relationship between cultural identity, place attachment, environmental policy, and the experience of injustice. An enormous, isolated, island continent situated within the Global South, Australia is a young nation with a Global North identity. Chaffing against its image as a modern, friendly country is Australia's ongoing history of settler colonialism (Veracini 2015), characterized by a longstanding subordinate position to Britain (and the United States), rampant mineral extraction by multinational corporations (Goodman and Worth 2008), a regressive stance on many global environmental and social issues (e.g. climate change, refugees), and a diabolical record of environmental degradation. Despite or perhaps because of the latter, "nature" has long been central to the Australian national identity (Smith 2011), whether as a spectacular but capricious opponent, a once-maligned but now beloved source of providence, or as a spiritual home and source of ancient wisdom. The latter mimics and celebrates Australian Aboriginal people's deep connectedness to nature (or more accurately, to Country, discussed below), and settlers' efforts to "indigenize" themselves as part of carving out a uniquely Australian (not British) identity (Wolfe 2006). Australian governments have a long history of undertaking massive landscape transforming projects, such as the Ord River Dam and irrigation scheme, the Snowy Mountain hydro-electric and irrigation scheme, (non-Aboriginal) broad-acre farming (Goodman and Worth 2008; Griffin 2003; Holmes and Mirmohamadi 2015), and large-scale mining. Although its role has significantly declined, agrarianism and associated notions of environmental conquest as nation-building remain strong within the Australian psyche.

In this chapter we review the relationship between this strong place identification and claims of environmental (in)justice. While historically not a major organizing principle or movement, applications and uses of EJ discourse in the Australian context are increasing. We begin by defining and arguing for the centrality of place attachment to conceptions of environmental justice. We then discuss three very different Australian cases where this approach is articulated: Aboriginal Australians' relationship to Country; coal and gas mining; and a fire event that brings together concerns about climate change and air pollution.

Environmental justice and attachment to place in Australia

Environmental justice has been one of the central organizing discourses of environmental and climate change movements over the last 30 years. In that time, social movements using environmental justice as an organizing theme have articulated a broad and expanding conception of justice (Schlosberg 2007; Schlosberg 2013). EJ is primarily seen as a concern for (in)equity in the distribution of environmental bads (and goods), an insistence on recognizing and redressing them where and when they occur, and a demand to avoid their expansion or replication. This focus began with classic studies of mal-distribution in the US (Bullard 1990; United Church of Christ Commission for Racial Justice 1987), and the demonstration of environmental inequity has since been replicated in many other places and discussed as an important reason for the geographic expansion of the EJ approach (Walker 2012). Australia, however, did not develop a self-identified EJ movement around these kinds of experiences (nor data repositories or environmental regulations) as happened in numerous other countries in the 1990s and 2000s. Historically, there have been similar histories of environmental inequality in many Australian towns and cities (Byrne and MacCallum 2013), but the documentation of the classic inequities approach to EJ is comparatively recent. Work has documented, for example, the inequitable distribution of environmental bads to poor and Indigenous communities around air pollution (Chakraborty and Green 2014), climate change impacts (Green et al. 2012; MacCallum et al. 2014), and the inequitable health impacts of coal mining (Colagiuri and Morrice 2015).

Importantly, the EJ discourse in Australia was first taken up by Aboriginal communities. Their concerns were not simply based in inequity, but more broadly on threats to land, country, resources and culture, concomitant with developmentalism and environmental degradation (for more, see Chapter 41). These concerns fit both an equity approach to understanding environmental injustice, and an expanded notion of EJ that has been developing over the past two decades. While equity has often been an initial issue, racism and other forms of cultural disrespect evince a concern with the politics of recognition – discrimination, disrespect, and stereotypes. In addition, exclusions that accompany inequity and disrespect have always been part of environmental justice discourse, fuelling concern for political participation, inclusion, and impact as a response to injustice. The demands for community voice, and participation in the review of development proposals and/or decision-making, have been constant (Keir et al. 2014). Finally, what movements have meant by the "justice" of environmental justice incorporates a range of basic capabilities and needs of individuals and communities. Environmental justice advocates have long talked about community health, good jobs, clean air and water, and the basic needs they require to live their lives – and about the environmental threats to those needs. Access to environmental goods is as important in this approach as avoiding exposure to environmental bads. This framework of EJ, encompassing equity, recognition, participation, and the provision and protection of capabilities (Schlosberg 2007), has been applied in numerous contexts.

What we want to argue here is that the Australian experience, at its core, often articulates another dimension of what justice in "environmental justice" means – a *relational* dimension, in particular *attachment to place*. Anguelovski (2013) argues that the combination of such place attachment and threats to that place – such as environmental decay – can create patterns of activism; in our cases, such threats to attachment to place are understood as environmental injustice. As Delaney (2016: 3) notes, "(In)justice is intrinsically social and relational in the sense that claims of injustice necessarily call into account inherently *social* states of affairs concerning contingent *social* arrangements – including socio-spatial arrangements." Attending to the quintessentially socio-spatial phenomena of some people's attachment to some places brings into play not only clear cases of "environmental bads" (e.g. the imposition of various forms of pollution) but the many more intangible losses of personally or locally valued "environmental goods," such as the loss of a local bird population, the clearance of a favourite patch of remnant vegetation, or the compulsory acquisition of a family farm to make way for infrastructure development. We argue for the overlapping and co-constructive quality of justice and place. While notions of distributional justice and equity have been muted themes in EJ controversies in Australia, there has been a major focus on the impacts of various practices on sacred or treasured places in the Australian landscape. Here, what is crucial is the relationship between environmental impacts on place and the experience, identity, and cultures of those living in these impacted places.

Places have material, inter-personal and symbolic dimensions. They are constituted of, and through, "a particular constellation of social relations . . . at a particular locus" (Massey 1997). They are highly relational in four ways. First, the concept of place attachment underlines the relationship between individuals and certain places; we can perceive unique or inherent qualities of some places —setting them apart from others, which can engender an emotional bond or affinity (attachment) with those places (Stedman 2003). Second, as Massey (1997) observes, place is relational because places are socially constituted in terms of both being subjective and being shaped by social relationships and hierarchies. People's identities and individual and collective characteristics relative to others (e.g. age, gender, class, race, sexual preference), relative levels of resource access, and lived experiences, all shape how they relate to any one place. In any one locality, some groups may be marginalized (e.g. for not being or for being "locals"), and thus they experience it as a very different sort of place than those who are privileged. Third, place is relational because it is shaped by other places and spaces. The voluntary and involuntary concentration of certain groups and processes in particular areas over time means that local places and experiences are inseparable from the more national, transnational and global processes that continually regenerate patterns of uneven development and divergent environmental quality.

This issue of environmental quality brings us to the final way in which place is relational, which is that places inscribe our bodies with location-based markers of environmental quality. The weather conditions we experience, the air we breathe, the food we eat, the water we drink and the green (or decidedly not green) spaces we walk through all impact our bodies with toxins, nutrients, pollens, viruses, bacteria, radioisotopes and the like. Places contribute to whether we are sick or healthy, not just via their physical qualities, but also via our emotional connection to them – whether a yearning, love, sadness or repulsion. For this reason, places play a key role in considerations of environmental (in)justice.

Our argument here is that EJ is, in part, about where and how we live in an environment; people's experience of and relationship to places is an important element of broader questions about environmental justice (or injustice). Here, justice hinges on a sense of a positive place attachment – and avoidance of negative impacts on place, such as pollution and threats to environmentally based, culturally valued practices. EJ also obviously depends on the relationships between social groups; while one aspect of environmental justice is the lack of a positive relationship between those who are impacted or marginalized and the place in which they are experiencing such disruption, a crucial component is the relationship between those marginalized and the elite who create, underpin, and sustain the experiences of injustice. The theme of attachment to place and to each other, and the threat to that attachment from elites or resource developers, local or far afield, ties together a wide range of EJ issues and movements across Australia, from Aboriginal people fighting uranium mines, to farmers blockading coal seam gas development, to communities fighting pollution.

Anguelovski (2014) identifies and addresses the importance of the creation of such attachment in neighbourhood revitalization movements. More theoretically, Groves (2015) has recently argued for the inclusion of a conception of attachment in the consideration of environmental justice – and the recognition that threats to our attachment to place will impact both that place and our identities. Like Anguelovski, he cites longstanding research that communities often frame activism around the degradation of place, the threat to complex socio-environmental links, and the reconstruction of attachment. Attachment to place is a process, a relationship, and a constitutive element of our identity and functioning, and EJ has long lamented – and acted upon – disruptions to that attachment. For Groves, connections or attachments between "place, identity, and agency" should be part of our understanding of a broad conception of environmental justice.

Part of the theoretical argument here is that positive attachment – both to other people and to place – is seen as a basic capability or need, essential for a fully functioning and flourishing life. In a capabilities approach (Nussbaum 2011; Sen 2001), justice occurs with the availability of all of the basic capabilities of human life. To deny, disrupt, or interrupt those capabilities by imposing environmental bads, removing valued environmental goods, and/or forcing one to move elsewhere is, then, the definition of an injustice. Going further, Albrecht (2005) argues that this kind of undermining of the capability of attachment to nonhuman places, including even the *threat of* loss of place, brings about "solastalgia" – a feeling of distress that comes "when your endemic sense of place is being violated" (see also Albrecht et al. 2007).

Our argument is that the disruption of attachment to place greatly deepens the injustice of the uneven distribution of environmental bads. To the extent that a vast array of areas is being gradually degraded by distal as well as proximate forces including climate change, this greatly opens out the scope of what is included within environmental injustice. Disruptions to place attachment are not just a case of the uneven distribution of environmental bads or loss of treasured environmental goods, but also involve procedural and cultural issues. The first is evident in how people's loss of place attachment is not fully considered a legitimate cost in conventional social impact assessment processes, which focus instead on quantifiable, economic costs. Although losses may be intense, they do not count in legal proceedings or other formal impact assessments. Second, this formal neglect of the significance of loss of place attachment intersects with issues of cultural misrecognition. Place attachment is especially core to the identity and wellbeing of certain groups such as Indigenous peoples or farm families. Rather, through the implicit application of a standard *Homo economicus* lens, judgments are made as to what a certain change in a local place should mean to a group, denying the unique and profound losses they may be experiencing (e.g. Durkalec et al. 2015).

Scale is also an important consideration, not only because it exposes how environmental bads are often displaced from one area to another (Delaney 2016), but because of how localness (or its lack) determines the legitimacy of one's voice. In many place-based issues, local residents (particularly property owners) are privileged; their claims are considered especially legitimate. But, importantly, concern for, and attachment to, place is reflected not only in local communities – such as those fighting against the impacts of coal or coal seam gas mining – but also from those distant from the impacted communities. "Outside" movement actors – from the cities, or from elsewhere – react to the potential impact on places that, while distant, nevertheless have strong environmental (the Great Barrier Reef) or cultural (bucolic farming regions) importance to them. There is thus a differentiation between what EJ means for local residents (some of whom may prioritize local jobs, and who have to live with environmental damage) and for people who care for, and are "attached" to, affected places but do not have to, or are not able to, live there. Often, these non-local concerns struggle for a legitimate stake in local issues, even when such "local" issues are clearly far from solely local – as with massive coal mines that will exacerbate global climate change, undermine prime farmland and food security, and potentially decimate national icons or important heritage sites.

In Australia, this denial of the legitimacy of non-local place attachment is illustrated by governments' recent efforts to alter the longstanding Environment Protection and Biodiversity Conservation Act in order to restrict the ability for environmental organizations to initiate legal action in response to major development proposals, such as mines, allowing only local residents a say. To the extent that EJ issues are boundary-crossing, and local groups and individuals struggle to launch legal proceedings or to articulate a general justice argument, there is a need to spatially unbound our sense of who is a "stakeholder" in any EJ issue, and to respect the link made between attachment, place, and identity by locals and nonlocals alike.

Finally, while we want to emphasize the important nature of place attachment, it is crucial to understand that "attachment" to place can also be a negative. In the history of environmental justice, it is an attachment to places that are damaged, poisoned, and/or dangerous, that defines many communities. Here, attachment is about the material or psychological dependence on a place and/or the inability to move away from it, generating the experience of being stuck, despite the location's negative effects. Whether it is toxic dumping in poor communities, border communities subject to air pollution, the children of poor families drinking contaminated water in Flint, Michigan, or marginalized and vulnerable people living along coastlines in danger of climate-induced flooding, the financial, physical, social and cultural inability to physically escape from environmental bads – and the attachment to threatened, damaged or damaging places – is part of the stressful experience of environmental injustice. We must see both the positive and the potentially negative experience of this relationship between attachment and place.

We turn now to our three examples of contemporary Australian environmental justice struggles. While current and interwoven, they replay to some degree historic land use conflicts. As explained below, the traditional Aboriginal owners of Australia were violently displaced by settler colonization (for agriculture, among other things). Agriculture in many regions is not only a source of numerous environmental problems but is itself now under direct and indirect attack from fossil fuel mining. And fossil fuel mining communities such as those in the La Trobe Valley now struggle to fight the direct and indirect impacts of both mining and its demise under economic, environmental and climate change pressure. All three cases involve competing place attachments and together they underline the complex justice issues that emerge as land claims, land uses, and their myriad feedbacks evolve and intersect.

Aboriginal Australians caring for Country

While environmental justice scholars have considered how multiple axes of difference shape lived experience, place attachment has very unique meanings for Aboriginal Australians. As Byrne and MacCallum (2013) have asserted, Aboriginal Australians share similar histories and experiences of dispossession, genocide and alienation to those which scholars have reported for other first nation peoples (e.g. Holifield 2012). This includes being disproportionately burdened by environmental harms (e.g. exposure to radiation from nuclear testing) and being spatially marginalized in towns and cities. Many cities still bear the traces of a segregationist colonial past, such as Brisbane, where "Boundary Street" in West End demarcates a physical boundary of Whiteness that Aboriginal people were once unable to cross (see also Taylor 2000). And until very recently, Aboriginal people were forcibly removed from their traditional homelands and kin, through State policies collectively referred to as the "stolen generations" (Prout 2009).

Scholars and activists have noted that for Aboriginal Australians, place attachment is a visceral experience. As one of the oldest cultures on the planet, Aboriginal Australians have developed a relationship of reciprocal exchange with the Australian landscape and its eco-systems, referred to as "Country." Citing Rose (1996), Wensing (2014) notes that Country is best described as a "place that gives and receives life." Aboriginal Australians have moral, cultural and spiritual obligations to Country, which they care for and nurture as if it were a family member. People worry for Country, fret for Country and mourn for Country (Wensing 2014). Obligations to Country include the requirement to perform ceremonies of nurturing and renewal. Indeed, Burgess et al. (2005) have argued that if these ceremonies do not occur, Country becomes ill (e.g. ecologically degraded), and Aboriginal Australians may suffer physical and mental illnesses.

Contemporary natural resource management (NRM) plans have appropriated this notion of caring for Country, but typically fail to recognize its real meaning for Aboriginal people and the socio-ecological relations entangled in Country (Lane and Corbett 2005). Grafted onto Western and scientific epistemologies, place-based relationships in NRM are diluted or lost because traditional knowledge is typically discounted or marginalized and Aboriginal custodians are sidelined. Resource management philosophies have also created tensions between environmentalists and Aboriginal custodians. Clashes have occurred around the complex relationship between the need to protect what White Australians regard as "special places" and Aboriginal Australians' right to self-determination. Environmental engagements with traditional owners can be fraught, especially when mining companies offer concessions that promise to deliver improvements in quality of life and livelihood previously denied to many Aboriginal people (Howlett et al. 2011). Conflict between environmentalists and tra-ditional owners has also occurred around efforts to protect "wild" rivers, to prevent land clearing, and to establish sites for the dumping of radioactive waste (Lane and Corbett 2005). Place-based environmental justice struggles in Australia, while often framed around lived experience, are not readily reduced to simple explanations (Coombes et al. 2012). Contestations between farming and coal mining are a good example.

Farmers fighting coal and gas mining

In Australia, movements organizing around coal and coal seam gas mining have thoroughly focused on impacts to the environments in which people's lives are immersed. The risks to land and water posed by mining are understood not only as risks to the environment, but also to the very relationship between community and place – and the survival of both. So a central focus of movement campaigns is on community attachment to place as part of both individual and collective identity and the functioning of place. Issues of cultural recognition and value, and of political participation and inclusion, are central and salient in these Australian cases.

The organization Lock the Gate is an important example of a recent movement adopt-ing a broad environmental justice discourse. The organization was founded in 2010 as a form of farmer resistance to the rapid, government-supported expansion of coal and coal seam gas mining, often against the will – and without the consideration – of local residents and communities. Unlike many farming communities faced with gas mining in the US, Australian communities allied with Lock the Gate are vibrant, fecund, agricultural communities – culturally and historically tied to farming. They see gas mining not as a new form of economic development, but as a threat to their own lands, ways of life, and the places and environments that support them.

The central aims articulated by Lock the Gate (undated) include the protection of water systems, agricultural lands, bushland, wetlands, and wildlife; the health of Australians; and Aboriginal and cultural heritage. All of these aims converge on the relationship between the functioning of environmental and human systems, communities' attachment to productive places, and recognition of the importance of this relationship. One of the organization's major campaigns was the "Call to Country" – a "call from the heart of this country, and the people that love it, to demand real action to restrict inappropriate coal and gas mining." Key to the campaign was a list of "places at risk," with a focus on iconic landscapes and rural communities. The main policy demands focus on the protection of these places and the productive relationship between these places and functioning communities.

The farmers and other activists at the heart of the Lock the Gate movement are thoroughly immersed in the relationship between environment, place, identity, and sustainability. The threat to that longstanding relationship and way of life is the reason farmers give for being some of the most outspoken critics of coal and coal seam gas mining. Contaminate the water, and you undermine the functioning of the environment; undermine that functioning, and you threaten community attachment to place. So it should not be surprising that we have farming communities worried about the way coal seam gas mining might, quite simply, contaminate the key element that makes their way of life, livelihoods, cultures, and their very attachment to place possible.

These Australian environmental activists are in the forefront of challenging our conception, and treatment, of the environments in which our everyday lives are immersed and on which we depend. It's not just about a faraway nature, and it's not only about getting polluted. It's about the threat to an environment that supports the functioning of a community, a threat to communities' attachment to place. Injustice is seen in the way that the very functioning of communities and their cultures are being threatened, that their attachment to place and environment is undermined.

One of the important implications of this focus on place attachment is that it has allowed coalitions and partnerships to develop that could not occur with a focus similar to traditional environmental campaigns. In the fight against the expansion of coal mining, and the impacts of coal seam gas, it is this focus on the intersection of environment, place, culture, and functioning communities that has allowed organizations like Greenpeace, Friends of the Earth, and the Environmental Defenders Office to work in alliances with farmers – and is why farmers and rural organizations are setting aside their longstanding distrust of environmental organizations that they have only seen previously as enemies. Such alliances and coalitions have only been enabled because the issues have redefined "environment" from the classic conception of a distinct natural world to one aligned with the environmental justice movement's longstanding definition of a place where we "live, work, and play."

It is crucial to note that the discourse of these movements is not only about such attachment and functioning, but also about the lack of recognition and value given to their ways of life, and the related political exclusion that comes with that. In one sense, people express that they, their communities, and their ways of life are simply valued less than the coal and gas that is taken out of the ground. They see longstanding, iconic ideals of farming in many of the most lush and productive areas of the country simply being dismissed as antiquated, or underproductive, in the face of the resource boom. And this lack of recognition is linked to anger that their agency as citizens, and their participation in the governance of their own communities, has been taken away. They see their governments as being in the pocket of mining companies – and various corruption cases have illustrated the problem (Australian Broadcasting Corporation 2013). Worse, communities see state and federal governments as actually creating barriers to public participation – for example, by defunding public legal aid organizations like the Environmental Defenders Office that help communities to participate in the political and legal processes, and by supporting laws that restrict both the public rights to challenge mining impacts in court (*Sydney Morning Herald* 2015) and to protest (*Sydney Morning Herald* 2016). Recognition and procedural justice remain central demands, part of the discourse of environmental justice, and linked to the threat of loss of place attachment, in these movements.

Coal power communities fighting for a future

Contrasting with the previous two cases, our third case study examines the environmental justice struggles of a regional community centred on an established but declining coal mining and power generation sector. The La Trobe Valley in the Gippsland region of Victoria was originally the beautiful forested home of the Gunai Kurnai people. Under settler colonialism it was turned into a successful dairy farming region, before its abundant shallow brown coal reserves were increasingly mined in large open cut operations to produce electricity for the state. The region's three mines are owned by and feed thermal coal power stations, collectively supplying about 85 per cent of Victoria's current electricity needs and selling electricity to other states. This urban dependence has dictated many decisions about the distant region, including the cancellation in 2009 and 2011 of the planned decommissioning of the most polluting plant, Hazelwood, which finally closed in 2017.

Of the region's current population of about 125,000, only a small proportion have permanent employment with the generators, though many others work as contractors or subcontractors in the mines and power plants (Snell et al. 2015). Employment in the coal power industry plummeted after the facilities were corporatized, privatized and fragmented in the mid-1980s, despite the efforts of "militant" trade union activities to save jobs (Barton and Fairbrother 2007; Weller 2012). The resultant period of economic recession and population decline left the region 'in search of future employment' (Barrett et al. 2012: 126). At the same time, one of the two towns in the region built to house mining workers – Yallourn – was demolished to make way for mine expansion (www.virtualyallourn.com). Today, economic pressures combined with environmental and climate change concerns mean that the entire future of the coal power sector in the region is in question, particularly given announcements by the owners of Hazelwood coal mine, Engie, that they want to abandon fossil fuel energy production, and close the mine and power plant (Australian Financial Review 2016). With few well-developed economic alternatives in the region (though see below), and many coal power sector workers poorly equipped to secure other jobs, job insecurity is a significant source of stress among workers (Snell et al. 2015). Overall, there is a fear that, as Lobao et al. (2016: 378) suggest about Appalachia in the US, the region will "become a national sacrifice zone in the movement towards cleaner energy." As discussed below, the challenge is how to transition to cleaner energy and advantage, not disadvantage, local populations.

The La Trobe Valley (LTV) case highlights two central environmental justice considerations. First, a major disadvantage for the region is its poor quality natural environment, notably the chronic and occasionally acute air pollution produced by the mines and power stations. Not only does this impact negatively on residents' health, it depresses house prices, limiting the degree to which residents can sell up and leave even if they wanted to.

Underpinning this negative place attachment is the way the region is broadly stigmatized for appearing out of step with modern standards, ways of life, and environmental consciousness (Tomaney and Somerville 2011), even as Melbourne remains heavily dependent upon its coal-fired power. Much of the escalating criticism of the regions' mines and power stations on environmental grounds tends to present the Valley as a whole in a negative light, which both obscures and adds to the challenges facing local people. Although the social justice implications of closing Hazelwood and the other stations are becoming better recognized, and the state has put in place a major economic recovery plan, opportunities for residents to enjoy formal, meaningful local involvement in discussions about the region's future remain limited (Pape et al. 2015). Overall, the LTV region has tended to be either broadly ridiculed and dismissed as an anachronism, or vilified as a symbol of Australia's broken climate change commitments, with local residents typically having little control over how their region is represented

Second, the LTV example underlines the growing salience of disasters and climate change in environmental justice matters. The case of the Hazelwood power plant and mine is especially instructive. Hazelwood was infamous for contributing enormous quantities of greenhouse gas emissions to the atmosphere, thanks to the young coal it used to burn and its ageing infrastructure. There was a certain perverse irony, then, when in February 2014 it was the site of a massive wild fire event, exacerbated by extreme fire weather and drought conditions. As fire spread in a disused, unrestored and exposed wall of the coal mine, becoming an inferno, local emergency response volunteers were constrained in their capacity to fight it thanks to problems at the site caused by corporate cost-cutting and priority being placed on protecting the power plant and ensuring the continued supply of electricity to Melbourne. In the end, the fire burnt for 45 days. During that time, volunteers and the local community lived and worked within a nauseating cloud of toxic, sticky ash and smoke. Two weeks into the event, the government belatedly called for residents to relocate on health grounds. But it was an evacuation order many locals did not have the financial or (then) physical capacity to follow, especially given their dependence on jobs in the mine and plant (Doig 2015).

The consequent public outcry quickly transformed the physical disaster into an ongoing political one. The state government was forced to hold a public inquiry into the situation. Although the inquiry's terms of reference framed the situation as a tragic emergency event, it inevitably reignited longstanding concern about the existence of Hazelwood, particularly in the context of climate change. It also brought into renewed focus the local LTV community and the myriad immediate challenges it faces. Assisted by others such as legal NGO Environmental Justice Australia, the community responded to the situation with a series of interventions that illustrate the strategies of antagonizing, reclaiming commons, and forging solidarity that Chatterton et al. (2013) identify at work in the climate justice movement. Locals demonstrated that amongst their negative place attachments lie many positive ones. Proud local voices and organizations have emerged. This includes the group Voices of the Valley, who successful called for a second public inquiry into the coal mine fire, which subsequently verified VoV's suspicions that – counter to the first inquiry's conclusions – the fire had been fatal, contributing to the deaths of some local people. Other positive initiatives include the Earthworker cooperative (http://earthworkercooperative.com.au/), which builds on the region's long-standing culture of trade unionism, manufacturing and energy production by offering workers a role in a unique owner–worker cooperative building solar hot water systems. In this way, the region is beginning to strengthen locals' positive attachments to the place while trying to reduce those negative ones borne of ill health and limited options.

Conclusions

This chapter has canvassed some of the distinctive environmental justice issues that arise with a focus on attachment to place, using contemporary Australian examples. There is clearly much more to say about the growing concern with regard to environmental justice in Australia. While all of these cases are about non-urban spaces, Australia is highly urbanized – a condition that both drives the resource extraction at the heart of the issues discussed above, and one that creates a range of its own environmental battles around housing, transportation, air quality, asbestos pollution, and more. In addition, an accounting of the broad and longstanding environmental injustices experienced and fought by Aboriginal communities would alone take volumes (see Chapter 41 for an overview).

We have offered here simply an introduction to the idea of attachment to place as an element of environmental justice, and have argued that it is particularly central to environmental justice considerations in Australia. While such attachment is a longstanding aspect of Aboriginal relationships with Country, we identify place attachment – and reactions to practices that undermine that attachment – in primarily white farming and mining communities as well.

In each of the cases presented here, people's understanding of and attachment to place reveals a tension between a celebration of valued places as vibrant, evolving, working landscapes, and a dislike of more revolutionary changes. For although, as Massey (1997: 30) notes, place is "an on-going production" of human relations, local people can experience "a (conscious) contradiction between a recognition of places as constantly changing, and a feeling of resistance to more modern activities that 'impact on' the nature of a place . . . ." This is a core theme in Australian environmental history, which is very much a story of sequential environmental disturbers and protectors. Moreover, there are small signs that things are changing in Australia, that new visions are starting to be taken seriously, that the social licence of the extractive industries is on the wane, and that both environmental justice and attachment to place are becoming central to a wide range of issues. We can see this in an interweaving of environmental justice and ecological justice concerns, spanning activities ranging from forestry and agriculture to resource extraction, and from concerns about the Great Barrier Reef and climate adaptation to wildlife conservation in urban brownfields. The Australian experience has much to offer future research on environmental (in)justice, and its particular relation to places and our attachment to them.

References

Albrecht, G. (2005). "'Solastalgia': A new concept in health and identity." *PAN: Philosophy Activism Nature*, vol. 3, pp. 41–55.

Albrecht, G., Sartore, G.-M., Connor, L., Higginbotham, N., Freeman, S., Kelly, B., et al. (2007). "Solastalgia: The distress caused by environmental change." *Australasian Psychiatry*, vol. 15(sup1), S95–S98.

Anguelovski, I. (2013). "From environmental trauma to safe haven: Place attachment and place remaking in three marginalized neighborhoods of Barcelona, Boston, and Havana." *City & Community*, vol. 12, pp. 211–237. doi: 10.1111/cico.12026

Anguelovski, I. (2014). *Neighborhood as refuge*. Cambridge, MA: MIT Press.

Australian Broadcasting Corporation (2013). *ICAC finds NSW coal corruption 'almost inevitable', makes sweeping recommendations* [online]. Australian Broadcasting Corporation. Available from: www.abc.net.au/news/2013-10-30/icac-finds-nsw-coal-corruption-27almost-invevitable27/5057174 [Accessed June 20 2016].

Australian Financial Review (2016). *Hazelwood closure could mark beginning of end for Victoria's brown coal* [online]. Available from: www.afr.com/business/energy/french-energy-giant-engie-mulls-closure-of-hazelwood-power-station-20160525-gp426a [Accessed June 20 2016].

Barrett, M., Maslyuk, S. and Pambudi, D. (2012). "Carbon tax: Economic impact on the Latrobe Valley." In Finch, N. (ed.), *Contemporary issues in mining: leading practice in Australia*. Basingstoke, Hampshire: Palgrave MacMillan, pp. 122–140.

Barton, R. and Fairbrother, P. (2007). "'We're here to make money; We're here to do business': Privatisation and questions for trade unions." *Competition & Change*, vol. 11(3), pp. 241–259.

Bullard, R. (1990). *Dumping in Dixie: race, class and environmental quality*. Boulder, CO: Westview Press.

Burgess, C., Johnston, F., Bowman, D. and Whitehead, P. (2005). "Healthy country: healthy people? Exploring the health benefits of Indigenous natural resource management." *Australian and New Zealand Journal of Public Health*, vol. 29(2), pp. 117–122.

Byrne, J. and MacCallum, D. (2013). "Bordering on neglect: 'Environmental justice' in Australian planning." *Australian Planner*, vol. 50(2), pp. 164–173.

Chakraborty, J. and Green, D. (2014). "Australia's first national level quantitative environmental justice assessment of industrial air pollution." *Environmental Research Letters*, vol. 9(4), 044010.

Chatterton, P., Featherstone, D. and Routledge, P. (2013). "Articulating climate justice in Copenhagen: Antagonism, the commons, and solidarity." *Antipode*, vol. 45(3), pp. 602–620.

Colagiuri, R. and Morrice, E. (2015). "Do coal-related health harms constitute a resource curse? A case study from Australia's Hunter Valley." *The Extractive Industries and Society*, vol. 2(2), pp. 252–263.

Coombes, B., Johnson, J.T. and Howitt, R. (2012). "Indigenous geographies I: Mere resource conflicts? The complexities in Indigenous land and environmental claims." *Progress in Human Geography*, vol. 36(6), pp. 810–821.

Delaney, D. (2016). "Legal geography II: Discerning injustice." *Progress in Human Geography*, vol. 40(2), pp. 267–274.

Doig, T. (2015). *The coal face*. Melbourne: Penguin.

Durkalec, A., Furgal, C., Skinner, M.W. and Sheldon, T. (2015). "Climate change influences on environment as a determinant of Indigenous health: Relationships to place, sea ice, and health in an Inuit community." *Social Science & Medicine*, vol. 136, pp. 17–26.

Goodman, J. and Worth, D. (2008). "The minerals boom and Australia's resource curse." *The Journal of Australian Political Economy*, vol. 61, pp. 201–219.

Green, D., Niall, S. and Morrison, J. (2012). "Bridging the gap between theory and practice in climate change vulnerability assessments for remote Indigenous communities in northern Australia." *Local Environment*, vol. 17(3), pp. 295–315.

Griffin, G. (2003). "Selling the snowy: The Snowy Mountains Scheme and national mythmaking." *Journal of Australian Studies*, vol. 27(79), pp. 39–49.

Groves, C. (2015). "The bomb in my backyard, the serpent in my house: Environmental justice, risk, and the colonisation of attachment." *Environmental Politics*, vol. 24(6), pp. 853–873.

Holifield, R. (2012). "Environmental justice as recognition and participation in risk assessment: Negotiating and translating health risk at a superfund site in Indian Country." *Annals of the Association of American Geographers*, vol. 102(3), pp. 591–613.

Holmes, K. and Mirmohamadi, K. (2015). "Howling wilderness and promised land: Imagining the Victorian mallee, 1840–1914." *Australian Historical Studies*, vol. 46(2), pp. 191–213.

Howlett, C., Seini, M., McCallum, D. and Osborne, N. (2011). "Neoliberalism, mineral development and Indigenous people: A framework for analysis." *Australian Geographer*, vol. 42(3), pp. 309–323.

Keir, L., Watts, R. and Inwood, S. (2014). "Environmental justice and citizen perceptions of a proposed electric transmission line." *Community Development*, vol. 45(2), pp. 107–120.

Lane, M.B. and Corbett, T. (2005). "The tyranny of localism: Indigenous participation in community-based environmental management." *Journal of Environmental Policy and Planning*, vol. 7(2), pp. 141–159.

Lobao, L., Zhou, M., Partridge, M. and Betz, M. (2016). "Poverty, place, and coal employment across Appalachia and the United States in a new economic era." *Rural Sociology*, vol. 81(3), pp. 343–386.

Lock the Gate, undated. *Our Call to Country* [online]. South Lismore, NSW: Lock the Gate. Available from: https://d3n8a8pro7vhmx.cloudfront.net/lockthegate/pages/2187/attachments/original/1439522455/LTG-CallToCountry-Brochure-web.pdf?1439522455 [Accessed June 20 2016].

MacCallum, D., Byrne, J. and Steele, W. (2014). "Whither justice? An analysis of local climate change responses from South East Queensland, Australia." *Environment and Planning C: Government and Policy*, vol. 32(1), pp. 70–92.

Massey, D. (1997). "A global sense of place." In Barnes, T. and Gregory, D. (eds.), *Reading human geography: The poetics and politics of inquiry*. London: Arnold, pp. 315–323.

Nussbaum, M.C. (2011). *Creating capabilities: The human development approach*. Cambridge, MA: Harvard University Press.

Pape, M., Fairbrother, P. and Snell, D. (2015). "Beyond the state: Shaping governance and development policy in an Australian region." *Regional Studies*, vol. 50(5), pp. 902–921.

Prout, S. (2009). "Security and belonging: Reconceptualising Aboriginal spatial mobilities in Yamatji country, Western Australia." *Mobilities*, vol. 4(2), pp. 177–202.

Rose, D.B. (1996). *Nourishing terrains: Australian Aboriginal views of landscape and wilderness*. Canberra: Australian Heritage Commission.

Schlosberg, D. (2007). *Defining environmental justice: Theories, movements, and nature*. Oxford: Oxford University Press.

Schlosberg, D. (2013). "Theorising environmental justice: The expanding sphere of a discourse." *Environmental Politics*, vol. 22(1), pp. 37–55.

Sen, A. (2001). *Development as freedom*. New York: Oxford University Press.

Smith, N. (2011). "Blood and soil: Nature, native and nation in the Australian imaginary." *Journal of Australian Studies*, vol. 35(1), pp. 1–18.

Snell, D., Schmitt, D., Glavas, A. and Bamberry, L. (2015). "Worker stress and the prospect of job loss in a fragmented organisation." *Qualitative Research in Organizations and Management: An International Journal*, vol. 10(1), pp. 61–81.

Stedman, R.C. (2003). "Is it really just a social construction?: The contribution of the physical environment to sense of place." *Society & Natural Resources*, vol. 16(8), pp. 671–685.

Sydney Morning Herald (2015). "Abbott government to change environmental laws in crackdown on 'vigilante' green groups." [online] Available from: www.smh.com.au/federal-politics/political-news/abbott-government-to-change-environment-laws-in-crackdown-on-vigilante-green-groups-20150818-gj1r4l.html [Accessed June 20 2016].

Sydney Morning Herald (2016). "Smaller penalties for CSG companies amid crackdown on protesters." [online] Available from: www.smh.com.au/nsw/smaller-penalties-for-csg-companies-amid-crackdown-on-protesters-20160307-gncbkk.html [Accessed June 20 2016].

Taylor, A. (2000). " 'The sun always shines in Perth': A post-colonial geography of identity, memory and place." *Australian Geographical Studies*, vol. 38(1), pp. 27–35.

Tomaney, J. and Somerville, M. (2011). "Climate change and regional identity in the Latrobe Valley , Victoria." *Australian Humanities Review*, vol. 49, pp. 29–47.

United Church of Christ Commission for Racial Justice (1987). *Toxic wastes and race in the United States: A national report on the racial and socio-economic characteristics of communities with hazardous waste sites*. New York: United Church of Christ.

Veracini, L. (2015). *The settler colonial present*. London: Palgrave Macmillan.

Walker, G. (2012). *Environmental justice: Concepts, evidence and politics*. London: Routledge.

Weller, S. (2012). "The regional dimensions of the 'transition to a low-carbon economy': The case of Australia's Latrobe Valley." *Regional Studies*, vol. 46(9), pp. 1261–1272.

Wensing, E. (2014). "Aboriginal and Torres Strait Islander peoples' relationships to Country." In Byrne, J., Sipe, N. and Dodson, J. (eds.), *Australian environmental planning: Challenges and future prospects*. London: Routledge.

Wolfe, P. (2006). "Settler colonialism and the elimination of the native." *Journal of Genocide Research*, vol. 8(4), pp. 387–409.

48

ENVIRONMENTAL JUSTICE IN SOUTH AND SOUTHEAST ASIA

Inequalities and struggles in rural and urban contexts

Pratyusha Basu

Environmental justice activism and scholarship in South and Southeast Asia (Figure 48.1) has focused on a wide range of rural and urban issues, including struggles for access to and control of land, water, and forests by Indigenous and peasant communities, and class-based inequities in access to environmental goods and exposure to environmental harms in urban metropolises. Urban–rural divisions in this region are linked to histories of colonialism, national development policies, and present-day flows of global capitalism. As part of colonial economies (mainly British in South Asia and Malaysia, French in mainland Southeast Asia, and Dutch in Indonesia), this region provided raw materials for European industrialization, while also serving as a market for European industrial products. Since raw materials were always cheaper than finished industrial products, a system of 'unequal exchange' characterized the colonial division of labour (McMichael 2012: 26–37).

Current rural–urban inequalities reflect the consequences of unequal exchange in two ways. First, as providers of raw materials, rural areas in the colonial period suffered loss of their environmental resources, and became increasingly tied to external needs rather than serving local populations. In the contemporary global economy, rural areas in the Global South often continue to play the same role. Second, in the colonial period, rural resources were channelled to European destinations through port cities, so that urbanization often occurred in coastal locations. The region reflects this historical pattern of uneven development where economic growth is concentrated in a few major cities, often national and regional capitals. Though urbanization is steadily increasing, many countries in the region continue to be majority rural (Table 48.1).

Peasant and Indigenous struggles occurring in the Global South are often characterized as an 'environmentalism of the poor' (Martinez-Alier 2002) since their aim is to protect small-scale rural livelihoods which require direct access to land, water, and forests (e.g. crop and livestock farming, fishing, gathering of non-timber forest products). An 'environmentalism of the rich' can also be identified in the region in the rise of middle-class environmental movements which have sought to remove low-income groups and their activities from urban areas (see Mawdsley 2004 for a study from India).

Indigenous communities in the region have often been disproportionately affected by dispossession and pollution due to resource extraction. Indigeneity in this region does not

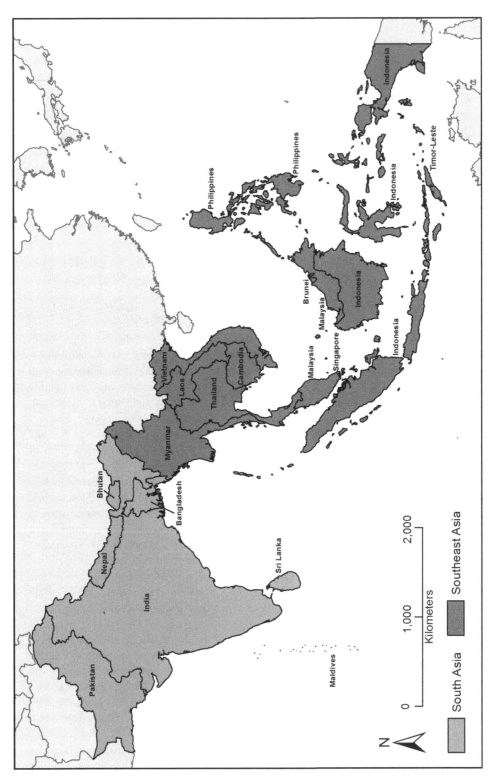

Figure 48.1   South and Southeast Asia

Table 48.1 Proportion of urban population in South and Southeast Asian countries

	Per cent urban population	
	1960	*2015*
SOUTH ASIA		
Bangladesh	5.1	34.3
Bhutan	3.6	38.6
India	17.9	32.7
Maldives	11.2	45.5
Nepal	3.5	18.6
Pakistan	22.1	38.8
Sri Lanka	16.4	18.4
SOUTHEAST ASIA		
Brunei Darussalam	43.4	77.2
Cambodia	10.3	20.7
Indonesia	14.6	53.7
Laos	7.9	38.6
Malaysia	26.6	74.7
Myanmar	19.2	34.1
Philippines	30.3	44.4
Singapore	100.0	100.0
Timor-Leste	10.1	32.8
Thailand	19.7	50.4
Vietnam	14.7	33.6

Source: World Bank 2016

always imply a verifiable claim to being original inhabitants, unlike Indigenous communities in the Americas and Australia which are Indigenous with respect to European populations. In many cases, Indigenous communities are characterized by dependence on forest-based livelihoods (Erni 2008; Karlsson 2003). Indigenous lands and forests were often usurped under colonial rule and brought under agricultural development (e.g. Galudra and Sirait 2009), and these dispossessions have continued under post-colonial governments for agricultural, industrial and infrastructural projects. Since Indigenous communities usually lack formal title to land, it becomes difficult to establish land ownership and seek compensation when faced with dispossession (Xanthaki 2003).

This chapter considers some prominent struggles against environmental injustices in rural and urban South and Southeast Asia, and the countries that comprise the region as defined in this chapter are depicted in Figure 48.1. The struggles discussed do not always explicitly identify environmental justice as their goal, but can be viewed as making claims similar to those of environmental justice movements (Walker 2012), since they draw attention to marginalized social groups which suffer the burdens of environmental pollution and dispossession (distributive environmental injustice), and challenge the lack of public participation in environmental decision-making (procedural environmental injustice) (see Chapter 9). Further, in academic scholarship and governmental and non–governmental reports, existing economic and social inequities and movements against them have often been analysed through the lens of environmental justice (e.g. Harding 2007; Moore and Pastakia 2007; Pangsapa 2015; Ravi Rajan 2014; Williams and Mawdsley 2006). This chapter also considers how environmental justice is becoming part of national environmental policy objectives in the region, though whether this will ultimately contribute to preventing and penalizing environmental injustices remains to be seen.

Environmental justice struggles in rural contexts

Environmental injustices in rural South and Southeast Asia have been opposed by social movements focusing on forests, dams, mining, chemical agriculture, and coastal areas. Access to forests has been a prominent site of conflict between corporate entities and local communities across the region. A much publicized struggle is the Chipko movement, which occurred in northern India in the 1970s (Guha 1989). In the Garhwal Himalayan region, currently the western part of the state of Uttarakhand, peasant communities had sought access to and control of surrounding forests through the formation of a *Gram Swarajya Sangh* (village self-rule association) in Dasholi village in 1964. When logging rights were allotted to a private company, protests took the form of people wrapping their bodies around tree trunks to prevent the cutting of trees, the most famous of these led by women in Reni village in 1974. The term 'Chipko' (or adhering to) alludes to this embracing of trees, a mode of protest that draws on longer histories and cultures of tree protection in India (see also Chapter 39).

As a consequence of the Chipko movement, the Indian government in 1980 declared a 15-year ban on tree felling in the Himalayan region, but deforestation remains a problem. However, Chipko did lead to a shift towards participatory management of forests (Sundar et al. 2001), and became the basis for other movements, including a political struggle for the formation of the state of Uttarakhand (Rangan 2000), opposition to the Tehri dam on the Bhagirathi river (Warner 2015), and protests against real estate development in the region. In its use of non-violent tactics and goal of village self-rule, the Chipko movement epitomized Gandhian values, while the active role of female activists in leadership and protests led to Chipko being considered an example of ecofeminism. As a movement which drew attention to the exclusion of peasant and Indigenous communities from both access to forest resources and decision-making about use of local forests, Chipko can also be considered to embody environmental justice principles.

Deforestation and its impacts on local communities is a significant issue in the tropical forest regions of Southeast Asia (Colfer et al. 2008; Contreras-Hermosilla and Fay 2005). While the Amazon region draws a large amount of attention as a critical environmental resource, Southeast Asia's tropical forests constitute an equally significant contributor to global carbon sequestration. These forests are under threat as multinational companies are increasingly moving to Southeast Asia in pursuit of low-cost forest resources. Currently, the major cause of deforestation in this region is the rise of oil palm plantations. Burning is a widespread and cheap method to clear forests, and the resultant air pollution covers the region in smoke or 'haze.' While shifting cultivation by small farmers has been blamed as a cause of the haze, it is more likely that the widespread air pollution in the region is linked to growth of large plantations. Oil palm plantations are an instance of 'land grabbing,' the process whereby large tracts of land are taken over by corporations and governments without taking local uses into consideration. The consequences of this deforestation are borne by Indigenous Dayak communities in Malaysia and Indonesia (mainly on the island of Borneo), as they lose the basis of their livelihoods and culture (Alcorn and Royo 2000). Dayak movements against deforestation have had limited success in a context where oil palm promises swift economic returns. Yet, these economic benefits are not shared by local communities – even as oil palm plantations provide some local jobs, these are low-wage jobs and accompanied by loss of small-scale farming land. Issues of distributive justice are thus raised by deforestation in Southeast Asia.

The construction of dams has been another key site for environmental struggles in the region. A prominent instance is the movement against the construction of dams on the Narmada River in central India (Dreze et al. 1997). Water control projects were crucial to enabling higher agricultural production in colonial South Asia, and the scale of dam building (in terms of both number and size of dams) has considerably expanded in the post-colonial period in this region, as well as across the Global South. The struggle against the Sardar Sarovar, the largest of the Narmada dams, began in 1985 when activists raised questions about the information available to local people about displacements due to the dam and their rights to compensation. In 1990, the Narmada Bachao Andolan (NBA) was established to oppose the construction of the Narmada dams. When the World Bank decided to withdraw funding from the Sardar Sarovar in 1992, it seemed that rights of displaced communities had begun to be acknowledged. However, the Supreme Court of India's judgment of 2000, which enabled the continuation of dam building, belied such expectations. Currently, the Andolan continues to oppose the Narmada dams, while also seeking adequate compensation for those displaced.

The Narmada movement has made environmental justice claims through focusing on both local consequences and transnational underpinnings of dam building. The NBA successfully drew tribal communities (who consider themselves *adivasi* or Indigenous inhabitants in the context of India) and small-scale farmers into an alliance against the Narmada dams, while also drawing on support from urban activists and international environmental organizations (Baviskar 1995). The movement revealed the nexus between international development agencies and multinational corporations, mainly the World Bank and companies in North America and Western Europe, which ensured that profits from dam construction accrued to corporations, while the burdens of displacement were borne by rural communities.

The Mekong river dams constitute another prominent example of the loss of small-scale uses of rivers in favour of large-scale projects which benefit richer farmers, urban water supply systems, and corporate interests (Middleton 2012). Dam construction here encompasses Laos, Cambodia, Thailand, Vietnam, and China, and concerns have been raised not only in terms of upstream displacements, but also downstream losses due to dams. In the Tonle Sap region of Vietnam, the decrease in fresh water from the river could lead to saltwater intrusion and corresponding loss of rice fields and freshwater fish species.

Environmental inequities associated with mining have also raised concerns across the region (e.g. TERI 2013 and Saha et al. 2011 provide studies from India; see also Chapter 30). A particularly well known instance of struggle against injustice occurred in the state of Odisha in eastern India, where mining of bauxite by the multinational Vedanta company in the Niyamgiri Hills was successfully opposed by the Indigenous Dongria Kondh community (Bera 2013). By preserving Niyamgiri Hills, Dongria Kondhs seek to protect both their cultural heritage and environmental resources – the Hills are considered an important deity, and bauxite mining will disrupt surface and groundwater sources. The Niyamgiri struggle is noteworthy in that the decision to reject mining was arrived at through a referendum conducted by *palli sabhas* (village councils) in 2013. This referendum was supported by India's Supreme Court and becomes a significant instance of local participation in environmental decision-making. Environmental justice concerns have also been raised in terms of gold mines in the Philippines and Indonesia, where children constitute a large part of the labour force and are especially susceptible to poisoning due to the continued use of mercury for processing (Paddock and Price 2013).

Chemical agriculture is another major contributor to environmental pollution in rural South and Southeast Asia. This region was one of the primary sites of the 'Green Revolution,' an agricultural development programme that led to an increase in productivity of staple food grains in the 1960s–90s through the adoption of hybrid seeds, increased use of chemical fertilizers and pesticides, and access to irrigation. The negative effects of agricultural chemicals on human health (e.g. reflected in increasing incidences of cancer) and soil and water quality has been one of the principal criticisms of the Green Revolution (e.g. Antole and Pingali 1994 provide a case study in the Philippines; Singh 2000 provides a case study in India; see also Chapter 33).

Coastal communities across the region are facing environmental injustices caused by climate change and industrial hazards (e.g. Nayak 2016). Altered temperature and precipitation regimes brought about by climate change will affect small farmers across the region, while rising sea levels affect coastal and island regions, including the Maldives and Bangladesh (Darlington 2014; Hossain et al. 2016; Sivakumar and Stefanski 2011). The clearing of mangroves makes coasts more vulnerable to sea level rise and flooding. In the 2004 Indian Ocean earthquake and tsunami, coastal areas which had extensive mangroves were relatively less impacted than those which had narrow strips of mangroves or no mangrove cover (Dahdouh-Guebas et al. 2005). One main reason for mangrove clearances is commercial shrimp farming, impelled in part by rising global demand for seafood, even as shrimp farming introduces salinity and harmful chemicals into the water (EJF 2004). The high stakes involved in commercial shrimp farming became visible in a 1999 protest by small-scale crop farmers in Bangladesh, which was attacked by shrimp farmers and led to the death of a female activist and injuries to many others. In terms of industrial hazards, the coasts of South Asia are favoured destinations for ship breaking due to the low cost of labour here. Ship breaking in Bangladesh, India, and Pakistan exposes men of various ages to hazardous materials (Bengali 2016).

The diversity of environmental injustices faced by rural communities makes them central in struggles for social and environmental justice in South and Southeast Asia. In many cases, these rural injustices can be linked to demands for resources emanating from urban regions, even as environmental injustices also characterize urban contexts. The next section shifts the focus to urban environmental injustices in South and Southeast Asia.

Environmental justice struggles in urban contexts

Some major facets of urban environmental injustice in the region include slums, industrial pollution, and vehicular air pollution. Slums are a major social and environmental justice issue in the region since slum dwellers often lack access to basic amenities, especially piped drinking water and sanitation facilities, and could be exposed to environmental harms due to the location and quality of their housing and occupations (e.g. Ahmed 2014; Choi 2016). In 2005, around 54 per cent of the urban population in South Asia (Bangladesh, India, Nepal, and Pakistan) resided in slums, while the corresponding figure for Southeast Asia (Cambodia, Indonesia, Laos, Myanmar, Philippines, Thailand, and Vietnam) was around 49 per cent (ESCAP 2011). Even as slums can be considered a 'creative' solution to lack of urban housing, slum residents usually lack secure tenure and face constant threats of eviction. The class antagonisms that constitute the city are revealed in programmes to remove slums. For city authorities, removal of slums is part of the process of making the city more attractive to investors. For real estate development, the proximity of slums to the city's economic centres makes them valuable areas for redevelopment. In some cases, as urban slums often house rural migrants from distant regions, sometimes from neighbouring countries, slum removal is also an aspect of anti-immigrant attitudes. Yet, slum residents are extremely valuable as a low-cost source of labour – for instance, working in the housing construction industry as well as in the homes of well-off urban residents. Thus, slum residents are contributing to urban economic development while remaining excluded from its benefits.

Industrial pollution is a matter of concern across the region. An infamous example of the consequences of industrial pollution was provided by the 1984 disaster in the city of Bhopal in central India (Hanna et al. 2005; Pariyadat and Shadaan 2014). An accident in a Union Carbide plant producing pesticides led to the leak of highly toxic methyl isocyanate as well as other chemicals into the air. Houses were densely packed around the factory and residents were not aware that they were living adjacent to a hazardous unit. In the immediate aftermath of the leak, residents were not provided any information on evacuation by governmental authorities. The informal quality of their dwellings did not offer protection, and due to prevailing wind conditions, residents moving away from the factory were actually following the flow of the gas leak (Delhi Science Forum 1985). The Bhopal gas tragedy is considered one of the worst industrial accidents in the world, and those affected by it continue to suffer from respiratory diseases with subsequent generations showing birth defects due to poisoning. The death toll from the disaster was estimated at around 15 000, but this does not take into account those who were injured and continue to suffer from the effects of the poisoning. Affected people were never adequately compensated either by the Indian government or by Union Carbide (now part of the Dow group). The Bhopal disaster exemplifies various facets of the environmental injustice faced by communities in the Global South, including lack of information about environmental dangers, manufacturing linked to multinational corporations which escape national and international regulatory and legal control, inability and unwillingness of governments to represent the interests of their own citizens, and lack of access to emergency and healthcare that would enable people to cope with the aftermath of disasters. As industrial-ization continues in the region, its burdens continue to be unevenly distributed. For example, petrochemical manufacturing in Map Ta Phut, Thailand's largest industrial estate, has affected the health of surrounding rural communities while also affecting coastal tourism through oil spills (Pangsapa 2015).

Another major problem in this region is outdoor air pollution. Increase of carbon monoxide, nitrogen dioxide, and fine particulate matter in the air has been linked to growth in automobile ownership in urban areas, as private automobile use is becoming an increas-ingly popular form of urban transport. The region's cities are also being targeted as new markets for mass and luxury automobiles, and have become hubs for automobile components manufacture and assembly. While exposure to air pollution encompasses the entire urban population, the ability to escape the consequences of air pollution through access to health-care or residence in less polluted neighbourhoods is not available to low-income social groups (e.g. Kathuria and Khan 2007; Kumar and Foster 2009; and Sabapathy et al. 2015 provide case studies from India). The rise of respiratory diseases among children and elderly populations is another example of the health disparities associated with air pollution (e.g. Ruchirawat et al. 2006 provide a study of Southeast Asia). Yet, efforts to curb air pollution can also be linked to environmental injustice. When polluting industries or transport vehicles are sought to be banned or removed from the city, often a key objective of middle-class environmental movements, the loss of jobs or costs associated with relocation often disproportionately affects low-income groups (e.g. Véron 2006).

While rural and urban environmental injustices have been considered separately in this chapter, processes and experiences of environmental injustices often connect across urban and rural areas. One major connection is through rural to urban migrations. When rural residents are displaced by resource extraction projects, or growth of cities into surrounding rural areas, they often move to urban areas and settle in slums. Subsequently, when slums are sought to be removed, rural residents face yet another round of displacement. In India, for instance, both the Chipko and Narmada movements have moved beyond their rural focus and become active in protesting the environmental degradation that accompanies urban growth in the Himalayas, and violent displacement of slum residents respectively.

Less developed rural and low-income urban communities face similar issues in terms of lack of access to adequate services, including electricity, water, and sanitation. In the absence of electricity, both urban and rural residents turn to wood and coal for cooking and heating. This burning of wood and coal contributes to indoor air pollution in low-income urban households as well as city-wide air pollution. In rural areas, the use of wood burning stoves has been identified as one of the main causes of indoor air pollution and respiratory illnesses in women and children (WHO 2006). Injustices related to water access are being faced in rural and urban South and Southeast Asia (e.g. Domènech et al. 2013; Luby 2008; Mehta et al. 2014; Venot and Clement 2013). Rivers often show high levels of pollution, especially near urbanized areas, due to dumping of industrial wastes and residential sewage (e.g. Kedzior 2014). Another major cause of water contamination is the presence of naturally occurring arsenic in ground water, which becomes a pressing issue in the absence of alternative sources of water (Buschmann et al. 2008; Hassan et al. 2005). Lack of piped domestic water and sanitation facilities affects women's work and health burdens in both rural and urban contexts. Water pipes and water treatment plants are often in need of upgradation to ensure water quality. While household-level water purification options are available, these are often not affordable for low-income groups. Various kinds of rural and urban injustices thus often combine to exacerbate environmental burdens borne by marginalized groups.

Governments, international development agencies and environmental justice

Partly in response to movements against environmental pollution and human displacement, and partly impelled by international development organizations, governments in South and Southeast Asia have begun to reform procedures associated with the implementation of development programmes and institute laws related to environmental protection (AJNE 2016; Harding 2007; Hofman 1998; Philippine Judicial Academy 2014; Tan 2004). Current policies of land acquisition, access to forests, and Indigenous rights are often continuations of colonial policies, so that their reform is long overdue. The utilization of environmental impact assessments (EIAs), establishment of 'green tribunals,' and public interest litigations (PILs) are prominent instances of the expansion of governmental and legal frameworks to address environmental injustice (e.g. Lima et al. 2015; Roque 1985; Sahu 2014). However, any reforms have to contend with the fact that achieving Western-style urbanization and consumption is a development priority for many countries in this region, and it remains doubtful whether concerns about pollution and dispossession will lead to any substantive change in economic and urban development policies.

International development and regional organizations, such as the Asian Development Bank (ADB), Association of Southeast Asian Nations (ASEAN), United Nations Environment Program (UNEP) and World Bank, have articulated the need to construct and strengthen environmental justice frameworks through environmental laws, and people's participation in development programmes. Conferences have been organized by the above mentioned organizations which seek to connect environmental justice to economic development and attention is being specifically directed towards the needs of the poor in this region (ADB 2012; UNEP 2013). In practice, such legal and participatory frameworks often become a means to cast a veneer of legality over continuing resource appropriation and environmental pollution. Though local communities and environmental organizations have often successfully challenged injustices through available legal mechanisms, the building and pursuing of a legal case is a very difficult process for rural and low-income communities that are already burdened with precarious livelihoods and lack of access to education and expertise. Further, international development organizations are currently in favour of private solutions to environmental problems – for instance, through insisting that low-income communities will benefit from the privatization of water supply systems (Sims 2015). In the context of neoliberalization, therefore, environmental access and ownership is increasingly shaped by market-led processes which exacerbate inequality rather than provide any lasting pathway to equity.

A major problem faced across South and Southeast Asia is the absence of adequate data to measure and analyse environmental injustice (e.g. lack of information in the context of India is discussed by Gokhale-Welch 2009, Kathuria 2009 and Shetty and Kumar 2016). Though the Census in individual countries of South and Southeast Asia provides valuable data on demographic characteristics, education, employment, and health, there is no corresponding governmental effort to collect data on rural displacement, loss of natural resources, and various forms of industrial pollution. Studies of displacement and loss of small-scale livelihoods often find it difficult to provide data about communities affected by these processes, since official records are non-existent or outdated. This is the case, for instance, with Indigenous communities whose use of forests or cultivation of land is often informal, posing problems when compensation for losses needs to be calculated and legally verified. Studies of pollution are often focused on the most prominent egregious cases, and the more widespread nature of everyday pollution remains undocumented. Social movements and environmental organizations have sought to combat this lack of information through their own data gathering practices (e.g. through Right to Information Acts, see Article 19 2015) and awareness campaigns (e.g. Hobson 2006). Overall, states in South and Southeast Asia have mainly been focused on procedural forms of justice, through the construction of legal forums, whereas social movements have struggled to establish both procedural and distributive forms of justice by challenging the inequalities embedded in processes of economic development.

Conclusion

This chapter has made two main arguments related to environmental justice in South and Southeast Asia. First, environmental justice in this region is a matter of vibrant people's struggles, rather than any concerted governmental efforts to construct and implement environmental laws. Social movements thus become an important component of understanding how environmental injustices are made visible and sought to be addressed in this region. Second, environmental justice struggles are shaped by the presence of rural–urban divisions. While this chapter has discussed rural and urban struggles separately, these often converge due to the fact that urban lifestyles are dependent on the continued extraction of rural resources and hence lead to rural dispossessions and migration, and lack of access to infrastructure and services affects low-income communities in both urban and rural areas.

A final point is that much remains to be studied and analysed about forms of environmental injustice in South and Southeast Asia. Future studies may consider the problems associated with evidence for environmental injustice given the absence of data sources. This has meant that environmental injustice becomes visible in terms of extraordinary struggles, rather than through everyday experiences of environmental pollution. More comprehensive portrayals of how environmental loss and pollution affect vulnerable and marginalized populations in this region – including rural and Indigenous communities, the urban poor, and women, children, and the elderly – thus remains an ongoing endeavour.

References

ADB (Asian Development Bank). (2012). *South Asia conference on environmental justice*. www.adb.org/publications/south-asia-conference-environmental-justice (accessed 23 August 2016).

Ahmed, A. (2014). "Implications of the environmental justice movement on redistributive urban politics: an example from megacity Dhaka, Bangladesh." In Leonard, L. and S. Kedzior, (eds.), *Occupy the earth: Global environmental movements*, Bingley, UK: Emerald Publishing, pp. 255–274.

AJNE (Asian Judges Network on Environment). (2016). *Home*. www.ajne.org (accessed 27 March 2017).

Alcorn, J., & Royo, A. (eds.) (2000). *Indigenous social movements and ecological resilience: Lessons from the Dayak of Indonesia*. Washington, DC: Biodiversity Support Program.

Antole J., & Pingali, P. (1994). "Pesticides, productivity, and farmer health: a Philippine case study." *American Journal of Agricultural Economics*, vol. 76, pp. 418–430.

Article 19. (2015). *Asia disclosed: A review of the right to information across Asia*. London. www.article19.org/resources.php/resource/38121/en/report:-the-right-to-know-across-asia (accessed 23 August 2016).

Bakker, L. (2009). "Community, *adat* authority and forest management in the hinterland of east Kalimantan." In McCarthy, J. and C. Warren (eds.), *Community, environment and local governance in Indonesia: Locating the Commonweal*, New York and London: Routledge, pp. 121–144.

Baviskar, A. (1995). *In the belly of the river: Tribal conflicts over development in the Narmada Valley*. Delhi: Oxford University Press.

Bengali, S. (2016). "Adult and underage workers risk their lives in Bangladesh's rising ship-breaking industry." *Los Angeles Times*, 9 March. www.latimes.com/world/asia/la-fg-bangladesh-ships-20160309-story.html (accessed 23 August 2016).

Bera, S. (2013). "Niyamgiri answers." *Down to Earth*, 31 August. www.downtoearth.org.in/coverage/niyamgiri-answers-41914 (accessed 23 August 2016).

Buschmann, J., Berg, M., Stengel, C., Sampson, M., Trang, P., & Viet, P. (2008). "Contamination of drinking water resources in the Mekong delta floodplains: arsenic and other trace metals pose serious health risks to population." *Environment International*, vol. 34, pp. 756–764.

Choi, N. (2016). "Metro Manila through the gentrification lens: disparities in urban planning and displacement risks." *Urban Studies*, vol. 53, pp. 577–592.

Colfer, C., Dahal, G., & Capistrano, D. (eds.) (2008). *Lessons from forest decentralization: Money, justice and the quest for good governance in Asia-Pacific*. London: Earthscan.

Contreras-Hermosilla, A., & Fay, C. (2005). *Strengthening forest management in Indonesia through land tenure reform: Issues and framework for action*. Washington DC: Forest Trends.

Dahdouh-Guebas, F., Jayatissa, L., Di Nitto, D., Bosire, J., Seen, D., & Koedam, N. (2005). "How effective were mangroves as a defence against the recent tsunami?" *Current Biology*, vol. 15, pp. 1337–1338.

Darlington, S. (2014). "Environmental justice in Thailand in the age of climate change." In Schuler, B. (ed.), *Environmental and climate change in South and Southeast Asia*, Leiden, The Netherlands: Brill, pp. 211–230.

Delhi Science Forum. (1985) "Bhopal gas tragedy." *Social Scientist*, vol. 3, pp. 32–53.

Domènech, L., March, H., & Saurí, D. (2013). "Contesting large-scale water supply projects at both ends of the pipe in Kathmandu and Melamchi Valleys, Nepal." *Geoforum*, vol. 47, pp. 22–31.

Dreze, J., Samson, M., & Singh, S. (eds.) (1997). *The Dam and the nation: Displacement and resettlement in the Narmada Valley*. Cambridge, UK: Cambridge University Press.

EJF (Environmental Justice Foundation). (2004). *Farming the sea, costing the earth: Why we must green the Blue Revolution*. London.

Erni, C. (ed.) (2008). *The concept of indigenous peoples in Asia: A Resource Book*. Copenhagen, Denmark: International Work Group for Indigenous Affairs (IWGIA).

ESCAP (United Nations Economic and Social Commission for Asia and the Pacific). (2011). "People – demographic trends – urbanization." *Statistical Yearbook for Asia and the Pacific 2011*. www.unescap.org/stat/data/syb2011/I-People/Urbanization.asp (accessed 23 August 2016).

Galudra, G., & Sirait, M. (2009). "A discourse on Dutch colonial forest policy and science in Indonesia at the beginning of the 20th century." *International Forestry Review*, vol. 11, pp. 524–533.

Gokhale-Welch, C. (2009). *Toxic release inventory for India: A discussion paper*. Chennai, India: Centre for Development Finance: Institute for Financial Management and Research. www.ifmrlead.org/wp- content/uploads/2015/OWC/TRI_Report_Finale.pdf (accessed 23 August 2016).

Guha, R. (1989). *The unquiet woods: Ecological change and peasant resistance in the Himalaya*. Berkeley, CA: University of California Press.

Hanna, B., Morehouse, W., & Sarangi, S. (2005). *The Bhopal reader: Remembering twenty years of the world's worst industrial disaster*. New York: Apex Press.

Harding, A. (ed.) (2007). *Access to environmental justice: A comparative study*. Leiden, The Netherlands: Martinus Nijhoff.

Hassan, M., Atkins, P., & Dunn, C. (2005). "Social implications of arsenic poisoning in Bangladesh." *Social Science and Medicine*, vol. 61, pp. 2201–2211.

Hobson, K. (2006). "Enacting environmental justice in Singapore: performative justice and the Green Volunteer Network." *Geoforum*, vol. 37, pp. 671–681.

Hofman, P. (1998). "Participation in Southeast Asian pollution control policies." In Coenen, F., D. Huitema and L. O'Toole (eds.), *Participation and the quality of environmental decision making*, Heidelberg, Germany: Springer, pp. 287–305.

Hossain, M., Dearing, J., Rahman, M., & Salehin, M. (2016). "Recent changes in ecosystem services and human well-being in the Bangladesh coastal zone." *Regional Environmental Change*, vol. 16, pp. 429–443.

Karlsson, B. (2003). "Anthropology and the 'indigenous slot:' claims to and debates about indigenous peoples' status in India." *Critique of Anthropology*, vol. 23, pp. 403–423.

Kathuria, V. (2009). "Public disclosures: using information to reduce pollution in developing countries." *Environment, Development and Sustainability*, vol. 11, pp. 955–970.

Kathuria, V., & Khan, N. (2007). "Vulnerability to air pollution: is there any inequity in exposure?" *Economic and Political Weekly*, vol. 42, pp. 3158–3165.

Kedzior, S. (2014). "Locating environmental knowledge in anti-pollution movements of northern India." In Leonard, L. and Kedzior, S. (eds.), *Occupy the earth: Global environmental movements*, Bingley, UK: Emerald Publishing, pp. 79–116.

Kumar, N., & Foster, A. (2009). "Air quality interventions and spatial dynamics of air pollution in Delhi and its surroundings." *International Journal of Environment and Waste Management*, vol. 4, pp. 85–111.

Lima, A. et al. (2015). *Environmental impact assessment systems in South Asia*. Paper presented at the International Association of Impact Assessment Conference, Florence, Italy. http://conferences.iaia. org/2015/Final-Papers/Sanchez-Triana,%20Ernesto%20-%20Environmental%20Impact%20 Assessment%20Systems%20in%20South%20Asia.pdf (accessed 23 August 2016).

Luby, S. (2008). "Water quality in South Asia." *Journal of Health, Population and Nutrition*, vol. 26, pp. 123–124.

Martinez-Alier, J. (2002). *The environmentalism of the poor*. Northampton, MA: Edward Elgar.

Mawdsley, E. (2004). "India's middle classes and the environment." *Development and Change*, vol. 35, pp. 79–103.

McMichael, P. (2012). *Development and social change*. Thousand Oaks, CA: Sage.

Mehta, V., Goswami, R., Kemp-Benedict, E., Muddu, S., & Malghan, D. (2014). "Metabolic urbanism and environmental justice: the water conundrum in Bangalore, India." *Environmental Justice*, vol. 7, pp. 130–137.

Middleton, C. (2012). "Transborder environmental justice in regional energy trade in mainland South-East Asia." *Austrian Journal of South-East Asian Studies*, vol. 5, pp. 292–315.

Moore, P., & Pastakia, F. (eds.) (2007). *Environmental justice and rural communities: Studies from India and Nepal*. Bangkok: IUCN.

Nayak, P., Armitage, D., & Andrachuk, M. (2016). "Power and politics of social-ecological regime shifts in the Chilika lagoon, India and Tam Giang lagoon, Vietnam." *Regional Environmental Change*, vol. 16, pp. 325–339.

Paddock, R., & Price, L. (2013). "Hunt for gold in Southeast Asia poisons child workers, environment." *Center for Investigative Reporting*. http://cironline.org/reports/hunt-gold-southeast-asia-poisons-child-workers-environment-5678 (accessed 23 August 2016).

Pangsapa, P. (2015). "Environmental justice and civil society: case studies from Southeast Asia." In Harris, P. and G. Lang (eds.) *Routledge handbook of environment and society in Asia*, New York and London: Routledge, pp. 36–52.

Pariyadat, R., & Shadaan, R. (2014). "Solidarity after Bhopal: building a transnational environmental justice movement." *Environmental Justice*, vol. 7, pp. 146–150.

Philippine Judicial Academy. (2014). *Citizens handbook of environmental justice.* www.ombudsman.gov. ph/UNDP4/wp-content/uploads/2013/02/s-HanBook-CC1.pdf (accessed 23 August 2016).

Rangan, H. (2000). *Of myths and movements: Rewriting Chipko into Himalayan history.* London: Verso.

Ravi Rajan, S. (2014). "A history of environmental justice in India." *Environmental Justice*, vol. 7, pp. 117–121.

Roque, C. (1985). "Environmental impact assessment in the Association of Southeast Asian Nations." *Environmental Impact Assessment Review*, vol. 5, pp. 257–263.

Ruchirawat, M., Navasumrit, P., Settachan, D., & Autrup, H. (2006). "Environmental impacts on children's health in Southeast Asia: genotoxic compounds in urban air." *Annals of the New York Academy of Sciences*, vol. 1076, pp. 678–690.

Sabapathy, A., Saksena, S., & Flachsbart, P. (2015). "Environmental justice in the context of commuters' exposure to CO and PM10 in Bangalore, India." *Journal of Exposure Science and Environmental Epidemiology*, vol. 25, pp. 200–207.

Saha, S., Pattanayak, S., Sills, E., & Singha, A. (2011). "Under-mining health: environmental justice and mining in India." *Health and Place*, vol. 17, pp. 140–148.

Sahu, G. (2014). *Environmental jurisprudence and the Supreme Court: Litigation, interpretation and implementation.* New Delhi: Orient BlackSwan.

Shetty, S., & Kumar, S. (2016). "Opaqueness of environmental information in India." *Economic and Political Weekly*, vol. 51. www.epw.in/journal/2016/30/commentary/opaqueness-environmental-information-india.html (last accessed 23 August 2016).

Sims, K. (2015). "The Asian Development Bank and the production of poverty: neoliberalism, technocratic modernization and land dispossession in the Greater Mekong Subregion." *Singapore Journal of Tropical Geography*, vol. 36, pp. 112–126.

Singh, R. (2000). "Environmental consequences of agricultural development: a case study from the Green Revolution state of Haryana, India." *Agriculture, Ecosystems and Environment*, vol. 82, pp. 97–103.

Sivakumar, M., & Stefanski, R. (2011). "Climate change in South Asia." In Lal, R., M. Sivakumar, M. Faiz, A. Mustafizur Rahman, and K. Islam. (eds.) *Climate change and food security in South Asia*, Heidelberg, Germany: Springer, pp. 13–30.

Sundar, N., Jeffery, R., & Thin, N. (eds.) (2001). *Branching out: Joint forest management in India.* New Delhi: Oxford University Press.

Tan, A. (2004). "Environmental laws and institutions in Southeast Asia: a review of recent developments." *Singapore Yearbook of International Law*, Vol. VIII, pp. 177–192.

TERI (The Energy Research Institute). (2013). *Equitable sharing of benefits arising from coal mining and power generation among resource rich states.* New Delhi.

UNEP (United Nations Environmental Program). (2013). *1st Asia and Pacific Colloquium on environmental rule of law.* www.unep.org/delc/worldcongress/WorkshopsEvents/AsiaPacificColloquium/tabid/132341/Default.aspx (accessed 23 August 2016).

Venot, J., & Clement, F. (2013). "Justice in development? An analysis of water interventions in the rural South." *Natural Resources Forum*, vol. 37, pp. 19–30.

Véron, R. (2006). "Remaking urban environments: the political ecology of air pollution in Delhi." *Environment and Planning A*, vol. 38, pp. 2093–2109.

Walker, G. (2012). *Environmental justice: Concepts, evidence and politics.* Oxon, UK: Routledge.

Warner, H. (2015). *The politics of dams: Developmental perspectives and social critique in modern India.* Oxford: Oxford University Press.

WHO (World Health Organization). (2006). *Fuel for life: Household energy and health.* www.who.int/indoorair/publications/fuelforlife/en/ (accessed 23 August 2016).

Williams, G., & Mawdsley, E. (2006). "Postcolonial environmental justice: government and governance in India." *Geoforum*, vol. 37, pp. 660–670.

World Bank. (2016). *Urban population (% of total).* http://data.worldbank.org/indicator/SP.URB. TOTL.IN.ZS?name_desc=false (accessed 23 August 2016).

Xanthaki, A. (2003). "Land rights of indigenous peoples in South-East Asia." *Melbourne Journal of International Law*, vol. 4. http://law.unimelb.edu.au/__data/assets/pdf_file/0007/1680361/Xanthaki.pdf (accessed 23 August 2016).

49

ENVIRONMENTAL JUSTICE IN A TRANSITIONAL AND TRANSBOUNDARY CONTEXT IN EAST ASIA

Mei-Fang Fan and Kuei-Tien Chou

Compared with the neoliberal regimes of Western society, in East Asia, faced with the liberalism of integrated markets and the authoritative expert politics of developmental states, the challenge for environmental justice is even greater. The opposition between authoritative politics and democratic governance is embedded in the social and political contexts of the regions. During the early modernization process in the 1970s, South Korea experienced authoritative bureaucracy (Han and Shim 2010). Despite Japan's defeat in World War II, the nation's militarism laid the foundation for an authoritative culture. In Taiwan, the KMT regime continued to fiercely suppress the rise of an opposition movement until the early 1980s. Under this regime, the developmental state was ruled with a mixture of bureaucracy and technological elitism. Technocrats guided national economic and technological development, forming the foundation of authoritative expert politics. With the political democratization of the 1980s, the political power consortia (chaebols, or big enterprise) in South Korea, Japan, and Taiwan gradually penetrated government, which weakened the power of the developmental state. The technocrats lost their autonomy in the decision-making process (Castells 1992; Weiss 1998), and were replaced by a regime of loosened control-oriented neoliberalism to further prioritize economic development (Kim 2007; Chu 2011; Shin 2012). Since China's economic reform of 1979 introduced capitalism, opened China to the rest of the world and modernized the nation, the policy of 'growth at any cost' has led to social unrest, a growing disparity between social classes, and environmental catastrophe (Joines 2012).

Minority and disadvantaged groups tend to suffer a disproportionate number of environmental burdens, particularly during rapid industrialization and urbanization. Government authorities have tended to adopt a 'least resistance' approach to decisions involving siting hazardous facilities, and residents of disadvantaged communities have often lacked adequate information on the related risks. A quest for environmental justice has unfolded democratization processes in East Asia's new democracies. Catalytic events have heightened public awareness of environmental degradation and trans-boundary health risks and injustices, and have aided people in fighting for the right to know and for a larger political space from which to influence decisions and institutional reform.

This chapter explores the unfolding of environmental justice issues in the context of democratic transition. It examines the complexity and dynamics of generating spaces of inequality, the development and significance of environmental justice campaigns, local activism against the development-oriented state, encroachment by strong business interests, and authoritative expert politics. This chapter highlights multiple forms of citizen activism for addressing environmental injustice. Collaboration among local residents, activists, and academia has challenged authoritative scientific regimes and regulatory policies and has contributed to knowledge production in a few situations. As new disparities develop at multiple scales, local and regional networks evolve to address common trans-boundary risks and fight for the well-being of future generations.

Environmental justice in Taiwan

Environmental justice was introduced in Taiwan by academia in the 1990s. Early research connected the environmental struggles in the United States with polluting activities and Indigenous struggles in Taiwan. Chi (2001) argued that Aboriginal tribes have suffered the most environmental injustice induced by Taiwan's capitalist expansion policy, as particularly indicated by the development of three national parks on historical Indigenous lands and hunting grounds, a nuclear waste facility on the Yami tribe's homeland, and proposed dam construction on the Rukai tribe's homeland. The Yami tribe's antinuclear waste movement is a manifestation of problems caused by distributional inequity, lack of recognition, and limited participation of tribesmen in decision-making, which are interwoven into political and social processes (Fan 2006a, 2006b). News media reported that a research team detected cobalt 60 and caesium 137 near the nuclear waste repository on Orchid Island in 2011. The Yami tribesmen were hired cheaply by a subcontractor to work in the nuclear repository and suffer from multiple risks and injustices, as well as from the problem of 'undone science' regarding the long-term impacts of nuclear waste on health and the environment. Antinuclear waste activists from various generations of tribesmen on Orchid Island have collaborated with expert activists, antinuclear organizations, and local activists from the nuclear power plants on the north coast of Taiwan to create hybrid antinuclear alliances; they took action to express their demands and challenged official regulatory standards and test methods (Fan 2016a).

The government announced that Da-Ren in Taitung County and Wang-An in Penghu County are favoured potential sites for final nuclear waste disposal in 2009. These two towns are relatively remote with comparatively sparse populations. Da-Ren is the homeland of the Pauwan tribe. The local opponents of nuclear waste have tended to perceive their situation as economically marginalized, politically excluded, and culturally patronized, and have charged that siting policies systematically and deliberately place vulnerable communities at a disadvantage. Nuclear waste siting policies must redress the broader pattern of the unjust treatment of disadvantaged groups (Huang et al. 2013).

Water diversion projects in Taiwan have also raised critical equity questions, such as 'What is the risk of harm to the environment within the basin of origin because of the water loss resulting from the transbasin water transfer?' When Typhoon Morakot hit Taiwan on 8 August 2009, neighbouring villages of the Tseng-Wen Reservoir Transbasin Water Diversion Project were severely affected by floods and mudslides, and local residents and environmental groups asserted that the water diversion project was responsible for the destruction of their villages. This case demonstrates the problems of uneven risks and catastrophic impacts on traditional ways of life; moreover, the tribes' unique cultural meanings and invisible cultural assets as well as their integrated relationship with nature are excluded from the EIA report, which lacks fully informed consent or substantial participation from local residents in the decision-making process (Fan 2016b).

Concerns are emerging regarding the notion that Indigenous minorities living in the vulnerable mountainous areas in Taiwan possibly suffer from harmful impacts and disproportionate losses and that they might become environmental refugees after encountering extreme climate change. Environmental and climate justice has provided an organizing frame for NGO coalitions that advocate policies for mitigation and low-carbon transition. The disaster of Typhoon Morakot attracted considerable attention to the vulnerabilities of the poor and Indigenous minorities living in the mountainous areas. Although the government maintained that the tragedy was caused by a 'natural disaster', local people tend to believe that it was caused by human activities. Moreover, the responsibility for the disaster has been ascribed to the government and its policies, such as the transbasin water transfer development project and forest management policies, failure regarding soil and water conservation, and a lack of comprehensive land-use policies and laws (Fan 2015).

Since the democratic transition of Taiwan in 1987, various democratic institutions have been established, and movements against environmental injustice have unfolded (e.g. the Environmental Impact Assessment Law was approved in 1994) (Tang et al. 2011). The movement against the expansion of science parks and electronic hazards in Taiwan has developed and adopted the concept of environmental justice during the last ten years. The controversies of the Hsinchu Science Park and former workers' lawsuits against RCA (Radio Corporation of America) have been collected in a book titled *Challenging the Chips: Labour Rights and Environmental Justice in the Global Electronics Industry*, which highlights that neighbouring residents and frontline workers suffer the most from electronic hazards (Chang et al. 2006). Campaigns against the electronic industry have protested against water contamination and environmental pollution, land expropriation and water division, and disproportionate health risks to local and nearby residents. A few Taiwanese academicians who used to study in the United States formed the Taiwan Environmental Action Network and invited the Silicon Valley Toxic Coalition to share their advocacy and organizing experiences. The movement has grown more robust with a central focus on distributive justice, the right to be recognized as stakeholders in corporations and political processes, and democratic decision-making and information transparency. The environmental organizations have established networks with farmers, fishermen, and local communities, and their advocating for tighter environmental regulations under the precautionary principle has led to the approval of the new version of the Toxic Chemical Substances Act (Chiu 2014).

Since President Ma took office and promoted the concept of environmental justice in 2008, environmental authorities have laid considerable emphasis on environmental justice as an essential policy principle and goal. For the government, environmental justice is particularly relevant for stipulating human rights protection as a part of basic rights in the Constitution and two conventions: International Covenant on Civil and Political Rights and International Covenant on Economic, Social, and Cultural Rights. Taiwan's Environmental Protection Agency (EPA) published 'Ten Lessons of Environmental Justice', introducing the concepts, international laws, and environmental justice cases of Taiwan and other countries, to raise public awareness and promote environmental education. The EPA tends to link environmental justice with the idea of sustainable homeland, pollution prevention, and carbon emission reduction. Moreover, the EPA considers several policy programmes and strategies associated with institutional constructions advocating sustainable development; resource recycling; and zero waste, energy saving, and mitigation measures as the main components of environmental justice (EPA 2013).

Nowadays environmental justice has been widely used in the press, environmental campaign leaflets, election advertisements, and government official announcements and policy commitments. Environmental NGOs and local activists often adopt an environmental justice framework, to appeal to public concerns about environmental destruction and disadvantaged situations of marginalized groups and to fight against developmental projects dominated by state and industry developers. Huang (2011) considers that policy declarations lacking clear definitions of environmental justice have caused strong critiques from NGOs when they encounter a major gap between the ideal and reality. The government must clearly define environmental justice and endeavour to construct institutions advocating environmental justice, as well as effectively utilize various policy tools for implementing policy goals.

Environmental justice in China

In China, environmental justice is framed in the broader context of human rights discourse. Because the government and industry appropriated the resources of socioeconomically disadvantaged communities for economic and national development, environmental inequality and human rights violations are inextricably interwoven, specifically under authoritarian regimes (Adeola 2000). Environmental injustice in China shows strong regional inequity. Developmental projects, such as coal mining, coal plants, and chemical plants, and intensive exploitation of natural resources in the surrounding provinces in Midwestern and Western China have caused irreversible environmental degradation (Guo 2008). Economically disadvantaged communities, who lie at the bottom of the society, have suffered social exclusion. Rural residents and specifically rural migrants are disproportionately exposed to industrial pollution, and townships with a high percentage of rural migrants are more likely to be exposed to high levels of air and water pollution (Ma 2010; Schoolman and Ma 2012).

Environmental justice discourses in China are mainly focused on the unequal distribution of environmental problems associated with toxic waste facility siting, chemical plants, and risks of developmental projects. Bell (2014) examined substantive, distributional, and procedural environmental justice in China. Although China has made relative progress in addressing its environmental concerns in recent years, substantive environmental justice problems associated with air, water, and soil pollutions and CO_2 emissions continue to exist. Rural migrants experience disproportionate environmental burdens after moving to periurban areas of cities. The *Hukou* system restricts rural migrants from accessing some public services, thus dissuading them from settling permanently and consequently preventing urban overcrowding. Rural migrants frequently move to polluted areas, where they find jobs and cheap accommodation. Bell (2014) indicated that environmental problems in China are highly mobile, generally moving from larger to smaller cities, from urban to rural areas, and from more to less developed areas of the country. Regarding procedural environmental justice, environmental legislation always functions in a top-down manner, and citizens rarely have opportunities to air their concerns. It is the elites who are most able to effectively engage in policy processes.

Although legal and institutional reforms have failed to improve the environmental situation in China, Balme (2014) reported that the collective action taken for environmental justice and litigation over the last decade has facilitated some improvements in environmental policymaking procedures. The Nu river mobilization successfully stopped a project that entailed construction of 13 dams. Barrington et al. (2012) illustrated various social and environmental forms of injustice concerning hydroelectric dams in the Mekong Basin, specifically its adverse ecological effects and the loss of livelihoods and homes. The authors indicated that engineers from various NGOs, such as Engineers Without Borders Austria, often work with affected communities to restore the ecological and social damages caused by the construction of large dams and suggest environmental justice polices.

Struggles against waste management facilities and incineration sites are also major components of environmental justice activism. An analysis of the anti-Liulitun incineration plant campaign in Beijing (Xie 2011) indicated that procedure, recognition, and participation are relevant aspects of environmental justice concerning China, specifically procedural justice. Residents claimed that the process of public consultation required by the Energy Information Administration (EIA) in China was 'not fairly conducted' and was 'lacking in justice'. The government was strongly condemned for its plan to build a highly risky incineration facility that may harm residents. In China, the aspect of recognition is often neglected during environmental policymaking. Formal procedures for considering various interests are mainly ineffective. The successful mobilization of urban citizens shows how certain social groups are more capable of getting their interests recognized.

Similarly, Liu et al. (2014) indicated that Mongolians have many long-established grievances, including the ecological destruction wrought by an unprecedented mining boom, which disproportionately benefits China at the cost of the environment, local residents' livelihood, and the rapid disappearance of Inner Mongolia's pastoral tradition. A public consultation by the Chinese EIA was not conducted for several small-scale illegal mines, and corrupt officials might have falsified information on many other mines. The authors argue that under the current political system in China, the government should adopt Sustainability Impact Assessments for promoting environmental justice while continuing the fight against corruption.

Environmental justice in South Korea

South Korean research on environmental justice began with environmental pollution. The concept of environmental justice was introduced by the government in South Korea after the 1990s. The Korean government has tried to include environmental justice in its laws and policies and has extensively discussed the concept. The Korean Ministry of Environment held several symposia on environmental justice, leading to the establishment of the Citizen's Movement for Environmental Justice (CMEJ) in 1992. In 1999, the CMEJ hosted the Environmental Conflict and Environmental Justice Forum, in which several case studies on environmental injustice, including a canal between Seoul and Incheon, the reclaimed land of Saemangum, the development of the city of Yongin without consideration of the environment, and the construction of the industrial complex in the Weechun region, were investigated from the perspective of environmental justice. South Korean discussions on environmental justice are concentrated on rural regions, where most environmentally contaminating facilities are located. Similar to the early UK environmental justice work initiated by Friends of the Earth in collaboration with NGOs and academics, environmental justice in South Korea was developed by academic elites rather than local people and communities (Lee 2009).

Bell (2014) reports that the substantive, distributional, and procedural aspects of environmental injustice continue in Korea. Substantive injustices include water pollution, inadequate housing, excessive use of natural resources, and weak environmental legislation. The endemic social inequality, especially when combined with geographical segregation on the basis of income, emphasizes distributional injustice. Contaminating facilities are often located in rural areas, and low-income residents are often disadvantaged during development and regeneration projects. In relation to procedural justice, freedom of expression is restricted, and participatory environmental decision-making is often tokenistic. The government considered growth and environmentalism compatible. However, green growth as a national development paradigm does not deliver environmental justice.

Lee (2009) studied a coal plant on the Yeongheung Island and showed that local low-income residents and fishers were exposed to potential environmental pollution because of the plant construction and that the construction was associated with national developmentalism, free trade, and neoliberalism for shaping local environmental issues. This study revealed that environmental injustice on a local scale is associated with regional development, coal energy demands on a national scale, and the political economy of resources on a global scale.

Moreover, Huh (2013) argued that the nuclear waste repository siting in Gyungju has many problems concerning environmental justice. The decision taken by the citizens to host the facility is attributable to economic depression. Compensation has been used to exploit the poor residents by buying them off cheaply. Another matter concerning environmental justice is the concentration of risky facilities in a limited area. Because most nuclear power stations are concentrated in the south east coastal area of the Korean peninsula, the construction of a nuclear waste repository in Gyungju will aggravate the situation.

In 2007, Samsung workers and family members established Supporters for the Health and Right of People (SHARP) for the benefit of semiconductor industry employees. In 2010, SHARP commenced civil litigation, suing the South Korean government and corporations for high-risk and long-term exposure of workers to toxic chemicals due to a failure to establish labour, health, and safety rights. In 2011, the courts agreed that a relationship exists between working at the semiconductor factory and various forms of cancer; subsequently, in 2012, the South Korean government's department of labour admitted the relationship between cases of cancer among Samsung female workers and their work environment.

Environmental justice in Japan

Literature on environmental justice in Japan is scant and recent. In Japan, environmental justice issues have been framed within the broader contexts of environmental pollution and environmental activism. Japan has suffered severe environmental pollution due to industrial growth. In the late 1950s, air, land, river, and sea contaminations caused by chemical plant and industrial wastes resulted in the 'Big Four' pollution incidents at Minamata Bay (Minamata disease), Yokkaichi City (Yokkaichi asthma), and the Jinzū (Toyama, *Tiai Itai* disease) and Agano (Niigata) rivers. Victims of the Big Four pollution diseases instigated civil law proceedings against the offending companies in 1967. Similar antipollution and antidevelopment movements have mobilized across the nation since the late 1960s; this is considered the rising 'wave of resistance' and a new stage in the democratization of Japan. By the early 1970s, the government had addressed many pollution problems and established strict regulatory standards. The Pollution Research Committee, as the most influential Japanese antipollution group, led the transformative movement for both environmental justice and democratic consolidation (Avenell 2012; Mori 2008).

High-tech pollution, including soil and groundwater contamination from the production sites of electronics, cell phones, and office equipment, was discovered in the early 1980s. The Taishi-cho Plant in Hyogo Prefecture was the first known case of groundwater contamination at the site of a semiconductor plant in Japan. The local government's position on the remediation was affected by the community's high dependence on Toshiba. Places similar to Taishi-cho existed and were economically dependent on one company, with no clear prospects for remediation and recovery, while other municipalities disclosed complete information and actively encouraged remediation plans. There are controversies about the responsibility of the polluter, the landowner, and the government authorities, as well as who will clean up and the degree of remediation needed (Yoshida 2006). Brownfield regeneration involves issues of equal opportunity and distribution of benefits and compensation between brownfield residents. A lack of meaningful local engagement in the brownfield regeneration process could cause unjust outcomes.

Ishimura and Takeuchi (2015) analysed the spatial distribution of disposal sites in Japan and showed that the existence of other waste-related facilities was associated with a high number of disposal sites for industrial waste per capita. The authors consider that companies might decide to place disposal sites in those areas where other waste-related facilities already exist because they expect less citizen conflict over such construction. Protest movements against large-scale development projects, such as dam construction, nuclear energy facilities, and so-called 'nuisance facilities', have been disparaged for 'regional egotism' or 'NIMBY movements' by any centre which tries to impose nuisance facilities onto the periphery. Japanese environmental studies have started debating whether to incorporate environmental justice into these movements (Aoki 2007).

The Fukushima nuclear accident in 2011 stimulated public debate on the fair distribution of risks related to energy in Japan. Using the Fukushima Daiichi nuclear accident as the preliminary case study, Shrader-Frechette (2012) suggested that prima facie evidence for environmental injustice arises not only because of toxic or polluting facilities located in communities with poor minorities but also because of racism and classism that cause disaster-related environmental injustice. Japanese environmental injustice victims include poor people and black or buraku nuclear workers and children. The Fukushima Daiichi disaster has awakened Japanese civil society, which has been depicted as comparatively weak and nonparticipatory. Citizens have stepped forward to engage in community-based science and challenge government authority (Aldrich 2012, 2013).

Transboundary risks in East Asia

This section focuses on the emerging significant transboundary risks and distributive justice issues in East Asia, including e-waste, nuclear crises, and air pollution, which involve local and regional concerns on a global scale. The transnational flows of e-waste have become major environmental justice concerns in East Asia, particularly in China, the largest importer of e-waste. A recently released export tracking study from the Basel Action Network (BNA) found that nearly one-third of the hundreds of low-value devices intended for recycling in the United States were being exported – almost exclusively – to Asia. The exported devices ended up in China, Hong Kong, Taiwan, Cambodia, Thailand, Kenya, and other countries. Most of the shipments were probably traded illegally under the laws of the transit or importing countries. The BNA observed that over the past decade the vast majority of e-waste from North America went to China (mostly to Guiyu in Guangdong Province). Most exportation went to Hong Kong's New Territories near the mainland border, known as the 'ground zero' for e-waste processing, because Hong Kong seems not to have enforced the strict Chinese ban on importing as mainland China has done (Basel Action Network 2016).

Exports of hazardous e-waste to developing countries have had negative impacts on green business development and sacrificed green jobs in industrial countries where the waste was created, while harming desperate workers and the environment in countries least able to contend with e-waste (Basel Action Network 2016). Ecological problems and health risks of exposure to hazardous materials have occurred unequally between industrial and developing countries, between Asian countries, and inside Asian nations with great socioeconomic differences. In 2002, the Asia-Pacific region overtook North America in semiconductor market shares. Pollution has moved from industrial countries to Asia in both the manufacturing and e-waste recycling phases. China imports e-waste from other Asian countries, particularly Japan, Taiwan, South Korea, Singapore, Thailand, and Malaysia. Hong Kong also transports its e-waste to nearby Guangdong Province. Research showed that residents in Guiyu, Taizhou, and other Chinese recycling centres have suffered from respiratory diseases at higher rates (Iles 2004).

Regarding the life cycles of electronic products from an environmental justice perspective, Iles (2004) argued that e-waste risks to unemployed labourers and the environment are generated through the intersections of global production systems, development pathways, and the failure of governments to introduce or enforce regulations. Efforts must be made to address the underlying causes of e-waste exports and imports, including poverty, a lack of industry accountability, and a weak regulatory system.

The Fukushima nuclear disaster on 11 March 2011 caused the release of radioactive materials into the local, regional, and global environment via air and seawater. Japan's nuclear disaster raised public concerns about whether radioactive materials have reached Taiwan and affected food safety and public health. The Taiwanese government reassured its people regarding safety from the threat of radioactive dust released from the Fukushima nuclear power plants, and Taiwan's Atomic Energy Council (AEC) and EPA have maintained that they had not detected any radioactive dust. However, many countries surrounding the Pacific Ocean, such as the United States, China, South Korea, the Philippines, and Vietnam, have reported the detection of radioactive dust. It was not until 31 March that the AEC reported detecting excessive radioactive iodine 131 in the air in response to an inquiry from academics and ENGOs (Chan and Chen 2011).

Kingston (2013) indicated that even after the disaster, Japan's government and nuclear industry continued to cover up the full extent of the risk caused by the emitted nuclear radiation; they did not enact policies related to radiation removal and lacked data transparency. In addition, they failed to develop practices for avoiding continued radiation damage to the victims. Japanese and international experts and scientists have suggested that spewing radiation has caused significant impacts on marine life. Radioactive substances have been carried everywhere by winds, rain, and ocean currents, entering the food chain through seaweed and seafood (The Watchers 2015). ENGOs and scientists have continued to focus on the transboundary risks and hazards.

The Fukushima accident has raised public awareness of the unequal distribution of environmental risks and generational justice and has incited a large-scale revival in the antinuclear movement not only in Japan but also in Taiwan and South Korea. There is an essential coupling between earthquake–tsunami-prone lands and nuclear power plants built along the coastlines and cooled by seawater in East Asian countries. According to a report by the *Asian Wall Street Journal*, many of the nuclear power plants in Japan and Taiwan, as well as China to a lesser extent, are on the 14 most dangerous list because of their proximity to geological faults and seas. It is worthwhile to rethink East Asian countries' more risky and unstable characteristics of nuclear technology in their historical, sociocultural, and geological contexts (Fu 2011).

Matsumoto (2013) believes that the Fukushima nuclear disaster was an inevitable structural disaster caused by long-term data cover-up. Chou (2004, 2008) reported that in Taiwan, technocrats and technological elites' narrow and empirical scientific risk assessments have created a culture of systematic neglect and risk cover-up for a long period, resulting in the public carrying the burden of high risk, and individualization of relatively vulnerable risks.

The government, nuclear industry, and academics have traditionally formed a nuclear village (Sugiman 2014; Aldrich 2013; Kingston 2013) or a tripartite complex (Matsumoto 2013) and have dominated Japan's nuclear power policy. Similar trends can also be seen in Taiwan, where a government-led nuclear power regime strongly connects industry personnel, officials, and academicians. In South Korea, such patronized association is even more extreme.

Similar to Japan's domestic public response to nuclear power facilities, South Korea's antinuclear activism successfully initiated the idea of discontinuing one of Seoul's nuclear plants and spread this initiative to other cities, whereas in Taiwan an even greater response was aroused from a previously slumbering antinuclear movement. In 2014, antinuclear groups temporarily stopped the commercial operation of the fourth nuclear power plant in Taiwan. In Japan, besides increasing nuclear risk knowledge, civil groups attempted to obstruct the rebooting of nuclear power in different regions through environmental litigation.

Air pollution caused by power plants, factories, and heavy industries has become an emerging crucial transboundary risk (Avenell 2012) and national and regional environmental justice issue in East Asia. Poor air quality and high $PM_{2.5}$ levels lead to child asthma and an increase in the probability of cancer and premature death. Coal burning is the major contributor of air pollution in Beijing and surrounding areas. Frequent air pollution crises have received substantial media attention, triggering public outcry for curbing coal burning in northwestern China. In recent years, the foul air pollution in China has contributed to some portions of Taiwan's air pollution. Taiwan will be affected by air pollution from China because China often encounters high atmospheric pressure masses in winter that prevent air pollutants from dispersing, and the high concentration of pollutants is then blown by seasonal winds toward Taiwan. Levels of $PM_{2.5}$ recorded at air quality monitoring stations in central and southern parts of Taiwan have become hazardous several times during the aforementioned period in winter. The sixth naphtha cracker complex in Yunlin County and the Taichung coal-fired power plant in Longjing are the two largest sources of air pollution in central Taiwan. Various civic groups protest against air pollution, claiming that the Taiwan EPA should be proactive in inspecting and fining the major parties responsible for air pollution, and citizens should have the right to fresh air and to exercise outdoors.

Recently, Japan, China, and South Korea issued a joint statement detailing their concerted efforts for resolving environmental issues, including reducing air pollution and protecting water quality and the maritime environment, while promoting multilevel collaboration involving municipalities, companies, and researchers. These countries agreed to exchange observation data on $PM_{2.5}$ and other particles in order to identify the travel routes of yellow dust.

Conclusions

E-waste exporting to China has shown various types of environmental injustice problems, both nationally and globally. Authoritative expert politics, which was not affected by political democratization, emerged in the late 1970s to form the path dependence of authoritative scientism in governance. Japan, South Korea, and Taiwan have sacrificed citizens' health and environmental rights in the name of strengthening national economic competitiveness and national security by geographical politics since the 1970s (Chou 2015a). Thus, a new robust imperialism–neoliberalism alliance of the developmental state has formed to deepen the special governance structure of environmental injustice within East Asian countries.

In other words, the compressed modernity (Chang 2010) or hidden, delayed risk society caused by authoritarian technocratic expert politics (Chou 2009) has structurally weakened the risk culture over a long period, leading to a distortion in environmental justice. However, in the past decade, with the wave of technological democracy, authoritative expert politics has increasingly been strongly challenged by a robust civil society, producing a new turning point in environmental governance (Chou 2015b).

Environmental justice campaigns in East Asian countries are primarily concerned with environmental 'bads': specifically, industrial and high-tech pollution that has caused cancer-related health problems among many workers and disadvantaged residents in remote areas. Future research could focus on assessing environmental 'goods', such as air quality, water, and safe foods, or on vulnerability to environmental and food risks. Moreover, future research should focus on the common substantive environmental justice issues and compare the experiences of East Asian countries with those of other regions, for example Singapore, which has suffered from the burning of Indonesia's rainforests, and on exploring how different civic groups organize and link their protests for rights, recognition, democracy, citizenship, sustainable development, and other activities. Such research can enrich literature and debate on environmental justice and contribute to deriving favourable solutions or transnational collaborations.

References

Adeola, F.O. (2000). "Cross-national environmental injustice and human rights issues: A review of evidence in the development world." *The American Behavioral Scientist*, vol. 43, no. 4, pp. 686–706.

Aldrich, D.P. (2012). "Post-crisis Japanese nuclear policy: From top-down directives to bottom-up activism." *Analysis from the East–West Center*, no. 103, pp. 1–12.

Aldrich, D.P. (2013). "Rethinking civil society–state relations in Japan after the Fukushima accident." *Polity*, vol. 45, no. 2, pp. 249–264.

Aoki, S. (2007). "Sociological perspectives on environmental justice: The case of a German residents' movement." In K. Ohbuchi (ed.), *Social Justice in Japan: Concepts, Theories and Paradigms*, Melbourne, Vic.: Trans Pacific Press, pp. 239–263.

Avenell, S. (2012). "From fearsome pollution to Fukushima: Environmental activism and the nuclear blind spot in contemporary Japan." *Environmental History*, vol. 17, no. 2, pp. 244–276.

Barrington, D.J., Dobbs, S. and Loden, D.I. (2012). "Social and environmental justice for communities of the Mekong River." *International Journal of Engineering, Social Justice, and Peace*, vol. 1, no. 1, pp. 31–49.

Basel Action Network (2016). *"Disconnect: Goodwill and Dell exporting the public's e-waste to developing countries."* Available at: www.ban.org/trash-transparency.

Bell, K. (2014). *Achieving Environmental Justice: A Cross-National Analysis*. Bristol, UK: Policy Press.

Balme, R. (2014). "Mobilising for environmental justice in China." *Asia Pacific Journal of Public Administration*, vol. 36, no. 3, pp. 173–184.

Castells, M. (1992). "Four Asian tigers with a dragon head: A comparative analysis." In R. Appelbaum and J. Henderson (eds.), *States and Development in the Asian Pacific Rim*, Newbury Park, CA: Sage, pp. 33–70.

Chan, C.C. and Chen, Y.M. (2011). "A Fukushima-like nuclear crisis in Taiwan or a nonnuclear Taiwan?" *East Asian Science, Technology and Society*, vol. 5, no. 3, pp. 403–407.

Chang, K.S. (2010). "The second modern condition? Compressed modernity as internalized reflexive cosmopolitization." *The British Journal of Sociology*, vol. 61, no. 3, pp. 444–464.

Chang, S.L., Chiu, H.M. and Tu, W.L. (2006). "Breaking the silicon silence: Giving voice to health and environmental impacts within Taiwan's Hsinchu Science Park." In T. Smith, D.A. Sonnenfeld and D.N. Pellow (eds.), *Challenging the Chip: Labor Rights and Environmental Justice in the Global Electronics Industry*, Philadelphia, PA: Temple University Press, pp. 170–180.

Chi, C.C. (2001). "Capitalist expansion and indigenous land rights: Emerging environmental justice issues in Taiwan." *The Asia Pacific Journal of Anthropology*, vol. 2, no. 2, pp. 135–153.

Chiu, H.M. (2014). "The movement against science park expansion and electronics hazards in Taiwan: A review from an environmental justice perspective." *China Perspectives*, no. 3, pp. 15–22.

Chou, K.T. (2004). "Monopolistic scientific rationality and submerged ecological and social rationality: A discussion of risk culture between local public, scientists, and the state." *Taiwan: A Radical Quarterly in Social Studies*, no. 56, pp. 1–63.

Chou, K.T. (2008). "Reflexive risk governance: A critical view of 'bring the state back in' in newly industrializing country." Paper presented at the 4S Annual Meeting, The Netherlands, 20–23 August, 2008.

Chou, K.T. (2009). "Reflexive risk governance in newly industrialized countries." *Development and Society*, vol. 43, no. 1, pp. 57–90.

Chou, K.T. (2015a). "Trans-boundary risk governance in East Asia." Paper presented at the World Congress of Risk Analysis, Singapore.

Chou, K.T. (2015b). "From anti-pollution to climate change risk movement: Reshaping civic epistemology," *Sustainability*, vol. 7, pp. 14574–14596.

Chu, W.W. (2011). "Democratization and economic development: The unsuccessful transformation of Taiwan's developmental state." *Taiwan: A Radical Quarterly in Social Studies*, vol. 84, pp. 243–288.

EPA. (2013). *Ten Lessons of Environmental Justice*. Taipei: EPA.

Fan, M.F. (2006a). "Environmental justice and nuclear waste conflicts in Taiwan." *Environmental Politics*, vol. 15, no. 3, pp. 417–434.

Fan, M.F. (2006b). "Nuclear waste facilities on tribal land: The Yami's struggles for environmental justice." *Local Environment: The International Journal of Justice and Sustainability*, vol. 11, no. 4, pp. 433–444.

Fan, M.F. (2015). "Disaster governance and community resilience: Reflections on Typhoon Morakot in Taiwan." *Journal of Environmental Planning and Management*, vol. 58, no. 1, pp. 24–38.

Fan, M.F. (2016a). "Whose risk, and whose regulatory and test standards? The controversy of nuclear waste on Orchid Island." Paper presented at the 5th International Conference on Law, Regulations and Public Policy, May 30–31, Singapore.

Fan, M.F. (2016b). "Environmental justice and the politics of risk: Water resource controversies in Taiwan." *Human Ecology*, vol. 44, no. 4, pp. 425–434.

Fu, D. (2011). "Introduction." An East Asian STS Panel Discussion on Japan's 3/11 and Fukushima Crises. *East Asian Science, Technology and Society*, vol. 5, pp. 377–379.

Guo, Y. (2008). "Environmental justice and rural issues in China." *Academic Forum*, vol. 7, pp. 38–41 (in Chinese).

Han, S.J. and Shim, Y.H. (2010). "Redefining second modernity for East Asia: A critical assessment." *The British Journal of Sociology*, vol. 61, pp. 465–488.

Hsu, A. (2015). "Seeing through the smog: China's air pollution change for East Asia." In P.G. Harris and G. Lang (eds.), *2015 Handbook of Environment and Society in Asia*. London: Routledge.

Huang, C.T. (2011). "American concept, Taiwanese interpretation: If environmental justice is the answer, what is the question?" *Open Public Administration Review*, vol. 22, pp. 217–250 (in Chinese).

Huang, G.C., Gray, T. and Bell, D. (2013). "Environmental justice of nuclear waste policy in Taiwan: Taipower, government, and local community." *Environment, Development and Sustainability*, vol. 15, no. 6, pp. 1555–1571.

Huh, Y. (2013). *Justice, democracy and the siting of nuclear waste repositories: The Buan and Gyungju cases of South Korea*. PhD thesis, Colorado: Colorado State University.

Iles, A. (2004). "Mapping environmental justice in technology flows: Computer waste impacts in Asia." *Global Environmental Politics*, vol. 4, no. 4, pp. 76–107.

Ishimura, Y. and Takeuchi, K. (2015). "Does conflict matter? Spatial distribution of disposal sites in Japan." *EPR-ASIA Discussion Paper Series*. Available at www.epr-asia.net/pdf/epradp001.pdf.

Joines, J. (2012). "Globalization of e-waste and the consequence of development: A case study of China." *Journal of Social Justice*, vol. 2. pp. 1–15.

Kim, Y.T. (2007). "The transformation of the East Asian states: From the developmental state to the market-oriented state." *Korean Social Science Journal*, vol. 34, no. 1, pp. 49–78.

Kingston, J. (2013). "Nuclear power politics in Japan, 2011–2013." *Asian Perspective*, vol. 37, pp. 501–521.

Lee, H. (2009). *The political ecology of environmental justice: Environmental struggle and injustice in the Yeongheung Island coal plant controversy*. PhD thesis, Florida: Florida State University.

Liu, L. (2010). "*Made in China*: Cancer villages." *Environment: Science and Policy for Sustainable Development*, vol. 52, no. 2, pp. 8–21.

Liu, L., Liu, J. and Zhang, Z. (2014). "Environmental justice and sustainability impact assessment: In search of solutions to ethnic conflicts caused by coal mining in Inner Mongolia, China." *Sustainability*, no. 6, pp. 8756–8774.

Ma, C.B. (2010). "Who bears the environmental burden in China: An analysis of the distribution of industrial pollution sources?" *Ecological Economics*, vol. 69, no. 9, pp. 1869–1876.

Matsumoto, M. (2013). "'Structural disaster' long before Fukushima: A hidden accident." *Development and Society*, vol. 42, no. 2, pp. 165–190.

Mo, W. (2012). "Samsung female worker was dead by breast cancer." *The Korean Officers Confirmed It as an Occupational Injury*, retrieved from: http://e-info.org.tw/node/82688

Mori, M. (2008). "Environmental pollution and biopolitics: The epistemological constitution of Japan's 1960s." *Geoforum*, vol. 39, no. 3, pp. 1466–1479.

Park, G.Y. and Bak, H.J. (2003). "Real concerns about GMO crops and food in Korea." Paper presented in Korea Conference on Innovative Science and Technology 2003, Jeju, Korea.

Schoolman, E.D. and Ma, C. (2012). "Class, migration, and environmental inequality: Pollution exposure in China's Jiangsu Province." *Ecological Economics*, no. 75, pp. 140–151.

Shin, K.Y. (2012). "The dilemmas of Korea's new democracy in an age of neoliberal globalization." *Third World Quarterly*, vol. 33, pp. 293–309.

Shrader-Frechette, K. (2012). "Nuclear catastrophe, disaster-related environmental injustice, and Fukushima, Japan: Prima-facie evidence for a Japanese 'Katrina'", *Environmental Justice*, vol. 5, no. 3, pp. 133–139.

Sugiman, T. (2014). "Lessons learned from the 2011 debacle of the Fukushima nuclear power plant." *Public Understanding of Science*, vol. 23, no. 3, pp. 524–267.

Tang, C.P., Tang, S.Y. and Chiu, C.Y. (2011). "Inclusion, identity, and environmental justice in new democracies: The politics of pollution remediation in Taiwan." *Comparative Politics*, vol. 43, no. 3, pp. 333–350.

The Watchers. (2015). "Fukushima world's radiation nightmare." Available at: http://thewatchers. adorraeli.com/2015/03/05/fukushima-worlds-radiation-nightmare/

Weiss, L. (1998). *The myth of the powerless state: Governing the economy in the global era*. Cambridge: Polity Press.

Xie, L. (2011). "Environmental justice in China's urban decision-making." *Taiwan in Comparative Perspective*, no. 3, pp. 160–179.

Yamaguchi, T. and Suda, F. (2010). "Changing social order and the quest for justification: GMO controversies in Japan." *Science, Technology, and Human Values*, vol. 35, pp. 382–407.

Yoshida, F. (2006). "High-tech pollution in Japan." In T. Smith, D.A. Sonnenfeld and D.N. Pellow (eds.), *Challenging the Chip: Labor Rights and Environmental Justice in the Global Electronics Industry*, Philadelphia, PA: Temple University Press, pp. 215–224.

50

ENVIRONMENTAL JUSTICE IN WESTERN EUROPE

Heike Köckler, Séverine Deguen, Andrea Ranzi,
Anders Melin and Gordon Walker

Introduction

Western Europe has a long tradition of dealing with interlinkages between social inequity, environmental quality and health outcomes on the one hand and the idea of a just city on the other hand. Especially during the periods of industrialization and urban growth in the nineteenth and twentieth centuries, these topics were of major concern (Mosse and Tugendreich 1994) and led to alternative concepts of more just cities that included environmental quality for the working class. Some of the most famous conceptual notions of this form were the Garden City Movement in the UK and elsewhere (Howard 1902) and the Athens Charta of Le Corbusier and others (CIAM 1933). The Athens Charta as a guiding principle for remaking cities was motivated, amongst others, by the idea of health promotion for all. Core elements of the Athens Charta are the overall goal of a functional city, a guaranteed minimum of solar exposure in all dwellings, access to parks and protection of dwellings from emissions. Most of these discourses were clouded by World War II and the rebuilding activities afterwards. Despite the overall goal of health promotion, the legacy of planning a functional city according to the Athens Charta contributes to many of the urban problems we still face today (Jacobs 1961).

Following the 1992 United Nations Earth Summit in Rio de Janeiro and related activities, sustainable development has become an active and influential discourse in Europe. As sustainable development calls for an integration of the social and environmental dimensions of development including the vision of a more just world, there was a clear resonance with the term 'environmental justice', which began to be used in the mid-1990s in different Western European countries. There was some flow of influence from the US, but also much reinterpretation into an agenda that connected to Europe-specific concerns and also considered the global dimensions of environmental justice (Stephens et al. 2001; Walker 2012; Diefenbacher 2001). Since then the discourse on environmental justice has been driven forward in Western Europe by relatively few people, academics mainly and some activists. Its status and meaning varies considerably from country to country, so that some distinctive national environmental justice discourses have emerged. As such a common identity of a Western European 'environmental justice community' does not exist. This might be due to both the mix of national identities and the ongoing evolution of European identity as a whole, including the expansion of the EU to Eastern Europe. Therefore in this chapter countries from Western Europe that are not part of the EU, such as Switzerland, are considered, while other EU-member countries, such as Poland and Hungary, are dealt with in Chapter 51 on Eastern and Central Europe.

In this chapter we aim to provide an overview of the variety of topics and methods that characterize the evolving set of environmental justice discourses in Western Europe. In acknowledging the variety of discourses we do not therefore judge whether a study, programme or any other kind of contribution is an environmental justice contribution or not. Instead we include studies and activities in this chapter that are stated as dealing with environmental justice by their protagonists. In broad terms, though, we see environmental justice as more than an analytical concept integrating social and environmental factors. It is about a positive vision for a more just world and it involves making claims about "how things ought to be" (normative), "how things are" (descriptive) and "why things are how they are" (explanatory) (Walker 2012: 40–41). Based on such claim-making, interventions can be derived in accordance with the vision of environmental justice that is put forward (Bolte et al. 2012).

In the first section of this chapter the range of discourses on environmental justice in Western Europe is introduced through an overview that focuses on the forms of concern and normativity that have been central to environmental justice claim-making. This is followed by an overview of similarities and differences in methodological approaches and results regarding distributional inequalities, based on selected studies. In the second section the role of spatial and environmental planning and the involvement of citizens in planning processes and access to information will be dealt with, focusing therefore on questions of procedural justice. Decision-making in this context will be focused on the specific context of implementation of the Aarhus Convention in the European Union. In the final section we take the dimension of future generations into account and discuss the relation between environmental justice and sustainability using the example of the energy sector and ways of dealing with nuclear power.

Discourses of environmental justice in Western Europe

In most Western European countries, such as the Netherlands, Germany, France and Italy, the environmental justice discourse has been initiated by people concerned with health inequalities, identifying the physical environment as one of the relevant predictors of disparities in health (see Chapter 26). To these actors the concept of environmental justice has been valued due to its multi-dimensional and intervention-oriented approach to providing new answers for the problem of increasing health inequality. Environmental justice therefore is mainly seen as an explanatory concept adding more determinants to explaining health inequalities. The overall normative sense of 'how things ought to be' is a distributional one of wanting to see health inequalities significantly reduced (Kruize et al. 2014; WHO 2010, 2012; Mielck and Bolte 2004).

In the UK in particular, there was a parallel and interconnected emergence of interest from the mid-1990s in applying notions of inequality and injustice to the environment. This included environmental groups, particularly Friends of the Earth (FoE), academics within geography, sociology, social policy, planning and health, and to some degree policy makers (Walker et al. 2003; Fairburn et al. 2005; Walker et al. 2007; Poustie 2004). In other European countries the discourse on environmental justice has also now extended into disciplines beyond public health and epidemiology. For example, in Germany urban and environmental planning now plays an active role in the discourse (Böhme and Bunzel 2014; Köckler 2014) as well as sociology (Elvers 2011), while in Sweden planning scientists and environmental lawyers shaped the environmental justice discourse from the very beginning (Ebbesson 2002; Bradley et al. 2008; Gunnarsson-Östling 2011). In France the first academic papers were from a political science perspective (Laurent 2014), whilst in Switzerland early research was from a sociological perspective (Diekmann and Meyer 2011). Publications and website materials on environmental justice are now available in an increasing number of countries, conferences on environmental justice have been held and environmental justice is more and more part of the scientific conferences of disciplinary associations.

As discussed in more detail below, in several countries, studies have been carried out that have identified spatial patterns that demonstrate forms of environmental distributional injustice, mainly through combining poverty and deprivation data with indicators of pollution, risk or environmental amenities. However, this has not always been the case, as some studies have found spatial patterns that do not show distributional injustice, or have produced contradictory results depending on methods applied.

As discourses of environmental justice in Western Europe are often closely connected to the sustainability discourse, they have included in some parts a global view, putting climate adaptation and mitigation and therefore energy issues firmly on the justice agenda. In several countries an older discourse on energy poverty is now discussed as an environmental or energy justice issue (see Chapter 31). For example, environmental justice has continued to be a significant campaigning theme for FoE in England, Wales and Scotland, and has been applied in various ways to both domestic and international issues. Climate justice has become particularly important as a theme for the mobilizations of many groups on climate change, and has featured particularly in opposition to fracking, new airport proposals and in debates around the major climate change negotiations. In parallel with political framings there has also been much new and continuing work on climate and energy justice themes (Bickerstaff et al. 2013).

The engagement with and role of government and civil society in relation to environmental justice has been quite differentiated across Europe. In the UK in the mid-1990s the then Labour government was particularly concerned with issues of inequality and exclusion, and there was an opportunity for those newly campaigning around environmental justice (such as FoE) to achieve political influence by tying into these concerns. This was particularly the case in Scotland, where recent devolution provided the context for environmental justice becoming a key theme for both the Scottish Government and FoE Scotland (Dunion 2003; Scandrett 2007). Governmental interest in environmental justice has, though, markedly declined in the UK in recent years, reflecting both changes in government and severe cuts to the funding and responsibilities of government departments and agencies.

In Germany and France the level of public and governmental interest is, in contrast, increasing. In Germany there are various public financed programmes dealing with environmental justice, especially in the field of urban planning (Böhme and Bunzel 2014). The Federal Environmental Agency has promoted environmental justice for several years by hosting conferences and funding projects, including some carried out by an environmental NGO, Deutsche Umwelthilfe. Other agencies started to become active in 2015. On the level of the federal states, Berlin and North-Rhine Westphalia are the most active. In Berlin an environmental justice analysis has been carried out and is integrated into the spatial planning system. In North-Rhine Westphalia environmental justice was mentioned in the coalition agreement from 2012 between the Social Democratic and Green parties, and it is intended to be followed into an action plan.

In Sweden within politics and grassroots movements, environmental justice has been increasingly discussed by actors such as WWF (World Wildlife Fund), Naturskyddsföreningen, other left wing oriented grassroot organizations and also, notably, addressed by some of the political parties such as the Left Party and the Green Party. Nonetheless, outside of these examples, in most Western European countries environmental justice is not at all significant within the agenda of parties or governments.

Therefore in Western Europe environmental justice is mainly driven by stakeholders in academic science and to some degree in public administration. In contrast to the US there has never been a strong citizen based environmental justice movement. This may be because the environmental movement in most of Western Europe remains the preserve largely of leaders and members who are white and middle class, rather than involving minority ethnic populations. However, a thorough systematic explanation remains to be undertaken.

Studies of distributional environmental injustice

Environmental justice has a clear spatial dimension. Here inequality is found in patterns of 'how things are', both in terms of the spatial distribution of access to environmental resources and exposure to environmental health determinants. Both, access and exposure, require a relevant coincidence in proximity and time between an individual and a certain state of the environment. Table 50.1 gives an overview of selected Western European studies on distributional inequality. These have been chosen to represent the work undertaken in different countries and to demonstrate the range of factors and methods that have been included.

The second column of Table 50.1, listing the factors which are included in the analyses, shows a common ground of combining social and environmental parameters. Some studies additionally integrate health outcome factors (Michelozzi et al. 2005; Fecht et al. 2015; Kihal-Talantikite et al. 2013a and b), whilst focusing on the relation between social status and specific environmental stressors, such as waste facilities (Martuzzi et al. 2010). For the social dimension different indices are applied. While in the UK the index of multiple

Table 50.1 A selection of studies on distributional justice in Western Europe

Place (source)	Factors included in analysis	Methods applied in analysis	Summary of results
France			
French metropolitan areas – Lille – Paris – Lyon – Marseille (Padilla et al. 2014)	– Air pollution: NO_2 – French socioeconomic index at the census block level	Spatial model: Generalized Additive Model account for spatial autocorrelation	The strength and direction of the association between deprivation and NO_2 estimates varied between cities: In Paris, census blocks with the higher social categories are exposed to higher mean concentrations of NO_2. In Lille and Marseille, the most deprived census blocks are the most exposed to NO_2. In Lyon, the census blocks in the middle social categories were more likely to have higher concentrations than in the lower social categories.

Place (source)	Factors included in analysis	Methods applied in analysis	Summary of results
Lyon metropolitan area (Kihal et al. 2013a)	– Green space – French socioeconomic index at the census block level	Spatial correlation	The census blocks with the highest greenness levels are found in the more wealthy areas, hence the spatial variations in the deprivation and greenness indexes have similar patterns.
Lyon metropolitan area (Kihal et al. 2013b)	– Noise – French socioeconomic index at the census block level	Spatial correlation GIS approach	Medium deprivation neighbourhoods had a slightly higher mean noise level. There is a significant difference of exposure to neighbourhood noise between the three deprivation categories.
Strasbourg metropolitan area (Havard et al. 2009)	– Air pollution: NO_2 – French socioeconomic index at the census block level	Ordinary least squares model and a simultaneous autoregressive model that controls for the spatial autocorrelation of data	The association between the deprivation index and NO_2 levels was positive and nonlinear: the midlevel deprivation areas were the most exposed.
Lyon metropolitan area (Lalloué et al. 2015)	– NO_2 annual concentrations, noise levels, proximity to green spaces, to industrial plants, to polluted sites and to road traffic – French socioeconomic index at the census block level	Integration of environmental factors in cumulative index. A Kruskal–Wallis test was performed to test the association between socioeconomic index and cumulative exposure categories.	Census blocks in categories with the lowest exposures to NO_2, noise or traffic are less deprived than those more exposed to NO_2 or pollutant industries.
Germany			
Berlin (BfS 2011)	– Social Urban Development Monitoring Index – Bio-climate – Noise – Ambient air (NOx, PM_{10} $PM_{2,5}$) – Open space	Spatial correlation for social index and environmental factors Summative approach to show multiple burdens	Deprived neighbourhoods contain a much higher percentage of areas with two, three or four environmental stressors. In contrast, areas without environmental stress are characterized by very high/high social development index values.

(continued)

Place (source)	Factors included in analysis	Methods applied in analysis	Summary of results
Hamburg (Raddatz and Mennis 2013)	– Toxic Release Facilities (PRTR) – Foreigners – Welfare recipients – Population density	Spatial correlation GIS zonal operation calculating mean distance Ordinary Least Squares regression	Toxic release facilities are disproportionately located within, and closer to, neighbourhoods with comparatively higher proportions of foreigners and the poor. The proportion of foreigners is higher in neighbourhoods with three and more facilities than the proportion of the welfare recipients.
Italy			
Italy (Martuzzi et al. 2010)	– Incinerators – Waste landfills – Socioeconomic index at area level (census block)	GIS distance from plants	Landfills and incinerators are predominantly located in areas with more deprived residents.
Rome (Cesaroni et al. 2012)	– Urban air pollution – Socioeconomic position at small-area level	Pollutant concentrations, population exposure and years of life gained	Evaluation of the environmental and health impact of two low-emission zones in Rome. These have been beneficial from a public health point of view for all socioeconomic groups of the population, but have also exacerbated social inequalities.
Bologna, Milan, Rome and Turin (Michelozzi et al. 2005)	– Mortality – Heat	Daily excess mortality as the difference between the number of deaths observed and the smoothed average	Heat waves (summer 2003). The greatest excess in mortality was registered in those with low socioeconomic status in Rome (+17.8%) and in those with lower education levels in Turin (+43%).
Netherlands/UK			
Netherlands / UK On different scales (from national down to city level) (Fecht et al. 2015)	– age (% children/ % 65 plus) – ethnicity – income support recipients – NO_2 – PM_{10}	Descriptive statistics and multivariate regression analysis on different spatial scales	Whether a neighbourhood is urban or not is one of the strongest determinants of environmental inequality in exposure to air pollution. Substantial inequalities in air pollution exposure exist for areas with high percentage of ethnic minorities, even when level of deprivation is taken into account.

Place (source)	Factors included in analysis	Methods applied in analysis	Summary of results
UK			
England (Walker et al. 2005)	– Index of Multiple Deprivation – Emissions from Integrated Pollution Control (IPC) Sites	Spatial coincidence and buffer analysis Concentration index	Deprived populations more likely to live near to IPC sites and to clusters of these sites. Also more likely to be near to sites with more hazardous and offensive pollutants.
England (Briggs et al. 2008)	– Five sets of pollutants in terms of proximity, emission intensity and environmental concentration – Index of multiple deprivation and its five domains	Bivariate and multivariate correlation Includes urban–rural distinctions	Finds weak environmental inequities associated with deprivation. Stronger for air pollution than other types of hazard, and for environmental concentrations rather than proximity to source or emissions. Associations are generally weak, subtle and complex.
Great Britain (Mitchell et al. 2015)	– Air pollutants (NO_2, PM_{10}, $PM_{2.5}$) – Townsend Deprivation Index for 2001 and 2011	Spatial association Longitudinal comparison of change over 10 year period	Improvement in GB's air quality has been substantial but unequal. NO_2 concentrations have fallen markedly, but the rate of improvement slower for the more deprived. PM_{10} concentrations have risen, and done so more quickly for the poor.
Scotland (Richardson et al. 2010)	– Municipal landfill sites and airborne emissions – Carstairs index of deprivation for 1981, 1991 and 2001	Exposure modelling Concentration indices Least square and logistic regression Longitudinal comparison over 30 years	Deprived areas disproportionately exposed to municipal landfills and have been since at least 1981. Area deprivation may have preceded disproportionate landfill siting to some extent, but landfill siting also preceded a relative increase in deprivation in exposed areas.

deprivation is available nationwide, in Germany the index used for the Berlin study is a city-specific approach. This is an example of limited comparability between the different studies. A range of environmental dimensions have been studied, with air quality (see Chapter 26) often the focus, including examining the distributional impact of policy interventions – such as research in Italy that evaluated the impact of the introduction of a low-emissions zone on social inequalities (Cesaroni et al. 2012). Especially in the field of noise and air quality, some of the relevant indicators are comparable, at least amongst EU member states, as they need to follow the same legal requirements. These data also have a fairly good availability, enabling Fecht et al. (2015) to compare associations between air pollution and socioeconomic characteristics of neighbourhoods in England and the Netherlands. Environmental goods, in particular green- or open space, are also dealt with in European studies, such as the Berlin and Lyon cases shown in Table 50.1.

Methodologically a variety of approaches have been applied. The Berlin study is a showcase for an assessment of multiple burdens, following a simple summative approach. Lalloué et al. (2015) applied a cumulative environmental index using a data mining approach (see also Chapter 22). The UK study by Briggs et al. (2008) is distinctive, in part, being concerned with identifying confounding factors affecting epidemiological study results.

The results of the studies are varied, as their short description in Table 50.1 shows. The French study on NO_2 shows differences in social inequality and NO_2 exposure between Paris and other French cities. The Berlin case shows that the most deprived people have to face more burdens than others. Fecht et al. (2015) show that inequalities in the UK and the Netherlands exhibit some regional variation; especially identifying urban areas as most affected by distributional inequality concerning air quality. The study by Richardson et al. (2010) is an example that goes beyond just the description of 'how things are', in trying to isolate the processes at work in generating inequalities. They examine alternative explanations, asking which came first: local deprivation or the siting of waste landfill sites?

While most of the studies have been undertaken for analytical reasons only, the environmental justice project in Berlin (BfS 2011) is an example of a study that has been conducted by the administration of Berlin Senate with the support of universities. It is meant to be actively used in local decision-making and has been integrated as a specific report in the legal land use zoning plan. Similarly the results of the study on French metropolitan areas (Padilla et al. 2014) were integrated in the final report of the second French national plan for environmental health.

The participatory dimension of environmental justice

Meaningful participation of all people in decision-making processes, or procedural environmental justice (see Chapter 9), is seen by different Western European authors as an important element of environmental justice (e.g. Ebbesson 2002; Bradley et al. 2008; Bolte et al. 2012; Köckler 2014). Todd and Zografos (2005: 485) provide the following definition in a paper on indicators for environmental justice in Scotland: "Procedural justice: this is concerned with how and by whom decisions are made, and encompasses participation and legitimacy as common concepts." In their work, which includes a participatory process to weigh between different indicators, it is significant that Scottish community environmental activists rated procedural justice higher than other dimensions of environmental justice.

The Western European urban and environmental planning system has a crucial influence on the distribution of land uses including the location and design of infrastructure (streets, landfills, housing, parks etc.). Correspondingly decisions in urban and environmental planning influence situations of distributional injustice (see Chapter 36). These decisions are prepared by the governmental administration and taken by local parliaments, legitimated through democratic elections, or by the administrative body, depending on the legal framework that is applied. In urban and environmental planning the participation of the public is implemented within different decision-making procedures. A number of studies in Western Europe have focused on procedural justice questions. For Sweden, Bradley et al. (2008) examine the relevance of the discursive dimensions of justice. They ask who is influential in a discourse defining justice for urban planning. Poustie (2004) explores, in a report for the Scottish Department of the Environment, the extent to which it can take account of environmental justice within its current legislative framework. Poustie identifies considerable rights of public participation in the domestic environmental law framework, but points out that there has been little research into their effectiveness from the perspective of those in communities disproportionately affected by pollution. Strelau and Köckler (2016) carried out a study on the perspectives on environmental justice of employees in German local environmental agencies. They came to the conclusion that environmental justice is hardly an issue for those environmental agencies being part of their investigation. One reason why those environmental agencies do not consider environmental justice as an issue is that within the general public there has been neither debate on the issue nor recognition of the concept of environmental justice. This can be linked to procedural injustice, because the social dimension of air quality in urban areas, which is a matter of environmental justice (see Table 50.1), was not integrated into the agencies' decision-making. Therefore no relevance was given to environmental justice by the environmental agencies considered in this analysis.

Across Europe the general rights of the public (individuals and their associations) were strengthened by the "Convention on Access to Information, Public Participation in Decision-Making and Access to Justice in Environmental Matters" (see also discussion in Chapter 9). This convention was signed by members of the United Nations Economic Commission for Europe in Aarhus, Denmark, in 1998 and is referred to as the Aarhus Convention. In 2003 the European Union adopted in two Directives core elements of the Aarhus Convention. They had to be implemented in the national law of the EU Member States by 2005. Also non EU-member states of Western Europe have adopted this convention: Norway in 2003, Iceland in 2011 and Switzerland in 2013. The potential of Aarhus for taking forward environmental justice was discussed from its very beginning (Ebbesson 2002).

Beyond doubt the implementation of the Aarhus Convention has led to improvements in the right to participation in decision-making processes, but the right to participation itself does not guarantee that this right will be used in practice by all people equally. People with fewer resources typically use the right to participation less than others (Köckler 2014). In a case study of the city of Bristol in England, Bell (2008) shows that though people are interested and active in environmental matters, their achievements are limited largely as a result of the asymmetry of power in environmental decision-making partnerships. This shows the complexity of procedural environmental justice: it is not only about the right to participate and the capabilities of participants, especially of deprived communities; it is also about the ability of those in power to integrate stakes in decision-making procedures to guarantee meaningful involvement for all.

Taking future generations into account

As already noted, in Western Europe there are often strong connections made between sustainability and environmental justice (see Chapter 14), reflecting wider calls for how these discourses should be brought together, including within the notion of 'just sustainability' (Agyeman and Evans 2004: 155–164). Justice between contemporary and future generations is a significant aspect of sustainability. *Our Common Future* defines sustainable development as "development that meets the needs of the present without compromising the ability of future generations to meet their own needs" (UNEP/WCED 1987). It can be argued therefore that intergenerational justice is an important part of environmental justice, although it is seldom recognized as such (see also Chapter 13 on human rights approaches).

Questions about what responsibility we have towards future generations have been especially important within debates about nuclear energy and nuclear waste management in Western Europe, as a consequence of the fact that nuclear waste will remain dangerous for future generations at least 100 000 years into the future and has to be stored safely (Rüdig 1990; Sovacool 2013) – see also Chapter 31. To provide specific examples, here we will focus on how the issue of intergenerational justice has been acknowledged within the debate on nuclear energy in two countries in Western Europe: Germany and Sweden. In Germany the first nuclear power station was opened in 1962 and during the next two decades over 30 additional power plants were built. Although the majority of the population accepted nuclear energy, there were also more critical voices. When the German Green Party entered the parliament for the first time in 1983, ethical discussions concerning nuclear energy became more common. One important theme was what responsibility we have to protect the planet for the sake of future generations (Schreurs 2014: 11–13).

The Fukushima accident in 2011 led to an increased concern about ethical issues, and German Chancellor Angela Merkel decided to temporarily shut down the seven oldest nuclear power plants and another one that had technical problems. She also established an ethics commission to analyse the ethical dimensions of Germany's energy production, which published its report on 30 May 2011. The fact that Merkel established a commission with a special focus on ethical issues is quite remarkable since the political discourse on environmental problems generally emphasizes technical and economic issues much more than ethical ones. Soon after the publication of the Commission's report the government decided to permanently shut down the eight nuclear power plants and to set out a schedule for the phase-out of the remaining nine, according to which the last nuclear power plants will be shut down in 2022 (Schreurs 2014). The report states that current generations have a clear responsibility to ensure future living conditions, in that whilst the use of nuclear energy may benefit us now, it will at the same time lead to risks for future generations (Ethics Commission for a Safe Energy Supply 2011).

In Sweden the first commercial reactor started operation in 1972, but political opinion soon switched against nuclear energy. In 1980 the Swedish parliament made the decision that the twelve reactors that were by then operating, under construction or planned should be allowed to continue working for their technical length of life, which was then assumed to be 25 years, but no more new stations should be built. The Swedish nuclear energy epoch should therefore be over by 2010. However, the decision to phase out nuclear energy by 2010 was cancelled in 2009, and at least so far, the Fukushima accident has not led to any significant shift in this policy (Kolare et al. 2016). Through these ongoing debates, questions about responsibilities to future generations have been an important theme within the Swedish mass media (Anshelm 2000). Moreover, a large number of official Swedish political documents on nuclear energy and nuclear waste management, such as private members' motions and public reports, mention the issue of intergenerational justice. A Swedish government report (SOU 2013) states, for example, that the fundamental moral principle that underlies the Swedish discussion of nuclear waste management is that each generation should have the right to decide for themselves which technology they want to use for managing nuclear waste, a principle that the report labels as the principle of "intergenerational autonomy". Future generations should have freedom of action. They should be given the same right as current generations to integrity, ethical freedom and responsibility (SOU 2013: 71).

Conclusion

Environmental justice in Western Europe can be characterized by several sub-discourses that are only rarely integrated. They extend from a focus on distributional analysis dealing with healthy living conditions, to concerns for procedural justice and into energy-related topics including energy poverty as well as the intergenerational ethical issues particularly raised by Europe's evolving and differentiated relationship with nuclear power. This variety reflects the different interests that there are in focusing on environmental justice (some more analytical, some more normative), disciplinary engagements and limitations and the specific social, political and geographical situations of the nations of Western Europe.

Studies on the social distribution of environmental goods and bads do show particular patterns of inequality, but simple generalizations across the European space are problematic. Distributional environmental justice analysis in Western Europe is a topic mainly focused on urban agglomerations and has been based on a variety of environmental and social indicators. Therefore the direct comparability of study results is not given. Participation in environmental decision-making has been the focus of analysis in many different European contexts, demonstrating that whilst there are established guarantees of participation within legal frameworks (through the Aarhus Convention in particular), these rights do not necessarily mean that opportunities will be acted on or be meaningful in practice, particularly for more marginal communities.

Environmental justice, as an explicit term, remains in Europe largely a topic for study, an analytical framework, and to some extent, it has also been adopted into policy and planning discourses and practices. Although there are examples of environmental groups taking up environmental justice as a campaigning theme in Western Europe, there is no integrated grassroots-based environmental justice movement. Typically communities who are affected by actual or expected negative consequences of environmental harm or access to environmental goods are not mobilizing under a collective justice discourse – except in relation to climate justice campaigning, where questions of global inequalities and responsibilities are predominant. However, as demonstrated by the intergenerational ethical questions that have been important to debates over nuclear power, this does not mean that matters of justice, fairness and equity in relation to the environment do not figure within European politics. From the local to the international scale normative justice concerns are part of environmental debate and the potential remains for them to become galvanized into a more explicit justice agenda, informed by the growing body of academic analysis of inequalities in distributional patterns and processes of decision-making.

To enhance the body of academic analysis on environmental justice in Western Europe we see the need for further research on a range of topics including: 1) the development of comparable cross-country study designs for analysing patterns of distributional justice; 2) analysis of dimensions of procedural injustice relating to the citizen engagement in different contexts across Western Europe, recognizing that these may be distinct from those in other regions in the world; 3) integration of justice into the guiding principles of environmental politics, learning in particular from the domains of energy and climate protection and adaptation where justice ideas have become more established; 4) examination of how the use of legal provisions does or does not support the making of environmental justice claims, given evidence that those treated unjustly are often not able to get effective access to the legal system; and 5) further analysis of cases where environmental injustice claims and discourses do not emerge where they might be expected to, seeking to understand the underlying dynamics and politics that are at work.

Acknowledgement

Thanks to Karolina Isaksson, Swedish National Road and Transport Research Institute (VTI), for detailed information on Sweden.

References

Agyeman, J. and Evans, B. (2004). "Just sustainability: the emerging discourse of environmental justice in Britain?" *The Geographical Journal*, vol. 170, no. 2, pp. 155–164.

Anshelm, J. (2000). *Mellan frälsning och domedag: Om kärnkraftens politiska idéhistoria i Sverige 1945–1999.* Stockholm: Brutus Östlings Bokförlag Symposion 2000.

Bell, K. (2008). "Achieving environmental justice in the United Kingdom: A case study of Lockleaze." *Environmental Justice*, vol. 1, no. 4, pp. 203–210.

BfS (Bundesamt für Strahlenschutz) (2011). *Special Issue II: Environmental Justice.* Federal Office for Radiation Protection (BfS), Federal Institute for Risk Assessment (BfR), Robert Koch Institute (RKI), Federal Environment Agency (UBA).

Bickerstaff, K., Walker, G. and Bulkeley, H. (eds.) (2013). *Energy Justice in a Changing Climate: Social Equity and Low Carbon Energy.* London: Zed.

Böhme, C. and Bunzel, A. (2014). *Umweltgerechtigkeit im städtischen Raum: Expertise, Instrumente zur Erhaltung und Schaffung von Umweltgerechtigkeit.* Berlin: Deutsches Institut für Urbanistik.

Bolte, G., Bunge, C., Hornberg, C., Köckler, H. and Mielck, A. (eds.) (2012). *Umweltgerechtigkeit: Chancengleichheit bei Umwelt und Gesundheit: Konzepte, Datenlage und Handlungsperspektiven.* 1st ed. Bern: Verlag Hans Huber.

Bradley, K., Gunnarsson-Östling, U. and Isaksson, K. (2008). "Exploring environmental justice in Sweden: How to improve planning for environmental sustainability and social equity in an 'eco-friendly' context". Justice, Equality & Sustainability, *MIT Journal of Planning*, vol. 8, pp. 68–81.

Briggs, D., Abellan, J.J. and Fecht, D. (2008). "Environmental inequity in England: Small area associations between socio-economic status and environmental pollution." *Social Science and Medicine*, vol. 67, no. 10, pp. 1612–1629.

Cesaroni, G., Boogaard, H., Jonkers, S., Porta, D., Badaloni, C., Cattani, G., et al. (2012). "Health benefits of traffic-related air pollution reduction in different socioeconomic groups: The effect of low-emission zoning in Rome." *Occupational and Environmental Medicine*, vol. 69, no. 2, pp. 133–139.

Congress Internationaux d'Architecture moderne (CIAM) (1933). *La Charte d'Athenes or The Athens Charter.* Trans J. Tyrwhitt. Paris, France: The Library of the Graduate School of Design, Harvard University, 1946.

Diefenbacher, H. (2001). *Gerechtigkeit und Nachhaltigkeit: Zum Verhältnis von Ethik und Ökonomie.* Darmstadt: WBG.

Diekmann, A. and Meyer, R. (2011). "Democratic smog? An empirical study on the correlation between social class and environmental pollution." *UMID* 2/2011, Special Issue Environmental Justice II, pp. 72–77.

Dunion, K. (2003). *Troublemakers: The Struggle for Environmental Justice in Scotland.* Edinburgh: Edinburgh University Press.

Ebbesson, J. (Ed.) (2002). *Access to Justice in Environmental Matters in the EU.* The Hague: Kluwer Law International.

Elvers, H.-D. (2011). "Umweltgerechtigkeit." In *Handbuch Umweltsoziologie*, ed. M. Groß. Wiesbaden: VS Verlag für Sozialwissenschaften. 464484.

Ethics Commission for a Safe Energy Supply (2011). *Germany's Energy Transition: A Collective Project for the Future.* Berlin.

Fairburn, J., Walker, G. and Smith, G. (2005). *Investigating Environmental Justice in Scotland: Links Between Measures of Environmental Quality and Social Deprivation.* Report UE4(03)01. Edinburgh: Scottish and Northern Ireland Forum for Environmental Research.

Fecht, D., Fischer, P., Fortunato, L., Hoek, G., de Hoogh, K., Marra, M., et al. (2015). "Associations between air pollution and socioeconomic characteristics, ethnicity and age profile of neighbourhoods in England and the Netherlands." *Environmental Pollution*, vol. 198, pp. 201–210.

Gunnarsson-Östling, U. (2011). *Just Sustainable Futures: Gender and Environmental Justice Considerations in Planning.* Stockholm: Royal Institute of Technology.

Havard, S., Deguen, S., Zmirou-Navier, D., Schillinger, C. and Bard, D. (2009). "Traffic-related air pollution and socioeconomic status: A spatial autocorrelation study to assess environmental equity on a small-area scale. " *Epidemiology.* vol. 20, no. 2, pp. 223–230.

Howard, E. (1902). *Garden Cities of Tomorrow.* London: Swan Sonnenschein & Co., Ltd.

Jacobs, J. (1961). *The Death and Life of Great American Cities.* New York: Random House.

Kihal-Talantikite, W., Padilla, C., Lalloué, B., Gelormini, M., Zmirou-Navier, D. and Deguen, S. (2013a). "Green space, social inequalities and neonatal mortality in France." *BMC Pregnancy and Childbirth.* October 20, vol. 13, p. 191.

Kihal-Talantikite, W., Padilla, C., Lalloué, B., Rougier, C., Defrance, J., Zmirou-Navier, D. and Deguen, S. (2013b). "An exploratory spatial analysis to assess the relationship between deprivation, noise and infant mortality: An ecological study." *Environmental Health.* Dec. 16, vol. 12, p. 109.

Köckler, H. (2014). "Environmental justice: Aspects and questions for planning procedures." *UVP Report,* vol. 28, no. 3&4, pp. 139–142.

Kolare, S., Vedung, E. and Lundberg, F.: "Kärnenergifrågan" in the webpage of Nationalencyklopedin (the National Encyclopedia). www.ne.se.proxy.mah.se/uppslagsverk/encyklopedi/lång/kärnenergifrågan (accessed 15 March 2016).

Kruize, H., Droomers, M., Van Kamp, I. and Ruijsbroek, A. (2014). "What causes environmental inequalities and related health effects? An analysis of evolving concepts." *International Journal of Environmental Research and Public Health,* vol. 11, no. 6, pp. 5807–5827.

Lalloué, B., Monnez, J.-M., Padilla, C., Kihal, W., Zmirou-Navier, D. and Deguen, S. (2015). "Data analysis techniques: A tool for cumulative exposure assessment." *Journal of Exposure Science and Environmental Epidemiology,* March-April, vol. 25, no. 2, pp. 222–230.

Laurent, É. (2014). "Environmental inequality in France: A theoretical, empirical and policy perspective", *Analyse & Kritik,* vol. 02/2014, pp. 251–262.

Martuzzi, M., Mitis, F. and Forastiere, F. (2010). "Inequalities, inequities, environmental justice in waste management and health", *European Journal of Public Health,* vol. 20, no. 1, pp. 21–26.

Michelozzi, P., de Donato, F., Bisanti, L., Russo, A., Cadum, E., DeMaria, M., et al. (2005). "The impact of the summer 2003 heat waves on mortality in four Italian cities." *Eurosurveillance,* vol. 10, no. 7, pp. 161–165.

Mielck, A. and Bolte, G. (eds.) (2004). *Umweltgerechtigkeit. Die soziale Verteilung von Umweltbelastungen.* Weinheim: Juventa.

Mitchell, G., Norman, P. and Mullin, K. (2015). "Who benefits from environmental policy? An environmental justice analysis of air quality change in Britain 2001–2011." *Environmental Research Letters,* vol. 10, no. 10, 105009.

Mosse, M. and Tugendreich, G. (1994). *Krankheit und Soziale Lage,* 4th edition. Freiburg im Breisgau: WiSoMed-Verlag.

Padilla, C., Kihal-Talantikite, W., Vieira, V., Rossello, P., Le Nir, G., Zmirou-Navier, D. and Deguen, S. (2014). "Air quality and social deprivation in four French metropolitan areas: A localized spatio-temporal environmental inequality analysis." *Environmental Research,* vol. 134, pp. 315–324.

Poustie, M. (2004). *Environmental Justice in SEPA's Environmental Protection Activities: A Report for the Scottish Environment Protection Agency.* Glasgow.

Raddatz, L. and Mennis, J. (2013). "Environmental justice in Hamburg, Germany", *The Professional Geographer,* vol. 65, no. 3, pp. 495–511.

Richardson, E.A., Shortt, N.K. and Mitchell, R.J. (2010). "The mechanism behind environmental inequality in Scotland: Which came first, the deprivation or the landfill?" *Environment and Planning A,* vol. 42, pp. 223–240.

Rüdig, W. (1990). *Anti-nuclear Movements: A World Survey of Opposition to Nuclear Energy.* Essex: Longman.

Scandrett, E. (2007). "Environmental justice in Scotland: Policy, pedagogy and praxis." *Environmental Research Letters,* vol. 2, no. 4, p. 045002.

Schreurs, M.A. (2014). "The ethics of nuclear energy: Germany's energy politics after Fukushima", *The Journal of Social Science,* vol. 77, pp. 9–29.

SOU (Statens Offentliga Utredningar) (2013). *Kunskapsläget på Kärnavfallsområdet 2013. Slutförvarsansökan under prövning: kompletteringskrav och framtidsalternativ.* Stockholm: SOU.

Sovacool, B.K. (2013). *Energy and Ethics: Justice and the Global Energy Challenge.* London: Palgrave Macmillan.

Stephens, C., Bullock, S. and Scott, A. (2001). *Environmental justice: Rights and means to a healthy environment for all.* Swindon: ESRC Global Environmental Change Programme.

Strelau, L. and Köckler, H. (2016). " 'It's optional, not mandatory': Environmental justice in local environmental agencies in Germany." *Local Environment,* vol. 21, no. 10, pp. 1215–1229.

Todd, H. and Zografos, C. (2005). "Justice for the environment: Developing a set of indicators of environmental justice for Scotland." *Environmental Values,* vol. 14, pp. 483–501.

United Nations Environment Programme (UNEP)/World Commission on Environment and Development (WCED) (1987) *Our Common Future*. Oxford: Oxford University Press, p. 43.

Walker, G. (2009). "Beyond distribution and proximity: Exploring the multiple spatialities of environmental justice." *Antipode*, vol. 41, no. 4, pp. 614–636.

Walker, G. (2012). *Environmental Justice: Concepts, Evidence and Politics*. Abingdon: Routledge.

Walker, G., Fairburn, J., Smith, G. and Mitchell, G. (2003). *Environmental Quality and Social Deprivation Phase II: National Analysis of Flood Hazard, IPC Industries and Air Quality*. Bristol, UK: Environment Agency.

Walker, G.P., Mitchell, G., Fairburn, J. and Smith, G. (2005). "Industrial pollution and social deprivation: Evidence and complexity in evaluating and responding to environmental inequality." *Local Environment*, vol. 10, no. 4, pp. 361–377.

Walker, G., Burningham, K., Fielding, J., Smith, G., Thrush, D. and Fay, H. (2007). *Addressing Environmental Inequalities: Flood Risk*. Science Report SC020061. Bristol: Environment Agency.

WHO (2010). "Environment and health risks: A review of the influence and effects of social inequalities", available at: www.euro.who.int/__data/assets/pdf_file/0003/78069/E93670.pdf (accessed 11 January 2012).

WHO (2012). "Environment health inequalities in Europe", available at: www.euro.who.int/__data/assets/pdf_file/0010/157969/e96194.pdf (accessed 10 March 2015).

51

ENVIRONMENTAL JUSTICE IN CENTRAL AND EASTERN EUROPE

Mobilization, stagnation and detraction

Tamara Steger, Richard Filčák and Krista Harper

Introduction

In 2007, residents of a small eastern Slovakian town called Trebišov were shocked to learn that private developers were planning to build a new thermal power plant on an abandoned sugar processing factory site just 200 meters from inhabited areas and 400 meters from the nearest school. The Slovak Ministry of the Environment conducted an Environmental Impact Assessment (EIA) in 2008; and, claiming manageable environmental impact, the construction permitting process was launched. Coal for the plant would be brought in from Ukraine, so why choose this predominantly agricultural-based town with a large population of poor people and Roma? Its operations would increase Slovakia's CO_2 emissions by 8 per cent and provide a limited number of jobs locally; but profits would go to the Czechoslovak Energy Company, a firm registered outside of the community. The town's people initiated legal action and set up an informal platform, called Trebišov Nahlas (Trebišov Aloud), supported by 3500 people during the construction permitting process. The group's subsequent petition against the construction of the plant gathered 9000 signatures. The petition's organizers engaged almost half of Trebišov's population but did not invite Romani residents to participate. The group organized protests on the grounds that the project would violate already approved development plans, while seriously and adversely affecting the environment and health of ordinary people living in close proximity to the site. Under pressure, the Košice county authorities denied the construction permit, and the project has been withdrawn.

Such cases of the struggle for environmental justice draw and inspire our attention and raise new (and old) questions in the effort to promote environmental justice in Central and Eastern Europe (CEE). The struggle inspires a story of meaningful civic engagement to promote environmental justice in part along class lines while, in the case of the petition process, still maintaining prejudice along ethnic lines against Roma. Romani communities are a specific population affected by acute environmental and health problems in everyday life (Steger 2007; Harper 2012). The Roma have lived in Central and Eastern Europe for six centuries, and they constitute the largest ethnic minority in many Central and Eastern European countries (Crowe 1996). Roma face systematic discrimination and social exclusion

(Ladányi and Szelényi 2001). In Hungary, researchers found that neighbourhoods in which the majority of residents are Romani are more exposed to environmental harms and have less access to public infrastructure than non-Romani neighbourhoods (Debrecen University School of Public Health 2004). Throughout Europe, Romani life expectancy is ten years lower than that of non-Roma (Doyle 2004). Despite the urgency of these issues, research on and policies targeting Romani populations in Central and Eastern Europe have traditionally not engaged Roma themselves in agenda setting. Roma are the region's largest ethnic minority, but they are not the only marginalized group. Recently, refugees from the Middle East, Africa, and South Asia have migrated to the region, and these new residents have also not been included in environmental decision-making.

We evoke this story to demonstrate the complexity of environmental justice in the context of CEE. Historically, under state socialism, environmentalism emerged first in the form of state-led nature conservation, but grassroots mobilizations in the region fuelled the development of environmental dissidence in the 1980s (Harper 2006; Snajdr 1998; Pavlinek and Pickles 2002). The fall of the Berlin Wall marked the beginning of post-socialist transformation characterized by mass privatization of land and industries, along with widening socioeconomic inequalities. Contemporary environmental justice mobilizations in Central and Eastern Europe grew out of this political and economic transition and reflect this distinctive trajectory. And, while clearly environmental injustice on the basis of class discrimination falls within this trajectory, prejudice along ethnic lines is accentuated as disillusionment with the free market system and democracy creates fertile ground for radical right wing politics (Wodak and Richardson 2012).

Environmental justice in CEE: from state socialism to post-socialist transformation

Central and Eastern Europe (CEE) is a diverse set of nation-states defined by their common post-Second World War history of membership in the Council for Mutual Economic Assistance (CMEA), participation in the Warsaw Treaty of Friendship, Cooperation and Mutual Assistance (the Warsaw Pact), and more generally by their location just east of the former Iron Curtain (with the exception of the former Yugoslavia and Albania). The region was characterized by state socialism based on a centrally planned economy[1] and rapid industrialization with varying degrees of political openness, and a common historical experience first with fascist and then Stalinist dictatorships. Despite some important differences, CEE (like the West) was driven by the principles of industrialism.

Challenges to CEE's economic competitiveness grew in the 1960s as globalization advanced, and states continued to support even the most economically inefficient enterprises (e.g. steelworks) while also working toward full employment. This came with a general acceptance of the environmental trade-offs, particularly pollution and ecological degradation associated with industrial production. The adverse environmental conditions in CEE began generating criticism that was tolerated to a certain extent by the ruling socialist parties in places such as Czechoslovakia, Hungary and Eastern Germany, and suppressed in others, like Romania.

In the period leading up to the fall of the Berlin Wall in 1989, environmental and political activists criticized the governments of the centrally planned economies in CEE for failing to deal with environmental problems, with significant implications for environmental justice. They called for access to information about exposure to environmental risks associated with Soviet-style industrialization involving extensive pollution and resource extraction. Demands for social equality and environmental justice were paramount, and in these struggles, "ordinary people" portrayed centralized economic policies and practices as bad for the environment and bad for people.[2]

The emerging environmental movements targeted what were seen as excesses of the centrally planned economy. They focused on the ways in which industrial growth disproportionately affected inhabitants, impoverishing them and imposing unequal environmental and health burdens. Citizens rallied against air and water pollution in the former Czechoslovakia (Kamieniecki 1993), chemical pollution in the former German Democratic Republic, dam construction in Latvia and mining in Estonia (Auer 1998), and there were anti-nuclear power protests against nuclear facilities at Ignalina in Lithuania (Dawson 1996) and Kozloduy in Bulgaria. In Hungary, and to a lesser extent in the former Czechoslovakia, citizens protested against the proposal to construct the Gabčíkovo-Nagymaros Dam on the Danube River and brought together tens of thousands of people expressing dissatisfaction with the socialist leadership and its centralized, irrational approach to industrialization.[3] Auer writes about the Estonian environmental movement and asserts that:

> From the ethnic Estonian perspective, the environmental crisis was precipitated by Soviet values and institutions that ethnic Estonians hated generally, including contaminated and wasted resources, the perceived theft of Estonia's natural riches, Russophones in the dominant employ of the republic's extractive industries, and Moscow's secretive and colonialist decision-making authority over Estonia's resource-based economy.
>
> *(Auer 1998: 660)*

While environmentalism was part of socialist regimes in the form of state-led nature protection units throughout CEE, the environmental movements that emerged leading up to the 1989 changes introduced a new discourse on environmental protection (Steger 2004a). Nature protectionists, while affiliated with the regimes, often aligned themselves with the more civic-based environmental activists for environmental protection. However, due to the perceived neutrality or relatively apolitical stance of nature protection efforts, political activists simultaneously used this "safe haven" for promoting political change (e.g. independence in the case of the Baltic states) under the guise of environmentalism. Still, interest in environmental protection was genuine, and it initially provided a relatively neutral arena for unification of otherwise deeply antagonistic social groups and political thinkers.

Environmental protest demonstrations and civic initiatives were increasingly politicized as the former regimes weakened over time. Major demonstrations took place in Teplice, a Northern Bohemian town, which preceded the Prague student protest march of 17 November 1989, which in turn precipitated the collapse of the socialist regime (Philip and Jehlička 2007). In Bulgaria, the first civil society association was established in 1988 – the Public Committee for the Ecological Protection of Rousse, a town susceptible to gas emissions from a Romanian chemical plant in Gurgevo, a Romanian-Bulgarian border town on the bank of the Danube.

Demands for free and open access to information (see Chapter 9) as well as the possibility to publish and share information fostered popular mobilization. Support for the Latvian environmental movement grew in large part because of a media campaign led by Dainis Ivans, a Latvian journalist and a member of the Latvian Writers' Union. Key to this campaign was a call for information on the environmental and social impacts of a proposed hydroelectric dam on the Daugava River. One of the first environmental movements to emerge in CEE was in Poland, and freedom of information was fundamental to their demands (Fisher 1993). The Solidarity movement had paved the way for the Polish Ecology Club (PEC), which enjoyed a brief period in which censorship of the press was lifted and a large support-ing membership was generated. Bulgaria's Ekoglasnost, the leading environmental movement organization of the time, made bold demands for transparency (Baumgartl 1995).

Beyond access to information, demands for public participation were also paramount in the articulation of environmental justice in the region. In the 1980s, the Bratislava branch of one of Slovakia's biggest associations, Slovenský zväz ochrancov prírody a krajiny (SZOPK) (the Slovak association of nature and countryside protectors), created an underground environ-mental movement and published "Bratislava Nahlas" [Bratislava Aloud]: "In our city, the basic conditions of life have become problematic. Contamination of the atmosphere is threatening the health of virtually all inhabitants, particularly the aged, the sick, and children." The group intertwined issues of environmental risks with the need for more meaningful political forums: "We expect that the public discussion . . . will not only articulate the interests of the citizens of Bratislava, but will mobilize their forces and renew the relationship of the citizens with respect to their city" (Glazer and Glazer 2014: 406–407).

In the last stages of the centrally planned socialist regimes, there were tendencies to search for a "third way" – a way of life that would be dominated by neither communism nor capi-talism. The "third way" discussion generated strategic thinking about oppositional choices. This line of thinking was, for instance, eminent among Czechoslovakian dissidents and environmentalists. Ivan Dejmal and Jozef Vavroušek published materials in the 1970s and 1980s about an alternative ecological society. Similar trends existed among Hungarian and East German environmentalists increasingly concerned about industrialization (Harper 2006; Bahro 1981).

As Western notions of capitalism and democracy flooded CEE, hopes for a third way were quickly replaced with a different kind of anticipation. Many believed that the "twin transi-tion" to capitalist democracy would result in efficiency (i.e. less waste and pollution) and more public participation (i.e. more pushback against ecologically destructive policies and develop-ments). In other words, pundits expected democratization and marketization to bring ecological modernization. Data from the European Environmental Agency and European Statistical Bureau clearly indicate that initially the level of pollution went down in practically all transforming countries as firms shut down (ESB 2008; EEA 1995, 2007, 2015). Democracy indeed opened space for access to information and public participation. However, this very same process of transformation also led to rising social inequalities within countries and localities across CEE.

Rapid political changes in CEE had several important implications from an environmental justice perspective. With the introduction of multiparty parliamentary systems and the advent of the first free elections after the political transition, the environment was on the agenda of every new party. New governmental institutions such as Ministries of the Environment were established. The Environment for Europe process and the EU enlargement project were launched. Many new environmental policies and legislative acts were adopted. Public opinion polls rated environment as a high priority in the early 1990s. Many CEE countries increasingly joined as signatories to the Aarhus Convention on access to information, public participation in decision-making and access to justice in environmental matters (see Chapters 9 and 50).

However, as the CEE economies were soon transformed by privatization and deregulation of CEE, under the auspices of the World Bank and Western governments, unemployment rose dramatically (Kabaj 2005; OECD 2004) along with other important social costs (Stark 1992; Burawoy and Krotov 1992; Burawoy 1996). Unskilled labour forces and marginalized groups (e.g. Roma) were devastated by this change. For example, between 1989 and 2010, the agricultural sector in Slovakia decreased from 250 000 workers to 56 000 (Filcak 2012a). As the employment sector declined, prices stagnated, while prices in rent, utilities, and water rose with liberalization and privatization (Eyal et al. 1998; Greskovits 1998; Hamm et al. 2012). The rising cost of living, furthermore, created demographic shifts in which people moved out of city dwellings and factory-owned dormitories into cheaper housing (and, in extreme cases, shantytowns) on the periphery.

The poorer rural regions of Eastern Poland, Slovakia, Bulgaria and Romania were especially hard hit by economic restructuring, resulting in a "double poverty" phenomenon (Ladányi and Szelényi 2001). This phenomenon took into consideration communities of place characterized by both poverty among the people and the general poverty of the rural landscape in which opportunities were minimal. While Roma have faced prejudices and discrimination for centuries, the post-1989 economic transformation led to significant deterioration of their economic status. They were usually among the first to lose jobs in industry or agriculture. This further increased the isolation of a subordinated group increasingly subject to a process of ghettoization (Filcak and Steger 2014).

Environmental injustices in post-socialist Central and Eastern Europe

While the initial transition brought hope and anticipation, the reality soon set in as a consumer-based society embedded in a free market economy took hold. Broad environmental coalitions, such as the Hungarian mobilization against the construction of the Gabčíkovo-Nagymaros dam or the activism of Bratislava Aloud soon disbanded. Some proponents became prominent advocates for neoliberal economic transformation and took on political positions in the new governments. Others joined non-governmental organizations or academia, sometimes eventually joining politics, or swung in and out of different sectors intermittently (i.e. following a "revolving door" pattern).

The institutionalization of environmental protection from both a governmental and a civil society perspective allowed for new seeds in environmental justice to be planted, but with time these seeds could hardly grow against the tides of global neoliberalizing trends in privatization, retraction of the state and shifts in employment coinciding with the marginalization of a reserve labour force, particularly among subordinated groups such as the Roma.

By the late 1990s, environmentalists across the region had articulated a post-socialist political ecology consisting of four branches of critique: "eco-colonialism," critiques of the emerging consumer society, democratic challenges to environmental risks, and the overall transformation of the green dissidence of the 1980s into a call for participatory citizenship and accountable political institutions (Harper 2006). These frames emphasized East–West economic inequalities between nations and citizens' access to environmental decision-making and public participation.

This initially widespread, diverse mobilization for environmental issues in Central and Eastern Europe was successful in establishing formalized advocacy and institutions. Governments' responsiveness to environmental issues and mobilizations varied from country to country, but almost all countries in the region adopted European Union environmental policies, established environmental ministries, and developed an independent civil sector of environmental organizations with professional expertise.

Distributional problems of environmental justice have proven more intractable. After the collapse of state socialism, CEE became a vulnerable repository for environmental risks and impacts from the West seeking low-regulation environments. Within this international political economy, attention must be given to the ways in which local socioeconomic inequalities make it politically possible to produce more pollution, "externalizing" environmental hazards to less favoured populations and sites (Ash et al. 2013; Filčák 2012c). Di Chiro (1996: 314) draws attention to the "colonial discourse of nature" in the characterization of the spaces of subordinate groups as "repositories of waste, garbage, vermin, disease, and depravity" which are challenged by the community-based counter discourse of the environmental justice movement.

When it comes to access to a healthy environment and basic environmental goods, the problem of unequal representation, distribution, and discrimination persists along class lines and particularly among subordinated groups such as the Roma. Roma in Europe have a shorter life expectancy than Europeans in general (EC 2009). The health problems associated with poor environmental conditions among the Roma include respiratory problems such as asthma, skin problems, and various illnesses related to exposure to toxics such as lead (Steger 2007). It is well known that social determinants of ill-health such as poor daily living conditions disproportionately negatively affect vulnerable groups (WHO 2013).

Contaminated sites are a legacy of past industrial and economic development in CEE related to post-war industrialization (see Chapter 25). Industrial enterprises, many formerly under state ownership, engaged in the overuse of cheap chemicals and wasteful technologies partially attributed to a lack of information, the low price of chemicals (external costs of production and consumption virtually did not exist) and the mismanagement and/or lack of effective opposition from the side of stakeholders. Transformation and privatization of the economy, that started most prominently with the political changes of 1989, created legal turmoil and, in many cases, company assets were grabbed as ownership went through several stages of selling and buying, eventually leading to bankruptcy and unclear liability for the environmental burden associated with the properties. Remediation costs subsequently were borne by the state, which ultimately claimed insufficient resources to carry out such costly work, or were left to people from the surrounding areas, who were exposed to health risks and a deteriorating environment.

Mining ventures have a long history of igniting mobilization for environmental justice in CEE (see Chapter 30). And the trail is paved with some success. The local inhabitants in Kremnica, Slovakia, were able to effectively block a gold mining project in one of the most historic cities in the country. Some other cases, like Rosia Montana in Romania, probably the biggest gold mining plan in CEE, are still open but have been challenged by effective opposition within the area (Velicu and Kaika 2015).

The exposure to environmental risks among Roma in CEE has been documented. For example, in Rudňany, Slovakia, there is a Roma settlement on an abandoned mine with new social housing constructed in the same area. In Ostrava, Czech Republic, the Romani people live in apartments located above an abandoned mine which emits methane. People living in proximity to polluted sites often have moved there drawn by higher wages offered by neighbouring industrial plants, or were moved there due to prejudice and pressure from non-Roma and the local authorities.

There are situations where infrastructure, such as water and sanitation, is provided to proximal residents while stopping right at the transition boundary to Roma settlements. In the Roma Fakulteta neighbourhood in Sofia, Bulgaria, for example, basic infrastructure is provided in the surrounding areas, but is cut short where the Roma neighbourhood begins. The site is additionally used as an illegal dump site for industrial waste. Field research in Slovakia conducted by Škobla and Filčák (2014) indicated that despite positive measures to install water pipelines, the problem of water access among Roma communities may not be solved, due to a lack of ability to pay and the existence of cheaper alternatives (i.e. transporting containers of water from free water sources). While the daily water consumption per capita in Roma communities connected to water pipelines is in many cases as low as 40 litres, the hygienic minimum is 80 litres.

Energy access as well as production impacts are also a challenge to environmental justice, not only in CEE, but in Western Europe and many other parts of the world. Lack of ability to pay for utilities is a global problem. In CEE, Rosalina Babourkova, for example, identified barriers to energy access in Bulgaria associated with the privatization of electricity utilities in two Roma settlements (2010). Extractive industries associated with energy production and access are global challenges with their fair share of environmental justice problems (see Chapter 31). Shale hydrocarbons extraction has raised arms in protest all over the world, and CEE is not excluded.

Roma and environmental justice in CEE: ecological recognition and distribution

The environmental movement that persisted after 1989 existed alongside the emergence of Romani civil rights activism and ethnic mobilizations in many countries in the region, but rarely did the two movements converge (see Chapter 3 on social movements). One of the founding civic organizations of the Hungarian Romani movement was the Anti-Ghetto Committee in Miskolc, which formed in early 1989 in response to the state socialist municipality's plan to demolish public housing units in the city centre and to relocate the largely Romani tenants to a neighbourhood in a swampy, flood-prone area on the edge of the city (Ladányi 1991). The Anti-Ghetto Committee organized demonstrations and a petition drive to protest against the plan as a case of discrimination in urban planning and housing policy, persuading the municipality to abandon the proposal. Invoking fairness in urban planning and the distribution of environmental vulnerability (in the form of flood risk and other hazards associated with the site), the Anti-Ghetto Committee's action deployed themes common to environmental justice mobilizations. However, Roma activism was largely contained in the "social issues" policy silo, apart from environmental issues.

In 2003, what was formerly known as the Central European University, Center for Environmental Policy and Law (CEPL), and is now known as the Environmental and Social Justice Action Research Group (ACT JUST), convened academics, activists, and environmental and human rights lawyers from Central and Eastern Europe (CEE) in two workshops supported by the European Union's PHARE programme and the Open Society Institute. In these workshops, the concept of environmental justice and its relevance in CEE was explored. The Coalition for Environmental Justice asserted that:

> An environmental injustice exists when members of a disadvantaged, ethnic, minority or other group suffer disproportionately at the local, regional (sub-national), or national levels from environmental risks or hazards, and/or suffer disproportionately from violations of fundamental human rights as a result of environmental factors, and/or are denied access to environmental investments, benefits, and/or natural resources, and/or are denied access to information; and/or participation in decision-making; and/or access to justice in environment-related matters.
>
> *(Steger 2007: 10)*

Two years later, the Trust for Mutual Understanding and the Roma Participation Program at the Open Society Institute supported reconvening the Coalition for Environmental Justice with environmental justice academics, activists, and lawyers from the US for the "Transatlantic Initiative to Promote Environmental Justice Workshop" at Central European University in Budapest, Hungary (Pellow et al. 2005). This was a ground-breaking effort to explore and expand mutual learning to promote environmental justice in the respective regions and globally.

These workshops and initiatives were the springboard for ongoing networking that allowed environmental and human rights activists to practise and contribute to their mutual work on environmental justice. The human rights organization in Ostrava, Czech Republic, called Life Together (*Vzájemne soužití*), worked with the International POPs Elimination Network (IPEN) to collect data on toxicity levels in an abandoned industrial site where Roma children played. Network contacts were consulted on future projects and challenges, for example between the European Roma Rights Organization in Budapest, Hungary, and lawyers in New Orleans, Louisiana, who were struggling in the aftermath of Katrina. Lawyers for Romani Criss, a Roma human rights organization in Romania, explored the plausibility of applying environmental laws to promote Roma human rights there. Global Action launched a letter-writing campaign to address the egregious environmental injustice related to lead exposure in internal displacement camps in Mitrovica/e, Kosovo, populated by Roma. In 2015, the Coalition's work was shared at the University of Alaska to explore similarities and opportunities for igniting environmental justice in native communities there in relationship to changes associated with climate change in particular (see Chapter 40).

In addition, researchers went on to articulate the issue of environmental justice in Central and Eastern Europe further by asserting a framework based on the European Union's policies and laws and international agreements and efforts (Antypas et al. 2008; Steger and Filčák 2008). Despite some progress in formal institutions, research revealed that environmental justice in CEE faced multiple challenges (Filčák and Steger 2014; Harper et al. 2009; Steger 2007; Filčák 2012b). Specific research on access to water among Roma communities was sponsored by the United Nations Development Programme in Slovakia and is the subject of ongoing research activities (Škobla and Filčák 2014).

Participatory action research partnerships between Roma communities, academics, and environmental organizations have documented environmental inequalities and sought policy solutions. In one such collaborative effort in northern Hungary, members of the Sajó River Association for Environment and Community Development (*Sajómenti Környezet- és Közösség Fejlésztök Egyesülete, SAKKF*) held community-based photography workshops to document environmental injustices in waste management, access to sewerage and water systems, fuel poverty, and other issues. Along with producing the "Sajó River Declaration on Environmental Justice," a team held photo exhibitions and facilitated discussions at the local, national, and international levels (Harper 2012).

More recently in 2015, ACTJUST convened a workshop on "Seeking Environmental Justice through Energy Poverty Alleviation among Roma" involving practitioners, researchers, and activists that mapped energy poverty among Roma communities in CEE with the purpose of exploring the development and application of decentralized energy systems to alleviate energy poverty in predominantly Roma communities.[4] Energy utility privatization and the rising cost of energy are among the factors excluding the poor from the grid. On the one hand, they pay for externalities of energy production in terms of exposure to air pollution and devastation of the countryside, while on the other hand, low wages and high unemployment prevent access to energy.

Energy poverty has resulted in the generation of alternative approaches particularly in poor and Roma communities. For example, some successful projects addressing home heating in Roma communities in Hungary have been launched. In an effort to simultaneously alleviate energy poverty and reduce illegal activity associated with the need for heat production (e.g. burning trash, illegal timber harvesting for fuelwood), biomass production projects were launched in two Roma communities in the villages of Bag and Told in Hungary. Involving civic entities (including the BAGázs Association) and local community members, these projects produced biomass briquettes using hand-operated pressure machines and materials gathered locally (Racz 2014).

Conclusions

This chapter has contextualized environmental justice in CEE under state socialism leading up to 1989, and during post-socialist transformation with EU enlargement and the introduction of multi-party parliamentary regimes, privatization, and marketization. The environmental movements that emerged during the period leading up to transition were characterized by horizontally connected environmental justice struggles to enhance participation in environmental decision-making, coinciding with advocacy for human rights. This widespread mobilization involved a transition in environmental discourse from mostly state-led nature protection to the introduction of environmental protection that clearly placed "people in the environment," while targeting the massive industrialization approaches of the socialist regimes.

After 1989, environmental justice in CEE entered a new phase with more opportunities for environmental justice from a vertical perspective. Formal state environmental institutions were expanded with the introduction of an EU environmental legal and policy framework. Opportunities for public participation increased through new legal structures and the development of a largely Western-financed, professionalized NGO sector. While EU enlargement benefited new CEE member states in terms of environmental policy and law, enforcement and financial support, it also paved the way for international capital, including investments related to the Transatlantic Free Trade Agreement. Many investors have been attracted to CEE as a region with untapped natural resources and politically quiescent locations for problematic investments (Castán Broto 2015).

This post-socialist phase has raised questions about environmental justice from the standpoint of a budding consumer society and the export of environmental risks from west to east reminiscent of "eco-colonialism." Additionally, post-socialist transformation ultimately challenged the horizontal diffusion of environmental justice (see Walker 2009), particularly in terms of discrimination along ethnic lines. Still, small initiatives for environmental justice did sprout and reverberate across communities within CEE. Despite inequalities in the distribution of environmental risks and access to basic environmental necessities such as water, sanitation, and energy across and within CEE countries, some innovative projects and efforts have been made to address these injustices at the local scale.

Almost 20 years after the collapse of the former Communist Party based regimes, Central and Eastern Europe has become part of the global market. Struggles for environmental justice that once confronted the former Soviet-style massive industrialization project are now instead confronted with a complex scenario of political and economic alliances at the national level alongside financialization mechanisms associated with domestic and international capital. As Grant (2001) points out, environmental problems are first and foremost political because they affect social groups differentially and impose different types of costs and burdens.

The Trebišov case at the opening of the chapter demonstrates a complicated pattern of mobilization for environmental justice in CEE in which communities struggling along class lines are pulled into ethnic prejudices characteristic of difficult times. Roma are particularly subject to environmental injustices, predominantly along ethnic lines in a growing right-wing political context. Efforts to promote environmental and social justice are subsumed or even distracted by such politics that divert class struggles to ethnic and social prejudice. Community success in preventing environmentally risky projects from coming to town may be only short-lived, as the case of Rosia Montana in Romania demonstrates.[5] Extractive industries have resources to carry the fight long term. Struggles for environmental justice require persistence and a commitment to direct democracy where people live, work and play, and the civic spirit is enjoined as a matter of everyday existence.

Notes

1 We use the term 'centrally planned socialism'. Calling the regimes communist is misleading, but socialism in turn does not capture the essence of these regimes, since the term 'socialism' can be used for democratic policy regimes.
2 Several environmental cases were instrumental for generating political support for the wave of revolution that preceded the collapse of centrally planned socialism.
3 It is important to note, however, that not all protesters were politically motivated as opponents of the government more broadly.
4 See www.ceu.edu/sites/default/files/attachment/event/14504/final-agendaconferenceseeking-environmental-justice-through-energy-poverty-alleviation-among-roma.pdf. This initiative was launched by Richard Robert Racz based on his thesis (2014) *Opportunities and Threats of Small Scale Decentralized Energy System: Biomass Briquette Production in Bag, Hungary*, Master's Thesis, Central European University, Budapest, Hungary.
5 Mining operations in Rosia Montana, while having a long history within the related communities, had stopped only briefly before a Canadian based corporation came to the region to extract remaining gold deposits.

References

Antypas, A., Cahn, C., Filčák, R., and Steger, T. (2008). 'Linking environmental protection, health, and human rights in the European Union: An argument in favor of environmental justice policy.' *Environmental Law and Management*, 20(1), 8-21.

Ash, M., Boyce, J.K., Chang, G., and Scharber, H. (2013). 'Is environmental justice good for white folks? Industrial air toxics exposure in urban America.' *Social Science Quarterly*, 94(3), 616–636.

Auer, M.R. (1998). 'Environmentalism and Estonia's independence movement.' *Nationalities Paper*, 26(4), 659–676.

Babourkova, R. (2010). 'The environmental justice implications of utility privatisation: The case of the electricity supply in Bulgaria's Roma settlements.' *International Journal of Urban Sustainable Development*, 2(1–2), 24–44.

Bahro, R. (1981). *The Alternative in Eastern Europe*. New York: Verso.

Bari, J. and the Photovoice Group of the Sajó River Association for Environment and Community Development (2007). *Sajó River Declaration*. Translated from Hungarian by K. Harper. Sájoszentpéter, Hungary, June 13.

Baumgartl, B. (1995). 'Green mobilisation against red politics: Environmentalists' contribution to Bulgaria's transition.' In Rüdig, W. (ed.) *Green Politics Three*. Edinburgh: Edinburgh University Press, pp. 154–191.

Burawoy, M. (1996). 'The State and economic involution: Russia through a China lens.' *World Development*, 24, 1105–1122.

Burawoy, M. and Krotov, P. (1992). 'The Soviet transition from socialism to capitalism: Worker control and economic bargaining.' *American Sociological Review*, 57, 16–38.

Castán Broto, V. (2015). 'Dwelling in a pollution landscape.' In Vaccaro, I., Harper, K., and Murray, D.S. (eds.) *Anthropology of Postindustrialism: Ethnographies of Disconnection*. New York: Routledge, pp. 91–112.

Crowe, D. (1996). *A History of the Gypsies of Eastern Europe and Russia*. New York: St. Martin's Griffin.

Dawson, J. (1996). *Eco-nationalism: Anti-nuclear Activism and National Identity in Russia, Lithuania, and Ukraine*. Durham, NC: Duke University Press.

Debrecen University School of Public Health (2004). *Telepek és telepszerű lakóhelyek felmérése*. (Survey of settlements and settlement-type residences). Debrecen, Hungary: Debrecen University School of Public Health.

Di Chiro, G. (1996). 'Nature as community: The convergence of environment and social justice.' In Cronon, W. (ed.) *Uncommon Ground: Rethinking the Human Place in Nature*. New York: W.W. Norton & Co., pp. 298–320.

Doyle, H. (2004). 'Improving access of Roma to health care through the Decade of Roma Inclusion.' *Roma Rights*, 3/4, 42–45.

European Commission (EC) (2009). *Health and the Roma Community, Analysis of the Situation in Europe. Bulgaria, Czech Republic, Greece, Portugal, Romania, Slovakia, Spain*. http://ec.europa.eu/justice/discrimination/files/roma_health_en.pdf

European Environment Agency (EEA) (1995). *Europe's Environment: The Dobris Assessment*. Copenhagen: EEA.

European Environment Agency (EEA) (2007). *Europe's Environment: The Fourth Assessment*. Copenhagen: EEA.

European Environment Agency (EEA) (2015). *SOER 2015 – The European Environment – State and Outlook 2015*. Copenhagen: EEA.

European Statistical Bureau (ESB) (2008). *Europe in Figures – Eurostat Yearbook 2008: Environment*. Brussels: ESB.

Eyal, G., Szelenyi, I., and Townsley, E. (1998). *Making Capitalism without Capitalists*. New York: Verso.

Filčák, R. (2012a). *Spoločnosť trhu a environmentálna politika: aktéri a konflikty* [Society of the market and environmental policy: actors and conflicts]. Bratislava: VEDA Publishing, 322.

Filčák, R. (2012b). 'Environmental justice and the Roma communities of eastern Slovakia: Settlements, land, entitlements, and the environmental risks.' *Czech Sociological Review*, (6), 1119–1147.

Filčák, R. (2012c). *Living beyond the Pale: Environmental Justice and Roma Minority*. Budapest: Central European University Press.

Filčák, R. and Steger, T. (2014). 'Ghettos in Slovakia: Confronting Roma social and environmental exclusion.' *Analyse and Kritik: A Journal for Social Philosophy*, 36(2), 229–250, Zurich: University of Zurich.

Fisher, D. (1993). 'The emergence of the environmental movement in Eastern Europe and its role in the revolutions of 1989.' In Jancar-Webster, B. (ed.) (1993) *Environmental Action in Eastern Europe: Responses to Crisis*. New York: M.E. Sharpe, Inc.

Glazer, M. and Glazer, P. (2014). 'On the trail of courageous behaviour.' In King, L. and McCarthy, D. (eds.) (2014) *Environmental Sociology: From Analysis to Action*. Maryland: Rowman and Littlefield.

Grant, W. (2001). 'Environmental policy and social exclusion.' *Journal of European Public Policy*, 8(1), 82–100.

Greskovits, B. (1998). *Political Economy of Protest and Patience: East European and Latin American Transformations Compared*. Budapest–New York: Central European University Press.

Hamm, P., King, L., and Stuckler, D. (2012). 'Mass privatization, state capacity, and economic growth in post-communist countries.' *American Sociological Review*, 77(2), 295–324.

Harper, K. (2006). *Wild Capitalism: Environmental Activism and Postsocialist Political Ecology in Hungary*. New York: East European Manuscripts/Columbia.

Harper, K. (2012). 'Visual interventions and the "crises in representation" in environmental anthropology: Researching environmental justice in a Hungarian Romani neighborhood', *Human Organization*, 71(3), 292–305.

Harper, K., Steger, T., and Filcak, R. (2009). 'Environmental justice and Roma communities in Central and Eastern Europe.' *Environmental Policy and Governance*, 19(1), 251–268.

Kabaj, M. (2005). *Ekonomia tworzenia i likwidacja mijsc pracy. dezaktywizacja polski?* Seria Studia i Monografie. IPiSS: Warsaw.

Kamieniecki, S. (1993). *Environmental Politics in the International Arena: Movements, Parties, Organizations, and Policy*. Albany, NY: State University of New York Press.

Ladányi, J. (1991). *A miskolci gettougy* [The Miskolc Ghetto Case]. *Valóság*, 34(4), 45–54.

Ladányi, J. and Szelényi, I. (2001). 'The social construction of Roma ethnicity in Bulgaria, Romania and Hungary during market transition.' *Review of Sociology*, 7(2), 79–89.

OECD (2004). *OECD Employment Outlook 2004*. Paris: Organisation for Economic Cooperation and Development.

Pavlinek, P. and Pickles, J. (2002). *Environmental Transitions: Transformation and Ecological Defense in Central and Eastern Europe*. London: Routledge.

Pellow, D., Steger, T., and McLain, R. (2005). *Proceedings from the Transatlantic Initiative to Promote Environmental Justice Workshop, Central European University, Budapest, Hungary*, October 27–30, 2005. Accessed online at http://archive.ceu.hu/publications/pellow/20015/42818

Philip, S. and Jehlička, P. (2007). 'Environmental movements in space-time: The Czech and Slovak Republics from Stalinism to post-socialism.' *Transactions of the Institute of British Geographers*, 32(3), 346–362.

Racz, R. (2014). *Opportunities and Threats of Small Scale Decentralized Energy System: Biomass Briquette Production in Bag, Hungary*. Master's Thesis. Central European University, Budapest, Hungary.

Škobla, D. and Filčák, R. (2014). 'Bariéry v prístupe k pitnej vode ako aspekt sociálneho vylúčenia rómskej populácie' [Barriers in access to potable water as an aspect of social exclusion of the Roma population]. *Sociológia*, 46(5), 620–637.

Snajdr, E. (1998). 'The children of the greens: New ecological activism in post-socialist Slovakia.' *Problems of Post-Communism*, 45(1), 54–62.

Stark, D. (1992). 'Path dependence and privatization strategies in East Central Europe.' *East European Politics and Societies*, 6, 17–51.

Steger, T. (2004a). *Environmentalism and Democracy in Hungary and Latvia*. PhD Dissertation. Syracuse, NY.

Steger, T. (ed.) (2007). *Making the Case for Environmental Justice in Central and Eastern Europe*. Central European University, Center for Environmental Policy and Law (CEPL), (Budapest, Hungary) and the Health and Environment Alliance (HEAL) (Brussels, Belgium).

Steger, T. and Filčák, R. (2008). 'Articulating the basis for promoting environmental justice in Central and Eastern Europe.' *Environmental Justice*, 1(1), 49–53.

Védegylet Egyesület (2010). *Környezeti igazságosság Magyarországon: Záró tanulmány* [Environmental justice in Hungary: Closing study]. Budapest: Védegylet. Accessed online on 24 December 2015 at http://kornyezetiigazsagossag.hu/zarotanulmany.pdf

Velicu, I. and Kaika, M. (2015). 'Undoing environmental justice: Re-imagining equality in the Rosia Montana anti-mining movement.' *Geoforum*. http://dx.doi.org/10.1016/j.geoforum.2015.10.012

Walker, G. (2009). 'Beyond distribution and proximity: Exploring the multiple spatialities of environmental justice.' *Antipode*, 41, 614–636.

Wodak, R. and Richardson, J.E. (2012). *Analysing Fascist Discourse: European Fascism in Talk and Text*. London: Routledge.

World Health Organization (WHO) (2013). *Review of Social Determinants and the Health Divide in the WHO European Region*. World Health Organization, Prepared by UCL, Institute of Health Equity. www.who.int/social_determinants/final_report/csdh_finalreport_2008.pdf

INDEX

The acronym EJ is used for environmental justice. Locators in *italics* refer to figures and those in **bold** to tables.

Printed in the United States
by Baker & Taylor Publisher Services